T0321902

Human Factors in Simple and Complex Systems, Third Edition

Human Factors in Simple and Complex Systems, Third Edition

By
Robert W. Proctor and Trisha Van Zandt

CRC Press
Taylor & Francis Group
Boca Raton London New York

CRC Press is an imprint of the
Taylor & Francis Group, an **informa** business

CRC Press
Taylor & Francis Group
6000 Broken Sound Parkway NW, Suite 300
Boca Raton, FL 33487-2742

© 2018 by Taylor & Francis Group, LLC
CRC Press is an imprint of Taylor & Francis Group, an Informa business

No claim to original U.S. Government works

ISBN 13: 13: 978-1-4822-2956-1 (hbk)

Library of Congress Cataloging-in-Publication Data

Names: Proctor, Robert W., author. | Van Zandt, Trisha, author.
Title: Human factors in simple and complex systems / by Robert W. Proctor and Trisha Van Zandt.
Description: Third edition. | Boca Raton : CRC Press, [2018] | Includes bibliographical references and index.
Identifiers: LCCN 2017020980| ISBN 9781482229561 (hardback : alk. paper) | ISBN 9781315156811 (ebook)
Subjects: LCSH: Human engineering. | Systems engineering.
Classification: LCC TA166 .P75 2018 | DDC 620.8/2--dc23
LC record available at https://lccn.loc.gov/2017020980

Visit the Taylor & Francis Web site at
http://www.taylorandfrancis.com

and the CRC Press Web site at
http://www.crcpress.com

Contents

PART I FOUNDATIONS OF HUMAN FACTORS

PART II   Perceptual Factors and Their Applications

PART III   Cognitive Factors and Their Applications

PART IV   Action Factors and Their Applications

PART V *Environmental Factors and Their Applications*

Preface to Third Edition

The first edition of *Human Factors in Simple and Complex Systems*, published in 1994, was prescient in highlighting the increasing role of humans in complex systems, an approach that has since come to be called *human–systems integration* (Boehm-Davis, Durso, & Lee, 2015; Booher, 2003a). We maintain this emphasis on complex systems in the present edition, while also acknowledging the need to systematically consider human factors in the simplest of systems and products. Many technological advances occurred in the 14 years prior to the second edition, published in 2008, and such advances have proceeded at an ever increasing pace from that time to the present. Mobile devices that allow access to the Internet from almost any place at any time have come into widespread use. Research and development teams can collaborate across the world instantaneously using software distributed across the Internet. Advanced display technologies have resulted in a major redesign of the air-traffic control system. All aspects of healthcare, in both medical and home settings, are becoming increasingly technological in nature. Using advancements made in robotics and communication technology, surgeons can now perform telesurgery, surgeries on patients in locations far from that of the surgeon. Intelligent vehicle systems that help avoid collisions, alert sleepy drivers, and assist drivers with navigation and parallel parking have become increasingly sophisticated. Advances such as these require significant input and evaluation from human factors specialists.

New technologies have continued to emerge that are now part of everyday life. Social networking services, including Facebook and Twitter, were in their initial phases: Twitter was founded in March, 2006, and Facebook opened registration to all computer users in September, 2006. Smartphones were in their infancy in 2008, with the first Android phone appearing that year and the first iPhone only a year earlier. Cell phone use while driving a vehicle has become an increasing concern because it places heavy demands on the physical and cognitive capabilities of the driver. Autonomous cars without human drivers are being allowed starting in the middle of 2018 in the state of California. In all of these examples, as well as for the multitude of technological advances that will occur in the years to come, usability issues are paramount. To maximize the effectiveness of any such technology, a host of human factors issues have to be addressed.

The third edition of *Human Factors in Simple and Complex Systems* is intended to update and expand our earlier editions taking into account the technological changes over the last 8 years. While technology may change, the foundation for understanding human performance remains the same. This foundation allows us to address issues of human factors and ergonomics in new and emerging technologies from a coherent, process-oriented perspective. As such, our goal continues to be to provide students with the knowledge necessary to understand the range of human factors issues that may be encountered in the design, implementation, and evaluation of products and systems. Our intent is to provide a foundation in the principles of human performance and a broad overview of the field of human factors for advanced undergraduate and graduate students.

Preface to Second Edition

Much has occurred during the 14 years since the first edition of *Human Factors in Simple and Complex Systems*. The World Wide Web, which was in its infancy in 1994, has become central to many aspects of life. Whereas use of the Internet was restricted primarily to e-mail messages between academic users, it is now part of the "information society." Without leaving home, we can order products and services of various types, engage in banking and other financial transactions, obtain information on almost any topic, and converse with other people from around the world.

One effect of this growth in use of the Internet is an increased concern with making the Internet accessible to all people, including the elderly, people with physical and mental disabilities, and people with limited education. Also, security of various forms—personal, national, etc.—has become a central issue. Despite this increased emphasis on security, new worms and viruses that can disrupt computer operations and destroy information arise regularly and cybercrimes, including identity theft and "phishing," are on the rise. Maximizing security of all types requires that people—end-users, system administrators, and security screeners—perform security-related tasks regularly and appropriately.

New technologies have emerged that are now part of everyday life. DVDs did not exist 14 years ago, yet today DVD players and recorders have surpassed VCRs as the most popular format for home entertainment systems. Cellular phones are everywhere, even where they should not be. Cell phone use while driving a vehicle has become a central concern because it places heavy demands on the physical and cognitive capabilities of the driver. Intelligent vehicle systems now help prevent collisions and death, and assist drivers with navigation and even parallel parking. In all of these examples, as well as for many other technological advances that will occur in the years to come, usability issues are paramount. To maximize the effectiveness of any such technology, a host of human factors issues have to be addressed.

The second edition of *Human Factors in Simple and Complex Systems* is intended to provide students with the knowledge necessary to understand the range of human factors issues that may be encountered in the design, implementation, and evaluation of products and systems. In writing this book, we had many of the same goals as in the first edition. Our intent is to provide a foundation in the principles of human performance and a broad overview of the field of human factors for advanced undergraduate and graduate students.

The book, organized around the human as an information-processing system, introduces students to a broad range of human factors topics. We emphasize throughout the text that there is a close relation between basic research=theory and application, and focus on methods, reasoning, and theories used by basic and applied researchers. The book provides an understanding of the variables that influence human performance and the ways that human factors experts draw upon this understanding. It also offers a framework of the research process in human factors and gives students an integrated view of the current state of our understanding of the factors that influence human performance and how these factors can be accommodated in system design.

The text fills the need for a textbook in human factors and ergonomics that bridges the gap between the conceptual and empirical foundations of the field. As Gavriel Salvendy noted in the foreword to the first edition, "The theoretical approach that it takes is in contrast to the 'cookbook' approach frequently seen in human factors, from which students get information about specific functions or attributes that can be applied to only a particular area. Instead, this book demonstrates a general approach to solving a broad range of system problems. It provides a long awaited and much needed coverage of the theoretical foundation on which the discipline of human factors is built."

Human Factors in Simple and Complex Systems is a complete human factors text that covers the full range of contemporary human factors and ergonomics. We wrote it for introductory courses at both the undergraduate and graduate level. Because the text is structured around human information processing, it could serve as well as a primary text for courses in applied cognition, omitting the chapters that cover topics not typically included in such courses.

The second edition differs from the first edition in three fundamental ways. First, we have updated and modified the textbook to reflect the current state of the field, with many new topics added to capture the tremendous changes in human factors and ergonomics that have taken place during the past decade. Moreover, we discuss concepts such as situation awareness that has come to be central in human factors but was not mentioned in the previous edition of the book. Second, we have provided a tighter integration of basic research and application throughout the text, strengthening the link between knowledge and the practice of human factors. This has resulted in reorganization of several sections and reduction in the number of chapters to 19. Third, we have made the writing more accessible. To break up the chapters, each chapter includes a separate box that discusses a topic of considerable current interest related to human interactions with computers and recent technology. In revising the text, we made a concerted effort not only to ensure clarity but also to convey the material in a straightforward and interesting manner.

Robert W. Proctor
Trisha Van Zandt

Authors

Robert W. Proctor is a distinguished professor of Psychological Sciences at Purdue University. He is a fellow of the American Psychological Association, the Association for Psychological Science, and the Human Factors and Ergonomics Society, and recipient of the Franklin V. Taylor Award for Outstanding Contributions in the Field of Applied Experimental/Engineering Psychology from Division 21 of the American Psychological Association in 2013. He is co-author of *Stimulus-Response Compatibility: Data, Theory and Application*, *Skill Acquisition & Training*, and coeditor of *Handbook of Human Factors in Web Design*.

Trisha Van Zandt is a professor of Psychology at The Ohio State University. She is a member of the Society for Mathematical Psychology, of which she was president in 2006–2007, and the American Statistical Association. She has received multiple research grants from the National Science Foundation and the Presidential Early Career Award for Scientists and Engineers in 1997. She is co-author of review chapters "Designs for and Analyses of Response Time Experiments" in the *Oxford Handbook of Quantitative Methods* and "Mathematical Psychology" in the *APA Handbook of Research Methods in Psychology*.

Part I

Foundations of Human Factors

1 Historical Foundations of Human Factors

Our interest in the design of machines for human use runs the full gamut of machine complexity—from the design of single instruments to the design of complete systems of machines which must be operated with some degree of coordination.

A. Chapanis, W. Garner, & C. Morgan
1949

INTRODUCTION

The quote with which we begin this chapter is from the first textbook devoted specifically to human factors, *Applied Experimental Psychology: Human Factors in Engineering Design,* by Alphonse Chapanis, Wendell Garner, and Clifford Morgan. Designing machines and systems, whether simple or complex, for human use was not only the central concern of their pioneering book but also the driving force for subsequent research on human factors and ergonomics over the past 69 years. The following quotation from the U.S. National Academy of Engineering in their report *The Engineer of 2020,* now more than 10 years old, captures the ever increasing importance of the role of human factors in the introduction of new technologies and products:

> Engineers and engineering will seek to optimize the benefits derived from a unified appreciation of the physical, psychological, and emotional interactions between information technology and humans. As engineers seek to create products to aid physical and other activities, the strong research base in physiology, ergonomics, and human interactions with computers will expand to include cognition, the processing of information, and physiological responses to electrical, mechanical, and optical stimulation. (2004, p. 14)

It is our purpose in this textbook to summarize much of what we know about human cognitive, physical, and social characteristics and to show how this knowledge can be brought to bear on the design of machines, tools, and systems that are easy and safe to use.

In everyday life, we interact constantly with instruments, machines, and other inanimate systems. These interactions range from turning on and off a light by means of a switch, to the operation of household appliances such as stoves and digital video recorders (DVRs), to the use of mobile smartphones and tablet computers, to the control of complex systems such as aircraft and spacecraft. In the simple case of the light switch, the interaction of a person with the switch, and those components controlled by the switch, forms a system. Every system has a purpose or a goal; the lighting system has the purpose of illuminating a dark room or extinguishing a light when it no longer is needed. The efficiency of the inanimate parts of this system, that is, the power supply, wiring, switch, and light bulb, in part determines whether the system goal can be met. For example, if the light bulb burns out, then illumination is no longer possible.

The ability of the lighting system and other systems to meet their goals also depends on the human components of the systems. For example, if a small person cannot reach the light switch, or an elderly person is not strong enough to operate the switch, then the light will not go on and the goal of illumination will not be met. Thus, the total efficiency of the system depends on both the performance of the inanimate component and the performance of the human component. A failure of either can lead to failure of the entire system.

ELECTRONIC AND DIGITAL EQUIPMENT

The things that modern electronic and digital equipment can do are amazing. However, how well these gadgets work (the extent to which they accomplish the goals intended by their designers) is often limited by the human component. As one example, the complicated features of video cassette recorders (VCRs) made them legendary targets of humor in the 1980s and 1990s. To make full use of a VCR, a person first had to connect the device correctly to the television and cable system or satellite dish system that provided the signal and then, if the VCR did not receive a time signal from the cable or satellite, accurately program its clock. When the person wanted to record a television program, she had to set the correct date, channel number, intended start and end times, and tape speed (SP, LP, or EP). If she made any mistakes along the way, the program she wanted would not be recorded. Either nothing happened, the wrong program was recorded, or the correct program was recorded for the wrong length of time (e.g., if she chose the wrong tape speed, say SP to record a 4 hour movie, she would get only the first 2 hours of the show). Because there were many points in this process at which users could get confused and make mistakes, and different VCRs embedded different command options under various menus and submenus in the interface, even someone who was relatively adept at programming recorders had problems, especially when trying to operate a machine with which he was unfamiliar.

Usability problems prevented most VCR owners from using their VCRs to their fullest capabilities (Pollack, 1990). In 1990, almost one-third of VCR owners reported that they had never even set the clock on the machine, which meant that they could never program the machine for recording at specific times. Usability problems with VCRs persisted for decades after their introduction in 1975.

Electronic technology continues to evolve. Instead of VCRs, we now have DVRs and DVR devices like TiVo and Roku. These products still require some programming, and, in many cases, they must be connected to other devices (such as a television set or a home Internet router) to perform their functions. This means that usability is still a major concern, even though we do not have to worry about setting their clocks any more.

You might be thinking right now that usability concerns only apply to older people, who may not be as familiar with technology as younger people. However, young adults who are more technologically sophisticated still have trouble with these kinds of devices. One of the authors of this textbook (Proctor) conducted, as a class project, a usability test of a modular bookshelf stereo with upper-level college students enrolled in a human factors class. Most of these students were unable to program the stereo's clock, even with the help of the manual. Another published study asked college students to use a VCR. Even after training, 20% of them thought that the VCR was set correctly when in fact it was not (Gray, 2000).

COMPUTER TECHNOLOGY

Perhaps nowhere is rapid change more evident than in the development and proliferation of computer technology (Bernstein, 2011; Rojas, 2001). The first generation of modern computers, introduced in the mid-1940s, was extremely large, slow, expensive, and available mainly for military purposes. For example, in 1944, the Harvard-IBM Automatic Sequence Controlled Calculator (ASCC, the first large-scale electric digital computer in the U.S.) was the length of half of an American football field and performed one calculation every 3–5 s. Programming the ASCC, which had nothing like an operating system or compiler, was not easy. Grace Hopper, the first programmer for the ASCC, had to punch machine instructions onto a paper tape, which she then fed into the computer. Despite its size, it could only execute simple routines. Hopper went on to develop one of the first compilers for a programming language, and in her later life, she championed standards testing for computers and programming languages.

The computers that came after the ASCC in the 1950s were considerably smaller but still filled a large room. These computers were more affordable and available to a wider range of users at businesses and universities. They were also easier to program, using assembly language, which allowed abbreviated programming codes. High-level programming languages such as COBOL and FORTRAN, which used English-like language instead of machine code, were developed, marking the beginning of the software industry. During this period, most computer programs were prepared on decks of cards, which the programmer then submitted to an operator. The operator inserted the deck into a machine called a *card reader* and, after a period of time, returned a paper printout of the run. Each line of code had to be typed onto a separate card using a keypunch. Everyone who wrote programs during this era (such as the authors of this textbook) remembers having to go through the tedious procedure of locating and correcting typographical errors on badly punched cards, dropping the sometimes huge deck of cards and hopelessly mixing them up, and receiving cryptic, indecipherable error messages when a program crashed.

In the late 1970s, after the development of the microprocessor, the first desktop-sized personal computers (PCs) became widely available. These included the Apple II, Commodore PET, IBM PC, and Radio Shack TRS-80. These machines changed the face of computing, making powerful computers available to everyone. However, a host of usability issues arose when computers, once accessible only by a small, highly trained group of users, became accessible by the general public. This forced the development of user-friendly operating system designs. For example, users interacted with the first PCs' operating systems through a text-based, command line interface. This clumsy and unfriendly interface restricted the PC market to the small number of users who wanted a PC badly enough to learn the operating system commands, but development of a "perceptual user interface" was underway at the Xerox Palo Alto Research Center (PARC). Only 7 years after Apple introduced the Apple II, they presented the Macintosh, the first PC to use a window-based graphical interface. Such interfaces are now an integral part of any computer system.

Interacting with a graphical interface requires the use of a pointing device to locate objects on the screen. The first computer "mouse" was developed by Douglas Engelbart in 1963 for an early computer collaboration system (see Chapter 15). He called it an "X-Y position indicator." His early design, shown in Figure 1.1, was later improved by Bill English at Xerox PARC for use with graphical interfaces. Like the graphical interface, the mouse eliminated some of the need for keyboarding during computer interaction.

Despite the improvements provided by graphical interfaces and the computer mouse, there are many usability issues yet to be resolved in human–computer interaction (HCI), and new issues appear as new functionality is introduced. For example, with a new piece of software, it is often hard to figure out what the different icons in its interface represent, or they may be easily confused with each other. One of the authors of this book, when he was not paying attention, occasionally clicked

FIGURE 1.1   The first computer mouse, developed by Douglas Engelbart after extensive usability testing.

unintentionally on the "paste" icon instead of the "save" icon for a popular word processor, because they were of similar appearance and in close proximity to each other. One very popular operating system required clicking on a button labeled "Start" when you wanted to shut the computer down, which most people found confusing. Moreover, like the old VCR problem, many software packages were very complex, and this complexity ensured that the vast majority of their users would be unable to use the software's full range of capabilities.

With the development of the Internet and the World Wide Web beginning around 1990, the individual PC became a common household accessory. It can function as a video game console, telephone (e-mail; voice messaging; instant messaging), digital video disc (DVD) player/recorder and DVR, stereo system, television, library, shopping mall, and so on. Usability studies form a large part of research into human factors relevant to interacting with the Web (Chen & Macredie, 2010; Vu & Proctor, 2011; Vu, Proctor, & Garcia, 2012). Although vast amounts of information are available on the Web, it is often difficult for users to find the information for which they are searching. Individual websites vary greatly in usability, with many being cluttered and difficult to comprehend. Starting with the introduction of the iPhone in 2007, most Americans now own smartphones, and many of those owners rely on those smartphones for access to the Web (Smith, 2015). A host of other usability issues are introduced by smartphones because of the devices' small display screens and restricted forms of user input (Rahmati et al., 2012). The preparation and structuring of content, as well as appropriate displays of information and input modes for a variety of devices, are important for the design of effective websites.

HEALTHCARE SYSTEMS

Over the past 15 years, healthcare has become a major research focus of human factors specialists. At the turn of the century, the Institute of Human Medicine published a report, *To Err Is Human: Building a Safer Health System* (Kohn, Corrigan, & Donaldson, 2000). This report pointed out that between 44,000 and 98,000 deaths annually were occurring as a result of human error during healthcare, and called for research clarifying the causes of the errors and a shift in focus of the healthcare system toward one of patient safety. This call has led to a torrent of human factors research on healthcare and the application of a human factors approach to the design of healthcare systems (Carayon & Xie, 2012; Carayon et al., 2014).

Interest in the role played by human factors in healthcare is sufficiently great that, starting in 2012, the Human Factors and Ergonomics Society has held an annual meeting, the International Symposium on Human Factors and Ergonomics in Health Care, which presents "the latest science in human factors as it applies to health-care delivery, medical and drug-delivery device design and health-care applications" (www.hfes.org/web/HFESMeetings/2015HealthCareSymposium.html). Among the technological advances in healthcare are electronic medical records, which allow the sharing of medical information among many parties, including possibly patients, but for which the interfaces must be usable by all potential users (Zarcadoolas, Vaughon, Czaja, Levy, & Rockoff, 2013).

CYBER SECURITY

In addition to usability, privacy and security on the Internet is a huge problem in general and for medical records in particular. Individuals and organizations want to keep some information open to the world while keeping other information secure and restricting its availability to authorized users only. An unsecured computer system or website can be damaged, either accidentally or intentionally, when an unauthorized person tampers with it, which may lead to severe financial damages. Careless system designers may make confidential information accessible to anyone who visits the site, which is what happened in the much-publicized 2014 hacking of Sony Pictures Entertainment in association with the motion picture *The Interview*. This cyber-attack resulted in the release of personal data and e-mail messages for thousands of then-current and former Sony employees. Lawsuits claim that the hacking was facilitated by Sony failing to provide an appropriate level of security for the information.

Increases in cyber security usually come at the cost of decreased usability (Proctor, Vu, & Schultz, 2009; Schultz, 2012). Most people, for a number of reasons, do not want to perform the additional tasks required to ensure a high degree of security. Other people try to make their data secure but fail to do so. So, questions remain about the best ways to ensure usability while maintaining security and privacy. The desire to allow people control over their private online data has resulted in the introduction of a new term, *human–data interaction*, to characterize the complex interactions between humans, online software agents, and data access (Mortier et al., 2014).

SERIOUS ACCIDENTS RESULTING FROM MAJOR SYSTEM FAILURES

Though some consumers' frustrations with their digital equipment may seem amusing, in some cases, great amounts of money and many human lives rely on the successful operation of systems. It is not difficult to find examples of incidents in which inadequate consideration of human factors contributed to serious accidents.

On January 28, 1986, the space shuttle *Challenger* exploded during launch, resulting in the death of its seven crew members. Design flaws, relaxed safety regulations, and a sequence of bad decisions led to the failed launch, the consequence of which was not only the loss of human lives but also a crippled space program and a substantial cost to the government and private industry. Even though the investigation of the *Challenger* disaster highlighted the problems within the National Aeronautics and Space Agency (NASA)'s culture and patterns of administration that led to the explosion, almost exactly the same kinds of mistakes and errors of judgment contributed to the deaths of seven astronauts on February 1, 2003, when the space shuttle *Columbia* broke up on re-entry to the earth's atmosphere.

The Columbia Accident Investigative Board concluded that the shuttle's heat shield was significantly damaged during liftoff when struck by a piece of foam from the booster. Although the cameras monitoring the launch clearly recorded the event, a quick decision was made by NASA officials that it did not threaten the safety of the shuttle. After all, they rationalized, foam had broken off during liftoff for other shuttle missions with no consequences. NASA engineers were not as cavalier, however. The engineers had worried for many years about these foam strikes. They requested (several times) that Space Shuttle Program managers coordinate with the U.S. Department of Defense to obtain images of the *Columbia* with military satellites so that they could determine the extent of the damage. These requests were ignored by Space Shuttle Program managers even after the engineers determined (through computer modeling and simulation) that damage must have occurred to the heat shield. The managers were suffering from the effects of a group behavior labeled "groupthink," in which it becomes very easy to ignore information when it is provided by people who are not part of the group.

NASA is not the only organization to have experienced disaster as a result of bad human factors. On March 28, 1979, a malfunction of a pressure valve triggered what ultimately became a core meltdown at the Three Mile Island nuclear power plant near Harrisburg, Pennsylvania. Although the emergency equipment functioned properly, poorly designed warning displays and control panels contributed to the escalation of a minor malfunction into the worst accident in the history of the U.S. nuclear power industry. This incident resulted in considerable "bad press" for the nuclear power industry as well as financial loss. The most devastating effect of the accident, however, was on U.S. citizens' attitudes toward nuclear power: Popular support for this alternative energy source dropped precipitously and has remained low ever since.

The Three Mile Island incident was a major impetus for the establishment of formal standards in nuclear plant design. Some of these standards attempted to remedy the obvious ergonomic flaws that led to the disaster. Other disasters have led to similar regulation and revision of design and safety guidelines. For example, in 1994, the *Estonia*, a Swedish car and passenger ferry, sank off the coast of Finland because the bow doors broke open and allowed water to pour into the hold. In what was regarded as one of the worst European maritime disasters since World War II, 852 lives were lost. The door locks failed because of poor design and lack of maintenance, and the crew failed to respond appropriately or quickly enough to the event. New safety guidelines were established for all European ferries after the disaster.

As with the *Challenger* and *Columbia* disasters, the increased attention to ergonomic issues on ferries did not prevent other disasters from occurring. After the *Estonia* disaster, the Norwegian ferry *Sleipner* ran into a rock and sank in November 1999, resulting in the loss of 16 lives; the crew was poorly trained, and few safety procedures existed.

Unfortunately, we can present still more examples. On October 31, 2000, a Singapore Airlines jumbo jet attempted to take off on a runway that was closed for construction, striking concrete barriers and construction equipment before catching fire. At least 81 people lost their lives as a consequence of this accident, which was due in part to poor placement of the barriers and inadequate signs. Yet another is the head-on collision, on September 12, 2008 in Chatsworth, California, of a commuter train and a freight train, which resulted in the loss of 25 lives and more than 100 injured people. The collision occurred when the commuter train ran a red signal and entered a section of track to which the freight train had been given access. Vision ahead was limited because of the curve in the track, which meant that the engineers of the two trains could not see each other until a few seconds before impact. Investigation showed that the signal was working properly and that the engineer of the commuter train was using his cellphone during the period in which the train passed the red light. Consequently, the National Transportation Safety Board (2010) determined, "The engineer failed to respond appropriately to a red signal at Control Point Topanga because he was engaged in text messaging at the time."

The *Challenger*, *Columbia*, Three Mile Island, and other disasters can be traced to errors in both the machine components and the human components of the systems. After reading this text, you should have a good understanding of how the errors that led to these incidents occurred and of steps that can be taken in the design and evaluation of systems to minimize the likelihood of their occurrence. You should also appreciate how and why human factors knowledge should be incorporated into the design of everything from simple products to complex systems with which people must interact.

WHAT IS HUMAN FACTORS AND ERGONOMICS?

When engineers design machines, they evaluate them in terms of their reliability, ease of operation or "usability," and error-free performance, among other things. Because the efficiency of a system depends on the performance of its operator as well as the adequacy of the machine, the operator and machine must be considered together as a single *human–machine system*. With this view, it then makes sense to analyze the performance capabilities of the human component in terms consistent with those used to describe the inanimate components of the system. For example, the reliability (the probability of successful performance) of human components can be evaluated in the same way as the reliability of machine components.

DEFINITION

The variables that govern the efficiency of the operator within a system fall under the topic of *human factors*: the study of those variables that influence the efficiency with which the human performer can interact with the inanimate components of a system to accomplish the system goals. This also is called *ergonomics*, and, in fact, the term *ergonomics* is more familiar than the term *human factors* outside of the U.S. and also to the general population within the U.S. (Dempsey, Wogalter, & Hancock, 2006). Other names include *human engineering*, *engineering psychology*, and, most recently, *human–systems integration*, to emphasize the many roles of humans in large-scale systems and systems of systems (Durso et al., 2015).

The "official" definition for the field of human factors, adopted in August, 2000, by the International Ergonomics Association and endorsed by the Human Factors and Ergonomics Society, is as follows:

> Ergonomics (or human factors) is the scientific discipline concerned with the understanding of inter-actions among humans and other elements of a system, and the profession that applies theory, prin-ciples, data, and other methods to design in order to optimize human well-being and overall system performance.

The field of human factors depends on basic research from relevant supporting sciences, applied research that is unique to the field, and application of the resulting data and principles to specific design problems. Human factors specialists thus are involved in research and the application of the data from that research to all phases of system development and evaluation.

Embodied in the definition of human factors is the importance of basic human capabilities, such as perceptual abilities, attention span, memory span, and physical limitations. The human factors specialist must know the limits of these capabilities and bring this knowledge to bear on the design of systems. For example, the placement of a light switch at an optimal height requires knowledge of the anthropometric constraints (i.e., the physical characteristics) of the population of intended users. For the switch to be used by people who are confined to wheelchairs as well as by people who are not, it should be placed at a height that allows easy operation of the switch by both groups. Similarly, the human factors specialist must consider people's perceptual, cognitive, and movement capabilities when designing information displays and controls, such as those found in automobiles, computer software packages, and microwave ovens. Only designs that accommodate and optimize the capabilities of the system's users will be able to maximize total system performance. Otherwise, the system performance will be reduced, and the system goals may not be met.

Barry Beith (2006, p. 2303), past president of the Human Factors and Ergonomics Society, said the following about what the human factors practitioner needs to know:

> HF/E is a field with very broad application. Essentially, any situation in which humans and technol-ogy interact is a focus for our field. Because of this breadth and diversity, the Basics are critical to success. By Basics, I am referring to the fundamental tools, techniques, skills, and knowledge that are the underpinnings of our discipline. Knowledge of human beings includes capabilities and limitations, behavioral and cultural stereotypes, anthropometric and biomechanical attributes, motor control, per-ception and sensation, cognitive abilities, and, most recently, emotional attributes addressed by affec-tive human factors.

The basics covered in this textbook will be important for your understanding of all the areas to which human factors and ergonomics analyses can be applied.

BASIC HUMAN PERFORMANCE

There is now a massive amount of scientific data on the limits of human capabilities. This research spans about 150 years and forms the core of the more general study of human performance. Specifically, the study of human performance involves analyses of the processes that underlie the acquisition, maintenance, transfer, and execution of skilled behavior (Healy & Bourne, 2012; Johnson & Proctor, 2017; Matthews, Davies, Westerman, & Stammers, 2000). The research iden-tifies factors that limit different aspects of a person's performance, analyzes complex tasks by breaking them into simpler components, and establishes estimates of basic human capabilities. With these data, we can predict how well people will be able to perform both simple and complex tasks.

Just as engineers analyze the machine component of a complex system in terms of its constituent subsystems (recall the wiring, switch, and light bulb of the lighting system), the human performance researcher analyzes the human component in terms of its subsystems. Before the light can be turned on, the human must perceive a need for light, decide on the appropriate action, and execute the action necessary to flip the switch. In contrast, the human factors specialist is concerned primarily with the interface between the human and machine components with the goal of making the communication of information between these two components as smooth and efficient as possible. In our lighting system example, this interface is embodied in the light switch, and the human factors issues involve the design and placement of the switch for optimal use. Thus, whereas the human performance researcher is interested in characterizing the processes within the human component, the human factors specialist is concerned with designing the human–machine interface to optimize achievement of the system goal.

In designing a system, we have much more freedom in how we specify the operating characteristics of the machine than in how we specify the characteristics of the human operator. That is, we can redesign and improve the machine components, but we shouldn't expect to be able to (or be permitted to) redesign and improve the operator. We can carefully screen and extensively train our operators before placing them in our system, but many limitations that characterize human performance cannot be overcome.

Because of this relative lack of freedom regarding the operator, it becomes imperative to know the constraints that human limitations impose on machine designs. Thus, the human factors specialist must consider basic human performance capabilities in order to wisely use the freedom that is available in the design of the machine component of the system.

HUMAN–MACHINE SYSTEMS AND DOMAINS OF SPECIALIZATION

Figures 1.2 and 1.3 summarize the domains of the design engineer, the human performance researcher, and the human factors specialist. Figure 1.2 shows a human–machine system: a person operating a microcomputer. The human–computer interface involves a video screen on which information is displayed visually by the microcomputer to communicate with the operator. This information is received by the operator through her visual sense. She processes this information and communicates back to the computer by pressing keys on the computer keyboard or moving the mouse. The computer then processes this information, and the sequence begins anew. The widespread use of microcomputers and other smart devices has forced the development of a branch of human factors that focuses exclusively on the problems involved in HCI (see Box 1.1).

Figure 1.3 shows a more abstract version of the human–computer system. In this abstraction, the similarity between the human and computer is clear. We can conceptualize each in terms of subsystems that are responsible for input, processing, and output, respectively. While the human receives input primarily through the visual system, the computer receives its input from the keyboard and other peripheral devices. The central processing unit in the computer is analogous to the cognitive capabilities of the human brain. Finally, the human produces output through overt physical responses, such as keypresses, whereas the computer exhibits its output on the display screen.

Figure 1.3 also shows the domains of the design engineer, the human performance researcher, and the human factors specialist. The design engineer is interested primarily in the subsystems of the machine and their interrelations. Similarly, the human performance expert studies the subsystems of the human and their interrelations. Finally, the human factors specialist is most concerned with the relations between the input and output subsystems of the human and machine components, or in other words, with the human–machine interface.

The final point to note from Figure 1.3 is that the entire human–machine system is embedded within the larger context of the work environment, which also influences the performance of the system. This influence can be measured for the machine component or the human component

FIGURE 1.2 Human–computer system.

as well as the interface. If the computer is in a very hot and humid environment, some of its components may be damaged and destroyed, leading to system failure. Similarly, extreme heat and humidity can adversely affect the computer user's performance, which may likewise lead to system failure.

The environment is not just those physical aspects of a workspace that might influence a person's performance. It also consists of those social and organizational variables that make work easier or harder to do. We use the term *macroergonomics* to describe the interactions between the organizational environment and the design and implementation of a system (Carayon, Kianfar, Li, & Wooldridge, 2015; Hendrick & Kleiner, 2002).

The total system performance depends on the operator, the machine, and the environment in which they are placed. Whereas the design engineer works exclusively in the domain of the machine, and the human performance researcher in the domain of the operator, the human factors specialist is concerned with the interrelations between machine, operator, and environment. In solving a particular problem involving human–machine interaction, the human factors specialist usually starts with a consideration of the capabilities of the operator. Human capabilities began to receive serious scientific scrutiny in the 19th century, well ahead of the technology with which the design engineer is now faced. This early research forms the foundation of contemporary human factors.

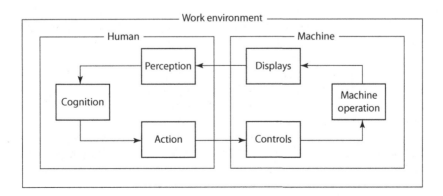

FIGURE 1.3 Representation of the human–machine system. The human and the machine are composed of subsystems operating within the larger environment.

BOX 1.1 HUMAN–COMPUTER INTERACTION

Human–computer interaction (HCI), also known as computer–human interaction (CHI), is the term used for an interdisciplinary field of study devoted to facilitating user interactions with computers. HCI has been a topic of burgeoning interest over the past 35+ years, during which time several organizations devoted to HCI have been established (e.g., the Special Interest Group on Computer–Human Interaction [SIGCHI] interest group of the Association for Computing Machinery [ACM]; see www.acm.org/sigchi/). HCI research is published not only in human factors journals but also in journals devoted specifically to HCI (e.g., *Human–Computer Interaction*, the *International Journal of Human–Computer Interaction*, and the *International Journal of Human-Computer Studies*). As well, there are textbooks (e.g., Kim, 2015; McKenzie, 2013), handbooks (Jacko, 2012; Weyers, Bowen, Dix, & Palanque, 2017), and encyclopedias (Soegaard & Dam, 2016) devoted exclusively to the topic of HCI.

Although some HCI experts regard it as a distinct field, it can be treated as a subfield of human factors, as we do in this text, because it represents the application of human factors principles and methods to the design of computer interfaces. HCI is fertile ground for human factors specialists because it involves the full range of cognitive, physical, and social issues of concern to the field (Carroll, 2003). These issues include everything from the properties of displays and data-entry devices, to the presentation of complex information in a way that minimizes mental workload and maximizes comprehension, to the design of groupware that will support team decisions and performance. More recently, there has been a push toward development of smart environments—offices, homes, businesses, classrooms, vehicles—in which adjustments and communications are made in response to input from sensors in the environment, people's preferences, and their actions (Hammer, Wißner, & André, 2015; Volpentesta, 2015).

The following are two examples of HCI considerations, one involving physical factors and the other involving cognitive factors. People who operate a keyboard for many hours on a daily basis are at risk for developing carpal tunnel syndrome, an injury that involves neural damage in the area of the wrist (Shiri & Falah-Hassani, 2015; see Chapter 16). The probability of developing carpal tunnel syndrome can be reduced by many factors, including the use of split or curved keyboards that allow the wrists to be kept straight, rather than bent. Thus, one concern of the human factors specialist is how to design the physical characteristics of the interface in such a way that bodily injuries can be avoided.

In addition to physical limitations, human factors professionals have to take into account the cognitive characteristics of targeted users. For example, most people have a limited ability to attend to and remember items such as where specific commands are located in a menu. The demands on memory can be minimized by the use of icons that convey specific functions of the interface that can be carried out by clicking on them. Icons can increase the speed with which functions are carried out, but if the icons are not recognizable, then more time is lost by executing the wrong functions.

Because computers are becoming increasingly involved in all aspects of life, HCI is studied in many specific application domains. These include educational software, computer games, mobile communication devices, and interfaces for vehicles of all types. With the rapid development of the Internet, one of the most active areas of HCI research in the past few years has been that associated with usability of the Internet and World Wide Web (Krug, 2014; Nielsen & Loranger, 2006; Vu & Proctor, 2011). Issues of concern include homepage design, designing for universal accessibility, e-commerce applications, Web services associated with health delivery, and conducting human research over the Web.

HISTORICAL ANTECEDENTS

The major impetus for the establishment of human factors as a discipline came from technological developments during World War II. As weapon and transport systems became increasingly sophisticated, great technological advances were also being made in factory automation and in equipment for common use. Through the difficulties encountered while operating such sophisticated equipment, the need for human factors analyses became evident. Human factors research was preceded by research in the areas of human performance psychology, industrial engineering, and human physiology. Thus, the historical overview that we present here will begin by establishing the groundwork within these areas that relate to human factors. The primary message you should take from this section is the general nature and tenor of work that provided an initial foundation for the field of human factors and not the details of this work, much of which is discussed more thoroughly in later chapters.

PSYCHOLOGY OF HUMAN PERFORMANCE

The study of human performance emphasizes basic human capabilities involved in perceiving and acting on information arriving through the senses. Research on human performance dates to the mid-19th century (Boring, 1942), with work on sensory psychophysics and the time to perform various mental operations being particularly relevant for human factors. Many of the concepts and methods these early pioneers developed to study human performance are still part of the modern human factors toolbox.

Sensory Psychophysics

Ernst Weber (b1795–d1878) and Gustav Fechner (b1801–d1887) founded the study of psychophysics and are considered to be the fathers of modern experimental psychology. Both Weber and Fechner investigated the sensory and perceptual capabilities of humans. Weber (1846/1978) examined people's ability to determine that two stimuli, such as two weights, differ in magnitude. The relation that he discovered has come to be known as *Weber's law*. This law can be expressed quantitatively as

$$\frac{\Delta I}{I} = K,$$

where:
- I is the intensity of a stimulus (say, a weight you are holding in your left hand),
- ΔI is the amount of change (difference in weight) between it and another stimulus (a weight you are holding in your right hand) that you need to be able to tell that the two stimuli differ in magnitude, and
- K is a constant.

Weber's law states that the absolute amount of change needed to perceive a difference in magnitude increases with intensity, whereas the relative amount remains constant. For example, the heavier a weight is, the greater the absolute increase must be to perceive another weight as heavier. Weber's law is still described in textbooks on sensation and perception (e.g., Goldstein, 2014) and provides a reasonable description for the detection of differences with many types of stimuli, except at extremely high or low physical intensities.

Fechner (1860/1966) formalized the methods that Weber used and constructed the first scales for relating psychological magnitude (e.g., loudness) to physical magnitude (e.g., amplitude). Fechner showed how Weber's law implies the following relationship between sensation and intensity:

$$S = K \log(I),$$

where:
  S is the magnitude of sensation,
  I is physical intensity,
  K is a constant, and
  log   is to any base.

This psychophysical function, relating physical intensity to the psychological sensation, is called *Fechner's law*. Like Weber's law, Fechner's law is presented in contemporary sensation and perception textbooks and still provokes theoretical inquiry concerning the relationship between what we perceive and the physical world (Steingrimsson, Luce, & Narens, 2006). The term *psychophysics* describes the research examining the basic sensory sensitivities, and both classical and contemporary psychophysical methods are described in Chapter 4.

Speed of Mental Processing

Fechner and Weber showed how characteristics of human performance could be revealed through controlled experimentation and, consequently, provided the impetus for the broad range of research on humans that followed. At approximately the same historical period, other scientists were making considerable advances in sensory physiology. One of the most notable was Hermann von Helmholtz (b1821–d1894), who made many scientific contributions that remain as central theoretical principles today.

One of Helmholtz's most important contributions was to establish a method for estimating the time for the transmission of a nerve impulse. He measured the difference in time between application of an electrical stimulus to a frog's nerve and the resulting muscle contraction, for two different points on the nerve. The measures indicated that the speed of transmission was approximately 27 m/s (Boring, 1942). The importance of this finding was to demonstrate that neural transmission is not instantaneous but takes measurable time.

Helmholtz's finding served as the basis for early research by Franciscus C. Donders (b1818–d1901), a pioneer in the field of ophthalmology. Donders (1868/1969) developed procedures called *chronometric methods*. He reasoned that, when performing a speeded reaction task, a person must make a series of judgments. He must first detect a stimulus (is something there?) and then identify it (what is it?). Then he may need to discriminate that stimulus from other stimuli (which one is it?). After these judgments, the observer selects the appropriate response to the stimulus (what response am I to make?).

Donders designed some simple tasks that differed in the combination of judgments required for each task. He then subtracted the time to perform one task from the time to perform another task that required one additional judgment. In this way, Donders estimated the time it took to make the judgments.

Donders' procedure is now called *subtractive logic*. The significance of subtractive logic is that it provided the foundation for the notion that mental processes can be isolated. This notion is the central tenet of *human information processing*—the approach that underlies most contemporary research on human performance. This approach assumes that cognition occurs through a series of operations performed on information originating from the senses. The conception of the human as an information-processing system is invaluable for the investigation of human factors issues, because it meets the requirement of allowing human and machine performance to be analyzed in terms of the same basic functions (Proctor & Vu, 2010). As Figure 1.4 shows, both humans and machines perform a sequence of operations on input from the environment that leads to an output of new information. Given this parallel between human and machine systems, it makes sense to organize our knowledge of human performance around the basic information-processing functions.

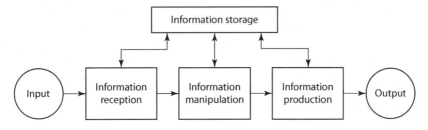

FIGURE 1.4 Information processing in humans and machines.

Wundt and the Study of Attention

The founding of psychology as a distinct discipline usually is dated from the establishment of the first laboratory devoted exclusively to psychological research by Wilhelm Wundt (b1832–d1920) in 1879. Wundt was a prolific researcher and teacher. Consequently, his views largely defined the field of psychology and the study of human performance in the late 1800s.

With respect to contemporary human performance psychology, Wundt's primary legacy is in his scientific philosophy. He promoted a deterministic approach to the study of mental life. He advocated the view that mental events play a causal role in human behavior. Wundt held that our mental representation of the world is a function of experience and the way our mind organizes that experience. Wundt and his students used a wide range of psychophysical and chronometric methods to investigate attentional processes in detail (de Freitas Araujo, 2016).

The topic of attention, as will be seen in Chapter 9, is of central concern for human factors. It attracted the interest of many other researchers in the late 19th and early 20th centuries. Perhaps the most influential views on the topic were those of William James (b1842–d1910). In his classic book *Principles of Psychology*, published in 1890, James devoted an entire chapter to attention. James states:

> Everyone knows what attention is. It is the taking possession by the mind, in clear and vivid form, of one out of what seems several simultaneously possible objects or trains of thought. Focalization, concentration, of consciousness are of its essence. It implies withdrawal from some things in order to deal effectively with others. (James, 1890/1950, pp. 403–404)

This quote captures many of the properties of attention that are still addressed in modern research, such as different types of attention, limitations in processing capacity, and the role of consciousness (see Johnson & Proctor, 2004; Nobre & Kastner, 2014).

Learning and Skill Acquisition

Whereas Wundt and others had tackled the experimental problem of isolating mental events, Hermann Ebbinghaus (b1850–d1909) was the first to apply experimental rigor successfully to the study of learning and memory. This contribution was important, because previously it had been thought that the quantitative study of such higher-level mental processes was not possible (Herbart, 1816/1891). In a lengthy series of experiments conducted on himself, Ebbinghaus (1885/1964) examined his ability to learn and retain lists of nonsense syllables. The procedures developed by Ebbinghaus, the quantitative methods of analysis that he employed, and the theoretical issues that he investigated provided the basis for a scientific investigation of higher mental function.

In a landmark study in the history of human performance, Bryan and Harter (1899) extended the topic of learning and memory to the investigation of skill acquisition. In their study of the learning of Morse code in telegraphy, Bryan and Harter determined many of the factors involved in the acquisition of what would later come to be called perceptual-motor skills. Using procedures and methods of analysis similar to those provided by Ebbinghaus, Bryan and Harter were able to contribute both to our basic understanding of skill learning and to our applied understanding how people learn to use a telegraph. By examining learning curves (plots of performance as a function of amount of practice), Bryan and Harter proposed that learning proceeds in a series of phases. Contemporary models of skill acquisition still rely on this notion of phases (e.g., Anderson, 1983; Anderson et al., 2004; see Chapter 12).

HUMAN PERFORMANCE IN APPLIED SETTINGS

Although we have dated the founding of human factors as a discipline to World War II, considerable applied work important to modern human factors was conducted prior to that time. Much of this work was oriented toward improving job performance and productivity, and at least two journals with a human factors emphasis were published prior to World War II.

Job Specialization and Productivity

Charles W. Babbage (b1792–d1871) wrote the book *Economy of Machinery and Manufactures* in 1832. In that book, he proposed methods for increasing the efficiency with which workers could perform their jobs. In the spirit of the modern factory assembly line, Babbage advocated job specialization, with the idea that this would enable a worker to become skilled and proficient at a limited range of tasks. Babbage also designed and planned two steam-powered machines: the Difference Engine, a special-purpose computer intended to perform differential equations, and the Analytical Engine, the first general-purpose programmable computer. One impetus for his work on computers was the desire to reduce the number of errors in scientific calculations.

We give credit to W. B. Jastrzębowski (1857) for being the first person to use the term *ergonomics*, in an article entitled "An Outline of Ergonomics, or the Science of Work Based upon the Truths Drawn from the Science of Nature" (Karwowski, 2006a; Koradecka, 2006). He distinguished useful work, which helps people, from harmful work, which hurts people, and emphasized the development of useful work practices.

Frederick W. Taylor (b1856–d1915) was one of the first people to systematically investigate human performance in applied settings. Taylor, who was an industrial engineer, examined worker productivity in industrial settings. He conducted one of the earliest human factors studies. As described by Gies (1991), Taylor was concerned that the workers at a steel plant used the same shovel for all shoveling tasks. By making careful scientific observations, he designed several different shovels and several different methods of shoveling that were appropriate for different materials.

Taylor is best remembered for developing a school of thought that is referred to as *scientific management* (Taylor, 1911/1967). He made three contributions to the enhancement of productivity in the workplace. The first contribution is known as *task analysis*, in which the components of a task are determined. One technique of task analysis is *time-and-motion study*. With this technique, a worker's movements are analyzed across time to determine the best way to perform a task. Taylor's second contribution was the concept of *pay for performance*. He suggested a "piecework" method of production, by which the amount of compensation to the worker is a function of the number of pieces completed. Taylor's third contribution involved *personnel selection*, or fitting the worker to the task. While personnel selection is still important, human factors emphasizes fitting the task to the worker. Although many of Taylor's contributions are now viewed as being dehumanizing and exploitative, Taylor's techniques were effective in improving human performance (i.e., increasing productivity). Moreover, time-and-motion study and other methods of task analysis still are used in contemporary human factors.

Frank Bunker Gilbreth (1868–1924) and Lillian Moller Gilbreth (1878–1972) are among the pioneers in the application of systematic analysis to human work. The Gilbreths developed an influential technique for decomposing motions during work into fundamental elements or "therbligs" (Gilbreth & Gilbreth, 1924). Frank Gilbreth's (1909) investigations of bricklaying and of operating-room procedures are two of the earliest and best-known examples of applied time-and-motion study.

Gilbreth, himself a former bricklayer, examined the number of movements made by brick-layers as a function of the locations of the workers' tools, the raw materials, and the structure being built. Similarly, he examined the interactions among members of surgical teams. The changes instituted by Gilbreth for bricklaying resulted in an impressive increase in productiv-ity; his analysis of operating-room procedures led to the development of contemporary surgical protocols. Lillian Gilbreth, who collaborated with Frank until his death, later extended this work on motion analysis to the performance of household tasks and designs for people with disabilities.

Another early contributor to the study of work was the psychologist Hugo Münsterberg (b1863–d1916). Though trained as an experimental psychologist, Münsterberg spent much of his career as an applied psychologist. In the book *Psychology and Industrial Efficiency*, Münsterberg (1913) examined work efficiency, personnel selection, and marketing techniques, among other topics.

Personnel selection grew from work on individual differences in abilities developed during World War I, with the use of intelligence tests to select personnel. Subsequently, many other tests of perfor-mance, aptitudes, interests, and so on were developed to select personnel to operate machines and to perform other jobs. While personnel selection can increase the quality of system performance, there are limits to how much performance can be improved through selection alone. A poorly designed system will not perform well even if the best personnel are selected. Thus, a major impetus to the development of the discipline of human factors was the need to improve the design of systems for human use.

Early Human Factors Journals

Two applied journals were published for brief periods in the first half of the 20th century that foreshadowed the future development of human factors. A journal called *Human Engineering* was published in the U.S. for four issues in 1911 and one issue in 1912 (Ross & Aines, 1960). In the first issue, the editor, Winthrop Talbot, described human engineering as follows:

> Its work is the study of physical and mental bases of efficiency in industry. Its purpose is to promote efficiency, not of machines but men and women, to decrease waste—especially human energy—and to discover and remove causes of avoidable and preventable friction, irritation or injury. (Quoted by Ross & Aines, 1960, p. 169)

From 1927 to 1937, a journal called *The Human Factor* was published by the National Institute of Industrial Psychology in England. The content and methodology of the journal are broader than those later associated with the field of human factors, with many of the articles focusing on voca-tional guidance and intelligence testing. However, the journal also included articles covering a range of issues in what were to become core areas of human factors. For example, the 1935 volume con-tained articles titled "The Psychology of Accidents," "A Note on Lighting for Inspection," "Attention Problems in the Judging of Literary Competitions," and "An Investigation in an Assembly Shop." The journal published a transcript of a radio speech given by Julian Huxley (b1887–d1975), the renowned biologist and popularizer of science. In this speech, Huxley differentiated what he called Industrial Health from Industrial Psychology:

> [Industrial Psychology], however, is something broader [than Industrial Health]. It, too, is dealing with the human factor in industry, but instead of dealing primarily with industrial disease and the prevention of ill-health, it sets itself the more positive task of finding out how to promote efficiency in all ways other than technical improvement of machinery and processes. To do this, it all the time stresses the necessity of not thinking of work in purely mechanical terms, but in terms of co-operation between a machine and a human organism. The machine works mechanically; the human organism does not, but has its own quite different ways of working, its own feelings, its fears and its ideals, which also must be studied if the co-operation is to be fruitful. (Huxley, 1934, p. 84)

Biomechanics and Physiology of Human Performance

Human performance also has been studied from the perspective of biomechanics and physiology (Oatis, 2016). The biomechanical analysis of human performance has its roots in the early theoretical work of Galileo and Newton, who helped to establish the laws of physics and mechanics. Giovanni Alphonso Borelli (b1608–d1678), a student of Galileo, brought together the disciplines of mathematics, physics, and anatomy in one of the earliest works on the mechanics of human performance (Borelli, 1679/1989).

Probably the most important contribution to biomechanical analysis in the area of work efficiency was by Jules Amar (b1879–d1935). In his book *The Human Motor*, Amar (1920) provided a comprehensive synthesis of the physiological and biomechanical principles related to industrial work. Amar's research initiated investigations into the application of biomechanical principles to work performance. The ideas of Amar and others were adopted by and applied to the emerging field of human factors.

Another major accomplishment was the development of procedures that allowed a dynamic assessment of human performance. In the latter part of the 19th century, Eadweard Muybridge (b1830–d1904) constructed an apparatus comprised of banks of cameras that allowed him to take pictures of animals and humans in action (e.g., Muybridge, 1955). Each series of pictures captured the biomechanical characteristics of complex action (see Figure 1.5). The pictures also could be viewed at a rapid presentation rate, with the result being a simulation of the actual movement.

Muybridge's work opened the door for a range of biomechanical analyses of dynamic human performances. In particular, the physiologist Etienne-Jules Marey (b1830–d1904) exploited related photographic techniques to decompose time and motion (Marey, 1902). Today, such analyses involve videotaping performances of human action that can then be evaluated. Modern camera-based systems, such as OPTOTRAK CERTUS® (see Figure 1.6), track small infrared sensors attached to a person's body to analyze movement kinematics in three dimensions.

FIGURE 1.5 A clothed man digging with a pickaxe. Photogravure after Eadweard Muybridge, 1887. By: Eadweard Muybridge and University of Pennsylvania.

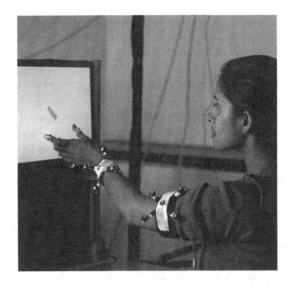

FIGURE 1.6 The Optotrak system for recording movement trajectories. The cameras pick up infrared light from infrared light-emitting diodes (IREDs) attached to the human subject. Computer analysis provides a three-dimensional account of each IRED at each sampling interval.

SUMMARY

A great deal of research conducted prior to the middle of the 20th century laid the foundation for the field of human factors. Psychologists developed research methods and theoretical views that allowed them to investigate various aspects of human performance; industrial engineers studied many aspects of human performance in job settings with an eye toward maximizing its efficiency; biomechanists and physiologists developed methods for examining physical and biological factors in human performance and principles relating those factors to work. Our coverage of these developments has of necessity been brief, but the important point is simply that without the prior work in these areas, human factors specialists would have had no starting point to address the applied design issues that became prominent in the latter half of the 20th century.

EMERGENCE OF THE HUMAN FACTORS PROFESSION

Although interest in basic human performance and applied human factors goes back to before the turn of the 20th century, a trend toward systematic investigation of human factors did not begin in earnest until the 1940s (Meister, 2006a). The technological advances brought about by World War II created a need for more practical research from the academic community. Moreover, basic research psychologists became involved in applied projects along with industrial and communications engineers. By the close of the war, psychologists were collaborating with engineers on the design of aircraft cockpits, radar scopes, and underwater sound detection devices, among other things.

Among the most significant developments for human factors was the founding in 1944 of the Medical Research Council Applied Psychology Unit in Great Britain (Reynolds & Tansey, 2003) and in 1945 of the Psychology Branch of the Aero Medical Laboratory at Wright Field in the U.S. The first director of the Applied Psychology Unit was Kenneth Craik (b1914–d1945), a leader in the use of computers to model human information processing, whose contributions to the field were cut short by his untimely death at age 31. The founding Head of the Psychology Branch, Paul M. Fitts (b1912–d1965), was a central figure in the development of human factors, who left a lasting legacy in many areas of research. Human factors and ergonomics remain prominent at Wright-Patterson Air Force Base in the 711th Human Performance Wing, particularly its Human Systems Integration

Directorate. Also, in 1946, Ross McFarland (1946) published *Human Factors in Air Transport Design*, the first book of which we are aware to use the term *human factors*. Around this period, some industries also began to research human factors, with Bell Laboratories establishing a laboratory devoted specifically to human factors in the late 1940s. The interdisciplinary efforts that were stimulated during and immediately after the years of the war provided the basis for the development of the human factors profession.

The year 1949 marked the publication of Chapanis et al.'s first general textbook on human factors, *Applied Experimental Psychology: Human Factors in Engineering Design*, from which we quoted to begin this book. Perhaps more important, the profession was formalized in England with the founding of the Human Research Group in 1949 (Stammers, 2006). In 1950, the group changed its name to the Ergonomics Research Society, subsequently shortened to the Ergonomics Society, and the term *ergonomics* came to be used in the European community to characterize the study of human–machine interactions. This term was intentionally chosen over the term *human engineering*, which was then popular in the U.S., because human engineering was associated primarily with the design-related activities of psychologists. According to Murrell (1969, p. 691), one of the founders of the Ergonomics Research Society, "In contradistinction the activities which were envisaged for the infant society would cover a much wider area, to embrace a broad spectrum of interests such as those of work physiology and gerontology." Reflecting the internationalization of human factors terminology and work, the Society changed its name again in 2009 to the Institute of Ergonomics and Human Factors, by which it is currently known.

Several years later, in 1957, the Ergonomics Society began publication of the first journal devoted to human factors, *Ergonomics*. In that same year, the Human Factors Society (changed in 1992 to the Human Factors and Ergonomics Society) was formed in the U.S., and the publication of their journal, *Human Factors*, began 1 year later. Also, the American Psychological Association established Division 21, Engineering Psychology, and in 1959, the International Ergonomics Association, a federation of human factors and ergonomics societies from around the world, was established.

CONTEMPORARY HUMAN FACTORS

From 1960 to 2000, the profession of human factors grew immensely. As one indicator of this growth, the membership of the Human Factors and Ergonomics Society increased from a few hundred people to more than 4500 by the late 1980s, which is approximately the current membership. The range of topics and issues investigated has grown as well.

Table 1.1 shows the composition of the Technical Groups of the Human Factors and Ergonomics Society in 2017. The topics covered by these groups indicate the broad range of issues now addressed by human factors specialists. Professional societies have been established in many other countries, and more specialized societies have developed. HCI alone is the focus of the Association for Computing Machinery's Special Interest Group on Computer–Human Interaction, the Software Psychology Society, and the IEEE Technical Committee on Computer and Display Ergonomics, as well as many others. Outside of computer science, the rapid growth of technology has made the human factors profession a key component in the development and design of equipment and machinery.

The close ties between human factors and the military have persisted since the years of World War II. The U.S. military incorporates human factors analyses into the design and evaluation of all military systems. All branches of the military have human factors research programs. These programs are administered by the Air Force Office of Scientific Research, the Office of Naval Research, and the Army Research Institute, among others.

Additionally, the military branches have programs to ensure that human factors principles are incorporated into the development of weapons and other military systems and equipment. These programs operated independently until 2009, when, as part of the National Defense Authorization

TABLE 1.1

Technical Groups of the Human Factors Society

Aerospace systems: Application of human factors to the development, design, certification, operation, and maintenance of human–machine systems in aviation and space environments.

Aging: Concerned with human factors appropriate to meeting the emerging needs of older people and special populations in a wide variety of life settings.

Augmented cognition: Concerned with fostering the development and application of real-time physiological and neurophysiological sensing technologies that can ascertain a human's cognitive state while interacting with computing-based systems … [to] enable efficient and effective system adaptation based on a user's dynamically changing cognitive state.

Children's issues: Consists of researchers, practitioners, manufacturers, policy makers, caregivers, and students interested in research, design, and application concerning human factors and ergonomics (HF/E) issues related to children's emerging development from birth to 18.

Cognitive engineering and decision making: Encourages research on human cognition and decision making and the application of this knowledge to the design of systems and training programs.

Communications: Concerned with all aspects of human-to-human communication, with special emphasis on communication mediated by technology.

Computer systems: Concerned with human factors in the design of computer systems. This includes the user-centered design of hardware, software, applications, documentation, work activities, and the work environment.

Education: Concerned with the education and training of human factors and ergonomics specialists.

Environmental design: Concerned with the relationship between human behavior and the designed environment … [including] ergonomics and macroergonomics aspects of design within home, office, and industrial environments.

Forensics: Application of human factors knowledge and techniques to "standards of care" and accountability established within the legislative, regulatory, and judicial systems.

Health care: Maximizing the contribution of human factors and ergonomics to medical system effectiveness and the quality of life of people who are functionally impaired.

Human performance modeling: Focuses on the development and application of predictive, reliable, and executable quantitative models of human performance.

Individual differences in performance: Interest in any of the wide range of personality and individual difference variables that are believed to mediate performance.

Internet: Interest in Internet technologies and related behavioral phenomena.

Macroergonomics: Focuses on organizational design and management issues in human factors and ergonomics as well as work system design and human–organization interface technology.

Occupational ergonomics: Application of ergonomics data and principles for improving safety, productivity, and quality of work in industry.

Perception and performance: Promotes the exchange of information concerning perception and its relation to human performance.

Product design: Dedicated to developing consumer products that are useful, usable, safe, and desirable … by applying the methods of human factors, consumer research, and industrial design.

Safety: Development and application of human factors technology as it relates to safety in all settings and attendant populations.

Surface transportation: Information, methodologies, and ideas related to the international surface transportation field.

System development: Integration of human factors/ergonomics into the development of systems.

Test and evaluation: All aspects of human factors and ergonomics as applied to the evaluation of systems.

Training: Information and interchange among people interested in training and training research.

Virtual environments: Human factors issues associated with human–virtual environment interaction.

Act, the U.S. Department of Defense was asked to establish the Human Systems Integration (HSI) initiative,

focused on the role of the human in the Department of Defense (DoD) acquisition process. The objective of HSI is to provide equal consideration of the human along with the hardware and software in the technical and technical management processes for engineering a system that will optimize total system performance and minimize total ownership costs. (Office of the Assistant Secretary of Defense, 2015)

The result of this initiative has been the *FY011 Department of Defense Human Systems Integration Management Plan* (2011), which distinguishes eight elements, including *Human Factors Engineering, Personnel, Manpower, Training, and Safety.*

The value of human factors analyses is also apparent in our everyday lives. For example, the automotive industry has devoted considerable attention to human factors in the design of automobiles (Gkikas, 2013). This attention has extended from the design of the automobile itself to the machinery used to make it. Similarly, modern office furniture has benefited significantly from human factors evaluations. Still, we often encounter equipment that is poorly designed for human use. The need for good human factors has become sufficiently obvious that manufacturers and advertising agencies now realize that it can become a selling point for their products. The makers of automobiles, furniture, ball-point pens, and the like advertise their products in terms of the specific ergonomic advantages that their products have over those of their competitors. If the present is any indication, the role of human factors will only increase in the future.

Standard human factors principles apply in space as well as on earth. For example, the factors contributing to a collision of the Russian supply spacecraft *Progress 234* with the *Mir* space station in 1997 included poor visual displays and operator fatigue resulting from sleep deprivation (Ellis, 2000). In addition, the unique conditions of extraterrestrial environments pose new constraints (Lewis, 1990). For example, in microgravity environments, a person's face will tend to become puffy. Furthermore, a person can be viewed from many more orientations (e.g., upside-down). Consequently, perception of the nonlinguistic cues provided by facial expressions likely is impaired, compromising face-to-face communication (Cohen, 2000). In recognition of the need to consider human factors in the design of space equipment, NASA published the first human factors design guide in 1987, updated and expanded in 2010 as the *Human Integration Design Handbook*, to be used by developers and designers to promote the integration of humans and equipment in space.

The plan to construct the International Space Station began in 1984. The U.S. committed NASA to developing the station in conjunction with space programs from other countries, including the European Space Agency and the Canadian Space Agency, among others. NASA quickly acknowledged the need to incorporate human factors into the design of all aspects of the space station (Space Station Human Productivity Study, 1985), and human factors engineering was granted an equal status with other disciplines in the system development process (Fitts, 2000). Human factors research and engineering played a role in designing the station to optimize the crew's quality of life, or habitability (Wise, 1986), as well as to optimize the crew's quality of work, or productivity (Gillan, Burns, Nicodemus, & Smith, 1986). Issues of concern included, for example, design of user interfaces for the equipment used by the astronauts to conduct scientific experiments (Neerincx, Ruijsendaal, & Wolff, 2001). Additional factors that have been investigated include how to achieve effective performance of multicultural crews (Kring, 2001), the dynamics of crew tension during the mission and their influence on performance (Sandal, 2001), and the effects of psychological and physiological adaptation on crew performance in emergency situations (Smart, 2001). Moving beyond the International Space Station to the extended duration that will be required for spaceflights to Mars and back, additional human factors issues emerge (Schneider et al., 2013).

SUMMARY

From its beginnings in the latter part of the 1940s, the field of human factors has flourished. Building on a scientific foundation of human performance from many disciplines, the field has developed from an initial focus primarily on military problems to concern with the design and evaluation of a broad range of simple and complex systems and products for human use. The importance of the contribution of human factors to industry, engineering, psychology, and the military cannot be overemphasized. When design decisions are made, failure to consider human factors can lead to waste of personnel and money, injury and discomfort, and loss of life. Consequently, consideration of human factors concerns at all phases of system development is of utmost importance.

In the remainder of the book, we elaborate on many of the themes introduced in this chapter. We will cover both the science of human factors, which focuses on establishing principles and guidelines through empirical research, and the profession, which emphasizes application and evaluation for specific design problems. We will describe the many types of research and knowledge that are of value to the human factors specialist and the many specific techniques that are available for predicting and evaluating many aspects of human factors.

RECOMMENDED READINGS

Casey, S. (1998). *Set Phasers on Stun: And Other True Tales of Design, Technology, and Human Error* (2nd ed.). Santa Barbara, CA: Aegean Publishing Company.

Chiles, J. R. (2001). *Inviting Disaster: Lessons from the Edge of Technology.* New York: Harper Business.

Norman, D. A. (2013). *The Design of Everyday Things* (revised and expanded edition). New York: Basic Books.

Perrow, C. (1999). *Normal Accidents* (updated edition). Princeton, NJ: Princeton University Press.

Rabinbach, A. (1990). *The Human Motor: Energy, Fatigue, and the Origins of Modernity.* New York: Basic Books.

Salas, E. & Maurino, D. (2010). *Human Factors in Aviation* (2nd ed.). San Diego, CA: Academic Press.

Vaughan, D. (1997). *The Challenger Launch Decision: Risky Technology, Culture, and Deviance at NASA.* Chicago, IL: University of Chicago Press.

Vicente, K. (2006). *The Human Factor: Revolutionizing the Way People Live with Technology.* New York: Routledge.

2 Research Methods in Human Factors

It may be said, fairly enough, that science progresses by the exposure of error and that in so far as an endeavor is scientific it is as ready to look for error within its own contentions as in those opposing it. In particular, it has to be stressed that observation, which plays so special a role in science, is not regarded as error-free.

W. M. O'Neil
1957

INTRODUCTION

In their book *Physics, the Human Adventure*, Holton and Brush (2000) state, "By far the largest part of the total research and development effort in science and engineering today is concerned, indirectly or directly, with human needs, relationships, health, and comforts" (p. 49). When we view science in this light, we see that human factors has a central place in contemporary science and engineering efforts. You can verify this by entering "role of human factors" into a Web search engine, which yields many entries covering aviation safety, healthcare, quality improvement, production networks, and scuba diving incidents, among others.

Human factors is an applied science. It relies on measurement of behavioral and physical variables in settings ranging from the laboratory to working human–machine systems. The human factors researcher must know the methods of science in general and the specific research methods that are available for conducting human factors research. The applied human factors specialist likewise must understand these methods, and their strengths and limitations, to be an effective consumer of available information and to be able to make wise decisions at all phases of the system development process.

Because it is an applied science, human factors involves a mix of basic and applied research. Basic and applied research can be classified using a 2×2 array, as shown in Figure 2.1, which was popularized by Stokes (1997). The rows identify a *Quest for Fundamental Understanding* (yes or no) and the columns *Considerations of Use* (yes or no; Stokes, 1997). The primary goal of basic research is to increase foundational knowledge on some topic, for example, attention, with no specific application of that knowledge in mind. In contrast, the primary goal of applied research is to solve practical real-world problems. The findings from basic research increase our scientific understanding, but perhaps with no obvious link to application, whereas the findings from applied research provide solutions to practical problems, but perhaps with little increase in scientific understanding. The emphasis on specific problems in applied research restricts its contributions to those problems and to existing technology. However, because new technologies continually arise to make older technologies obsolete, new problems continually crop up. This means that we need basic research that generates knowledge transcending particular applications so we can address new problems effectively as they occur.

Consistent with this point, the influence of basic and applied research on system development happens at very different times (Adams, 1972). If we look at the important research events leading to a system innovation, the immediately preceding events come primarily from applied research, whereas the longer-term contributions arise from basic research. In other words, basic research provides a foundation of conceptual and methodological tools that can subsequently be used to resolve specific applied problems.

Considerations of use?

	Yes	No
Yes	Use-inspired basic research	Pure basic research
No	Pure applied research	

Quest for fundamental understanding?

FIGURE 2.1 Two dimensions of scientific research

Consider, for example, human attention (see Chapter 9). Contemporary research on attention is usually dated to experiments by Mackworth (1948; see Murray, 1974), who worked with airborne operators during World War II. These operators were required to monitor radar screens for hours at a time while searching for enemy submarines. Tasks like this are called *vigilance tasks.* Basic research on attention initiated by the issue of vigilance and related applied problems led to the development of theories that conceptualized attention as one or more limited-capacity resources (see Chapter 9), which resulted in the development of methods to measure mental workload that are widely used today in human factors (Young, Brookhuis, Wickens, & Hancock, 2015).

Applied research identifies issues of human performance that need to be addressed in a particular setting and provides a criterion for meaningful research. Again, using attention as an example, the scientific study of attention has benefited considerably from investigations of applied problems such as those involved in display design. This interplay between basic and applied research is the foundation of many sciences, including human factors. As Alan Baddeley, a noted memory researcher, said, "Sometimes choosing what appears to be a practical problem actually can have considerable theoretical impact" (quoted in Reynolds & Tansey, 2003, p. 48), and vice versa.

In addition to pure basic and applied research, Stokes' (1997) taxonomy includes a third type of research, called *use-inspired basic research* (Stokes, 1997). This research is driven by both considerations of use and a quest for fundamental understanding. This type of research is particularly productive because it involves conducting basic research in the pursuit of applied problems (Gallaway, 2007).

Use-inspired basic research is valuable in human factors, but it is difficult to conduct. Because so many different researchers collaborate in a system design, problems of communication can result from different ways of talking about and solving problems, as well as conflict from different goals. However, when successful, as in the case of the work of Thomas Landauer and colleagues on Latent Semantic Analysis (see Chapter 10), the theoretical and applied contributions can be profound (Evangelopoulos, 2013), demonstrating that "fundamental research and work on solutions to practical human problems can live together in interesting and useful ways" (Streeter, Laham, Dumais, & Rothkopf, 2005, p. 2). The point to remember is that pure basic, use-inspired basic, and pure applied research are complementary, and all contribute to the discipline of human factors (Proctor & Vu, 2011).

An understanding of human factors research requires an understanding of scientific methodology, research methods, and measurement. The purpose of this chapter is to outline the primary features of scientific methodology and to present the general research and statistical methods used in the investigation of human performance. We assume that if you have not already taken formal courses in research methods and statistics, then you will be taking them shortly. The outline we present in this chapter reviews the essential tools you need to formulate and critically evaluate human factors studies. It also provides different perspectives on how to think about problems unique to human factors. Techniques specific to particular areas of human factors, discussed in later chapters, build from the concepts introduced in this chapter.

DISTINGUISHING FEATURES OF SCIENCE

The definition of human factors provided in Chapter 1 emphasized that it is a scientific discipline. Consequently, to understand the field, you need to appreciate what it means to take a scientific approach. But what is science? Any definition of science will always fall short, because science is not a thing. Rather, it is a process—a way of learning about the world. This process involves making informal observations, forming alternative hypotheses about the nature of things, and testing these hypotheses empirically.

Of course, science is not the only way to learn about the world. For example, as a student of human factors, you might undertake the task of designing a keyboard entry device for a new computer system. In designing the keyboard, you could appeal to a number of sources for information to determine the most effective design. You could consult an established authority, perhaps the instructor of your course. Or, you could examine various keyboards already available to determine the traditional wisdom. You might even design the keyboard on the basis of your personal experience and ideas about an optimal design. Each of these approaches can provide valuable information. If you were to take a scientific approach, however, these approaches would serve as the starting points of your search, rather than as ends in and of themselves.

Why are sources of information like authority, tradition, or personal experience insufficient for designing a keyboard entry device? Imagine that a fellow classmate has been given the same assignment. Although you both may consult the same sources of information to complete the project, you may not interpret each of these sources in the same way. Therefore, it is quite likely that you would arrive at different keyboard designs.

You then have to decide which keyboard is best. This is where the methods of science take over. Science provides systematic ways to resolve this question. In fact, using a scientific approach, not only could you test to determine which keyboard is best and whether either is better than existing keyboards, but you could also discover those specific attributes that make one design better than another and why they do. You could both resolve this specific design issue and make a contribution to understanding human factors in keyboard design.

FOUNDATIONS OF SCIENCE

Science is based on *empiricism*. Empiricism means pursuing knowledge by observation. This observation can range from uncontrolled, direct observations within natural settings, to tightly controlled experiments in artificial settings. For example, if we are interested in the performance of operators in the control room of a nuclear power plant, we can record and analyze their activities during work, conduct specific exercises on a simulator, test the operators' ability to identify alternative displays, and so on. The key point behind the principle of empiricism is that statements are evaluated on the basis of observable events. Thus, science provides objective criteria for evaluating the truth value of alternative statements.

Science is distinguished from other ways of acquiring knowledge because it is self-correcting. Empiricism provides the mechanism for self-correction: We continually test our scientific statements with observations. When reliable observations deviate systematically from our predictions and explanations, we revise the scientific statements. Thus, the observations provide feedback that allows correction of error, as O'Neil (1957) emphasizes in the quote with which this chapter begins. Science therefore operates as a closed-loop system of the type described in Chapter 3.

The self-correcting characteristic of science ensures that new knowledge will be dependable and will help advance our understanding of the world. Consequently, scientists accept any statement tentatively, with the degree of acceptance of a particular scientific statement being a direct function of the amount of evidence in support of it. They constantly test the validity of scientific statements, and such tests are open to public observation and scrutiny. These self-correcting aspects of science are embodied in what is called the *scientific method*. What scientists do for the most part, then, is to systematically apply the scientific method.

SCIENTIFIC METHOD

The scientific method is a logical approach to obtaining answers to questions. This approach is often equated with the steps by which hypotheses are generated and tested, beginning from general observations in the world and ending with a detailed, documented statement of the factors that give rise to observed phenomena. Figure 2.2 shows the steps involved in hypothesis testing.

The scientific enterprise begins with curiosity about the cause of some observed phenomenon. For example, you might wonder why people who use cell phones when they drive tend to get into more accidents. To answer this question, you need to phrase the question in a way that allows it to become a problem that can be investigated. Often this involves deciding which behavioral measures will reflect the problem. For example, the time it takes a driver to brake in response to an obstacle in the road might be a behavioral measure that reflects the amount of attention devoted to driving. High levels of attention might lead to fast braking times, and low levels of attention might lead to slower braking times. You might suspect that cell phone use reduces the amount of attention devoted to driving, increasing the time it takes for drivers to respond to changing road conditions. This is a *hypothesis*, a tentative causal statement about the relations among the factors involved in the occurrence of the phenomenon.

The hypothesis serves as the statement that is to be assessed by research. Once you have formulated a hypothesis, you can begin to make other observations to test the hypothesis, compare it with alternative hypotheses, and increase your understanding of the phenomenon. For example, if attentional factors are responsible for accidents during cell phone use, then maybe other kinds of distractions also lead to accidents. You could perform experiments comparing response times during cell phone use with response times during conversations with passengers. You will continually refine and modify your hypothesis, or even reject it, based on the results of your experiments, until you have a complete understanding of the phenomenon.

Note that the hypothesis is tested by conducting experiments designed to confirm or disconfirm its predictions. We do not test the hypothesis directly; rather, we test the relations between measurable and observable variables predicted by the hypothesis. The viability of the hypothesis is determined by how appropriately it captures the relations among the factors of interest compared with alternative hypotheses. For example, you might compare the hypothesis that lack of attention during cell phone use causes accidents with a hypothesis that states that one-handed driving during cell phone use causes accidents. You could simultaneously test both the attentional and the one-handed driving hypothesis in experiments to determine which hypothesis makes better predictions. As always, you will then apply the information about which hypothesis is better back to the original problem.

Finally, the last step of the scientific method is telling other researchers what you have learned. The hypotheses you tested and the data you collected, together with your interpretation of the results, must be written up and distributed to the scientific community. You do this with conference presentations, journal articles, technical reports, and books. It is at this point that your new information

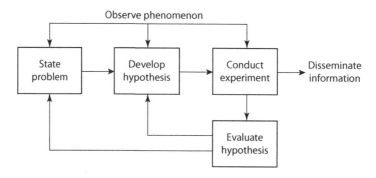

FIGURE 2.2 Steps in hypothesis testing.

becomes part of the scientific knowledge base. Without this last step, the most important characteristic of the scientific method would be missing: the characteristic of self-correction. It is possible that you made a mistake somewhere, maybe in the design of your experiments or how you analyzed your data, so that your conclusions are not valid. Making your work available to the entire scientific community lets other researchers examine it closely, attempt to replicate it, and come up with their own contributions to the problem. If you made a mistake somewhere, they will find it and correct it.

Our depiction of the hypothesis testing process is by necessity oversimplified. A hypothesis is not tested in isolation. Many other factors, some known and some unknown, also act to influence the results of an experiment (see Proctor & Capaldi, 2006). For example, a study may be conducted in more than one room, and slight differences in the way that light falls on the materials or in the placement of equipment around the testing area could affect the scientist's measurements. Or, the measuring instrument may not be appropriate for testing the hypothesis, it may not be sensitive enough, or it may be improperly calibrated. Therefore, failure to confirm a hypothesis may be due to one or more unknown factors, rather than to the inadequacy of the hypothesis itself. Also, the predictions drawn from the hypothesis may not be valid. It is possible that by emphasizing some originally neglected aspect of the hypothesis, a somewhat different correct prediction may follow. So, even when an experiment fails to support a hypothesis, rejecting that hypothesis immediately is not always the best thing to do. A lot of good science has resulted when researchers have resolutely held on to a supposedly disconfirmed hypothesis.

GOALS OF SCIENCE

Because of the emphasis on data collection, it is easy to assume that science is basically a fact-gathering activity. However, this is far from the case. The goals of science are explanation, prediction, and control. The vehicle for achieving these goals is *theory*. According to Kerlinger and Lee (2000, p. 11),

> A theory is a set of interrelated constructs (concepts), definitions, and propositions that present a systematic view of phenomena by specifying relations among variables, with the purpose of explaining and predicting the phenomena.

A scientific theory about a particular problem is closely related to the empirical evidence collected on that problem. The theory that we accept at any point in time will be the one that provides the best explanation of existing findings (Haig, 2009; Koslowski, 2012). Thus, whereas the layperson tends to conceive of a theory as an armchair speculation, a scientific theory provides instead a detailed, specific organization of existing knowledge.

Good theories explain findings that previously resisted explanation (Brush, 1989). However, not only do good theories provide an explanation of established data, but they must also produce new predictions that can be tested empirically. These new predictions advance science further. Finally, the theory states precisely which factors are important for a particular problem and what the relationships are between those factors. Consequently, it allows control over the phenomenon by giving an understanding of the conditions under which it occurs.

Theory is beneficial not only to the basic researcher who seeks to understand the underlying nature of things, but also to the applied human factors researcher and practitioner. Theory offers at least four benefits to the practitioner (Kantowitz, 1989, p. 1060). It

1. Enables sensible interpolation to a specific real-world problem when there are no data;
2. Provides quantitative predictions of the type desired by engineers and designers;
3. Allows the practitioner to recognize relations between problems that seem unrelated on the surface; and
4. Can be used cheaply and efficiently to aid system design.

Thus, an understanding of human performance in terms of theoretical concepts is essential for the effective application of existing knowledge to human factors problems.

Most important, science is an approach to thinking about problems and acquiring knowledge. This point is captured clearly by the late, well-known astronomer and popularizer of science, Carl Edward Sagan (b1934–d1996), who said:

> Science is much more than a body of knowledge. *It is a way of thinking.* This is central to its success. Science invites us to let the facts in even when they don't conform to our preconceptions. It counsels us to carry alternative hypotheses in our heads and see which ones best match the facts. It urges on us a fine balance between no-holds-barred openness to new ideas, however heretical, and the most rigorous skeptical scrutiny of everything—new ideas *and* established wisdom. *We need wide appreciation of this kind of thinking. It works.* (Sagan, 1990, p. 265; italics ours)

MEASUREMENT

The problem that leads a researcher to conduct a study has two consequences. First, it defines a domain of interest. In human factors, the domain of interest is usually human performance within a system context. Within this domain, we *operationally define* the objects or events under study in terms of the physical characteristics that we intend to measure. For example, in the preceding section, we operationally defined an increase in attention as a reduction in braking time during driving. Second, it defines the conditions under which we can make useful measurements. If we are interested in the effects of cell phone use on driving performance, we will need to observe people using cell phones in realistic driving conditions or simulated driving scenarios, not people listening to radios while mowing the lawn.

Because a person's performance and the conditions under which we observe it can vary, we define the conditions and performance measures collectively as *variables*. That is, a variable is any event that can change, whether we change it ourselves or whether we observe a change caused by something else. We refine the initial question into a researchable form; we establish the primary variables of interest and begin to structure the types of research designs that we will use.

There are several ways to classify variables. Most research in human factors involves the measurement of behavioral variables. *Behavioral variables* are any overt, observable behaviors. They can range from simple key presses to the complex responses required to pilot a jet aircraft. Behavior never occurs in isolation, but always in the context of a collection of stimulus events. Stimuli that can have an effect on the behavior of an organism are referred to as *stimulus variables*. Stimulus variables can range from a simple buzzer indicating a desired response, to the complex auditory and visual messages received by an air-traffic controller. Most research in human factors is concerned with the influence of stimulus variables on behavioral variables.

Another way to classify variables is in terms of whether they are manipulated or measured. An *independent variable* is one that is manipulated by the researcher. Most often, the manipulation is made for stimulus variables, such as the level of illumination. We manipulate independent variables to determine their effects on other variables, which are referred to as *dependent variables*. Dependent variables usually are behavioral variables. These are sometimes called *criterion variables* (Sanders & McCormick, 1993) and can be grouped into measures that reflect performance (such as speed and force of responding), physiological indexes (such as heart rate and EEG recordings), or subjective responses (such as preferences and estimates of effort). The distinction between independent and dependent variables forms a cornerstone of the true experiment, because it allows us to establish causal relations.

In nonexperimental or descriptive research, we can make no distinction between independent and dependent variables. We can make only tentative statements about causal relations. For example, uncontrolled observation of consumer behavior in a grocery store is nonexperimental. We do not identify or manipulate independent variables, and we might not measure a dependent variable either. In other kinds of descriptive research (in particular, differential research, which attempts to understand how people differ from each other), we can label independent and dependent variables. However, in differential research, the independent variable is a subject variable, that is, a property of the subjects, such as height, weight, gender, age, and personality classification. We cannot manipulate subject variables. For example, we can't take a group of children and raise half of them with healthy diets and half with unhealthy diets; that would be unethical. We can, however, find children with healthy and unhealthy diets and measure differences between them. Because we can't manipulate or control subject variables, we can't assign cause and effect relationships to the different variables. We will talk more about descriptive research later in this chapter.

RELIABILITY, VALIDITY, AND CONTROL

Two of the most important concepts in human factors and ergonomics research are those of reliability and validity (Kanis, 2014). *Reliability* refers to the consistency of measurements. Measurements are said to be reliable if we get similar values when we measure the same thing more than once. Experimental results are reliable if we can replicate them under similar conditions. For example, if you give a test to the same group of people at two different times, the test is said to have high "test-retest" reliability if the scores for each person are similar for the two administrations of the test. Another way to think of reliability is that any measure, call it $X_{observed}$, has two parts. One part is "true" (X_{true}) and the other part is random error. The two parts are added together to give the final measure: $X_{observed} = X_{true} + error$. The larger the true part is relative to the error part, the higher the reliability.

Validity refers to the degree to which an experiment, a procedure, or a measurement does what it is supposed to do. However, the idea of validity is complex, and we can talk about validity in many ways (see, for example, Jacko, Yi, Sainfort, & McClelland, 2012). A variable is said to have high *face validity* if it seems on the surface to measure what it is supposed to be measuring. For example, measuring the number of mistakes that a pilot makes during simulated flight of a new aircraft to evaluate the quality of that aircraft's cockpit design has high face validity. A variable is said to have *construct validity* if it is a true measure of the construct it represents. For example, the time to perform a task decreases systematically with practice, indicating that response time is a valid measure of learning. One task of a researcher is to optimize validity.

Three types of validity are particularly important for human factors research: ecological, internal, and external validity. A research setting is said to have high *ecological validity* when it closely resembles the situations and tasks of concern in everyday life and low ecological validity when it does not. As an example, you might be studying the effects of attention on driving performance. You could design an experiment in which road-like conditions are presented on a computer screen and the driver performs by pressing keys on the computer keyboard. This experiment would have low ecological validity. Or, you could design an experiment in a simulator that had every appearance of a real car, and the driver could perform real driving tasks. This experiment would have higher ecological validity. Applied human factors research typically strives for relatively high ecological validity, whereas basic research on human performance often does not.

A study is said to have high *internal validity* if the relations observed in it can be attributed with a high degree of confidence to the variables of interest and low internal validity if they cannot. Typically, laboratory experiments yield the highest level of internal validity because of the level of control we can exert on the variables of interest. Ecological validity is often sacrificed for internal validity. It is easier to obtain internal validity in a low ecologically valid setting than in a high ecologically valid setting. This is because settings with high ecological validity often have more uncontrolled situational variables, such as the number of elements in the visual field, the actions available to the performer, and so forth.

A study is said to have high *external validity* if its results, or the principles derived from the results, can be generalized to a variety of other settings, and low external validity if they cannot. It is important to understand that high ecological validity does not ensure high external validity; nor does high internal validity necessarily mean that the findings cannot also be high in external validity (Fincannon, Keebler, & Jentsch, 2014). That is, even research conducted in ecologically valid settings may yield results that are only obtained in the same setting, and research conducted in the lab can yield principles that can be generalized broadly.

Remember that empirical observations are the central facet of scientific research. In human factors, these typically are observations of behavior. This observed behavior can be influenced by a multitude of variables, not all of which are known to the researcher. Moreover, some of the variables may be of interest to the researcher, whereas others are extraneous. If the extraneous variables have a sufficiently great effect, they can cloud the effects of the variables of interest. For example, investigating the performance of workers under new arrangements of assembly line stations may be confounded by wage reductions instituted at the time the study began. If productivity decreases after we rearrange the line, is this because of the changes we made or is it because the workers are angry about their cut in pay? In short, extraneous variables threaten the validity of the research, because the observed effects may not be due to what we think. Consequently, an essential aspect of research is the reduction of the influence of these variables.

Control procedures are the systematic methods that a researcher uses to reduce the influence of extraneous variables that threaten the validity of the research (Proctor, Capaldi, & Vu, 2003). It is important to realize that extraneous variables may exert their influence on both the subjects and the researcher. For instance, the way that the experimenter records and classifies behavioral events can be affected by her bias, or how she wants the experiment to turn out. If the research is to be internally valid, the measurement of the dependent variable must be an accurate reflection of the influence of the independent variable.

RESEARCH METHODS

Because of the wide range of issues investigated in the study of human factors, there is no single research method preferred for all problems. There are many research techniques in the behavioral sciences, each suited for a different type of investigation. These techniques allow you to ask a lot of different questions, but they differ in the degree to which you can be confident in the answers. The degree of confidence depends on the relative control that you have over the various factors involved in the situation you are investigating. Thus, procedures range from the observation and reporting of phenomena in natural settings to tightly controlled laboratory experiments.

No single method will be most appropriate for answering all types of questions, and for many purposes, a converging set of findings from several methods will provide the strongest evidence for a hypothesis. Research that relies on a variety of methods is called *multimethod* or *mixed methods* research (Hesse-Biber & Johnson, 2015). The multimethod approach is particularly relevant for targeting specific human factors issues because these issues have their genesis in the use of systems and products in real-world contexts, and the knowledge gained from more controlled methods must be linked back to evaluations of usability and usefulness for the end users.

In this section, we outline some of the more commonly used methods in human factors research. In each case, we describe the strengths and weaknesses of each method, including when and where it can be used, the nature of questions that can be asked and the answers that will be provided, and the type of statistical rigor that can be applied.

Descriptive Methods

We can use a scientific approach to asking questions about the world even for situations in which true experiments are not possible. Such situations typically arise when you are not able to exercise any control over the events under investigation. Some experts in human factors (e.g., Kanis, 2002; Meister, 1985) place considerable emphasis on these descriptive methods, because the ultimate concern of human factors is the operational system, which by its nature is complex and not subject to precisely controlled investigation. When aspects of these systems are studied in controlled laboratory situations, the research often loses its relevance to the original system. This is because the constraints imposed by tight experimental control will make the task environment differ in potentially significant ways from the real-world setting. As we noted earlier, the extent to which a research setting emulates the real-world setting is called *ecological validity*. In this section, we summarize descriptive methods that preserve ecological validity.

Archival Data

One source of data for human factors researchers is archival data, that is, preexisting data that have been collected for some other purpose (Bisantz & Drury, 2005; Stewart, 2012). Such data may come from injury or incident reports (often required by law) and records produced by people in the performance of their jobs. The massive connectivity of devices like smartphones and cloud-based voice services has resulted in new sources of information about what people are doing in everyday life activities. These sources include social media interactions such as blog entries and product reviews, health records, online search entries, logs of phone usage, and so on. This massive set of archival data that is available is often referred to as "big data," and a lot of attention is being devoted to developing special methods for making such data widely available and for analyzing the data sets, while protecting the privacy of users (Fan, Han, & Liu, 2014).

Archival data may be useful for developing hypotheses to test experimentally and to obtain important information about an operational system. In addition, they can be used to look for evidence that a phenomenon established primarily in laboratory research generalizes to the real world. Consider, for example, the "dilution effect." In laboratory studies, when people are asked to make a decision about something, say, a medical diagnosis, pieces of information or cues (like a patient's symptoms) are provided by the experimenter. A cue is "diagnostic" when it provides valuable information about the correct choice. The dilution effect occurs when some cues are not diagnostic. People tend to pay less attention to diagnostic cues when nondiagnostic cues are present.

One experiment used archival data to look for dilution effects in financial auditing (Waller & Zimbelman, 2003). Auditors have to decide whether mistakes in financial reporting are simply mistakes or evidence of fraud. Some patterns of reporting are diagnostic of fraud, but there are a great many cues to consider in any financial document. If the dilution effect holds for auditors making real audits, they should pay less attention to the diagnostic cues as the number of nondiagnostic cues increases. Data from 215 real audits conducted by an auditing firm confirmed that the auditors experienced dilution. This is real-world evidence of an effect previously found only in the laboratory.

Because the researcher has no control over the collection of archival data, those data must be used with extreme caution. For example, for injury reports, only some injuries may be reported, leading to an underestimation of the injury rate in a particular setting. Similarly, operating records may not contain many important details. Bisantz and Drury (2005) aptly summarize the situation as follows: "Archival data represent a valuable resource for human factors research and analysis, but are full of hidden traps for the unwary" (p. 65).

Naturalistic Observation and Ethnographic Methods

The greatest ecological validity arises when a researcher observes behaviors in naturalistic, or field, settings. When conducting *naturalistic observation*, the researcher is a passive observer of behavior. His intent is to be nonreactive and nonintrusive, so that the individuals under observation are free to behave with virtually no constraints. In human factors, one role of observational research is to characterize the way that people perform their work in real-world, functioning systems (Bisantz & Drury, 2005). For example, a human factors analysis of task performance could begin with observation of the task within the work environment itself. Sometimes the researcher will obtain a complete narrative record (i.e., a faithful reproduction of the behavior as it originally occurred). Often this will be in the form of a videotaped or audiotaped record that he can examine later for behaviors of interest.

Observation can be casual or formal. Casual observation is used most often at the earliest stages of research, to gather initial impressions about what is important and make decisions about how best to study it. Casual observation also provides an opportunity for researchers to see how users typically interact with a system. For example, through casual observation the researcher may note that an employee skips a checklist of procedures unless a supervisor is present. In later stages of research, most naturalistic measures are made through formal observation. Formal observations rely on a system of procedures developed by the researcher. When a certain set of behaviors is of interest, the researcher typically will record only those events that correspond to those behaviors. A checklist can be used to record the presence or absence of these specific behaviors. Measures of the frequency or duration of the behaviors then can be derived. Also, behaviors can be rated in terms of their amount, duration, or quality.

Observational measurement methods also vary in several other ways (Meister, 1985):

1. The observations can be recorded at the time the observation is made or later.
2. The content and amount of detail in the observations can vary.
3. The length of time during which observations are made can be short or long.
4. Observations can vary in terms of the amount of inference, or degree of interpretation, that is required to classify events into the measurement categories.

In conducting observational research, the investigator must develop a taxonomy of the behaviors that she will observe, decide on a strategy for observation, establish the reliability and validity of the taxonomy and strategy, and organize the resulting data in a way that makes sense (Sackett, Ruppenthal, & Gluck, 1978). In deciding on a behavioral taxonomy, the investigator determines whether the measurements are to be molecular or molar. Molecular measurements are defined in terms of specific actions, whereas molar measurements are more abstract and are defined according to function or outcome. For example, the number of times a particular lever is used on a control panel would be a molecular measurement, and the number of products completed on an assembly line would be a molar measurement.

One significant methodological advance in recent years has been the development of sophisticated computerized systems, such as *The Observer*® (Noldus Information Technology, Wageningen, Netherlands), for the collection and management of observational data, as well as for purposes of analysis. These systems can record many aspects of behavior, including the activities in which a person is engaged, their postures, movements, and positions while engaged in these activities, and so on. With the advent of laptop computers, tablets, and smartphones, this form of data collection has become particularly valuable for observations conducted in field settings. However, even using software like *The Observer*, analysis of observational data can be very laborious (Stanton et al., 2013).

One of the most important considerations in observational research is observer reliability. If an observer is unreliable, unreliable data will be produced. Reliable observations require well-constructed measurement scales and well-trained observers. Typically, we establish reliability in observational research by using more than one observer. We can then calculate some measure of the agreement between the observers, such as the percentage agreement between two observers,

$$\frac{\text{Number of times two observers agree}}{\text{Number of opportunities to agree}} \times 100.$$

High observer reliability provides assurance that the measurements are accurate, but not necessarily that they are valid. We can use videotape to check the reliability of observers, as well as to provide a permanent record of the behavior for future reference.

Observational procedures are useful when there is not much data available on a topic. This kind of research often serves as a basis for hypotheses that can be tested later with experimental methods. It is also probably the most efficient way of doing research if we are interested in behavior in real-world settings. For example, observational procedures are very important for evaluating how users interact with a product in their natural surroundings (Beyer & Holtzblatt, 1998) or engage in distracting activities when operating a vehicle (Sullman, 2012). The general weakness of observational methods is that they do not provide a firm basis for drawing inferences about the cause of the observed behavior, because we can exert no systematic control over important variables.

An *ethnographic study* is a type of observational method that comes from the discipline of anthropology (Murchison, 2010). With respect to human factors, the method is used to understand the users' culture and their work environments, and to identify the artifacts they use to accomplish goals. Unlike most observational techniques, ethnography relies on participant observation; that is, the observer actively participates with the users in the environment of interest. The ethnographic researcher spends time with the people being studied, with the intent of understanding them and their activities in the context of interest. Ethnographic research aims to represent the participant's perspective and understanding of the phenomenon of interest; the context in which the phenomenon occurs is seen by the ethnographer as being as important as the phenomenon itself.

Possible drawbacks of ethnographic studies include their heavy reliance on subjective interpretations of the ethnographer and the fact that the participants' behaviors could be influenced by the ethnographer. That is, the subjectivity that a researcher attempts to control in more traditional observational research is embraced in the ethnographic approach, rendering the results suspect from a scientific perspective. Also, ethnographic studies typically take a long time to conduct, because the researcher is trying to understand the group so well that she "becomes" a member of the group.

Within human factors, ethnographic methods may be used in product development, where the ultimate goal is to bring the group of potential users to life so that engineers have a sense of who the users of a product or system are, develop empathy for them, and ultimately develop products to fit their needs (Fulton-Suri, 1999). For example, Paay (2008) advocates the use of ethnographic methods in the design of interfaces for mobile information systems, suggesting that ethnographic data can be used as the basis for design sketches and subsequent mock-ups of proposed interfaces. Based on a study conducted using ethnographic methods to study teams performing software engineering projects, Karn and Cowling (2006) conclude, "Initial findings indicate that ethnographic methods are a valuable weapon to have in one's arsenal when carrying out research into human factors" (p. 495).

Surveys and Questionnaires

Sometimes the best way to begin addressing a problem is by asking the people at work what they think. This information is invaluable, because the operators of a particular system will be familiar with it in a way that an outsider could not be. The questioning can be done informally, but often we will need to construct more formal surveys or administer questionnaires (Charlton, 2002). Questionnaires or surveys are particularly useful when you want to elicit information from a large group of users and the issues of concern are relatively simple. By using a carefully designed set of questions, you can obtain a succinct summary of the issues, and determine probable relations among variables. The benefits of questionnaires include being able to obtain information from different user populations and getting information that is relatively easy to code. However, the types of questions asked affect the validity of the questionnaire, and the return rate is typically low.

A questionnaire must be well constructed, but even the simplest questionnaire can be difficult to construct. As with any other measurements, the data that you obtained will only be as good as your measurement device. There are several steps involved in preparing a questionnaire (Harris, 2014; Shaughnessy, Zechmeister, & Zechmeister, 2014). First, decide on the information that you want the questionnaire to provide and prepare a plan for developing it. Then, decide whether to use a new questionnaire or one that is already available. If the latter, make sure that the questions are clear and unambiguous and that they are ordered such that earlier questions will not bias answers to later ones. After initial questionnaire development, revise and pretest the questionnaire with a small sample to fine-tune the final questionnaire.

Something else you must consider is the form of the responses, which can be open-ended or closed-ended. Open-ended questions allow the respondent to answer freely, which provides a lot of unstructured data that can be difficult to analyze. Multiple-choice questions are a common form of closed-ended questions that provide a limited number of distinct options and are easy to analyze. Another common response format is the rating scale. This allows the respondent to indicate a strength of preference or agreement with the statement being rated. For example, if you ask someone to rate the amount of workload imposed by a particular task, the possible responses could be *very low, low, moderate, high,* and *very high.*

We can use questionnaires in conjunction with other research methods to obtain demographic data such as age, gender, and ethnicity, and other background information. The results of surveys and questionnaires can be summarized by descriptive statistics of the type presented later in the chapter. As with all nonexperimental methods, we should be hesitant about inferring causal relations from questionnaire data. However, they can provide good starting points from which experimental investigations can proceed.

Interviews and Focus Groups

We can conduct structured and unstructured interviews with operators and users at any phase of the research process and for a variety of purposes (Sinclair, 2005). Structured interviews usually present a set of predetermined questions in a fixed order to all interviewees. The questions used in a structured interview should follow the same development procedure as used for questionnaires to avoid problems with misleading, poorly worded, and ambiguous questions. Unstructured interviews typically have few prepared questions, and the interviewer asks questions flexibly, with the responses of the interviewee used in part to direct the interview. Most interview techniques are "semi-structured"; that is, they have an intermediate degree of structure.

Focus groups usually consist of 5–10 users who are brought together in a session to discuss different issues about the features of a system, product, or service (Krueger & Casey, 2015). The group is directed by a moderator, who is in charge of keeping the group on track and getting all users involved in the discussion. Focus groups are good for getting information on different aspects of the system and allowing users to react to ideas presented by other users. Focus groups have the same disadvantages as questionnaires and interviews, the biggest being that what users say may not truly reflect what they do. For a number of reasons, users may not be able to articulate all the steps in a task they perform or knowledge they possess. Also, a single talkative individual can dominate or influence the discussion. Focus groups are good for determining high-level goals, such as generating a list of functions or features for a product. They do not work well for discovering specific usability problems in a product.

Diaries and Studies of Log Files

The purpose of a diary is to record and evaluate actions of a user over a period of time (Rieman, 1996). The user records events related to the task or product under study in a diary, as well as thoughts and insights regarding those events. The researcher can also provide a camera so that the user can take pictures that complement the diary. A video diary can be obtained if the user wears a wireless video camera to record work activities. Although user diaries can provide detailed information about the problem under study, it is important that the diary-keeping not be invasive or difficult to implement, or users will not keep detailed records. One additional, negative factor is the tendency of users to delay entering task information in the diary until a "convenient time." This reduces the amount of information recorded and may result in a loss of some of the insights that users have as they perform the task.

An alternative to diaries is a log file. With a log file, users' actions are recorded as they interact with a system, and these actions are analyzed to determine trends and predict behavior. Log files can be obtained relatively easily for any activities involving computer use, for example, users' retrieval of files on their personal computers (Fitchett & Cockburn, 2015). A large amount of data can be collected from a variety of users, and the data collection process does not interfere with how each user would normally interact with the system. The drawback of log files is that irrelevant or incorrect data may be logged and important behaviors may not be logged. Furthermore, the data do not reflect any of the potentially relevant cognitive processes in which the users were engaged when performing the logged actions.

CORRELATIONAL AND DIFFERENTIAL RESEARCH

We have been discussing the techniques of naturalistic observation, in which the researcher makes no attempt to control the environment. Correlational research gives slightly more control than naturalistic observation. In correlational designs, we must decide ahead of time which behavioral variables we will measure. Typically, we choose these variables on the basis of some hypothesis about how they relate to each other. After we make the measurements, we use statistical procedures to evaluate how the variables change together, or covary. In the simplest case, we measure two variables to determine the degree of relationship between them. We can determine the extent of this relationship by calculating a correlation coefficient, as described in the statistical methods section. Correlational research can be conducted using any of the descriptive methods discussed in the present section.

The value of correlational procedures is that they enable prediction of future events based on the strength of the relationship between observed variables. That is, if we establish that a reliable relationship exists between two variables, we can predict with some accuracy the value of one when we know the value of the other. Suppose, for example, that the number of accidents attributable to operator error increases as the total number of hours on duty increases. We can use this correlation to predict the likelihood of an operator making an error given the amount of time spent on duty. Then, we could use this information to determine the optimal shift length.

We cannot say that one variable (like time on shift) causes another variable (error rate) to increase or decrease just because we observe that the factors covary. This restriction is due in part to the fact that uncontrolled, intervening variables may influence the correlation. Thus, while correlational procedures provide predictive power, they contribute relatively little to an understanding of the causal variables involved in the phenomena. For instance, although errors increase with time on duty, some other variable, such as boredom or fatigue, may be involved.

Research on differences between people is called *differential research*. Szalma (2009) indicates that it is important to consider individual differences in usability and human factors studies. Differential research examines the relations among variables for groups of people who share a common characteristic, such as high versus low intelligence or a difference in personality traits. Often, the distinction between groups serves as the basis for the choice of independent variables. For example, the performance of a group of young adults may be compared with that of a group of elderly people. The distinction between these two groups (age) is the subject variable. As we discussed above, subject variables are not true experimental variables because, by their very nature, they preclude the random assignment of people to groups and we can't manipulate them. This means that there may be many unknown and uncontrolled variables covarying along with the designated subject variable.

How well a differential study provides insight into a phenomenon depends on the strength of the relation between the subject variable and the phenomenon of interest. Differential research has the additional benefit of allowing the use of more sophisticated statistical methods than are possible with the other research methods. But, as with the other nonexperimental designs described thus far, causal inferences are risky. Therefore, even if a phenomenon does covary with the subject variable, you cannot make a causal statement.

EXPERIMENTAL METHODS

True experiments have three defining features:

1. They test a hypothesis that makes a causal statement about the relation among variables.
2. We compare a dependent measure at no fewer than two levels of an independent variable.
3. By randomly assigning people to experimental conditions, we make sure that the effects of many potentially confounding factors are distributed equally across the conditions.

Which particular independent and dependent variables we examine will depend on the hypotheses under consideration. With a random assignment, each person has an equal probability of being assigned to any condition. Random assignment ensures that there can be no systematic influence from extraneous factors, such as education or socioeconomic status, on the dependent variables. Consequently, we can attribute differences among treatment conditions solely to the manipulation of the independent variable. As such, we can make a causal statement about the relation between the independent and dependent variables.

An alternative to random assignment is to perform *stratified sampling*, whereby individuals are assigned to groups in such a way that the proportions of different subject variables, such as age and gender, are the same as the proportions of those variables in the population of interest. We can't make causal statements about the relations between the dependent variables and these subject variables, but we increase the validity of the experiment by making sure our groups look as much like the population as possible.

Because of the restricted nature of laboratory experiments, well-designed experiments have high internal validity. As we noted earlier, this strict control can result in low ecological validity, because the controlled experimental situation is far removed from the real-world environment. Not too surprisingly, whereas experiments examining basic human capabilities, such as vision, tend to use artificial environments, experiments examining applied human factors issues often use simulated environments (e.g., performance in a driving simulator) or field settings (e.g., performance of actual driving).

Between-Subject Designs

In between-subject designs, two or more groups of people are tested, and each group receives only one of the treatment conditions of the independent variable. Subjects in such experiments are usually assigned to each condition randomly. Because subjects are randomly assigned, the groups are equivalent (within chance limits) on the basis of preexisting variables. Thus, any reliable performance difference should be a function of the independent variable.

In cases where we know that a subject variable is correlated with the dependent measure, we can use a matching design. There are several alternative matching procedures, but the general idea behind all of them is to equate all experimental groups in terms of the subject variable. For example, suppose you want to compare two methods for loading crates onto a truck and determine which method is best. The company by which you are employed has 20% female dock workers and 80% male. You will assign half the workers to one method and the other half to the other method. Because physical strength is strongly correlated with sex, if you use a strictly random assignment of workers to groups, one group might contain a higher percentage of females than the other and thus be less strong on the average. Consequently, any differences in performance between the two groups might not be due to the loading methods. A better way is to match the groups in terms of the percentages of males and females in each group, so that the physical strength in each group is approximately equal.

Matching designs allow the systematic distribution of subject variables across treatment conditions. Whereas random assignment ensures that there is no systematic difference in the makeup of the groups prior to the experiment, the matching procedure gives you the added confidence that you have spread a known extraneous factor (strength, in the example above) equally across the treatment conditions.

Within-Subject Designs

Random assignment, stratified sampling, and matching are ways by which we try to make groups equivalent. Another way to equate different groups is to use the same subjects in each one. That is, each person is tested in all conditions, and serves as his or her own control. This increases the sensitivity of the design, making it more likely that small differences in the treatment conditions will be detected. It also substantially reduces the number of people who must be tested.

Within-subject designs have two major drawbacks. First, *carryover effects* may occur, in which previously received treatment conditions influence a subject's performance on subsequent conditions. Second, *practice* or *fatigue effects* may occur, regardless of the particular treatment orders. We can use various *counterbalancing procedures* to minimize these problems. Such procedures equate and/or distribute the order of treatments in various ways to minimize their impact. For example, if subjects are tested under both conditions A and B, we can counterbalance order by testing half with condition A followed by condition B, and half in the reverse order. Again, we would use random assignment, this time involving the assignment of people to the two orders. Although within-subject designs are useful for many situations, we can't use them when a person's participation in one condition precludes participation in another.

Complex Designs

In most cases, we will manipulate more than one independent variable in an experiment. We can use any combination of between-, matched-, and within-subject designs with these manipulations. Such complex experiments enable the researcher to determine whether the variables have interactive effects on the dependent measure. That is, does the manipulation of one variable exert the same effect regardless of the presence of the other variable? If so, the effects of the variables are independent; if not, the variables are said to interact. We will discuss examples of such interactions later in the chapter. Examination of interactions among variables is important, because many variables operate simultaneously in the real world. Moreover, patterns of interaction and noninteraction can be used to infer the structure of the processes that underlie the performance of a task.

SUMMARY

Whenever possible, we must use experimental designs to answer scientific questions. That is, if we desire a precise understanding of the causal nature of phenomena, experimental designs are necessary. However, this does not deny the importance of descriptive methods for the human factors specialist. Such methods provide important information about real-world systems that cannot be obtained from controlled experiments. They are useful not only in contributing to the human factors knowledge base, but also during the system design process, where they can be used to obtain quick information about user characteristics and usability that is helpful to designers (see Box 2.1). Also, in situations for which the specialist needs only to predict behavior without understanding of the causal mechanisms, descriptive procedures are useful. In short, because of the distinct strengths and weaknesses of the experimental and descriptive methods, a blend of both is necessary in human factors research. This point will be illustrated by an example in the last section of the chapter.

As the preceding paragraph suggests, it is important to keep in mind that the goals for basic and applied research are very different. Consequently, the studies you conduct from each perspective will be quite distinct. For basic research, your primary concern is internal validity, and you will often use artificial environments that allow you to control variables that are confounded in

BOX 2.1 USABILITY EVALUATION AND USER EXPERIENCE

Human–computer interaction (HCI) and human factors specialists working in industry want to ensure that the products and systems developed by their companies are "user friendly." The primary goal of this type of research is not to contribute to the scientific knowledge base but to see that the product or system is usable. Usability research is often conducted to meet short deadlines, and often, early in the design process, controlled experiments cannot be conducted. Because controlled experiments are impossible to perform, there are several other methods that human factors specialists can use to assess usability (see Vu, Zhu, & Proctor, 2011), and these can be classified as either inspection-based or user-based.

Usability inspection methods are techniques used by software developers and human factors professionals to assess a system's usability without testing users (Cockton, Woolrych, Hornbæk, & Frøkjær, 2012). Among the most well-known of these methods are *heuristic evaluation* and *cognitive walkthrough*. For a heuristic evaluation, one or more usability experts determine whether the software, website, or product under consideration conforms to established guidelines such as "minimize user memory load." Because different evaluators tend to find different problems, usually 5–7 evaluators are needed to find most of the usability problems. For a cognitive walkthrough, the usability expert interacts with the system by performing tasks that a user would typically perform and evaluates the system from the users' perspective. For each of the steps required to achieve a certain goal, such as pasting a picture into text and adjusting its size and location appropriately, the evaluator tries to answer questions about whether the required action will be intuitive to the user. Both the heuristic evaluation and cognitive walkthrough depend on the availability of experts to make the evaluations and on the use of appropriate heuristics or tasks in the evaluation process.

User-based evaluations are usually called *usability tests* (Dumas & Fox, 2012). Designers perform these tests in a usability lab that is intended to mimic the environment in which the system will be used. The designers must be sure to select users for a usability study who are representative of the target user group. The designers record the users' behavior as they perform certain tasks. Often an observation room adjoins the test room, separated from it by a one-way mirror, through which the designer can observe the user performing the tasks. Some measures that the designer might record include standard behavioral measures such as the time to perform a task and the number of tasks completed successfully.

The designers will also make a continuous video/audio recording of the session in case they need to do further analysis of the user performance. Furthermore, verbal reports obtained from users during the test and follow-up interviews or questionnaires given to users after the test can provide additional information regarding user preferences of the product. The design of the usability test is typically quite simple, and designers will test only a small number of subjects (5 or more), since the goal is to obtain quick, useful information for design purposes and not to add to the scientific database.

Other methods used in usability evaluation include ethnographic methods and diary studies of the type described in the text, as well as interviews, focus groups, and questionnaires (Volk, Pappas, & Wang, 2011). These methods are used to understand how users interact with the product and to obtain their preferences regarding product characteristics. The methods generally provide a large amount of qualitative data that may be valuable to the design team. However, a lot of time and effort is required to collect and analyze data using these methods, and extensive interpretation of the users' data is necessary.

User experience is a broader concept, which includes usability but also is affected by other aspects of the product or system. According to Nielsen and Norman (2016), "The first requirement for an exemplary user experience is to meet the exact needs of the customer, without fuss or bother. Next comes simplicity and elegance that produce products that are a joy to own, a joy to use." Note the emphasis here on creating a pleasurable experience, which means that the user's emotional reaction is considered to be one of the most important dimensions (Jokinen, 2015). Apple, Inc. have the reputation of emphasizing the overall user experience in their product designs, taking a system perspective on value to the customer (Pynnönen, Ritala, & Hallika, 2011). The aim is to fulfill the wants and needs of the customer, and consumers can assume that Apple's products and services have been designed with this aim in mind. The usability of a device's interface (e.g., that of an iPhone) is treated along with the usability of the service–user interface (e.g., for iTunes) as a system, so that the entire experience of interacting with the service through the device will be positive for users. Trust established with users regarding the pleasurable user experience results in many who stand in line to buy a new product when it is introduced.

User experience is difficult to measure, but studies typically use subjective ratings as their primary data. For example, Cyr, Head, and Larios (2010) compared the user experience of an electronics e-commerce site displayed using three distinct website color schemes (blue, gray, and yellow). Subjects from Canada, Germany, and Japan rated their disagreement/agreement (on a 1–5 scale) with statements about the pleasantness of the color scheme, trust in the website, and overall satisfaction with the site. The yellow scheme was rated as less appealing than the other two color schemes, and these ratings also correlated highly with the ratings of overall satisfaction with the website. The assumption is that users who are less satisfied with their experience of a website or product are less likely to use or purchase that product in the future.

real-world environments. You will use the results of the research to test predictions derived from theories about underlying processing mechanisms, and your concern should be that the resulting theory will generalize to situations to which it is intended to apply. For applied research, however, your emphasis is on ecological validity, and your research setting should resemble the real-world environment of interest. Your results will let you make predictions about the real-world behavior of interest and generalize your findings to the real-world environment. Both basic and applied research are essential to the field of human factors and ergonomics.

STATISTICAL METHODS

The results obtained from both descriptive and experimental research typically consist of many numerical measurements for the variables of interest. These results must be organized and analyzed if they are to make sense to the researcher and to others who wish to make use of the information. We use statistical methods to perform these functions. There are two types of statistical procedures: descriptive and inferential. For each kind of statistics, we have to distinguish between samples and populations. A sample is a set of measurements made on a subset of people from a larger population. Rarely do we have the resources to measure an entire population, so we depend on samples to tell us what the population characteristics are.

DESCRIPTIVE STATISTICS

As implied by the name, *descriptive statistics* describe or summarize the results of research. One concept that is fundamental to descriptive statistics is that of the *frequency distribution*. When we obtain many measurements of a variable, we can organize and plot the frequencies of the observed values. For example, if we have a group of people estimate the mental workload imposed by a task on a scale of 1–7, we can record the number of people who responded with each value. This record of the frequency with which each score occurred is a frequency

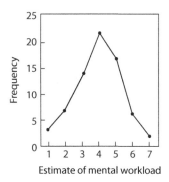 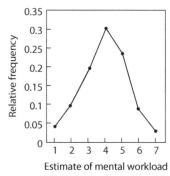

FIGURE 2.3 Frequency and relative frequency polygons.

distribution. A frequency distribution often is plotted in the form of a frequency polygon, as is shown in Figure 2.3. A *relative frequency distribution*, also shown in the figure, displays the same plot on the scale of the proportion (or percentage) of times that each score was observed. We can describe a score in terms of its *percentile rank* in the distribution. A percentile is a point on a measurement scale below which a specified percentage of scores falls. The percentile rank is the percentage of scores that falls below that percentile. We use percentile ranks for, among other things, creating tables of anthropometric data and applying these data in the design of equipment for human use.

Central Tendency and Variability

Although a distribution helps to organize the data from research, other summary values may convey the crucial information more succinctly. Typically, measures of central tendency and variability are the primary descriptive statistics reported in research articles. Measures of central tendency indicate the middle or representative score for the distribution. Most studies present their results in the form of measures of central tendency, such as means or medians. The *arithmetic mean* is obtained by adding up all of the score values and dividing by the total number of scores. If X represents the variable of interest, then the mean X is given by

$$\overline{X} = \frac{1}{n}\sum_{i=1}^{n}X_i,$$

where X_i refers to the ith of n scores. The mean is an estimate of the population mean μ. The *median* is the score which 50% of the distribution falls below and 50% above, or, in other words, the score with the percentile rank of 50%. The mean and median values are equivalent for a symmetric distribution, but not otherwise. If there are extreme low or high scores, the median may be a better estimate of central tendency, because it is sensitive only to the ordinal properties of the scores and not their magnitudes.

One other measure of central tendency is the *mode*, the most frequently occurring score. In most cases, the mode will not be very useful. However, for qualitative (non-numeric) variables, it is the only meaningful measure of central tendency. The mode is also used to classify the shapes of distributions. For example, a distribution is said to be unimodal if it has only one mode and bimodal if it has two.

Measures of variability provide indications of the dispersion of the individual scores about the measure of central tendency. In other words, most scores may be close to the most typical score, or they may be widely dispersed. Measures of variability increase as the amount of dispersion increases. The most widely used measures of variability are the *variance* and the *standard deviation*. The formula for the variance of a sample of scores is

$$S_X^2 = \frac{1}{n-1} \sum_{i=1}^{n} \left(X_i - \overline{X} \right)^2,$$

where:

S_X^2 represents the sample variance,

X represents the mean, and

n represents the number of scores or observations in the sample.

This statistic is an estimate of the population variance σ^2. Another name for the variance is *mean squared deviation*, which emphasizes that the variance reflects the average of the squared deviations of each individual score from the mean.

The sample standard deviation is obtained by taking the square root of the sample variance:

$$S_X = \sqrt{\frac{1}{n-1} \sum_{i-1}^{n} \left(X_i - \overline{X} \right)^2}.$$

The advantage of the standard deviation for descriptive purposes is that it exists in the same scale as the original measurements and thus gives a measure of how different the scores are from the mean. Note that both the variance and the standard deviation are always positive numbers.

In many situations, the population from which a sample was taken will approximate a normal (Gaussian) curve (unimodal, bell-shaped and symmetric; see Figure 2.4). If measurements come from a normal distribution, we can describe them by their population mean μ and standard deviation σ. Because there are an infinite number of normal distributions (since there are an infinite number of values for μ and σ), we frequently transform such distributions into the standard normal distribution, which has mean 0 and standard deviation 1. Measurements from the standard normal distribution give the value of the variable X in terms of the number of standard deviation units it is from the mean. Such measurements are called *z-scores*, and we compute the z-scores of a sample as

$$Z_i = \frac{X_i - \overline{X}}{S_X}.$$

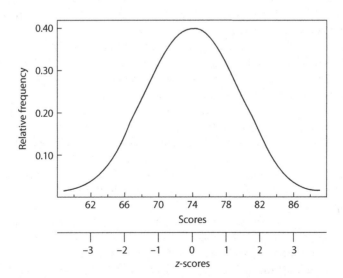

FIGURE 2.4 A normal distribution with $X = 74$ and $S = 4$, and corresponding z-scores.

The score Z_i is the z-score corresponding to the ith observation in the original scale of the variable. z-scores are useful because it is more informative to know that $Z_i = 1.0$, meaning one standard deviation above the mean, than to know that $X_i = 80$, where we don't know what the mean and standard deviation are.

We can transform any sample of measurements into z-scores, even if the sample was taken from a population that is not normally distributed. However, for the resulting sample of z-scores to represent a sample from a standard normal population, the measurements X must be normally distributed. Because any normally distributed variable X can be transformed into z-scores, the relative frequencies and percentile ranks of normally distributed variables are presented in z tables (see Appendix I).

As an example, in 1985, the thumb-tip reach, or distance from thumb to shoulder, for U.S. women has a mean of 74.30 cm and a standard deviation of 4.01 cm. For a woman with a thumb-tip reach of 70.00 cm, the z-score would be

$$\frac{70.00 - 74.30}{4.01} = -1.07,$$

or 1.07 standard deviation units below the mean. Using the table in Appendix I, we can determine that this individual has a percentile rank of 14%. In other words, 14% of women have a shorter thumb-tip reach than this person.

Correlation Coefficient

In correlational research, a common descriptive statistic is the Pearson product-moment correlation r:

$$r = \frac{S_{XY}}{S_X S_Y},$$

where X and Y are the two variables and

$$S_{XY} = \frac{1}{n-1} \sum_{i-1}^{n} \left(X_i - \overline{X} \right) \left(Y_i - \overline{Y} \right)$$

is the covariance between X and Y. The covariance between two variables is a measure of the degree to which changes in one variable correspond to changes in another. The correlation coefficient is an indicator of the degree of *linear* relationship between two variables.

The coefficient r is always between -1.0 and $+1.0$. When X and Y are uncorrelated, r will equal 0. When r is 0, there is no linear relationship between X and Y. When they are perfectly correlated, r will equal $+1.0$ or -1.0, and the values of one variable can be related to the values of the other variable by a straight line. A positive correlation means that, as values of X increase, so do values of Y; a negative correlation means that, as values of X increase, those of Y decrease. Figure 2.5 provides illustrations of data for several values of r. Note that X and Y may be related to each other and still be uncorrelated. Because r only measures linear relationships, if X and Y are nonlinearly related, say, $Y = X^2$, the correlation may be zero. Another useful statistic is r^2, which gives the proportion of total variance that can be traced to the covariance of the two variables. It is often said that r^2 reflects the amount of variance "explained" by the linear relationship.

We can illustrate the use of the correlation coefficient, as well as means and standard deviations, with an example. Lovasik, Matthews, and Kergoat (1989) investigated the effect of foreground and background color on performance of a visual detection task, in which people were asked to search a computer display for a target symbol during a 4-h period. The observers' response times to find the target in each display were measured as a function of the amount of

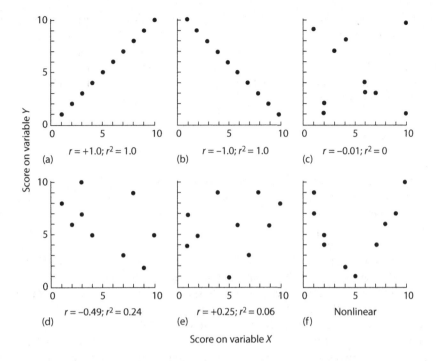

FIGURE 2.5 Scatterplots for different correlations between X and Y variables.

time they had already spent on the task and the foreground and background colors of the display. Table 2.1 presents the search times under all time-on-task and color conditions for each observer. These search times are the percentage of the observers' search times after half an hour on the task.

In Table 2.1, we compute the mean percent search times for each of the three color conditions. Note that a mean of 100% means there was no improvement from the initial search time, and that

TABLE 2.1

Percent Target Search Time Relative to the First Half-Hour of Time on Task (in Hours)

Observation (N)	Time on Task (T)	Red/Black (R)	Blue/Black (B)	Red/Green (RG)
1	0.5	100	100	100
2	1.0	97	93	102
3	1.5	95	94	98
4	2.0	90	89	99
5	2.5	91	88	100
6	3.0	94	90	98
7	3.5	91	87	97
8	4.0	90	83	98
$\sum\limits_{i=1}^{8}$	18.0	748	734	792
	$\overline{T} = \dfrac{18.0}{8}$	$\overline{R} = \dfrac{748}{8}$	$\overline{B} = \dfrac{734}{8}$	$\overline{RG} = \dfrac{792}{8}$
	$= 2.25$	$= 93.50$	$= 91.75$	$= 99.00$

the smaller the mean is, the greater the improvement. Observers showed virtually no improvement in the red/green condition, and the most improvement in the blue/black condition. We compute the variance of the target search time in Table 2.2, as well as the correlation between search time and time-on-task. The red/green condition has the smallest variance. The correlations between time-on-task and search time are negative for all display conditions, indicating that as time-on-task increased, search time decreased.

INFERENTIAL STATISTICS

Probability

Inferential statistics rely on the concepts of probability theory. We speak of events as having some probability of occurring when we are uncertain of the outcome. Probabilities are measurements that assign a number from 0.0 to 1.0 to an event, according to how likely that event is to occur. Probability represents the proportion of times that the event occurs relative to all other possible events. Thus, a *probability distribution* is a relative frequency distribution over the entire set of possible events. If an event has a probability of 0.0, then it has no chance of occurring. If an event has a probability of 1.0, then we know with absolute certainty that it will occur. If an event has probability 0.5, then approximately half of the outcomes will consist of that event. For example, the event *heads* that could be observed on the toss of a fair coin has probability 0.5. Thus, if we flip the coin many times, we would expect to observe heads approximately half of the time.

The combined outcome of two separate events is called a *joint event*. For example, if we rolled two dice and observed two dots on the first die, event 2, and four dots on the second die, event 4, the outcome (2, 4) is a joint event. Just as we can observe a relative frequency distribution over all possible outcomes of a single event, we can observe a joint relative frequency over all possible outcomes of a joint event. In other words, we can calculate the proportion of times that each possible joint event occurs.

The two events that make up the joint event may or may not depend on each other. If knowing the outcome of the first event provides no information about what might happen for the second event, then the two events are independent. This implies that, for independent events A and B, the probability of the joint event A and B, written $A \cap B$, is

$$P(A \cap B) = P(A) P(B).$$

That is, the joint probability can be written as the product of the marginal probabilities of the individual events. It is possible, however, that the outcome of the first event influences the outcome of the second, in which case the two events are dependent. For dependent events, the probability that event B occurs given that event A occurred is

$$P(B \mid A) = \frac{P(A \cap B)}{P(A)}.$$

Probabilities are important, because a researcher tests a sample selected from a larger population. However, the researcher is not interested in reaching conclusions about the specific sample, but about the population from which it was drawn. Yet, because the entire population was not measured, there is a probability of some measurement error. For example, if you are interested in determining the mean height of adult males for a particular population, you might measure the height for a sample of 200 males. Each height that you observe has some probability of occurring in the population, which has an average around 5 ft 10 in. You could compute the sample mean and use it as an estimate of this population mean. However, it is only an estimate. It is possible, although not likely, that all 200 men in the sample would be taller than 6 ft. Thus, the sample mean would be greater than 6 ft, even though the population mean is closer to 5 ft 10 in. This is an instance of sampling

TABLE 2.2
Computations of Variance and Correlation Coefficients for the Data in Table 2.2

N	$T-\bar{T}$	$(T-\bar{T})^2$	$R-\bar{R}$	$(R-\bar{R})^2$	$B-\bar{B}$	$(B-\bar{B})^2$	$RG-\overline{RG}$	$(RG-\overline{RG})^2$	$(T-\bar{T})(R-\bar{R})$	$(T-\bar{T})(B-\bar{B})$	$(T-\bar{T})(RG-\overline{RG})$
1	-1.75	3.06	6.50	42.25	8.25	68.06	1.00	1.00	-11.38	-14.44	-1.75
2	-1.25	1.56	3.50	12.25	1.25	1.56	3.00	9.00	-4.38	-1.56	-3.75
3	-0.75	0.56	0.50	0.25	2.25	5.06	-1.00	1.00	-0.38	-1.69	0.75
4	-0.25	0.06	-4.50	20.25	-2.75	7.56	0.00	0.00	1.12	0.69	0
5	0.25	0.06	-5.50	30.25	-3.75	14.06	1.00	1.00	-1.38	-0.94	0.25
6	0.75	0.56	-3.50	12.25	-1.75	3.06	-1.00	1.00	-2.62	-1.31	-0.75
7	1.25	1.56	-6.50	42.25	-4.75	22.56	-2.00	4.00	-8.12	-5.94	-2.50
8	1.75	3.06	-0.50	0.25	1.25	1.56	-1.00	1.00	-0.88	2.19	-1.75
$\sum_{i=1}^{8}$		10.48		160.00		123.48		18.00	-28.00	-23.00	-9.50
S^2		$\frac{10.48}{7}=1.50$		$\frac{160.00}{7}=22.22$		$\frac{123.48}{7}=17.64$		$\frac{18.00}{7}=2.57$			
r									$\frac{-28.02}{(10.48)(160.00)}=-0.68$	$\frac{-23.02}{(10.48)(123.48)}=-0.64$	$\frac{-9.50}{(10.48)(18.00)}=-0.69$

error. Generally, increases in the sample size will lead to more precise estimates of the population characteristic of interest.

Probability theory will be important in later chapters. The concepts of probability theory are central to reliability analysis (Chapter 3), signal-detection theory (Chapter 4), and decision theory (Chapter 11). They also are pivotal to simulations of human and system performance.

Statistical Hypothesis Testing

Typically, in conducting experiments or other types of research, the researcher wants to test hypotheses. In the simplest type of experiment, two groups of subjects may be tested. One group receives the experimental treatment and the other, referred to as the control group, does not. The concern is whether the difference between the two groups results in a change in a dependent variable. Usually, we will compute the sample means for this variable for each group. We can compare these means to determine whether they differ. However, because these sample means are only estimates of the population means and are subject to sampling error, just examining these values will not tell us whether the treatment had an effect. In other words, we need a way to decide how much of a difference between the sample means we need to see before we conclude that the difference reflects a "real" difference in the population means.

Inferential statistics provide a way to answer this question. In our two-group experiment, we begin by formulating a *null hypothesis*. The null hypothesis is that the treatment had no effect or, in other words, that the population means for the two conditions do not differ. The inferential test determines the probability that the observed mean difference could have been due solely to chance, given some estimate of error variance and assuming that the null hypothesis is true. It is important to note that the researcher does not know whether or not the null hypothesis is true. Based on the probabilistic evidence provided by the statistical test, the researcher must decide whether to accept or reject the null hypothesis. The combination of two possible states of the world (null hypothesis true or false) with two possible decisions (null hypothesis true or false) yields a 2 by 2 matrix of outcomes, two of which are correct (see Table 2.3).

As the table shows, two distinct types of errors can occur when trying to infer characteristics of the population from the sample. The first of these is called a *type I error*. This type of error occurs when the sample mean difference is sufficiently large that the null hypothesis is rejected, but in fact, the population mean difference is zero. The second, called a *type II error*, occurs when the population means are different, but the observed sample means are sufficiently similar that the null hypothesis is accepted. The researcher, given some knowledge about the distribution of the population means, must select an acceptable probability for a type I error by deciding the point at which the null hypothesis will be rejected. This probability is called the α-level.

Traditionally, the rule has been to conclude that the experimental manipulation had a reliable or *significant* effect if the difference between the sample means is so large that the probability of obtaining a difference that large or larger is less than 0.05, if the null hypothesis were true. The null hypothesis is not rejected if this probability is greater than 0.05. If a 0.01 α-level were used instead, the probability of a type I error would be less, but the probability of a type II error would increase. In other words, the criterion level adopted by the researcher affects the relative likelihood of each error occurring (Lieberman & Cunningham, 2009).

TABLE 2.3
Statistical Decision Making

Null Hypothesis	Decision	
	True	**False**
True	Correct Acceptance $(1-\alpha)$	Type I Error (α)
False	Type II Error (β)	Correct Rejection $(1-\beta)$

Whenever differences between conditions are reported for experiments described in this book, it can be assumed that the difference was significant at the 0.05 level. Remember that a significant difference is meaningful only in a well-designed study. That is, a significant inferential test suggests that something other than chance apparently was operating to distinguish the two groups, not what that something was. Also, failing to reject the null hypothesis does not necessarily mean that the independent variable had no effect. The failure to show a reliable difference could reflect only large measurement error or, in other words, low experimental power.

One widely used inferential test that is important to issues discussed later in the text is the analysis of variance (ANOVA). One strength of the ANOVA is that it can be applied to experiments with complex designs. The ANOVA allows us to evaluate the interactions between different independent variables. In the two-variable case, an ANOVA will tell not only whether each variable had an overall effect on the dependent variable (i.e., a main effect), but also whether the two independent variables together had an interactive effect. Figure 2.6 shows some example patterns of interaction and noninteraction. Nonparallel lines indicate interactive effects, whereas parallel lines indicate independent effects. When an experiment includes more than two independent variables, we can evaluate complex interaction patterns among all of the variables in the same way.

As an example, alcohol and barbiturates are known to have interactive effects on behavior. If A and B represent the level of alcohol and barbiturates consumed, respectively, let A_0 indicate no alcohol and B_0 no barbiturates, and A_1 and B_1 some fixed dosage of each. Figure 2.7 shows an interaction of these variables on driving performance, where the dependent variable is the variance of the distance from the center of the lane in feet squared. As you can see, performance under the combined effect of both drugs is considerably worse than with either drug alone.

A STUDY EVALUATING HUMAN FACTORS DESIGN

To make some of the issues regarding research and statistical issues concrete, we will discuss in detail a specific study that, although conducted several decades ago, still provides an excellent example of how various types of research methods can be brought to bear on solving a specific

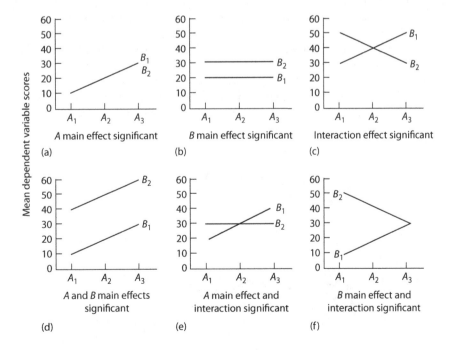

FIGURE 2.6 Example patterns of interaction and noninteraction.

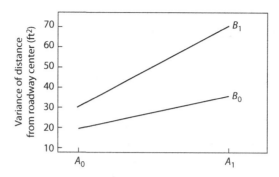

FIGURE 2.7 Interactive effects of alcohol (a) and barbiturates (b) on driving performance.

human factors problem. Marras and Kroemer (1980) conducted a laboratory and field study on design factors crucial to the efficient operation of boating distress signals. To begin, Marras and Kroemer surveyed the distress signals that were available on the market and categorized them according to the steps that were required to identify, unpack, and operate them. After they categorized the signals, they conducted a preliminary study in which naive and experienced boaters were videotaped while operating the signals. They used the performance of these individuals with the various signals to identify the design variables of interest, such as form, labeling, size, and so on.

The next stage in the study was a series of laboratory tests in which they used each identified design variable as an independent variable in an experiment. For example, one test investigated how different shapes affected the identification of the distress-signal device. They painted three differently shaped flares (one that fully complied with proposed human factors regulations, one that partially complied, and one that did not comply) red and left them unlabeled. The person's task was to pick up the flare from among five other devices, and the researchers measured the time it took them to do so. They found that people selected the flare that complied with the human factors guidelines faster than the flare that did not.

The final stage of the study took place in a natural environment. They took their participants by boat to an island where they boarded a rubber raft that was rigged to deflate. They told the boaters that the purpose of the study was to rate the visibility of a display shown from shore. They placed one of two types of hand-held flares on board the raft (see Figure 2.8). They pushed the raft onto the lake, and within 2 min it began to deflate, prompting the boater to use the distress signal. Then they pulled the participant back to shore, and repeated the procedure with the second device. Table 2.4 shows the total times required to unpack and operate the two types of flares for 20 people.

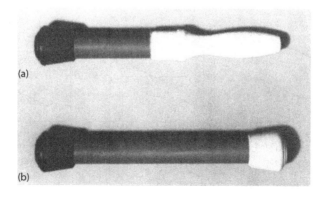

FIGURE 2.8 The two types of hand-held flares tested.

TABLE 2.4
Results of On-Water Experiments: Performance Times in Minutes

Subject	Flare A	Flare B
1	0.21	1.46[a]
2	0.29[a]	0.87
3	0.10	0.64[a]
4	0.23[a]	0.92
5	0.24	0.81[a]
6	0.43[a]	0.36
7	0.42	1.42[a]
8	0.16	2.32[a]
9	0.11	0.80[a]
10	0.10[a]	2.23
11	0.25[a]	0.72
12	0.22	0.98[a]
13	0.57[a]	1.39
14	0.44	2.53[a]
15	0.35[a]	0.73
16	0.67[a]	1.84
17	0.20	1.87[a]
18	0.27[a]	2.45
19	0.26	2.14[a]
20	0.26[a]	1.04
Mean	0.289	1.376
Standard Deviation	0.152	0.687

[a]First trial.

For flare A, which had a handle that clearly distinguished the grip end of the flare, the mean performance time was 0.289 min. For flare B, which did not have a clearly distinguished grip, the mean performance time was 1.376 min. An analysis of variance showed that the difference was significant. In other words, the grip reliably reduced total performance time.

One nice characteristic of this study is that it shows the systematic progression from descriptive survey and observational methods, to controlled laboratory experiments, and back to a field experiment conducted in an ecologically valid setting. Thus, the research coupled the ecological strengths of the observational and field methods with the internal strengths of the experimental method. In this way, the study provides an ideal example of how different research methods can be combined to address a specific human factors problem.

You should be able to determine from the description of the field experiment that it used a within-subjects design, because each person was tested in both flare conditions. Therefore, the order in which the flares were used is an extraneous variable that could potentially confound the results. For example, if flare A had been tested first and flare B second for all people, then the disadvantage for flare B could have been due to fatigue, or some other order effect, such as the absence of fear in the second test. To control for such order effects, Marras and Kroemer (1980) counterbalanced the order in which the two devices were tested. Half of the people were tested with flare A first, and half were tested with flare B first.

Even though order was counterbalanced, you still might question the appropriateness of the within-subject design. One reason for conducting the field experiment was for boaters to perform in a real emergency. After the test with the first flare, it is very likely that the boaters became aware of the true purpose of the experiment. No evidence is apparent in the data that performance was affected greatly by any such awareness, so it may not be a problem. This potential drawback of the within-subject design was offset by the benefits that fewer people had to be tested and extraneous variables, such as boating experience, were equated for the two conditions. This example of the relative merits of within- and between-subject designs indicates that there are no hard and fast rules for making research design decisions.

The changes in smoke signal and flare design that Marras and Kroemer (1980) recommended on the basis of their study were subsequently instituted. In 1987, Kroemer remarked,

> But nowadays, if you go into the store and buy those marine emergency signals, they are of different colors, they are of different designs, and they work fine. Nobody ever said anything but I have always considered this one of the very satisfying results. We may have saved some people this way. This really made me happy. We never got any formal reply but we know it was being used—*that's the best.* (quoted in Rogers, 1987, p. 3)

SUMMARY

The present chapter has provided an outline of the basics needed to understand research in human factors. Human factors is an applied science. Thus, scientific reasoning guides the activities of the researcher and the designer. Because science relies on empirical observation, the methods by which we make these observations are of fundamental importance. Furthermore, we must understand the statistics used to summarize and evaluate these observations.

The scientific approach requires the continuous development and refinement of theory based on our observations. The observations provide the basic facts, and the theories help to explain why they are so. Good theories not only explain but also make predictions about new situations and allow us to optimize performance in those situations.

The research methods used to obtain data vary according to how tightly controlled the setting is and to how much the setting approximates the environment of interest. Laboratory experiments typically are highly controlled but have low ecological validity, whereas descriptive methods tend to be relatively uncontrolled but have high ecological validity. Thus, the choice between the various experimental and descriptive methods depends on the goals of the study, and often some combination of methods provides the best understanding of a human factors issue.

Data are summarized and evaluated by means of descriptive and inferential statistics. Proper conclusions require the use of appropriate statistical analyses. Probability and statistics underlie many of the more sophisticated analyses of human performance that will be encountered later in the text, as well as those of human reliability, which will be covered in the next chapter.

RECOMMENDED READINGS

Bechtel, W. (1988). *Philosophy of Science: An Overview for Cognitive Science.* Hillsdale, NJ: Lawrence Erlbaum Associates.

Chalmers, A. F. (2014). *What Is This Thing Called Science?* (4th ed.). Indianapolis, IN: Hackett.

Creswell, J. W. (2013). *Introduction to Qualitative Research Methods* (4th ed.). Thousand Oaks, CA: Sage.

Haig, B. D. (2014). *Investigating the Psychological World: Scientific Method in the Behavioral Sciences.* Cambridge, MA: MIT Press.

Meister, D. (1985). *Behavioral Analysis and Measurement Methods.* New York: Wiley.

Pagano, R. R. (2010). *Understanding Statistics in the Behavioral Sciences* (9th ed.). Belmont, CA: Cengage Learning.

Proctor, R. W., & Capaldi, E. J. (2006). *Why Science Matters: Understanding the Methods of Psychological Research.* Malden, MA: Blackwell.

Rubin, J., & Chisnell, D. (2008). *Handbook of Usability Testing: How to Plan, Design, and Conduct Effective Tests.* Hoboken, NJ: Wiley.

Stanovich, K. E. (2013). *How to Think Straight about Psychology* (10th ed.). Boston, MA: Pearson.

3 Reliability and Human Error in Systems

One afternoon in the early 1970s, I was boiling a kettle for tea. The teapot … was waiting open-topped on the kitchen surface. At that moment, the cat – a very noisy Burmese – turned up at the nearby kitchen door, howling to be fed … . I opened a tin of cat food, dug in a spoon and dolloped a large spoonful of cat food into the teapot.

James Reason
2013

INTRODUCTION

We all make slips and mistakes like that made by Prof. Reason, an expert in the area of human error. Sometimes these have humorous outcomes and sometimes more serious consequences, as described in the subtitle of his book, *From Little Slips to Big Disasters*. For example, while driving, it is easy to get distracted. You might be talking to a friend on a cell phone or trying to operate the infotainment system and so fail to see that the car in front of you has stopped. You might just crash into the car in front without braking, or you might not have time enough to complete the braking action to prevent a crash. Allowing your attention to be taken from the road for whatever reason was an error, and errors often have adverse consequences.

In July 2001, 6-year-old Michael Colombini was in the Westchester Medical Center, New York, for a magnetic resonance imaging (MRI) exam following surgery. An MRI machine consists of a very large and powerful magnet. Michael was placed in the middle of this magnet for his scan. Regulations about what can and cannot be brought into MRI exam rooms are very explicit; even paperclips are not allowed because they will be drawn into the center of the magnet at high speed when the machine is turned on. Nonetheless, someone brought an oxygen tank into the exam room. The heavy tank was drawn into the center of the magnet, and Michael died of blunt force trauma to the head.

When medical errors occur, like the one that cost Michael Colombini his life, they are usually human errors. As described in Chapter 1, in the year 2000, the number of deaths in the U.S. resulting from medical errors was estimated to be between 44,000 and 100,000 people (Kohn, Corrigan, & Donaldson, 2000). In response to Kohn et al.'s report, *To Err is Human: Building a Safer Health System*, the U.S. President launched a series of initiatives to boost patient safety, with the goal of reducing preventable medical errors by 50% by 2005. Among the initiatives was a requirement that all hospitals participating in the government's Medicare program be required to institute error reduction programs and support for research into medical errors. This was accompanied by calls for the medical profession to implement a human factors approach to human error similar to that employed by the aviation industry (Crew Resource Management, or CRM; see, for example, Leape, 1994) and to integrate information about the sources of human error into the medical curriculum (Glavin & Maran, 2003).

Despite the increased awareness of the importance of reducing medical error, Leape and Berwick (2005) concluded that progress had been "frustratingly slow" (p. 2385), with deaths from medical errors reduced only slightly by 2005. Recent estimates of the number of error-related deaths put the figure now at 400,000 per year, considerably greater than the estimate in 2000 (James, 2013; McCann, 2014), suggesting that medical error is now the third leading cause of death in the United States. This high rate of medical error contributes to the U.S.'s low health system ranking among those of other developed countries (Davis, Stremikis, Squires, & Schoen, 2014). Pleas for additional research and education on medical human factors have been made, for example, in the areas of surgical teams (Kurmann et al., 2012) and resuscitation teams (Norris & Lockey, 2012). One area that is particularly in need of examination is the design and use of information technology for electronic health records (EHRs), which may be based on outdated or poorly designed software. Design deficiencies and poor usability of EHRs are directly responsible for a number of high-profile medical errors (Sawchuk, Linville, Cornish et al., 2014).

As we discussed in Chapter 1, systems can be small (like the lighting system) or large (like the many people, equipment, policies, and procedures that compose a U.S. hospital). Within each system we can identify one or more operators, people in charge of using a machine, implementing a policy, or performing a task, who help guide the system toward achieving its goal. A primary mission of the human factors specialist is to minimize human error and so to maximize system performance. This requires the specialist to identify the tasks performed by the operator and determine possible sources of error. This information must then be incorporated into the design of the system, if performance is to be optimized. Because large-scale systems may involve many people, a term that describes this process is *human–systems integration* (Durso, Boehm-Davis, & Lee, 2015), to emphasize the idea that system design involves much more than just consideration of the individual human in the system. Before considering ways that the likelihood of human error can be evaluated, we must first consider the system concept and its role in human factors.

CENTRAL CONCEPT IN HUMAN FACTORS: THE SYSTEM

A human–machine system

> "is a system that involves an interaction between people and other system components, such as hardware, software, tasks, environments, and work structures. The system may be simple, such as a human interacting with a hand tool, or it may be complex, such as an aviation system or a physician interacting with a complex computer display that is providing information about the status of a patient"

> **(Czaja & Nair, 2012, p. 38).**

A system operates for the purpose of achieving a goal. A hospital operates to cure disease and repair injury. An automobile operates to move people from one place to another. As human factors specialists, we believe that the application of behavioral principles to the design of systems will lead to improved functioning of those systems and will increase our abilities to achieve our goals. Indeed, the U.S. National Academy of Engineering indicated the importance of the systems approach to engineering in general, stating, "Contemporary challenges—from biomedical devices to complex manufacturing designs to large systems of networked devices—increasingly require a systems perspective" (2005, p. 10).

The system approach has its basis in *systems engineering*, which is "an engineering approach that provides an understanding of the interaction of individual parts that operate in concert with one another to accomplish a task or purpose" (Cloutier, Baldwin, & Bone, 2015, p. 1). Systems engineering emphasizes the overall goals of the system or product under development during the design process (Kossiakoff, Sweet, Seymour, & Biemer, 2011). Beginning with the identification of an operational need, designers determine the requirements of the system, which in turn results in a system concept. Designers implement this concept in a system architecture, dividing the system into optimized subsystems and components. For example, a hospital might plan a cancer research center, including state-of-the-art diagnostic devices (like an MRI machine), treatment facilities, counseling, and hospice care. Each of these separate components of the cancer center can be treated as a subsystem, tested and optimized, and then integrated into the overall system, which in turn is evaluated and tested. The result is the final cancer research center, which, hopefully, satisfies the system goals throughout its life cycle.

Systems engineering (as well as systems management) does not focus specifically on the human component of the system (Folds, 2015). This is the domain of the human factors specialist, who is concerned with optimizing the human subsystems, primarily through the design of human–machine interfaces, individual and team training materials, and so forth that promote effective human use. System analyses applied to the human component provide the basis for evaluating human reliability and error, as well as for the design recommendations intended to minimize errors. They also provide the basis for the safety assessment of existing technological systems, such as nuclear power plants (Cacciabue, 1997). Several implications of the system concept are important for evaluating human reliability and error (e.g., Bailey, 1996). These include the operator, the goals and structure of the system, its inputs and outputs, and the larger environment in which it is placed.

IMPLICATIONS OF THE SYSTEM CONCEPT

Several implications of the system concept are important for evaluating human reliability and error (e.g., Bailey, 1996). These include the operator, the goals and structure of the system, its inputs and outputs, and the larger environment in which it is placed.

The Operator Is Part of a Human–Machine System

We must evaluate human performance in applied settings in terms of the whole system. That is, we must consider the specific system performing in the operational environment and study human performance in relation to the system.

The System Goals Take Precedence over Everything Else

Systems are developed to achieve certain goals. If these goals are not achieved, the system has failed. Therefore, evaluations of all aspects of a system, including human performance, must occur with respect to the system goals. The objective of the design process is to satisfy the system goals in the best way possible.

Systems Are Hierarchical

A system can be broken down into smaller subsystems, which in turn can be broken down into components, subcomponents, and parts, or it can be conceived as a component in a more encompassing "system of systems." Higher levels in the system hierarchy represent system functions (i.e., what the system or subsystem is to accomplish), whereas lower levels represent specific physical components or parts. A human–machine system can be broken into human and machine subsystems, and the human subsystem can be characterized as having subgoals that must be satisfied for the overriding system goals to be met. In this case, the components and parts represent the strategies and elementary mental and physical acts required to perform certain tasks. We can construct a hierarchy of goals and subsystems by considering components within both the human and machine subsystems. Consequently, we can evaluate each subsystem relative to a specific subgoal, as well as to the higher-level goals within the system.

Systems and Their Components Have Inputs and Outputs

We can identify the inputs and outputs of each subsystem. The human factors specialist is particularly concerned with the input to the human from the machine and the actions that the human performs on the machine. Because the human subsystem can be broken down into its constituent subprocesses, we are also interested in the nature of the inputs and outputs from these subprocesses and how errors can occur.

A System Has Structure

The components of a system are organized and structured in a way that achieves a goal. This structure provides the system with its own special properties. In other words, the whole operating system has properties that emerge from those of its parts. By analyzing the performance of each component within the context of the system structure, the performance of the overall system can be controlled, predicted, and/or improved. To emphasize the emergent properties of a whole complex system, advocates of an approach called *cognitive work analysis* like to conceive of the entire system, humans and machines alike, as a single intelligent cognitive system rather than as separate human and machine subsystems (Sanderson, 2003).

Deficiencies in System Performance Are Due to Inadequacies of System Design or System Components

The performance of a system is determined by the nature of the system components and their interactions with each other. If the system design is appropriate for achieving certain goals, we must attribute system failures to the failure of one or more system components.

A System Operates within a Larger Environment

The system itself cannot be understood without reference to the larger physical and social environment in which it is embedded. If we fail to consider this environment in system design and evaluation, we will make an inadequate assessment of the system. Although it is easy to say that there is a distinction between the system and its environment, the boundary between them is not always clearly defined, just as the boundaries between subsystems are not always clearly defined. For example, a data-management expert works at a computer workstation in the immediate environment of his or her office, but this office resides in the environment created by the policies and guidelines mandated by his or her employer.

SYSTEM VARIABLES

A system consists of all the machinery, procedures, and operators carrying out those procedures, which work to fulfill the system goal. There are two kinds of systems: mission-oriented and service-oriented (Meister, 1991). Mission-oriented systems subordinate the needs of their personnel to the goal of the mission (Guo, Wang, Guo, & Si, 2013). These systems, such as weapon and transport systems, are common in the military, and failure of the system means that the mission must be terminated. Service-oriented systems cater to personnel, clients, or users (Chang, 2010). The service provider delivers some product or process to the client in a mutually agreed-upon transaction. Service systems include supermarkets, Internet providers, and offices.

Most systems fall between the extremes of mission and service orientations and involve components of both. For example, an automobile assembly plant has a mission component, that is, the goal of building a functional vehicle. However, it also has a service component in that the vehicle is being built for a consumer. Furthermore, assembly line workers, whose welfare is of concern to the system designers, build the vehicle. The company must service these workers to fulfill its mission to build automobiles.

The variables that define a system's properties, such as the size, speed, and complexity of the system, in part determine the requirements of the operator necessary for efficient operation of the system. Following Meister (1989), we can talk about two types of system variables. One type describes the functioning of the physical system and its components, whereas the other type describes the performance of individual and team operators. Table 3.1 lists some variables of each type.

Physical System Variables

Physical systems are distinguished by their organization and complexity. Complexity is a function of the number and arrangement of subsystems. How many subsystems operate at any one time, which subsystems receive inputs from and direct outputs to the other subsystems, and the ways that the subsystems or components are connected, all contribute to system complexity.

The organization and complexity of the system determine interdependencies among subsystems. Subsystems that depend on others for their input and those that must make use of a common resource pool to operate are interdependent. For interdependent subsystems, the operation of one subsystem directly influences the operation of another because it provides inputs and uses resources required by another subsystem.

TABLE 3.1

System Variables Identified by Meister (1989)

Physical System Variables

1. Number of subsystems
2. Complexity and organization of the system
3. Number and type of interdependencies within the system
4. Nature and availability of required resources
5. Functions and tasks performed by the system
6. Requirements imposed on the system
7. Number and specificity of goals
8. Nature of system output
9. Number and nature of information feedback mechanisms
10. System attributes: for example, determinate/indeterminate, sensitive/insensitive
11. Nature of the operational environment in which the system functions

Operator Variables

1. Functions and tasks performed
2. Personnel aptitude for tasks performed
3. Amount and appropriateness of training
4. Amount of personnel experience and skill
5. Presence or absence of reward and motivation
6. Fatigue or stress condition
7. The physical environment for individual or team functioning
8. Requirements imposed on the individual or team
9. Size of the team
10. Number and type of interdependencies within the team
11. The relationship between individual/team and other subsystems

An important characteristic of a system has to do with feedback. Feedback refers to input or information flow traveling backward in the system. Different systems may have different kinds of feedback mechanisms, and often more than one. Feedback usually provides information about the difference between the actual and the desired state of the system. Positive feedback is added to the system input and keeps the state of the system changing in its present direction. Such systems are usually unstable, because positive information flow can amplify error instead of correcting it. The alternative to positive feedback is negative feedback, which is subtracted from the system input. It is often beneficial for a system to include negative feedback mechanisms.

Suppose, for instance, that a system's goal is to produce premixed concrete. A certain amount of concrete requires some amount of water for mixing. If too much water is added, sand can be introduced to the mixture to dry it. A negative feedback loop would monitor the water content of the mixture, and this information would be used to direct the addition of more water or more sand until the appropriate mix had been achieved.

Systems that make use of feedback are called *closed-loop systems* (see Figure 3.1b). In contrast, systems that do not use feedback are referred to as *open-loop systems* (see Figure 3.1a). Closed-loop systems that use negative feedback are error correcting because the output is continuously monitored. In contrast, open-loop systems have no such error-detection mechanisms. In complex systems, there may be many feedback loops at the different hierarchical levels of the system.

The goals, functions, organization, and complexity of a system determine its attributes. As one example, a system can be relatively sensitive or insensitive to deviations in inputs and outputs. A small change in airflow probably will not affect the systems in a typical office building, but it might be devastating for the systems in a chemical processing plant. Also, systems can be determinate or

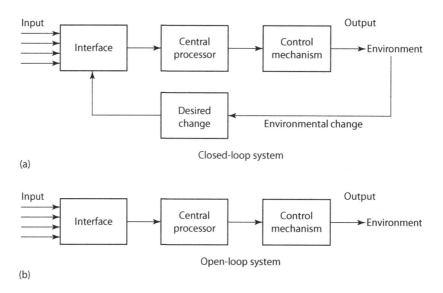

FIGURE 3.1   (a) Closed- and (b) open-loop systems.

indeterminate. Determinate systems are highly proceduralized. Operators follow specific protocols and have little flexibility in their actions. Indeterminate systems are not as highly proceduralized, and there is a wide range of activities in which the operators can engage. Also, in indeterminate systems, the operator's response might be based on ambiguous input, with little feedback.

Finally, systems operate in environments that may be friendly or unfriendly. Adverse conditions, such as heat, wind, and sand, take their toll on system components. For the system to operate effectively, the components must be able to withstand the environmental conditions around them.

Operator Variables

The requirements for system operators depend on the functions and tasks that must be performed for effective operation of the system. To perform these tasks, operators must meet certain aptitude and training requirements. For example, fighter pilots are selected according to aptitude profiles and physical characteristics, and they also must receive extensive training.

Performance is also affected by motivation, fatigue, and stress. Depending on the levels of these factors, a person's performance can vary from good to bad. Consider the problem of medical errors. Before July 2003, hospital residents routinely worked more than 80 hours per week in shifts that frequently exceeded 30 hours (see Lamberg, 2002). Because the guidelines differed at different hospitals, the true workload was unknown. There are many reasons why such demanding schedules are required of doctors-in-training, one being that the long duty shifts allow young doctors to observe the course of an illness or trauma from start to finish.

However, sleep deprivation contributes to medical mistakes (Landrigan et al., 2004). A number of consumer advocacy and physicians' groups petitioned the U.S. Occupational Health and Safety Administration (OSHA) to establish national limits on work hours for resident physicians in the early 2000s. Facing pending legislation, the U.S. Accreditation Council for Graduate Medical Education (ACGME) set new rules, which took effect in July 2003 and were updated in 2011, that restrict the number of resident work hours to no more than 80 a week, in shifts no longer than 30 hours (16 hours for first-year residents), and provide at least one day off in seven.

In the year after the ACGME's new duty-hour standards, however, over 80% of interns reported that their hospitals were not complying with the standards, and they had been obliged to work more than 80 hours per week. Over 67% of interns reported that their shifts had been longer than 30 hours, and these violations had occurred during one or more months over the year (Landrigan et al., 2006).

A more recent study after implementation of the 2011 standards shows some improvement, but more than half of the residents and program directors still reported violating the standards (Drolet, Schwede, Bishop, & Fischer, 2013). It is still difficult to determine the extent to which teaching hospitals are complying with the new standards: Residents are reluctant to report violations, and they often choose to work longer hours than they are scheduled to care for their patients (Gajilan, 2006). Regardless, residents still make many more mistakes at the end of an extended shift, such as stabbing themselves with needles or cutting themselves with scalpels (Ayas et al., 2006), and their risk of being involved in auto accidents traveling home after their shift is greatly increased (Barger et al., 2005).

The demands placed on an individual will vary across different physical environments even in the absence of complicating factors such as stress and fatigue. Variables like temperature, humidity, noise level, illumination, and so on may exert their effects through increasing stress and fatigue (see Chapter 17). Also, when several people must work together to operate a system, team factors become important. The size of the team and the interrelations among the various team members influence the efficiency with which the team operates and, hence, the efficiency with which the system operates.

SUMMARY

The system concept, as developed in the field of systems engineering, is fundamental to the discipline of human factors. This is exemplified by the terms *human–machine system* and *human–system integration*. We must think about a system in terms of both its physical and mechanical variables and its individual and team operator variables. We must evaluate the performance of the operators with respect to the functioning of the entire system. The assumptions and implications of the system concept dictate the way that researchers and designers approach applied problems. The system concept is the basis for reliability analysis, which we consider later in the chapter, as well as for the information-processing approach to human performance, which is discussed in Chapter 4.

HUMAN ERROR

On the night of November 24, 2014, a private jet attempting to take off from Biggin Hill Airport, U.K., ran off the runway and was damaged beyond repair. The final accident report summarized the incident as follows:

> The aircraft lined up for takeoff in conditions of reduced visibility. The crew believed that the lights they could see ahead were runway centreline lights when they were actually runway edge lights. The aircraft began its takeoff run but ran off the paved surface and onto grass. The commander closed the thrust levers to reject the takeoff.
>
> Information available to the pilots allowed them to develop an incorrect mental model of their route from the holding point to the runway. Environmental cues indicating that the aircraft was in the wrong position for takeoff were not strong enough to alert the pilots to the fact that they had lost situational awareness. (AAIB Bulletin, 12/2015)

This is just one example of human error, which is often invoked as a contributing factor to a disaster. When human error is said to be a contributing factor, it means that something that the operator or user did or did not do played a role in the mishap. In this case, the report identified contributing human factors relating to visual sensory processing (poor visibility) and comprehension of the information (incorrect mental model and lack of situational awareness), topics that we cover in later chapters.

Because human error has multiple sources and can be examined from multiple perspectives, it is difficult to define. One useful definition is that a human error occurs when an action is taken that was "not intended by the actor; not desired by a set of rules or an external observer; or that led the task or system outside its acceptable limits" (Senders & Moray, 1991, p. 25). Therefore, we see that whether an action is considered to be an error is determined by the goals of the operator and of the system. In some situations a slow or sloppy control action may not qualify as error, but in other situations it might.

For example, in normal flight, displacements of an aircraft a few meters above or below an intended altitude are not crucial and would not be classified as errors. However, in stunt flying, and when several planes are flying in formation, slight deviations in altitude and timing can be fatal. In 1988, at a U.S. air base in Ramstein, West Germany, three Italian Air Force jets collided in one of the worst-ever air show disasters. A total of 70 people, including the three pilots, were killed when one of the planes collided with two others and crashed into a crowd of spectators. The collision occurred when the jets were executing a maneuver in which one jet was to cross immediately above five jets flying in formation. A former member of the flying team concluded, "Either the soloist was too low or the group was too high … In these situations a difference of a meter can upset calculations … [This deviation could have been caused by] a sudden turbulence, illness or so many other things" (1988, UPI wire story). In this case, the system failed because of a very slight altitude error.

The principal consideration of the human factors specialist is with system malfunctions that involve the operator. Although we typically refer to such errors as human errors, they frequently are attributable to the design of the human–machine interface and/or the training provided to the operator (Peters & Peters, 2006). Thus, the failure of a technological system often begins with its design. The system design can put the user in situations for which success cannot be expected. We restrict the term *operator error* to refer to those system failures that are due entirely to the human and the term *design error* to refer to those human errors that are due to the system design.

WHY HUMAN ERROR OCCURS

There are several viewpoints about what causes human error (Wiegmann & Shappell, 2001). From one perspective, human error can be traced to inadequacies of the system design: Because the system involves humans and machines operating within a work environment, the human is rarely the sole cause of an error. Inadequacies of system design fall into three groups (Park, 1987): task complexity, error-likely situations, and individual differences.

Task complexity becomes an issue when task requirements exceed human capacity limits. As we will see in later chapters, people have limited capacities for perceiving, attending, remembering, calculating, and so on. Errors are likely to occur when the task requirements exceed these basic capacity limitations. An error-likely situation is a general situational characteristic that predisposes people to make errors. It includes factors such as inadequate workspace, inadequate training procedures, and poor supervision. Finally, individual differences, which we talked about in Chapter 2, are the attributes of a person, such as abilities and attitudes, which in part determine how well he or she can perform the task (Joe & Boring, 2014). Some important individual differences are susceptibility to stress and inexperience, which can produce as much as a tenfold increase in human error probability (Miller & Swain, 1987).

A second view about the causes of error is oriented around the cognitive processing required to perform a task (Manchi, Gowda, & Hanspal, 2013). One assumption of cognitive models (described more fully in Chapter 4) is that, in the brain, information progresses through a series of processing stages from perception to initiation and control of action. Errors occur when one or more of these intervening processes produce an incorrect output. For example, if a person misperceives a display indicator, the bad information will propagate through the person's cognitive system and lead to actions that result in an error because decisions are based on this information.

A third view of human errors, popular within the context of aviation, borrows from an aeromedical perspective (Raymond & Moser, 1995), that is, one involving the medical aspects of physiological and psychological disorders associated with flight. From this view, errors can be attributed to an underlying physiological condition. This approach emphasizes the role of physiological status in affecting human performance. This perspective has been responsible for much of the attention devoted to the factors of fatigue and emotional stress, and how they are influenced by work schedules and shift rotations.

Two final views emphasize group interactions and their effects on human error with an emphasis on psychosocial and organizational perspectives (Dekker, 2005; Perrow, 1999). The psychosocial perspective looks at performance as a function of interactions among many people. This is particularly relevant for commercial aviation, where there are several members of the flight crew, each with different duties, air-traffic controllers with whom they must communicate, and the flight attendants who interact with both passengers and crew. In addition, ground crews supervise the loading and fueling of the aircraft, and maintenance personnel work on maintaining the aircraft in good condition. The psychosocial perspective emphasizes that errors occur when communications among the group members break down.

The organizational perspective (Drews, 2012), which emphasizes the roles played by managers, supervisors, and other people in an organizational hierarchy, is important in industrial settings. The risky, and ultimately fatal, decision to launch the space shuttle *Challenger* on a cold morning, despite concern expressed by engineers that the O-rings would not properly seal the joints at low temperatures, is one of the most well-known incidents in which social and organizational dynamics were significant contributors to a disaster. Such errors are often called *management errors* (Taylor, 2016).

Error Taxonomies

It is useful to discuss human error with a taxonomy, a scheme for categorizing different kinds of errors. There are many useful error taxonomies (Stanton, 2006a; Stanton, & Salmon, 2009). Some refer to the type of action taken or not taken, others to particular operational procedures, and still others to the location of the error in the human information-processing system. We will describe the taxonomies of action, failure, processing, and intentional classification, and the circumstances under which each is most appropriate.

Action Classification

Some errors can be traced directly to an operator's action or inaction (Meister & Rabideau, 1965). An *error of omission* is made when the operator fails to perform a required action. For example, a worker in a chemical waste disposal plant may omit the step of opening a valve in the response sequence to a specific emergency. This omission might be in relation to a single task (failing to open the valve) within a more complicated procedure, or an entire procedure (failing to respond to an emergency). An *error of commission* occurs when an action is performed, but it is inappropriate. In this case, the worker may close the valve instead of opening it.

We can further subdivide commission errors into timing errors, sequence errors, selection errors, and quantitative errors. A *timing error* occurs when a person performs an action too early or too late (e.g., the worker opened the valve but too late for it to do any good). A *sequence error* occurs when she performs steps in the wrong order (e.g., she opened the valve but before waste had been diverted to that valve). A *selection error* occurs when she manipulates the wrong control (e.g., she opened a valve but it was the wrong one). Finally, a *quantitative error* occurs when she makes too little or too much of the appropriate control manipulation (e.g., she opened the valve but not wide enough).

Failure Classification

An error may or may not inevitably lead to a system failure. This is the distinction between recoverable and nonrecoverable errors. Recoverable errors are ones that can potentially be corrected and their consequences minimized. In contrast, nonrecoverable errors are those for which system failure is inescapable. Human errors are most serious when they are nonrecoverable. Recoverable errors require feedback to make the operator aware of the error and the actions he should take to recover the system. All systems should be designed to provide feedback to the operator so that errors are recoverable.

Human-initiated system failures can arise because of operating, design, assembly, or installation/maintenance errors (Meister, 1971). An *operating error* occurs when a machine is not operated according to the correct procedures. A *design error* can occur when the system designer creates an error-likely situation by failing to consider human tendencies or limitations. An *assembly* or *manufacturing* error arises when a product is misassembled or faulty, and an *installation* or *maintenance* error occurs when machines are either installed or maintained improperly.

In 1989, British Midlands flight 092 from London to Belfast reported vibration and a smell of fire in the cockpit—signs of an engine malfunction. Although the malfunction occurred in the Number 1 engine, the crew throttled back the Number 2 engine and tried to make an emergency landing using only the malfunctioning engine. During the landing approach, the Number 1 engine lost power, resulting in a crash 900 meters short of the runway. Forty-seven of the 126 passengers and crew were killed. You might think that advances in cockpit design since the 1980s would have eliminated the possibility of erroneously shutting down the functioning engine instead a malfunctioning one, but a similar accident occurred in February, 2015, when a TransAsia Airways flight from Taipei that had lost power in one engine crashed after the pilot exclaimed, "Wow, pulled back the wrong side throttle" (quoted in Hung & Govindasamy, 2015).

With regard to the British Midlands flight, at first, investigators speculated that there had been a maintenance error: Possibly the fire-warning panel of the Boeing aircraft was miswired to indicate that the wrong engine was on fire. The investigation revealed that in fact the panel had not been miswired. However, the tragedy led to inspections of other Boeing aircraft, which revealed 78 instances on 74 aircraft of miswiring in the systems designed to indicate and extinguish fires (Fitzgerald, 1989). To avoid future wiring errors during assembly and maintenance, Boeing redesigned the panel wiring connectors so that each would be a unique size and miswiring would be impossible.

Processing Classification

We can also classify errors according to their locus within the human information-processing system (Berliner, Angell, & Shearer, 1964; see Table 3.2). *Perceptual errors* are those attributable to sensory and perceptual processes. *Mediational errors* reflect the cognitive processes that translate between perception and action. *Communication errors* involve inaccurate transmission of information between members of a team. *Motor errors* are those that are due to the selection and execution of physical responses. Table 3.2 lists specific behaviors for which errors of each type can occur. In

TABLE 3.2

Berliner's Processing Classification of Tasks

Processes	Activities	Example Behaviors
Perceptual	Searching for and receiving information	detect; inspect; observe; read; receive; scan; survey
Mediational	Identifying objects, actions, and events	discriminate; identify; locate
	Information processing	calculate; categorize; compute; encode; interpolate; itemize; tabulate; transfer
Communication	Problem solving and decision making	analyze; choose; compare; estimate; predict; plan
		advise; answer; communicate; direct; indicate; inform; instruct; request; transmit
Motor	Simple, discrete tasks	activate; close; connect; disconnect; hold; join; lower; move; press; raise; set
	Complex, continuous tasks	align; regulate; synchronize; track; transport

subsequent chapters, we will elaborate the details of the human information-processing system and spell out in more detail the specific sources of errors within it.

Rasmussen (1982) developed an information-processing failure taxonomy that distinguishes six types of failures: stimulus detection, system diagnosis (a decision about a problem), goal setting, strategy selection, procedure adoption, and action. An analysis of approximately 2000 U.S. Naval aviation accidents using Rasmussen's and other taxonomies concluded that major accidents had a different cognitive basis than minor ones (Wiegmann & Shappell, 1997). Major accidents were associated with judgment errors like decision making, goal setting, or strategy selection, whereas minor accidents were associated more frequently with procedural and response-execution errors. This study illustrates how a more detailed analysis may help to resolve issues about the nature of errors; in this case whether the causes of major accidents are fundamentally similar to or different from the causes of minor accidents.

Intention Classification

We can classify errors as slips or mistakes, according to whether or not someone performed the action that she intended. A *slip* is a failure in execution of action, whereas a *mistake* arises from errors in planning of action. Reason (1990) related the distinction between slips and mistakes to another taxonomy of behavior modes developed by Rasmussen (1986, 1987). According to this taxonomy, an operator is in a *skill-based* mode of behavior when performing routine, highly over-learned procedures. When situations arise that are relatively unique, the operator switches to a *rule-based* mode, where her performance is based on recollection of previously learned rules, or a *knowledge-based* mode, where performance is based on problem solving. Reason attributes slips to the skill-based mode and mistakes to either misapplication of rules or suboptimal problem solving.

Consider an operator in a nuclear power plant who intended to close pump discharge valves A and E but instead closed valves B and C inadvertently. This is a slip. If the operator used the wrong procedure to depressurize the coolant system, this is a mistake (Reason, 1990). For a slip, the deviation from the intended action often provides the operator with immediate feedback about the error. For example, if you have both mayonnaise and pickle jars on the counter when making a sandwich and intend to open the mayonnaise jar to spread mayonnaise on your sandwich, you will notice your error quickly if you slip and open the pickle jar instead. You do not get this kind of feedback when you make a mistake, because your immediate feedback is that the action you performed was executed correctly. It is the intended action that is incorrect, and so the error is more difficult to detect. Consequently, mistakes are more serious than slips. We can also identify a third category of errors, *lapses*, which involve memory failures such as losing track of your place in an action sequence (Reason, 1990).

There are three major categories of slips (Norman, 1981): faulty formation of an action plan, faulty activation of an action schema, and faulty triggering of an action schema. An action schema is an organized body of knowledge that can direct the flow of motor activity. We will talk more about action schemata in Chapter 14. For now, it is only important to understand that before an action is performed, it must be planned or programmed, and this is what an action schema does. Well-practiced or familiar actions may come from an action schema.

The faulty formation of an action plan is often caused by ambiguous or misleading situations. Slips resulting from poor action plans can be either mode errors, due to the misidentification of a situation, or description errors, for which the action plan is ambiguous or incomplete. Mode errors can occur when instruments, like a digital watch, have several display modes. You can misinterpret a display (e.g., reading the dial as the current time when the watch was displaying stopwatch time), and perform an action (e.g., turning off the oven) that would have been appropriate had the display been in a different mode.

The second category of slip, faulty activation of action schemas, is responsible for such errors as failing to make an intended stop at the grocery store while driving home from work. The highly overlearned responses that take you home are activated in place of the less common responses that take you to the grocery store. The third kind of slip, faulty triggering of one of several activated schemas, arises when a schema is triggered at the incorrect time or not at all. Common forms of such errors occur in speech. For example, "spoonerisms" are phrases in which words or syllables are interchanged. You might say, "You have tasted the whole worm" instead of "You have wasted the whole term," for example.

In contrast to slips, we can attribute mistakes to the basic processes involved in planning (Reason, 1987). First, all of the information a person needs to act correctly may not be part of the information she uses in the planning processes. The information she does use, selected according to a number of factors such as attention or experience, will then include only a small amount of potentially relevant information or none at all. Second, the mental operations she engages to plan an action are subject to biases, such as paying too much attention to vivid information, a simplified view of how facts are related, and so on. Third, once she formulates a plan, or sequence of action schemas, it will be resistant to modification or change; she may become overconfident and neglect to consider alternative action plans. Various sources of bias can lead to inadequate information on which to base the choice of action, unrealistic goals, inadequate assessment of consequences, and overconfidence in the formulated plan.

We can find an application of the slips/mistakes taxonomy in a study of human error in nursing care (Narumi et al., 1999). Records of reported accidents and incidents in a cardiac ward from August, 1996 to January, 1998 showed that 75 errors caused patients discomfort, and these were split about evenly between skill-based slips (36) and rule-based mistakes (35), with the remaining 4 errors being knowledge-based slips. Of 12 life-threatening errors, 11 were rule-based mistakes. The 12th error was a skill-based slip. There were only four errors involving procedural matters, with three due to skill-based slips and one a knowledge-based error. Note that, as for Wiegmann and Shappell's (1997) study of human error in aviation, major errors involved decisions (in this case, predominantly rule-based mistakes) and minor errors tended to be of a more procedural nature (action slips).

We can distinguish errors (slips, mistakes, and lapses) from violations, which involve disregard for the laws and rules that are to be followed (Reason, 1990; Wiegmann et al., 2005). *Routine violations* are those that occur on a regular basis, such as exceeding the speed limit when driving on the highway. They may be tolerated or encouraged by organizations or individuals in authority, as would be the case if they adopted a policy of not ticketing a driver for speeding unless the vehicle's speed was more than 10 miles per hour above the speed limit. As this example suggests, routine violations can be managed to some extent by authorities adopting appropriate policies. *Exceptional violations* are those that do not occur on a regular basis, such as driving recklessly in an attempt to get to the office of an overnight postal service before it closes for the day. Exceptional violations tend to be less predictable and more difficult to handle than routine violations.

Errors and violations both are unsafe acts performed by operators. Reason (1990) also distinguished three higher levels of human failure: organizational influences, unsafe supervision, and preconditions for unsafe acts. The Human Factors Analysis and Classification System (HFACS; Wiegmann & Shappell, 2003) provides a comprehensive framework for human error, distinguishing 19 categories of causal factors across the 4 different levels (see Figure 3.2). At the highest level are organizational influences, which include the organizational climate and process, and how resources are managed. These may lead to unsafe supervision, including inadequate supervision and violations on the supervisor's part, planning inappropriate operations, and failing to correct problems. Unsafe supervision may result in preconditions for unsafe acts, which can be partitioned into factors involving the physical and technical environments, conditions of operators (adverse mental and physiological states, as well as physical and mental limitations), and personnel factors (crew resource management and personnel readiness). The unsafe acts that may then occur are classified in a manner similar to the errors and violations described previously, but with a slightly different distinction made among the error categories.

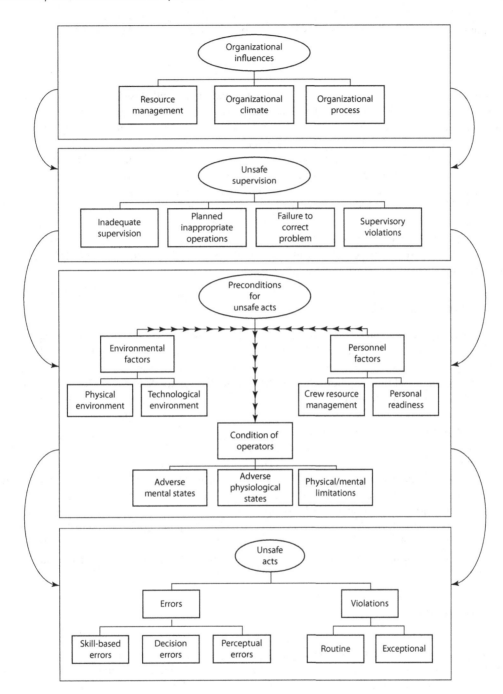

FIGURE 3.2   The human factors analysis and classification system framework.

The strength of HFACS is that it incorporates organizational, psychosocial, aeromedical, and cognitive approaches to human error within a single framework. HFACS provides a valuable tool for analyzing human error associated with general and commercial aviation (Wiegmann et al., 2005), military aviation (Li & Harris, 2005), remotely piloted aircraft (Tvaryanas, Thompson, & Constable, 2006), train accidents (Reinach & Viale, 2006), and surgical operating-room procedures (ElBardissi, Wiegmann, Dearani, Daly, & Sundt, 2007). Persons trained in classifying errors within HFACS show high interrater reliabilities for classifying errors into the 4 different tier categories and lower, but still good, reliabilities for classifying errors into the specific 19 categories encapsulated by the respective tiers (Ergai et al., 2016).

Summary

The four error taxonomies (action, failure, processing, and intentional classification) capture different aspects of human performance, and each has different uses. The action and failure classifications have been used with success to analyze human reliability in complex systems, but they categorize errors only at a superficial level. That is, errors that are considered to be instances of the same action category may have quite different cognitive bases. The processing and intentional classifications are "deeper" in the sense that they identify underlying causal mechanisms within the human operator, but they require us to make more assumptions about how people process information than do the action and failure classifications. Because the processing and intentional classifications focus on the root causes of the errors, they have the potential to be of greater ultimate use than the classifications based on surface error properties. HFACS, which incorporates these latter classifications within a context of organizational, psychosocial, and aeromedical factors, provides the best framework to date for comprehensively analyzing human error in complex systems.

RELIABILITY ANALYSIS

When a system performs reliably, it completes its intended function satisfactorily. The discipline of reliability engineering began to develop in the 1950s (Birolini, 2014). The central tenet of reliability engineering is that the total system reliability can be determined from the reliabilities of the individual components and their configuration in the system (O'Connor & Kleyner, 2012). Early texts (e.g., Bazovsky, 1961) and comprehensive works (e.g., Barlow & Proschan, 1965) provided quantitative bases for reliability analysis by combining the mathematical tools of probability analysis with the organizational tools of system analysis.

The successful application of reliability analyses to hardware systems led human factors specialists to apply similar logic to human reliability. The discipline of reliability engineering and its close relative, probabilistic risk assessment (PRA; Bahr, 2017), which estimates risks associated with failure, has shown increasing recognition of the importance of including estimates of human performance reliability as part of an overall reliability analysis of a complex system such as a nuclear power plant (Dhillon, 2009; La Sala, 1998). This is because human error is a contributing factor in the majority of serious incidents involving any complex system. In the sections that follow, we describe the basics of reliability analysis in general and then explain human reliability analysis in more detail.

SYSTEM RELIABILITY

Although it would be nice if constructed systems functioned well forever, they do not. The term *reliability* is used to characterize the dependability of performance for a system, subsystem, or component. We define reliability as "the probability that an item will operate adequately for a specified period of time in its intended application" (Park, 1987, p. 149). For any analysis of reliability to be meaningful, we need to know exactly what system performance constitutes "adequate" operation. The decision about what constitutes adequate operation will depend on what the system is supposed to accomplish.

There are three categories of failure for hardware systems: operating, standby, and on-demand failures (Dougherty & Fragola, 1988). An operating failure is one that occurs for continuously operated equipment; a standby failure is when a piece of equipment that is normally dormant malfunctions when it is required to operate; a demand failure is one that occurs in periodically operated equipment. At the time that this chapter was written, one of the authors was experiencing a building-wide air conditioning system failure. This failure was not an operating failure, because the air conditioning never came on. If the air conditioning had started working and then failed, we would have called the situation an operating failure. The people in charge of maintaining the air conditioning system argued that it was an on-demand failure: Although the system was adequately maintained during the winter, they claimed it could not be turned on when the weather became unseasonably warm. The building staff, on the other hand, having experienced intermittent operating failures of the same system during the previous warm season, argued that it was a standby failure: Poor maintenance of an already unreliable system resulted in a failure over the winter months when the system was not in operation.

A successful analysis of system reliability requires that we first determine an appropriate taxonomy of component failures. After this determination, we must estimate the reliabilities for each of the system components. The reliability of a component is the probability that it does not fail. Thus, the reliability r is equal to $1-p$, where p is the probability of component failure. When we know or can estimate the reliabilities of individual components, we can derive the overall system reliability by developing a mathematical model of the system using principles of probability. For these methods, we usually rely on empirical estimates of the probability p, or how frequently a particular system component has been observed to fail in the past.

When determining system reliability, a distinction between components arranged in *series* and in *parallel* becomes important (Dhillon, 1999). In many systems, components are arranged such that they all must operate appropriately if the system is to perform its function. In such systems, the components are in series (see Figure 3.3). When independent components are arranged in series, the system reliability is the product of the individual probabilities. For example, if two components, each with a reliability of 0.9, must both operate for successful system performance, then the reliability of the system is $0.9 \times 0.9 = 0.81$. More generally,

$$R = (r_1) \times (r_2) \times ... (r_n) = \prod_{i=1}^{n} r_i,$$

where:
  r_i is the reliability of the ith component, and
  R is the system reliability.

Remember two things about the reliability of a series of components. First, adding another component in series always decreases the system reliability unless the added component's reliability is 1.0 (see Figure 3.4). Second, a single component with low reliability will lower the system reliability considerably. For example, if three components in series each has a reliability of 0.95, the system reliability is 0.90. However, if we replace one of these components with a component whose reliability is 0.20, the system reliability drops to 0.18. In a serial system, the reliability can only be as great as that of the least reliable component.

Another way to arrange components is to have two or more perform the same function. Successful performance of the system requires that only one of the components operate appropriately. In other words, the additional components provide redundancy to guard against system

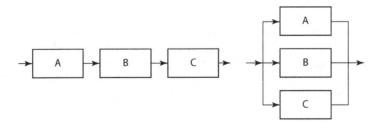

FIGURE 3.3 Examples of serial (left) and parallel (right) systems.

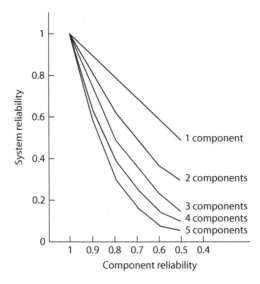

FIGURE 3.4 Reliability of a serial system as a function of number of task components and the reliability of each component.

failure. When components are arranged in this manner, they are *parallel* (see Figure 3.2). For a simple parallel system in which all components are equally reliable,

$$R = \left[1 - (1-r)^n \right],$$

where:

r is the reliability of each individual component, and
n is the number of components arranged in parallel.

In this case, we compute overall system reliability by calculating the probability that at least one component remains functional.

The formula for the reliability of a parallel system can be generalized to situations in which the components do not have equal reliabilities. In this case,

$$R = 1 - \left[(1-r_1)(1-r_2)...(1-r_n) \right] = 1 - \prod_{i=1}^{n} (1-r_i),$$

where r_i is the reliability of the ith component. When i groups of n parallel components with equal reliabilities are arranged in series,

$$R = \left[1 - (1-r)^n \right]^i.$$

More generally, the number of components within each group need not be the same, and the reliabilities for each component within a group need not be equal. We find the total system reliability by considering each of n subsystems of parallel components in turn. Let c_i be the number of components operating in parallel in the ith group, and let r_{ji} be the reliability of the jth component in the ith group (see Figure 3.5). The reliability for the ith subsystem is

$$R_i = 1 - \prod_{j=1}^{c_i} (1-r_{ji}).$$

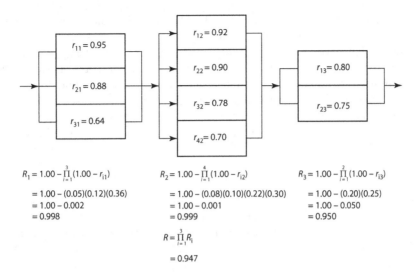

$$R_1 = 1.00 - \prod_{i=1}^{3}(1.00 - r_{i1})$$

$$= 1.00 - (0.05)(0.12)(0.36)$$
$$= 1.00 - 0.002$$
$$= 0.998$$

$$R_2 = 1.00 - \prod_{i=1}^{4}(1.00 - r_{i2})$$

$$= 1.00 - (0.08)(0.10)(0.22)(0.30)$$
$$= 1.00 - 0.001$$
$$= 0.999$$

$$R_3 = 1.00 - \prod_{i=1}^{2}(1.00 - r_{i3})$$

$$= 1.00 - (0.20)(0.25)$$
$$= 1.00 - 0.050$$
$$= 0.950$$

$$R = \prod_{i=1}^{3} R_i$$

$$= 0.947$$

FIGURE 3.5 Computing the reliability of a series of parallel subsystems.

Total system reliability, then, is the reliability of the series of parallel subsystems:

$$R = \prod_{k=1}^{n} R_k.$$

Whereas in serial systems the addition of another component dramatically decreases system reliability, in parallel systems it increases system reliability. It is clear from the expression for R_i that, as the number of parallel components increases, the reliability tends to 1.0. As an illustration, the system reliability for five parallel components each with a reliability of 0.20 is $1.0 - (1.0 - 0.20)^5 = 0.67$. When ten components each with a reliability of 0.20 are arranged in parallel, the system reliability is $1.0 - (1.0 - 0.20)^{10} = 0.89$. This makes sense if you think of all the components in a parallel system as "backup units." The more backup units you have, the greater the probability will be that the system will continue to function even if a unit goes bad.

Some effects on a system, such as the heat caused by a fire, are sudden. Other environmental processes, such as the effect of water on underwater equipment, affect the reliability of the system continuously over time. Consequently, we use two types of reliability measures. For demand or shock-dependent failures, $r = P[S < \text{capacity of the object}]$. That is, reliability is defined as the probability that the level of shock S does not exceed the capacity of the equipment to withstand the shock during the equipment's operation. For time-dependent failures, $r(t) = P[T > t]$, where T is the time of the first failure. In other words, reliability for time-dependent processes is defined as the probability that the first failure occurs after time t. When we have to consider many components simultaneously, as within the context of a large system, time-dependent reliability analysis can be extremely difficult.

We will talk a lot about *models* in this text. A model is an abstract, simplified, usually mathematical representation of a system. The model has parameters that represent physical (measurable) features of the system, such as operating time or failure probabilities, and the structure of the model determines how predictions about system performance are computed. Later in this chapter, and later in this book, we will talk about models that represent the human information-processing system. Such models do not always represent how information is processed very accurately, and sometimes it is very difficult to interpret their parameters. However, the power of these models is in the way they simplify very complex systems and allow us to make predictions about what the system is going to do.

There is considerable debate about whether the focus of reliability analysis should be on empirically based quantitative models of system architecture, like the serial and parallel models we have described in this section, or on "physics-of-failure" models (Czichos, 2013). Physics-of-failure models are concerned with identifying and modeling the physical causes of failure, and advocates of this approach have argued that reliability predictions using it can be more accurate than those derived from empirical estimates of failure probabilities. As we shall see in the next section, the human reliability literature has been marked by a similar debate between models that focus on the reliabilities of observable actions and models that focus on the cognitive processes that underlie these actions.

Human Reliability

We can apply procedures similar to those used to determine the reliability of inanimate systems to the evaluation of human reliability in human–machine systems (Spurgin, 2010). In fact, to perform a probabilistic safety analysis of complex systems such as nuclear power plants, we must provide estimates of human error probabilities as well as machine reliabilities, since the system reliability is to a considerable extent dependent on the operators' performance. Human reliability analysis thus involves quantitative predictions of operator error probability and of successful system performance, although there has been increasing interest in understanding the causes of possible errors as well (e.g., Hollnagel, 1998; Kim, 2001).

Operator error probability is defined as the number of errors made (e) divided by the number of opportunities for such errors (O; e.g., Bubb, 2005):

$$P\left(\text{operator error}\right) = \frac{e}{O}.$$

Human reliability thus is 1 − P(operator error). Just as we can classify hardware failures as time-dependent and time-independent, we can also classify operator errors.

We can carry out a human reliability analysis for both normal and abnormal operating conditions. Any such analysis begins with a task analysis that identifies the tasks performed by humans and their relation to the overall system goals (see Box 3.1). During normal operation, a person might perform the following important activities (Whittingham, 1988): routine control (maintaining a system variable, such as temperature, within an acceptable range of values); preventive and corrective maintenance; calibration and testing of equipment; restoration of service after maintenance; and inspection. In such situations, errors of omission and commission occur as discrete events within the sequence of a person's activity. These errors may not be noticed or have any consequence until abnormal operating conditions arise. Under abnormal operating conditions, the person recognizes and detects fault conditions, diagnoses problems and makes decisions, and takes actions to recover the system. Although action-oriented errors of omission and commission still can occur during recovery, perceptual and cognitive errors become more likely.

Human reliability analyses are based on either *computational methods*, which analyze errors and their probabilities, or *Monte Carlo methods*, which simulate performance on the basis of a system model (Boff & Lincoln, 1988). The steps for performing such analyses are shown in Figure 3.6. As in any system/task analysis, the first step for both methods involves a description of the system: that is, its components and their functions. For the computational method, after we describe the system, we identify potential errors for each task that must be performed and estimate the likelihoods and consequences of each error. We then use these error probabilities to compute the likelihood that the operator accomplishes her or his tasks appropriately and the probability of success for the entire system. Error probabilities can come from many sources, described later; they must be accurate if the computed probabilities for successful performance of the operator and the system are to be meaningful.

BOX 3.1 TASK ANALYSIS

A first step in human reliability analysis is to perform a task analysis. Such an analysis examines in detail the nature of each component task, physical or cognitive, that a person must perform to attain a system goal, and the interrelations among these component tasks. Task analysis is also a starting point in general for many other human factors concerns, including the design of interfaces and the development of training routines. A fundamental idea behind task analysis of any type is that tasks are performed to achieve specific goals. This emphasis on task and system goals is consistent with the importance placed on system goals in systems engineering, and it allows the task analysis to focus on ways to structure the task to achieve those goals.

As we discussed in Chapter 1, Taylor (1911) and Gilbreth (1909) developed the first task analysis methods. They analyzed physical tasks in terms of motion elements and estimated the time to perform the whole task by adding together the time for each individual motion element. In so doing, Taylor and Gilbreth could redesign tasks to maximize the speed and efficiency with which they could be performed. Taylor and Gilbreth's approaches focused primarily on physical work and, consequently, were applicable primarily to repetitive physical tasks of the type performed on an assembly line. During the century that has passed since their pioneering efforts, the nature of work has changed, and, consequently, many different task analysis methods have been developed to reflect these changes (Diaper & Stanton, 2004; Strybel, 2011).

One of the most widely used task analysis methods is *hierarchical task analysis* (Annett, 2004; Stanton, 2006b). In hierarchical task analysis, the analyst uses observations and interviews to infer the goals and subgoals for a task; the operations, or actions, that a person must perform in order to achieve those goals; and the plans that specify the relations among the component operations. The end result is a diagram specifying the structure of the task. An example of a hierarchical task analysis for a simple task, selecting an item from a pop-up menu in a computer application, is shown in Figure B3.1 (Schweickert, Fisher, & Proctor, 2003). This diagram shows the goal (selecting an item), three elementary operations (search the menu, move the cursor, and double click), and the plan specifying the order of these operations. Of course, the diagrams for most tasks will be considerably more complex than this.

One of the major changes in jobs and tasks with increasingly sophisticated technology is an increase in cognitive demands and a decrease in physical demands in many work environments. Consideration of cognitive demands is the primary concern for computer interface design, which is the target of much current work on task analysis (Diaper & Stanton, 2004). Consider a website, for example. The information that needs to be available at the site may be quite complex and varied in nature, and different visitors to the site may have different

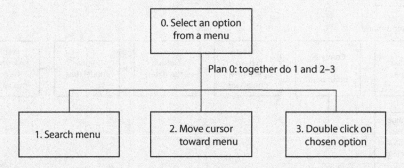

FIGURE B3.1 Example hierarchical task analysis for selecting an item from a pop-up menu.

goals. Task analyses must evaluate the goals that users have in accessing this information, the strategies they employ in searching for the information, how to structure the information to allow users to be able to achieve their goals, and the best ways to display this information to maximize the efficiency of the search process (Strybel, 2011).

The term *cognitive task analysis* refers to techniques that analyze the cognitive activity of the user or operator, rather than the user's observable physical actions (May & Barnard, 2004; Schraagan, Chipman, & Shalin, 2000). The most widely used analysis method of this type, which was developed explicitly for human–computer interaction, is the GOMS model and its variants (John, 2003), described in more detail in Chapter 19. GOMS stands for goals, operators, methods, and selection rules. With a GOMS analysis, a task is described in terms of goals and subgoals, and methods are the ways in which the task can be carried out. A method specifies a sequence of mental and physical operators; when more than one method exists for achieving the task goal, a selection rule is used to choose which is employed. A GOMS model can predict the time to perform a task by estimating the time for each of the individual operations that must be performed in order to accomplish the task goal.

For the Monte Carlo method, the second step is to model the system in terms of task interrelations. At this stage, we must make decisions about the random behavior of task times (e.g., are they normally distributed?) and select success probabilities to simulate the operations of the human and the system. We repeat the simulation many times; each time, it either succeeds or fails in accomplishing its task. The reliability of the human or system is the proportion of times that the task is completed in these simulations.

The computational and Monte Carlo methods are similar in many respects, but each has its own strengths and weaknesses. For example, if the computational method is to be accurate, we must perform detailed analyses of the types of errors that can occur, as well as their probabilities

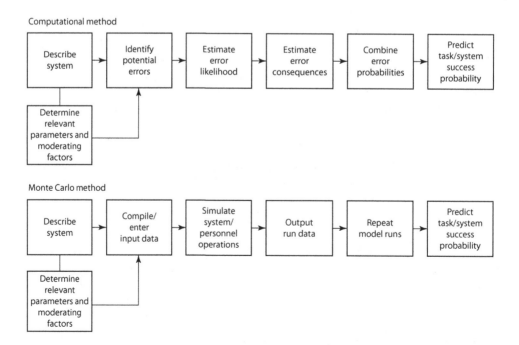

FIGURE 3.6   Computational and Monte Carlo methods of conducting human reliability analysis.

and consequences. The Monte Carlo method, in turn, requires us to develop accurate models of the system.

There are many ways to perform a human reliability analysis. One review summarized 35 techniques that had either direct or potential application to the field of healthcare (Lyons, Adams, Woloshynowych, & Vincent, 2004). Kirwan (1994) provides a more detailed review of several major techniques available for the quantification of human error probabilities and discusses guidelines for the selection and use of techniques. There is a difference between first- and second-generation techniques (Hollnagel, 1998; Kim, 2001), although they overlap somewhat. First-generation techniques closely follow those of a traditional reliability analysis by analyzing human task activities instead of machine operations. They typically emphasize observable actions, such as errors of commission and omission, and place little emphasis on the cognitive processing underlying the errors. The second-generation techniques are much more cognitive in nature. We will provide detailed examples of two first-generation techniques, one that uses the Monte Carlo method (the stochastic modeling technique) and another that uses the computational method (THERP), and two associated more recent relatives of these (SHERPA and TAFEI). We will then describe three representative second-generation techniques (HCR, ATHEANA, and CREAM).

Technique for Human Error Rate Prediction (THERP)

THERP, developed in the early 1960s, is one of the oldest and most widely used of the computational methods for human reliability analysis (Swain & Guttman, 1983). It was designed initially to determine human reliability in the assembly of bombs at a military facility, and it subsequently has been the basis of reliability analyses for industry and nuclear facilities (Bubb, 2005).

The reliability analyst using THERP proceeds through a series of steps (Miller & Swain, 1987):

1. Determine the system failures that could arise from human errors.
2. Identify and analyze the tasks performed by the personnel in relation to the system functions of interest.
3. Estimate the relevant human error probabilities.
4. Integrate the human reliability analysis with a system reliability analysis to determine the effects of human errors on the system performance.
5. Recommend changes to the system to increase the reliability, and then evaluate these changes.

The most important steps in THERP are the third and fourth. These involve determining the probability that an operation will result in an error and the probability that a human error will lead to system failure. Such probabilities can be estimated from a THERP data base (Swain & Guttmann, 1983) or from any other data, such as simulator data, that may be relevant.

Figure 3.7 depicts these probabilities in an event tree diagram. In this figure, a is the probability of successful performance of task 1, and A is the probability of unsuccessful performance. Similarly, b and B are the probabilities for successful and unsuccessful performance of task 2. The first branch of the tree thus distinguishes the probability of performing or not performing task 1. The second level of branches involves the probabilities of performing or not performing task 2 successfully, depending on the performance of task 1. If the two tasks are independent (see Chapter 2), then the probability of completing task 2 is b and of not completing it is B. If we know the probability values for the individual component tasks, we can compute the probability of any particular combination of performance or nonperformance of the tasks, as well as the overall likelihood for total system failure resulting from human error.

As an example, suppose that we need to perform a THERP analysis for a worker's tasks at one station on an assembly line for portable radios. The final assembly of the radio requires that the electronic components be placed in a plastic case. To do this successfully, the worker must bend a wire for the volume control to the underside of the circuit board and snap the two halves of the case together. If the worker fails to wrap the wire around the board, the wire may be damaged when he

FIGURE 3.7 Task/event tree diagram.

closes the case. He might also crack the case during assembly. The probability that he positions the wire correctly is 0.85, and the probability that he does not crack the case is 0.90. Figure 3.8 illustrates the event tree for these tasks. The probability that the radio is assembled correctly is 0.765. The benefit of the THERP analysis in this example is that weaknesses in the procedure, such as the relatively high probability of poorly placing the wire, can be identified and eliminated to increase the final probability of correct assembly.

Though THERP compares favorably to other human reliability assessment techniques for quantifying errors (Kirwan, 1988), the THERP error categorization procedure relies on the action classification described earlier; that is, on errors of omission and commission. This focus is problematic (Hollnagel, 2000). Because THERP relies on an event tree (Figure 3.8), we see each step in a sequence of actions as either a success or a failure. Categorizing errors in this way is independent of the human information processes that produce the specific errors. More recent techniques, such as the Human Cognitive Reliability (HCR) model discussed later, place more emphasis on the processing basis of errors. The importance of THERP cannot be overstated, though, as Boring (2012)

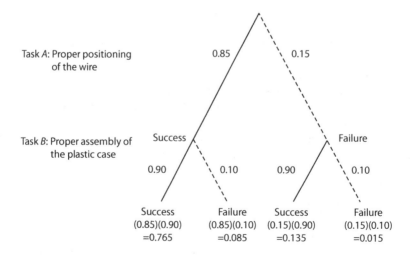

FIGURE 3.8 Event tree diagram for the assembly of portable radios.

notes: "All subsequent HRA [human reliability analysis] methods are derived as a refinement of THERP or as an attempt to address perceived shortcomings with the original technique" (p. 10).

Stochastic Modeling Technique

An example of the Monte Carlo method of human reliability analysis is the stochastic modeling technique developed by Siegel and Wolf (1969). The technique is intended to determine whether an average person can complete all tasks in some allotted time, and to identify the points in the processing sequence at which the system may overload its operators (Park, 1987). It has been applied in complex situations, such as landing an aircraft on a carrier, in which there are many subtasks that the operator must execute properly. The model uses estimates of the following information:

1. The mean time to perform a particular subtask; the average variability (standard deviation) in performance time for a representative operator;
2. The probability that the subtask will be performed successfully;
3. An indication of how essential successful performance of the subtask is to completion of the task;
4. The subtask that is to be performed next, which may differ as a function of whether or not the initial subtask is performed successfully.

We make three calculations based on these data for each subtask (Park, 1987). First, urgency and stress conditions are calculated according to the subtasks to be performed by the operator in the remaining time. Second, a specific execution time for the subtask is selected by randomly sampling from an appropriate distribution of response times. Finally, whether the subtask was performed correctly is determined by random sampling using the probabilities for successful and unsuccessful performance.

The stochastic modeling technique is used to predict the efficiency of the operator within the entire system based on the simulated performance of each subtask. This technique has been applied with reasonable success to a variety of systems. Moreover, it has been incorporated into measures of total system performance.

Systematic Human Error Reduction and Prediction Approach (SHERPA) and Task Analysis for Error Identification (TAFEI)

SHERPA (Embrey, 1986; Stanton, 2013) and TAFEI (Stanton & Baber, 2005) are related methods that can be used easily to predict human errors when a person is interacting with a device. The first step for both is a hierarchical task analysis (see Box 3.1) that decomposes work activities into a hierarchy of goals, operations to achieve the goals, and plans for executing these operations in an appropriate sequence. The resulting task hierarchy provides the basis for determining possible errors and their relative likelihood.

To use SHERPA, the reliability analyst takes each operation at the lowest level of the task hierarchy and classifies it as one of five types: action, retrieval, checking, selection, or information communication. For each operation, he must identify several possible error modes. For example, an action error may be one of mistiming the action, or a checking error may be one of omitting the check operation. He then considers the consequences of each error, and for each, whether the operator could take any recovery action. He will assign a "probability" of low if the error is unlikely to ever occur, medium if it occurs on occasion, and high if it occurs frequently. He also designates each error as critical (if it would lead to physical damage or personal injury) or not critical. In the last step, the analyst provides strategies for error reduction. The structured procedure and error taxonomy makes SHERPA relatively easy to perform, but the analysis does not consider cognitive bases of errors.

To use TAFEI, after first performing the hierarchical task analysis, the analyst constructs state space diagrams that represent a sequence of states through which the device can pass until it reaches its goal. For each state in the sequence, he will indicate links to other system states to represent the possible actions that can be taken to move the system from the present state to another state. He then enters this information into a transition matrix that shows the possible transitions from different current states to other states. The matrix records legal transitions as well as illegal, error transitions. This procedure results in design solutions that make it impossible for a user to make illegal transitions. TAFEI and SHERPA, when used in combination, will allow the analyst to make very accurate reliability predictions.

Human Cognitive Reliability Model

First-generation models such as the stochastic modeling technique and THERP are primarily concerned with predicting whether humans will succeed or fail at performing various tasks and subtasks. Second-generation models are more concerned with what the operator will do. The HCR model, developed by Hannaman, Spurgin, and Lukic (1985), is one of the earliest second-generation models because of its emphasis on human cognitive processes. The approach was developed to model the performance of an industrial plant crew during an accident sequence. Because the time to respond with appropriate control actions is limited in such situations, the model provides a way to estimate the probability of time-dependent operator failures (nonresponses). The input parameters to the model are of three types: category of cognitive behavior, median response time, and environmental factors that shape performance.

As with all the other techniques, the human reliability analyst first identifies the tasks the crew must perform. Then, she must determine the category of cognitive process required for each task. HCR uses the categories from Rasmussen's (1986, 1987) taxonomy described earlier: skill-based, rule-based, and knowledge-based behavior. Recall that skill-based behavior represents the performance of routine, overlearned activities, whereas rule-based and knowledge-based behaviors are not as automatic. Rule-based behavior is guided by a rule or procedure that has been learned in training, and knowledge-based behavior occurs when the situation is unfamiliar (see earlier discussion in this chapter). HCR is based on the idea that the median time to perform a task will increase as the cognitive process changes from skill-based to rule-based to knowledge-based behavior.

The analyst estimates the median response times for a crew to perform its required tasks from a human performance data source, some of which are described in the next section. She then modifies these times by incorporating performance-shaping environmental factors such as level of stress, arrangement of equipment, and so on. She must also evaluate response times according to the time available to perform the task, so providing a basis for deciding whether the crew will complete the required tasks in the available time.

The most important part of the HCR model is a set of normalized time-reliability curves, one for each mode of cognitive processing (see Figure 3.9). These curves estimate the probability of a nonresponse at any point in time. The normalized time T_N is

$$T_N = \frac{T_A}{T_M},$$

where T_A is the actual and T_M is the median time to perform the task. The analyst uses these normalized curves to generate nonresponse probabilities at various times after an emergency in the system develops.

The HCR model was developed and evaluated within the context of operation of nuclear power plants and focuses mainly on the temporal aspects of crew performance. Many of its fundamental hypotheses have been at least partially verified (Worledge, Joksimovich, & Spurgin, 1988), leading Whittingham (1988) to propose that a combination of the HCR and THERP models should provide a good predictor of human reliability. An application of these two models can be found in a report that quantified improvements in human reliability for a nuclear power plant (Ko, Wu, &

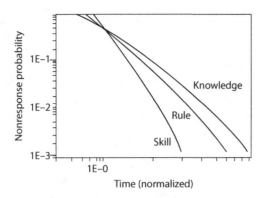

FIGURE 3.9 Normalized crew nonresponse curves for skill-, rule-, and knowledge-based cognitive processing.

Lee, 2006). The plant had implemented a severe accident management guidance program, which provided operators with structured guidance for responding to an emergency condition. The analysis was conducted to show that the implementation of the structured guidance program changed the operators' behavior mode from knowledge based (i.e., problem solving) to rule based (following the rules of the program). This made it more likely that operators would complete necessary tasks within the time limits.

A Technique for Human Error ANAlysis (ATHEANA)

Another model representative of a second-generation technique is ATHEANA (USNRC, 2000). As for a typical probabilistic reliability analysis, ATHEANA begins by identifying possible human failure events from accident scenarios. The analyst describes these events by enumerating the unsafe actions (errors of omission or commission) of the operators, and then characterizing them further using Reason's (1990) distinctions between slips, lapses, mistakes, and violations of regulations. The model combines environmental factors and plant conditions affecting the likelihood of human errors in error-forcing contexts; that is, situations in which an error is likely. The descriptions of these error-forcing contexts may lead to better identification of possible human errors and where they are most likely to occur in a task sequence. The final result of an ATHEANA analysis is a quantitative estimate of the conditional probability of an unsafe action as a function of the error-forcing context in the situation under study.

ATHEANA is very detailed and explicit. Most importantly, after an accident, the reliability expert can identify particular errors of commission resulting from an error-forcing context. However, it has several limitations (Dougherty, 1997; Kim, 2001). Because it is a variant of probabilistic reliability analysis, it suffers from the many of the shortcomings associated with this. As one example, ATHEANA continues to make a distinction between errors of commission and omission, which, as we noted earlier, is linked to probabilistic reliability analysis and is independent of the cognitive basis for the errors. Another shortcoming concerns the model's emphasis on an error-forcing context, which might imply that a particular situation may allow no chance for success. Because this context is used as a substitute for the many factors that influence human cognition and performance in a task, it may be more profitable to develop more detailed models of cognitive reliability, as in the next method we consider.

Cognitive Reliability and Error Analysis Method (CREAM)

CREAM (Hollnagel, 1998) takes a *cognitive engineering* perspective, according to which the human–machine system is conceptualized as a joint cognitive system, and human behavior is shaped by the context of the organization and the technological environment in which it resides.

After a task analysis, CREAM requires an assessment of the conditions under which the task is commonly performed. Some of these conditions might include the availability of procedures and plans to the operator, the available time for the task, when the task is performed, and the quality of collaboration among members of the crew. Given the context in which a task is performed, the reliability analyst then develops a profile to identify the cognitive demands of the task. The analyst describes these demands using the cognitive functions of observation, interpretation, planning, and execution. Then, for each task component, he assesses what kinds of strategies or control modes are used by the operators to complete the task.

CREAM considers four possible control modes: strategic, tactical, opportunistic, or scrambled. For the strategic mode, a person's action choices are guided by strategies derived from the global context; for the tactical mode, her performance is based on a procedure or rule; for the opportunistic mode, salient features of the context determine the next action; for the scrambled mode, the choice of the next action is unpredictable. The reliability analysis is completed when the reliability expert identifies what cognitive function failures are most likely to occur and computes the cognitive failure probabilities for the task elements and for the task as a whole.

CREAM is a detailed method for quantifying human error in terms of the operator's cognitive processes. CREAM's method is more systematic and clear than that of ATHEANA, and it allows the analyst to perform both predictive and retrospective analyses using the same principles (Kim, 2001). One of its limitations is that it does not explicitly take into consideration how people might recover from erroneous actions: All errors are assumed to be nonrecoverable. This means that CREAM will tend to underestimate human reliability in many situations.

Human Performance Data Sources

Human reliability analysis requires that we explicitly specify estimates of human performance for various tasks and subtasks. Such estimates include the probability of correct performance, reaction time, and so on. Figure 3.10 shows several possible sources for useful performance estimates. The best estimates come from empirical data directly relevant to the task to be analyzed. Such data may come from laboratory studies, from research conducted on trainers and simulators, or from actual system operation. Data like these are summarized in data banks (such as *Human Reliability Data Bank for Nuclear Power Plant Operators*, Topmiller, Eckel, & Kozinsky, 1982, and the *Engineering Data Compendium: Human Perception and Performance*, Boff & Lincoln, 1988) and handbooks (such as the *Handbook of Human Factors and Ergonomics,* Salvendy, 2012), with more detailed descriptions presented in the original research reports. The primary limitation of these data sources is that the most commonly used data come from laboratory studies that are typically conducted under restricted, artificial conditions; generalization to more complex systems thus should be made with caution. Moreover, the amount of data available in any data bank is limited.

Simulators provide another source of data for complex systems, such as chemical waste disposal plants, for which a failure can be hazardous (Collier, Ludvigsen, & Svengren, 2004). The simulator can create specific accident sequences to analyze the performance of the personnel in such circumstances without endangering the system or its operators. This permits the analyst to measure the response accuracy and latency to critical events, as well as the possibility for using interviews to obtain information from the operators about the displays and indicators to which they attended and how they made decisions (Dougherty & Fragola, 1988, p. 50).

Another way to estimate human error probability parameters is from computer simulations or mathematical models of human performance (Yoshikawa & Wu, 1999). An accurate model can provide objective probability estimates for situations for which direct empirical data are not available. A final option is to ask experts and obtain their opinions about the probabilities of specific errors. However, information obtained in this way is highly subjective, and so you should interpret it cautiously.

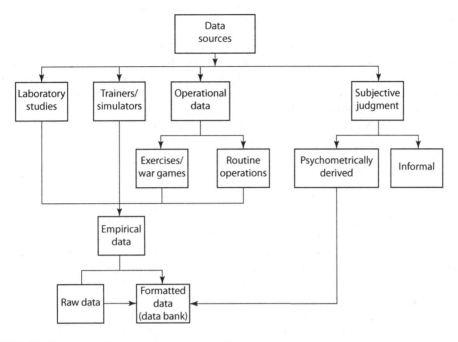

FIGURE 3.10 Human performance data sources and outputs.

PROBABILISTIC RISK ANALYSIS

In complex systems, the risks associated with various system failures are assessed as part of a reliability analysis. Risk refers to events that might cause harm, such as a nuclear power plant releasing radioactive steam into the atmosphere. A *risk analysis*, therefore, considers not only the reliability of the system but also the risks that accompany specific failures, such as monetary loss and loss of life. Probabilistic risk analysis, the methods of which were developed and applied primarily within the nuclear power industry, involves decomposing the risk of concern into smaller elements for which the probabilities of failure can be quantified (Bedford & Cooke, 2001). These probabilities then are used to estimate the overall risk, with the goal of establishing that the system is safe and to identify the weakest links (Paté-Cornell, 2002).

The human risk analysis of a complex system like a nuclear plant includes the following goals:

1. Represent the plant's risk contribution from its people and their supporting materials, such as procedures;
2. Provide a basis upon which plant managers may make modifications to the plant while optimizing risk reduction and enhancing human factors; and
3. Assist the training of plant operators and maintenance personnel, particularly in contingencies, emergency response, and risk prevention (Dougherty & Fragola, 1988, p. 74).

The nuclear power industry uses probabilistic risk analysis methods to identify plant vulnerabilities, justify additional safety requirements, assist in designing maintenance routines, and support the decision-making process during routine and emergency procedures (Zamanali, 1998).

The focus of a reliability analysis is on the successful operation of the system, and so we look at the system environment in terms of its influence on system performance. In contrast, the focus of a risk analysis is to evaluate the influence of system failures on the environment. Maximization of system reliability and minimization of system risk require that we conduct risk and reliability analyses and address design concerns at all phases of system development and implementation.

SUMMARY

The operator is part of a human–machine system. Consequently, the system concept plays a central role in human factors. We must examine the contribution of the operator from within the context of the system. The performance of a system depends on many variables, some unique to the mechanical aspects of the system and some unique to the human aspects of the system. We can find still more variables in the system environment.

Errors by a system's operator can result in system failure. A fundamental goal of human factors is to minimize risk while maximizing system reliability. This requires that the human factors expert perform an analysis of the sources of potential human errors and an evaluation of their consequences for overall system performance. The expert can use several alternative classifications for types of errors for this purpose.

We estimate system reliability from the reliabilities of the system's components and the structure of the system. Reliability analysis can successfully predict the reliability of machines. Human reliability analysis is based on the assumption that the performance of the operator can be analyzed using similar methods. Human and machine reliability analyses can be combined to predict the overall performance of the human–machine system and the overall risk associated with its operation.

A theme we will repeat frequently in this book is that optimal system design requires us to consider human factors at every stage of the system development or design process. This means we must consider the potential for different types of human errors at every stage of the system development process. By incorporating known behavioral principles into system design and evaluating design alternatives, the human factors specialist ensures that the system can be operated safely and efficiently.

RECOMMENDED READINGS

Birolini, A. (2014). *Reliability Engineering: Theory and Practice* (7th ed.). Berlin: Springer.

Dhillon, B. S. (2009). *Human Reliability, Error, and Human Factors in Engineering Maintenance with Reference to Aviation and Power Generation*. Boca Raton, FL: CRC Press.

Hollnagel, E. (1998). *Cognitive Reliability and Error Analysis Method*. London: Elsevier.

Gertman, D. I., & Blackman, H. S. (1994). *Human Reliability & Safety Analysis Handbook*. New York: John Wiley.

Kirwan, B. (1994). *A Guide to Practical Human Reliability Assessment*. London: Taylor & Francis.

Kossiakoff , A., Sweet, W. N., Seymour, S., & Biemer, S. M. (2011). *Systems Engineering: Principles and Practice* (2nd ed.). Hoboken, NJ: John Wiley.

Reason, J. (1990). *Human Error*. Cambridge: Cambridge University Press.

Reason, J. (2013). *A Life in Error: From Little Slips to Big Disasters*. Burlington, VT: Ashgate.

Senders, J. W., & Moray, N. P. (1991). *Human Error: Cause, Prediction, and Reduction*. Hillsdale, NJ: Lawrence Erlbaum.

Spurgin, A. (2010). *Human Reliability Assessment: Theory and Practice*. Boca Raton, FL: CRC Press.

4 Human Information Processing

Information processing lies at the heart of human performance. In a plethora of situations in which humans interact with systems, the operator must perceive information, transform that information into different forms, take actions on the basis of the perceived and transformed information, and process the feedback from that action, assessing its effect on the environment.

C. D. Wickens & C. M. Carswell
2012

INTRODUCTION

The human information-processing approach to studying behavior characterizes the human as a communication system that receives input from the environment, acts on that input, and then outputs a response back to the environment. We use the information-processing approach to develop models that describe the flow of information in the human, in much the same way that system engineers use models to describe information flow in mechanical systems. The similarity between the human information-processing and system perspectives is not coincidental; the human information-processing approach arose from the contact that psychologists had with industrial and communication engineers during World War II.

Information-processing concepts have been influenced by information theory, control theory, and computer science (Posner, 1986). However, experimental studies of human performance provide the empirical base for the approach. An information processing account of performance describes how inputs to the perceptual system are coded for use in cognition, how these codes are used in different cognitive subsystems, the organization of these subsystems, and mechanisms by which responses are made. Diagrams of hypothesized processing subsystems can identify the mental operations that take place in the processing of various types of information as well as the specific control strategies adopted to perform the tasks.

Figure 4.1 shows a simple example of an information-processing model. This model explains human performance in a variety of tasks in which responses are made to visually presented stimuli (Townsend & Roos, 1973). The model consists of a set of distinct subsystems that intervene between the presentation of an array of visual symbols and the execution of a physical response to the array. The model includes perceptual subsystems (the visual form system), cognitive subsystems (the long-term memory components, the limited-capacity translator, and the acoustic form system), and action subsystems (the response-selection and response-execution systems). The flow of information through the system is depicted by the arrows. In this example, information is passed between stages and subsystems.

Engineers can look inside a machine to figure out how it works. However, the human factors expert cannot look inside a person's head to examine the various subsystems that underlie performance. Instead, he must infer how cognitive processing occurs on the basis of behavioral and physiological data. There are many models that he can consider, which differ in the number and arrangement of processing subsystems. The subsystems can be arranged serially, so that information flows through them one at a time, or in parallel, so that they can operate simultaneously. Complex models can be hybrids that are composed of both serial and parallel subsystems. In addition to the arrangement and nature of the proposed subsystems, the models also must address the processing cost (time and effort) associated with each subsystem.

Using these kinds of models, we can make predictions about how good human performance will be under different stimuli and environmental conditions. We evaluate the usefulness of any model by comparing its predictions with experimental data. The models that are most consistent with the

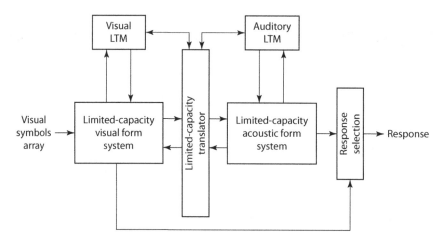

FIGURE 4.1   A model for performance of visual information processing.

data are more credible than alternative models. However, credible models must do more than simply explain a limited set of behavioral data. They must also be consistent with other behavioral phenomena and with what we know about human neurophysiology. In keeping with the scientific method, we will revise and replace models as we gather additional data.

The importance of the information-processing approach for human factors is that it describes both the operator and the machine in similar terms (Posner, 1986). A common vocabulary makes it easier to treat the operator and the machine as an integrated human–machine system. For example, consider the problem of a decision-support system in an industrial control setting (Rasmussen, 1986). The system assists operators during supervisory tasks and emergency management by providing information about the most appropriate courses of action in a given circumstance. Whether or not system performance is optimal will depend on the way the system presents information about the machine and the way the operator is asked to respond. The more useful and consistent this information is, the better the operator will be able to perform. Models of human information processing are prerequisites for the conceptual design of such systems, because these models can help determine what information is important and how best to present it (McBride & Schmorrow, 2005; Rasmussen, 1986). In human–computer interaction in particular, information-processing models have resulted in solutions for a range of issues (Proctor & Vu, 2012). Card, Moran, and Newell (1983, p. 13) note, "It is natural for an applied psychology of human-computer interaction to be based theoretically on information-processing psychology."

Because "human society has become an information processing society" (Sträter, 2005), the information-processing approach provides a convenient framework for understanding and organizing a wide variety of human performance problems. It gives a basis for analyzing the components of a task in terms of their demands on perceptual, cognitive, and action processes. In this chapter, we will introduce the basic concepts and tools of analysis that are used in the study of human performance.

A THREE-STAGE MODEL

Figure 4.2 presents a general model of information processing that distinguishes three stages intervening between the presentation of a stimulus and the execution of a subsequent response. Early processes associated with perception and stimulus identification are in the perceptual stage. Following this stage are intermediate processes involved with decision making and thought: the cognitive stage. Information from this cognitive stage is used in the final action stage to select, prepare, and control the movements necessary to effect a response.

The three-stage model provides an effective organizational tool, which we will use in this book. Keep in mind that the model fails to capture preparatory processes that occur prior to the

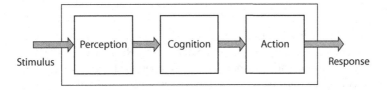

FIGURE 4.2   Three stages of human information processing.

presentation of a stimulus (i.e., how a person anticipates or gets set to perform a particular task), as well as the intimate and cyclical relation between action and perception (e.g., Knoblich, 2006). Researchers who study human performance are interested in determining to which processing stage experimental findings should be attributed, and in characterizing information flow within the system. However, the boundaries between perception, cognition, and action are not as clearly defined as is implied by the model. It is not always clear whether a change in performance should be attributed to the perceptual, cognitive, or action stage. Once a specific change in performance can be clearly attributed to a stage, detailed models of the processing within that stage can be developed.

Perceptual Stage

The perceptual stage includes processes that operate from the stimulation of the sensory organs (e.g., Wolfe et al., 2015). Some of this processing might occur without the person even becoming aware of it through the processes involved in the detection, discrimination, and identification of the stimulation. For example, a visual display produces or reflects patterns of light energy that are absorbed by photoreceptors in the eye (see Chapter 5). This triggers a neural signal that travels to areas of the brain devoted to filtering the signal and extracting information contained in it, such as shape, color, or movement. The ability of the brain to extract information from the signal depends on the quality of the sensory input. This quality is determined by, among other things, the clarity and duration of the display.

If the display is not clear, much of the information will be lost. When a film projector is poorly focused, you cannot see the details of the picture. Similarly, a poorly tuned television picture can be snowy and blurred. Displays that are presented very briefly or that must be examined very quickly, such as road signs and some computer error messages, do not allow the extraction of much information during the time that they are available. Such degradations of input to the sensory system restrict the amount of information that can be extracted and, thus, will restrict performance.

Cognitive Stage

After the perceptual stage has extracted enough information from a display to allow the stimulus to be identified or classified, processes begin to operate with the goal of determining the appropriate action or response. These processes might include the retrieval of information from memory, comparisons among displayed items, comparison between these items and the information in memory, arithmetic operations, and decision making (e.g., Groome & Eyesenk, 2016). The cognitive stage imposes its own constraints on performance. For example, people are not generally very good at paying attention to more than one source of information or performing complicated calculations in their heads.

Errors in performance may arise from these and a number of other cognitive limitations. We often characterize cognitive limitations in terms of cognitive resources: If there are few available resources to devote to a task, then task performance may suffer. One of our goals as human factors specialists is to identify the cognitive resources necessary for the performance of a task and systematically remove limitations associated with these resources. This may require additional information displays, redesigning machine interfaces, or even redesigning the task itself.

ACTION STAGE

Following the perceptual and cognitive stages of processing, an overt response (if required) is selected, programmed, and executed (e.g., Schneider, 2015). Response selection is the problem of choosing which of several alternative responses is most appropriate under the circumstances. After a response is selected, it then must be translated into a set of neuromuscular commands. These commands control the specific limbs or effectors that are involved in making the response, including their direction, velocity, and relative timing.

Selection of the appropriate response and specification of the parameters of the movement take time. We usually see that the time required to begin a movement increases as the difficulty of response selection and movement complexity increase (Henry & Rogers, 1960). The action stage therefore imposes its own limitations on performance, just as the cognitive stage imposes limitations. There are also physical limitations that must be considered: An operator cannot press an emergency button at the same time as she is using both hands to close a valve, for example. Action stage limitations can result in errors in performance, such as the failure of a movement to terminate accurately at its intended destination.

HUMAN INFORMATION PROCESSING AND THE THREE-STAGE MODEL

The three-stage model is a general framework that we are using to organize much of what we know about human capabilities. It enables us to examine performance in terms of the characteristics and limitations of the three stages. This simple classification of human information processing allows a more detailed examination of the processing subsystems within each stage. For example, Figure 4.3 shows how each of the stages can be further partitioned into subsystems whose properties can then be analyzed. Box 4.1 describes more general cognitive architectures and the computational models we can develop from them based on detailed specifications of the human information-processing system.

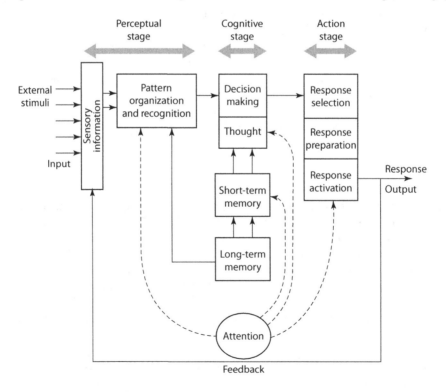

FIGURE 4.3 An elaborated model of human information processing.

BOX 4.1   COMPUTATIONAL MODELS AND COGNITIVE ARCHITECTURES

In this book, you will encounter a vast array of data and theories about human information processing and cognition. Our understanding of human–computer interaction (HCI), and how we solve applied problems in HCI and other areas of human factors, depends on these data and theories. One result of this work is the development of cognitive architectures that allow computational modeling of performance for a variety of different tasks and task environments.

Cognitive architectures are "broad theories of human cognition, based on a wide selection of human experimental data, that generally are implemented as computer simulations" (Byrne, 2015, p. 353). Although there are a variety of mathematical and computational models to explain performance in specific task domains, cognitive architectures emphasize "broad theory." That is, they are intended to provide theories that integrate and unify findings from a variety of domains through computational models. The architecture specifies in detail how human information processing operates but requires researchers and designers to provide specific information on which it operates before it can model a particular task. Cognitive architectures were developed initially by academic researchers interested primarily in basic theoretical issues in cognition, but they are now used extensively by practitioners in HCI and human factors.

One of the values of cognitive architectures is that they can provide quantitative predictions for a variety of measures, such as performance time, error rates, and learning rates. In contrast to goals, operators, methods, and selection rules (GOMS) models of the type described in Box 3.1, you can program a computational model to actually execute the processes involved in task performance upon the occurrence of a stimulus event. Consequently, they can be used as cognitive models for simulated virtual worlds (Jones et al., 1999) and as a model of a learner's knowledge state in educational tutoring systems (Anderson, Douglass, & Qin, 2005).

The most popular cognitive architectures are ACT-R (Adaptive Control of Thought–Rational; Anderson et al., 2004; Borst & Anderson, 2016), Soar (States, Operators, and Results; Lehman, Laird, & Rosenbloom, 1998; Peebles, Derbinsky, & Laird, 2013), and EPIC (Executive Process Interactive Control; Kieras & Meyer, 1997; Kieras, Wakefield, Thompson, Iyer, & Simpson, 2016). All of these architectures are classified as production systems in that they are based on production rules. A production rule is an "if-then" statement: If a set of conditions is satisfied, a mental or physical act is produced. These architectures provide depictions of the entire human information-processing system, although they differ in the details of the architecture and the level of detail at which the productions operate.

As an example, we will briefly describe EPIC (Kieras & Meyer, 1997). Figure B4.1 shows the overall structure of the EPIC architecture. Information enters auditory, visual, and tactile processors, and it is then passed on to working memory. Working memory is part of the cognitive processor, which is a production system. In addition to productions for implementing task knowledge, executive knowledge for coordinating the various aspects of task performance is represented in productions; hence the emphasis on "executive process" in EPIC's name. Oculomotor, vocal, and manual processors control the responses selected by the production system. A salient feature of EPIC is that all of the processors operate in parallel with each other, there is no limit to the number of operations that the cognitive processor can perform in parallel, and executive and task productions can execute in parallel. When you use EPIC to model the performance of a particular task, you will need to specify parameters for how long it will take each processor to carry out its operations. We will elaborate on EPIC and the other architectures later in the book.

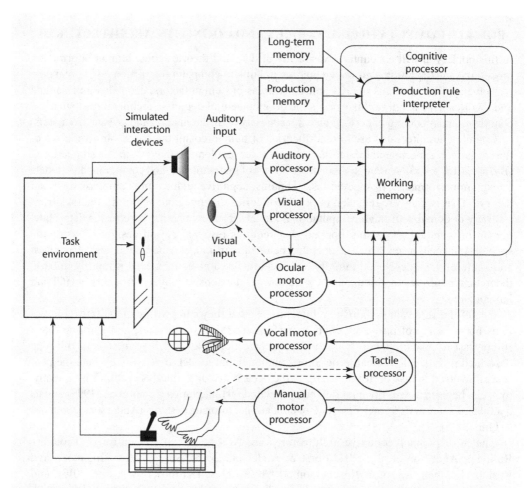

FIGURE B4.1   Structure of the EPIC architecture.

There are three kinds of processing limitations that can cause processing errors at each stage: data, resource, and structural limitations (Norman & Bobrow, 1975). *Data-limited processing* takes place when the information input to a stage is degraded or imperfect, such as when a visual stimulus is only briefly flashed or when speech signals are presented in a noisy environment. *Resource-limited processing* occurs if the system is not powerful enough to perform the operations required for a task efficiently, such as the memory resources required to remember a long-distance phone number until it is dialed. *Structurally limited processing* arises from an inability of one system to perform several operations at once. Structural limitations can appear at any stage of processing, but the most obvious effects occur in the action stage when two competing movements must be performed simultaneously with a single limb.

Although Norman and Bobrow's distinction between different kinds of processing limitations may not accurately characterize the ways that cognitive processing can be realized in the human brain, it is a useful taxonomy. With this taxonomy, it becomes easier to determine whether performance limitations are due to problems in the way that information is being delivered to the operator, the kinds of information being used for a task, or components of the task itself. Furthermore, although it is easiest to see data limitations in the perceptual stage, resource limitations in the cognitive stage, and structural limitations in the action stage, it is important to remember that all three of these limitations may appear at any stage of processing.

PSYCHOLOGICAL REPRESENTATION OF THE PHYSICAL WORLD

Viewing the human being as an information-processing system is not a new idea in human factors and psychology. We can trace similar ideas as far back as the work of Ernst Weber and Gustav Fechner in the 1800s, discussed in Chapter 1, and even farther. The information-processing view brings with it a number of questions, two of which we address in this section. These are (1) what are the limits of the senses to sensory stimulation, and (2) how do changes in stimulus intensity relate to changes in sensory experience? Researchers who concern themselves with answering these kinds of questions are called *psychophysicists*. Many psychophysical techniques have been developed to measure sensory experience (e.g., Kingdom & Prins, 2010; Szalma & Hancock, 2015), and these techniques are a valuable part of every competent human factors specialist's toolbox.

Most psychophysical techniques rely on the frequency with which certain responses are made under different stimulus conditions. For example, we might be concerned with the number of times a radiologist sees a shadow on the X-ray of someone's lung under different lighting conditions. The frequency of times a shadow is reported can be used to infer properties of the sensory experience provided by a particular X-ray, lighting scheme, and so on.

The psychophysical methods that we discuss here provide precise answers to questions about detectability, discriminability, and perceived magnitude for almost any conceivable kind of stimulus. *Detectability* refers to the absolute limits of the sensory systems to provide information that a stimulus is present. *Discriminability* involves the ability to determine that two stimuli differ from each other. Discovering the relation between perceived magnitude and physical magnitude is referred to as *psychophysical scaling*.

Most of what we know about the dynamic ranges of the human senses and the precise sensory effects of various physical variables, such as the frequency of an auditory stimulus, was discovered using psychophysical techniques. These techniques also have been used to investigate applied issues in many areas, although perhaps not as widely as is warranted. The importance of using psychophysical methods to study applied problems was stressed by Uttal and Gibb (2001), who conducted psychophysical investigations of night-vision goggles. They concluded that their work supports the general thesis that "classical psychophysics provides an important means of understanding complex visual behavior, one that is often overlooked by both engineers and users" (p. 134). It is important to realize that basic psychophysical techniques can be used by human factors specialists to solve specific problems relating to optimal design.

CLASSICAL METHODS FOR DETECTION AND DISCRIMINATION

The most important concept in classical psychophysics is that of the threshold. An *absolute threshold* is the smallest amount of intensity a person needs to detect a stimulus (VandenBos, 2015). A *difference threshold* is the smallest amount of difference a person needs to perceive two stimuli as different. The goal of the classical psychophysical methods is to measure these thresholds accurately.

The definition of a threshold suggests fixed values below which stimuli cannot be detected or differences discriminated, and above which stimuli are always perfectly detected. That is, the relation between physical intensity and detectability should be a step function (illustrated by the dashed line in Figure 4.4). However, psychophysical studies always show a range of stimulus values over which an observer will detect a stimulus or discriminate between two stimuli only some percentage of the time. Thus, the typical psychophysical function is an S-shaped curve (as shown by the points in Figure 4.4).

Fechner developed the classical methods we use for measuring thresholds. Although many modifications to his procedures have been made over the years, the methods in this tradition still follow closely the steps that he outlined. These methods require us to make many measurements in carefully defined conditions, and we estimate the threshold from the resulting distribution of responses. Two of the most important methods are the method of limits and the method of constant stimuli (Kingdom & Prins, 2010).

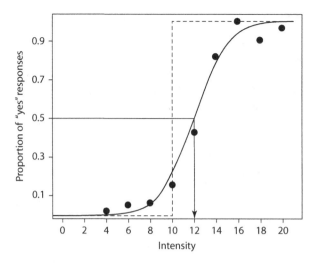

FIGURE 4.4   Proportion of detection responses as a function of stimulus intensity for idealized (dashed) and actual (points) observers.

To find an absolute threshold using the *method of limits*, we present stimulus intensities that bracket the threshold in a succession of small increments to an observer. For instance, to determine an observer's threshold for detecting light, we could work with a lamp that can be adjusted from almost-zero intensity to the intensity of a 30-watt light bulb. We could start with an ascending sequence of intensities, beginning with the lowest intensity (almost zero) and gradually increasing the intensity. At each intensity level, we ask the observer whether he sees the light. We keep increasing the intensity until the observer reports that he has seen the light. We then compute the average of the intensities for that trial and the immediately preceding trial and call that the threshold intensity for the series. It is important to also present a descending series of trials, where we start with the highest intensity and decrease the intensity on each step. We must repeat this procedure several times, with half of the series being ascending and the other half being descending. For each series, we will obtain a value for the threshold. The average of all of the thresholds that we compute for a particular observer is defined as the absolute threshold for that observer.

To find difference thresholds with the method of limits, we present two stimuli on each trial. The goal is to determine the smallest difference between the two stimuli that the observer can detect. If we are still concerned with light intensities, we would need two lamps. The intensity of one lamp remains constant, and we call that intensity the *standard*. The intensity of the other lamp varies, and we call that intensity the *comparison*. On each trial, the observer must say whether the comparison stimulus is less than, greater than, or equal to the standard stimulus. We increment and decrement the intensity of the comparison in exactly the same way as we just described for determining the absolute threshold, and measure the intensities at which the observer's response changes from less than to equal or from greater than to equal. These intensities define two thresholds, one for ascending and one for descending sequences. We could average the two thresholds to get an overall difference threshold, but it is common to find that the two thresholds obtained in this way are actually quite different, so averaging might not provide an accurate threshold value.

The second method is the *method of constant stimuli*. In contrast to the ascending and descending series of intensities presented in the method of limits, we present different intensities in random order. If we are still working with light intensities, the observer would report whether the light was seen on each trial, just as before. For each intensity, we compute the number of times that the light was detected, and the threshold intensity is defined as that intensity for which the light was detected 50% of the time. Figure 4.4 shows how the data might look for this procedure.

It is also easy to determine difference thresholds using the method of constant stimuli. Using two lamps, we again hold the intensity of the standard constant, but the intensity of the comparison varies randomly from trial to trial. The difference threshold is defined as the comparison intensity for which the observer reports 50% "greater than" and 50% "less than" responses.

The methods of limits and constant stimuli, as well as many others, have been used successfully since Fechner's work to obtain basic, empirical information about the characteristics of each of the senses. It is important to understand that the measurement of a threshold in an isolated condition is not very informative. Rather, it is the changes in the threshold under varying conditions that provide critical information about sensory limitations. For example, the light intensity threshold we just described will depend on the color of the light (see Chapter 5). Also, the difference threshold that we compute will depend on the intensity of the standard: The higher the intensity of the standard, the higher the difference threshold will be. For auditory stimuli, the detection threshold is a function of the stimulus frequency; very high-pitched and very low-pitched sounds are more difficult to hear. Findings like these do not only reveal basic characteristics of the visual and auditory systems, but also provide information for the human factors specialist about how visual and auditory information should be presented to ensure that a person can perceive it.

In some situations, we might want to insure that people can't perceive something. For example, Shang and Bishop (2000) evaluated the impact of introducing a transmission tower or oil refinery tanks into landscape settings. They edited photographs of landscapes to include these (ugly) structures and used psychophysical methods to obtain difference thresholds for detection, recognition, and visual impact (degradation of the views caused by the structure). They showed that the size of the structure and its contrast with the surroundings determined all three threshold types. The authors recommended extension of threshold measurement to assess the aesthetic effects of other types of changes to the environment, such as billboards and clearcuts.

Despite their utility, there are several problems with the threshold concept and the methods used to evaluate it. Most serious is the fact that the measured value for the threshold, which is assumed to reflect sensory sensitivity, may be affected by the observer's desire to say *yes* or *no*. Some evidence for this is the common finding that the difference threshold is not the same for ascending and descending sequences. Consider the extreme case, where a person decides to respond *yes* on all trials. In these circumstances, no threshold can be computed, and we have no way of knowing whether she actually detected anything at all. We could address this problem by inserting some catch trials on which no stimulus is presented. If she responds *yes* on those trials, her data could be thrown out. However, catch trials will not pick up response biases that are less extreme.

SIGNAL-DETECTION METHODS AND THEORY

The problem with the classical methods is that threshold measurements are subjective. That is, we have to take the observer's word that she detected the stimulus. This is like taking an exam that consists of the instructor asking you whether or not you know the material and giving you an *A* if you say "yes." A more objective measure of how much you know requires that that your "yes" response be verified somehow. An objective test can be used that requires you to distinguish between true and false statements. In this case, the instructor can evaluate your knowledge of the material by the extent to which you correctly respond *yes* to true statements and *no* to false statements. Signal-detection methods are much like an objective test, in that the observer is required to discriminate trials on which the stimulus is present from trials on which it is not.

Methods

In the terminology of signal detection (Green & Swets, 1966; MacMillan & Creelman, 2005), noise trials refer to those on which a stimulus is not present, and signal-plus-noise trials (or signal trials) refer to those on which a stimulus is present. In a typical signal-detection experiment, we select a single stimulus intensity and use it for a series of trials. For example, the stimulus may be a tone of a particular frequency and intensity presented in a background of auditory noise. On some trials we present only the noise, whereas on other trials we present both the signal and the noise. The listener must respond *yes* or *no* depending on whether he heard the tone. The crucial distinction between the signal-detection methods and the classical methods is that the listener's sensitivity to the stimulus can be calibrated by taking into account the responses made when the stimulus is not present.

Table 4.1 shows the four combinations made up of the two possible states of the world (signal, noise) and two responses (yes, no). A *hit* occurs when the observer responds *yes* on signal trials, a *false alarm* when the response is *yes* on noise trials, a *miss* when the response is *no* on signal trials, and a *correct rejection* when the response is *no* on noise trials. Because the proportions of misses and correct rejections can be determined from the proportions of hits and false alarms, signal-detection analyses typically focus only on the hit and false-alarm rates. You should note that this 2 × 2 classification of states of the world and responses is equivalent to the classification of true and false null hypotheses in inferential statistics (see Table 4.2); optimizing human performance in terms of hits and false alarms is the same as minimizing Type I and Type II errors, respectively. The key to understanding signal-detection theory is to realize that it is just a variant of the statistical model for hypothesis testing.

A person's sensitivity to the stimulus is good if his hit rate is high and his false-alarm rate low. This means that he makes mostly *yes* responses when the signal is present and mostly *no* responses when it is not. Conversely, his sensitivity is poor if the hit and false-alarm rates are similar, so that he responded *yes* about as often as he responded *no* regardless of whether or not a signal was presented. We can define several quantitative measures of sensitivity from the hit and false-alarm rates, but they are all based on this general idea.

Sometimes a person might tend to make more responses of one type than the other regardless of whether the signal is present. We call this kind of behavior a *response bias*. If we present the same number of signal trials as noise trials, an unbiased observer should respond *yes* and *no* about equally often. If an observer responds *yes* on 75% of all the trials, this would indicate that he has a bias to respond *yes*. If *no* responses are more frequent, this would indicate that he has a bias to respond *no*. As with sensitivity, there are several ways that we could quantitatively measure response bias, but they are all based on this general idea.

Theory

Signal-detection theory provides a framework for interpreting the results from detection experiments. In contrast to the notion of a fixed threshold, signal-detection theory assumes that the sensory evidence for the signal can be represented on a continuum. Even when the noise alone is presented, some amount of evidence will be registered to suggest the presence of the signal. Moreover, this amount will vary from trial to trial, meaning that there will be more evidence at some times than at others to suggest the presence of the signal. For example, when detecting an auditory signal in noise, the amount of energy contained in the frequencies around that of the signal frequency will vary from trial to trial due to the statistical properties of the noise-generation process. Even when no physical noise is present, variability is introduced by the sensory registration process, neural transmission, and so on. Usually, we assume that the effects of noise can be characterized by a normal distribution.

TABLE 4.1
Classifications of Signal and Response Combinations in a Signal-Detection Experiment

Response	State of the World	
	Signal	Noise
"Yes" (present)	Hit	False alarm
"No" (absent)	Miss	Correct rejection

Noise alone will tend to produce levels of sensory evidence that are on average lower than the levels of evidence produced when a signal is present. Figure 4.5 shows two normal distributions of evidence, the noise distribution having a smaller mean μ_N than the signal distribution, which has a mean of μ_{S+N}. In detection experiments, it is usually not easy to distinguish between signal and noise trials. This fact is captured by the overlap between the two distributions in Figure 4.5. Sometimes noise will look like signal, and sometimes signal will look like noise. The response an observer makes on any trial will depend on some criterion value of evidence that she selects. If the evidence for the presence of the signal exceeds this criterion value, then she will respond "yes," the signal is present; otherwise, she will respond "no."

Detectability and Bias

We have explained in general terms what we mean by sensitivity (or detectability) and response bias. Using the framework shown in Figure 4.5, we now have the tools needed to construct quantitative measurements of detectability and bias. In signal-detection theory, the detectability of the stimulus is reflected in the difference between the means of the signal and noise distributions. When the means are identical, the two distributions are perfectly superimposed, and there is no way to discriminate signals from noise. As the mean for the signal distribution shifts away from the mean of the noise distribution in the direction of more evidence, the signal becomes increasingly detectable. Thus, the most commonly used measure of detectability is

$$d' = \frac{\mu_{S+N} - \mu_N}{\sigma},$$

where:

   d' is detectability,

   μ_{S+N} is the mean of the signal + noise distribution,

   μ_N is the mean of the noise distribution, and

   σ is the standard deviation of both distributions.

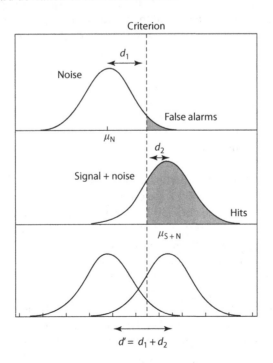

FIGURE 4.5 Signal and noise distributions of sensory evidence illustrating determination of d'.

The d' statistic is the standardized distance between the means of the two distributions. The placement of the criterion reflects the observer's bias to say *yes* or *no*. If the signal trials and noise trials are equally likely, an unbiased criterion setting would be at the evidence value for which the two distributions are equal in height (i.e., the value for which the likelihood of the evidence coming from the signal distribution is equal to the likelihood of it coming from the noise distribution). We call an observer "conservative" if she requires very strong evidence that a signal is present. A conservatively biased criterion placement would be farther to the right than the point at which the two distributions cross (see Figure 4.5). We call her "liberal" if she does not require much evidence to decide that a signal is present. A liberally biased criterion would be farther to the left than the point at which the two distributions cross. The bias is designated by the Greek letter β. This value is defined as

$$\beta = \frac{f_S(C)}{f_N(C)},$$

where:

   C is the criterion, and
   f_S and f_N are the heights of the signal and noise distributions, respectively.

If $\beta = 1.0$, then the observer is unbiased. If β is greater than 1.0, then the observer is conservative, and if it is less than 1.0, the observer is liberal.

It is easy to compute both d' and β from the standard normal table (Appendix I). To compute d', we must find the distances of μ_{S+N} and μ_N from the criterion. The location of the criterion with respect to the noise distribution is conveyed by the false-alarm rate, which reflects the proportion of the distribution that falls beyond the criterion (see Figure 4.5). Likewise, the location of the criterion with respect to the signal distribution is conveyed by the hit rate. We can use the standard normal table to find the z-scores corresponding to different hit and false-alarm rates. The distance from the mean of the noise distribution to the criterion is given by the z-score of $(1 - \text{false-alarm rate})$, and the distance from the mean of the signal distribution is given by the z-score of the hit rate. The distance between the means of the two distributions, or d', is the sum of these scores:

$$d' = z(H) + z(1 - FA)$$

where:

   H is the proportion of hits, and
   FA is the proportion of false alarms.

Suppose we perform a detection experiment and observe that the proportion of hits is 0.80 and the proportion of false alarms is 0.10. Referring to the standard normal table, the point on the abscissa corresponding to an area of 0.80 is $z(0.80) = 0.84$, and the point on the abscissa corresponding to an area of $1.0 - 0.10 = 0.90$ is $z(0.90) = 1.28$. Thus, d' is equal to $0.84 + 1.28$, or 2.12. Because a d' of 0.0 corresponds to chance performance (i.e., hits and false alarms are equally likely) and a d' of 2.33 to nearly perfect performance (i.e., the probability of a hit is very close to one, and the probability of a false alarm is very close to zero), the value of 2.12 can be interpreted as good discriminability.

The bias in the criterion setting can be found by obtaining the height of the signal distribution at the criterion and dividing it by the height of the noise distribution. This is expressed by the formula

$$\beta = \exp\left\{-\frac{1}{2}\left(z(H)^2 - z(1 - FA)^2\right)\right\}.$$

For this example, β is equal to 1.59. The observer in this example shows a conservative bias because β is greater than 1.0.

Changes in Criterion

The importance of signal-detection methods and theory is that they allow measurement of detectability independently of the response criterion. In other words, d' should not be influenced by whether the observer is biased or unbiased. This aspect of the theory is captured in plots of *receiver operating characteristic* (ROC) curves (see Figure 4.6). For such curves, the hit rate is plotted as a function of the false-alarm rate. If performance is at chance ($d' = 0.0$), the ROC curve is a straight line along the positive diagonal. As d' increases, the curve pulls up and to the left. A given ROC curve thus represents a single detectability value, and the different points along it reflect possible combinations of hit and false-alarm rates that can occur as the response criterion varies.

How can the response criterion be varied? One way is through instructions. Observers will adopt a higher criterion if we instruct them to respond *yes* only when they are sure that the signal was present than if we instruct them to respond *yes* when they think that there is any chance at all that the signal was present. Similarly, if we introduce payoffs that differentially favor particular outcomes over others, they will adjust their criterion accordingly. For instance, if an observer is rewarded with $1.00 for every hit and receives no reward for correct rejections, he will adjust his criterion downward so that he can make a lot of "signal" responses. Finally, we can vary the probabilities of signal trials $p(S)$ and noise trials $p(N)$. If we present mostly signal trials, the observer will lower his response criterion, whereas if we present mostly noise trials he will raise his criterion. As predicted by signal-detection theory, manipulations of these variables typically have little or no effect on measures of detectability.

When signal and noise are equally likely, and the payoff matrix is symmetric and does not favor a particular response, then the observer's optimal strategy is to set β equal to 1.0. When the relative frequencies of the signal and noise trials are different, or when the payoff matrix is asymmetric, the optimal criterion will not necessarily equal 1.0. To maximize the payoff, an ideal observer should set the criterion at the point where $\beta = \beta_{\text{opt}}$, and

$$\beta_{\text{opt}} = \frac{p(N)}{p(S)} \times \frac{\text{value}(CR) - \text{cost}(FA)}{\text{value}(H) - \text{cost}(M)},$$

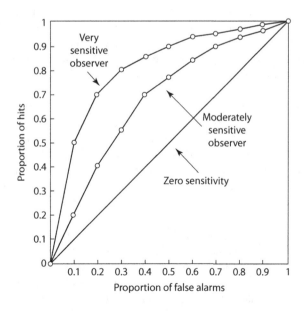

FIGURE 4.6 ROC curves showing the possible hit and false-alarm rates for different discriminabilities.

where, *CR* and *M* indicate correct rejections and misses, respectively, and value and cost indicate the amount the observer gains for a correct response and loses for an incorrect response (Gescheider, 1997). We can compare the criterion set by the observer with that of the ideal observer to determine the extent to which performance deviates from optimal.

Applications

Although signal-detection theory was developed from basic, sensory detection experiments, it is applicable to virtually any situation in which a person must make binary classifications based on stimuli that are not perfectly discriminable. As one example, signal-detection theory has been applied to problems in radiology (Boutis, Pecaric, Seeto, & Pusic, 2010). Radiologists are required to determine whether shadows on X-ray films are indicative of disease (signal) or merely reflect differences in human physiology (noise). The accuracy of a radiologist's judgment rarely exceeds 70% (Lusted, 1971). In one study, emergency room physicians at a Montreal hospital were accurate only 70.4% of the time in diagnosing pneumonia from a child's chest X-rays (Lynch, 2000, cited in Murray, 2000). There are now many alternative medical imaging systems that can be used instead of the old X-ray radiograph, including positron-emission tomography (PET), computed tomography (CT), and magnetic resonance imaging (MRI) scanning. By examining changes in d' or other measures of sensitivity, we can evaluate different imaging systems in terms of the improvement in detectability of different pathologies (Barrett & Swindell, 1981; Swets & Pickett, 1982).

With some thought, you may be able to think of other situations where signal-detection techniques can be used to great benefit. The techniques have been applied to such problems as pain perception, recognition memory, vigilance, fault diagnosis, allocation of mental resources, and changes in perceptual performance with age (Gescheider, 1997).

PSYCHOPHYSICAL SCALING

In psychophysical scaling, our concern is with developing scales of psychological quantities that, in many cases, can be mapped to physical scales (Marks & Gescheider, 2002). For instance, we may be interested in measuring different sounds according to how loud they seem to be. Two general categories of scaling procedures for examining psychological experience can be distinguished: indirect and direct. Indirect scaling procedures derive the quantitative scale indirectly from a listener's performance at discriminating stimuli. To develop a scale of loudness, we would not ask the listener to judge loudness. Instead, we would ask the listener to discriminate sounds of different intensities. In contrast, if we used a direct scaling procedure, we would ask the listener to rate her perceived loudness of each sound. Our loudness scale would be based on her reported numerical estimation of loudness.

Fechner was probably the first person to develop an indirect psychophysical scale. Remember from Chapter 1 that he constructed scales from absolute and difference thresholds. The absolute threshold provided the zero point on the psychological scale: The intensity at which a stimulus is just detected provides the smallest possible value on the psychophysical scale. He used this intensity as a standard to determine a difference threshold, so the "just-noticeable-difference" between the zero point and the next highest detected intensity provided the next point on the scale. Then he used this new stimulus intensity as a standard to find the next difference threshold, and so on.

Notice that Fechner made the assumption that the increase in an observer's psychological experience was equal for all the points on the scale. That is, the amount by which experience increased at the very low end of the scale close to the absolute threshold was the same as the amount by which experience increased at the very high end of the scale. Remember, too, that the increase in physical intensity required to detect a change (jnd) increases as the intensity of the standard increases, so that the change in intensity at the high end of the scale is very much larger than the change in intensity at the low end of the scale. Therefore, the function that describes the relation of level of psychological experience to physical intensity (the "psychophysical function") is negatively accelerated and, as discussed in Chapter 1, usually well described by a logarithmic function.

Direct scaling procedures have a history of use that roughly parallels that of indirect procedures. As early as 1872, scales were derived from direct measurements (Plateau, 1872). However, the major impetus for direct procedures came from Stevens (1975). Stevens popularized the procedure of magnitude estimation, in which observers rate stimuli on the basis of their apparent intensity. The experimenter assigns a value to a standard stimulus, for example, the number 10, and the observers are then asked to rate the magnitude of other stimuli in proportion to the standard. So, if a stimulus seems twice as intense as the standard, an observer might give it a rating of 20.

With these and other direct methods, the resulting psychophysical scale does not appear to be logarithmic. Instead, the scales appear to follow a power function

$$S = aI^n,$$

where:

S is (reported) sensory experience,
a is a constant,
I is physical intensity, and
n is an exponent that varies for different sensory continua.

This relationship between physical intensity and psychological magnitude is *Stevens' law*. Figure 4.7 shows the functions for three different kinds of stimuli. One is an electric shock, which varies in voltage; one is the length of a line, which varies in millimeters; one is a light, which varies in luminance. For the experience associated with perceiving lines of different lengths, the exponent of Stevens' law is approximately 1.0, so the psychophysical function is linear. For painful stimuli like the electric shock, the exponent is greater than 1.0, and the psychophysical function is convex. This means that perceived magnitude increases at a more rapid rate than physical magnitude. For the light stimulus, the exponent is less than 1.0, and the psychophysical function is concave. This means that perceived magnitude increases less rapidly than physical magnitude. Table 4.2 shows the exponents for a variety of sensory continua.

Psychophysical scaling methods are useful in a variety of applied problems. For example, the field of environmental psychophysics uses modified psychophysical techniques to measure a person's perceived magnitude of stimuli occurring in the living environment. Such analyses can be particularly useful

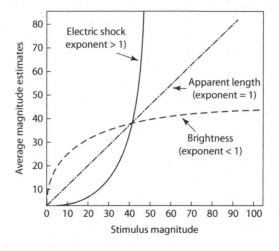

FIGURE 4.7   Power-function scales for three stimulus dimensions.

TABLE 4.2

Representative Exponents of the Power Functions Relating Sensation Magnitude to Stimulus Magnitude (Based on Stevens, 1961)

Continuum	Exponent	Stimulus Conditions
Loudness	0.6	Both ears
Brightness	0.33	5° target (dark-adapted eye)
Brightness	0.5	Point source (dark-adapted eye)
Lightness	1.2	Gray papers
Smell	0.55	Coffee odor
Taste	0.8	Saccharine
Taste	1.3	Sucrose
Taste	1.3	Salt
Temperature	1.0	Cold (on arm)
Temperature	1.6	Warmth (on arm)
Vibration	0.95	60 Hz (on finger)
Duration	1.1	White noise stimulus
Finger span	1.3	Thickness of wood blocks
Pressure on palm	1.1	Static force on skin
Heaviness	1.45	Lifted weights
Force of handgrip	1.7	Precision hand dynamometer
Electric shock	3.5	60 Hz (through fingers)

Source: Stevens, S. S. (1961). To Honor Fechner and Repeal His Law. *Science*, 133(3446), 80–86.

for evaluating the psychological magnitude of noxious stimuli such as high noise levels or odorous pollution. Berglund (1991) and her colleagues developed a procedure that obtains scale values for environmental stimuli from magnitude estimation judgments. With this procedure, the estimates from controlled laboratory studies are used to standardize the judgments people make to environmental stimuli.

Consider, for example, the odor that arises around hog farms (Berglund et al., 1974). They spread wet manure on a field in different ways and, after different amounts of time and from different distances, asked people to judge the magnitude of the odor. The same people provided magnitude estimates for several concentrations of pyridine (which has a pungent odor), and the researchers used these latter estimates to convert the estimates of odor strength for the manure to a master scale on which the scale values for each person were comparable. This sort of analysis provides valuable information about factors that reduce the perceived magnitude of noxious environmental stimuli, in this case how manure is spread and how far the hog farm is away from the people affected by it.

Psychophysical scaling methods have been applied to problems in manual lifting tasks (Snook, 1999). People are able to judge relatively accurately the highest acceptable workload they could maintain for a given period of time (say, an 8-hour work day) based on their perceived exertion under different physiological and biomechanical stresses. So, a package handler's estimate of the effort required to load packages of a particular weight at a particular rate under particular temperature conditions can be used to establish limits on acceptable materials handling procedures. Comprehensive manual handling guidelines, including the National Institute for Occupational Safety and Health (NIOSH) lifting equation (Lu, Waters, Krieg, & Werren, 2014), described in Chapter 17, have been developed based primarily on studies using variants of the psychophysical scaling methods described in this section.

For a final example, Kvälseth (1980) proposed that magnitude estimation could be profitably applied to evaluation of the factors influencing the implementation of ergonomics programs in industry. He asked employees of several firms to estimate the importance of 21 factors for the implementation of an appropriate ergonomics program in their company. They rated one factor first (say, accident rates) and then rated all other factors according to the ratio of importance relative to the first (e.g., if the second factor was twice as important as the first, they were to assign it a numerical value twice that of the first). Kvälseth's procedure demonstrated that, surprisingly, the two factors judged to be most important were management's perception of the need for ergonomic implementation and management's knowledge of the potential benefits of having satisfactory working conditions and environment. This result was surprising, because these factors were perceived as more than twice as important as the factor that was ranked 12th: the extent of work accidents and incidents of damage to health in the firm.

CHRONOMETRIC METHODS

The rise of the information-processing approach coincided with increased use of reaction time and related chronometric measures to explore and evaluate human performance (Lachman, Lachman, & Butterfield, 1979; Medina, Wong, Díaz, & Colonius, 2015). In a reaction-time task, a person is asked to make a response to a stimulus as quickly as possible. Whereas in the psychophysical approach, response frequency was the dependent variable upon which performance was evaluated, in the chronometric approach we look at changes in the reaction times under different response conditions.

There are three types of reaction-time tasks. In *simple reaction time*, a single response is made whenever any stimulus event occurs. That is, the response can be executed as soon as a stimulus event (e.g., the appearance of a letter) is detected. It does not matter what the stimulus is. A *go–no go reaction time* is obtained for situations in which a single response is to be executed to only some subset of the possible stimulus events. For example, the task may involve responding when the letter *A* occurs but not when the letter *B* occurs. Thus, the go–no go task requires discrimination among possible stimuli. Finally, *choice reaction time* refers to situations in which more than one response can be made, and the correct response depends on the stimulus that occurs. Using the preceding example, this would correspond to designating one response for the letter *A* and another response for the letter *B*. Thus, the choice task requires not only deciding what each stimulus is, but also that the correct response be selected for each stimulus.

SUBTRACTIVE LOGIC

Donders (1868/1969) used the three kinds of reaction tasks in what has come to be called *subtractive logic*. Figure 4.8 illustrates this logic. Recall from Chapter 1 that Donders wanted to measure the time that it took to perform each unique component of each reaction task. Donders assumed that the simple reaction (a type *A* reaction in his terminology) involved only the time to detect the stimulus and execute the response. The go–no go reaction (type *C*) required an additional process of identification of the stimulus, and the choice reaction (type *B*) included still another process, response selection. Donders argued that the time for the identification process could be found by subtracting the type *A* reaction time from the type *C*. Similarly, the difference between *B* and *C* should be the time for the response-selection process.

Subtractive logic is a way of estimating the time required for particular mental operations in many different kinds of tasks. The general idea is that whenever a task variation can be conceived as involving all the processes of another task, plus something else, the difference in reaction time for the two tasks can be taken as reflecting the time to perform the "something else." One of the clearest applications of the subtractive logic appears in studies of mental rotation. In such tasks, two geometric forms are presented that must be judged as *same* or *different*. One form is rotated relative to the other, either in depth or in the picture plane (see Figure 4.9). For *same* responses, reaction time is a linearly increasing function of the amount of rotation. This linear function has been interpreted as indicating that people mentally rotate one of the stimuli into the same orientation as the other before making the *same–different* judgment. The rate of mental rotation can then be estimated from the slope of the function. For the conditions shown in Figure 4.9, each additional deviation of 20°

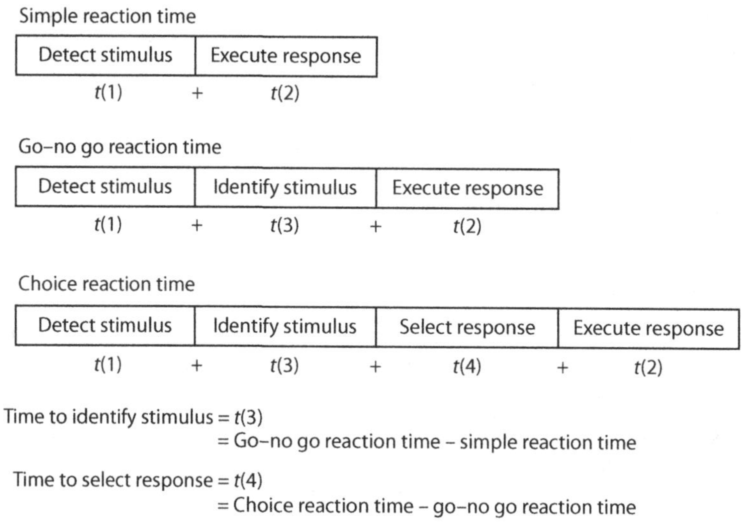

Simple reaction time

Detect stimulus	Execute response
$t(1)$ +	$t(2)$

Go–no go reaction time

Detect stimulus	Identify stimulus	Execute response
$t(1)$ +	$t(3)$ +	$t(2)$

Choice reaction time

Detect stimulus	Identify stimulus	Select response	Execute response
$t(1)$ +	$t(3)$ +	$t(4)$ +	$t(2)$

Time to identify stimulus = $t(3)$
= Go–no go reaction time – simple reaction time

Time to select response = $t(4)$
= Choice reaction time – go–no go reaction time

FIGURE 4.8 The subtractive logic applied to simple (*A*-reaction), go-no go (*C*-reaction), and choice (*B*-reaction) reaction times.

between the orientations of the two stimuli adds approximately 400 ms to the reaction time, and the rotation time is roughly 20 ms/degree. This is an example of the subtractive logic in that the judgments are assumed to involve the same processes except for rotation. Thus, the difference between the reaction times to pairs rotated 20° and pairs with no rotation is assumed to reflect the time to rotate the forms into alignment.

ADDITIVE-FACTORS LOGIC

Another popular method uses an *additive-factors logic* (Sternberg, 1969). The importance of the additive-factors logic is that it is a technique for identifying the underlying processing stages. Thus, whereas the subtractive logic requires that you assume what the processes are and then estimate their times, additive-factors logic provides evidence about how these processes are organized.

In the additive-factors approach, we assume that processing occurs in a series of discrete stages. Each stage runs to completion before providing input to the next stage. If an experimental variable affects the duration of one processing stage, differences in reaction time will reflect the relative duration of this stage. For example, if a stimulus-encoding stage is slowed by degrading the stimulus, then the time for this processing stage will increase, but the durations of the other stages should be unaffected. Importantly, if a second experimental variable affects a different stage, such as response selection, that variable will influence only the duration of that stage. Because the two variables independently affect different stages, their effects on reaction time should be additive. That is, when an analysis of variance is performed, there should be no interaction between the two variables. If the variables interact, then they must be affecting the same stage (Schweickert, Fisher, & Goldstein, 2010).

The basic idea behind additive-factors logic is that through careful selection of variables, it should be possible to determine the underlying processing stages from the patterns of interactions and additive effects that are obtained. Sternberg (1969) applied this logic to examinations of memory search tasks, in which people are given a memory set of items (letters, digits, or the like), followed by a target item. The people must decide whether or not the target is in the memory set. Sternberg was able to show that the size of the memory set has additive effects with variables that should influence target identification, response selection, and response execution. From these findings, he argued for the existence of a stage of processing in which the memory set is searched for a target, and that this stage was arranged serially and was independent of all other processing stages.

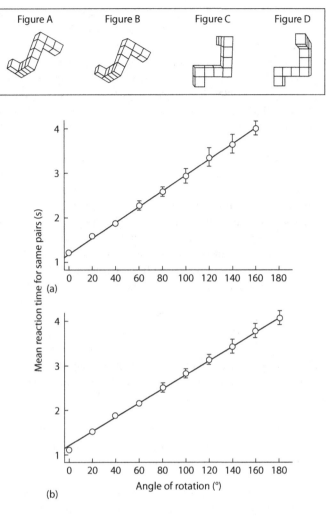

FIGURE 4.9   Mental rotation stimuli (upper panel) and results (lower panel) for rotation (a) in the picture plane and (b) in depth.

We can find fault with both the additive-factors and subtractive logics, for several reasons (Pachella, 1974). An assumption underlying both the subtractive and additive-factors logics is that human information processing occurs in a series of discrete stages, each with constant output. Because of the highly parallel nature of the brain, this assumption is an oversimplification that is difficult to justify in many circumstances. Another limitation of these approaches is that they rely on analyses of reaction time only, without consideration of error rates. This can make it difficult to apply these techniques to real human performance problems, because people make errors in most situations, and it is possible to trade speed of responding for accuracy, as described in the next section. Despite these and other limitations, the additive-factors and subtractive methods have proved to be robust and useful (Sanders, 1998).

CONTINUOUS INFORMATION ACCUMULATION

In recent years, researchers have advocated more continuous models of information processing in which many operations are performed simultaneously (Heathcote & Hayes, 2012). Information is not transmitted in chunks or discrete packets, as in the subtractive/additive-factors logics, but instead flows through the processing system like water soaking through a sponge. An important aspect of these kinds of theories is that partial information about the response that should be made to a particular stimulus can begin to arrive at the response-selection stage very early in processing, resulting in "priming" or the partial activation of different responses (Eriksen & Schultz, 1979; McClelland, 1979; Servant, White, Montagnini, & Burle, 2015). This idea has found a good deal of empirical support. Coles et al. (1985) demonstrated empirically that responses show partial activation during processing of the stimulus information in some circumstances. Neurophysiological evidence also suggests that information accumulates gradually over time until a response is made (Schall & Thompson, 1999).

In reaction-time tasks, because people are trying to make responses quickly, those responses are sometimes wrong. This fact, along with the evidence that responses can be partially activated or primed, has led to the development of processing models in which the state of the human processing system changes continuously over time. Such models account for the relation between speed and accuracy through changes in response criteria and rates of accumulation.

One way that we can characterize gradual accumulation of information is with a *random walk* (Klauer, 2014; see Figure 4.10, top panel). Suppose that a listener's task is to decide which of two letters (A or B) was presented over headphones. At the time that a letter (suppose it is "A") is read to the listener, evidence begins to accumulate toward one response or the other. When this evidence reaches a critical amount, shown as the dashed lines in Figure 4.10, a response can be made. If evidence reaches the top boundary, marked "A" in Figure 4.10, an "A" response is made. If evidence reaches the bottom boundary, marked "B," a "B" response can be made. The time required to accumulate the evidence required to reach one of the two boundaries determines the reaction time.

Suppose now that we tell the listener that he must respond much more quickly. This means that he will not be able to use as much information, because it takes time to accumulate it. So he sets his criteria closer together, as shown in the bottom panel of Figure 4.10. Less information is now required for each response, and, as a consequence, responses can be made very quickly. However, chance variations in the accumulation process make it more likely that the response will be wrong. As shown in the bottom panel of Figure 4.10, the decreased criteria result in an erroneous "B" response because of a brief negative "blip" in the accumulation process.

Accumulation models like the random walk are the only models that naturally explain the relation between speed and accuracy. Because the models can explain a wide range of phenomena involving both speed and accuracy, more detailed quantitative accounts of specific aspects of human performance are often of this type (Ratcliff, Smith, Brown, & McKoon, 2016). Another closely related family of models that is well suited for modeling information accumulation assumes the simultaneous activation of many processes. These models are called *artificial neural networks*, and they have been widely applied to problems in human performance, as well as any number of medical and industrial situations requiring diagnosis and classification (O'Reilly & Munakata, 2003; Rumelhart & McClelland, 1986).

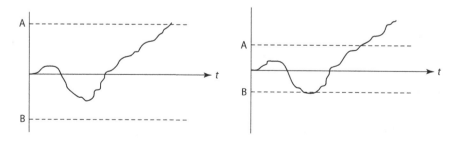

FIGURE 4.10 The random walk model and its relation to speed and accuracy. Information accumulates to the criterion level for response A or B. The relation between the right and left panels demonstrates a tradeoff between speed and accuracy.

In such networks, information processing takes place through the interactions of many elementary units that resemble neurons in the human brain. Units are arranged in layers, with excitatory and inhibitory connections extending to other layers. The information required to perform a particular task is distributed across the network in the form of a pattern of activation: Some units are turned on and others are turned off. Network models have been useful in robotics and machine pattern recognition. In psychology, they can provide an intriguing, possibly more neurophysiologically valid, alternative to traditional information-processing models. We can describe the dynamic behavior of many kinds of artificial networks by accumulation processes like the random walk, which makes the accumulator models doubly useful.

PSYCHOPHYSIOLOGICAL MEASURES

Psychophysiological methods measure physiological responses that are reliably correlated with certain psychological events. These kinds of measurements can greatly enhance the interpretation of human performance provided by chronometric methods (Luck, 2014; Rugg & Coles, 1995). One very important technique is the measurement of event-related potentials (ERPs) that appear on an electroencephalogram (EEG).

The living brain exhibits fluctuations in voltage produced by the electrochemical reactions of neurons. These voltage fluctuations are "brain waves," which are measured by an EEG. Technicians place electrodes precisely on a person's scalp, over particular brain areas, and the EEG continuously records the voltage from these areas over time. If the person sees a stimulus that requires a response, characteristic voltage patterns, ERPs, appear on the EEG. The "event" in the ERP is the stimulus. The "potential" is the change in the voltage observed in a particular location at a particular time.

An ERP can be positive or negative, depending on the direction of the voltage change. They are also identified by the time at which they are observed after the presentation of a stimulus. One frequently measured ERP is the P300, a positive fluctuation that appears approximately 300 ms after a stimulus is presented. We observe the P300 when a target stimulus is presented in a stream of irrelevant stimuli, and so it has been associated with processes involving recognition and attention.

An ERP like the P300 is a physiological response to a certain kind of stimulus. Reliable physiological indices like this are invaluable for studying human behavior and testing different theories about how information is processed. By examining how these measures change for different tasks and where in the brain they are produced, we have learned a great deal about how the brain functions. We can also use these kinds of measures to pinpoint how specific tasks are performed and determine how performance can be improved.

The EEG has been used to study performance for many years, but it has its shortcomings. The brain waves measured by the EEG are like the ripples on a pond: If we throw a tire into the pond, we will reliably detect a change at some point on the surface of the pond. Each time we throw in a tire, we will record approximately the same change at approximately the same time. But if we don't know for certain what was thrown into the pond and exactly where, the measurement of the ripple will only be able to suggest what happened and where. More recent technology has given us virtual windows on the brain by mapping where changes in neural activity occur. Methods such as PET and functional magnetic resonance imaging (fMRI) measure changes in the amount of oxygenated blood being used by different parts of the brain (Huettel, Song, & McCarthy, 2014). Neural activity requires oxygen, and the most active regions of the brain will require the most oxygen. Mapping where in the brain the oxygen is being used for a particular task provides a precise map of where processing is localized. Compared with the EEG, these methods have very high spatial resolution.

However, unlike the EEG, PET and fMRI have very low temporal resolution. Returning to the pond analogy, these methods tell us exactly where something was thrown into the pond but they are unable to tell us when. The EEG can tell us precisely when something is happening but not what it is. The problem is that PET and fMRI depend on the flow of blood into different brain areas: It takes at least a few seconds for blood to move around to where it's needed. It also takes some time for the oxygen to be extracted, which is the activity that PET and fMRI record. To test theories about mental processing that occurs in a matter of milliseconds, imaging studies require sometimes elaborate control conditions. Relying heavily on logic similar to Donders' subtractive logic, control conditions are devised that include all the information-processing steps except the one of interest (Poldrack, 2010). The patterns of blood flow during the control conditions are subtracted from the patterns of blood flow during the task of interest. The result is an image of the brain in which the activity is concentrated in those areas responsible for executing the task.

Like the other methods we have presented in this chapter, psychophysiological measures come with a number of problems that make interpretation of results difficult. However, they are invaluable tools for determining brain function and the specific kinds of processing that take place during reaction tasks. Although the human factors specialist may not always have access to the equipment necessary to record psychophysiological measures of human performance, basic research with these techniques provides an important foundation for applied work. Moreover, work on cognitive neuroscience is being integrated closely with human factors issues in the emerging approach that is called *neuroergonomics* (Johnson & Proctor, 2013) and in the study of *augmented cognition* (Stanney, Winslow, Hale, & Schmorrow, 2015). The goal of this work is to monitor neurophysiological indexes of mental and physical functions to adapt interfaces and work demands dynamically to the changing states of the person being monitored.

SUMMARY

The human information-processing approach views the human as a system through which information flows. As with any other system, we can analyze human performance in terms of subsystem components and the performance of those components. We infer the nature and organization of these subsystems from behavioral measures, such as response accuracy and reaction time, collected from people when they perform different tasks. General distinctions among perceptual, cognitive, and action subsystems provide a framework for organizing our basic knowledge of human performance and relating this knowledge to applied human factors issues.

There are many specific methods for analyzing the human information-processing system. We use response accuracy, collected using classical threshold techniques and signal-detection methods, to evaluate basic sensory sensitivities and response biases. We use reaction times and psychophysiological measures to clarify the nature of the underlying processing stages. We can use continuous models of information processing to characterize the relations between speed and accuracy of performance across many task situations.

The chapters in this book report many studies on human performance. The data upon which certain theories are based and recommendations for optimizing performance are made were collected using the methods described in this chapter. When you read about these studies, you will notice that we do not usually provide specific details about the experimental methods that were used. However, you should be able to determine such things as whether the reported data are thresholds, whether a conclusion is based on additive-factors logic, or which methods would be most appropriate in that particular situation. Because the distinction between perception, cognition, and action subsystems provides a convenient way to organize our knowledge of human performance, the next three sections of the book will examine each of these subsystems in turn. In the final section, we will discuss the influence of the physical and social environment on human information processing.

RECOMMENDED READINGS

Eyesenk, M. W., & Keane, M. T. (2010). *Cognitive Psychology: A Student's Handbook* (6th ed.). New York: Psychology Press.

Gazzaniga, M. S., Ivry, R. B., & Mangun, G. R. (2014). *Cognitive Neuroscience* (4th ed.). New York: W. W. Norton.

Gescheider, G. A. (1997). *Psychophysics: The Fundamentals* (3rd ed.). Mahwah, NJ: Erlbaum.

Lachman, R., Lachman, J. L., & Butterfield, E. C. (1979). *Cognitive Psychology and Information Processing: An Introduction*. Hillsdale, NJ: Erlbaum.

Luce, R. D. (1986). *Response Times: Their Role in Inferring Elementary Mental Organization*. New York: Oxford University Press.

MacMillan, N. A., & Creelman, C. D. (2005). *Detection Theory: A User's Guide*. Mahwah, NJ: Erlbaum.

Marks, L. E., & Gescheider, G. A. (2002). Psychophysical scaling. In H. Pashler & J. Wixted (Eds.), *Stevens' Handbook of Experimental Psychology* (3rd ed.), 4: *Methodology in Experimental Psychology* (pp. 91–138). Hoboken, NJ: Wiley.

Ratcliff, R., Smith, P. L., Brown, S. D., & McKoon, G. (2016). Diffusion decision model: Current issues and history. *Trends in cognitive sciences*, 20, 260–281.

Rosenbaum, D. A. (2014). *It's a Jungle in There: How Competition and Cooperation in the Brain Shape the Mind*. New York: Oxford University Press.

Part II

Perceptual Factors and Their Applications

5 Visual Perception

The information that we have about the visual world, and our perceptions of objects and visual events in the world, depend only indirectly upon the state of that world. They depend directly upon the nature of the images formed on the backs of our eyeballs, and these images are different in many important ways from the world itself.

T. N. Cornsweet
1970

INTRODUCTION

For an organism to operate effectively within any environment, natural or artificial, it must be able to get information about that environment using its senses. Likewise, the organism must be able to act on that information and to perceive the effect of its action on the environment. How people perceive environmental information is of great importance in human factors. Because the performance of a person in any human–machine system will be limited by the quality of the information he perceives, we are always concerned about how to display information in ways that are easily perceptible.

We must understand the basic principles of sensory processing and the characteristics of the different sensory systems when considering the design of displays, controls, signs, and other components of the human–machine interface (e.g., Proctor & Proctor, 2012). A good display will take advantage of those features of stimulation that the sensory systems can most readily transmit to higher-level brain processes. In this chapter we give an overview of the visual system and the phenomena of visual perception, with an emphasis on characteristics that are most important for human factors.

It is important to distinguish between the effect of a stimulus on a person's sensory system and her perceptual experience of the stimulus. If someone looks at a light, the sensory effect of the intensity of the light (determined by the number of photons from the light falling on the retina) is quite different from the perceived brightness of the light. If the light is turned on in a dark room, he may perceive it as being very bright indeed. But if that same light is turned on outside on a sunny day, he may perceive it as being very dim. In the discussion to follow, we will distinguish between sensory and perceptual effects. By the end of this chapter, you should be able to determine whether a phenomenon is sensory or perceptual.

Before we discuss the visual system specifically, we will consider some general properties of sensory systems (Møler, 2014). In this section, we describe the basic "cabling" of the nervous system and how information is coded by this system for processing by the brain.

PROPERTIES OF SENSORY SYSTEMS

Sensation begins when a physical stimulus makes contact with the "receptors" of a sensory system. Receptors are specialized cells that are sensitive to certain kinds of physical energy in the environment. For example, the light emitted by a lamp takes the form of photons that strike the receptor cells lining the back of the eye. A sound takes the form of a change in air pressure that causes vibrations of the tiny bones in the middle ear. These vibrations result in movement of receptor cells located in the inner ear.

Receptor cells transform physical energy into neural signals. Highly structured neural pathways carry these signals to the brain. A pathway is a cable of sorts: a chain of specialized cells called neurons that produce tiny electrical currents. Neural pathways act as filters, sorting and refining incoming information according to specific characteristics, such as color, shape, or intensity. There is no simple, passive transmission of information from the receptors to the brain. Information processing begins at the moment a sensation begins.

The pathways' first stop within the brain is the thalamus, a walnut-sized lump of tissue located in the center of the brain (Sherman & Guillery, 2013). One of the functions of the thalamus is to serve as a kind of switching station, sending neural signals to the appropriate areas in the cortex for further processing. The cortex is the outermost surface of the brain, a wrinkled layer of highly interconnected neurons only a few millimeters thick. Different areas on the cortex are highly structured and sensitive to specific kinds of stimulation. When a neural signal reaches the cortex it is very refined, and the operations that take place in the cortex refine it even more.

Any single neuron has a baseline level of activity (firing or spike rate) and receives excitatory and inhibitory input from many other neurons. An increase in activity for neurons with excitatory input will increase the firing rate of the neuron, whereas an increase in activity for those with inhibitory inputs will decrease the firing rate. These inputs determine the specific features of stimulation (e.g., color) to which the neuron is sensitive. The neurons in the cortex respond to very complicated aspects of a stimulus (e.g., specific shapes) as a result of inputs from lower-level neurons responsive to simpler aspects of the stimulus (e.g., lines and spots). The cortical neurons send very complicated signals to many different brain areas responsible for motor coordination, memory, emotion, and so forth. By the time a signal is completely processed, it has probably traveled through most of the major areas of the brain.

THE VISUAL SENSORY SYSTEM

Consider the vast amount of visual information received by your brain when you open your eyes in the morning. Somehow, all the patterns of light and movement that fall on your retinas organize themselves into a representation of the world. What kinds of stimulation do you need to be able to get out of bed? What kinds of stimulation can you ignore? It turns out that a good deal of stimulation is simply ignored. Our perceptions of the world are made up of a very limited amount of the information that we actually receive.

Take as an example the visual information received by the brain while you drive (Castro, 2009). Certain features of the world are important for your task, such as the location of the roadway, the yellow line down the center of the road, the locations of other vehicles on the road, and so forth. Rarely do you perceive the shapes of clouds or the color of the sky, even though this information is impinging on your senses.

The visual system is unique among our senses in that it provides us with information about where objects are in our environment without requiring that we actually touch those objects. This gives us many abilities, including the ability to reach for an object and also to avoid objects we do not wish to touch. The ability to negotiate around objects in an environment without actually touching anything is the basis for our ability to guide moving vehicles, as well as our own movement.

Vision lets us read written information in books, magazines, and newspapers, not to mention signs or television images. The visual modality is the most common and most reliable format for transmitting information from a machine to an operator in a human–machine interface. For example, all of the dials and gauges on an automobile dashboard require that the driver can see them to obtain the information they convey. In this and in all cases of visual sensation, information is conveyed by photons projected or reflected into the eye (Schwartz, 2010).

All light is electromagnetic radiation, which travels from a source in waves of small particles at a speed of 3.0×10^8 m/s (Hecht, 2016). A particular kind of electromagnetic radiation, such as a radio-station signal, an X-ray, or visible light, is defined by its range of wavelengths within the electromagnetic spectrum. Visible light, the range of wavelengths to which the human eye is sensitive, is a tiny range within the entire electromagnetic spectrum. The light that reaches the eye can be characterized as waves of photons that are either emitted by or reflected from objects in the environment. The intensity of a light is determined by the number of photons it produces. The color of a light is determined by its wavelength.

The range of wavelengths to which humans are sensitive runs from approximately 380 to 760 billionths of a meter, or nanometers (nm). Long wavelengths are perceived as red, whereas short wavelengths are perceived as violet (see Figure 5.1). Most colors that we experience are not composed of a single wavelength, but are mixtures of many different wavelengths. White light, for example, is composed of approximately equal amounts of all the different wavelengths.

When photons enter the eye, they are absorbed by the layer of receptor cells that line the back of the eye (Remington, 2012). These cells are located in the retina (see the next section), which acts like a curved piece of photosensitive paper. Each individual receptor cell contains a photopigment that is chemically reactive to photons of different wavelengths. When struck by a photon, the receptor cell generates an electrochemical signal that is passed to the nerve cells in the retina. A visual image results from a complex pattern of light that falls on the retina. For that pattern to be interpretable as an image, the light waves must be focused in the same way that an image must be focused through the lens of a camera.

THE FOCUSING SYSTEM

A schematic diagram of the eye is shown in Figure 5.2. Light is projected by a source or reflected from a surface into the eye. It enters the eye through the transparent front covering, called the cornea, and passes through the pupil, which varies in size. The light then is directed through the lens and focused on the retina. When the eyes are not moving and a person is attempting to look at something specific, we say that the eyes are fixated, or that an object has been fixated by the eyes. The location in space of a fixated object is the point of fixation. Once an object is fixated, its image must be brought into focus.

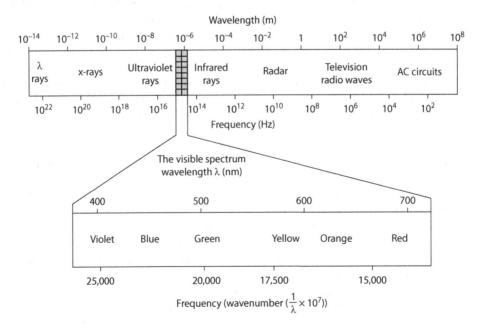

FIGURE 5.1   The visual spectrum, as located within the electromagnetic spectrum.

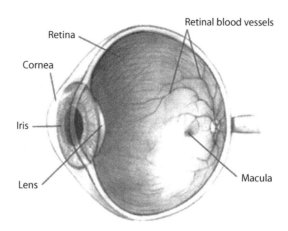

FIGURE 5.2   The human eye and optic nerve.

Cornea and Lens

Most of the focusing power of the eye comes from the cornea and the lens (Bowling, 2016). The shape of the cornea does most of the work by strongly bending the light before it even gets into the eye. After the light passes through the pupil, it must then pass through the lens. The lens is a transparent, gelatinous structure that makes fine adjustments, depending on the distance of the object that is being fixated and brought into focus.

More power (bending) is needed to focus an image when an object is close than when it is far away. The lens provides this additional power through the process of *accommodation*. During accommodation, the lens changes its shape (see Figure 5.3). When a fixated object is approximately 3 m or further away, the lens is relatively flat. The distance at which the lens no longer accommodates is called the *far point*. As the distance to the object decreases from the far point, tiny muscles attached to the lens relax, decreasing their tug on the lens and allowing the lens to become progressively more spherical. The more spherical the lens, the more the light is bent. Accommodation has a *near point*, which is approximately 20 cm in young adults. For objects closer than the near point, further increases in power are impossible and the image degrades.

It takes time and, sometimes, noticeable effort to accommodate to a change in an image's distance. For example, while driving, taking your eyes from the roadway to look at your speedometer requires a change from the far point of accommodation to a nearer point to bring the speedometer into focus. Accommodative changes in young adults are usually accomplished within about 900 ms after an object appears in the field of vision (Campbell & Westheimer, 1960).

The accommodative process is influenced by the amount of light in the environment. Accommodation is different in full light than in darkness. In darkness, the muscles of the lens are in a resting state, or *dark focus*. This is a point of accommodation somewhere between the near and far points (Andre, 2003; Andre & Owens, 1999). The dark focus point differs among different people and is affected by a number of different factors, such as the distance of a prior sustained focus and the position of the eyes (Hofstetter, Griffin, Berman, & Everson, 2000). On average, the dark focus is less than 1 m. The distance between a person's dark focus and the distance of an object to be accommodated is called the *lag of accommodation*. Eye strain can be caused by the placement of displays at distances that require continuous changes in accommodation, even for very small accommodative lags. This is especially true when conditions such as a low level of ambient lighting or the nature of the task encourage fixation at the dark focus (Patterson, Winterbottom, & Pierce,

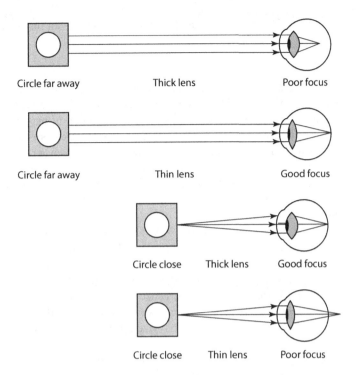

FIGURE 5.3 The process of accommodation.

2006). The constant tug of the eye muscles required to maintain focus when the lens muscles drift toward dark focus is the cause of the strain.

Pupil

The pupil is the hole in the middle of the iris. The iris, the colored part of your eye, is a doughnut-shaped muscle that controls how much light enters the eye. When the pupil dilates, it increases in size up to a maximum of 8 mm in diameter, and more light is allowed in. When the pupil contracts, it decreases in size to a minimum of 2 mm, and less light is allowed in. The amount of light entering the eye when the pupil is completely dilated is about 16 times the amount of light entering the eye when the pupil is completely contracted.

Dilation and contraction have been thought to be mostly reflexive behaviors determined by the amount of light falling on the eye (Watson & Yellott, 2012). However, recent evidence suggests that "the pupillary light response is far more than the low-level reflex that it was historically thought to be" (Mathôt & Van der Stigchel, 2015). Rather, it is affected by cognitive factors, including whether or not you are consciously aware of the stimulus, whether you are attending to the stimulus, and its apparent brightness. The size of the pupil also varies with a person's state of arousal, with an increase in arousal level resulting in dilation (Bradley, Miccoli, Escrig, & Lang, 2008).

The size of the pupil determines the *depth of field* of a fixated image (Marcos, Moreno, & Navarro, 1999). Suppose that you fixate an object some distance away, and so its image is clearly in focus. For some distance in front of the object and for some distance behind it, other objects in the image will also be clearly in focus. The total distance in depth for which objects in a scene are in clear focus is the depth of field. When the pupil is small, depth of field is greater than when the pupil is large. Consequently, for situations in which the pupil is large, such as when illumination is low, accommodation must be more precise (Randle, 1988) and there is an increased likelihood of eye strain.

Vergence

Another factor in focusing that occurs as a function of the distance of a fixated object is the degree of *vergence* of the two eyes (Morahan, Meehan, Patterson, & Hughes, 1998). Take a moment right now and look at the end of your nose. You should find that, in trying to bring your nose into focus, you crossed your eyes. Your eyes rotated toward each other. If you look from the end of your nose to an object some feet away, your eyes rotate away from each other. Vergence refers to the degree of rotation of the eyes inward or outward required to cause the light from a fixated object to fall on the central regions (the foveas) of the left and right eyes (see Figure 5.4). The vergence process allows the images from the two eyes to be fused and seen as a single object.

We can talk about the *line of sight* for each eye: a line drawn from the center of the back of each eye outward to the point of fixation in the world. When the point of fixation changes from far to near, the eyes turn inward and the lines of sight intersect at the point of fixation. Conversely, when the point of fixation changes from near to far, the eyes diverge and the lines of sight become almost parallel. Beyond fixated distances of approximately 6 m, the lines of sight remain parallel and there is no further divergence. The near point of convergence is approximately 5 cm; at this distance, objects become blurred if they are moved any closer. Look again at the end of your nose. Although you can probably fixate it easily enough, unless you have a very long nose you will not be able to bring it into clear focus.

Vergence is controlled by muscles that are attached to the outer surface of the eye. There is a reflexive connection between these muscles and the muscles that attach to the lens. This means that accommodation will change when eye position changes (Schowengerdt & Seibel, 2004). Remember that the muscles attached to the lens that control accommodation have a resting state, the dark focus. Similarly, the muscles controlling the degree of vergence have a resting state, which is measured as the degree of vergence assumed in the absence of light. This state is called *dark vergence* (Jaschinski, Jainta, Hoormann, & Walper, 2007). The angle formed by the lines of sight in dark vergence is somewhere between the angles formed by vergence on near and far objects (Owens & Leibowitz, 1983).

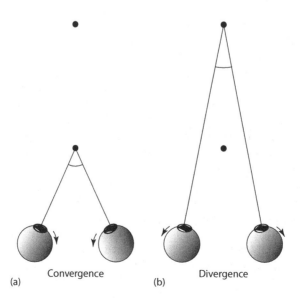

FIGURE 5.4 Vergence angle at near and far points.

Different people have very different dark vergence angles (Jaschinski et al., 2007), and these individual differences can affect the performance of visual inspection tasks. In one study, students performed a task for more than half an hour in which they inspected enlarged images of contact lenses for defects, at viewing distances of 20 and 60 m (Jebaraj, Tyrrell, & Gramopadhye, 1999). The students took more than twice as long to inspect the lenses at the near viewing distance as at the far distance, and reported greater visual fatigue, even though the sizes of the images on the retina were equal (see our discussion of visual angle later in this chapter). This effect was correlated with dark vergence distance but not with dark focus. In another study, people sat 20 cm from a video display terminal and searched the screen for a target letter among distractors (Best, Littleton, Gramopadhye, & Tyrrell, 1996). People with near dark vergence angles performed the search task faster than those with far dark vergence angles. Because everyone was positioned close to the display, this outcome suggests that the performance of visual inspection tasks is best when the difference between the viewing distance and the dark vergence posture is minimized.

Focusing Problems

Accurate perception depends on the proper functioning of the focusing system, which includes the cornea, pupil, and lens. The most common flaw in this system is in the shape of the eye (Naess, 2001). An eye that is too long or too short can result in an inability to focus an image on the receptors, regardless of the amount of accommodation attempted. In other words, the receptors are not at a position where an image can be focused (see Figure 5.5). The main purpose of glasses and contact lenses is to provide the extra focusing power necessary to correct this problem.

For nearsightedness, or *myopia*, the eye is too long, resulting in a focal point that is in front of the receptors when the lens is relaxed. For farsightedness, or *hyperopia*, the eye is too short, resulting in a focal point that is behind the receptors when the lens is fully flexed. As people become older, the speed and extent of their accommodation decrease continually. With age, the lens becomes harder and less responsive to the pulls of the muscles in the eye, so accommodative ability decreases and essentially all people become hyperopic. This condition is called *presbyopia*, or old-sightedness.

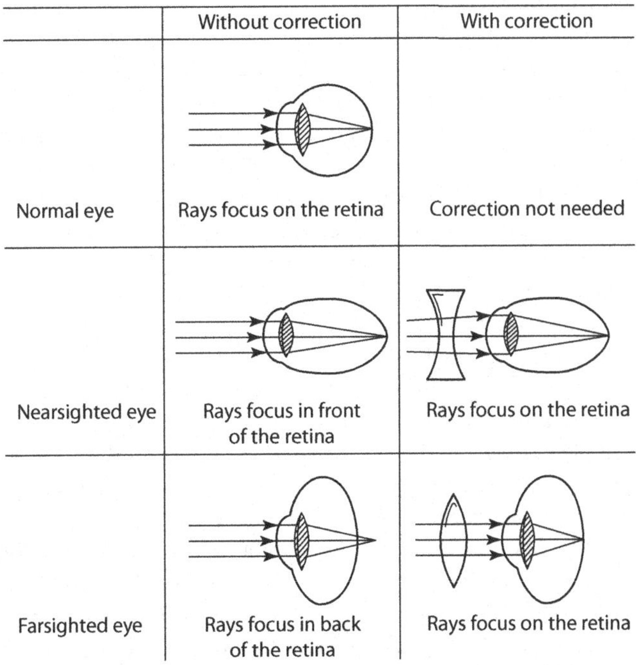

	Without correction	With correction
Normal eye	Rays focus on the retina	Correction not needed
Nearsighted eye	Rays focus in front of the retina	Rays focus on the retina
Farsighted eye	Rays focus in back of the retina	Rays focus on the retina

FIGURE 5.5 Focusing for normal, nearsighted, and farsighted eyes.

The near point can increase from as close as 10 cm for 20-year-olds to as far as 100 cm by age 60. Presbyopia can be corrected with reading glasses or bifocals, which typically are not prescribed until age 45 years or older. A person can have perfect vision in all other respects but still need reading glasses to compensate for the decreased accommodative ability of the lens.

A problem similar to presbyopia is *accommodative excess*, which results in either accommodative insufficiency or accommodative infacility. These disorders are tied to the muscles that control accommodation. These muscles can spasm, reducing accommodative ability and greatly increasing the time required for accommodative adjustments. Accommodative insufficiency is sometimes called early presbyopia, and results in an inability to properly adjust accommodation for close objects. Accommodative infacility refers to difficulty changing from near to far focus (and vice versa), resulting in poor accommodation and significantly slowed accommodation times. Sometimes accommodative excess can be improved by making changes to a person's corrective lenses, but it can also be treated with vision therapy, a program of exercises designed to reduce the tendency of the lens muscles to spasm.

Eye discomfort or eye strain is usually caused by fatigue of accommodative and vergence muscles. This sort of discomfort is particularly problematic for people who engage in a lot of close work or spend a lot of time at a computer monitor. Displays that are close to the viewer require both more vergence and more accommodation, and if fixation on such displays is required for an entire workday, the eye muscles can (not surprisingly) get very tired.

We mentioned that the amount of eye strain experienced from close visual work varies as a function of individuals' dark vergence and dark focus postures. People with far dark vergence angles report experiencing more visual fatigue after prolonged near work than do people with close vergence angles (Owens & Wolf-Kelly, 1987; Tyrrell & Leibowitz, 1990). Similarly, people whose dark focus point is further show more visual fatigue during close work than those with a nearer focus point. For people using a visual display screen, those people with more distant dark foci have the greatest visual fatigue when viewing the screen from 50 cm. However, at a viewing distance of 100 cm, people with longer dark foci experience no more or less fatigue than anyone else (Jaschinski-Kruza, 1991). Also, people with far dark foci tend to position themselves further from a visual display screen than do people with near dark foci, perhaps in an attempt to reduce vergence effort (Heuer, Hollendiek, Kroger, & Romer, 1989).

Working at a computer monitor induces another kind of eye strain that seems to be due to the accommodation muscles. Text on a computer monitor is different from printed text. Whereas printed text has sharp edges, the text on a computer monitor is sharp in the middle but has blurry edges because of the way the light fades out around the edge of an image on the screen. The combination of in- and out-of-focus signals that the eyes receive causes the state of accommodation to drift to the dark focus. This means that a person who spends a long time reading text on a computer monitor must continuously work to keep the text on the screen in focus. The tug-of-war between near and dark focus can cause significant discomfort.

Finally, a person can have problems focusing because he has an *astigmatism*. This problem is similar to myopia and hyperopia, which are caused by problems in the shape of the eye. For astigmatism, the problem is due to irregularities in the shape of the cornea. These irregularities cause light to be bent asymmetrically as it passes through the cornea. This means that contours in certain orientations will be in clear focus on the retina whereas those in other orientations will not. No matter how much the eye accommodates, some parts of the image will always be blurred. As with myopia and hyperopia, astigmatism can be corrected by glasses.

The cornea and lens must be transparent to allow light to pass into the eye. Injury to the eye and disease can cloud these organs and interfere with vision. The cornea can be scarred, which results in decreased acuity and an increase in the scattering of light. This can cause the perception of halos around light sources, especially at night. Another common problem is *cataracts*, which are hard, cloudy areas in the lens that usually occur with age. Seventy-five percent of people over 65 have cataracts, although in most cases the cataracts are not serious enough to interfere with the person's activities. Sometimes surgical intervention is necessary to correct major corneal and lens problems.

Summary

Research on accommodation, vergence, and other aspects of the focusing system plays an important role in human factors. The focusing system determines the quality of the image that is received by the eyes and limits the extent of visual detail that can be resolved. The system is also susceptible to fatigue, which can be debilitating for an operator. One of the most interesting things about the focusing system is that the degree of accommodation and vergence varies systematically with the distance of fixated objects. Therefore, the state of the focusing system can provide information about how far away an object is and how big it is. This means that an operator's judgments about distant objects will be influenced by the position and focus of his eyes. When accurate judgments about distant objects are critically important, such as while driving a car or piloting a plane, designers must take into account how vergence and accommodation will be influenced by the displays in the vehicle, the operator's position relative to those displays, the light environment in which the displays will most likely be viewed, and how judgments about objects outside of the vehicle will be influenced by all of these other factors.

THE RETINA

In a healthy eye, visual images are focused on the *retina*, which is the organ that lines the back of the eye (Ryan et al., 2013). The retina contains a layer of receptor cells, as well as two other layers of nerve cells that perform the first simple transformations of the retinal image into a neural signal. Most people are surprised to learn that the receptor cells are located behind the nerve cell layers, so a lot of the light that enters the eye never reaches the receptors at all. The light must penetrate these other layers first, as well as the blood supply that supports the retina. Consequently, only about half of the light energy that reaches the eye has an effect on the photoreceptors, which initiate the visual sensory signal.

Photoreceptors

The retina contains two types of receptors, *rods* and *cones* (Packer & Williams, 2003). The receptors of both types are like little pieces of pH-paper. At the end of each photoreceptor there is a little bit of photosensitive pigment. These photopigments absorb photons of light, which results in the photoreceptor being "bleached" and changing color. This change initiates a neural signal.

Rods and cones respond to different things. While cones are responsive to different colors, rods are not. All rods have the same kind of photopigment, whereas there are three types of cones, each with different photopigments. The four photopigments are most sensitive to light of different wavelengths (rods, 500 nm; short-wavelength cones, 440 nm or bluish; middle-wavelength cones, 540 nm or greenish; and long-wavelength cones, 565 nm or reddish), but each responds at least a little bit to light falling within fairly broad ranges of wavelength. There are many more rods (approximately 90 million) than cones (approximately 4–5 million; Packer & Williams, 2003).

The most important part of the retina is the *fovea*, which is a region about the size of a pinhead that falls directly in the line of sight. Its total area, relative to the area of the entire retina, is very tiny (approximately 1.25° diameter). There are only cones in the center of the fovea. Both rods and cones are found outside of the fovea, but there the rods greatly outnumber the cones. As we will discuss in more detail later, the rod system is responsible for vision in dim light (*scotopic* vision), whereas the cone system is responsible for vision in bright light (*photopic* vision). The cone system is responsible for color vision and perception of detail. The rod system is unable to provide any information about color and fine detail, but it is much more sensitive than the cone system in that rods can detect tiny amounts of light that cones cannot.

Another landmark of the retina is the *blind spot*. The blind spot is a region located on the nasal side of the retina that is approximately 2–3 times as large as the fovea. It is the point at which the fibers that make up the optic nerve leave the eye, and so there are no receptor cells here. Consequently, any visual stimulus that falls entirely on the blind spot will not be seen. You can "see" the blind spot for yourself. Close one eye, hold a pencil upright at arm's length and look at the eraser. Now, keeping the pencil at arm's length, move the pencil slowly away from your nose without moving your eyes. After you have moved the pencil about a foot, you should notice that the top part of the pencil has disappeared. At this point, you can move the pencil around a little bit, and watch (without moving your eyes!) the eraser pop in and out of view. As you move the pencil around in this way, you will realize that the blind spot is actually relatively large: there is a great empty hole in your visual field.

Although the blind spot is rather large, we rarely even notice it. One reason for this is that the region of the image that falls on one eye's blind spot falls on a part of the retina for the other eye that contains receptors. However, even when we look at the world with only one eye, the blind spot is rarely evident. If a pattern falls across the blind spot, it is perceived as continuous and complete under most circumstances (Baek, Cha, & Chong, 2012; Kawabata, 1984). Pick up your pencil again and find your blind spot. If the pencil is long enough, you should be able to move it so that the top of the pencil sticks above the blind spot and your hand holding the other end is below. You will probably notice that it becomes much harder to see the blind spot in the middle of the pencil: the pencil tends to look whole. This is the first example of an important principle that will recur in our discussion of perception: *the perceptual system fills in missing information* (Ramachandran, 1992).

Neural Layers

After the receptor cells respond to the presence of a photon, they send a signal to the nerve cells in the retina (Lennie, 2003). An important characteristic of these nerve cells is that they are extensively connected to each other, so that light falling on one area in the retina may have at least some small effect on the way that the nerve cells respond in another area of the retina. These "lateral" connections ("lateral" here meaning connections within the same layer of tissue) are responsible for some of the different characteristics of rods and cones, as well as several interesting visual illusions.

A phenomenon called Mach bands, which can be seen in Figure 5.6, is thought to be due to interactions between nerve cells in the retina (e.g., Keil, 2006). The figure shows a graduated sequence of gray bars ranging in lightness from light to dark. Although the bars themselves are of uniform

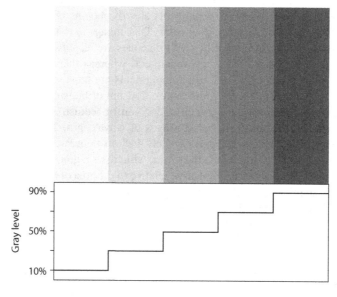

FIGURE 5.6 Mach bands.

intensity, darker and lighter bands are perceived at the boundaries of transition from one region to another. These bands arise from cells responding to brighter areas decreasing the activity of cells in a nearby darker area, and cells in darker areas tending to increase the activity of cells in a nearby lighter area. In short, the light and dark bands that are perceived are not actually present in the physical stimulus. They are induced by the competing activities of retinal neurons.

The difference in sensory characteristics between rods and cones also comes from the neural structure of the retina. While there are approximately 95 million rods and cones on the retina, these receptors are connected to only approximately 6 million nerve cells. Consequently, the signals from many receptors are pooled onto a single nerve cell. Approximately 120 rods converge on a single nerve cell, whereas only approximately 6 cones on average converge on a single nerve cell.

The relatively small amount of convergence in the cone system allows the accurate perception of details. Because light falling on different cones tends to be sent to different nerve cells, the spatial details of the retinal image are faithfully reproduced in the neurons that convey this information. However, each of the cones in the system must absorb its own photons for the image to be complete. In contrast, the relatively large amount of convergence in the rod system results in a loss of fine detail. Because information from many spatial locations is sent to a single cell, that cell cannot "know" anything about where on the retina the signal it receives originated. However, because so many rods converge on that cell, photons falling on only a few of them are sufficient to produce a signal. So, while the cone system needs a lot of light to function well, the rod system needs very little.

The sensory pathways subsequent to the receptors are specialized to process distinct characteristics of the stimulation along at least three parallel streams: the parvocellular, magnocellular, and koniocellular streams (Percival, Martin, & Grünert, 2013). Because we don't yet understand the function of the cells in the koniocellular stream in sensation and perception, we will restrict our consideration to the first two streams. The cells in the parvocellular stream (p cells) have small cell bodies, exhibit a sustained response in the presence of light stimuli (i.e., they continue to fire as long as light is falling on the retina), are concentrated around the fovea, show sensitivity to color, have a slow transmission speed, and have high spatial resolution but low temporal resolution. In contrast, those in the magnocellular stream (m cells) have large cell bodies, show a transient response to light stimuli (i.e., these neurons produce an initial burst of firing when light falls on the retina but the response rate decreases gradually as long as the light remains on), are distributed evenly across the retina, are broadband (i.e., not sensitive to color), show a fast speed of transmission, and have low spatial resolution and high temporal resolution (i.e., sensitivity to movement).

These properties have led researchers to speculate that the parvocellular stream is important for the perception of pattern and form, whereas the magnocellular stream is important for the perception of motion and change (McAnany & Alexander, 2008). As we will see, the distinction between the parvocellular and magnocellular streams extends through the primary visual cortex and forms the basis for two systems involving many areas of the brain that perform parallel analyses of pattern and location information.

Retinal Structure and Acuity

As we have now noted, the structure of the retina determines many characteristics of perception. One of the most important of these is the ability to perceive detail as a function of retinal location. This ability to resolve detail is called *acuity*. An example of a task that requires good visual acuity is detecting a small gap between two lines. If the gap is very small, the two lines may look like a single line, but when it is larger it is easier to see two lines. Figure 5.7 shows how acuity is highest at the fovea and decreases sharply as the image is moved further into the periphery. The acuity function is similar to the distributions of cone receptors and p cells across the retina, suggesting that it may be determined by the parvocellular system. The smaller degree of convergence for these cells results in better acuity.

It is not surprising that acuity varies with ambient light levels. Under photopic viewing conditions, the cone system is doing a lot of the work, and so acuity is great. Under scotopic viewing

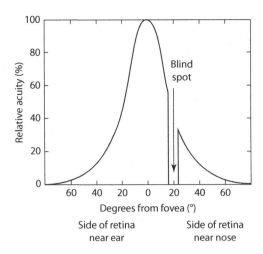

FIGURE 5.7　Acuity as a function of retinal location.

conditions, only the rods are operating. Because the rod system has a much higher degree of convergence than the cone system, acuity is much worse. Fine detail cannot be discriminated in the dark, only in full light.

The acuity function is relevant to many human factors problems. For example, in determining the design of instrument faces and where instruments should be located on an instrument panel, the human factors specialist must take into account where images will fall on the retina and the level of detail that must be resolved by the operator. Gauges and dials and so forth in the peripheral field of view will need to be larger, with less fine detail, than those located in the center of the visual field. Some electronic displays use gaze-contingent multiresolution (Reingold, Loschky, McConkie, & Stampe, 2003). These displays adjust so that the part of the display at which the operator is looking has higher resolution than the rest of the display (see Box 5.1). We will discuss other factors that affect acuity later in the chapter.

VISUAL PATHWAYS

Once the optic nerve leaves the eye, visual signals become progressively more refined. The optic nerve splits, sending half of the information from each eye to one half of the brain and the other half to the other. Information about objects located in the right visual field first goes to the left half of the brain, while information about objects in the left visual field goes to the right half of the brain. These separate signals are put back together later in processing. After passing through a region of the thalamus called the lateral geniculate nucleus (LGN), where the parvocellular and magnocellular pathways are kept distinct and all neurons are monocular (i.e., they respond to light at only one eye), the next stop for visual information is the primary visual cortex. Put your hand on the back of your neck: the bump from your skull that hits the top edge of your hand is approximately where the visual cortex is located.

Visual Cortex

The *visual cortex* is highly structured (Hubel and Wiesel, 1979). It consists of several layers that contain approximately 10^8 neurons. The cortical neurons are spatiotopic, which means that if one cell responds to stimulation at one area on the retina, cells very close to it will be responsive to stimulation at nearby locations on the retina. The magnocellular and parvocellular inputs from the LGN have their effects on neurons in distinct layers of the visual cortex.

BOX 5.1 GAZE-CONTINGENT MULTIRESOLUTIONAL DISPLAYS

Many human–computer interaction tasks require users to search for information on large screens and monitors. Overall, high-resolution displays are more desirable than low-resolution displays, because it is easier to identify objects using high-resolution displays. However, high-resolution displays are expensive, and their processing requirements may exceed the processing capacity and/or transmission bandwidth for some computer systems or networks. Consequently, a high-resolution display is not always practical or feasible.

Gaze-contingent multiresolutional displays (GCMRDs) take into account the fact that high resolution is only useful for foveal vision and is wasted to a large extent in peripheral vision because of low acuity in the periphery (see Reingold, Loschky, McConkie, & Stampe, 2003). For a GCMRD, only a limited region of the display is presented in high resolution, and this region corresponds to the area falling in or about the fovea. The user viewing the display wears an eye-tracker, which monitors where he is looking, and the display updates in real time, presenting in high resolution only the area that is currently fixated.

How tasks and displays are designed has consequences for perception and performance when GCMRDs are used. A considerable amount of human factors research can be conducted to investigate these consequences. Consider, for example, the basic idea behind GCRMDs: the user only needs high resolution in central vision and can use peripheral cues just as effectively whether they are of low or high resolution. Loschky and McConkie (2002) investigated this hypothesis by manipulating the size of the high-resolution area to see how performance of searching for a target object in a scene was affected. If low resolution is just as good as high resolution in the periphery, then the size of the high-resolution area should not have any effect on response times. Contrary to this hypothesis, a smaller high-resolution area actually led to longer search times than a larger area. The reason for this seems to be due to eye movements. Because the distance between successive eye movements was shorter for smaller high-resolution areas, observers made more fixations until the target was found. Apparently, people had to make more eye movements with the smaller high-resolution area because the target object did not stand out well when it was in the low-resolution region.

Although GCMRDs may be useful, Loschky and McConkie's (2002) results suggest that performance of tasks that require visual search may not be as good with such displays as with high-resolution displays. However, their study used images with only two levels of resolution: high and low. There was a sharp boundary between the low- and high-resolution areas of each image. Because the decrease in a person's visual acuity from the fovea to the periphery is continuous, it may be that little decrement in performance would occur with only a small high-resolution area if the image instead used a gradient of resolutions from high to low, so that there was a more gradual drop-off in resolution better matching these changes in visual acuity.

Loschky, McConkie, Yang, and Miller (2005) obtained evidence consistent with this hypothesis. The people in their experiments viewed high-resolution displays of scenes. Occasionally, for a single fixation, GCMRD versions with decreasing resolution from fixation to periphery appeared. The person was to push a button as fast as possible when he or she detected blur. When the decrease in resolution from fixation to periphery was slight, and less than the limits imposed by the retina, people did not detect any blur. The image looked normal to them. Moreover, across different amounts of decrease in resolution, the blur detection results and eye fixation durations were predicted accurately by a model in which contrast sensitivity decreases gradually from fovea to periphery.

GCMRDs can be used as a research tool to investigate other human factors issues. One example is to measure the useful field of view (UFOV), which is the region from which a person can obtain visual information during a single fixation. The UFOV is influenced by many factors, including the moment-to-moment cognitive load imposed on the person. Because a restricted UFOV may lead to increased crash risk when driving, Gaspar et al. (2016) devised a task to measure the UFOV dynamically for people in a driving simulator. For this purpose, they used a gaze-contingent task in which a participant had to periodically discriminate the orientation of a briefly presented grating. The performance of this task yielded a good measure of transient changes in the UFOV as a function of cognitive load, leading Gaspar et al. to conclude: "The GC-UFOV paradigm developed and tested in this study is a novel and effective tool for studying transient changes in the UFOV due to cognitive load in the context of complex real-world tasks such as simulated driving" (p. 630).

The cells in the cortex are distinctive in terms of the kinds of information to which they respond. The earliest, most fundamental cortical cells have circular center-surround receptive fields. This means that they respond most strongly to single spots of light on preferred locations on the retina, and they tend to fire less when light is presented around that location. However, other cells are more complex. Simple cells respond best to bars or lines of specific orientations. Complex cells also respond optimally to bars of a given orientation, but primarily when the bar moves across the visual field in a particular direction. A subclass of simple and complex cells will not fire if the stimulus is longer than the receptive field length. Therefore, the cortical cells may be responsible for signaling the presence or absence of specific features in visual scenes.

There are many interesting effects in visual perception that have their origins in the orientation-sensitive cells in the visual cortex. If you look at a vertical line over a field of tilted lines, it will appear to be tilted in the opposite direction from the field (see Figure 5.8a), a phenomenon called *tilt contrast* (Tolhurst & Thompson, 1975). A related effect occurs after you fixate for a while on a field of slightly tilted lines (see Figure 5.8b). If you look at a field of vertical lines after staring at the tilted field, the vertical lines will appear to be tilted in the opposite direction from the field (Magnussen & Kurtenbach, 1980). Tilt aftereffect and tilt contrast are due to the interactions between the neurons in the visual cortex (see, e.g., Bednar & Miikkulainen, 2000; Schwartz, Sejnowski, & Dayan, 2009), just as Mach bands are due to interactions between the cells in the retina.

Another perceptual phenomenon attributable to the orientation-sensitive cells of the visual cortex is called the *oblique effect*. People are much better at detecting and identifying horizontally or vertically oriented lines than lines of any oblique orientation. This effect seems to be due to a larger proportion of neurons in the visual cortex that are sensitive to vertical and horizontal orientations. Not as many cortical neurons are sensitive to oblique orientations (Gentaz & Tschopp, 2002). Because more neurons are devoted to horizontal and vertical orientations, these orientations can be detected and identified more easily.

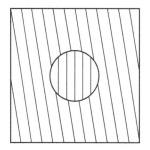 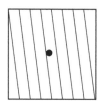 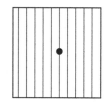

FIGURE 5.8 (a) Tilt contrast and (b) the tilt aftereffect.

Dorsal and Ventral Streams

The primary visual cortex is just the first of more than 30 cortical areas involved in the processing of visual information (Frishman, 2001). This information is processed in two streams, called the dorsal (top) and ventral (bottom) streams, which are sometimes called the "where" and "what" streams, respectively. The dorsal stream receives much of its input from the magnocellular pathway and seems to be involved primarily in perception of spatial location and motion and in the control of actions. In contrast, the ventral stream receives both parvocellular and magnocellular input and is important for perception of forms, objects, and colors. Because "what" and "where" are analyzed by distinct systems, we should expect to find situations in which people make "what" and "where" decisions better, depending on which pathway is used.

For example, Barber (1990) examined performance measures of short-range air defense weapon operators in simulated combat together with measurements of the operators' basic visual perception abilities. The combat task involved detection of aircraft, identification of the aircraft as friendly or hostile, aiming the guns at the hostile aircraft, and tracking the hostile aircraft with the gun system once an initial fix was obtained. He correlated how well each operator performed each component of the task with the operators' scores on simple visual perception tasks. This analysis suggested that the dorsal system helps to control detection and acquisition, and the ventral system helps to control detection and identification. Barber proposed that a third subsystem, which receives both magnocellular and parvocellular input, helps to control identification and tracking.

In other studies, Leibowitz and his associates (Leibowitz, 1996; Leibowitz & Owens, 1986; Leibowitz & Post, 1982; Leibowitz, Post, Brandt, & Dichgans, 1982) examined how the different visual pathways influence night driving performance. To do this, they have classified perceptual tasks according to whether they require "focal" processing (primarily in the central visual field, requiring the ventral system) or "ambient" processing (across the entire visual field, requiring the dorsal system). They hypothesize that focal processing is required for object recognition and that ambient processing is required for locomotion and orientation in space. For driving, the focal mode is involved in the identification of road signs and objects in the environment, whereas the ambient mode directs guidance of the vehicle.

Night driving fatality rates are three to four times higher than daytime rates (after adjusting for the fewer dark hours in a day). Why should an accident at night be more likely to lead to a fatality than an accident during the day? Leibowitz and his associates suggest that this occurs because the focal system is adversely affected under low illumination levels but the ambient system is not. Because the focal system does not function well in the dark, drivers can't recognize objects as easily or accurately. However, because the ambient system is relatively unaffected, drivers can steer vehicles as easily at night as during the day. Moreover, most objects that require recognition, for example road signs and dashboard instruments, are illuminated or highly reflective. Consequently, drivers underestimate the extent to which focal perception is impaired and do not reduce speed accordingly.

An impairment of the focal system becomes obvious only when a non-illuminated obstacle, such as a parked car, fallen tree, or pedestrian, appears in the road. Drivers may take much more time to identify these objects than they have to be able to stop safely. In many cases, drivers report not even seeing an obstacle before their accidents. Leibowitz and Owens (1986) suggest that night accidents might be reduced if drivers were educated about the selective impairment of recognition vision at night.

VISUAL PERCEPTION

As we mentioned earlier, while you drive your car from home to work, a lot of different visual stimuli impinge on your eyes, but you actually perceive very few of those stimuli and use information about these few to make driving decisions. Furthermore, it is not the physical properties of the important stimuli, such as intensity or wavelength, on which our decisions are based, but rather, their corresponding perceptual properties, like brightness or color. While some perceptual properties of visual stimuli correspond directly to physical properties of the stimulus, some do not, and others might arise under more than one set of conditions. However, all (visual) perceptual phenomena can be traced to the structure of the visual sensory system, whether directly or indirectly.

In the rest of this chapter we discuss the basic properties of visual perception, beginning with brightness and acuity, and how our perceptions of brightness and acuities depend on the environment in which perception is taking place.

BRIGHTNESS

Automobile drivers must share the road with other wheeled vehicles, including bicycles and motorcycles. For various reasons, drivers often "don't see" these other vehicles, and because of the smaller and less protective nature of a motorcycle or a bicycle, the riders of these alternative vehicles can be seriously injured or killed in accidents with cars. For this reason, several U.S. states mandate that motorcycle headlamps shall be turned on even in the daylight hours, in an attempt to increase motorcycle visibility.

Assuming that the problem with car–motorcycle accidents really is one of visibility, is such a law effective? For the headlamp to increase the visibility of the motorcycle, we might consider whether the headlamp increases the perceived *brightness* of the motorcycle. The primary physical determinant of brightness is the intensity of the energy produced by a light source (*luminance*). The physical measurement of light energy is called *radiometry*, with radiant intensity being the measure of total energy. The measurement process that specifies light energy in terms of its effectiveness for vision is called *photometry*. Photometry involves a conversion of radiant intensity to units of luminance by weighting a light's radiance according to the visual system's sensitivity to it. Different conversion functions, corresponding to the distinct spectral sensitivity curves (see below), are used to specify luminance in candelas per square meter (see also Chapter 17).

On a warm, sunny day (the best sort of day for riding a motorcycle), a shiny new motorcycle reflects a lot of sunlight from its chrome and highly polished surfaces. It looks bright already, even without turning on the headlamp. We can determine perceived brightness by measuring the intensity of the light reflecting from its surface. The relation between brightness and light intensity is generally described well by a power function

$$B = aI^{0.33},$$

where:

B is brightness,
I is the physical intensity of the light, and
a is a constant, which might be different for different automobile drivers (see Chapter 4).

Although the power function relationship is useful in a theoretical sense, pragmatically, different people will judge the same physical intensity to be of different levels of brightness. The bril scale is a way to quantify brightness that measures everyone's perceived brightness on the same scale (Stevens, 1975). To understand the bril scale, it is important to understand the concept of a decibel. You have probably heard the term "decibel" used before in relation to noise levels, but it can be used for any perceptual effect related to stimulus intensity. A decibel (dB) is a unit of physical intensity that is defined as

$$\log_{10} \frac{I}{S},$$

where S is the intensity of some standard stimulus. Notice, then, that a measurement in decibels is entirely dependent on the intensity S. For the bril scale, one bril is the brightness of a white light that is 40 dB above a person's absolute threshold for detecting light. So, for one bril, S is the intensity of a white light at absolute threshold.

One question we might ask about motorcycle headlamps is whether turning them on increases the brightness noticeably. We already know from earlier chapters that as luminance increases, greater changes are needed to produce equivalent changes in brightness. This means that we might not perceive much, if any, change in brightness between a motorcycle with its headlamp turned off and one with its headlamp turned on if the sun is already making the motorcycle appear very bright. However, on a cloudy day, when the sunlight reflecting off the motorcycle is greatly reduced, turning on the headlamp may greatly increase perceived brightness, even though the headlamp has exactly the same physical intensity on sunny and cloudy days.

It is clear, then, that although brightness is primarily a function of stimulus intensity, it is influenced by many other factors (Fiorentini, 2003). Although in some situations keeping the motorcycle headlamp on might not increase perceived brightness, in others it might, so perhaps the headlamp law is not such a bad idea. There are other factors that can contribute to perceived brightness, and we turn to a discussion of them now. Of particular importance are the state of adaptation of the observer and the wavelength of the perceived light, the duration of the light, and its contrast with the background illumination.

Dark and Light Adaptation

We discussed earlier in this chapter the major differences between rods and cones. One important factor is convergence: because many rods converge on a single nerve cell, rods are much better at detecting small amounts of light than cones are. Light presented on the periphery of the visual field appears brighter for this reason. You can verify this prediction by looking at the night sky. Find a dim star and look at it both directly and out of the corner of your eye. Sometimes you can even see a star peripherally that you cannot see while looking straight at it.

The differences between rods and cones are responsible for the phenomenon of *dark adaptation* (Reuter, 2011). When you first enter a dark room, it is very difficult to see anything at all. However, during the first few minutes in the dark, your ability to see improves substantially and then levels off (see Figure 5.9). After about 8 minutes, you will experience another improvement, and your ability to see will again get better and continue to do so until approximately 45 minutes have passed since

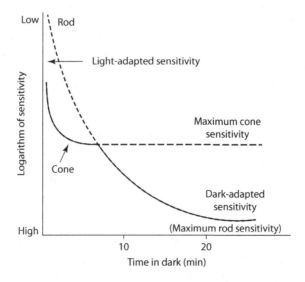

FIGURE 5.9 Dark adaptation function, which shows sensitivity to light as a function of time in the dark.

you entered the room. At this time, your sensitivity to light is close to 100,000 times greater than it was when you entered the room!

Why does this happen? When you enter the dark, many of your photoreceptors, both rods and cones, are bleached. They have absorbed photons from the lighted environment from which you came and have yet to regenerate new photopigment. The cones regenerate their pigment the fastest, resulting in the first improvements in your ability to see. At the end of 3 minutes, the cones have finished regenerating their pigment, but remember that cones are not very good for seeing in the dark. For your vision to improve, you have to wait a little longer for the rods to start helping out. After about 8 minutes in the dark, the rods catch up with the cones, and your ability to see again starts to improve. The remainder of the increase in sensitivity as time progresses is due to the rods continuing to regenerate photopigment.

We can do experiments that can verify that rods and cones are responsible for the different stages of dark adaptation. For instance, we could take you into a completely dark room and then only allow little bits of light energy to fall on your fovea (where there are no rods) and measure your sensitivity to this light over time. We would find that your ability to see never improves much after the first 3 minutes in the dark. Also, there are some people who have no cones (called rod monochromats). If we brought one of these people into the dark room, she would remain relatively blind until 8 minutes had passed, when her rods had regenerated enough photopigment for her vision to begin to improve.

The opposite of dark adaptation is light adaptation. Light adaptation happens after you come out of the dark room. If you have dark adapted, it is usually uncomfortable to return to a fully lighted environment. This discomfort arises because your eyes are far more sensitive to light. In fact, if we were to measure your ability to detect small amounts of light, your threshold would be very low if you were dark adapted. After you return to a lighted environment, your threshold begins to increase. After about 10 minutes, your threshold will have stabilized to a relatively high level, meaning that you can no longer detect the small lights that you could see while dark adapted (Hood & Finkelstein, 1986). The reason for your increased threshold has to do with the number of bleached photoreceptors in your retina. The more light that enters your eye, the more photoreceptors get bleached. Bleached photoreceptors do not respond to light, so sensitivity decreases.

For any environment, light or dark, your eyes will adapt. Light adaptation is a concern in night driving, where a driver needs to dark adapt to maintain maximal sensitivity to light. If the light intensity created by a driver's own headlights close to his vehicle is too high, the driver's eyes will light adapt, and he will not be as able to see objects farther in front of the car (Rice, 2005). However, the brighter and wider the headlight beam pattern can be farther from the vehicle, the better the driver will be able to see (Tiesler-Wittig, Postma, & Springer, 2005). However, this increased distance and intensity must be accomplished without creating too much glare for other drivers.

In some situations, changes in the environment will force rapid changes in adaptation level. Again, we find an example of this in driving. Highway tunnels, however well lit, force drivers to change from one level of adaptation, that required by the environment outside of the tunnel, to another, that required by the tunnel lighting. Problems in light sensitivity are particularly severe just upon entering a tunnel and just after leaving a tunnel. During daytime driving, the roadway will appear very dark at the tunnel entrance and very bright at its exit. Brighter lights placed at the beginning and end of a tunnel provide more gradual changes of illumination and less visual impairment (Oyama, 1987).

Spectral Sensitivity

The different photoreceptors have different spectral sensitivities. Figure 5.10 shows how the photopigments for both rods and cones are broadly tuned, stretching between 100 and 200 nm, depending on the photopigment. The peak for the combined absorption spectra of the three cone photopigments is around 560 nm, whereas the peak rod absorbance is approximately 500 nm.

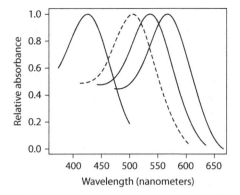

FIGURE 5.10 Pigment absorption spectra for rod (dotted line) and cone (solid lines) photopigments.

Recall that rods are important for scotopic viewing conditions and cones are important for photopic viewing conditions. The wavelength difference in peak sensitivities for rods and cones can be seen in spectral sensitivity curves (see Figure 5.11). These curves are the absolute thresholds for the detection of a light as a function of its wavelength. The sensitivity curve for photopic vision is similar to that of the combined cone photopigment absorption curve, and the sensitivity curve for scotopic vision is similar to that of the rod photopigment absorption curve. These curves indicate that at either photopic or scotopic levels of illumination, sensitivity to light energy varies across the spectrum.

One interesting thing to notice is that rods are not sensitive to low-intensity, long-wavelength (red) light. In the presence of red light, only long-wavelength cones will be bleached. A dark-adapted person can enter a room lit with red light and remain dark adapted. There are many situations where someone might want to see and still preserve his or her state of dark adaptation. Astronomers might need to read charts without losing their ability to see dim objects through a telescope. Military personnel on night missions may need to read maps or perform other tasks while preserving dark adaptation. This need has led to the design of low-intensity red flashlights, red-lit cockpits and control rooms, red finger-lights for map reading, and red dials and gauges in vehicle control systems.

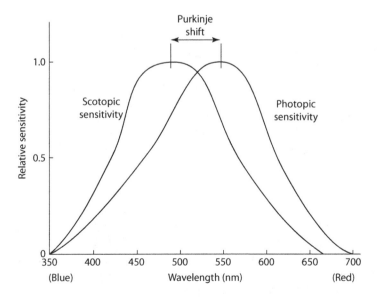

FIGURE 5.11 Scotopic and photopic spectral sensitivity functions illustrating the Purkinje shift.

The Purkinje Shift

Look again at Figure 5.11. The difference between the rod and cone peak sensitivities is the source of a perceptual effect called the *Purkinje shift* (Wolfe & Ali, 2015), a subtle perceptual effect that you may never have noticed. This shift refers to the fact that when two light sources, one short and one long wavelength, appear equally bright under photopic conditions, they will not do so under scotopic conditions. The short-wavelength light will look brighter than the long-wavelength light under conditions of dark adaptation because of the contribution of the rods. The long-wavelength light will look dimmer. You can notice the Purkinje shift at dusk. During the day, red and yellow objects may appear brighter than green and yellow objects. As day fades into night and the rod system takes over, blue and green objects will tend to appear brighter than red and yellow objects.

Temporal and Spatial Summation

Brightness is also influenced by how long a light is on and how big it is. For lights that are on only very briefly (100 ms or less), brightness is a function of both the intensity and the exposure time. This relation, known as Bloch's law (Gorea, 2015), is

$$T \times I = C,$$

where:

    T is the exposure time,
    I is light intensity, and
    C is a constant brightness.

In other words, a 100 ms light that is half the intensity of a 50 ms light will appear equally bright, because the energy of both lights over a 100 ms period is equal. For these flickering or very short-duration lights, it is the total amount of light energy during the presentation period that determines brightness.

The area or size of a light also affects its detectability and brightness. For very small areas of approximately 10 min of visual angle (discussed later in this chapter), Ricco's law states that

$$A \times I = C,$$

where A is area. For larger stimuli, Piper's law states that

$$\sqrt{A \times I} = C.$$

Whereas Bloch's law describes how light energy is summed over time, Ricco's and Piper's laws describe how light energy is summed over space (Khuu & Kalloniatis, 2015). Spatial summation occurs because of the convergence in the rod system. Perception of brightness is less influenced by stimulus size or area in the fovea than in the periphery (Lie, 1980), because the degree of convergence is much greater in the periphery.

LIGHTNESS

The amount of light reflected from an illuminated surface is a function of both the level of illumination and the degree to which the surface reflects light. While brightness is the perceptual attribute associated with overall light intensity, the term *lightness* refers to the perceptual attribute associated with reflectance (Gilchrist, 2006). Lightness describes how dark or light an object appears on a scale from black to white. Black surfaces have low reflectance and absorb most of the light that falls on them, whereas white surfaces have high reflectance and reflect most of the light that falls on them.

Lightness is very different from brightness in a number of ways. For instance, brightness is a function of intensity: As intensity increases, brightness increases. Consider, however, the reflectance of two surfaces under two levels of illumination. Under high levels of illumination, both surfaces will reflect much more light energy than under low levels of illumination, yet their relative lightness will tend to remain the same (Soranzo, Galmonte, & Agostini, 2009). This phenomenon is called *lightness constancy*.

For example, pieces of white and black paper will look white and black whether viewed inside or outside in the sun. Because the intensity of illumination outside typically is greater than inside, the black paper may actually be reflecting more light outside than the white paper does inside. The perception of lightness is tied to the reflectance properties of the objects rather than the absolute amount of light reflected from them.

Lightness contrast refers to the fact that the perceived lightness of an object is affected by the intensity of surrounding areas (e.g., Soranzo, Lugrin, & Wilson, 2013). The key difference to note between lightness contrast and lightness constancy is that the former occurs when only the intensity from surrounding regions is changed, whereas the latter occurs when the intensity of illumination across the entire visual field is changed. Figure 5.12 shows lightness contrast, because the center squares of constant intensity appear progressively darker as the surround becomes lighter.

In general, two stimuli will appear equally light when the intensity ratio of each stimulus to its respective surround is the same (Wallach, 1972). Lightness contrast in Figure 5.12 arises because this ratio is changed. While the center square is of a constant gray level, the gray level of the surround is changed. Gelb (1929) demonstrated the importance of the contrast ratio. He suspended a black disc in black surroundings. A hidden light source projected light only onto the disc. In this situation, the black disc looked white. In terms of the constant-ratio rule, the conditions for constancy were violated because the disc had a source of illumination that its background did not. However, when he placed a small piece of white paper next to the black disc so that it also was illuminated by the hidden light source, the black disc then looked black.

Gilchrist (1977) made one of the most compelling demonstrations of this type. He arranged a situation in which a white card was seen as white or dark gray, depending on the card's apparent position in space. As shown in Figure 5.13, the observer looked at three cards through a peephole. Two of the cards (a white test card and a black card) were in a front chamber that was dimly illuminated, while the third card (also white) was in a back chamber that was brightly illuminated. By changing the shape of the white test card, Gilchrist made the third card look as if it was either behind (panel a in Figure 5.13) or in front of (panel c) the other two cards.

Observers judged the lightness of the test card (the white card in the front room). When the test card looked as if it was in the front room with the black card, it was seen as white. However, when the shape of the card made it look as if it was in the back room with the more brightly illuminated white card, it was seen as almost black. This phenomenon suggests that the perception of illumination is important to lightness. If the test card really had been in the brightly lit back chamber, it would have been reflecting less light than the white card and so appeared dark. Apparently, the perceptual system uses "logic" like this to compute lightness. Thus, even the basic aspects of sensory experience of the type covered in this chapter are subject to computations performed by higher-level brain processes.

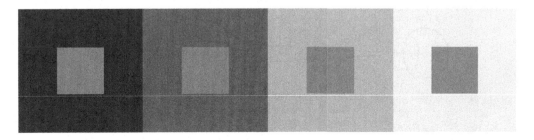

FIGURE 5.12 Lightness contrast.

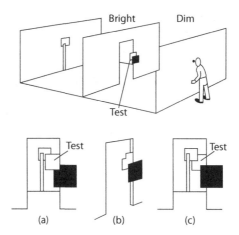

FIGURE 5.13 Gilchrist's apparatus with the test stimulus constructed to appear in the front room (a) or in the back room (b and c).

SPATIAL AND TEMPORAL RESOLUTION

Acuity

We discussed acuity in general terms earlier in this chapter. To discriminate objects in the visual field, differences between regions of different intensity must be resolved. More formally, acuity can be measured by finding the minimum *visual angle* for a detail that can be resolved. Visual angle is a measure of stimulus size that does not depend on distance: it is a measure of the size of the retinal image. Because of this property, it is the most commonly used measure of stimulus size. As shown in Figure 5.14, the size of the retinal image is a function of the size of the object and its distance from the observer. The visual angle is given by

$$\alpha = \tan^{-1}\left(S/D\right),$$

where:
- S is the size of the object, and
- D is the viewing distance, in equivalent units.

There are several types of acuity. *Identification acuity* can be measured by the use of a Snellen eye chart, which is the chart you've probably seen in your doctor's office. It consists of rows of letters that become progressively smaller. Acuity is determined by the smallest letters that the observer

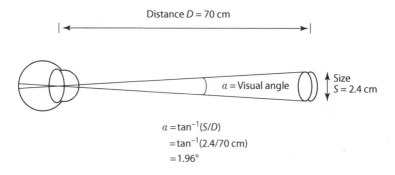

FIGURE 5.14   Visual angle of a quarter viewed at 70 cm.

can identify. Identification acuity is often specified in terms of the distance at which the person could identify letters that an observer with normal vision could identify at a standard distance. In the U.S., this standard distance is 20 ft (6.1 m); thus, a person with 20/20 vision can identify letters at 20 ft that a normal observer could from 20 ft away, whereas a person with 20/40 vision can only identify letters that a normal observer could from as far as 40 ft away.

Other kinds of acuity are vernier and resolution acuity. *Vernier acuity* is based on a person's ability to discriminate between a broken line and an unbroken line (Westheimer, 2005). *Resolution acuity* is a measure of the person's ability to distinguish multiple bars (or gratings) from a single area of the same average intensity (Chui, Yap, Chan, & Thibos, 2005).

Acuity varies as a function of many of the factors that influence brightness. As we already discussed, acuity decreases as the location of a shape is moved out from the fovea to the periphery. This decrease is even more drastic if irrelevant random shapes are presented nearby (Mackworth, 1965) or if a person's attention is focused on other stimuli in the center of the visual field (Williams, 1985).

Acuity is better under photopic viewing conditions than under scotopic conditions. Like Bloch's law for brightness, acuity is a function of time and contrast for durations of up to 300 ms (Kahneman, Norman, & Kubovy, 1967). In other words, for shapes presented for less than 300 ms, we can increase a person's acuity by increasing the contrast (difference in intensity between light and dark regions) or by increasing exposure duration.

Usually, we think of acuity as the ability to resolve detail in static displays, or images that do not change over time. However, motion can affect acuity. *Dynamic acuity* is measured when there is relative motion between an object and the observer (Miskewicz-Zastrow, Bishop, Zastrow, Cuevas, & Rainey, 2015). Typically, dynamic acuity is poorer than static acuity (Morgan, Watt, & McKee, 1983; Scialfa et al., 1988), but they are highly correlated. That is, a person with good static acuity will probably have good dynamic acuity. Both types of acuity decline with age, although the decline is greater for dynamic acuity.

Acuity is an important consideration for any task that requires processing of detailed visual information, like driving. All states in the U.S. require that applicants for driver's licenses pass identification acuity examinations. These examinations are tests of static identification acuity under high levels of illumination, and typically a minimum acuity of 20/40 is required for a license to drive without corrective lenses. Because driving involves dynamic vision, and night driving occurs under low levels of illumination, the traditional driving acuity test does not measure acuity under anything like actual driving conditions. You should not be surprised to learn, then, that dynamic acuity predicts driving performance better than static acuity measures (Sheedy & Bailey, 1993). A study of young adult drivers showed high correlations between dynamic visual acuity and identification of highway signs under dynamic viewing conditions, suggesting that at least part of the relation between dynamic visual acuity measures and driving performance may be due to the ease and accuracy with which highway signs can be read (Long & Kearns, 1996).

The standard acuity test seems particularly inappropriate for older drivers. People over the age of 65 show little deficiency on the standard acuity test, but show significant impairment relative to younger drivers when static acuity is measured under low illumination (Sturr, Kline, & Taub, 1990). Moreover, the elderly report specific problems with dynamic vision, such as difficulty reading signs on passing buses (Kosnik, Sekuler, & Kline, 1990), which correlates with the larger problem of reduced dynamic acuity (Scialfa et al., 1988). To provide assessment of visual ability for driving, Sturr et al. (1990) have recommended a battery of acuity tests involving static and dynamic situations under high and low levels of illumination.

Spatial Sensitivity
Another way to view acuity is in terms of spatial contrast sensitivity, or sensitivity to fluctuations between light and dark areas. The spatial distribution of light in a visual scene is a complex pattern that can be analyzed according to how quickly the fluctuations between light and dark occur. Parts of a scene may fluctuate very quickly whereas other parts may fluctuate more slowly. These

FIGURE 5.15   Sine-wave gratings of (a) low, (b) medium, and (c) high spatial frequencies.

different parts have different spatial frequencies. It turns out that people are not equally sensitive to all spatial frequencies.

This fact can be seen in the human *contrast sensitivity function*. This function reflects the ability of a person to discriminate between a sine-wave grating and a homogeneous field of equal average illumination. A sine-wave grating is a series of alternating light and dark bars that, in contrast to a square-wave grating, are fuzzy at the edges (see Figure 5.15). High-frequency gratings are composed of many fine bars per unit area, whereas low-frequency gratings are composed of few wide bars per unit area. We can measure a threshold for contrast detection by finding the lowest amount of contrast between the light and dark bars necessary for the observer to discriminate a grating from a homogeneous field.

The contrast sensitivity function for an adult (see Figure 5.16) shows that we are sensitive to spatial frequencies as high as 40 cycles per degree of visual angle. Sensitivity is greatest in the region of 3–5 cycles per degree and decreases sharply as spatial frequencies become lower or higher. The visual system is less sensitive to very low spatial frequencies and high spatial frequencies than to intermediate ones. Because the high frequencies convey the fine details of an image, this means that under low levels of illumination, such as those involved in driving at night, we will not be able to see details well.

The contrast sensitivity function specifies how both size and contrast limit perception, whereas standard visual acuity tests measure only size factors. Ginsburg, Evans, Sekuler, and Harp (1982) compared the ability of standard acuity measures and contrast sensitivity functions to predict how

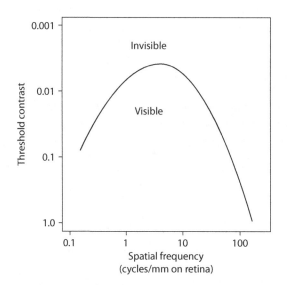

FIGURE 5.16   Spatial contrast sensitivity function for an adult.

well pilots could see objects under conditions of reduced visibility (like twilight or fog). Pilots flew simulated missions and then landed. On half of the landings, an object blocked the runway, and the landing had to be aborted. Their results indicated that the pilots who saw the object at the greatest distance were those with the highest contrast sensitivity.

Contrast sensitivity to intermediate and low spatial frequencies can predict how well people see stop signs at night (Evans & Ginsburg, 1982) and recognize faces (Harmon & Julesz, 1973). Such results suggest that measurement of contrast sensitivity may be useful in screening applicants for jobs that require a lot of visually oriented tasks. Measurement of contrast sensitivity under dynamic viewing conditions may provide a better overall indicator of a person's visual functioning than any other acuity measure, and it may allow improved prediction of performance at driving and other dynamic visual-motor tasks (Long & Zavod, 2002). Unfortunately, getting measurements of both static and dynamic contrast sensitivity is expensive and time-consuming relative to measurements of simple static acuity, and so evaluation of contrast sensitivity is often neglected in favor of the cheap and easy Snellen eye chart.

Temporal Sensitivity

Lights that flicker and flash are present everywhere in our environment. In some cases, as with train-crossing signals, the flicker provides an important message that drivers need to see. In other cases, such as video display screens, flicker is undesirable. The visibility of a continuously flickering light is determined by a person's *critical flicker frequency* (CFF), or the highest rate at which flicker can be perceived (Brown, 1965; Davranche & Pichon, 2005). With high-luminance stimuli of relatively large size, like a computer monitor, the CFF can be as high as 60 Hz. It is lower for stimuli of lower luminance and smaller size. Many other factors, such as retinal location, influence the CFF.

Video displays or sources of illumination that are intended to be seen as continuous should be well above the CFF, whereas displays intended to be seen as intermittent should be well below it. For example, fluorescent lamps flicker continuously at 120 Hz (cycles per second), a rate that is sufficiently high for new lamps that flicker is not detectable. However, because the CFF decreases for lower luminance, when a lamp needs to be replaced, the decrease in luminance of the old lamp makes the flicker visible.

Using methods similar to those used to determine spatial contrast sensitivity, we can ask how well an observer can distinguish between a light whose luminance level increases and decreases sinusoidally (flickers) and a constant light. This ability depends on the overall intensity of the light (Watson, 1986). We can measure temporal contrast sensitivity for different temporal frequencies and luminance levels and plot it as a temporal contrast sensitivity function (see Figure 5.17). As with spatial contrast sensitivity, temporal contrast sensitivity increases with temporal frequency to an intermediate value (around 8 Hz for a bright light), then decreases with further increases in temporal frequency up to the CFF of about 60 Hz (de Lange, 1958). The form of the function is influenced by factors such as the intensity of ambient lighting and the spatial configuration of the light relative to its background.

Masking

When two visual images are presented to the retina in close spatial or temporal proximity to each other, the perception of one of those images (which we will call the mask) can interfere with the perception of the other (the target). Such interference, called *masking*, occurs in a variety of situations (Breitmeyer & Öğmen, 2006). We refer to simultaneous masking when the target and mask occur simultaneously, while we use the terms *forward masking* and *backward masking* to describe perceptual problems that arise when the mask precedes or follows the target, respectively. With forward and backward masking, the amount of time between the onsets of the two stimuli will determine the degree of masking.

We can distinguish at least three broad categories of masking situations for which the location of the mask on the retina overlaps the location of the target stimulus (Breitmeyer &

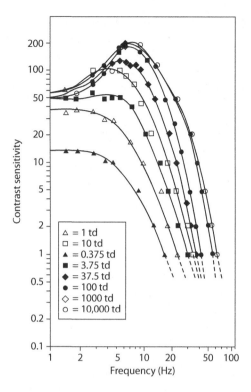

FIGURE 5.17 Temporal contrast sensitivity functions for an adult at several background intensities (in Trolands, td, where 1 td = 1 candle/m² luminance × 1 mm² pupil area).

Öğmen, 2000). For homogeneous light masking, the mask is an equiluminant patch of light flashed before, during or after a target; for structure masking, the mask shares many features with the target; for visual noise masking, the mask consists of a clump of random contours. The effects of these types of masking are usually most pronounced when the target and mask are presented together, as is the case under conditions of glare, where the light energy from a glare source (the sun reflected off a driver's rearview mirror, the headlamps of an oncoming car) decreases the contrast of an image (the view of the roadway in front of your car). As the time between the target and the masking image increases, the masking effect decreases. Part of this effect appears to be due to temporal integration. That is, even when presented separately, the light energy of both the target and the mask is at least partially summed, reducing the visibility of the target.

A fourth type of masking occurs when the target and mask do not overlap. This is called *metacontrast* or *lateral masking*. The magnitude of metacontrast masking decreases as the spatial separation between the stimuli increases. Moreover, when the target and mask are of approximately equal luminance, the largest masking effect occurs when the target precedes the mask by 50–100 ms. Theories of metacontrast focus on lateral connections in the visual system, attributing the effect to the different properties of the parvocellular and magnocellular systems (e.g., Breitmeyer & Ganz, 1976) or to the temporal dynamics of the neurons involved in the formation of boundaries and contours (Francis, 2000). However, it also is possible that metacontrast reflects, at least in part, higher-level attention and decision processes (e.g., Shelley-Tremblay & Mack, 1999). Regardless of the causes for visual masking, the human factors specialist needs to be aware that masking can occur when an operator must process multiple visual stimuli in close spatial and temporal proximity.

SUMMARY

Machines in the environment communicate with humans by displaying information in at least one sensory modality, most often vision. The first step toward the optimization of information displays is to understand the sensitivities and characteristics of these sensory input processes. Because the senses are not equally sensitive to all aspects of stimulation, a good display must be based on those aspects that will be readily perceived. For example, if a display is intended for use under low levels of illumination, it makes no sense to use color coding, because the user's cones will not be responsive in the dark.

The human factors specialist needs to know the properties of the physical environment to which the visual sensory receptors are sensitive, the nature of the process involved in the conversion of the physical energy into a neural signal, and the way in which the signal is analyzed in the sensory pathways. It is also important for her to know how vision is limited, so that the displays she designs can compensate for those limitations.

Although the visual sensory system constrains what can be perceived, perception involves more than a passive registration of the results of the sensory analyses. Perception is often characterized as a highly constructive process in which the sensory input serves as the basis for the construction of our perceptual experience. In the next chapter, we will examine the factors that influence the way in which we organize and perceive the world around us.

RECOMMENDED READINGS

Boff, K. R., Kaufman, L., and Thomas, J. P. (Eds.) (1986). *Handbook of Perception and Human Performance: Volume 1, Sensory Processes and Perception*. New York: Wiley.

Coren, S., Ward, L. M., and Enns, J. T. (2004). *Sensation and Perception* (6th ed.). San Diego, CA: Harcourt Brace.

Goldstein, E. R. (Ed.) (2001). *Blackwell Handbook of Perception*. Malden, MA: Blackwell.

Goldstein, E. R. (2016). *Sensation and Perception* (10th ed.). Belmont, CA: Wadsworth.

Wade, N. J., and Swanston, M. T. (2013). *Visual Perception: An Introduction* (3rd ed.). New York: Psychology Press.

Wolfe, J. M., Kluender, K. R., Levi, D. M., Bartoshuk, L. M., Herz, R. S., Klatzky, R. L., Lederman, S. J., & Merfeld, D. M. (2015). *Sensation and Perception* (4th ed.). Sunderland, MA: Sinauer Associates.

6 Perception of Objects in the World

The study of perception consists ... of attempts to explain why things appear as they do.

J. Hochberg
1988

INTRODUCTION

In the last chapter we introduced the visual system and some of the perceptual effects that can arise from the way that the visual system is put together. We continue this discussion in this chapter, emphasizing now the more complicated aspects of perceptual experience. Whereas previously we talked about how intense a perceptual experience is (in terms of brightness or lightness), now we focus on less quantifiable experiences such as color or shape.

You know something now about the basic signals that the brain uses to construct a perception. An amazing phenomenal characteristic of perception is how automatically and effortlessly a meaningful, organized world is perceived given these very simple neural signals. From a two-dimensional (2D) array of light energy, somehow we are able to determine how its pieces go together to form objects, where those objects are located in three dimensions, and whether changes in position of the image on the retina are due to movements of objects in the environment or to our own movements.

To drive a car down a highway, fly a plane, or even just walk across a room, a person must accurately perceive the locations of objects in either 2D or 3D space. Information presented on gauges, indicators, and signs must be not only detected but also identified and interpreted correctly. Consequently, the design of control panels, workstations, or other environments often relies on information about how people perceive objects around them. Design engineers must recognize how people perceive color and depth, organize the visual world into objects, and recognize patterns. These are the topics that we cover in the present chapter.

COLOR PERCEPTION

In daylight, most people see a world that consists of objects in a range of colors. Color is a fundamental part of our emotional and social lives (Davis, 2000). In art, color is used to convey many emotions. The color of your wardrobe tells others what kind of person you are. Using color, we can discriminate between good and bad foods or decide if someone is healthy or sick. Color plays a crucial role in helping us acquire knowledge about the world. Among other things, it aids in localizing and identifying objects.

At the most basic level, color is determined by the wavelength of light reflected from or emitted by an object (Malacara, 2011; Ohta and Robertson, 2005). Long-wavelength light tends to be seen as red and short-wavelength light tends to be seen as blue. But your experience of blue may be very different from your best friend's experience of blue. As with brightness, the perception of color is psychological, whereas wavelength distinctions are physical. This means that other factors, such as ambient lighting and background color, influence the perception of color.

Color Mixing

Most colors that we see in the environment are not spectral colors. That is, they are not composed of light of a single wavelength. Rather, they are mixtures of light of different wavelengths. We call colors from these mixtures *nonspectral colors*. Nonspectral colors differ from spectral colors in their degree of saturation, or color purity. By definition, spectral colors, consisting of a single wavelength, are pure, or completely saturated. Nonspectral colors are not completely saturated.

There are two ways to mix colors. First, imagine the colors that result when you mix two buckets of paint together. Paint contains different pigments that reflect light of different wavelengths. Mixtures of pigments result in what is called a *subtractive* color mixture. Next, imagine shining each of two light sources through a gel of a different color, like the lighting systems on a theatrical stage. If the gels placed in front of the light sources are of different colors, then when those two light sources are focused on the same location, their combination is an *additive* color mixture. Most of the rules of color mixing that you can recall (e.g., "blue plus yellow makes green") refer to subtractive color mixtures. Because of the different pigments that color different substances, it is less easy to predict the results of a subtractive color mixture than an additive one.

What happens when light of two wavelengths is mixed additively? It depends on the specific wavelengths and the relative amounts of each. In some cases, a color may look very different from its components. For example, if long-wavelength (red) light and middle-wavelength (yellow) light are mixed in approximately equal amounts, the color of the combination will be orange. If the middle-wavelength component is increased, then the mixture will appear more yellowish. Combinations of other spectral light sources may yield no color. For example, if a short-wavelength (blue) light and an upper-middle-wavelength (yellow) light are mixed in approximately equal amounts, the resulting combination will have no hue. More generally, we can reconstruct any hue (with any saturation) as an additive mixture of three primary colors (one long, one middle, and one short wavelength).

A color system that describes the dimensions of hue and saturation is the *color circle* (see Figure 6.1). Isaac Newton created the color circle by imagining the spectrum curved around the outside of a circle. He connected the low (red) and high (blue) wavelengths with nonspectral purples. Thus, the outer boundary of the color circle corresponds to the monochromatic or spectral colors plus the highly saturated purples. The center of the circle is neutral (white or gray). If we draw a diagonal from the center to a point on the rim, the hue for any point on this line corresponds to the hue at the rim. The saturation increases as the point shifts from the center to the rim.

We can estimate the appearance of any mixture of two spectral colors from the color circle by first drawing the chord that connects the points for the spectral colors. The point corresponding to

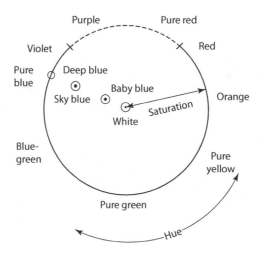

FIGURE 6.1 The color circle.

the mixture falls on this chord, with the specific location determined by the relative amounts of the two colors. If the two are mixed in equal percentages, the mixture will be located at the midpoint of the chord. The hue that corresponds to this mixture will be the one at the rim at that particular angle, and the saturation will be indicated by the distance from the rim.

A more sophisticated color mixing system is the one developed in 1931 by the Commission Internationale de l'Eclairage (CIE; the International Color Commission; Oleari, 2016). This system incorporates the fact that any color can be described as a mixture of three primaries. The CIE system uses a triangular "chromaticity" space (see Figure 6.2). In this system, a color is specified by its location in the space according to its values on three imaginary primaries, called X, Y, and Z. These primaries correspond to long-, medium-, and short-wavelength light, respectively. The coordinates in the chromaticity space are determined by finding the proportions of the color mixture that are X and Y:

$$x = X / (X + Y + Z);$$

$$y = Y / (X + Y + Z).$$

Because $x + y + z = 1.0$, z is determined when x and y are known, and we can diagram the space in terms of the x and y values.

TRICHROMATIC THEORY

The fact that any hue can be matched with a combination of three primary colors is evidence, recognized as early as the 1800s, for the view that human color vision is trichromatic (Helmholtz, 1852; Young, 1802; see Mollon, 2003). *Trichromatic color theory* proposes that there are three

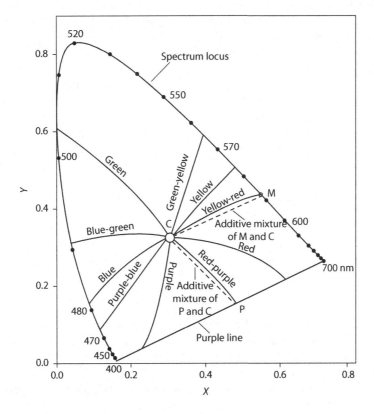

FIGURE 6.2 The CIE color space.

types of photoreceptors, corresponding to blue, green, and red, that determine our color perception. According to trichromatic theory, the relative activity of the three photoreceptors determines the color that a person perceives.

As trichromatic theory predicted, there are three types of cones with distinct photopigments. Color information is coded by the cones in terms of the relative sensitivities of the pigments. For example, a light source of 500 nm will affect all three cone types, with the middle-wavelength cones being affected the most, the short-wavelength cones the least, and the long-wavelength cones an intermediate amount (see Figure 5.11). Because each color is signaled by the relative levels of activity in the three cone systems, any spectral color can be matched with a combination of three primary colors.

Because there is only one rod photopigment, which is sensitive to a range of wavelengths across the visual spectrum, there is no way to determine whether a high level of rod activity is being caused by high-intensity light of a wavelength to which the photopigment is not very sensitive or by lower-intensity light of a wavelength to which the photopigment is more sensitive. This means that it is the relative levels of activity within the three cone subsystems that allow the perception of color.

Approximately 200 million people worldwide are color blind, or, more accurately, have a congenital color vision deficiency (Machado, Oliveira, & Fernandes, 2009). Men are more likely to have deficient color vision, with as many as 8% of men affected but only 0.5% of women (Simunovic, 2010). Deficient color vision is characterized by how many primary colors a person needs to match any hue. We say that someone is color blind when they need fewer than three primaries to match any color. Most color blind individuals are dichromats: they have *dichromatic vision*, meaning that two colors may look the same to a color blind individual that look different to a person with normal trichromatic vision (a trichromat). These people are usually missing one of the three types of cone photopigments, although the total number of cones is similar to that of a normal trichromat (Cicerone & Nerger, 1989). The most common form of dichromatic color vision is deuteranopia, which is attributed to a malfunction of the green cone system. Deuteranopes have difficulty distinguishing between red and green, although research findings have shown that they have a richer color experience than one might expect and use the words "red" and "green" consistently to label their color percepts (Wachtler, Dohrmann, & Hertel, 2004).

Some people still need three primaries to match all spectral colors but are said to have anomalous color vision, because the color matches that they make are not the same as those made by normal trichromats (Frane, 2015). Finally, there are rare individuals (monochromats) with no cones or only one type of cone who have monochromatic vision, as we mentioned briefly in Chapter 5.

Commercial products that use color filters, often in the form of a tinted contact lens, have been developed for use by people with color blindness to try to reduce their color confusions (Simunovic, 2010). A red-green color blind individual wears a red lens monocularly, which passes light primarily in the long-wavelength region of the spectrum. The basic idea is that a green color will look relatively darker in the filtered image than in the unfiltered image at the other eye, whereas a red color will not, providing a cue to help differentiate red from green. Unfortunately, the benefits of such filters are limited (Sharpe & Jägle, 2001). In fact, they may have serious side effects: They reduce luminance (since some light is filtered out), which may be particularly harmful for other aspects of vision at night, and impair depth perception (by altering binocular cues; see later in this chapter).

OPPONENT PROCESS THEORY

Although human color vision is based on trichromatic physiology, there are some characteristics of color perception that seem to be due to the way that the signals from cone cells interact in the retina. We mentioned already that if equal amounts of blue and yellow light are mixed additively, the result is an absence of any hue, a white or gray. The same effect occurs when red and green are mixed additively. Also, no colors seem to be combinations of either blue and yellow or red and green. For example, although orange seems to be a combination of red and yellow, there is no color corresponding to a combination of red and green.

The relations of red with green and blue with yellow show up in other ways, too. If you fixate on a yellow (or red) patch of color briefly (a procedure known as adaptation) and then look at a gray or white surface, you will see an afterimage that is blue (or green). Similarly, if you look at a neutral gray patch surrounded by a region of one of the colors, the gray region will take on the hue of the complementary color. A gray square surrounded by blue will take on a yellow hue, and a gray square surrounded by red will take on a green hue. Therefore, in addition to the three primary colors red, green and blue, yellow appears to be a fourth basic color.

These phenomena led Ewald Hering to develop the *opponent process theory* of color vision in the 1800s. He proposed that neural pathways linked blue and yellow together and red and green together. Within each of these pathways, one or the other color could be signaled, but not both at the same time. Neurophysiological evidence for such opponent coding was obtained initially from the retina of a goldfish (Svaetchin, 1956) and later in the neural pathways of rhesus monkeys (DeMonasterio, 1978; De Valois & De Valois, 1980). The nature of the cells in these pathways is such that, for example, red light will increase their firing rate and green light will decrease it. Other cells respond similarly for blue and yellow light.

There are a number of other perceptual phenomena that support the idea of opponent color processes. Many phenomena depend on the orientation of the stimulus (linking color perception to processing in the visual cortex), direction of motion, spatial frequency, and so on. We can explain most of these phenomena by the fact that the initial sensory coding of color is trichromatic and that these color codes are wired into an opponent-process arrangement that pairs red with green and blue with yellow (see, e.g., Chichilnisky & Wandell, 1999). By the time the color signal reaches the visual cortex, color is evidently coded along with other basic features of the visual scene.

Human Factors Issues

Most of the environments that we negotiate every day contain important information conveyed by color. Traffic signals, display screens, and mechanical equipment of all types are designed under the assumption that everyone can see and understand color-coded messages. For most color blind people, this bias is not too much of a concern. After all, the stop light is red, but it is also always at the top of the traffic signal, so it does not matter very much if one out of every ten male drivers cannot tell the difference between red and green.

However, there are other situations where color perception is more important. Commercial pilots, for example, must have good color vision so that they can quickly and accurately perceive the many displays in a cockpit. Electricians must be able to distinguish wiring of different colors, because the colors indicate which wires are "hot," and also (for more complex electronics) which wires connect to which components. Paint and dye manufacturing processes require trained operators who can distinguish between different pigments and the colors of the products being produced. Therefore, although most color blind individuals do not perceive themselves as being disabled in any way, color blindness can limit their performance in some circumstances.

The human factors engineer must anticipate the high probability of color blindness in the population and, when possible, reduce the possibility of human error due to confusion. The best way to do this is to use dimensions other than color to distinguish signals, buttons, commands, or conditions on a graph (Frane, 2015; MacDonald, 1999). The redundant coding of location and color for traffic lights described above is an example of using more than one dimension. This guideline is followed inconsistently, as illustrated by the fact that the standard default coding in current Web browsers uses redundant coding for links (blue color and underlined) but only color to distinguish sites that have recently been visited from ones that have not.

PERCEPTUAL ORGANIZATION

Our perceptual experience is not one of color patches and blobs, but one of objects of different colors at specific locations around us. The perceptual world we experience is constructed; the senses provide rough cues, for example, similarities and differences of color, that are used to evaluate hypotheses about the state of the world, but it is these hypotheses themselves that constitute perception. A good example involves the blind spot, which we discussed in Chapter 5. Sensory input is not received from the part of the image that falls on the blind spot, yet no hole is perceived in the visual field. Rather, the field is perceived as complete. The blind spot is filled in on the basis of sensory evidence provided by other parts of the image. In the rest of this chapter we will discuss how the perceptual system operates to construct a percept.

Perceptual organization is how the brain determines what pieces in the visual field go together (Kimchi, Behrmann, & Olson, 2003), or "the process by which we apprehend particular relationships among potentially separate stimulus elements (e.g., parts, features, dimensions)" (Boff & Lincoln, 1988, p. 1238). A widely held view around the beginning of the 20th century was that complex perceptions are simply additive combinations of basic sensory elements. A square, for example, is just a combination of horizontal and vertical lines. However, a group of German psychologists known as the Gestalt psychologists demonstrated that perceptual organization is more complicated than this. Complex patterns of elementary features show properties that emerge from the configuration of features that could not be predicted from the features alone (Koffka, 1935).

A clear demonstration of this point was made by Max Wertheimer in 1912 with a phenomenon of apparent movement that is called *stroboscopic motion* (Wade & Heller, 2003). Two lights are arranged in a row. If the left light alone is presented briefly, it looks like a single light turning on and off in the left location. Similarly, if the right light alone is presented briefly, then it looks like a single light turning on and off in the right location. Based on these elementary features, when the left and right lights are presented in succession, the perception should be that the left light comes on and goes off, and then the right light comes on and goes off. However, if the left and right lights are presented one after the other fairly quickly, the two lights now look like a single light moving from left to right. This apparent movement is the emergent property that cannot be predicted on the basis of the elementary features.

Figure and Ground

One of the most fundamental tasks the perceptual system must perform is the organization into *figure and ground* (Wagemans et al., 2012). Visual scenes are effortlessly perceived as objects against a background. Sometimes, however, the visual system can be fooled when the figure–ground arrangement is ambiguous. For the images shown in Figure 6.3, each part of the display can be seen as either figure or ground. Figure–ground ambiguity can produce problems with perception of signs, as for the one shown in Figure 6.4.

Examples like those shown in Figure 6.3 illustrate some major distinctions between objects classified as figure and those classified as ground. The figure is more salient than the ground and appears to be in front of it; contours usually seem to belong to the figure; and the figure seems to be an object, whereas the ground does not. Six principles of figure–ground organization are summarized in Table 6.1 and illustrated in Figure 6.3. The cues for distinguishing figure from ground include symmetry, area, and convexity. In addition, lower regions of a figure tend to be seen as figure more than upper regions (Vecera, Vogel, & Woodman, 2002). Images, scenes, and displays that violate the principles of figure–ground organization will have ambiguous figure–ground organizations and may be misperceived.

Grouping Principles

Probably more important for display design are the principles of *Gestalt grouping* (Gillam, 2001; Wagemans et al., 2012), which are illustrated in Figure 6.5. This figure demonstrates the principles of proximity, similarity, continuity, and closure. The principle of proximity is that elements close together in space tend to be perceived as a group. Similarity refers to the fact that similar elements (in terms of color, form, or orientation) tend to be grouped together perceptually. The principle of continuity is embodied in the phenomenon that points connected in straight or smoothly curving lines

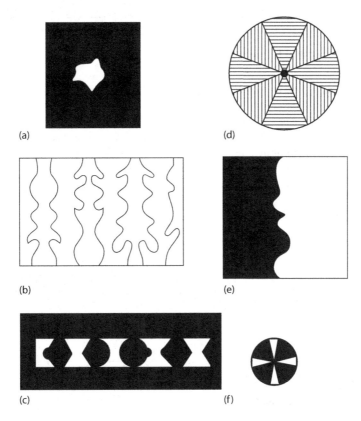

FIGURE 6.3   Factors that determine figure–ground organization: (a) surroundedness; (b) symmetry; (c) convexity; (d) orientation; (e) lightness or contrast; and (f) area.

FIGURE 6.4   Road sign intended to depict "no left turn."

tend to be seen as belonging together. Closure refers to a tendency for open curves to be perceived as complete forms. Finally, an important principle called *common fate*, which is not shown in the figure, is that elements that are moving in a common direction at a common speed are grouped together.

Figure 6.6 shows a very complicated arrangement of displays in an interior view of a simulation of the cockpit of the now-decommissioned space shuttle *Atlantis*. This simulator is a faithful

TABLE 6.1
Principles of Figure-Ground Organization

Principle	Description
Surroundedness	A surrounded region tends to be seen as figure while the surrounding region is seen as ground
Symmetry	A region with symmetry is perceived as figure in preference to a region that is not symmetric
Convexity	Convex contours are seen as figure in preference to concave contours
Orientation	A region oriented horizontally or vertically is seen as figure in preference to one that is not
Lightness or contrast	A region that contrasts more with the overall surround is preferred as figure over one that does not
Area	A region that occupies less area is preferred as figure

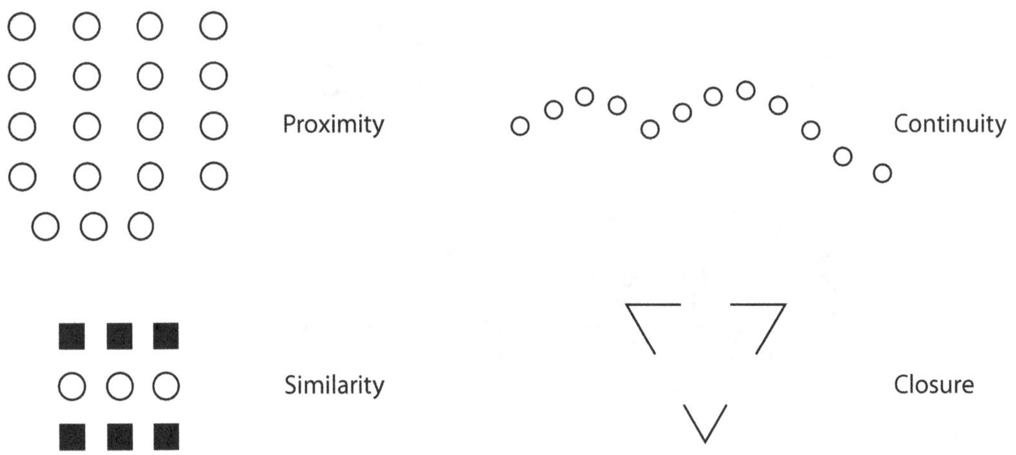

FIGURE 6.5 The Gestalt organizational principles of proximity, similarity, continuity, and closure.

reproduction of the real *Atlantis*. Several Gestalt principles are evident in the design of the cockpit. First, displays and controls with common functions are placed close to each other, and the principle of proximity assures that they are perceived as a group. This is particularly obvious for the controls and indicators in the upper center of the cockpit. The principles of proximity and similarity organize the linear gauges below the ceiling panel into three groups. The digital LEDs to the right of the array of gauges use both proximity and continuity to form perceptual groups.

There are two ways that grouping can be artificially induced by the inclusion of extra contours (Rock & Palmer, 1990). Dials or gauges that share a common function can be grouped within an explicit boundary on the display panel or connected by explicit lines (see Figure 6.7). Rock and Palmer call these methods of grouping *common region* and *connectedness*, respectively. They seem to be particularly useful ways to ensure that dials are grouped by the observer in the manner intended. Returning to the cockpit of the *Atlantis*, you can see several places where the cockpit designers exploited these principles to ensure the groupings of similar displays.

Wickens and Andre (1990) demonstrated that when a task (like landing the shuttle) requires integration across display elements, organizational factors have different effects on performance than when the task requires focused attention on a single display element. The task that they used involved three dials that might be found in an aircraft cockpit, indicating air speed, bank, and flaps. Pilots either estimated the likelihood of a stall (a task that required integrating the information from all three dials) or indicated the reading from one of the three dials (the task that required focused attention on a single dial).

Spatial proximity of the dials had no effect in Wickens and Andre's (1990) experiments. However, they found that performance for focused attention was better when display elements were of different colors than when they were all the same color. In contrast, integration performance was best

FIGURE 6.6   Interior of a simulator of the cockpit of the space shuttle *Atlantis*.

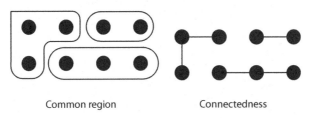

Common region   Connectedness

FIGURE 6.7   Displays grouped by common region and connectedness.

when all display elements were the same color. Wickens and Andre also experimented with displays that combined the information given by the three elements into a single rectangular object, the position and area of which were determined by air speed, bank, and flaps. They concluded that the usefulness of such an integrated display depends upon how well the emergent feature (the rectangle) conveys task-relevant information.

Another feature of displays that determines perceptual organization is the orientation of different components in the display. People are particularly sensitive to the orientation of stimuli (e.g., Beck, 1966). When forms must be discriminated that are the same except for orientation (e.g., upright Ts from tilted Ts), responses are fast and accurate. However, when pieces of the stimuli are all oriented in the same direction, for example, upright Ts from backward Ls (see Figure 6.8), it is much harder to discriminate between them.

An example where grouping by orientation can be useful is shown in Figure 6.9. This figure shows two example display panels for which *check reading* is required. In check reading, panels of gauges or dials must each be checked to determine whether they all register normal operating values. The bottom of Figure 6.9 shows a configuration where the normal settings are indicated by pointers at the same orientation, whereas the top shows a configuration where they differ. Because orientation is a fundamental organizing feature, it is much easier to tell from the bottom display than the top that one dial is deviating from normal (Mital & Ramanan, 1985; White, Warrick, & Grether, 1953). With the bottom arrangement, the dial that deviates from the vertical setting would "pop out" and the determination that a problem existed would be made rapidly and easily.

More generally, the identification of information in displays will be faster and more accurate when the organization of the display is such that critical elements are segregated from the distracting elements. For example, when observers must indicate whether an F or a T is included in a display that has noise elements composed of features from both letters (see Figure 6.10), they are slower to respond if the critical letter is "hidden" among the distractors by good continuity, as in Figure 6.10b, or proximity (Banks & Prinzmetal, 1976; Prinzmetal & Banks, 1977).

When designing pages for the World Wide Web, organizing the page in a manner consistent with the Gestalt grouping principles can facilitate a visitor's perception of the information on the page. Because of the difficulty of evaluating the overall organizational "goodness" of Web pages based on the various individual principles, Hsiao and Chu (2006) developed a mathematical model based on five grouping principles: Proximity; Similarity; Continuity; Symmetry; Closure. Web-page designers use a seven-point scale (from very bad to very good) to rate the extent to which each of these principles is used on a Web page for (1) layout of graphics, (2) arrangement of text, and (3)

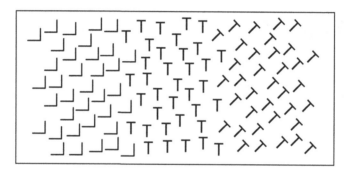

FIGURE 6.8   Example of orientation as an organizing feature.

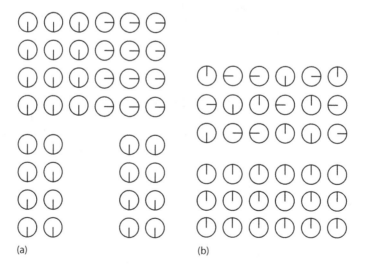

(a)   (b)

FIGURE 6.9   Displays grouped by proximity and similarity (a), and display groups with similar and dissimilar orientations (b).

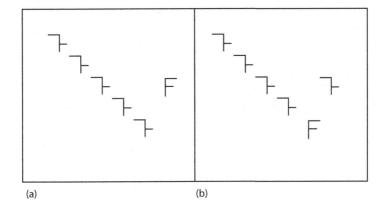

(a)   (b)

FIGURE 6.10 Example stimuli used to illustrate how good continuation influences target (F) identification when it is grouped (a) separately from and (b) together with distractors.

optimal use of colors. The model generates a value from 0 to 1 from these ratings, with a higher value for a Web page indicating more effective use of the Gestalt principles. This measure can be used by Web-page designers to evaluate whether a page is organized well visually, which should be correlated with the ease with which the content of the page can be comprehended and navigated by users. Though developed specifically for Web-page design, the method may also be useful for visual interface design more generally.

In summary, we can use Gestalt organizational principles to help determine how visual displays will be perceived and the ease with which specific information can be extracted from them. A good display design will use these principles to cause the necessary information to "pop out." Similarly, if we wish to obscure an object, as in camouflage, the object can be colored or patterned in such a manner that the parts will blend into the background.

DEPTH PERCEPTION

One of the most amazing things that our visual system does is transform the 2D image that falls on the retina into a complex 3D scene, where objects fall behind other objects in depth. As a first guess, you might think that our ability to see depth is a function of binocular cues associated with having two eyes. This is, in fact, part of the story. However, by closing one eye, you can see that it is not the entire story. Depth can still be perceived to some extent when the world is viewed with only a single eye.

The visual system uses a number of simple cues to construct depth (Howard, 2002, 2012; Proffitt & Caudek, 2013), and most of them are summarized in Figure 6.11. Notice that while many of them are derived from the retinal image, some come from the movement of the eyes. Many depth cues are monocular, explaining why a person can see depth with a single eye. In fact, depth perception from monocular cues is so accurate that the ability of pilots to land aircraft is not degraded by patching one eye (Grosslight, Fletcher, Masterton, & Hagen, 1978), nor is the ability of young adults to drive a car (Wood & Troutbeck, 1994). Another study examined the driving practices of monocular and binocular truck drivers, and found that monocular drivers were just as safe as binocular drivers (McKnight, Shinar, & Hilburn, 1991).

The extent to which the cues outlined in Figure 6.11 contribute to the perception of a 3D image is something to be considered when designing displays for virtual environments of the type used in simulators (see Box 6.1). The view outside the simulator window shown in Figure 6.6 is an artificial scene constructed using simple depth cues. How people use these cues to perceive depth is a basic problem that has been the focus of a great deal of study. We will now discuss each type of

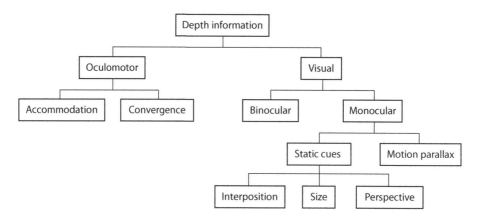

FIGURE 6.11   A hierarchical arrangement of the cues to depth.

cue, oculomotor and visual, and explain how the visual system uses these cues in the perceptual organization of depth.

OCULOMOTOR CUES

Oculomotor depth cues are provided proprioceptively. Proprioception is the ability to feel what your muscles are doing and where your limbs are positioned. The position of the muscles of the eye can also be perceived proprioceptively. We have already discussed (in Chapter 5) how the muscles of the eye work and even how abusing these tiny muscles can lead to eye strain and fatigue. The two motions that these muscles accomplish are accommodation and vergence, and the states of accommodation and vergence are two oculomotor cues to depth.

Recall that accommodation refers to automatic adjustments of the lens that occur to maintain a focused image on the retina, and vergence refers to the degree to which the eyes are turned inward to maintain fixation on an object. The information about the position of the muscles controlling the degree of vergence and accommodation could potentially be used as feedback in the visual system to help determine information about depth. Because the extent of both accommodation and vergence depends on the distance of the fixated object from the observer, high levels of accommodation and vergence signal that an object is close to the observer, whereas the information that the eye muscles are relatively relaxed signals that an object is farther from the observer.

Accommodation only varies for stimuli that are between approximately 20 and 300 cm from the observer. This means that proprioceptive information about accommodation could only be useful for objects that are very close. Vergence varies for objects up to 600 cm from the observer, so proprioceptive information about vergence is potentially useful over a wider range of distances than accommodation.

Morrison and Whiteside (1984) performed an interesting experiment to determine how important vergence and accommodation were to the perception of depth. They asked observers to guess how far away a light source was. They did this in such a way that in some situations the observers' degree of vergence was held constant and accommodation varied with distance, but in other situations accommodation was held constant while vergence varied with distance. They determined that changes in vergence were useful for making accurate distance estimates over a range of several meters, but changes in accommodation were not. Mon-Williams and Tresilian (1999, 2000) reached a similar conclusion that vergence plays a significant role in near-space perception and that "accommodation is almost certain to play no direct role in distance perception under normal viewing conditions" (Mon-Williams & Tresilian, 2000, p. 402).

We should take note of one important factor in Morrison and Whiteside's (1984) experiment. The light was presented very briefly, too briefly for the observers to actually make the necessary vergence changes, and so the proprioceptive information provided by vergence posture could

BOX 6.1   THREE-DIMENSIONAL DISPLAYS

A standard computer display screen is 2D, and many of the displays presented on them are also 2D. For example, the 2D start screen for any version of the Windows operating system contains any number of 2D icons displayed at various locations on the screen. Also, a word processor used for preparing and editing documents displays part of a page of the document, framed by toolbars that contain icons for various operations and (sometimes) rulers that specify horizontal distance from the left side and vertical distance from the top. One reason why the 2D display works well for icon selection is because the input device used for selection, typically a computer mouse, operates in two dimensions. Similarly, the 2D display for text editing is deliberately representative of the paper on which a copy can be printed or on which one can choose to write or type instead.

However, our interactions with the world and knowledge of the relations among objects involve the third dimension of depth. For example, an air-traffic controller must be able to comprehend the locations and flight paths of many aircraft within the flight environment that he is controlling. Likewise, the operator of a telerobotic system must be able to manipulate the movements of a remotely controlled robot in three dimensions. In situations like these, a person's performance may benefit from 3D displays. Depth can be represented on a two-dimensional screen using many of the monocular cues described in this chapter. Static monocular cues can be used to provide depth information, as in many Windows icons intended to represent objects. One common icon depicts a document contained in a folder, an image that uses interposition as a cue to depth. Some monocular cues can also be used to create more complex perspective displays of 3D relationships. The perception of depth can be particularly compelling when movement is introduced to the display. Using specialized goggles, stereoscopic views can be created by presenting different images to the two eyes, resulting in an even more compelling experience of depth. These kinds of tricks are used in gaming and virtual reality software.

3D displays are aesthetically and intuitively appealing because they depict shapes of objects and the relations among them in a realistic way. However, there are problems with 3D displays that may limit their effectiveness. Rendering a 3D image on a 2D screen means that some information has been lost. This loss can introduce ambiguity about the location of objects along lines of sight, and distortions of distances and angles, making it difficult to determine exactly where an object is supposed to be. One way to overcome the effects of these ambiguities and distortions is (for certain tasks) to use a multiple-view 2D display instead of a single 3D display.

Park and Woldstad (2000) examined a person's performance of a simulated telerobotic task, where the goal was to use a Spaceball® 2003 3D controller (a sphere that responds to pressure in the appropriate direction and that has buttons for specific operations such as picking up) to pick up an object and place it in a rack. They provided the people performing the task with either a multiple-view 2D, monocular 3D, or stereoscopic 3D display of the work area. The multiple-view 2D display consisted of two rows of three displays each: force-torque display, plan-view, right side–view, left side–view, front-view, and task status display.

People performed the task best when using the multiple-view 2D display. When visual enhancement cues (e.g., reference lines extending from the face of the gripper to the object to be grasped) were added to the 3D displays, the performance differences were eliminated, but the 3D displays still produced no better performance than the multiple-view 2D display.

It seems a bit surprising that the 3D displays do not result in better performance than the multiple-view 2D display. However, this finding has been replicated (St. John, Cowen, Smallman, & Oonk, 2001). Observers asked to make position judgments about two objects or two natural terrain locations performed better with multiple-view 2D displays than 3D displays. However, when the task required identifying the shapes of a block figure or terrain, they did better with 3D displays. The advantage of the multiple-view 2D displays in relative position judgment (which, it should be noted, was also an important component of Park and Woldstad's, 2000, simulated telerobotic task) is due to the fact that those displays minimize distortion and ambiguity. The advantage for 3D perspective displays in understanding shape and layout is due to the fact that the three dimensions are integrated in the display, rather than requiring the user to expend effort to integrate them mentally. The 3D displays also allow the rendering of extra depth cues and the depiction of hidden features, both of which can aid in shape identification. Therefore, it should not be too surprising that a recent review concluded that stereoscopic 3D displays are most useful for tasks that require manipulation of objects or locating, identifying, and categorizing objects (McIntire, Havig, & Geiselman, 2014).

One use of 3D is in the area of virtual environment, or virtual reality, displays. In virtual reality, the goal is not just to depict the 3D environment accurately, but also to have the user experience a strong sense of "presence," that is, of actually being in the environment. Because vision is only one sensory modality involved in virtual environments, we will delay discussion of them until Box 7.1 in the next chapter.

not have been the source of the information used to make distance estimates. Morrison and Whiteside proposed instead that the observers relaxed into the constant dark vergence posture (see Chapter 5) and used other cues, like binocular disparity (see later), as a source of information about depth. This finding suggests that although in some cases vergence cues may contribute directly to depth perception, in others their contribution may occur indirectly through joint effects with other cues.

MONOCULAR VISUAL CUES

The monocular visual cues sometimes are called *pictorial cues*, because they convey impressions of depth in a still photograph. Artists use these cues to portray depth in paintings. Figure 6.12 illustrates several of these cues.

The top panel (a) in Figure 6.12 shows a complex 3D scene. The scene seems to consist of a rectangular object lying flat on a field, with three monoliths to the left and two monoliths to the right. The separate components of the scene are unpacked in the bottom panel (b). The changes in the texture gradient of the field aid in the perception that the field recedes in depth toward a horizon. The changes in size of the three monoliths provides a relative size cue, which makes them appear to be three equally sized monoliths placed at different distances from the observer. The linear perspective implied by the unequal angles of the quadrangle in the foreground suggests that a flat rectangle recedes into the distance.

Probably the most important cue, that of interposition, is based on the fact that a near object will block the view of a more distant one if they are in the same line of vision. For the two monoliths on the right, the view of one monolith is partially obscured by the other, suggesting that the obscured monolith is farther away than the other. Interposition also contributes to the perceived locations of the other objects in Figure 6.12 because of the way the receding field is obscured by each piece.

Interposition can be very compelling. Edward Collier's painting *Quod Libet*, shown in Figure 6.13, relies heavily on interposition to portray a collage of 3D objects. In this painting, Collier also makes very clever use of the attached shadow cue, which we discuss below. This type of painting is referred to as a "trompe l'oeil," a French phrase that means "fool the eye." An artist's expert use of pictorial

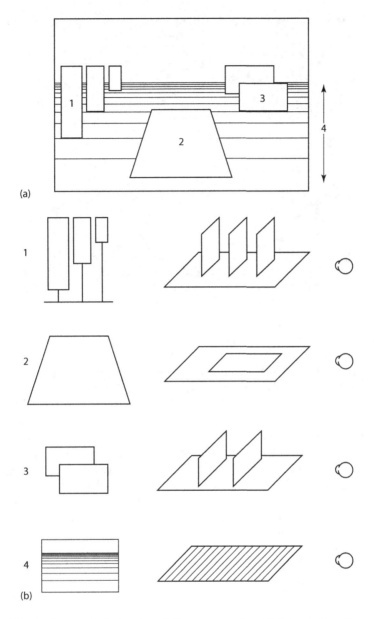

(a)

(b)

FIGURE 6.12 The depth cues of relative size (1), linear perspective (2), interposition (3), and texture gradient (4) in the complete image (a) and in isolation (b).

depth cues can sometimes, as in *Quod Libet*, give the impression that the images in a painting are real and not rendered.

Another important source of information about depth is the size of the perceived objects. Size cues can be of two sorts. First, an object, like a coffee cup, might have a familiar size that you have learned through experience. If you sense an image of a tiny coffee cup, you might conclude that the cup is very far away. Beginning around the late 1970s, the size of the average car began to decrease. In 1985, when larger cars perhaps were still more familiar than small cars, small cars tended to be involved in accidents more frequently than large cars (Eberts and MacMillan, 1985). One reason for this is that the smaller visual image of the smaller cars made a less familiar small car look like

FIGURE 6.13   Edward Collier, *Quod Libet* (1701).

a more familiar large car far away. This meant that the smaller cars were routinely judged to be farther away than they really were, resulting in a higher accident rate.

Second, the image of the object has a retinal size, referring to the area on the retina taken up by the image. This cue depends on the idea of a visual angle, which we discussed in Chapter 5. For an object of constant size, like a quarter, the closer it is to you, the larger the size of the retinal image will be. Thus, the relative size of images within the visual field can be used as a cue to distance.

Perspective is another important cue to depth. There are two types of perspective: aerial and linear. We saw an example of linear perspective in Figure 6.12. More formally, linear perspective refers to the fact that parallel lines receding into depth converge to a point in an image. This is true not only for visible lines, but also for the relations among objects that can be captured by invisible lines (see Figure 6.14).

Aerial perspective refers to interference in an image produced by particles in the air. The farther away an object is, the more opportunity there is for some of the light from it to be scattered and absorbed. This causes the image from a faraway object to be bluer than images from nearby objects and not as sharply defined. The blue coloration comes from the fact that short-wavelength blue light scatters more than longer-wavelength light.

Linear perspective and relative size are combined in texture gradients (see Figure 6.15 and also Figure 6.12). A gradient is characterized by parts of a texture's surface that become smaller and more densely packed as they recede in depth. A systematic texture gradient specifies the depth relations of the surface. If the texture is constant, it must be from an object facing the observer directly in the frontal plane (Figure 6.15, panel a). If the texture changes systematically, it indicates a surface that recedes in depth. The rate of change specifies the angle of the surface. The faster the texture increases in density, the more perpendicular to the observer the surface is.

The attached shadow cue is based on the location of shadows in a picture (Ramachandran, 1988; see Figure 6.16). Regions with shadows at the bottom tend to be perceived as elevated. Regions with shadows at the top tend to be perceived as depressed into the surface. These perceptions are what we expect to see when the light source projects from above, as is typically the case. The light source in Collier's *Quod Libet* is from above, so all of the objects shaded from below tend to project forward from the surface of the painting. For instance, the sheaves of paper curl outward because of the shadows he painted below each curl.

In situations where the light on an image projects from below, the attached shadow cue can be misleading. Take another look at Figure 6.16, but this time turn the book upside-down. The bubbles

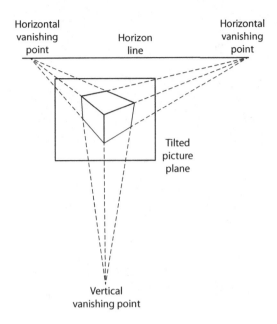

FIGURE 6.14 Vanishing points for linear perspective.

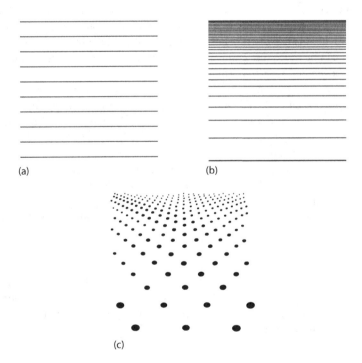

FIGURE 6.15 Texture gradients for surfaces (a) parallel to the frontal plane and (b and c) receding in depth.

that popped out when the image was upright should now appear to be depressions. This happens because we tend to see the figure with the light source coming from above no matter what the orientation of the figure is. If the light source is actually from below, what appear to be bubbles in the upright image are actually depressions (and vice versa).

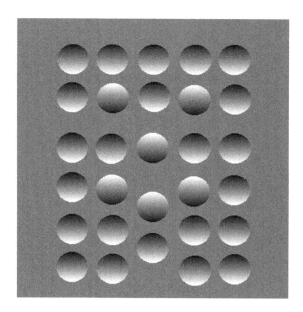

FIGURE 6.16   The attached shadow cue.

All the monocular cues described to this point are available to a stationary observer. It is because the observer cannot move relative to the objects in an image that sometimes our perceptions can be fooled. For instance, there is usually one best place to look at a trompe l'oeil painting. If you move around, the illusion can be much less compelling. This means that some information about depth is conveyed through movement. One important movement-based cue is called *motion parallax* (Ono & Wade, 2005). If you are a passenger in a car, fixate on an object to the side of the car, such as a cow. Objects in the foreground, like telephone poles or fence posts, will appear to move backward, whereas objects in the background, like trees or other cows, will seem to move forward in your direction. Also, the closer an object is to you, the faster its position in the visual field will change. The fence posts will travel by very rapidly, but the trees in the background will move very slowly. Similar movement cues can be produced on a smaller scale by turning your head while looking at an image.

Motion parallax is perceived when an observer is moving along beside an image. Motion also provides depth information when you move straight ahead. The movement of objects as you look straight ahead is called *optical flow*, which can convey information about how fast you are moving and how your position is changing relative to those of environmental objects. For example, as you drive down the road, the retinal images of trees on the roadside expand and move outward to the edges of the retina (see Figure 6.17). When the relation between the speed of your movement and the rate of the optical flow pattern changes, the perception of speed is altered. This is apparent if you watch from the window of an airplane taking off. As the plane leaves the ground and altitude increases, the size of the objects in the image decreases, the optical flow changes, and the speed at which the plane is moving seems to decrease.

BINOCULAR VISUAL CUES

Although you can see depth relatively well with only one eye, you can perceive depth relations more accurately with two. This is most obvious when comparing the perception of depth obtained from a 2D picture or movie with that provided by 3D, or stereoscopic, pictures and movies. Stereoscopic pictures mimic the binocular depth information that would be available from a real 3D scene. People can perform most tasks that involve depth information much more rapidly and accurately when using both eyes (Sheedy, Bailey, Burl, & Bass, 1986). For example, surgeons' perceptual-motor

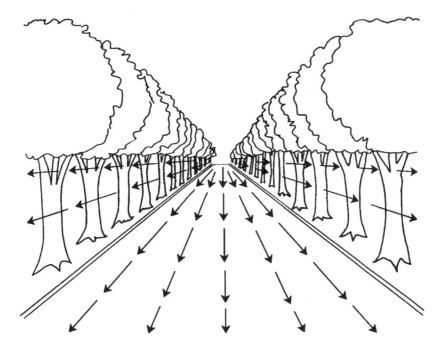

FIGURE 6.17   The optical flow of a roadway image for a driver moving straight ahead.

performance during operations is worse with image-guided surgical procedures, such as laparoscopic surgery, than with standard procedures, in part because of degraded depth perception caused by the elimination of binocular cues (DeLucia, Mather, Griswold, & Mitra, 2006).

The cues for binocular depth perception arise from *binocular disparity*: each eye receives a slightly different image of the world because of the eyes' different locations. The two images are merged through the process of fusion. When you fixate on an object, the image from the fixated area falls on the fovea of each eye. An imaginary, curved plane (like the wall of a cylinder) can be drawn through the fixated object, and the images from any objects located on this plane will fall at the same locations on each retina. This curved plane is called the *horopter* (see Figure 6.18). Objects in front of or behind the horopter will have retinal images that fall on different points in the two retinas.

Objects that are further away than the point of fixation will have uncrossed disparity, whereas those closer than fixation will have crossed disparity. The amount of disparity depends on the distance of the object from the horopter, and the direction of disparity indicates whether an object is in front of or behind the horopter. Thus, disparity provides accurate information about depth relative to the fixated object.

Stereoscopic pictures take advantage of binocular disparity to create an impression of depth. A camera takes two pictures at a separation that corresponds to the distance between the eyes. A stereoscope presents these disparate images to each respective eye. The red and green or polarized lenses used for 3D movies accomplish the same purpose. The lenses allow each eye to see a different image. A similar effect occurs while viewing random-dot stereograms (Julesz, 1971), pairs of pictures in which the right stereogram is created by shifting a pattern of dots slightly from the locations in the left stereogram (see Figure 6.19). This perception of objects in depth takes place in the absence of visible contours. "Magic Eye®" posters, called *autostereograms*, produce the perception of 3D images in the same way but in a single picture (see Figure 6.20). This happens when you fixate at a point in front of or behind the picture plane, which then allows each eye to see a different image (Ninio, 2007). We don't yet understand how the visual system determines what dots or part of an image go together to compute these depth relations in random-dot stereograms.

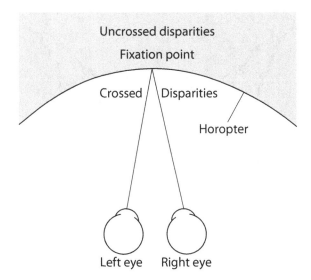

FIGURE 6.18 The horopter, with crossed and uncrossed disparity regions indicated.

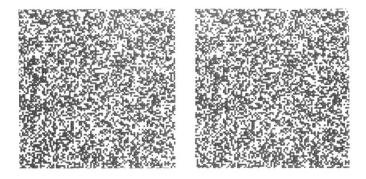

FIGURE 6.19 A random-dot stereogram in which the left and right images are identical except for a central square region that is displaced slightly in one image.

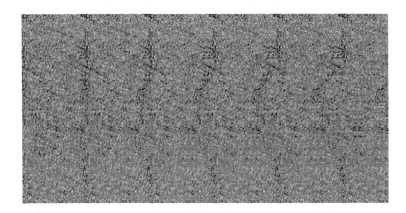

FIGURE 6.20 Random-dot autostereogram.

SIZE AND SHAPE CONSTANCY

Depth perception is closely related to the phenomena of *size constancy* and *shape constancy* (Walsh & Kulikowski, 1998). These refer to the fact that we tend to see an object as having a constant size and shape, regardless of the size of its retinal image (which changes with distance) and the shape of its retinal image (which changes with slant). The relationship between these constancies and depth perception is captured by the size-distance and shape-slant invariance hypotheses (Epstein, Park, & Casey, 1961). The size-distance hypothesis states that perceived size depends on estimated distance; the shape-slant hypothesis states that perceived shape is a function of estimated slant. The strongest evidence supporting these relations is that size and shape constancy are not very strong when depth cues are eliminated. Without depth cues, there is no way to estimate the distance and slant of an object (Holway & Boring, 1941).

ILLUSIONS OF SIZE AND DIRECTION

In most situations, the Gestalt organizational principles and depth cues we have discussed contribute to an unambiguous, accurate percept of objects in 3D space. However, many illusions occur that attest to the fallibility of perception. Figures 6.21 and 6.22 show several such illusions of size and direction.

Figure 6.21 illustrates five size illusions. In each panel, there are two lines or circles that you should compare. For instance, in the Müller-Lyer illusion (panel a), which of the two horizontal lines is longer? Because of the contours at the end of each line, the left line appears to be longer than the right. However, the two lines are exactly the same size (measure them to convince yourself). In each of the panels in Figure 6.21, the forms to be compared are exactly the same size.

Figure 6.21 shows several illusions of direction. In each panel, perfectly straight or parallel lines appear bent or curved. For instance, the Poggendorff illusion (panel a) shows a straight line running behind two parallel vertical lines. Although the line is perfectly straight, the upper part of the line does not seem to continue from the bottom of the line: it looks offset by at least a small amount. Using a straight edge, convince yourself that the line is really straight. In each of the panels in Figure 6.22, the presence of irrelevant contours causes distortions of linearity and shape.

There are many reasons why these illusions occur (Coren and Girgus, 1978; Robinson, 1998). These include inaccurate perception of depth, displacement of contours, and inaccurate eye movements, among others. Consider, for example, the Ponzo illusion (see Figure 6.21, panel b). The defining feature of this illusion is the two vertical lines that converge toward the top of the figure. Although the two horizontal lines are exactly the same length, the top line appears to be slightly longer than the bottom line. Recall from Figure 6.12 that vertical lines, like these, that converge at the top suggest (through linear perspective) a recession into the distance. If this depth cue is applied here, where it shouldn't be, the horizontal

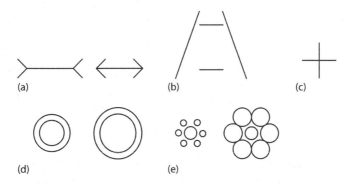

FIGURE 6.21   Illusions of size: (a) the Müller-Lyer illusion; (b) the Ponzo illusion; (c) the vertical-horizontal illusion; (d) a variation of the Delboeuf illusion; and (e) the Ebbinghaus illusion.

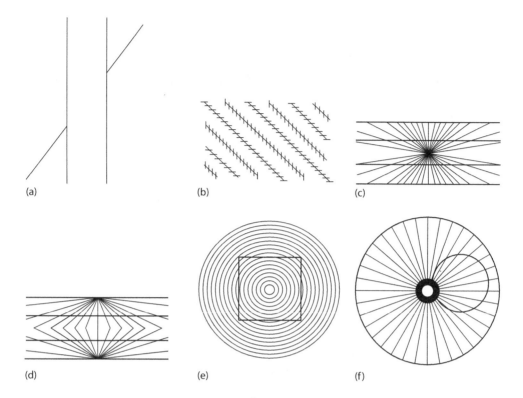

FIGURE 6.22 Illusions of direction: (a) the Poggendorff illusion; (b) the Zöllner illusion; (c) the Hering illusion; (d) the Wundt illusion; (e) the Ehrenstein illusion; and (f) the Orbison illusion.

line located higher in the display is further away than the one located lower in the display. Now, the retinal images of these two lines are exactly the same. Therefore, if the top one is further away than the bottom one, it must be longer than the bottom one. Hence, the top line is perceived as longer than the bottom.

Note that a similar illusion can be seen in the two monoliths illustrating interposition in Figure 6.12. The occluded monolith appears more distant than the monolith in front of it, and this distance is exaggerated by the receding texture gradient of the field. However, the two monoliths are exactly the same size (measure them). That is why, in panel a, the monolith in the back appears larger than the monolith in the front.

These seemingly artificial illusions can create real-world problems. Coren and Girgus (1978) describe a collision between two commercial aircraft that were approaching the New York City area at 11,000 and 10,000 ft, respectively. At the time, clouds were protruding above a height of 10,000 ft, forming an upward-sloping bar of white against the blue sky. The crew of the lower aircraft misperceived the planes to be on a collision course and increased their altitude quickly. The two aircraft then collided at approximately 11,000 ft. The U.S. Civil Aeronautics Board attributed the misjudgment of altitude to a naturally occurring variant of the Poggendorff illusion (see Figure 6.22a) created by the upward-sloping contours of the cloud tops. The clouds gave the illusion that the two flight paths were aligned even though they were not, and the altitude correction brought the lower plane into a collision course with the upper plane.

A recurring problem for pilots flying at night occurs when landing under "black hole" conditions in which only runway lights are visible. In such situations, pilots tend to fly lower approaches than normal, with the consequence that a relatively high proportion of night flying accidents involve crashes short of the runway. Experiments have shown that the low approaches arise from overestimates of approach angles due to the insufficiency of the available depth cues, like motion parallax and linear perspective (Mertens & Lewis, 1981, 1982). Because the pilot must evaluate the few cues provided by the runway lights according to some familiar standard, he or she will tend to make lower approaches when the runway has a larger ratio of length to width than that of a familiar runway with which the pilot has had recent experience (Mertens & Lewis, 1981).

PERCEPTION OF MOTION

Not only do we perceive a structured, meaningful world, but we see it composed of distinct objects, some stationary and others moving in various directions at different rates of speed. How is motion perceived? An initial answer that you might think of is that changes in displacement of an image on the retina are detected. However, how can the visual system determine when an object moves versus when the observer moves? Changes in retinal location can be due either to movement of objects in the environment or to the observer's movement. How the perceptual system resolves the locus of movement constitutes the primary problem of motion perception.

Object Motion

Motion perception can be thought of in terms of two separate kinds of systems (Gregory, 2015). The image-retina system responds to changes in retinal position, whereas the eye-head system takes into account the motion from our eye and head movements. The image-retina system is very sensitive. People are good at discriminating movement as a function of changes in retinal position. Movement can be seen if a small dot moves against a stationary background at speeds as low as 0.2° of visual angle per second. (From a 1 m viewing distance, 0.2° corresponds approximately to 3 mm.) Sensitivity to movement is even greater if a stationary visual reference point is present (Palmer, 1986). In such situations, tiny changes of as little as 0.03° of visual angle (approximately 0.5 mm at 1 m distance) per second produce a perception of movement.

Displacement of a retinal image does not necessarily mean that an object is moving, because the displacement may be due to movement of the observer. However, if an object is moving, we might track that object by moving our eyes. Such eye movements are called *smooth-pursuit movements*. During smooth-pursuit movements, the image remains on the fovea but we perceive that the object is moving. This sort of motion perception is due to the eye-head movement system.

Two theories have been proposed to explain how the eye-head system can tell the difference between an observer's own movements and movement of objects in the world (Bridgeman, 1995). Sherrington (1906) proposed what is often called *inflow theory*. According to this theory, feedback from the muscles that control eye movements is monitored by the brain. The change in the position of the eyes is then subtracted from the shift in location of the image on the retina. In contrast, *outflow theory*, proposed by Helmholtz (1867), states that the motor signal sent to the eyes is monitored instead. A copy of this outgoing signal, which is called a *corollary discharge*, is used to cancel the resulting movement of the image on the retina.

Research on motion perception has tended to favor outflow theory over inflow theory. Helmholtz noticed that if you press (gently!) on your eyelid while looking at an object, the object appears to move. In this situation, your eye is moving because you have moved it with your finger, not by moving the eye muscles. Because the muscles have not moved, there is no corollary discharge from them. According to outflow theory, this discharge must be subtracted from the movement of the retinal image; without the discharge, the retinal movement cannot be corrected, and the object appears to move.

One prediction of outflow theory is that if the muscles of the eye provide a corollary discharge but the retinal image remains fixed, motion of an object should also be perceived. This prediction has been confirmed (Bridgeman & Delgado, 1984; Stark & Bridgeman, 1984). Imagine a situation where pressure is applied to your eye, as with your finger, but you use the eye muscles to prevent the eye from moving. A corollary discharge will occur, but the retinal image will remain fixed, and the object appears to move. More complicated experiments have been performed using curare to temporarily paralyze an observer. When the observer tries to move his or her eyes (which do not actually move), the scene appears to move to a new position (Matin, Picoult, Stevens, Edwards, & McArthur, 1982; Stevens, Emerson, Gerstein, Kallos, Neufeld, Nicholas, & Rosenquist, 1976).

Induced Motion

Although we are very good at perceiving very small movements against a stationary background, stationary backgrounds can also lead to illusions of movement. In such illusions, movement is attributed to the wrong parts of the scene. One example of this is called the *waterfall effect*, which can take many forms. If you stare closely at a waterfall, a downward-moving pattern of water against a stationary background of rocks, you may experience the perception that the water stops falling and the rocks begin moving upward. You can also experience a waterfall effect while watching clouds pass over the moon at night. Often, the moon appears to move and the clouds remain still.

Motion illusions are easy to reproduce in a laboratory setting by presenting observers with a test patch of stationary texture and surrounding it with a downward-drifting inducing texture. When the test and inducing objects are in close spatial proximity, the effect is called *motion contrast*. When the test and inducing objects are spatially separated, the phenomenon is called *induced motion*. Induced motion can be demonstrated when one of two stimuli is larger than and encloses another. If the larger stimulus moves, at least part of the movement is attributed to the smaller enclosed stimulus. The enclosing figure serves as a frame of reference relative to which the smaller stimulus is displaced (Mack, 1986).

Apparent Motion

We usually perceive the movement of retinal images as smooth, continuous movement of objects through a visual scene. However, discrete jumps of a retinal image can produce the same perception of smooth movement. We introduced this phenomenon, called *apparent motion*, when discussing Gestalt organization. Apparent motion is the basis for the perceived movement of lights on a theater marquee, as well as for the movement perceived in motion pictures and on television. The fact that we perceive smooth movement from motion pictures conveys the power of apparent motion.

We know a lot about when apparent motion will be perceived from experiments conducted with very simple displays, such as the two lights used to illustrate stroboscopic motion discussed earlier. Two factors determine the extent to which apparent motion will be perceived: the distance and the time between successive retinal images. Apparent motion can be obtained over distances as large as 18°, and the interval that provides the strongest impression of apparent motion depends on the distance. As the degree of spatial separation increases, the strongest impression of apparent motion will be given by interval durations that are longer and longer.

Our current understanding of apparent motion is that there are two processes involved. A short-range process is responsible for computing motion over very small distances (15 min of visual angle or less) and rapid presentations (100 ms or less). Another long-range process operates across large distances (tens of degrees of separation) and over time intervals of up to 500 ms. Whereas the short-range process is probably a very low-level visual effect, the long-range process appears to involve more complex inferential operations.

PATTERN RECOGNITION

Up to this point, we have been discussing how our perceptual system uses different kinds of visual information to construct a coherent picture of the world. Another important job that the perceptual system must perform is the recognition of familiar patterns in the world. In other words, we have to be able to identify what we see. This process is called *pattern recognition*.

Because pattern recognition seems to be a skill that is fundamental to almost every other cognitive process a person might engage, it has been the focus of a tremendous amount of basic research. Many experiments have examined performance in a task called "visual search," which requires observers to decide if a predetermined target item is present in a visual display. Earlier in this chapter, we talked about how grouping of display elements can help make a target letter "F" more or less easy to find (Figure 6.9). This is an example of a visual search task. Knowing how people perform this task is critical to the good design of certain displays and task environments.

The idea that objects in a visual scene can be taken apart in terms of their basic "features" is again an important concept in understanding pattern recognition (Treisman, 1986). Visual search that is based on a search for primitive features such as color or shape can be performed very rapidly and accurately: it is very easy to find the one green object in a display containing red objects, no matter how many red objects there might be. However, if a target is defined by a combination of more than one primitive feature, and those features are shared by other objects in the display, the time to determine whether the target is present is determined by the number of nontarget objects in the array (array size). Whereas search for a single primitive feature is rapid and effortless, search for conjunctions of features requires attention and is effortful.

This basic fact of pattern recognition in visual search has implications for the design of computer interfaces. For menu navigation, highlighting subsets of options by presenting them in a distinct color should shorten the time for users to search the display. This has been confirmed in several studies (Fisher, Coury, Tengs, & Duffy, 1989; Fisher & Tan, 1989). Users are faster when a target is in the highlighted set and slower when it is not or when no highlighting is used. Moreover, even if the target is not always guaranteed to be in the highlighted set, the benefit of highlighting is greater when the probability is high that the target will be in the highlighted set than when the probability is low.

Another characteristic of primitive features that is important for display design is that of integral and separable dimensions (Garner, 1974). Whereas a feature is a primitive characteristic of an object, a dimension is the set of all possible features of a certain type that an object might have. For example, one feature of an object might be that it is red. The dimension that we might be interested in is an object's color, whether red, green or blue.

Dimensions are said to be *integral* if it is not possible to specify a value on one feature dimension without also specifying the value on the other dimension. For example, the hue and brightness of a colored object are integral feature dimensions. If dimensional combinations can exist independently of one another, they are called *separable*. For example, color and form are separable dimensions. You can pay attention to each of two separable feature dimensions independently, but you cannot do so for two integral dimensions. Thus, if a judgment about an object requires information from one of its dimensions, that judgment can be made faster and more accurately if the dimensions of the object are separable. On the other hand, if a judgment requires information from all of an object's dimensions, that judgment will be easier if the dimensions are integral. Another way to think about integrality of dimensions is in terms of correlations between object features. If a set of objects has correlated dimensions, a specific value on one dimension always occurs in the presence of a specific value on the other dimension.

Another kind of dimension is called a *configural dimension*. Configural dimensions interact in such a way that new emergent features are created (Pomerantz, 1981). Emergent features can either facilitate or interfere with pattern recognition, as Figure 6.23 shows. This figure shows an array for a visual search task, in which the target to be detected is the line slanting downward (in the lower right-hand corner). For both the top and bottom panels in the figure, exactly the same set of contextual features is added to each object in the array. Because the same features are added to each object, these features alone do not provide any help in recognizing the target object. However, when we examine the final configuration of the objects after the new features are added, we see in the top array that the contextual features have enhanced the differences between the objects, and the time to recognize the target (RT) is greatly reduced. In the bottom row, the contextual features have obscured the target, and RT is greatly increased.

Up to this point, our discussion of pattern recognition has focused on the analysis of elementary features of sensory input. This analysis alone does not determine what we perceive. Expectancies induced by the context of an object also affect what we perceive. Figure 6.24 shows a famous example of the influence of context. This figure shows two words, "CAT" and "THE." We easily perceive the letter in the middle of CAT to be an A and the letter in the middle of the word THE to be an H. However, the character that appears in the middle of each word is ambiguous: it is neither

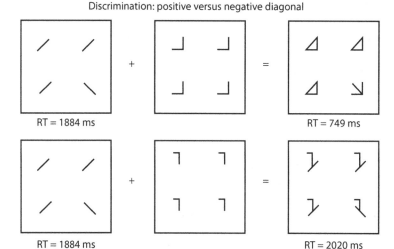

Discrimination: positive versus negative diagonal

RT = 1884 ms　　　　RT = 749 ms

RT = 1884 ms　　　　RT = 2020 ms

FIGURE 6.23 The additional configural context that facilitates (top row) and impedes (bottom row) performance.

TAE CAT

FIGURE 6.24 The effect of context on perception. The same symbol is seen as the letter H in THE and CAT.

an A nor an H, and it is in fact identical for both words. The context provided by the surrounding letters determines whether we recognize an A or an H.

Similar expectancy effects occur for objects in the world. Biederman, Glass, and Stacy (1973) presented organized and jumbled pictures and had subjects search for specific objects within these pictures. They presumed that the jumbled picture would not allow the viewers to use their expectations to aid in searching for the object. Consistently with this hypothesis, search times for coherent scenes were much faster than those for jumbled scenes. Biederman et al. also examined the influence of probable and improbable objects within the coherent and jumbled scenes. They found that for both kinds of pictures it was much easier to determine that an improbable object was not present than to determine that a probable object was not present. This finding indicates that observers develop expectations about objects that are possible in a scene with a particular theme. What we perceive thus is influenced by our expectancies as well as by the information provided by the senses.

The influence of expectations is critical when objects fall into the peripheral visual field (Biederman et al., 1981). It is difficult to detect an unexpected object in the periphery, particularly when it is small. The rate at which targets are missed in visual search increases to 70% as the location of an unexpected object shifts from the fovea to 4° in the periphery. The miss rate for a peripheral object is reduced approximately by half when the object is expected.

SUMMARY

Perception involves considerably more than just passive transmission of information from the sensory receptors. The perceived environment is constructed around cues provided by many sensory sources. These cues allow both 2D and 3D organization of visual, auditory, and tactile information as well as pattern recognition. The cues are comprised of encoded relations among stimulus items, such as orientation, depth, and context.

Because perception is constructed, misperceptions can occur if cues are false or misleading, or if the display is inconsistent with what is expected. It is important, therefore, to display information in ways that minimize perceptual ambiguities and conform to the expectancies of the observer. In Chapters 5 and 6, we have concentrated on visual perception because of its importance to human factors and the large amount of research conducted on the visual sense. The next chapter discusses auditory perception and, to a lesser extent, the senses of taste, smell, and touch.

RECOMMENDED READINGS

Cutting, J. E. (1986). *Perception with an Eye to Motion*. Cambridge, MA: MIT Press.

Ninio, J. (2007). The science and craft of autostereograms. *Spatial Vision*, *21*, 185–200.

Palmer, S. E. (1999). *Vision Science: Photons to Phenomenology*. Cambridge, MA: MIT Press.

Palmer, S. E. (2003). Visual perception of objects. In A. F. Healy and R. W. Proctor (Eds.), *Experimental Psychology* (pp. 179–211). Volume 4 in I. B. Weiner (Editor-in-Chief) *Handbook of Psychology*. Hoboken, NJ: Wiley.

Snowden, R., Thompson, P., and Troscianko, T. (2012). *Basic Vision: An Introduction to Visual Perception* (revised ed.). Oxford, UK: Oxford University Press.

Wagemans, J. (Ed.) (2015). *The Oxford Handbook of Perceptual Organization*. Oxford, UK: Oxford University Press.

Wagemans, J., Elder, J. H., Kubovy, M., Palmer, S. E., Peterson, M. A., Singh, M., and von der Heydt, R. (2012). A century of Gestalt psychology in visual perception: I. Perceptual grouping and figure–ground organization. *Psychological Bulletin*, *138*, 1172–1217.

7 Hearing, Proprioception, and the Chemical Senses

Our lives are multi-sensory and our interactions vary from the bold to subtle.

F. Gemperle et al.,
2001

INTRODUCTION

While vision is essential for navigational tasks like driving, we use our other senses too. The information from these other senses may be important for navigation itself or for other activities being performed at the same time, like listening to the radio or engaging in conversation. For driving, auditory stimuli may convey significant navigational information: A driver may be alerted to the fact that she has drifted onto the shoulder of the road in part by the difference in sound that is made by the tires, or she may detect an unusual sound associated with a mechanical problem in the engine. The skin senses provide feedback about such things as whether the temperature is adjusted appropriately in the vehicle and where her hands are positioned at any moment in time. Vibration and noise produced by rumble bars warn the driver to slow down for a potential hazard. Information about speed and acceleration is provided by the vestibular sense, although the driver typically is not aware of this fact. Smell and taste play less of a direct role in driving, although a smell may alert the driver to the fact that a mechanical failure is imminent. Taste can become a distracting factor when the driver is eating and driving at the same time.

As this example illustrates, all of the senses provide input that we use to perceive the world around us. Consequently, information in many sensory modalities can be used by design engineers in appropriate circumstances to transmit different critical signals to machine operators during the performance of a task. Therefore, human factors specialists need to be aware of the basic properties of each of these other sensory systems and the perceptual properties and phenomena associated with them. We will devote the majority of this chapter to the sense of hearing and auditory perception, because audition is second only to vision in its relevance to human factors and ergonomics. In the remainder of the chapter, we will provide brief descriptions of the attributes of the remaining senses.

HEARING

The sense of hearing plays an important role in the communication of information (Yost, 2013). Sound provides us with information about such things as the location of objects and the speed and directions of their approach. In the example above, sound can not only inform the driver that the car's engine is malfunctioning, but it also can indicate that the turn signal has been left on, a tire is flat, the car has struck an object, the heater fan is running, and so on. One common use of auditory signals is to alert the operator to potential problems. For instance, most cars will signal an open door, an unfastened seatbelt, or keys left in the ignition with a chime. Auditory signals are also used for signaling emergency situations: Smoke alarms signal potential fires with a loud aversive noise, and alarms sound in aircraft cockpits when potentially serious situations arise, such as when a dangerous altitude is reached or when another plane is too close.

A benefit of auditory signals is that they can be detected and perceived regardless of where they are located relative to a person. This is not true of visual signals, which must be in a person's line of sight. Auditory signals are also more attention-demanding than visual signals. Furthermore, virtually all of our direct communications with other people occur through speech by way of the sense of hearing. Speech-based messages are common in human–machine systems. For example, automated people-movers, such as those used in larger airports, use speech messages to warn passengers that the doors are closing and the vehicle is about to move.

To understand how the human brain processes auditory signals, we need to know how the sense of hearing operates. This means we need to understand the nature of sound and the anatomy and functioning of the auditory sensory system. As with visual perception, it is also important to know what specific characteristics of a sound signal are perceived and how we react to them.

SOUND AND THE AUDITORY SENSORY SYSTEM

Sound Stimuli

Sound begins with a mechanical disturbance that produces vibrations. Banging two pots together, for example, produces vibrations that are transmitted outward from the pots in all directions, through collisions between molecules in the air, at a speed of 340 m/s. Consider a tuning fork, a simple metal object consisting of a pair of upright parallel prongs mounted on a handle. When the fork is struck by tapping it against a hard surface, a simple, pure sound is produced. This sound originates in the oscillating motion of the prongs. As the prongs move outward, they push the air molecules before them. This produces a small increase in air pressure (compression) that reaches a peak when the prongs attain their maximal outward displacement. As the prongs move inward, the air pressure decreases (rarefaction) and reaches a minimum when they attain their maximum inward displacement. These repeated cycles of compression and rarefaction produce a sound wave, which moves through the air at a speed of 340 m/s.

We can measure the changes in air pressure at a fixed distance from the tuning fork. If we plot these changes over time, we see that they follow a sinusoidal pattern (see Figure 7.1). The sinusoid can be characterized in several ways. First, we can consider its frequency (F), period (T), or wavelength (λ). Frequency is defined as the number of complete cycles that occur in 1 second, or hertz (Hz). For example, a 1 Hz tone goes through exactly one compression/rarefaction cycle per second; a 1 kHz tone goes through 1000 cycles/s. The perceived pitch of the tone is closely related to the frequency. Higher-frequency tones are perceived to have higher pitches than lower-frequency tones. The period T of the wave form is the duration of a single cycle and is the inverse of the frequency F. The sound's wavelength λ is the distance between two adjacent peaks. It is calculated from the frequency and the speed of sound, c, as

$$\lambda = \frac{c}{F}.$$

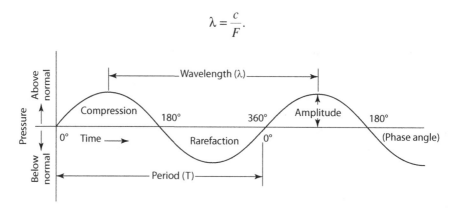

FIGURE 7.1 Simple sound wave.

Second, we can consider the sound wave's amplitude, which is specified in terms of pressure or intensity. High-intensity tones are perceived to be louder than tones of lower intensity. The sound pressure is a function of the difference between the maximal and minimal pressures shown in Figure 7.1, the upper and lower peaks of the sinusoid. However, the pressure changes occur very rapidly over time. Therefore, pressure usually is measured frequently within some interval of time. Suppose, for example, we measure the changes in air pressure $p(t_1), p(t_2), \ldots, p(t_n)$ at times $t_1, t_2, \ldots, t_n$. These changes occur relative to a background pressure p_0 (what the air pressure would be if there were no sound). The sound pressure is then specified as the root mean square (RMS) deviation from the static pressure, or

$$\sqrt{\frac{\sum_{i=1}^{n}\left(p(t_i) - p_0\right)^2}{n}}.$$

Intensity is closely related to RMS pressure and is specified in units of watts per square meter. For the tuning fork, amplitude is determined by the distance over which the fork moves. Striking the fork forcefully produces high-amplitude movement and high-amplitude sound waves, whereas striking it less forcefully results in lower-amplitude movements and sound waves. The amplitude or intensity of the sound also depends on the distance from the sound source. The further the point of measurement is from the source, the lower the intensity will be. Intensity follows an *inverse square law*: measured intensity is proportional to one over the square of the distance from the source.

The tone produced by a tuning fork is "pure": the changes in air pressure produced follow a perfect sine wave. Rarely do we encounter pure tones. Typically, sound waves are much more complex, and those from many sources are mixed into a single complex wave. However, any sound wave, from that produced by a jet when taking off to that of a person singing, can be described as a weighted sum of pure sinusoidal tones. The procedure by which a complex sound is decomposed into pure tones is called *Fourier analysis* (Kammler, 2007).

Waveforms that repeat themselves, such as the simple sinusoid, are called *periodic*. A complex, periodic tone, such as that produced by a musical instrument, has a fundamental frequency, f_0, that is the inverse of the period, T_0:

$$f_0 = \frac{1}{T_0}.$$

Such waveforms also contain *harmonics* that are integer multiples of the fundamental frequency. Aperiodic complex waveforms that have randomly varying amplitudes across a range of frequencies are sometimes called *noise*. There are several kinds of noise. "White" noise has an equal average intensity for all component frequencies. "Wideband" noise has frequencies across most or all of the auditory spectrum, and "narrowband" noise has only a restricted range of frequencies.

Outer and Middle Ear

The human ear serves as a receiver for sound waves (see Figure 7.2). Sound is collected by the *pinna*, the scoop-shaped outer part of the ear. The pinna amplifies or attenuates sounds, particularly at high frequencies, and plays a significant role in sound localization. The pinna funnels sound into the *auditory canal*, which isolates the sensitive structures of the middle and inner ears from the outside world, thus reducing the likelihood of injury. This canal has a resonant frequency of 3–5 kHz (Shaw, 1974), which means that sounds with frequencies in this range, such as normal speech, receive a boost in amplitude. At the far end of the auditory canal is the eardrum, or *tympanic membrane*, which vibrates when sound-pressure waves strike it. In other words, if the sound wave is a 1 kHz tone, then the eardrum vibrates at 1000 cycles per second. Perforation of the eardrum results in scar tissue and thickening of the membrane, which reduces its sensitivity to vibrations. This in

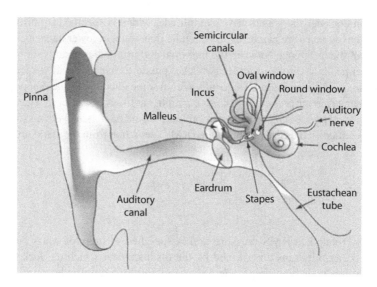

FIGURE 7.2   The ear.

turn leads to a decreased ability to detect tones, particularly those of high and middle frequencies (Anthony & Harrison, 1972).

The eardrum separates the outer ear from the middle ear. The function of the middle ear is similar to that of the eardrum: the transmission of vibration further (deeper) into the structures of the auditory system. The middle ear passes the vibrations of the eardrum to a much smaller membrane, the *oval window*, which provides entry into the inner ear. The transmission between these two membranes occurs by means of three bones, which collectively are called the *ossicles*. Individually, these bones are the malleus (hammer), incus (anvil) and stapes (stirrup), in reference to their appearance. The malleus is attached to the center of the eardrum, and the footplate of the stapes lies on the oval window, with the incus connecting the two. Movement of the eardrum thus causes movement of the three bones, which ultimately results in the oval window vibrating in a pattern similar to that of the eardrum. Typically, the role of the ossicles is described as one of impedance matching. The inner ear is filled with fluid. If the eardrum were given the task of transferring vibrations directly to this fluid, the increase in density from air to fluid would damp the incoming sound waves, reducing their amplitude. The transmission of the waves through the eardrum and ossicles to the smaller area of the oval window serves to amplify the wave, so that the change in medium from air to fluid occurs efficiently.

The middle ear is connected to the throat by the Eustachian tube. This tube maintains the air pressure within the middle ear at the level of the outside atmospheric pressure, which is necessary for the middle ear system to function properly. Discomfort and difficulty in hearing are often experienced when a plane changes altitude, because the air pressure of the middle ear has yet to adjust to the new atmospheric pressure. Yawning opens the Eustachian tubes and allows the pressure to equalize. If you have ever had to use ear drops, you were probably able to taste the medicine in the back of your mouth after putting them in. This happens when small amounts of the medicine permeate the eardrum and filter through the Eustachian tube.

Finally, the middle ear contains small muscles connected to the eardrum and to the stapes, which together produce the *acoustic reflex* in the presence of loud sounds (Fletcher & Riopelle, 1960; Schlauch, 2004). This reflex reduces the sound vibrations sent from the outer ear to the inner ear by making the eardrum and ossicles difficult to move; thus, the inner ear is protected from potentially damaging sounds. For people with intact middle ear structures and normal hearing, the acoustic reflex kicks in for sounds of about 85 decibels (Olsen, Rasmussen, Nielsen, & Borgkvist, 1999), but this will differ from person to person. The acoustic reflex requires about 20 ms to stabilize the ossicles, and it attenuates primarily low-frequency sounds. Thus, the reflex does not provide protection from sound with rapid onset (e.g., a gunshot) or from intense high-frequency sounds.

One function of the acoustic reflex may be to reduce a person's sensitivity to his/her own voice, because the reflex is triggered prior to and during speech (as well as chewing; Schlauch, 2004). Because the low-frequency components of speech can often mask the high-frequency components, selective attenuation of the low-frequency components probably improves speech perception.

Inner Ear

After sound vibrations travel through the middle ear, they reach the inner ear by way of the oval window. The inner ear contains several structures, but the one of importance to hearing is the *cochlea* (Young, 2007). The cochlea is a fluid-filled, coiled cavity that contains the auditory sensory receptors (see Figure 7.3). It is partitioned into three chambers: the vestibular canal, the cochlear duct, and the tympanic canal. All three of the chambers are filled with fluid. The cochlear duct is completely separate from the vestibular and tympanic canals and contains a different fluid. The latter two canals are connected by a pinhead-size opening at the apex of the cochlea, allowing fluid to pass between them. The oval window, which vibrates in response to the stapes, is at the base of the vestibular canal, and the round window is at the base of the tympanic canal. Together, these two windows allow pressure to be distributed within the cochlea.

The membrane that separates the cochlear duct from the tympanic canal is called the *basilar membrane*. The role of the basilar membrane in hearing is similar to the role of the retina in vision. Sounds transmitted through the middle ear cause movement at the base of the basilar membrane that spreads to the apex (Békésy, 1960). However, because the width and thickness of the basilar membrane vary along its length, the magnitude of vibration of the membrane will not be equal across its entire extent. Low-frequency tones produce the greatest movement at the end farthest away from the oval window. As the tone increases in frequency, the peak displacement shifts progressively toward the oval window. Thus, tones of different frequency will cause maximal displacement at different places on the basilar membrane.

The sensory receptors are rows of hair cells that run the length of the basilar membrane (see Figure 7.4). These hair cells have cilia sticking up into the fluid in the cochlear duct, with the tops of some of the cilia touching the tectorial membrane. There are two groups of hair cells: inner and outer. There are approximately 3500 inner hair cells, which are in a single row along the basilar membrane. In contrast, there are approximately 25,000 outer hair cells lined up in three to five rows. The bending of the cilia for both types of hair cells initiates a neural signal.

How do the cilia bend? The presence of a sound wave in the ear ultimately leads to motion (waves) within the fluid of the inner ear. Because the basilar membrane is flexible, it shows a wave motion consistent with the waves in the fluid. However, the tectorial membrane moves only slightly and in directions opposite to that of the basilar membrane. These two opposing actions generate fluid streams across the tops of the hair cells, causing the cilia to bend. When the cilia bend, an electrical change is triggered within them.

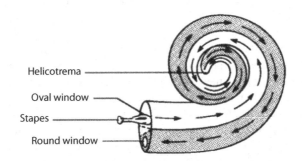

FIGURE 7.3 Schematic diagram of the cochlea.

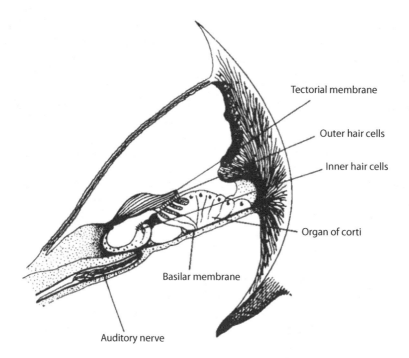

FIGURE 7.4   Cross-section of the cochlea.

We do not know as much about the basilar membrane as we do about the retina. It seems as though the inner hair cells provide the detailed information about the nature of the auditory stimulus. However, the exact role of the outer hair cells has been a puzzle for many years, because most of them connect to inner hair cells. The outer hair cells both vibrate in response to an alternating electric field and create their own electric fields. Thus, they both generate a field and respond to it, as a positive feedback system. This system may serve to make the inner hair cells more sensitive to different sound frequencies, allowing better detectability of differences between sounds.

The vibrations of the outer hair cells are the basis of a strange phenomenon called *otoacoustic emissions* (Dar & Hall, 2012). That is, not only does the basilar membrane register and transmit external sound waves; it produces sound waves of its own. What role (if any) these emissions play in hearing is still under investigation. However, their presence is completely normal, and measurement of the emissions provides a basis for evaluating basilar membrane function, especially in newborns.

The Auditory Pathways

The electrical activity of the hair cells in the cochlea causes the release of transmitter substances at their bases. These substances act on the receptor sites of neurons that make up the auditory nerve. There are approximately 30,000 neurons, 90% of which are devoted to the inner hair cells. Thus, although there are many more outer hair cells than inner hair cells, fewer neurons are devoted to the outer hair cells in the auditory nerve.

The neurons that make up the auditory nerve have preferred or characteristic frequencies. Each neuron fires maximally to a particular frequency, and less so to frequencies that deviate from it. The characteristic frequency is presumed to result from where a particular hair cell is located on the basilar membrane. A particular frequency will result in one point on the basilar membrane being displaced the most, and the cells at this point will respond most strongly. The neuron upon which these cells converge will have a characteristic frequency roughly equal to the frequency to which the cells respond most strongly. The entire sensitivity curve for a neuron is called a *frequency tuning curve* (see Figure 7.5). Thus, the auditory nerve is composed of a set of neurons that have frequency tuning curves with distinct characteristic frequencies.

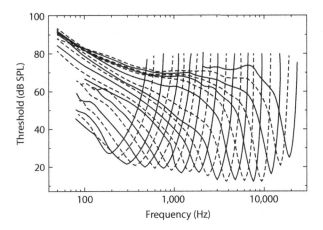

FIGURE 7.5   Frequency tuning curves for auditory neurons.

Neurons with similar characteristic frequencies are located near each other in the auditory nerve, a property called *tonotopic coding*. As for color vision, the specific frequency of an auditory stimulus must be conveyed by the overall pattern of activity within the set of neurons and not just which neuron is responding the most strongly. One important characteristic of auditory neurons is how they respond to continuous stimulation. If a sound remains "on" and a particular frequency is continuously transmitted to the basilar membrane, resulting in a continuous stimulation of particular hair cells, the level of neural activity declines. This phenomenon is known as *adaptation*.

The activity of an auditory neuron in response to a tone can be suppressed by the presence of a second tone. This phenomenon, called *two-tone suppression*, occurs when the frequency of the second tone falls just outside the tuning curve for the neuron. This suppression is thought to reflect the responsiveness of the basilar membrane (Pickles, 1988) and likely plays a role in the psychological phenomenon of auditory masking, discussed later in this chapter.

Unlike the optic nerve, the auditory nerve projects to several small neural structures before reaching the thalamus. These structures process several important parts of the auditory signal, such as interaural *time differences* (the time differences between signals to the two ears) and interaural *intensity differences* (the intensity differences between the two ears). This allows the extraction of spatial information, or where a sound is coming from. Some parts of the pathway perform complex analyses of the frequencies of the sound, in essence taking apart the different frequencies of a sound wave that the neurons from the basilar membrane put together. Eventually, the auditory signal reaches the thalamus, which then sends the signal on to the auditory cortex, located in the brain's temporal lobes. The temporal lobes are located in front of the occipital lobe, approximately at the level of each ear.

All of the neurons in the structures to which the auditory nerves project have neurons that show tonotopic coding similar to the spatiotopic coding shown for vision. A neuron that responds optimally to a given frequency will be located in close proximity to neurons that respond optimally to similar frequencies. Neurons in these structures are also sensitive to complex patterns in the signals. There are some neurons that respond to tone onsets, other neurons that respond to tone onsets and then continue to respond after a brief pause, and some neurons that respond with repeated bursts interspersed with pauses. The thalamus contains cells similar to the center/surround cells in vision. These cells respond optimally to energy in a certain frequency range; energy in surrounding ranges decreases the firing rate.

The auditory cortex also exhibits a tonotopic organization (Palmer, 1995). Further, many cortical cells respond to relatively simple features of stimulation. They show on responses, off responses, or on-off responses. Other cells respond to more complex sounds, such as bursts of noise or clicks. One type of cell is called a *frequency sweep detector*. It responds only to changes in frequency that occur in specific directions (higher or lower) within a limited frequency range. In short, as in the visual cortex, the neurons of the auditory cortex are specialized for extracting important features of stimulation.

Summary

An auditory stimulus, such as a change in air pressure, initiates a complex sequence of events that leads to the perception of sound. Physical vibrations of the eardrum, ossicles, and oval window produce a wave motion in the fluid of the inner ear. This wave motion causes neural signals through the bending of cilia on the basilar membrane. The auditory information is transmitted along pathways in which the neurons respond to different frequencies and other acoustic features. As with vision, the processing of the sensory signal performed by the auditory system provides the basis for auditory perception.

PERCEPTION OF BASIC PROPERTIES

As with vision, some attributes of auditory perception correspond relatively closely to the influence of sound on the sensory system, whereas others do not. Because the receptor cells in the auditory system are sensitive to the amplitude and frequency of sound, the perception of loudness and pitch is closely linked to the structure of the auditory sensory system. We will examine these properties, as well as some other qualitative characteristics of auditory perception, in the present section.

Loudness

The quantitative dimension of auditory perception is loudness (Schlauch, 2004). As with perceived brightness and physical luminance, loudness is psychological and is correlated with the physical dimension of intensity. Using magnitude estimation procedures, Stevens (1975) found that the perception of loudness is best described by a power function

$$L = aI^{0.6},$$

where:
- L is the loudness,
- I is the physical intensity of the sound, and
- a is a constant.

On the basis of this function, Stevens derived a scale for measuring loudness in which the unit is called the *sone*. One sone is the loudness of a 1000 Hz stimulus at 40 dB intensity. Figure 7.6 shows the sone scale and some representative sounds. This scale is used in human factors to describe the relative loudness of noise in different contexts. For example, we would use the sone scale to describe the noisiness of the interiors of different automobiles.

The loudness of a tone also can be influenced by its frequency. This fact is captured by *equal loudness contours*, which are shown in Figure 7.7. These contours are obtained by presenting a standard, 1000 Hz tone at a given intensity level, and then adjusting tones of other frequencies so that they sound equally loud. The equal loudness contours illustrate several important points. First, to sound equally loud, tones of different frequencies must be adjusted to different intensity levels. An alternative way of stating this relation is that if tones of different frequencies are presented at identical intensities, they will have different loudnesses. Second, tones in the range of 3–4 kHz are most easily detected, because they do not have to be as intense as tones outside this range to sound as loud. Third, low-frequency tones below approximately 200 Hz are hardest to detect. Finally, these differences in loudness across the frequencies progressively diminish as intensity increases.

One consequence of the relationship between intensity and loudness is that music recorded at high intensity levels will sound different when played back at low levels. Most obviously, the low-frequency bass will "drop out." Although it is present to the same relative extent in the sound wave, the lower intensity levels make it difficult to hear the low-frequency energy in the signal.

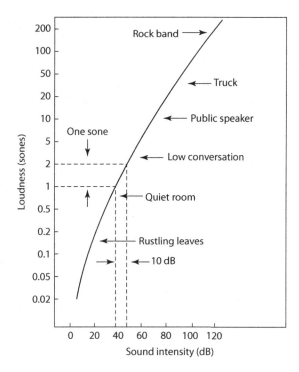

FIGURE 7.6   The sone scale of loudness.

Many high-fidelity amplifiers produced in the 1970s and 1980s have a "loudness" switch to compensate for this change. When it is switched on, the low and high frequencies are increased in intensity, making the recording sound more normal if played at a low intensity level.

As with visual intensity, we must calibrate the measurement of sound intensity for purposes of human hearing according to the human's sensitivity to different frequencies. Sound-pressure level L_p for a particular sound p typically is specified in decibels as

$$L_p = 20 \log_{10} \left(\frac{p}{p_r} \right),$$

where p_r is a reference pressure of 20 micropascals (1 µPa = 1 N/m²). The reference pressure of 20 µPa is approximately the smallest change in air pressure needed for a young adult to detect the presence of a tone. The resulting measure in decibels is a good characterization of the effective intensity of the sound at moderate energy levels.

Recall from our discussion of brightness how the visual system responds over space and time. The auditory system behaves similarly. If a sound is played for a brief period of time (200 ms), temporal summation occurs. The perceived loudness will be a function not only of the intensity of the tone but also of how long it was presented. Longer tones will be perceived as louder than shorter tones. However, the auditory system adapts to continuously presented tones, which means that the loudness of the tone diminishes over time. Finally, the loudness of complex tones is affected by the bandwidth, or the range of frequencies (e.g., 950–1050 Hz) that are included in the tone. As the bandwidth is increased, with the overall intensity held constant, loudness is not affected until a *critical bandwidth* is reached. Beyond this point, loudness increases as higher and lower frequencies are added to the complex tone.

Whether a sound can be heard depends on other sounds in the environment. If a sound is audible by itself but not in the presence of other sounds, the other sounds are said to mask it. As for vision,

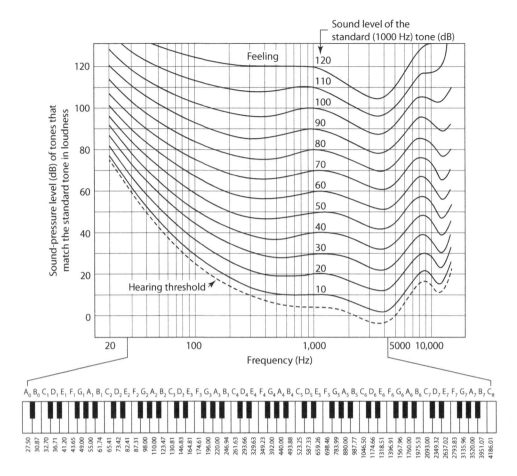

FIGURE 7.7 Equal loudness contours.

simultaneous masking occurs when the stimulus tone and the mask occur simultaneously. As the intensity of a mask is increased, the greater the stimulus intensity must be for the stimulus to be detected. The largest masking effect occurs when the stimulus and the mask are of the same or similar frequencies. If the stimulus is of lower frequency than the mask, the masking effect is very small and the tone can be easily detected (Zwicker, 1958). When the stimulus is of higher frequency than the mask, the masking effect is much greater (see Figure 7.8). The asymmetric masking effect on tones of higher and lower frequency is thought to be due to the pattern of movement on the basilar membrane.

Consider the design problem involved in constructing a warning signal (say, a fire alarm) for use in a relatively noisy environment (say, a factory floor). To avoid problems involved in adaptation, the alarm should be intermittent rather than continuous. It should also have a fairly large bandwidth so that its perceived loudness is maximized. Finally, its frequency should be on average lower than the frequency of the noise in the environment, to avoid the alarm being masked by the background. Human factors specialists need to remember that low-frequency tones will be less susceptible to masking in most environments, but that low-frequency noise may affect the perceptibility of a wide range of tones of higher frequency.

There are wide ranges of individual differences in the ability to hear, more so than in the ability to see. For instance, cigarette smokers are less sensitive to high-frequency tones than nonsmokers (Mehrparvar et al., 2015), and people who regularly take large doses of aspirin can suffer from 10–40 dB temporary decreases in sensitivity, often accompanied by tinnitus—a high-pitched

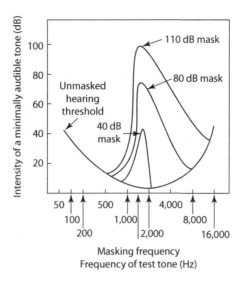

FIGURE 7.8 Thresholds for a tone stimulus in the presence of a narrow band of masking noise centered at 1200 Hz.

ringing in the ears (Stolzberg , Salvi, & Allman, 2012). These effects arise from changes in blood flow to the inner ear induced by nicotine and aspirin.

The audible range of frequencies decreases progressively across a person's life. A young person can hear tones as low as 20 Hz and as high as 20 kHz. As a person ages, the ability to hear high-frequency tones is lost. By age 30, most people are unable to hear frequencies above 15 kHz. By age 50, the upper limit is 12 kHz, and by age 70, it is 6 kHz. Substantial loss also occurs for frequencies as low as 2 kHz.

The hearing loss at high frequencies for the elderly means that they will not perform well when perception of high-frequency tones is required. Many land-line telephones use electronic bell or beeper ringers in comparison to the old-style mechanical bell. The frequency spectrum of the electronic bell ringer spans 315–20 kHz and that of the electronic beeper ringer 1.6–20 kHz. In contrast, the mechanical bell covers a range of 80–20 kHz. Berkowitz and Casali (1990) presented people of various ages with these three ringer types. The electronic beeper was virtually inaudible for older people and was easily masked by noise. This outcome is hardly surprising, given that the beeper includes virtually no acoustic energy in the range at which the elderly are most sensitive. The electronic bell actually was more audible to the elderly than the mechanical bell, apparently because it has a concentration of energy around 1 kHz.

Pitch

Pitch is the qualitative attribute of hearing that is the equivalent of hue in vision, and, as mentioned earlier, it is determined primarily by the frequency of the auditory stimulus (Schmuckler, 2004). However, just as the loudness of a sound can be affected by variables other than intensity, the pitch can be influenced by variables other than frequency. For example, *equal pitch contours* can be constructed by varying the intensity of stimuli of a given frequency and judging their pitches. As shown in Figure 7.9, below approximately 3 kHz, pitch decreases with increasing intensity, whereas above 3 kHz it increases.

Pitch also can be influenced by the duration of a tone. At durations of less than 10 ms, any pure tone will be heard as a click. As tone duration is increased up to approximately 250 ms, tone quality is improved. This results in an increase in ability to discriminate between pitches of longer-duration tones.

Two theories of pitch perception were developed in the 1800s, and contemporary research indicates that both are needed to explain pitch-perception phenomena. The first theory, proposed by

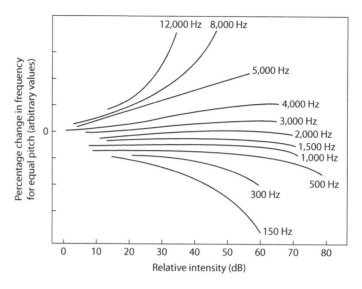

FIGURE 7.9   Equal pitch contours.

Rutherford, is called *frequency theory*. This theory suggests that the basilar membrane vibrates at the frequency of the auditory stimulus. This frequency of vibration of the basilar membrane then is transformed into a pattern of neural firing at the same frequency. Thus, a 1 kHz tone would cause the basilar membrane to vibrate at a 1 kHz rate, which in turn would cause neurons to respond at this frequency.

The alternative theory is *place theory*, which was first proposed by Helmholtz. He noticed that the basilar membrane was a triangular shape and suggested that it consisted of a series of resonators of decreasing length that respond to different frequencies. Thus, the frequency of the tone would affect a particular place on the basilar membrane; the activity of the receptors at this location then would send a signal along the particular neurons that received input from that place.

Place theory received new life from the work of George von Békésy (1960). Physiological investigations revealed that the basilar membrane does not act as a series of resonators, as Helmholtz had proposed. However, by observing the action of the basilar membrane in the ears of guinea pigs, von Békésy established that different frequency tones produce traveling waves that have maximal displacement at distinct locations on the basilar membrane. As we saw already, displacements for low-frequency tones are at the wide end of the basilar membrane, away from the oval window. As the frequency of the tone is increased, the location of peak displacement shifts progressively toward the oval window. The term *traveling wave* comes from the fact that the action of the basilar membrane corresponds to the action of a rope that is secured at one end and shaken at the other; a wave ripples down the basilar membrane (see Figure 7.10).

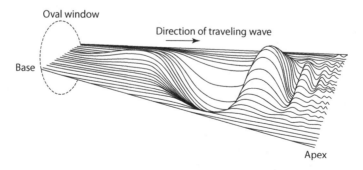

FIGURE 7.10   Traveling wave motion of the basilar membrane.

A major problem with traveling wave theory is that the peak displacement for all tones below approximately 500 Hz is at the far end of the basilar membrane. Thus, place coding does not seem possible for these frequencies. Information about where on the basilar membrane the displacement occurs is the same for all low-frequency tones. Because tones below 4 kHz can be accommodated by frequency theory, the widely accepted view is that place coding holds for tones that exceed 500 Hz, whereas frequency coding holds for tones of less than 4 kHz. Thus, between 500 and 4 kHz, both place and frequency coding must signal pitch.

Timbre, Consonance, and Dissonance

When the same musical note is played by different instruments, it does not sound identical, because of their different resonance properties. This qualitative aspect of auditory perception that can occur even for sounds of equivalent loudness and pitch is called *timbre* (Plomp, 2002). Timbre is determined by many factors, with one being the relative strengths of the harmonics in the sound wave. Figure 7.11 shows the frequency spectra for a note with a fundamental frequency of 196 Hz played on a bassoon, guitar, alto saxophone, and violin. The pitch, which is determined by the fundamental frequency, sounds the same. However, the relative amounts of energy at the different harmonic frequencies vary across the instruments; these different spectral patterns are one reason why the tones have distinct timbres. Timbre also is influenced by the time course of the buildup and decay of sound at the beginning and end of the tone.

Consonance and *dissonance* refer to the degree of pleasantness of combinations of two or more tones. When pure tones are combined, the relative dissonance is a function of the critical bandwidth. Tones within the critical band sound dissonant, whereas tones separated by more than the critical band sound consonant. Within the critical band, a small separation in frequencies leads to the perception of beats, or oscillations of loudness, and the pitch sounds like one of a frequency intermediate to the two-component frequencies. Outside of the range that produces beats, the tone is perceived to be rough. For more complex musical tones, the harmonics are also important in determining whether a combination is heard as consonant or dissonant.

PERCEPTION OF HIGHER-LEVEL PROPERTIES

Our range of auditory perceptions is nearly as rich as that of our visual perceptions. We can perceive complex patterns, determine the locations of stimuli in the environment, and recognize speech with

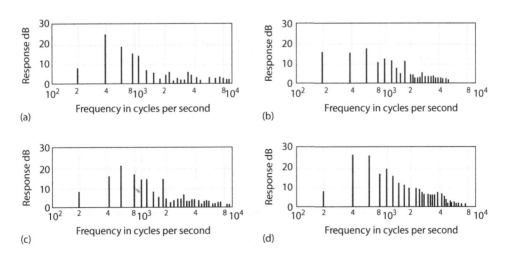

FIGURE 7.11 The harmonic structure of a tone with a fundamental frequency of 196 Hz, played with a (a) bassoon, (b) guitar, (c) alto saxophone, and (d) violin.

proficiency. Consequently, it is important to know how auditory perception is influenced by organizational factors, spatial cues, and the features of speech stimuli.

Perceptual Organization

Although the principles of perceptual organization have been studied most thoroughly for vision, they also apply to the other senses. For audition, the principles of proximity and similarity are important. Temporal proximity of tones is more important than spatial proximity. Tones that follow each other closely in time will tend to be perceived as belonging together.

Similarity is determined primarily by the pitch of the tones. Tones with similar pitches tend to be grouped together perceptually. This point was illustrated by Heise and Miller (1951). They played a sequence of tones for which the frequency increased linearly. Listeners heard a tone in the middle as part of the sequence even if its frequency was a bit higher than it should have been at that location in the series. However, when the frequency of this tone became too deviant, listeners heard it as an isolated tone against the background of the increasing sequence.

In music, a rapid alternation between high- and low-frequency notes leads to the perception of two distinct melodies, one pitched high and the other pitched low. This perceptual effect is called *auditory stream segregation*. Bregman and Rudnicky (1975) showed that how tones are organized can influence how they are perceived. They presented listeners with two standard tones, A and B, that differed in frequency (see Figure 7.12). The listeners' task was to determine which tone occurred first. When the tones were presented by themselves, performance was good, but when the tones were preceded and followed by an extra "flanking" tone of lower frequency, performance was poor. However, if additional tones of the same frequency as the flanking tones were presented before and after the flanking tones, performance again was good. Apparently, the additional tones caused the flanking tones to be segregated into one auditory stream and the standard tones into another stream. Consequently, the standard tones did not "get lost" among the flankers, and their order was easy to perceive.

Recall what happens to retinal images when they fall across the blind spot. The visual system "fills in" the missing information as best it can. A similar effect is observed in audition. When a tone is interrupted for a brief interval by presentation of wideband noise (i.e., noise with frequencies across a broad range of the auditory spectrum), the auditory system fills in the tone across the interval, and an illusion of continuity is produced. Bregman, Colantonio, and Ahad (1999) showed that several variables had similar effects on both illusory continuity and the perception of a single auditory stream, suggesting that both illusory continuity and streaming depend on an early process in perception that links together the parts of a sequence that have similar frequencies.

Sound Localization

Although audition is not primarily a spatial sense, our ability to locate sounds in space is still fairly good. There are a number of important cues the auditory system uses to locate sound (Blauert, 1997).

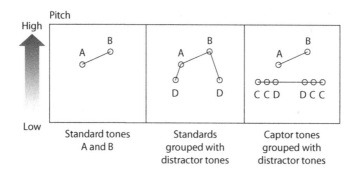

FIGURE 7.12 Auditory streaming: Standards (Left); Grouped with distractor tones (Center); Distractor tones grouped with captor tones (Right).

For example, as a train moves toward you and then past, there is a systematic transformation in the frequency of the sound pattern that results in a shift of pitch corresponding to the change in location (the Doppler effect). Most research on spatial perception in hearing has been on sound localization along the azimuth, the horizontal plane, which we will consider here.

Our ability to localize briefly presented sounds is relatively good. In a typical experiment on sound localization, the listener will be blindfolded. Sounds then are presented from various locations around him. In most cases, he will be able to identify the location of the sound along the horizontal plane accurately.

Accurate sound localization depends on having two ears. There are two different sources of information: *interaural intensity differences* and *interaural time differences*, which are analyzed by different neural mechanisms (Marsalek & Kofranek, 2004). The relative intensity at each ear varies systematically as the location of a sound is moved from front to back. When the sound is at the front of the listener, the intensity at each ear is the same. As the location changes progressively toward the right side, the intensity at the right ear relative to the left increases, with the difference reaching a maximum when the sound is directly to the right. As the location is moved behind the listener, the difference shifts back toward zero.

What causes the intensity differences? The answer to this question is that the head produces a sound shadow, much like the way that a large rock produces a "dead spot" behind it in the flow of a stream. This sound shadow is only significant for frequencies above approximately 2 kHz. The reason for this limitation is that the head is too small relative to the length of the low-frequency sound waves to cause a disturbance. In the natural world, sounds originating from different locations have unique combinations of spectral cues and interaural timing and intensity cues, which seem to be brought into alignment by neurons in the auditory pathway (Slee & Young, 2014). These cues provide information that can be used to localize the sounds.

Interaural time differences show a similar pattern of change to the intensity differences when the origin of a tone is moved. In contrast to the intensity differences, the time differences are most effective for low-frequency tones below 1000 Hz. It seems that while interaural time differences are used to localize low-frequency sounds, interaural intensity differences are used to localize high-frequency sounds. Not surprisingly, localization accuracy is worst for tones in the 1–2 kHz range, in which neither intensity nor time differences provide good spatial cues.

You should note that both time and intensity cues are ambiguous, in that two different locations on each side (one toward the front, the other toward the back) produce similar time and intensity relations. Some monaural information distinguishing the locations comes from different "coloration" of the sounds (distortion of high frequencies) provided by the pinna (Van Wanrooij & Van Opstal, 2005). Also, head movements provide dynamic changes that allow a sound to be localized more accurately (Makous & Middlebrooks, 1990). When head movements are restricted, the most common types of errors people make in localization are front to back reversals.

Anything that decreases the intensity of an auditory signal reaching the ears should decrease the localization accuracy, particularly if it alters the relative timing and intensity relations at the two ears. Caelli and Porter (1980) had listeners sit in a car and judge the direction from which a siren sounded. Localization accuracy was poor when all windows were rolled up, and people made a lot of front-back reversals. Accuracy was even worse when the driver-side window was rolled down, as we might expect because of the alteration of the relative-intensity cue.

Vertical sound localization (elevation) is less accurate than localization in the horizontal plane, primarily because it cannot be based on interaural differences. The torso, head, and pinna all modify the acoustic signal, providing complex spectral cues that provide information about vertical location (Van Wanrooij & Van Opstal, 2005). However, these cues are not as strong as the binaural timing and intensity cues, so judging the distance of sounds vertically is difficult. It relies on the intensity of the sound and on reflections of sound waves from nearby objects. A sound source of constant intensity will be perceived as louder when it is near to the listener than when it is farther away.

Three-dimensional auditory displays are used to enhance cockpit displays in military aircraft. Accurate localization is important for these kinds of displays. To synthesize a sound's location in space, the display designer must first measure the changes that the ear and head make to a sound wave for many sound source positions relative to the body (Langendijk & Bronkhorst, 2000). The designer can incorporate these changes in digital filters that are then used to simulate sounds from different locations. The bandwidth of these signals determines how useful they are. King and Oldfield (1997) noted that the communication systems in most military aircraft are fairly narrow, while broadband signals encompassing frequencies from 0 to at least 13 kHz are required for listeners to be able to localize signals accurately. This means that auditory cockpit displays must be broadband if they are to serve their purpose.

Speech Perception

To perceive speech, we must be able to recognize and identify complex auditory patterns (Pisoni & Remez, 2005). Usually we process speech patterns quickly and effortlessly. As with most perceptual processes, the ease with which we perceive speech does not reflect the complexity of the problem that must be solved by the speech pattern-recognition system.

The basic unit for speech is the *phoneme*, which is the smallest speech segment that, when changed, will alter the meaning of a word. Figure 7.13 shows the phonemes for English, which correspond to the vowel and consonant sounds. Because a change in the phoneme results in the perception of a different utterance, people must be able to identify parts of speech at the level of the phonemes. Research on speech and auditory perception has concentrated largely on the process of identification.

Figure 7.14 illustrates a *speech spectrogram* for a short speech utterance. The abscissa of the figure is time, and the ordinate is the frequency in the sound. The dark regions at any point in time show that the acoustic signal includes energy at those frequencies. Most of the energy is contained in distinct, horizontal bands of frequencies, which are called *formants*. The formants represent vowel sounds. The initial consonant phonemes correspond to formant transitions (or changes) that occur early in the signal. The problem faced by researchers is to identify the parts of the acoustic signal that represent the presence of specific phonemes.

Identifying phonemes may start with looking for invariant acoustic cues; that is, aspects of the acoustic signal that uniquely accompany particular phonemes in all speech contexts. However, if we examine a wide variety of speech spectrograms, there are no obvious invariant cues.

	Major consonants and vowels of English and their phonetic symbols		
Consonants		**Vowels**	
p pea	θ thigh	i beet	o go
b beet	ð thy	ɪ bit	ɔ ought
m man	s see	e ate	a dot
t toy	ʒ measure	ɛ bet	ə sofa
d dog	tʃ chip	æ bat	ɝ urn
n neat	dʒ jet	u boot	ai bite
k kill	l lap	U put	aU out
g good	r rope	ʌ but	ɔɪ toy
f foot	y year	ɒ odd	oU own
ç huge	w wet		
h hot	ŋ sing		
v vote	z zip		
ʍ when	ʃ show		

FIGURE 7.13   Phonemes of the English language.

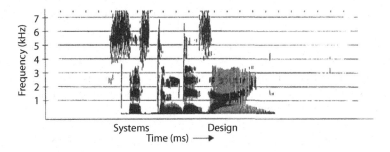

FIGURE 7.14 A speech spectrogram for the words "systems design."

Figure 7.15 illustrates this point by showing a schematic spectrogram used to produce artificial speech corresponding to the utterances *dee* and *do*. Because the vowel phonemes are different, it is not surprising that the formants for the two utterances differ. However, while the consonant phoneme is the same in the context of the two vowels, the formant transitions are not. The transition for the higher-frequency formant rises in the acoustic signal for *dee* but falls in the signal for *do*.

Because of this and other examples of acoustic variability for phonemes in the speech spectrogram, phoneme perception must occur in other ways that do not rely on invariant cues. One hypothesis is that phoneme perception is a function not only of the acoustic signal but also of the way that the sound is produced; for example, the proprioceptive feedback (see below) that would be provided by the muscles of the mouth, tongue, and throat if the phoneme were spoken (see Fowler & Galantucci, 2005; Mattingly & Studdert-Kennedy, 1991).

One important phenomenon in speech perception is *categorical perception* (Altmann et al., 2014). We can illustrate this phenomenon using the sounds *da* and *ta*, which differ primarily in terms of voice onset time (i.e., the delay between when the utterance begins and the vocal cords start vibrating). For *da*, the voice onset time is approximately 17 ms, and for *ta*, the voice onset time is approximately 91 ms. With artificial speech, this onset time can be varied between 17 and 91 ms. The question is, what do listeners perceive for these intermediate onset times? The answer is that the stimuli are heard as either *da* or *ta*, with a relatively sharp boundary occurring at an intermediate onset. Moreover, stimuli on the same side of the boundary sound exactly alike even when their voicing onsets differ. In other words, people do not hear the physical differences between the stimuli; the stimuli are strictly categorized as *da* or *ta*.

The amount we can learn about speech perception by investigating how people process phonemes is limited. Other research on speech perception has focused on more natural conversational speech. The problems we have to solve for conversational speech are much more complex than those for phoneme perception. In conversational speech, there are no physical boundaries between

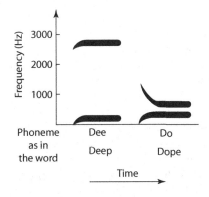

FIGURE 7.15 Artificial speech spectrograms for "dee" and "do."

words. Any boundaries that we think we hear are imposed by our perceptual systems. Additionally, people do not enunciate words clearly in conversational speech. If we record a stream of conversation and then listen to an individual word from that stream, it will be difficult to identify. The context in which a word is embedded determines how we perceive it. When this context is ambiguous, the word may be confused for one that sounds similar. In 1990, a passenger on an interstate bus shouted, "There is a bum in the bathroom!" However, the bus driver mistakenly heard the utterance as, "There is a bomb in the bathroom!" Consequently, the bus was stopped, state troopers were called, the highway was blocked off, and the bus was searched by a bomb-sniffing dog. The transient in the bathroom was charged with misdemeanor theft for avoiding the price of a ticket (Associated Press wire story, 1990b).

Because of the complexity of conversational speech, speech perception depends heavily on *semantic* and *syntactic context*. This is illustrated in a classic experiment by Miller and Isard (1963). They had listeners repeat aloud strings of words as they heard them. The strings were either (1) normal sentences (e.g., Bears steal honey from the hive), (2) semantically anomalous but grammatically correct sentences (e.g., Bears shoot work on the country), or (3) ungrammatical strings (e.g., across bears eyes work the kill). The percentage of complete strings listeners repeated accurately was lowest for the ungrammatical strings (56%). The semantically anomalous but syntactically correct sentences were repeated with higher accuracy (79%), indicating that consistency with grammatical rules benefits perception. Moreover, performance was even better for the meaningful sentences (89%), indicating that semantic context is also important for perception.

One intriguing context effect in speech perception is the phonemic restoration effect. Warren (1970) had people listen to the passage, "The state governors met with their respective legislatures convening in the capital city," and replaced the first *s* in "legislatures" with a cough. No one noticed that the *s* was missing or could identify the location of the cough. This effect also occurred when the context prior to the word was ambiguous, and the phoneme had to be determined from subsequent words. Once again, the phoneme was constructed by the listeners' perceptual systems on the basis of the sentence context. For restoration of linguistic as well as nonlinguistic auditory stimuli, not only must the context provide sufficient cues but the interpolated sound must also be in the frequency range for which the replaced sound could potentially be masked (Bashford, Riener, & Warren, 1992). The findings obtained with conversational speech indicate that a listener's expectancies influence speech perception, much as they do visual pattern recognition.

THE VESTIBULAR SYSTEM

The sensory receptors for the *vestibular sense* are located within the membranous labyrinth of the inner ear (Goldberg et al., 2012). This sense allows us to feel the movements of our bodies. It also contributes to our ability to control the position of our eyes when we move our heads and to maintain an upright posture. The vestibular organ is comprised of three structures: the utricle and saccule (collectively called the otolith organs) and the semicircular canals. Within the otolith organs and the semicircular canals are receptor cells very similar to those found on the basilar membrane (Lackner & DiZio, 2005). Displacement of cilia within these organs results in a neural signal.

The otolith organs are lined with hair cells whose cilia are embedded in a gelatinous liquid that contains "ear stones." When you tilt your head, the stones move through the liquid. This movement displaces the cilia and results in information about the direction of gravity and linear acceleration. These signals are used to control posture. The semicircular canals are located in three roughly orthogonal planes. When you turn your head, a relative motion between the fluid and the canal is created. This results in a shearing action on the hair cells within each canal. These receptors provide information about rotary and angular acceleration.

The system functions together with vision and proprioception to assist in the control of movements. The movements you make because of signals from your vestibular system are mostly involuntary. For example, when you fixate your eyes on an object and then turn your head, your fixation is maintained by a vestibulo-ocular reflex, which moves your eyes in the opposite direction from your head movement. Similar reflexive movements of your eyes to counter body motion help you stabilize your gaze in other situations. The vestibular system also plays a role in maintaining posture and balance. Reduced efficiency of the vestibular system is one factor that contributes to falls in older adults.

When a person is subjected to unfamiliar motion or vibratory patterns, she may experience illusions of visual and auditory localization and of perceived self-orientation. Unfamiliar motion may also result in motion sickness (Lackner & DiZio, 2005), possibly due to a mismatch between vestibular cues and cues from other senses. The vestibular sense is implicated in motion sickness because people with loss of the vestibular system exhibit little if any motion sickness (Paillard et al., 2013). Motion sickness often occurs for persons riding in moving vehicles (aircraft, boats, cars, trains, and so on), but it is also prevalent in simulated environments (where the simulated vehicle is not actually moving) and in virtual environments (Harm, 2002; see Box 7.1). Symptoms include headache, eye strain, nausea, and vomiting. The severity of the symptoms varies as a function of individual differences in susceptibility and the magnitude of the motion.

The vestibular system contributes to human abilities during manned space flight (Young, 2000). Vestibular cues are different from those an astronaut usually experiences during all phases of the space flight, from launch to orbit to landing. When in space, gravity is absent, causing the astronaut's vestibular reactions to be altered. These altered reactions lead to spatial disorientation and motion sickness, which typically begins during the first few hours in orbit and may last up to 3 days. Space motion sickness reduces the amount of work that the astronaut can accomplish in the first few days in space. The work schedules during this period must reflect this reduced capacity to work. The astronaut may also experience spatial orientation illusions. Disorientation and difficulty with postural control may persist throughout a space mission.

One possible solution to the longer-term problems of a zero-gravity environment is to provide artificial gravity by rotating the space vehicle. However, this will also produce unusual vestibular stimulation associated with the rotation. In sum, there are a variety of human factors issues associated with the vestibular reaction to all aspects of the flight sequence.

THE SOMESTHETIC SYSTEM

While driving your car, you may reach for the gearshift without looking in its direction. The sensation provided when your hand comes into contact with the knob tells you when you have grasped the shift. Also, you can operate the gearshift by moving it through the various settings without looking at it. The information that allows you to identify the gearshift and its settings is provided by the *somesthetic senses*. These include the sense of touch as well as the senses of pressure, vibration, temperature, pain, and proprioception.

Sensory System

Most sensory receptors of the somesthetic system are located in the skin, which consists of two parts. The epidermis is formed by layers of dead cells on top of a single layer of living cells. The dermis is an inner layer in which most of the nerve endings reside. These nerve endings are of a variety of types. Some respond primarily to pressure stimuli, whereas others are particularly responsive to pain stimuli. These nerves respond to mechanical, temperature, or electrical stimulation by generating an action potential that is transmitted along axonal fibers to the brain.

The nerve pathways can be organized according to two major principles (Coren, Ward, & Enns, 2004): the type of nerve fiber and the place of termination of the pathway in the cortex. Fibers differ in terms of the stimulus type to which they are most responsive, whether they are slow- or fast-adapting, and whether their receptive fields are small or large. The receptive fields have the same center–surround type of organization as in the visual system.

BOX 7.1 VIRTUAL ENVIRONMENTS

"Virtual reality is a medium with tremendous potential. The ability to be transported to other places, to be fully immersed in experiences, and to feel like you're really there—present—opens up previously unimagined ways to interact and communicate" (Parisi, 2016). Virtual reality (VR) or virtual environments (VEs) are complex human–computer interfaces (Stanney, 2002). As captured by the opening quote, a VE is designed to immerse the user in a "world" that changes and reacts to the user's actions in the same way as it would if the user were really there. No one completely agrees on what constitutes a VE, but among its most basic properties is a three-dimensional (3D) visual display. Not all 3D displays are VE displays. Wann and Mon-Williams (1996) argue, "The term VE/VR should be used to describe systems which support salient perceptual criteria (e.g., head motion parallax, binocular vision) such that the user is able to perceive the computer generated image as structured in depth" (p. 835). In other words, a VE system should produce perceptual changes as the user moves that correspond to those that would occur in the physical world. Another property of a VE system is that it must permit the user to interact directly with the environment through manipulation of objects.

VE designers often strive for a strong sense of presence, which is the user's experience of being in the VE instead of the physical environment in which she is actually located. The experience of presence is a function of involvement, or the extent to which the user's attention is focused on the activities in the VE, and immersion, or the degree to which the VE envelopes the user (Sun, Li, Zhu, & Hsiao, 2015). Many factors influence presence, including the realism of the visual display, the ease with which the user can interact with the VE, the extent to which the user has control over her own actions, the quality of the VE hardware and software, and beliefs about whether the VE environment represents a digitized live environment or not (Bouchard et al., 2012).

While VE designers are greatly concerned about how to display 3D information, the goal in a VE system is to incorporate all of the senses in such a way that the sensory experience provided by the VE is indistinguishable from that of the physical environment it represents. Because audition is almost as important as vision for interacting with the world, 3D auditory displays are often included in VE to add realism and increase the experience of presence (Ruotolo et al., 2012). Realistic spatial localization can be achieved through headphones using filters based on head-related transfer functions, which specify how acoustic signals at different points in space are transformed as they travel to the ears. Although not as widely used, haptic displays can be added to allow the user to "feel" manipulated objects and receive force feedback when operating a virtual device (Reiner & Hecht, 2009), and an acceleration system can be incorporated that mimics the effects of body acceleration on the vestibular system (Maeda, Ando, & Sugimoto, 2005).

Because of technological limitations, not all of the sensory changes resulting from interacting with a VE will occur in the exact time and manner that they would in the physical world, and this will result in some degree of conflict between the sensory systems. As a consequence, VE users often experience a form of motion sickness called "cybersickness" (Rebenitsch & Owen, 2016). More than 80% of VE users experience some level of cybersickness, ranging from minimal symptoms to nausea (Stanney, 2003). During VE exposure, physiological adaptation will occur that tends to reduce the symptoms, but this adaptation may produce aftereffects such as disturbances in balance and hand–eye coordination. VE users must be cautious to minimize any potential health and safety risks.

The feasibility of developing VE systems for various purposes continues to increase. Just in the past 5 years, systems such as the Oculus Rift and Google Cardboard have opened the way to low-cost devices through which VE can be delivered (Riva & Wiederhold, 2015).

Areas of application include the engineering design process, medical training, simulation of novel environments and infrequent emergency events, team training, and scientific visualization, as well as, of course, gaming. One of the benefits of VEs is that they are not constrained by the rules that govern our interactions with the physical world (Wann & Mon-Williams, 1996). Many of the most useful applications of VEs will take advantage of this fact.

The nerve fibers follow two major pathways. The first is called the *medial lemniscus*. The fibers in it conduct information quickly and receive their inputs primarily from fibers with corpuscles. This pathway ascends the back portion of the spinal cord on the same side of the body as the receptors that feed into it. At the brainstem, most of the fibers cross to the other side of the body. The pathway continues until it reaches the somatosensory cortex. The fibers in this system respond primarily to touch and movement. The second pathway is the *spinothalamic pathway*. The fibers in it conduct information slowly in comparison to the lemniscal fibers. This pathway ascends to the brain on the opposite side of the body from which the fibers terminate, passes through several areas in the brain, and ends up in the somatosensory cortex. The spinothalamic pathway carries information about pain, temperature, and touch.

The somatosensory cortex is organized much like the visual cortex. It consists of two main parts, each with distinct layers. The organization is spatiotopic in that stimulation of two adjacent areas on the skin will result in neural activity in two adjacent areas of the cortex. Areas of the skin to which we are more sensitive have relatively larger areas of representation in the somatosensory cortex (see Figure 7.16). Cells respond to features of stimulation, such as movement of ridges across the skin.

Receptors located within muscle tendons and joints, as well as the skin, provide information about the position of our limbs. This information is called *proprioception* and, when related to movement, *kinesthesis*. It plays a fundamental role in the coordination and control of bodily movement. The input for proprioception comes from several types of receptors. Touch receptors lie deep in layers of tissue beneath the skin. Stretch receptors attached to muscle fibers respond to the stretching of the muscles. Golgi tendon organs sensitive to muscle tension are attached to the tendons that connect the muscles to bones. Joint receptors are located in joints and provide information about joint angle. The neurons that carry the information for proprioception travel to the brain by way of the same two pathways as for touch. They also project into the same general area of the somatosensory cortex.

PERCEPTION OF TOUCH

A sensation of touch can be evoked from anywhere on the body. Absolute thresholds for touch vary across the body, with the lowest thresholds being in the face (see Figure 7.17). Vibrating stimuli are easier to detect than punctate stimuli. Using psychophysical methods, we can obtain *two-point thresholds* by asking people to determine when two simultaneous points of stimulation are perceived as two distinct points. The threshold, measured as spatial distance between the two points, provides an indication of the accuracy with which points on the skin can be localized. The two-point thresholds across the body show a function similar to absolute thresholds for touch. The primary difference between the two-point and absolute touch thresholds is that the fingers, rather than the face, have the lowest two-point threshold.

For vibratory stimuli, we can measure threshold and above-threshold magnitudes of sensation as a function of the frequency of vibration. Figure 7.18 shows that equal-sensation contours are of the same general nature as the equal loudness contours found for different auditory frequencies. People show the best sensitivity to vibration for frequencies in the region of 200–400 Hz.

In addition to direct tactile stimulation, we also receive such stimulation indirectly by using tools, by wearing gloves, or with other interposing materials. Operators can use tools to detect

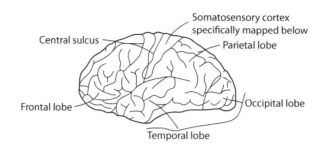

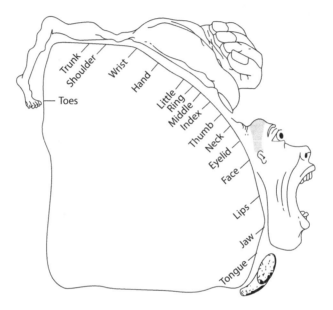

FIGURE 7.16 The somatosensory cortex.

faults in manufactured goods in situations for which direct touch might be injurious. For example, a worker might be asked to examine the edges of panes of glass for flaws. This is not something we would want him to do with his bare hands, but instead with some kind of instrument. Kleiner, Drury, and Christopher (1987) investigated factors that influence the ease with which people can perform such indirect fault detection. They found that the probability of detecting a fault increased as the size of the fault increased and that sensitivity decreased as the instrument tip increased in diameter.

One of the most important distinctions we can make in tactile perception is between *passive touch* and *active touch*. In passive touch, the skin is stationary and an external pressure stimulus is applied to it. This is the type of procedure we would use to obtain the absolute and two-point thresholds discussed above. In active touch, the person contacts the stimulus by moving the skin. This corresponds more to what we do when we grasp an object and try to identify it.

Gibson (1950) emphasized that passive touch results in the perception of pressure on the skin, whereas active touch results in the perception of the object touched. Although we don't yet understand the reasons for these quite distinct perceptual experiences, probably the most important factor is that active touch is purposive. That is, you manipulate an object for the purpose of identifying or using it; you use expectancies about the object to encode the relations among the sequence of

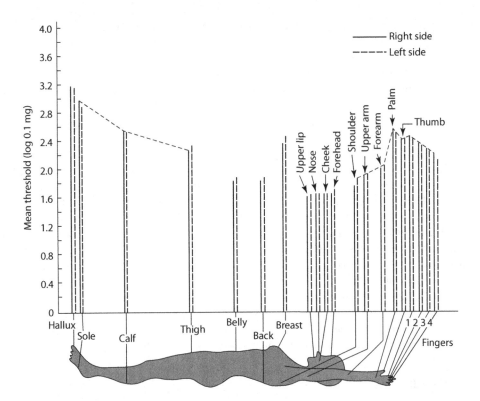

FIGURE 7.17 Absolute thresholds across the body.

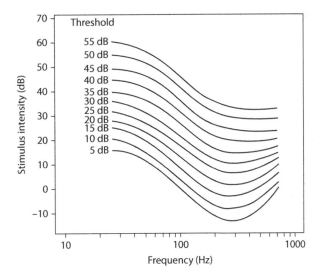

FIGURE 7.18 Equal-sensation contours for vibratory stimuli.

sensations. The important point for human factors is that movement of stimuli across the skin is required for accurate perception. This is similar to the role of movement in visual perception that was demonstrated by the phenomenon of biological motion.

We know that people can read from tactual input as well as from visual input. Because the tactual sense is not as sensitive to spatial details as vision, the patterns that represent text must be larger and more distinct than printed text. The Braille alphabet is one system that meets this requirement. However, a trained Braille reader can read a maximum of about 100 words per minute, whereas an average visual reader can read at 250–300 words per minute. The slower reading speed with Braille reflects lower tactile than visual acuity, and, because of this, Braille characters must be sufficiently wide and spaced far enough apart that the reader only perceives a single letter at a time.

Nontext material can also be tactile. We can present graphical material using raised surfaces. Consider, for example, a graph of the Dow-Jones Industrial Average over the past year. The axes and their labels, as well as the fluctuating value of the Dow, can all be depicted as raised lines. One question we might have is whether the graph should include a grid to aid in localizing points, the better to determine what the Dow was at any specific point in time. The answer depends on the information that must be identified from the graph. For questions about position, like the value of the Dow, people do better with a grid than without, whereas for questions about overall configuration, like whether the Dow is rising or falling, people do better without the grid (Aldrich & Parkin, 1987; Lederman & Campbell, 1982). Another example of tactile information comes from navigational symbols on walking paths (Courtney and Chow, 2001). People with reduced vision might benefit from foot-discriminable symbols used on the paths. Even wearing shoes, people can accurately identify up to 10 different symbols using their feet.

Tactual devices are used in medical simulators to train doctors in "hands on" operating techniques, in air and space vehicles to present information to pilots under conditions of high gravity where vision is limited, and in virtual environments to help people navigate.

PERCEPTION OF TEMPERATURE AND PAIN

We measure temperature sensitivity by applying thermal stimuli to the skin. Temperature thresholds show nearly perfect temporal summation for 0.5–1.0 s (Stevens, Okulicz, & Marks, 1973) and spatial summation over large areas (Kenshalo, 1972). This means that if you press a heated flat surface on your skin, it will feel hotter than if you just press the edge. You will adapt to thermal stimuli over a period of several minutes. You will be able to tell where hot and cold stimuli are applied to your body, but not very precisely.

There is a lot of research on the neurophysiological and psychological basis of pain (Cervero, 2012). Pain can come from extreme environmental conditions, such as loud noise or cold temperatures, which signal the risk of physical harm if the conditions persist for any length of time. Pain may also be beneficial: it signals that you should minimize your activity, for example, stop walking on a sprained ankle, and so promote healing (Coren et al., 2004).

Pain perception and its measurement are an important component of physical ergonomics. Lower back pain and pain associated with upper extremity disorders are common, and the costs of such pain to individuals and their employers are high (Feuerstein, Huang, & Pransky, 1999; Garofolo & Polatin, 1999). How to prevent or minimize physical pain from injury and cumulative trauma is the topic of Chapter 17.

Pain receptors are of two kinds: free nerve endings, located throughout the body, and nerve endings called Schwann cells, located in the outer part of the skin, that terminate in a sheath. The pain receptors and the fibers to which they are attached respond only to high-intensity stimuli. The fibers connect mainly with the spinothalamic pathway. Studies of pain have used many devices designed to apply extreme mechanical, thermal, chemical, and electrocutaneous stimuli to all parts of the body. As Sherrick and Cholewiak (1986) describe, "The full array of devices and bodily loci employed would bring a smile to the lips of the Marquis de Sade and a shudder of anticipation to the Graf van Sacher-Masoch" (pp. 12–39).

Sensitivity to pain varies across the body, being lowest for the tip of the nose, sole of the foot, and ball of the thumb, and highest for the back of the knee, bend of the elbow, and neck region

(Geldard, 1972). In contrast to touch and temperature, pain thresholds show little temporal or spatial summation. Pain perception does, however, show adaptation during prolonged stimulation. In other words, you will get used to pain over time.

THE CHEMICAL SYSTEMS

Taste and smell are the chemical senses, because the stimulus is molecules of substances in the mouth and nose (Di Lorenzo & Youngentob, 2013). Taste and smell are important both for aesthetics and for survival. Things that taste or smell bad are often poisonous. Thus, taste and smell provide us with important information about objects and substances in our environment.

For example, a Web page devoted to distinguishing forged 1938 German postcards from authentic ones recommends a smell test: "You smell old cards, no kidding! Try to smell one of your old cards and a new one. You surely will smell a difference—genuine old cards most often smell like [something] coming from the attic" (Forgery Warning I, Oct. 1999). In some situations, we can add smelly substances to other odorless, potentially harmful substances to convey warnings. For example, mercaptans, a family of strong-smelling chemical compounds, are added to natural gas, which is naturally odorless. The mercaptans make it easy for anyone to detect a gas leak. Mercaptans are produced naturally by skunks to ward off potential predators.

The physical stimulus for taste is a substance dissolved in saliva. The dissolved substance affects receptors that are located on the tongue and throat. The receptors for taste are groups of cells called *taste buds*. Each taste bud is made up of several receptor cells arranged close together. These cells are continually developing and have a lifespan of only a few days. The receptor mechanisms for taste are thought to be contained in projections from the top end of each cell that lie near an opening on the surface of the tongue called a *taste pore*.

There seem to be at least four basic taste qualities—sweet, salty, sour, and bitter—with umami, the taste evoked by monosodium glutamate (MSG), being a possible fifth. These tastes are related to the molecular structure of the substances that produce them. While all areas of the tongue respond to all the taste qualities, sensitivity for each quality depends on the location on the tongue. We don't yet know much about how the molecules affect the sensory receptors to initiate a neural signal. Fibers from the taste buds make up three large nerves that go to several nuclei, including a thalamic center, before projecting to a primary area near the somatosensory cortex. A secondary cortical area is located in the anterior temporal lobe.

We can smell substances that are volatile; that is, they can evaporate. Air currents carry the molecules to our nose, where they affect smell receptors. The receptor cells are located in a region of the nasal cavity called the *olfactory epithelium*. Each receptor cell has an extension, called an *olfactory rod*, which goes to the surface of the epithelium. The olfactory rod contains a knob near its end, from which hairlike structures, *olfactory cilia*, protrude. These cilia are most likely the receptor elements. Like the taste receptors, smell receptors have a limited lifespan. They function for about 4–8 weeks. The axons from the smell receptors make up the olfactory nerve, which goes to the olfactory bulb at the front of the brain. The primary route from the olfactory bulb to the cortex is called the *lateral olfactory tract*.

Smell and taste are closely related. You can demonstrate this for yourself by holding your nose and tasting various foods. This interaction between smell and taste explains in part why food tastes "off" when you have a cold. The activity of tasters for products such as alcoholic beverages emphasizes the relation between taste and smell. One company that makes whiskey uses tasters in the sensory evaluation department and the quality control department (Associated Press wire story, 1990a). In the first department, expert tasters oversee the progress of aged whiskey during the 3–5-year aging process. Tasters in the quality control department perform tests during the blending process and after bottling. Although the tasters sample the whiskey in their mouths, their judgments are based primarily on the smell of the whiskey. As one taster says,

"When I am tasting, I really do it by aroma. It's just like the wine tasters. You take it up, you swish it around, you smell, and you can tell almost everything like that. I taste it after that just to sort of reinforce my first opinion" (quoted in Balthazar, 1998).

SUMMARY

Perception involves input from a variety of senses in addition to vision. The sense of hearing provides critical information about many events in the world. We can discriminate several intensities and frequencies, as well as more complex properties of sound. As with vision, the auditory scene is constructed around cues provided by many sensory sources, including timing, intensity, and context. The vestibular sense is closely related to the auditory sense and provides us with information about our orientation and spatial relation to the external environment. The skin senses provide the basis for perception of touch, temperature, and pain. The chemical senses allow us to perceive different tastes and smells.

In Chapters 5 through 7 we have emphasized that perception is constructed. One consequence of this is that we may experience errors in perception if the cues surrounding a stimulus are false or misleading, or if the stimulus within its context is inconsistent with what we expect. It is important, therefore, to display information in ways that minimize perceptual ambiguities and that conform to the expectancies of the observer. In the next chapter, we discuss how information can be displayed to optimize the accuracy of perception.

RECOMMENDED READINGS

Dalton, P. (2002). Olfaction. In H. Pashler & S. Yantis (Eds.), *Stevens' Handbook of Experimental Psychology* (3rd ed.), *Vol. 1: Sensation and Perception* (pp. 691–746). New York: Wiley.

Halpern, B. P. (2002). Taste. In H. Pashler & S. Yantis (Eds.), *Stevens' Handbook of Experimental Psychology* (3rd ed.), *Vol. 1: Sensation and Perception*. (pp. 653–690). New York: Wiley.

Klatzky, R. L., & Lederman, S. J. (2013). Touch. In A. F. Healy & R. W. Proctor (Eds.), *Experimental Psychology* (pp. 152–178). Volume 4 in I. B. Weiner (Editor-in-Chief) *Handbook of Psychology* (2nd ed.). Hoboken, NJ: John Wiley.

Krueger, L. (Ed). (1996). *Pain and Touch. Handbook of Perception and Cognition* (2nd ed.). San Diego, CA: Academic.

Plack, C. J. (2005). *The Sense of Hearing*. Mahwah, NJ: Erlbaum.

Stoffregen, T. A., Draper, M. H., Kennedy, R. S., & Compton, D. (2002). Vestibular adaptation and aftereffects. In K. M. Stanney (Ed.), *Handbook of Virtual Environments: Design, Implementation, and Applications. Human Factors and Ergonomics* (pp. 773–790). Mahwah, NJ: Erlbaum.

Warren, R. M. (1999). *Auditory Perception: A New Analysis and Synthesis*. New York: Cambridge University Press.

Yost, W. A. (2013). Audition. In A. F. Healy & R. W. Proctor (Eds.), *Experimental Psychology* (pp. 120–151). Volume 4 in I. B. Weiner (Editor-in-Chief) *Handbook of Psychology* (2nd ed.). New York: John Wiley.

8 The Display of Visual, Auditory, and Tactual Information

The information was presented in a manner to confuse operators.

Report of the President's Commission on the Accident at Three Mile Island

INTRODUCTION

Information displays are part of the background of everyday life. Animated billboards, plasma television screens, stock tickers, and giant digital signs are common sights in most large cities. From its outset, the discipline of human factors has been concerned with the best way to present information. The most important principle to remember is that the display should convey the intended information in as simple and unambiguous a manner as possible. For a wide range of applications, human factors experts have asked what sensory modality is best (e.g., visual or auditory), how much information is required, and how that information should be coded.

For more complex human–machine interfaces, such as the cockpit of an airplane or the control room of a nuclear power plant, well-designed displays ensure the safe and effective operation of the system. However, display design considerations are equally important in other less complex and critical situations. For example, the increased use of visual display terminals that has accompanied the development of computer workstations and microcomputers has led to concern about the optimal designs for such displays. Instructional labels and the signs used in public facilities can vary in the efficiency with which they communicate vital information to passers-by. New display technologies provide an increasingly broad array of display options, and with each technology, unique human factors issues emerge.

In the present chapter, we examine issues to consider in display design, with particular emphasis on relating design guidelines to the principles of human perception. The chapter focuses primarily on visual and auditory displays, because the vast majority of displays use these senses. Tactual displays are used for limited purposes, such as for controls that must be identified by "feel" and for conveying spatially distributed information to the blind, and smell and taste displays are rarely used.

The issue of display modality (particularly visual or auditory) can often be resolved by considering the message that the display is intended to convey. Is the message long or short? Simple or complex? What action will the receiver of the message be required to take? In what kind of environment will the receiver be acting? Table 8.1 presents general guidelines for determining whether the visual or auditory modality is most appropriate for a particular message. These guidelines are based on the distinct properties of the two senses, as well as on the characteristics of the environment in which the display will be used.

If the environment is noisy, or the auditory system is overburdened by other auditory information, auditory messages may be masked and difficult to perceive. In such situations a visual display will usually be most effective. When the visual field is cluttered, visually displayed information may be difficult to perceive, and so auditory displays may be more appropriate. A visual display must be located in the field of view if it is to be seen, whereas the exact location of an auditory display in relation to the person is usually unimportant. Therefore, the position and movements of the person partially determine the best modality for information presentation.

TABLE 8.1

When to Use Auditory or Visual Displays

Use auditory presentation if:

1. The message is simple
2. The message is short
3. The message will not be referred to later
4. The message deals with events in time
5. The message calls for immediate action
6. The visual system of the person is overburdened.
7. The receiving location is too bright or dark-adaptation integrity is necessary
8. The person's job requires continual motion

Use visual presentation if:

1. The message is complex
2. The message is long
3. The message will be referred to later
4. The message deals with location in space
5. The message calls for immediate action
6. The auditory system of the person is overburdened
7. The receiving location is too noisy
8. The person's job allows remaining in one position

Because we can make spatial discriminations most accurately with vision, spatial information is best conveyed through visual displays. Likewise, because temporal organization is a primary attribute of auditory perception, temporal information is best conveyed through auditory displays. Auditory information must be integrated over time, which provides the basis for the recommendation that auditory messages should be simple, short, and not needed for later operations. Finally, auditory signals attract attention more readily than visual signals and should be used when immediate action is required.

Consider, for example, the problems you would encounter trying to get a message to an auto assembly line worker. Suppose this message says that a metal press is miscalibrated and components coming off the press are misshapen. An auto assembly line is a noisy place, and the assembly line workers, most of whom should be wearing ear protection of some sort, are in constant motion. The miscalibrated press requires immediate action: It must be turned off and recalibrated. From Table 8.1, we can see that some features of this situation can be addressed through a visual display and others through an auditory display. If the auto workers are wearing ear protection, an auditory display will have to be very loud to be perceived. However, because the workers are in constant motion, a spatially fixed message, such as a warning light blinking on the press, may not be detected rapidly enough to prevent faulty components from entering the assembly line. Because the message is short and simple ("recalibrate me") and requires immediate action, an auditory alarm perceptible through ear protection may be the most appropriate way to convey the message.

Many individuals have impaired vision or hearing. Consequently, it may be a good idea to use more than one display modality when possible. For example, you may have encountered crosswalk signals that convey the traditional visual "walk/don't walk" message to sighted pedestrians along with an auditory "chirp" or spoken message "walk sign is on" for the visually impaired.

We can find another example in a technology that permits patients to "hear" drug labels for their prescription medications. If a person is unable to read a pill bottle, she may take the wrong pill at the wrong time, take the wrong dose, or be unaware of facts such as that the drug may cause drowsiness. She may also require the pharmacist to repeat back for her the information that is already on the label. It will be difficult for her to call in requests for prescription refills. To remedy these problems, some prescription labels contain an embedded microchip that converts the information into a speech message generated by a voice synthesizer (Spektor, Nikolic, Lekakh, & Gaynes, 2015). As with most human factors innovations, the talking labels system has several potential benefits, including less time spent by pharmacists on reading-related problems and improved safety for their patients. Moreover, because many visually impaired people are elderly, the labels can help improve the quality of life for these individuals by allowing them to live independently or in assisted-living arrangements rather than in a nursing home.

VISUAL DISPLAYS

One of the first applications of human factors was in the design of aircraft display panels for the military (Green, Self, & Ellifritt, 1995). Engineers devoted substantial effort to determining the optimal arrangement of instruments on the display panel as well as the most effective ways to present the information within each instrument. Considerably more research has been conducted since that early work, resulting in an extensive data base on the optimal design of visual displays. One of the most fundamental distinctions that we can make is between *static* and *dynamic displays*. Static displays are fixed and do not change, like road signs, signs marking exits in buildings, or labels on equipment. Dynamic displays change over time and include such instruments as speedometers, pressure gauges, and altimeters. Displays such as changeable electronic message signs along the highway, on which a series of discrete alphanumeric messages is flashed, fall somewhere in between.

Some displays can render complex system or environmental changes virtually instantaneously—as soon as they are detected. These displays are commonly used to convey complex dynamic patterns of information. For example, though television weather maps used to show only a static depiction of the locations of storms and fronts, now we see dynamic displays that convey the direction and speed with which the storms and fronts are moving. For the operation of complex systems such as process control plants, we can display dynamic information about the system at various levels of abstraction in multiple displays (see Box 8.1).

Dynamic displays are usually much more complex than static displays. However, dynamic displays often have many static features, like the tick marks and digits on the dial of a speedometer. On a weather map, county and state boundaries and town labels are static against the dynamic, moving pattern of an approaching storm front. Consequently, in the discussion to follow, we will present issues involved in the design of first static displays and then dynamic displays.

STATIC DISPLAYS

Effectiveness of Displays

Several factors must be considered when designing a good static display or sign (Helander, 1987). Table 8.2 shows some principles that enhance the effectiveness of visual displays. The first two principles, *conspicuity* and *visibility*, are perhaps the most important. Consider a road sign or a billboard. Conspicuity refers to how well the sign attracts attention, whereas visibility refers to how well the sign can be seen. An inconspicuous and invisible sign is not going to do a very good job of conveying its message.

Conspicuity and visibility will be determined by where a sign is placed, how well it attracts attention, and the environmental conditions in which it is found. For instance, we know that visual acuity and color sensitivity decrease as a stimulus moves out further into the periphery of the visual field (see Chapters 5 and 6). This suggests that we should put a display or sign where people are likely to be looking, or design it to attract attention so that people will be compelled to look at it. Furthermore, if the sign will be placed in conditions of darkness, bright light,

BOX 8.1   ECOLOGICAL INTERFACE DESIGN

Ecological interface design (EID), developed by Vicente and Rasmussen (1992), is a popular approach to designing computer interfaces for complex work domains (such as nuclear power plant control rooms). This approach originates from the fact that, although operators of complex human–machine systems respond for the most part to routine events for which they have acquired considerable skill, on some occasions they must respond to unfamiliar events, in some cases ones that were anticipated and in other cases ones that were not (Torenvliet & Vicente, 2006).

The EID approach relies on two conceptual tools. The first is known as the *abstraction hierarchy* (Rasmussen, 1985). The idea here is that any work domain can be described at different levels of abstraction. For process control, these are (1) the functional purpose of the system, (2) the abstract function of the system (the intended causal structure), (3) the generalized function (the basic functions of the system), (4) the physical function (the components and their interconnections), and (5) the physical form of the system. Under "abnormal" circumstances, the interface should convey the goal structure and relations at these different levels of abstraction, allowing the operator to consider the system at different levels of detail.

The second conceptual tool is Rasmussen's (1983) taxonomy, introduced in Chapter 3, which distinguishes skill-based, rule-based, and knowledge-based modes of behavior. Skill-based behavior is the mode that characterizes a skilled or expert operator engaging in routine activities. Through extensive experience, the operator has acquired highly automatized perception-action procedures that rely primarily on pattern recognition. The skill-based mode thus is less effortful than the other modes, and skilled operators will typically prefer it. An implication of this fact is that an interface should be designed to allow the operator to be in a skill-based behavior mode whenever possible. However, because even skilled operators will need to rely on one of the two higher-level modes of behavior in many cases where they are confronted with complex or novel problems, interfaces should be designed to support these modes as well.

EID consists of three prescriptive principles that are intended to match the display properties with the appropriate level of control. At the skill-based level, interfaces should be designed to take advantage of the operators' highly overlearned procedures by allowing their behavior to be directly guided by low-level perceptual properties of the interface, with the structure of the displayed information matching the structure of the movements that are to be made. In other words, the operator should be able to look at the display and, with little effort, know what it is signaling and what actions to take. The rule-based mode of behavior depends on retrieval of an appropriate rule in response to a cue, which then allows selection of the correct action. Here, the EID principle is to provide a consistent mapping between the constraints of the work domain and the cues provided by the interface.

The knowledge-based mode of cognitive control, with its emphasis on problem solving, is the most effortful and error-prone. The work domain should be represented in the form of an abstraction hierarchy, depicting the processes in the system at different levels, as described above. Designing the interface around a system hierarchy presumably provides the operator with an externalized mental model that supports her problem-solving efforts.

Vicente (2002) evaluated progress in the evaluation and implementation of EID. He concluded that interfaces designed according to the EID principles can improve performance in comparison with those based on more traditional design approaches currently used in industry, but this improvement is primarily for situations involving complex problem solving; that is, those that require knowledge-based behavior. Vicente concluded that evidence indicates that the benefits of EID arise from the functional information provided by the interface in support of higher-level control and the greater reliance on visuospatial displays rather than on textual displays. EID has been successfully applied in a variety of domains, including some nuclear power plant applications, a neonatal intensive care unit, and hypertext information retrieval (Bennett & Flach, 2011; Chery & Vicente, 2006), leading to the identification of new information requirements for interfaces.

inclement weather, and so forth, it will be necessary to ensure that it will be visible under those conditions. For example, a road sign should be as visible during rain and fog, and at night, as it is on a sunny day.

Visibility and conspicuity are important for emergency vehicles. A 2009 U.S. Federal Emergency Management Agency report (FEMA, 2009) emphasized that materials that reflect light back toward the source significantly increase the nighttime conspicuity and visibility of fire engines and other emergency vehicles. The fact that these vehicles move through traffic at high speed makes emergency driving particularly hazardous, and the number of accidents in which fire trucks are involved is disproportionately high, even during the day (Solomon & King, 1997). This high accident rate is in part because of the fact that the color red, which is the color of the majority of fire trucks, is not very visible or conspicuous. While there were few, if any, red vehicles on the road before 1950, allowing a red fire engine to stand out, such is not the case now. Also, our visual systems are very insensitive to the long-wavelength (red) region of the visual spectrum at night and relatively insensitive during daylight, and red cannot be detected very far into the periphery of the visual field (see Chapter 5). Moreover, color blind people have difficulty identifying the color red.

TABLE 8.2

Principles That Enhance the Effectiveness of Visual Displays

Conspicuity

The sign should attract attention and be located where people will be looking. Three main factors determine the amount of attention people devote to a sign: prominence, novelty, and relevance.

Visibility

The sign or the label should be visible under all expected viewing conditions, including day and night viewing, bright sunlight, and so forth.

Legibility

Legibility may be optimized by enhancing the contrast ratio of the characters against the background, and by using type fonts that are easy to read.

Intelligibility

Make clear what the hazard is and what may happen if the warning is ignored. Use as few words as possible, avoiding acronyms and abbreviations. Tell the operator exactly what to do.

Emphasis

The most important words should be emphasized. For example, a sign might emphasize the word "danger" by using larger characters and borderlines.

Standardization

Use standard words and symbols whenever they exist. Although many existing standards may not follow these recommendations, they are usually well established and it might be confusing to introduce new symbols.

Maintainability

Materials must be chosen that resist the aging and wear due to sunlight, rain, cleaning detergents, soil, vandalism, and so forth.

You may have seen lime-yellow emergency vehicles in your community. The human photopic sensitivity function shows that people are maximally sensitive to lime-yellow (see Chapter 5). This means that this color is distinguishable from most backgrounds even in rural areas. Solomon and King analyzed the accident rates for red and lime-yellow fire trucks in the city of Dallas, Texas, in 1997, when they were used in roughly equal numbers. The lime-green trucks were involved in significantly fewer accidents than the red trucks. Because the lime-green color is easier to detect than red, drivers of other vehicles had more time to take evasive actions to avoid the approaching fire truck.

One successful human factors analysis involving visibility and conspicuity is that of the centered, high-mounted brake light required by law in the U.S. on automobiles since the 1986 model year and on light trucks since the 1993 model year. Several studies field-tested different configurations of brake lights on cabs and company vehicles, and showed that rear-end collisions were reduced significantly for vehicles that had the high, central brake light. Vehicles that were involved in rear-end collisions had less damage.

The reduction in accidents and damage occurred because the center-mounted brake light is more conspicuous than brake lights at other locations, because it is located directly in the line of sight (Malone, 1986). Ricardo Martinez, then administrator of the U.S. National Highway Traffic Safety Administration (NHTSA), praised the lights in 1998, saying, "The center high-mounted stop lamp is an excellent example of a device that provides significant safety benefits at a fraction of its cost to consumers" (U. S. Department of Transportation, 1998). By one estimate, the high-mounted, centered brake lights prevent 194,000–239,000 accidents, 58,000–70,000 nonfatal injuries, and $655 million in property damage each year in the U.S. alone (Kahane, 1998).

Conspicuity is also a problem for other kinds of vehicles. As we noted in Chapter 5, motorcycles are not very conspicuous or visible under all driving conditions. Increasing the conspicuity of these vehicles will decrease accidents. For motorcycles, daytime conspicuity is better when the headlamp is on and when the cyclist wears a fluorescent vest and helmet cover (Mitsopoulos-Rubens & Lenné, 2012; Sivak, 1987). Because many other vehicles often travel with running lights, ways to make the motorcycle more conspicuous relative to these other vehicles need to be considered. Cavallo and colleagues proposed that the motorcycle headlamp be colored yellow to make motorcycles stand out (Cavallo & Pinto, 2012; Pinto, Cavallo, & Saint-Pierre, 2014).

Another vehicle that lacks conspicuity at night is the tractor-trailer rig, especially the trailer. The conspicuity of the trailer can be increased by the use of reflectorized materials and running lights. Pedestrians also are not conspicuous under conditions of poor visibility (Langham & Moberly, 2003). Reflectorized materials on shoes and clothing increase the conspicuity of pedestrians at night (Sayer & Mefford, 2004). A lack of conspicuity is also the cause of the high number of fatal accidents for jet skis, because of their relatively small size and unusual patterns of movement, and similar steps can be taken to increase their conspicuity (Milligan & Tennant, 1997).

To determine whether something is more or less conspicuous after changing something like its reflective materials or the color of its lights or paint, we must measure conspicuity. A simple estimate of conspicuity can be obtained by measuring the effects of irrelevant surrounding visual information on a person's ability to perceive or identify an object. The farther a person's gaze can be diverted away from the object while it can still be perceived, the greater is its conspicuity (Wertheim, 2010; Wertheim, Hooge, & Smeets, 2011).

The remaining principles in Table 8.2 deal with more fundamental properties of a display, including what it's made of and how it looks. An important principle is *legibility*, the ease with which the symbols and letters that are present in the display can be discerned. Thus, legibility is closely related to visual acuity (see Chapter 5) and so is influenced by such factors and the stroke width of the lines comprising letters and other unfilled forms (Woodson, Tillman, & Tillman, 1992). Legibility for images on both older cathode ray tube (CRT) displays and thin film transistor liquid crystal displays (LCDs) increases as pixel density increases, because higher pixel density allows higher resolution.

One factor influencing legibility is the contrast ratio between figures on the display and their background (see Chapter 17). As a general rule, the higher the contrast ratio, the better the legibility. Contrast is determined by the amount of light reflected by the figures on the display. The amount of light reflected by red, blue, and green pigments is usually more than is reflected by black and less than by white, so the contrast ratio of black-on-white is highest. This means that black characters against a white background will usually be more legible than when characters are red, blue, or green against a white background.

As with conspicuity, we can increase the legibility of a display in daylight by using fluorescent colors and at night by using reflective materials. These materials are important for maximizing *legibility distance*, the distance at which a person can read the display (Dewar, 2006). For traffic signs and signals, legibility distance has to be great enough that a driver has enough time to read the signs and respond to them. Fluorescent traffic signs are legible at farther distances in daylight than nonfluorescent signs of the same color (Schnell, Bentley, & Hayes, 2001). At night, fully reflectorized license plates are more legible than nonreflectorized plates (Sivak, 1987). There are published guidelines that specify the reflectance values needed for legible highway signs. For instance, fully reflectorized signs should have a figure–ground contrast ratio of 12:1 (Sivak & Olson, 1985).

Readability is another important quality of a visual display, and this quality incorporates the principles of intelligibility, emphasis, and standardization. A readable display allows people to recognize information quickly and accurately, particularly when the display is composed of alphanumeric characters. The message on the display should be simple and direct. Key words, such as WARNING or DANGER, should "pop out" with large letters or a distinct color. The display should use standardized symbols and words, rather than symbols and words that may be unfamiliar or confusing.

The message in the display should be unambiguous, a feature that is related to *intelligibility*. For example, consider the airport sign illustrated in Figure 8.1. In this sign, it is unclear which set of gates goes with which arrow. To eliminate the ambiguity, we would redesign it so that the gates and their arrows are grouped together.

In summary, a good sign will be conspicuous, have legible characters, and convey a readable, interpretable message. For practical purposes, *maintainability* also is important. Signs must be constructed of materials that will withstand soil, mistreatment, and weather, and maintain high levels of conspicuity, legibility, and readability.

Alphanumeric Displays

An alphanumeric display is any display that uses words, letters, or numbers to convey information. Such displays are everywhere, from the buttons in the elevator, to road signs and warning labels, to the text in books, magazines, instruction manuals, newspapers, and documents available on the Web. They are the most widely used and important kind of display we encounter, but they have some drawbacks. For instance, some letters and digits share many features and are easily confused with each other. Also, the use of words or phrases in displays means that the person for whom the sign or display is intended must be literate in the language of the text.

The contrast ratio plays an important role in the legibility and readability of alphanumeric displays, as we mentioned above. Another important role is played by the stroke width of the lines that make up the alphanumeric forms. For black letters on a white background, under good illumination, the optimal stroke width-to-height ratio is from 1:6 to 1:8. For white letters on a black background, it is from 1:8 to 1:10.

Because the contrast ratio is the same for black-on-white and white-on-black displays, it is not obvious why the optimal stroke width-to-height ratio should be different. Thinner lines for white-on-black text are required because, in general, it is more difficult to read white-on-black text than black-on-white text. The difficulty is caused by a phenomenon known as *radiation*, or *sparkle*, in which the details of the white characters tend to "bleed" together.

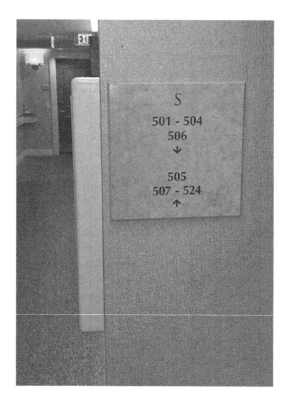

FIGURE 8.1 A sign for which the direction of the rooms on the floor to the left or right is ambiguous.

Another factor that influences legibility and readability is the size of the characters. Smaller characters will usually be more difficult to read than larger characters, but the maximum character size will be limited by the size of the display. The optimal character size will depend on factors such as viewing distance and ambient lighting. We can partly overcome some of the adverse effects of smaller characters by increasing the contrast ratio. Similarly, we can overcome the adverse effects of low contrast ratio by increasing the size of the characters (Snyder & Taylor, 1979).

You encounter printed material in a variety of forms, and these materials may use any of thousands of different type fonts, which may differ in their legibility (Chaparro, Shaikh, & Chaparro, 2006). There are four kinds of fonts: serif, sans serif, script, and those that do not fit into the other three categories. Most of what you read uses a serif or sans serif font, and these kinds of fonts are usually appropriate in many conditions, although some will be more legible than others. Serif fonts, which have few embellishments, typically are used for text. (The text you are reading now is using a serif font.) It may be easier to segregate words with serif fonts, and different letters may be easier to identify (Craig, 1980). However, there is no difference in reading speed for serif and sans serif fonts (Akhmadeeva, Tukhvatullin, & Veytsman, 2012; Arditi & Cho, 2005). When we consider font types for use on CRT and LCD computer monitors, we also need to consider point size, screen resolution, and monitor size (Kingery & Furuta, 1997). A lot of trial and error goes into determining the best way to present characters on these displays.

A font called Clearview was developed specifically to improve the legibility and readability of road signs (Garvey, Pietruch, & Meeker, 1998). The standard road sign font used in the U.S., called Highway Gothic, was established many years ago, before the development of high-reflectance materials that are used for road signs today. Consequently, Highway Gothic suffers from the radiation phenomenon we just discussed: At night, when signs are illuminated by bright headlights, the letters tend to "fill in" and legibility is poor (see Figure 8.2, right column). The newer Clearview font has a narrower stroke width than standard road sign fonts so that problems with radiation are reduced. In addition, whereas most older road signs use only uppercase letters, Clearview uses lowercase letters for all but the first letter. People recognize words better with lowercase letters, because the "word envelope" provides a cue about what the word is that is not available when all letters are uppercase. For example, the word *Blue* has a different global shape than *Bird*, but BLUE has the same shape as BIRD.

The development of Clearview proceeded in a series of stages. Several versions were created, and the final font (see Figure 8.2) was selected after an iterative design process involving field testing and laboratory studies. People recognize words at a 16% greater distance with the Clearview font than with traditional highway fonts, which at 55 miles per hour translates into an additional 2 seconds to read the sign. Two Pennsylvania road signs are shown in Figure 8.2. The sign on the bottom row uses the Clearview font, whereas the sign on the top row uses the traditional font.

Clearview was adopted in the U.S. as an interim highway standard in 2004. However, in January 2016, the U.S. Federal Highway Administration returned to Highway Gothic as the only approved font (Capps, 2016). The stated reason for the reversal was that research has not confirmed the benefit of the Clearview font for positive contrast signs and, more importantly, has shown that the font is inferior to Highway Gothic for negative contrast (black-on-white) signs at night (Holick, Chrysler, Park, & Carlson, 2006).

There are four basic characteristics of alphanumeric display formats that influence the ability of an observer to read or interpret the display: overall density, local density, grouping, and layout complexity (Tullis, 1983). Overall display density is the number of characters shown over the total area of the display (compare Figure 8.3a with Figures 8.3b,c). Local density is the density in the region immediately surrounding a character (compare Figure 8.3b with Figure 8.3c). Grouping is related to the Gestalt organizational principles we discussed in Chapter 6 (see Figure 8.3d). Layout complexity is the extent to which the layout is predictable.

Tullis (1986) developed computer-based methods of analysis to aid in the quantitative evaluation of alternative display formats. He concluded that, for best readability, overall display density should be as low as possible, with local density at an intermediate level. This reduces lateral masking between display characters and increases the ease with which a reader can locate information in the display. Grouping display elements will improve readability as long as the groups are appropriate, but there is a tradeoff between grouping and layout complexity. More groups mean higher complexity, and increased layout complexity could mean decreased readability.

One experiment looked at the effects of grouping and complexity for graphical user interface screens (Parush, Nadir, & Shtub, 1998). They showed computer users a dialogue box that required them to select an appropriate action from among several alternatives. The alternatives were grouped with frames (see Figure 6.7 in Chapter 6). At the same time, they varied complexity by aligning the alternatives differently within each frame. Dialogue boxes with grouping frames and lower

FIGURE 8.2 Irradiation of Standard Highway font versus Clearview font.

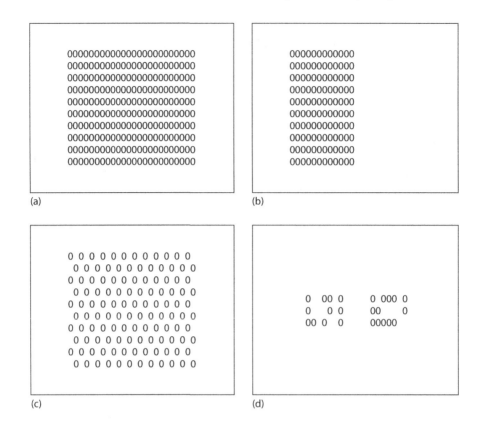

FIGURE 8.3 Four examples of different display densities and grouping: (a) overall density = 100%, local density = 81%; (b) overall density = 50%, local density = 72%; (c) overall density = 50%, local density = 39%; and (d) grouping into two sets.

complexity (where alternatives within a frame were aligned) produced faster performance than those without grouping frames and higher complexity.

While an alphanumeric display is often a sign or other static representation, we can also use electronic alphanumeric displays to present information that may change over time. Such displays might be used in industrial applications as machine status displays and message boards. There are many dynamic alphanumeric displays in cars, such as electronic clocks and radio-station settings; even the speedometers in some models are dynamic alphanumeric displays. Electronic LCDs need a light source to form letters and symbols. This means that the perceptibility of electronic displays will depend on the ambient lighting level.

Consider two examples of automatic luminance control. The first is the luminance of the electronic displays on the dashboard of a car (such as the numbers on the electronic clock). In many cars, when you turn the headlights on, these displays will dim. This design feature assumes that you will not turn on the headlights unless it is dark. If it is dark, your vision will be dark adapted and so you will be more sensitive to light than you would be in full daylight. The change in the intensity of the clock display as a function of whether the headlights are on is a very simple display adjustment. A second example of automatic luminance control is used by your smartphone. Many smartphones have a light sensor in the bezel next to the screen. When the ambient light intensity is low, the display luminance will be low. When the ambient light intensity is high, the display luminance will increase to compensate.

Automatic luminance controls are also found in some airplane cockpits (Gallimore & Stouffer, 2001). Many cockpit instruments are combined into single electronic multifunction displays. The dynamic flight environment means that different displays will require different settings, and these requirements may vary over time. The pilot cannot be expected to waste time and effort manually adjusting the luminance settings of each of the displays. In fact, to avoid such efforts, pilots will often set the luminance of the displays to the maximum value and leave them there. This practice may create problems with display visibility and legibility. Commercial aircraft now have automatic luminance control systems, but military aircraft do not. In military aircraft, there are many more variables that determine the optimal display luminance, including the type of mission, different kinds of head-up displays (HUDs), and the use of night-vision goggles.

Symbolic Displays

Symbols, sometimes called *pictographs*, are often effective for conveying information (Wogalter, Silver, Leonard, & Zaikina, 2006). They are most useful for concrete objects that can be easily drawn. It is more difficult to develop an effective symbol for abstract or complex concepts. For instance, think about how you might design a symbol indicating "exit" without using the word exit or any other text. Because effective symbols directly depict the concepts that they represent, a person does not need to know any particular language to understand the message. Hence, symbolic displays are used extensively in facilities such as airports and train stations, where many travelers may not be familiar with the local language. For the same reasons, manufacturers of exported products prefer to use symbols to depict usage instructions and warning messages.

A symbolic display must be identifiable and understandable if it is to be effective. People must reliably be able to recognize the depicted object or concept and to determine the referent of the sign. A Canadian study investigated how well people could interpret road signs that used pictograms (Smiley, MacGregor, Dewar, & Blamey, 1998). They asked people to read highway tourist signs from highways in Ontario, Canada, but they only gave them as much time to read them as a driver traveling at 50 miles per hour (80 km/h) would have. Then they asked them to interpret the signs. The pictographs on the signs increased the number of errors people made, because there were several pictographs that they couldn't understand.

Even if a person can recognize the concept depicted by a pictograph, there is no guarantee that she will be able to comprehend the display's message. One study examined the information symbols for different services within a hospital, such as orthopedics, dentistry, and so on (Zwaga, 1989). While people could easily recognize certain symbols, they misunderstood their referents. Figure 8.4 shows a symbol used to designate the orthopedics clinic. Although the symbol can be recognized as a leg in a plaster cast, most people misinterpreted the referent to be the "plaster room." In contrast, everyone recognized and comprehended the tooth symbol for dentistry shown in Figure 8.5.

One example of the use of symbols and pictographs involves the process of screening potential blood donors who may have come into contact with the HIV virus that causes acquired immune deficiency syndrome (AIDS). In developing pamphlets to communicate to someone whether they fall into the high-risk category and therefore should not donate blood, Wicklund and Loring (1990) proposed that the information be portrayed symbolically to reach people of low literacy. The concepts that need to be communicated, such as "do not give blood if you are a man who has had sex with another man even once since 1977," are very abstract. Consequently, Wicklund and Loring examined the effectiveness with which the intended message could be communicated with alternative symbol designs. Figure 8.6 shows the designs that they evaluated for this concept. Of these symbols, the only one that was rated as very effective is D.

The symbols in Figure 8.7 are abstract pictographs. For some displays, such as pamphlets, we can use representational pictographs that involve more detailed line drawings. Representational pictographs are less ambiguous than abstract pictographs. Wicklund and Loring concluded that information about high-risk behavior is conveyed best by representational pictographs that show interpersonal relationships unambiguously, accompanied by a limited amount of text (see Figure 8.7).

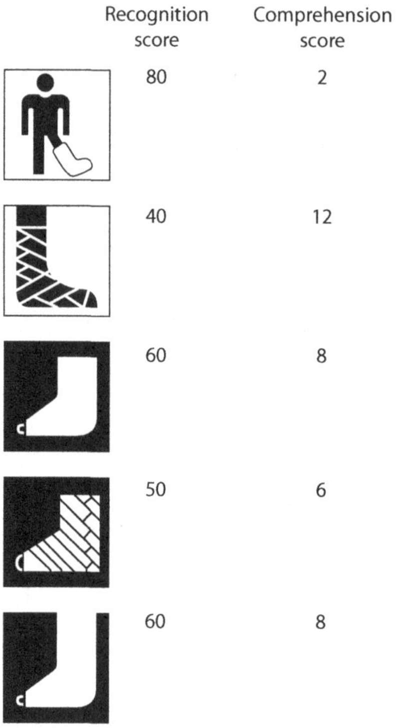

	Recognition score	Comprehension score
	80	2
	40	12
	60	8
	50	6
	60	8

FIGURE 8.4   Recognition and comprehension scores for orthopedics symbols.

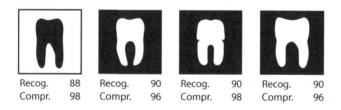

Recog. 88	Recog. 90	Recog. 90	Recog. 90
Compr. 98	Compr. 96	Compr. 98	Compr. 96

FIGURE 8.5   Recognition and comprehension scores for dentistry symbols.

As we have now seen in several contexts, the speed and accuracy with which people can identify symbolic displays are influenced by Gestalt organizational principles. Easterby (1967, 1970) provided examples of how symbolic codes can be made more easily interpretable by designing them to be consistent with general organizational principles such as figure–ground, symmetry, closure, and continuity (see Figure 8.8). A clear figure–ground distinction helps eliminate ambiguity about the important elements of the display. Simple and symmetric symbols will enhance readability. Closed, solid figures are easier to interpret than (potentially) more complex open figures. Figure contours should be smooth and continuous, unless discontinuity contributes to the information that is to be conveyed. Easterby's examples illustrate that subtle changes in display design can affect the way in which the display is organized perceptually and, hence, the overall effectiveness of the display.

One issue that confronts the human factors specialist is whether to use an alphanumeric display or a symbolic display. For example, should highway signs be verbal, or should they

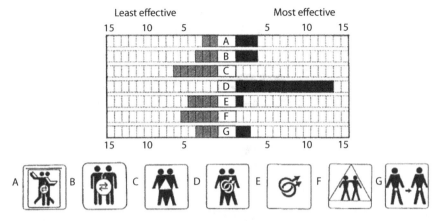

FIGURE 8.6 Effectiveness ratings for alternative pictographs indicating the same concept.

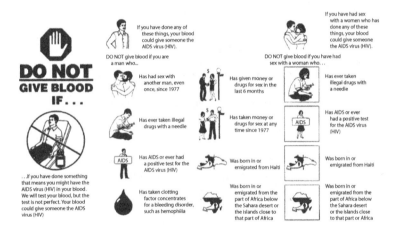

FIGURE 8.7 Prototype AIDS prevention pamphlet for men.

be symbolic? An advantage of verbal signs is that reading is a highly overlearned process for literate people fluent with the language to be used, and no new relationships between symbols and concepts need to be learned. However, there are some disadvantages, including the amount of cognitive effort required to interpret the display. Because symbolic displays can depict the intended information directly, less processing should be required than for verbal displays.

It is generally true that people can interpret signs faster when they are symbolic than when they are verbal (Ells & Dewar, 1979). This difference is most pronounced under degraded viewing conditions. When displays are difficult to see, people are much faster with symbolic displays. This may be because verbal messages require more complex visual patterns that will be less legible and readable under poor viewing conditions. Legibility and readability are not critical for symbolic codes.

In some situations, a message may be conveyed using both symbols and words. One study asked drivers to rate the urgency of a message displayed on a computer by symbols alone, the action to

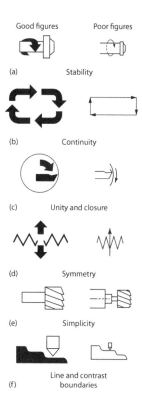

FIGURE 8.8 Principles of figure/ground stability, continuity, figure unity and closure, symmetry, simplicity, and line and contrast boundaries for symbolic codes.

be taken (e.g., "coolant low" or "fill up soon"), or both (Baber & Wankling, 1992). The drivers rated the symbol plus action message as having the highest urgency, which suggests that additional text information with symbols for in-car warnings may be most effective for getting the drivers' attention.

Coding Dimensions

Some information may be conveyed in ways that are neither verbal nor pictorial. We can sometimes arbitrarily assign different display features to code objects or concepts. Such codes can be based on alphanumeric forms, nonalphanumeric forms, colors, sizes, flash rates, and any of a number of other different dimensions. Road signs in the U.S. use color to convey information: green signs are informational, brown signs indicate sites of historic or recreational interest, blue signs signal the availability of services like hotels and gas stations, yellow signs are warnings, and white signs are regulatory. Although the appropriateness of a specific coding dimension depends on the particular task, we can provide some general guidelines (see Table 8.3).

Recall that an absolute judgment refers to the classification of a stimulus when several options are available (e.g., deciding that a signal is "high" when the options are high, medium, or low). If a stimulus varies along a single dimension (e.g., pitch or hue), the number of such stimuli between which people will be able to discriminate reliably is limited to between five and seven (see Chapter 4). Thus, we will need to keep the number of values on a particular coding dimension small if we require people to make absolute judgments about them. The number of items that people can accurately distinguish is greater for multidimensional stimuli or when people can make relative judgments (or compare one item directly with another).

TABLE 8.3

Comparison of Coding Methods

Code	Number of Code Steps[a]		Evaluation	Comments
	Maximum	Recommended		
Color				
Lights	10	3	Good	Location time short. Little space required. Good for qualitative coding. Larger alphabets can be achieved by combining saturation and brightness with the color code. Ambient illumination not a critical factor.
Surfaces	50	9	Good	Same as above, except ambient illumination must be controlled. Has broad application.
Shapes				
Numerals and letters	Unlimited		Fair	Location time longer than for color or pictorial shapes. Requires good resolution. Useful for quantitative and qualitative coding. Certain symbols easily confused.
Geometric	15	5	Fair	Memory required to decode. Requires good resolution.
Pictorial	30	10	Good	Allows direct association for decoding. Requires good resolution. Good for qualitative coding only.
Magnitude				
Area	6	3	Fair	Requires large symbol space. Location time good.
Length	6	3	Fair	Requires large symbol space. Good for limited applications.
Brightness	4	2	Poor	Interferes with other signals.
Visual number	6	4	Fair	Ambient illumination must be controlled.
Frequency	4	2	Poor	Requires large symbol space. Limited application. Distracting. Has merit when attention is demanded.
Stereo depth	4	2	Poor	Limited population of users. Difficult to instrument.
Angle of inclination	24	12	Good	Good for limited application. Recommended for quantitative code only.
Compound codes	Unlimited		Good	Provides for large alphabets for complex information. Allows compounding of qualitative and quantitative codes.

[a] The maximum number assumes a high training and use level of the code. Also, a 5% error in decoding must be expected. The recommended number assumes operational conditions and a need for high accuracy.

Color Coding

Color coding of information, like U.S. road signs, can be very effective (Christ, 1975), particularly when the color for a particular object or concept is unique. Consider a person's medications as an example. For people who require different eye-drop medications, confusing them may have serious health consequences. To avoid confusion, the American Academy of Ophthalmology adopted a uniform color-coding system for eye-drop dispenser caps. In this system, the caps are tan for anti-infectives, pink for anti-inflammatories/steroids, yellow for beta-blockers, and so on. The uniform code parallels the red for stop, green for go, and yellow for caution codes of traffic lights, so the dispenser caps convey the amount of risk or hazard associated with the medication within the bottle (Trudel, Murray, Kim, & Chen, 2015).

When a task requires searching for items or counting the number of items of a given type, the benefit of color coding by type increases as the display density increases. Imagine, for example, trying to determine how many apples there are in a large basket of apples and oranges. This is a much easier task than determining how many tangerines there are in a large basket of tangerines and oranges. The relation between color and display density holds because the time to search for one colored item (an apple) among those of other colors (oranges) is unaffected by the number of items of the other color, as long as all of them can be seen at once (see Chapter 6).

Color coding is an important tool that helps people read maps. In particular, color can distinguish between different levels and kinds of information. Yeh and Wickens (2001) asked people to answer questions about a battle that required them to use different kinds of information from an electronic map, such as where tanks were located or where ground troops were deployed. Using different colors to classify different kinds of information (e.g., people in red, terrain in green, roads in blue, and so on) enhanced people's abilities to access the information they needed. They were better able to segregate different parts of the map display, extract the information relevant to their task, and ignore the clutter caused by other information.

Shape Coding

Shape is a particularly valuable way of representing information, because people can distinguish between a very large number of geometric shapes. Shapes are not constrained by the rule that we use no more than seven different stimuli, because they can vary along more than one dimension (e.g., area, height, and width). However, some shapes are more discriminable than others, so we have to be careful about the shapes that we use. For instance, circles and triangles are more easily discriminated than circles and ellipses.

Shape discriminability is influenced by several factors (Easterby, 1970). Triangles and ellipses are best discriminated by their areas, and rectangles and diamonds by their maximum dimensions (e.g., height or width). More complicated shapes, such as stars and crosses, are best discriminated by their perimeters. Other coding dimensions, such as the size of forms, number of forms, angle of inclination, and brightness, have more limited uses (see Grether & Baker, 1972).

Combination Codes

We have a lot of data about different types of coding, combinations of codes, and the circumstances in which each will be most effective. Figure 8.9 shows five codes, including shape, configuration, and color, used in one study to represent information in various sections of a map (Hitt, 1961). People scanned the display and identified, localized, counted, compared, or verified the locations of different targets. Figure 8.10 presents the number of correct responses they made per minute as a function of the code symbol used. People performed best with numeral and color codes, but configuration codes were not as effective. However, the performance differences between code sets disappear with practice (Christ & Corso, 1983). This means that if your goal is to improve people's long-term performance, the code sets you choose probably will not matter in the long run.

DYNAMIC DISPLAYS

Analog and Digital Displays

For dynamic displays, information is conveyed by movement within the display. That is, the operator must be able to perceive changes in the state of the system as the display changes. Figure 8.11 shows several types of dynamic displays. These displays are either analog or digital. Digital displays present information in alphanumeric form. Analog displays have a continuous scale and a pointer. The position of the pointer indicates the momentary value on the scale.

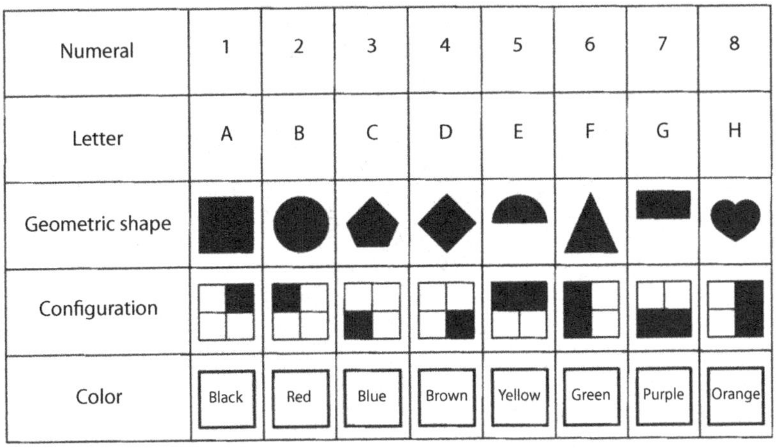

Numeral	1	2	3	4	5	6	7	8
Letter	A	B	C	D	E	F	G	H
Geometric shape	■	●	⬟	◆	⌓	▲	▬	♥
Configuration								
Color	Black	Red	Blue	Brown	Yellow	Green	Purple	Orange

FIGURE 8.9 Code symbols used in mapping used by Hitt (1961).

There are two ways that analog displays can be designed, and these determine the behavior of the scale and the pointer, and the shape of the display. A display can have a moving pointer and a fixed scale, or a fixed pointer and a moving scale. The speedometer in most cars has a moving pointer and a fixed scale. The needle moves against a fixed background of numerals. In contrast, most bathroom scales have a fixed pointer and a moving scale. The numeric dial turns under a fixed pointer. The shape of the display can be circular (like a speedometer), linear (like a thermometer), or semicircular (like a voltmeter).

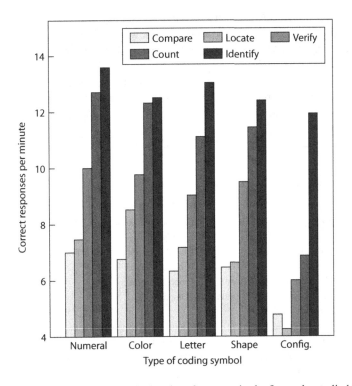

FIGURE 8.10 Relation between coding method and performance in the five tasks studied by Hitt (1961).

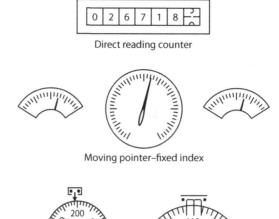

FIGURE 8.11 Digital, moving pointer, and fixed pointer dynamic displays.

One major issue in the design of a dynamic display is whether the display should be analog or digital (see Table 8.4). The best display type will not be the same in all situations, because analog and digital displays differ in how efficiently they can convey different kinds of information. Digital displays convey exact numerical values well. However, they are difficult to read when measurements are changing rapidly. Also, it is harder to see trends in the measurements, such as whether your car is accelerating or decelerating. Analog displays convey spatial information and trends efficiently but do not provide precise values. From these properties, we can determine a general rule for digital displays: they are more appropriate for devices such as clocks or thermometers in which the measurements are not changing rapidly.

But even this simple guideline depends on the kind of response that a person needs to make to the display. For example, people can report clock time much faster with digital displays than with analog clock faces (Miller & Penningroth, 1997). However, this is only true for terse responses ("two

---

TABLE 8.4

Choice of Display Indicator as a Function of Task

Use of Display	Type of Task	Display Typically Used for	Type of Display Preferred
Quantitative reading	Exact numerical value	Time from a clock, rpm from tachometer	Counter
Qualitative reading	Trend, rate of change	Rising temperature, ship off course	Moving pointer
Check reading	Verifying numerical value	Process control	Moving pointer
Setting to desired value	Setting target bearing, setting course	Compass	Counter or moving pointer
Tracking	Continuous adjustment of desired value	Following moving target with cross hair	Moving pointer
Spatial orientation	Judging position and movement	Navigation aids	Moving pointer or moving scale

thirty-seven") or responses given as minutes after the hour ("thirty-seven minutes after two"). For responses of minutes before the hour ("twenty-three minutes before three"), there is no difference in how long it takes to read the two display types, because people have to compute the time remaining in the hour by subtracting the minutes in the digital display. So even a task that requires reading precise values will not have an obvious "best" display type, and a designer must consider how the displayed information maps onto the task requirements.

Analog displays can also be representational (Hegarty, 2011). This means that rather than a scale and pointer, the display presents a direct depiction of the system state. Tasks that require spatial processing often benefit from the use of analog, representational displays. Schwartz and Howell (1985) conducted a simulated hurricane-tracking task in which they presented historical information about the previous and current positions of a hurricane to observers. Observers watched the display until they were ready to make a decision about whether the hurricane would hit a city. The observers made earlier and better decisions when the position of the hurricane was displayed graphically in a representational, analog display rather than numerically, particularly when their decisions had to be made under time pressure.

Moving pointer–fixed scale displays are very common and will usually be easiest for people to use. This is in part because the stationarity of the scale markers and labels makes them easier to read. When the display shows changes in the system that occur in direct response to an operator's manual control movements, a moving pointer display will often be the most appropriate. The choice between circular and linear displays may be arbitrary; there is little difference in the ease with which circular and linear arrays can be read (Adams, 1967). However, circular arrays do not require as much space as linear arrays, and they are simpler to construct.

There are some other issues that we must consider when designing an analog display. Labels or symbols used to mark the scales must be legible. We have to make decisions about the scale units and how to mark them, as well as the type of pointer to use. Scale progressions are easier to read if they are marked in factors of 10 (e.g., 10, 20, 30, ...; 100, 200, 300, ...) than by some other values (e.g., 1, 7, 13; Whitehurst, 1982). On a unit scale, major markers can indicate each multiple of ten units (10, 20, 30, ...), with minor markers designating each single unit. The major markers should be distinct, and we commonly do this by making them longer or wider than the minor markers. If we know that people will be using the display under low illumination, the markers will need to be wider than if they are using the display under normal illumination. The increased width compensates in part for the decreased acuity of the operator under scotopic viewing conditions. The tip of the pointer should meet the smallest scale markers, and we must angle or color it so that people will not confuse its tip with the marker to which it points.

Display Arrangements

In some situations, a display panel will be a complex arrangement of many dials and signal lights. In such situations, the human factors specialist needs to be sensitive not only to the factors that influence the perceptibility of information within each of the individual dials, but also to the overall organization of the display. As we discussed in Chapter 6, Gestalt organizational principles can be used to group dials with related functions.

Another factor that the human factors specialist must consider is the rapid decrease in visual acuity outside of the fovea. This means that the operator can only see a small region of the display panel clearly at any point in time. One design principle that arises from this limitation is that of *frequency of use*. We must locate the most frequently used and important displays close to central vision under normal viewing conditions.

The limited acuity across much of the retina means that eye and head movements are required to see several displays clearly. Because eye movements take time, the farther apart two displays are located, the longer it will take a person to redirect his gaze from one display to another. Thus, a second design principle is to locate displays according to their *sequence of use*. That is, if there are fixed sequences in which people must scan displays, we should arrange the displays in that sequence. Even when there is no fixed sequence, different displays usually have different functions, and so we should group them according to these functions.

A technique we can use to assist in the design of display configurations is *link analysis* (Stanton et al., 2013). A link is a connection between a pair of items, in this case display elements, indicating a certain relation between them. For display configurations, links represent the percentage of eye movements shifting from one display to another. We should design display configurations so that the distance between displays with high-value links is shorter than the distance between displays with low-value links. Also, we should locate displays that are examined most frequently close to the line of sight.

There are four steps in a link analysis of display arrangements (Cullinane, 1977). First, we must prepare a diagram that shows the interactions between the display components. Second, we must examine all relations between the displays and establish link values in terms of the frequency of eye movements between the displays. Third, we develop an initial link diagram in which the displays are rearranged so that the most frequently used displays are located in close proximity in the central visual field. Finally, we refine the diagram we created in the first step to make the final layout. There is a computer application for performing link analysis that incorporates these four steps and allows easy application of link analysis to systems with many elements and links (Zhao, Hignett, & Mansfield, 2014).

Link analysis has been around for a long time, but it still is an important step in determining how display panels should be arranged. Fitts, Jones, and Milton (1950) performed a link analysis of the scanning patterns of pilots during aircraft instrument landings. They recorded the eye movements of each pilot during approaches to the runway using the standard instrument arrangement shown in Figure 8.12. The highest link value (29% of all eye movements) was between the cross-pointer altitude indicator and the directional gyro. They also found that the number of fixations per minute was greatest for these two display elements. Thus, an improved display arrangement would place the cross-pointer and directional gyro adjacent to each other in the central part of the panel.

Later, Dingus (1995) used link analysis to evaluate the impact of navigation aids (such as global positioning system (GPS) map displays) on the eye-scanning behavior of drivers. He found that under all driving conditions, drivers spent a constant, relatively small percentage of time devoted to scanning instruments, mirrors, and signs/landmarks. Attending to navigation aids therefore reduced the time drivers were able to devote to forward, left, and right roadway scanning, which may increase the likelihood of collisions. Voice displays reduced the visual attention demands of

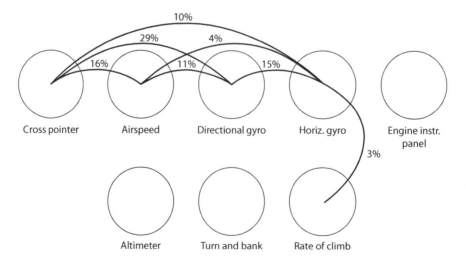

FIGURE 8.12 Links among dials in an airplane control panel.

the navigation aid and allowed the driver more time to scan the roadway than when the display was visual. Also, the drivers spent less time scanning the navigation aid and more time scanning the roadway locations when the aid automatically planned the route, but more time scanning the aid when it required the drivers to plan the route while en route to their destination.

Another application of link analysis is to analyze work environments for ambulances and hospitals: One study found that access to equipment and consumables for paramedics from their preferred seat in an ambulance was suboptimal (Ferreira & Hignett, 2005), and another used link analysis to show that the proposed design of a clinical healthcare department did not provide the optimal layout for efficient clinical activities by the staff (Lu & Hignett, 2009). All of the studies in this section indicate that link analysis can provide insight into the way that new tasks or possible system configurations will impact operator behavior.

Motion Interpretability

Many vehicles use representational displays to convey information about the movement of the vehicle as it is controlled by its driver. In such situations, what is the best way to represent the vehicle's motion? That is, what frame of reference should the display use? Should it portray the external world as moving around a stationary vehicle, or should it portray the vehicle as moving through a stationary world? This issue arises with the attitude displays used in aircraft, which indicate the orientation of the plane with respect to the horizon.

Figure 8.13 depicts three possible types of attitude displays. The inside-out display shows the plane's attitude by changing the line that marks the horizon. In other words, the horizon marker corresponds to the orientation of the actual horizon that the pilot would see while looking out. In contrast, the outside-in display holds the horizon constant and varies the tilt of the aircraft indicator. This display portrays the attitude of the plane that an observer from the outside would see.

The inside-out display has the advantage that it is compatible with the view seen by the pilot; the disadvantage is that it is incompatible with the control action that should be taken to return the aircraft to level (see Chapter 13). That is, it might look as if the pilot should turn the control counterclockwise to bring the horizon line to horizontal, when in fact she should do the opposite. When a display and an action are not well matched, as in this case, we say that they have poor compatibility. However, while the outside-in display has good display-control compatibility, it does not correspond with the view of the world that the pilot sees. Which display is best?

Cohen, Otakeno, Previc, and Ercoline (2001) compared performance with inside-out and outside-in attitude displays in helmet-mounted displays (see later) for pilots and nonpilots. Nonpilots performed better on a simple navigational task with the display-control compatible outside-in display than with the inside-out display. In contrast, the pilots performed equally well with both displays, but they expressed a preference for the more common inside-out display. This difference between pilots and nonpilots suggests that the pilots' experience with the inside-out display allowed them to adapt, at least to some extent, to its less compatible display-control relation.

Another type of display, the frequency-separated display, combines the advantages of the inside-out and outside-in displays (Beringer, Williges, & Roscoe, 1975). This display acts like an inside-out

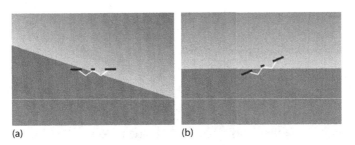

(a)   (b)

FIGURE 8.13   (a) Inside-out and (b) outside-in attitude displays.

display when the pilot does not often adjust the plane's attitude, but it changes to an outside-in display when the pilot makes frequent adjustments. Thus, when the pilot makes rapid control actions, the display is compatible with these actions and reduces the number of response reversals; when the pilot is not making such actions, the display corresponds to the pilot's view of the world.

Professional pilots performed better with the inside-out and frequency-separated displays than with the outside-in display when engaged in recovery from unknown attitudes (during which the frequency-separated display acts like an inside-out display; Beringer et al., 1975). However, when the pilot had to respond to a change in attitude (position of the aircraft relative to the horizon) during tracking, their performance was better with a frequency-separated display.

The extent of a pilot's experience with these different kinds of displays is important. Under the reunification of Germany in 1990, the German Air Force merged equipment from the former West and East Germany (Pongratz, Vaic, Reinecke, Ercoline, & Cohen, 1999). West German pilots in the Air Force were used to flying jets with inside-out display indicators. However, the East German Soviet-constructed MIG-29 jets used a mixed outside-in display representation for bank and inside-out display representation for pitch. The lack of familiarity of some pilots with the East German displays resulted in problems of spatial disorientation when they flew the MIG-29 in adverse conditions, including conditions of low visibility and high gravity forces. Simulator training with the new displays is one way to deal with this problem.

OTHER DISPLAYS

Head-up Displays

Military pilots often fly under very different conditions from commercial airline pilots. Military pilots may occasionally be required to take evasive action, fly in close formation with other planes, or engage another plane in aerial combat. Sometimes, then, the few seconds that a military pilot spends looking at his or her instrument panel instead of the scene outside the cockpit windshield may have life or death consequences. To allow the pilot to continuously monitor events as they unfold outside the cockpit, the military developed the HUD. An HUD is a virtual imaging display of collimated light images projected onto the windshield in front of the pilot (see Figure 8.14; Crawford & Neal, 2006). HUDs were introduced into fighter aircraft in the 1960s, and by the 1970s, all fighter aircraft in the U.S. were equipped with HUDs. Since then, HUDs have been installed on some commercial aircraft, automobiles, and video games (Caroux & Isbister, 2016).

FIGURE 8.14 A head-up display.

The HUD display is intended to minimize eye movements, shifts of attention, and changes of accommodation and vergence during flight navigation. Because the display is superimposed on the view through the windshield, the pilot can monitor critical information from the display along with the visual information from outside during rapid or delicate maneuvers. The collimated light image is intended to allow the pilot's accommodation to remain at optical infinity, as if the pilot's gaze were fixated on a distant object like the ones outside the cockpit window.

HUDs offer numerous advantages over panel-mounted displays (Ercoline, 2000), the majority of which are due to the reduced number of attentional shifts the pilot needs to make to navigate the aircraft. Some HUDs incorporate flight-path markers and acceleration cues. A flight-path marker is a representation of the projection of the aircraft though space that allows the pilot to directly see where the aircraft is heading. An acceleration cue allows the pilot to immediately detect sudden changes in airspeed. These features have significantly improved pilot performance during delicate maneuvers. For example, a study conducted with pilots in a flight simulator showed that under conditions of restricted visibility, an HUD with a flight-path marker reduced lateral error during landing (Goteman, Smith, & Dekker, 2007).

Unfortunately, there are several drawbacks to HUDs. Most prominent among these is the fact that pilots become (spatially) disoriented more frequently when using a HUD than when using traditional instrumentation. Pilots occasionally report being unable to determine the position of the plane relative to the earth. This can have deadly consequences: between 1980 and 1985, the U.S. Air Force lost 54 planes in clear-weather "controlled flights into the terrain" (crashes) when the pilots were using HUDs (McNaughton, 1985). One factor contributing to this problem is that the display elements move rapidly during turbulence. Another factor is that HUD attitude information is easy to misinterpret. However, many of these crashes may be due to problems of accommodation (Roscoe, 1987).

The use of a collimated virtual image does not guarantee that the pilot's eyes are accommodated for distant viewing at optical infinity. In fact, many pilots tend to fixate at a distance about an arm's length away when they view the objects in an HUD, a point closer to dark accommodation than optical infinity (see Chapter 5). This *positive misaccommodation* causes objects in the visual field to appear smaller and more distant than they actually are. This in turn causes distant objects to appear more distant and items below the line of sight, such as the runway, to appear higher than they really are.

Another problem is visual clutter. Because all HUD display elements are the same color (green), the displays rely more on alphanumeric codes, and so too many symbols appear in the displays. These symbols may occlude important visual information outside the cockpit. Even when clutter is minimized, when a pilot's attention is focused on the HUD, he or she may fail to observe critical events, such as an aircraft moving onto the runway on which the plane is landing (e.g., Foyle, Dowell, & Hooey, 2001). As we will see in Chapter 9, this phenomenon of "inattentional blindness" is relatively common (Simons, 2000).

Despite the problems involved with HUDs, they are useful tools. Although pilots report a tendency toward disorientation when using such displays, they do not consider the problem sufficiently severe to warrant discontinuing their use (Newman, 1987). In fact, the tendency toward disorientation decreases with training and/or better integration of the HUD into the cockpit. Ercoline (2000) emphasizes that most, though not all, of the problems with HUDs are ones that can be corrected.

Although HUDs have been used primarily in aircraft, they began to be implemented in automobiles in the 1980s (Dellis, 1988). Automotive designers anticipate that with HUDs, drivers will be able to maintain their eyes on the road more than with dashboard displays, and their need to shift accommodation from near to far distances will be reduced. However, Tufano (1997) notes that the potential risks associated with HUD use in automobiles have not received sufficient consideration.

Some studies showed that drivers are no faster at reading information from an HUD when the display's focal distance increases beyond about 2 m (about the distance of the edge of the hood). Accordingly, designers have adopted the 2 m focal distance rather than focusing the HUD images at optical infinity. Tufano argues that this will exacerbate problems of positive misaccommodation of the type that occur in aircraft displays. A focal distance of 2 m will demand that the driver's eyes be focused at a closer distance than the real objects outside of the car, leading to misjudgments of object size and distance. "Cognitive capture," a phenomenon in which the driver's attention is inappropriately directed to the HUD, may also lead to the driver's failure to respond to unexpected obstacles. This is especially true for objects that appear in the visual periphery, which are harder to detect when more than four symbols are visible on the HUD (Burnett & Donkor, 2012).

Helmet-Mounted Displays

Helmet-mounted displays (HMDs) serve purposes similar to those of HUDs. The displays are used to present alphanumeric, scenic, and symbolic images that enhance the capability of pilots flying military aircraft. As with HUDs, they allow the pilot to obtain critical flight information without taking his or her attention from the scene outside of the aircraft (Houck, 1991). One of the principal benefits of HMDs in comparison to HUDs is that the pilot can be looking outside in any direction and still see the images on the display. The primary barriers to the use of HMDs have been their excessive weight and bulk. However, the development of miniature CRTs and micro LCDs, together with improved graphics processing, has made the use of HMDs more practical (see Figure 8.15). Today's HMDs can weigh less than a pound, but the total weight depends on their capabilities.

With an HMD, the image on a miniature CRT or LCD is reflected off a beam splitter into the eye. The display can be provided only to one eye, leaving the other eye with an unobstructed view outside the aircraft, or by using a transparent projection system, it can be provided to both eyes. It can provide the pilot with cues to help determine the flight path of his/her own aircraft as well as the flight path of an adversary.

Single-view HMDs are often used with thermal imaging systems (Rash, Verona, & Crowley, 1990). Thermal systems contain sensors that detect infrared radiation emitted by objects in the field of view. They can assist the pilot's ability to perform effectively at night and during adverse weather. A helmet-mounted thermal imaging system is used on the Apache AH-64 attack helicopter. For the pilot, a sensor mounted on the nose of the aircraft provides an image of the external environment. This is coupled with displays indicating speed, heading, altitude, and so on.

Several special problems arise from the single-view HMD. The field of view can be reduced to as little as 20°, although some models allow up to 50°. This restricted field of view limits performance and requires the pilot to make more head movements than would be necessary without the display. These head movements lead to problems of disorientation at night and, because the helmet can be heavy, will produce increased muscular fatigue and persisting neck pain (Ang & Harms-Ringdahl, 2006).

Training can reduce the problems associated with increased head movements. Seagull and Gopher (1997) showed that helicopter pilots who received training to produce head movements while performing a visual scanning task in a simulator did better on subsequent flight tests than

FIGURE 8.15 A helmet-mounted display (HMD) unit.

those who did not receive the training. The trained pilots learned to increase their head movements, whereas the untrained pilots spontaneously reduced theirs.

Another way to deal with the problem of head movements is to expand the display to a wider field of view. Rogers, Asbury, and Haworth (2001) showed that performance by virtually any measure was better with a 99° field of view than with a more typical 40°.

A unique problem with single-view HMDs arises from the placement of the sensor at the nose of the aircraft. The optical flow of the display will correspond to the flow that would occur not from the pilot's vantage point, but rather, from the vantage of someone sitting where the sensor is. Consequently, different images of apparent motion, motion parallax, and distance are presented to each eye. Other possible limitations arise from the elimination of binocular disparity as a depth cue, and the potential for binocular rivalry (inattention to the scene at one eye) between the direct view of the environment at the unobstructed eye and the image presented through the HMD. Improper positioning and instability of the helmet can also lead to confusing images. Despite these problems, single-view HMDs can significantly enhance pilot performance.

Binocular, stereoscopic HMDs find wide use in military applications. They are also popular for creating immersive virtual reality environments for applications such as remote handling of hazardous materials and games for the amusement industry (Shibata, 2002). For these applications, a helmet or pair of goggles is worn, and the system presents separate images to each eye. The sensation of immersion depends on the system's ability to reproduce the retinal images that a person would get under normal stereoscopic viewing conditions.

One problem in the use of stereoscopic three-dimensional displays is that there may be a mismatch between the accommodation and vergence distance, creating a conflict between these depth cues that may contribute to a user's visual fatigue. Another problem is that because the images vary along several dimensions, such as magnification and vertical offset, any differences between the left and right displays on these dimensions may create discomfort (Meltzer & Moffitt, 1997). Also, "cybersickness" can result due to a mismatch of the visual cues for motion with vestibular cues (see Box 7.1). Although this is a problem for simulated environments in general, it is particularly problematic for virtual reality environments because of the lag between when the head moves and when the image is updated. There are many human factors issues still to be addressed before the potential of binocular HMDs can be fully realized.

Warning Signals and Labels

We can use visual displays for presenting warning information, either as general alert or warning signals or as warning labels. There are three types of warning signals: warnings, cautions, and advisories (Meyer, 2006). A warning signal evokes immediate attention and should require an immediate response, a caution signal evokes immediate attention and requires a relatively rapid response, and an advisory signal evokes general awareness of a marginal condition. Alarms can be classified into these categories by considering the consequences of the event being signaled, how rapidly these consequences could occur, the worst outcome that would arise if the signal were ignored, the time required to correct the problem, and how fast the system recovers.

The display design should maximize the detectability of high-priority alerting signals. For visual signals, this means presenting them as near to the operator's line of sight as possible, as well as making them sufficiently large (at least 1° of visual angle) and bright (twice as bright as other displays on the panel). Because flashing stimuli are more readily detected, the signal should flash against a steady-state background. Legends should be sufficiently large to be readable. Everyone has prior experience with red warning signals and amber cautionary signals, such as traffic lights, and so we reserve these colors for warning and advisory signals, respectively. In cases where this relationship does not hold or is reversed, responses to the signals will tend to be slower and less accurate.

In some cases, we will be forced to locate visual alerting signals in the periphery of the visual field. In such situations, we can use a centrally located master signal to indicate the onset of one of several alerting signals, which will improve the time and accuracy of responding to the alerting signal itself (Siegel & Crain, 1960). The readily detectable master signal alerts the operator to the presence of a specific alarm signal, which must then be located. The use of more than one master signal is unnecessary and may cause confusion.

Many of the issues we discussed previously for static displays apply to the design of warning labels, such as those found on most electric appliances. Warnings are more effective if they describe the consequences of noncompliance, have wide, colorful borders, are short and to the point, salient and relevant to the users' goals, and are presented near the hazard (Parsons, Seminara, & Wogalter, 1999). Warnings are less likely to be effective if the users are already familiar with the object or product to which the warning is affixed, as compared with when the object is unfamiliar. Similarly, when users do not perceive a risk associated with the use of an object, warning labels will be less effective. The designer of the warning label must try to overcome these obstacles, and many of the display design principles that we have touched on in this chapter will be relevant for the design of effective labels.

AUDITORY DISPLAYS

We use auditory displays primarily to convey simple information at low rates of transmission. In fact, one of the foremost uses of auditory displays is as emergency alarms and warning signals (Walker & Kramer, 2006). When we need to transmit more complicated information auditorily, we usually use speech. There are other types of auditory displays, such as auditory icons (representational sounds with stereotypical meanings) and earcons (recognizable sequences of tones to provide information, such as to signal the arrival of an e-mail message), that we can use in some situations (Altinsoy & Hempel, 2011).

WARNING AND ALARM SIGNALS

Auditory warning and alarm signals must be detectable within the normal operating environment, and the information conveyed by the signal should be easily communicated to the operator. For detectability, the concept of the *masked threshold* is important (Haas & Edworthy, 2006; Sorkin, 1987). The difference between the masked and absolute thresholds (see Chapter 7) is that the masked threshold is determined relative to some level of background noise, whereas the absolute threshold is determined in the absence of noise. Because warning signals are often presented in a noisy environment, our concern must be with the masked threshold in that particular environment. To measure this threshold, an observer is presented with two bursts of noise (usually over headphones), one of which contains a signal. He or she must then indicate in which noise burst (the first or second) the signal was contained. The masked threshold is defined as the signal intensity level required for 75% correct selection of the noise burst (50% is chance).

Several guidelines can be used to determine the optimal level for auditory signals (Sorkin, 1987). To ensure high detectability, the intensity of the signal should be well above threshold. An intensity 6–10 dB above the masked threshold will usually be needed at a minimum. As you might expect from Weber's law (see Chapter 1), the increase above masked threshold will need to be larger for high noise levels than for low noise levels. Emergency vehicle sirens often go unheard by drivers because the siren intensity is not sufficiently above the level of the background noise (Miller & Beaton, 1994). Several factors contribute to this problem of detectability. The intensity of the siren cannot exceed certain limits, in order to prevent ear damage to anyone who might be close to the siren. Because intensity is inversely proportional to distance, detectability will drop off rapidly even with only moderate distances from the siren. Detectability will also be attenuated considerably by the body of the vehicle. Given the many sources of background noise within the car (CD player, air conditioner fan, and so on), not to mention innovations in automobile soundproofing, it should not be too surprising that in many situations the siren intensity will not be much, if at all, above the masked threshold.

If a rapid response is required to a warning signal, then the intensity should be at least 15–16 dB above the masked threshold. An overly loud signal can interfere with speech communication and be generally disruptive, so the intensity of an auditory warning signal in most cases should not exceed 30 dB above the masked threshold (Patterson, 1982). Antin, Lauretta, and Wolf (1991) investigated levels of intensity for auditory warning tones under different driving conditions. Each warning signal required a response from the driver. They measured masked thresholds relative to three background noise conditions: quiet (56 km/h on a smooth road); loud (89 km/h on a rough road); and radio (56 km/h on a smooth road with the radio on). They then determined the tone intensity that allowed 95% detection for each noise condition. For the quiet noise condition, a warning tone 8.70 dB above the masked threshold was required on average. For the loud and radio noise conditions, 17.50 and 16.99 dB increases in the warning tone above the respective thresholds were required. Drivers indicated that they preferred even louder tones, perhaps to ensure that they would hear and react quickly to them.

Auditory signals can differ in terms of their distributions of energy across the frequency spectrum, which affects how the signal is perceived (Patterson, 1982). The fundamental frequency of a warning signal should be between 150 and 1000 Hz, because low-frequency tones are less susceptible to masking. Furthermore, the signal should have at least three other harmonic frequency components. This maximizes the number of distinct signals that we can generate and stabilizes the pitch and sound quality under various masking conditions. Signals with harmonically regular frequency components are better than ones with inharmonic components, because their pitches will be perceived as relatively constant in different auditory environments. These additional components should be in the range from 1 to 4 kHz, for which human sensitivity is high. If the signal is dynamic, that is, changing with the state of the environment, then a listener's attention can be "grabbed" by including rapid glides (changes) in the signal's fundamental frequency.

The temporal form and pattern of the auditory signal are also important. Because the auditory system integrates energy across time (see Chapter 7), the minimum duration for a signal should be 100 ms. Brief signals are useful for environments where verbal communication is important, such as in the cockpit of an aircraft, and when temporal patterning is used to code information. Rapid onset rates will sound instantaneous to the listener and may produce a startle response. Thus, we should design signals with gradual onsets and offsets over a period of approximately 25 ms.

We can use temporal coding of information for the pattern of the signals. For example, we might use a rapid intermittent signal for high-priority messages and a slower intermittent signal for low-priority messages (Patterson, 1982). A pattern called the "temporal three" signal was adopted as a requirement for fire alarms in the U.S. by the National Fire Protection Association in 1993, and it has since become both a national and an international standard (Richardson, 2003). The pattern consists of three 0.5 s bursts, separated by 0.5 s off-intervals, followed by a 1.5 s off-interval before the next cycle begins. The temporal code was implemented because it can be signaled by an audible tone of any frequency and because the pattern will be the same in various auditory environments. This is not true for the pitch of a frequency code signal.

Different auditory warnings signal problems of different urgency. Warning signals that produce the highest ratings of perceived urgency and the fastest response times are those with high frequency, high intensity, and shorter intervals between pulses (Haas & Edworthy, 1996). We can scale perceived urgency from pitch, speed, repetition rate, inharmonicity, and warning length, using magnitude estimation procedures (see Chapter 4; Hellier & Edworthy, 1999). For example, the power-law scale of perceived urgency for warning length in milliseconds is

$$\text{Perceived urgency} = 1.65(\text{Warning length})^{0.49}.$$

The scales for each parameter (pitch, speed, etc.) have different exponents. Compared to the warning length exponent of 0.49, speed has the largest exponent with 1.35. This somewhat large exponent (greater than 1) means that a relatively small increase in warning speed will produce a large change in perceived urgency. All these scales mean that we can select warning signals for an application and match them to the relative urgency of the events they signal, so that perceived urgency will be highest for those situations in which quick reactions to the signals are most necessary.

For some purposes, it may be better to use auditory icons rather than warning signals. For example, a warning icon signaling that a collision between two vehicles is imminent might consist of the sound of screeching tires and breaking glass. Belz, Robinson, and Casali (1999) compared the effectiveness of conventional auditory warnings and auditory icons to signal impending collision information to commercial truck drivers. The conventional auditory warnings were 350-ms pulses of four concurrent tones of 500, 1000, 2000, and 3000 Hz (forward collision) and a "sawtooth" waveform (side collision), whereas the auditory icons were the sound of a tire skidding (front-end collision) and a long horn honk (impending side collision). Drivers' braking response times were shorter and collisions fewer with the auditory icons than with the conventional auditory signals. The icons were more effective in part because they were more easily identified.

Another kind of auditory signal is a *likelihood alarm*. A likelihood alarm warns of an impending event, but sounds different depending on how likely the event is. Usually, an automated monitoring system computes the probability of the event, and the alarm sounds according to the computed probability. So, for example, a monitoring system might present proximity warnings for objects on a factory floor. When the system determines that there is a small (less than 5%) chance of a collision, it is silent. When the system determines that there is a moderate (20%–40%) chance of a collision, it will sound a warning signal of moderate urgency. When the system determines that there is a high (greater than 80%) chance of a collision, it will sound an urgent warning signal. Likelihood alarms can improve people's ability to process information and respond to alarms, because the alarms help them allocate attention across several tasks, and they can easily integrate the information provided by such alarms into their decisions (Sorkin, Kantowitz, & Kantowitz, 1988).

In very complex systems, emergencies can result in too many auditory warning signals (Edworthy & Hellier, 2006). During the Three Mile Island nuclear reactor incident, over 100 auditory signals sounded during the critical period of system failure, exceeding the capability of the operators to process the information that was being provided. Similarly, some aircraft can deliver as many as 30 possible auditory signals. Not only does an excessive number of possible alarms increase confusion, but it also increases the likelihood of a false alarm, which may cause a pilot or plant operator to ignore alarms. In complex systems, designers should restrict the number of high-priority warning signals to five or six. One or two additional signals, called *attensons* (Patterson, 1982), can be used to signal lower-priority conditions that the operator can then diagnose from a speech or computer display.

Another problem in the use of auditory alarms is that operators tend to disable them. This has been a contributing factor in several significant air and rail accidents (Sorkin, 1989). Operators are most likely to disable aversive alarms with high false-alarm rates. The disruptive and annoying signal frequently sounds when there is usually no need to take action. Turning off the alarm system is a natural way for the operator to avoid distress. Designers of alarms and warning signals must avoid the use of disruptive auditory signals when false-alarm rates are high.

THREE-DIMENSIONAL DISPLAYS

Although audition is not primarily spatial, auditory cues can provide spatial information. Such cues can direct an operator's attention to a particular location without requiring a change in visual fixation. These kinds of cues are used in displays presented to fighter pilots. Auditory localization cues give information about the locations of threats and targets, reduce some of the visual clutter in the cockpit and decrease visual overload, and can significantly improve fighter pilots' performance.

Localization cues are given over headphones by introducing interaural intensity and time differences that mimic those that occur naturally, providing the illusion of sounds emanating from different locations in three-dimensional space (see Chapter 7). Displays of this type are called *dichotic displays* (Shilling & Shinn-Cunninghman, 2002). They are relatively easy to implement but are limited in providing information only about the lateral location of a sound. To present dichotic information about location effectively, intensity and time differences must be adjusted to correspond with the orientation of the pilot's head. For example, if a sound is localized in front of the pilot, the intensity at the left ear relative to the right should increase when the pilot's head turns to the right (Sorkin, Wightman, Kistler, & Elvers, 1989). Because people are best able to localize noise (as compared with speech or pure tones; Valencia & Agnew, 1990), the stimuli presented on dichotic displays should contain a broad band of frequencies.

Virtual reality applications require dichotic displays to ensure that the user experiences immersion. There are a variety of signal-processing techniques system designers can use to produce stereo signals that contain most of the cues present in the real world (Shilling & Shinn-Cunningham, 2002). The auditory displays produced using these techniques are very realistic, producing sounds that are localized in space around the listener and difficult to distinguish from a (real) free-field presentation. However, quantifying each sound's location in advance is a time-consuming and information-intensive process. Consequently, the sounds often are actually measured at only a single distance only a few times, and sound amplitude is manipulated by calibrated filters in an attempt to represent the other distances. Also, even though different people will perceive the same simulated sound to be at different locations, most applications assume that everyone is the same.

Speech Displays

Speech messages are common for the transmission of auditory information. When designing a speech display, the designer has a choice between natural and artificially generated speech (Stanton, 2006c). Digitized natural speech is easier to comprehend than synthesized speech, but it requires more data storage and cannot be generated as needed (Baldwin, 2012). Regardless of the kind of display the designer chooses, the voice must be intelligible. For natural speech, intelligibility depends primarily on the frequencies between about 750 and 3000 Hz. Intelligibility is affected by several other factors (Boff & Lincoln, 1988): the type of material being spoken, speech filtering, the presence of visual cues, and the presence of noise. Speech intelligibility is better for structured material, such as sentences, than for unstructured material, primarily because of the redundancy provided by the structure. We presented evidence in Chapter 7 that grammatically and semantically correct sentences are perceived more accurately than strings of unrelated words. Similarly, identification thresholds for single words vary as a function of the number of syllables, the phonetic content, and the stress pattern.

Redundancy can be provided not only through structure in the speech signal but also by visual information. The lip-reading cues provided by simultaneously displaying a visual image of a person speaking with the speech signal can increase intelligibility. Even when auditory and visual displays are unintelligible when presented alone, we can often make the speech intelligible by combining the displays and presenting them simultaneously.

For reproducing speech, because a full speech signal is very large and complex, we need to know what frequencies can be filtered, or deleted, from the auditory signal without degrading intelligibility. Because most acoustic energy for speech is in the range between 750 and 3000 Hz, intelligibility decreases most when frequencies within this range are filtered. Frequencies lower or higher than this range can be deleted without having much effect. The ability to perceive speech accurately

TABLE 8.5

Methods for Reducing the Masking of Speech by Noise

1. Increase the redundancy of speech.
2. Increase the level of the speech relative to the level of the noise.
3. Utter the speech with moderate (vs. high or low) vocal force.
4. Peak-clip the speech signal and reamplify to original levels.
5. Exclude noise at the microphone by using a throat microphone, pressure-gradient microphone, or noise shield.
6. Provide intra-aural cuing by presenting the speech out of phase in the two ears.
7. Use earplugs when noise levels are high.

declines with age, particularly after age 60; this effect of aging is much greater for speech that has been degraded than for speech under optimal conditions (Bergman et al., 1976).

As with nonspeech auditory signals, the human factors specialist must be concerned with the intelligibility of speech within the specific environment in which a speech display is to be used. When speech is presented over a noisy background, its intelligibility will be reduced. The extent of this reduction depends on the signal-to-noise intensity ratio, the amount of overlap between the frequency components of speech and noise, and other factors. Table 8.5 lists methods for reducing the masking of speech by noise.

The first two methods, increasing the redundancy of the speech message and increasing the signal-to-noise ratio, should not need elaboration. The recommendation to utter the speech with moderate vocal force arises from the fact that low-intensity speech will be "lost" in the noise, whereas high-intensity speech is less intelligible than moderate-intensity speech, regardless of the noise level. Peak clipping is performed by setting a maximum amplitude for the sound wave, then clipping any signal that exceeds that amplitude to the maximal value. Peak clipping the speech signal, then reamplifying it to the original intensity level, will produce a relative increase in the intensities for the frequencies of lower amplitude in the original signal (see Figure 8.16). These lower-amplitude frequencies convey the information about consonants and typically are the limiting factors in speech perception. Thus, a reamplified peak-clipped signal will be more intelligible than an unclipped signal of the same average intensity.

Eliminating noise at the microphone minimizes noise effects at the point of transmission. Presenting the speech and noise out of phase at the two ears assists the listener in localizing the speech and noise signals, thus improving intelligibility. Earplugs can contribute to speech intelligibility under conditions of high-intensity noise by reducing the sound intensity to levels at which the ear is not overloaded. The best earplugs for improving speech are ones that do not filter frequencies below 4 kHz. Thus, the intensity of the speech signal is not reduced much, whereas that of the noise is.

There are several ways to estimate the intelligibility of speech in noise, including the *articulation index* (Kryter & Williams, 1965; Webster & Klumpp, 1963). There are two methods for calculating this index: the 20-band method and the weighted one-third octave band. For the 20-band method, we measure the intensity levels of speech and noise for each of 20 frequency bands that contribute equally to speech intelligibility. We then normalize the average of the differences between the speech and noise levels in each band to yield an articulation-index value between 0.0 and 1.0. An articulation index of 0.0 means that the speech will be unintelligible, whereas an index of 1.0 means virtually perfect intelligibility.

The weighted one-third octave band method is easier to compute but is less precise. Table 8.6 shows a worksheet for computing the articulation index with this method. The five computational steps are as follows (Kryter, 1972). First, determine the peak intensity level of the speech signal for each of the 15 one-third octave bands shown in Table 8.6. Then, do the same for the steady noise

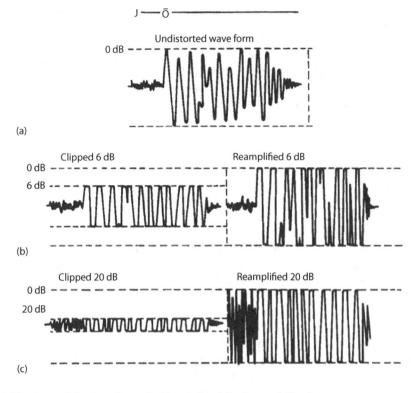

FIGURE 8.16 Reamplification of a peak-clipped signal for the word "Joe."

TABLE 8.6
Worksheet for One-Third Octave Band Method

One-third octave band (Hz)	Center frequency (Hz)	Speech peak to noise peak difference (dB)	Weight	Peak difference × weight
180–224	200	_____	0.0004	_____
224–280	250	_____	0.0010	_____
280–355	315	_____	0.0010	_____
355–450	400	_____	0.0014	_____
450–560	500	_____	0.0014	_____
560–710	630	_____	0.0020	_____
710–900	800	_____	0.0020	_____
900–1120	1000	_____	0.0024	_____
1120–1400	1250	_____	0.0030	_____
1400–1790	1600	_____	0.0037	_____
1790–2240	2000	_____	0.0038	_____
2240–2800	2500	_____	0.0034	_____
2800–3530	3150	_____	0.0034	_____
3530–4480	4000	_____	0.0024	_____
4480–5600	5000	_____	0.0020	_____

that reaches the ear. Third, find the difference between the speech peak and noise levels for each band. Assign the value 30 to differences of 30 dB or more, and 0.0 to negative differences (noise more intense than speech peak). Fourth, multiply each difference by the appropriate weighting factors. These weights are based on the relative importance of the respective frequency bands for speech perception. Fifth and finally, add these weighted values to obtain the articulation index. You can interpret this index in the same way as the one you obtained using the 20-band method.

The articulation index is a good predictor of recognition accuracy under a wide range of conditions (Wilde & Humes, 1990). Under 21 conditions differing in the type of noise (wideband nonspeech or speech), type of hearing protection (unprotected, earplugs, or earmuffs), and signal-to-noise ratio (three levels), the articulation index can accurately predict the percentages of words recognized both by normal listeners and by listeners with high-frequency sensorineural hearing loss. Thus, the articulation index is useful for predicting performance not only under optimal conditions but also under conditions in which hearing protection is provided or listeners are deaf or hard of hearing.

One limitation of the articulation index and related measures is that it is based on the assumption that the different frequency bands contribute to intelligibility additively and independently. However, the instantaneous levels within adjacent levels during speech show a high positive correlation, rather than being independent, meaning that the information provided in the bands is redundant. The Speech Transmission Index (an index of speech intelligibility similar to the articulation index) makes more accurate predictions when these dependencies between octave bands are accounted for (Steeneken & Houtgast, 1999).

Additional considerations arise when we decide to use artificial speech. The artificial speech signal is not as redundant as natural speech. Consequently, speech perception is disrupted more by background noise and the removal of context (Luce, Feustal, & Pisoni, 1983; Pisoni, 1982). Low-quality speech synthesis may be adequate if there are only a small number of messages or if context information is provided in advance (Marics & Williges, 1988), but higher-quality speech generation may be needed when the messages are unrestricted. Because more effort is required to perceive artificial speech, there is poorer retention of the information that is presented (Luce et al., 1983; Thomas, Gilson, Ziulkowski, & Gibbons, 1989). However, an advantage of artificial speech is that the system designer has considerable control over speech parameters. Thus, the voice can be generated to suit the particular task environment.

Aesthetic considerations are as important as performance considerations when evaluating synthesized speech. A voice evokes a more emotional response than a light or a tone. If the voice is unpleasant or the message is one that the listener does not want to hear, then the voice can have an irritating effect. This point is illustrated most succinctly by the short-lived use of speech messages, such as "Your seatbelt is not fastened," in automobiles in the early 1980s. Ratings of the usefulness of a synthetic voice are not valid indicators of its effectiveness: these ratings reflect its perceived pleasantness and not the rater's performance (Rosson & Mellon, 1985).

The acoustic properties of words can influence their effectiveness as warning signals. Hellier et al. (2002) showed that for both spoken words and artificial speech with similar acoustic properties, words spoken in an "urgent" manner (spoken louder, at a higher frequency, and with a broader frequency range) were rated as more urgent than the same words spoken in a "nonurgent" manner. Urgency ratings also were affected by word meaning, with "deadly" rated as of much greater urgency than "note," for example, regardless of the manner in which it was spoken. To convey the strongest urgency with a spoken warning, a word with an "urgent" connotation, such as "danger," should be spoken with an urgent emphasis.

TACTILE DISPLAYS

The tactile sense is important in situations for which spatial information is required but vision is not possible or is overloaded. Controls often are coded to be distinguishable by touch, because the operator will not be able to look at them. Similarly, tactual information is crucial for people who must work in a dark environment and for people with visual impairments.

Tactile displays are not good for alerting signals, because they tend to be disruptive. (Imagine being poked by something while concentrating intently on solving a problem.) However, if you decide to use such a display, the stimulation should be vibratory to maximize detectability. Many smartphone applications use vibratory displays to signal the arrival of a message or other information. The amplitude of the vibration should be detectable on the specific region of the body to which it will be delivered. Sensitivity is greatest on the hands and the soles of the feet. The stick shaker in an aircraft is a vibratory display that warns the pilot that a stall is imminent by vibrating the control column.

We can code tactile stimuli for identification according to physical dimensions in the same way as visual and auditory stimuli. Most important are the dimensions of shape and texture, although size and location also can be used. Figure 8.17 presents a standard set of controls distinguishable by touch that have been adopted by the military for use in aircraft.

Tactile stimulation also can be used to supplement the visual and auditory systems in conditions where they are overloaded. For example, Jagacinski, Miller, and Gilson (1979) compared people's performance using a tactile display with that using a visual display in a system control task. The tactile display was a variable-height slide on the control handle, which indicated the direction and magnitude of error between the actual and desired control settings. People's performance was poorer overall with the tactile displays, but in some conditions performance approximated that obtained with visual displays. Tactile displays show promise as a mode for replacing visual displays in many applications, including driving navigation systems (Tan, Lim, & Traylor, 2000). An array of "tactors" embedded in the back of the driver's seat can provide directional information by providing rapid, successive stimulation across locations of the array in a particular direction.

Tactile displays can also replace visual displays for the visually impaired. The most commonly used tactile displays involve Braille. Visual displays of such things as floor numbers in elevators often appear with Braille characters embossed on their surfaces. Another tactile display that has received widespread use is the Optacon (*opt*ical-to-*t*actile *con*verter), which was developed as a

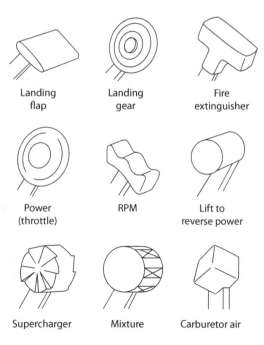

Landing
flap

Landing
gear

Fire
extinguisher

Power
(throttle)

RPM

Lift to
reverse power

Supercharger

Mixture

Carburetor air

FIGURE 8.17   A standard set of aircraft controls distinguishable by touch.

reading aid for the blind. To use the Optacon, an individual places an index finger on an array of 6×24 vibrotactile stimulators that vibrate at 230 Hz. She then passes a light-sensitive probe over the text or other pattern that is to be examined. The scanning of the probe produces a spatially corresponding scanning pattern of activation on the vibrotactile display. Skilled Optacon readers can read text at up to 60–80 words/minute.

Although stimulation of a single finger works relatively well for reading, it doesn't support three-dimensional "virtual reality" tactual displays. These virtual displays have been "largely inaccessible to visually impaired people" (Lee, Bahn, and Nam, 2014, p. 892), because a much higher quality of haptic rendering is required for someone to rely entirely on the tactile modality than is needed when using tactile stimulation together with visual and other inputs in a multimodal virtual reality system. Many researchers are working toward developing devices that will allow visually impaired people to be able to experience a three-dimensional virtual world through the sense of touch and audition (Chebat, Maidenbaum, & Amedi, 2015; Kristjánsson et al., 2016; Sevilla, 2006).

Tactile displays can also aid speech perception for the deaf and for people who are hearing impaired. The adequacy of the tactile sense for speech perception is illustrated by a "natural" method, called Tadoma, in which a deaf-blind person places a hand on the face and neck of the speaker. With this method, people can become relatively proficient at recognizing speech (Reed et al., 1985), although it is not used much today.

Several synthetic devices have been developed for tactile communication of speech. These devices convey characteristics of the speech signal through arrays of tactile stimulators. The devices have three characteristics in common (Reed et al., 1989): (1) reliance on variations in location of stimulation to convey information; (2) stimulation of only the skin receptors and not the proprioceptive receptors; and (3) all of the elements in the stimulating array are identical. Several devices are available commercially. The Tactaid VII uses seven vibrators worn 2–4 in. apart. It presents coded speech information by the place of vibration, movement of vibration, strength of vibration, and duration of vibration. The Tactaid 2000 uses tactile cues specifically intended to allow differentiation of high-frequency speech sounds, which are difficult to hear. A single vibrator provides information about frequencies lower than 2000 Hz, and five vibrators provide information from the range of 2000 to 8000 Hz. The Tickle Talker differs from the Tactaid devices in extracting the fundamental frequency and second formant from the speech signal for presentation on the skin. With this device, electrical pulses are transmitted through eight rings, one on each finger, excluding thumbs. Changes in the fundamental frequency affect the perceived roughness of the stimulation, while second formant frequency is represented by the location stimulated (front/back of different fingers). None of these tactile devices are currently commercially available, in part because cochlear implants for the severely hearing impaired were found to improve speech perception more.

SUMMARY

Displays of information are used for many different purposes. The central message of this chapter is that not all displays are created equal. Performance in simple and complex human–machine systems can be affected drastically by display design. The choice of an appropriate display involves consideration of the characteristics of the sensory modality for which it is intended. Such characteristics include the temporal, spatial, and absolute sensitivities of the modality. These factors interact with the operating environment, the purpose of the system, the nature of the information that is to be communicated, and the capabilities of the population that will be using the display.

For static visual displays, alphanumeric stimuli and symbols can be used effectively. The display must be visible and conspicuous, and the elements should be legible and intelligible. Many additional factors come into play for dynamic displays, including choice of digital or analog format, and for analog displays intended to represent the physical environment, the best frame of reference for representing the environment. For display panels, the placement of the displays can determine how easily an operator will be able to process the information needed to carry out various tasks. Many similar factors must be considered for displays designed for other sensory modalities, though there are unique considerations for each modality.

This chapter has focused primarily on perceptual factors that affect the ease with which a display can be used. However, because the ultimate purpose of a display is to convey information to the observer, performance will be affected by cognitive factors as well. These factors are the topic of the next section of the book.

RECOMMENDED READINGS

Easterby, R., & Zwaga, H. (Eds.) (1984). *Information Design*. New York: Wiley.

Lehto, M. R., & Miller, J. D. (1986). *Warnings (Vol. 1: Fundamentals, Design, and Evaluation Methodologies)*. Ann Arbor, MI: Fuller Technical Publications.

Stanney, K. M. (Ed.) (2002). *Handbook of Virtual Environments: Design, Implementation, and Applications*. Mahwah, NJ: Erlbaum.

Wogalter, M. S. (Ed.) (2006). *Handbook of Warnings*. Mahwah, NJ: Erlbaum.

Zwaga, H. J. G., Boersema, T., & Hoonhout, H. C. M. (Eds.) (1999). *Visual Information for Everyday Use*. London: Taylor & Francis.

Part III

Cognitive Factors and
Their Applications

9 Attention and the Assessment of Mental Workload

The theoretical importance of attention can be seen at two different levels. First as one of the three main limits on human information processing … Second, attentional properties underlie many other psychological phenomena … The applied importance of attention is also manifest in several ways … The dangers of distracted drivers, the attentional overload of making sense of massive data bases, the rapid attention switching required in the electronic workplace, [and] the success or failure of unreliable alarms to capture attention … are all examples.

C. D. Wickens & J. S. McCarley
2008

INTRODUCTION

Driving a car is a complex task composed of many subtasks, each of which must be performed at the appropriate times and within certain speed and accuracy requirements. For example, you must decide where you want to go and the route you want to take to get there, and then you have to navigate the car toward your intended destination. You must steer the car so that it stays in the desired lane, and use the gas and brake pedals to maintain proper speed. You have to see, read, and comprehend the information signs located along the roadway and modify your driving accordingly. You may find that you need to change the settings of entertainment and air conditioning systems within the passenger compartment, and operate the turn signal and windshield wipers. All the while, you must monitor the environment continuously for unexpected events, such as obstacles appearing in the roadway or approaching emergency vehicles that require rapid reactions on your part. Although it is not part of the task of driving, you may frequently engage in conversations with other passengers or use your cell phone.

Because of the many perceptual, cognitive, and motor demands that driving imposes on a driver, the task of driving incorporates almost all of the aspects of attention that will be of concern in this chapter. It should come as no surprise that a lot of applied research on attention is devoted to studying the performance of drivers of land, air, and water vehicles under different cognitive demands.

Historically, what we call "attention" has been of interest since Aristotle's time. Research on attention began in earnest in the last half of the 19th century and the early 20th century, as indicated by the quote from William James in Chapter 1. Much of this early work focused on the role of attention in determining the contents of conscious experience. In part because of attention's reliance on unseen mental events, and in part because of a lack of theoretical concepts for depicting the mechanisms of attention, the study of attention received less emphasis during the period from 1910 to 1950. However, research on attention never ceased entirely during that period, and several important contributions to our present understanding were made (Johnson & Proctor, 2004, Chapter 1; Lovie, 1983).

Our ability to attend to stimuli is limited, and where we direct attention will determine how well we perceive, remember, and act on information. Information or objects that do not receive attention usually fall outside of our awareness and, hence, have little influence on our performance. Thus, an information display important for a task (like the fuel gauge while driving) may not be exploited if the system operator is not attending to it. However, when a single highly practiced response has been given to a stimulus many times in the past, attention is not needed for accurate or fast execution of the response. This means that highly familiar but irrelevant stimuli may interfere with and draw attention away from relevant stimuli that require attention. These and other attentional factors determine an operator's level of performance for any assigned task.

There are two kinds of attention. *Selective attention* determines our ability to focus on certain sources of information and ignore others: for example, you may often find yourself at a party or in a classroom where more than one person is talking at once, yet you are able to listen to what only one speaker is saying. *Divided attention* determines our ability to do more than one thing at once, such as driving a car while simultaneously carrying on a conversation. No matter what kind of attention is being used to perform a task, to understand the conditions that make people perform better or worse, we need to know the amount of *mental effort* that people are expending to perform the task. We call a task that requires considerable mental effort "attention demanding." We also need to know what kind of *executive control* is being used. Executive control refers to the strategies that a person adopts in different task environments to control the flow of information and task performance.

The concept of mental effort is closely related to that of *mental workload*, which is an estimate of the cognitive demands of an operator's duties. Many techniques for measuring and predicting workload in applied settings have been developed, based on the methods and concepts derived from basic research on attention. In this chapter we describe alternative models of attention, consider different aspects of attention in detail, and examine techniques for assessing mental workload.

MODELS OF ATTENTION

Several useful models of attention have been developed over the years, each of which has generated research that has enhanced our understanding of attention. Because each model is concerned with explaining some different aspect of attention, it is important that you note exactly what aspect a model is focused on so that you can appreciate the situations that are appropriately characterized by each model.

Figure 9.1 shows a hierarchical classification of attention models. The first distinction between models separates bottleneck from resource models. Bottleneck models specify a particular stage in the information-processing sequence where the amount of information to which we can attend is limited. In contrast, resource models view attention as a limited-capacity resource that can be allocated to one or more tasks, rather than as a fixed bottleneck. For bottleneck models, performance gets worse as the amount of information stuck at the bottleneck increases. For resource models, performance gets worse as the amount of resources decreases.

We can make further distinctions within each of these categories. Bottleneck models can be referred to as either "early-selection" or "late-selection," depending on where the bottleneck is placed in the information-processing sequence (closer to perception or closer to the response). Resource models can be distinguished by the number of resource pools that are used to perform a

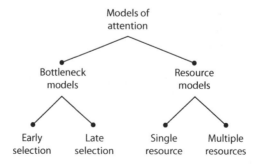

FIGURE 9.1 Hierarchical classification of attention models.

task: Is there only a single resource or are multiple resources involved? Both the bottleneck and the resource model share the view that all human information processing is limited in capacity.

One last class of models attempts to explain human performance without hypothesizing any capacity limitations. These models, called *executive control models*, view decrements in performance as a consequence of the need to coordinate and control various aspects of human information processing. We describe the characteristics of bottleneck, resource, and executive control models in turn, along with the experimental evidence that supports each one.

BOTTLENECK MODELS

Filter Theory

After the resurgence of interest in the topic of attention, the first detailed model of attention, called *filter theory*, was proposed by Broadbent (1958). The filter theory is an early-selection model in which stimuli enter a central processing channel one at a time to be identified. Extraneous or unwanted messages are filtered out early, prior to this identification stage. The filter can be adjusted on the basis of relatively gross physical characteristics, such as spatial location or vocal pitch, to allow information from only one source of input to enter the identification stage (see Figure 9.2).

Broadbent (1958) proposed this particular model because it was consistent with what was known about attention at that time. Many attentional studies were conducted in the 1950s, primarily with auditory stimuli. Probably the best known of these studies is Cherry's (1953) investigation of the "cocktail party" phenomenon, so called because many different conversations occur simultaneously at a cocktail party. Cherry presented listeners with several simultaneous auditory messages. The listeners' task was to repeat word for word ("shadow") one of the messages while ignoring the others, much like the situation you would find yourself in at a cocktail party (although you probably wouldn't want to repeat back the conversation word for word). Listeners were able to do this as long as the messages were physically distinct in some way. For example, when messages were presented through headphones to the left and right ears, listeners could shadow the message in the left ear while ignoring the right, or vice versa.

Not only were listeners able to selectively attend to one of the messages; they also showed little awareness of the unattended message other than its gross physical characteristics (e.g., whether the message was spoken by a male or a female). In one study (Moray, 1959), listeners had no memory

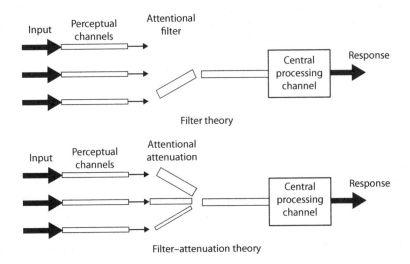

FIGURE 9.2   Filter and attenuation theories of attention.

of words that had been repeated even up to 35 times in the unattended message! In another study, Treisman (1964a) switched the unattended message from English to French after the first few words and then switched back before the end of the message. Fewer than half of the listeners were aware that the language had changed. Consistently with the filter theory, these selective-attention experiments suggested that the unattended message is filtered before the stage at which it is identified.

Another critical experiment was performed by Broadbent (1958) using what is called a *split-span technique*. He presented listeners with pairs of words simultaneously, one to each ear, at a rapid rate. The listeners' task was to report back as many words as they could in any order. The listeners tended to report the words by ear; that is, all of the words presented to the left ear were reported in order of occurrence, followed by any that could be remembered from the right ear, or vice versa. Because the messages presented to both ears required attention, this finding suggests that the message from one ear was blocked from identification until the items from the other ear had been identified, again consistent with the filter theory.

The filter theory nicely captures the most basic phenomena of attention: *It is difficult to attend to more than one message at a time, and little is remembered about an unattended message.* Consequently, filter theory remains one of the most useful theories of attention to this day for the design and evaluation of human–machine systems. For example, Moray (1993) recommended, for design purposes, "Broadbent's original Filter Theory, which is probably both necessary and sufficient to guide the efforts of the designer" (p. 111).

However, most researchers have concluded that filter theory cannot be entirely correct (Bronkhorst, 2015). As usually happens when a theory of human behavior is sufficiently specific to be falsifiable, evidence accumulated that is inconsistent with the filter theory. For example, Moray (1959) found that 33% of listeners were aware that their own name had occurred in the unattended message, a finding replicated under more stringent conditions by Wood and Cowan (1995b). Also, Treisman (1960) showed that, when prose passages were switched between the two ears, listeners continued to shadow the same passage after it had been switched to the "wrong ear." So, the context provided by the earlier part of the message was sufficient to direct the listener's attention to the wrong ear. These and other studies indicate that the content of unattended messages is identified at least in some circumstances, a fact that the filter theory cannot explain.

Attenuation and Late-Selection Theories

Treisman (1964b) attempted to reconcile the filter theory with these conflicting findings. She proposed a *filter-attenuation model* in which an early filter served only to attenuate the signal of an unattended message rather than to block it entirely (see Figure 9.2). This could explain why the filter sometimes seemed to "leak," as in the two examples described in the previous paragraph. That is, an attenuated message would not be identified under normal conditions, but the message could be identified if familiarity (e.g., as for your name) or context (e.g., as for the prior words in a prose passage) sufficiently lowered the identification threshold of the message. Although the filter-attenuation model is more consistent with the experimental findings than the original filter theory, it is not as easily testable.

An alternative approach to address the problems of the filter theory was to move the filter to later in the processing sequence, after identification had occurred. Deutsch and Deutsch (1963) and Norman (1968) argued that all messages are identified but decay rapidly if not selected or attended. There is some evidence that supports this *late-selection model*. For example, Lewis (1970) presented listeners with a list of five words in one ear at a rapid rate. The listeners shadowed the list, while an unattended word was presented in their other ear. Listeners were not able to recall the unattended word, but its meaning affected their response times to pronounce the simultaneously presented, attended word. Response time was slowed when the unattended word was a synonym of the one being shadowed, which is evidence that the meaning of the unattended word interfered with the pronunciation of the attended word.

One possible resolution to the debate over whether selection is early or late is that the locus of selection varies as a function of the specific task demands. Johnston and Heinz (1978) suggested that as the information-processing system shifts from an early-selection to a late-selection model, more information is gathered from irrelevant sources, requiring a greater amount of effort to focus on a relevant source.

A hybrid early- and late-selection model of attention incorporating this idea is called *load theory* (Lavie, Hirst, Fockert, & Viding, 2004). In load theory, whether selection is early or late will depend on the perceptual load (e.g., the number of stimuli and the rate at which they are presented). When perceptual load is high, selection is shifted to early in the process, and irrelevant stimuli are not identified. When perceptual load is low, selection can be delayed until later in the processing sequence. In this situation, both irrelevant and relevant stimuli are identified. Load theory is a fruitful area of inquiry, because "Distraction is an issue in almost every occupation and activity, so the potential benefits of load theory applications are numerous" (Murphy, Groeger, & Greene, 2016, p. 20).

RESOURCE MODELS

The difficulty of pinpointing the locus of a single bottleneck of attention led some researchers to take a different approach and develop resource models of attention. Instead of focusing on a specific location in the information-processing sequence where attentional limitations arise, resource models postulate that attentional limitations arise because a limited capacity of resources is available for mental activity. Performance suffers when resource demand exceeds the supply.

Unitary-Resource Models

Unitary-resource models were proposed by several authors in the early 1970s. The best known is that of Kahneman (1973), which is illustrated in Figure 9.3. According to his model, attention

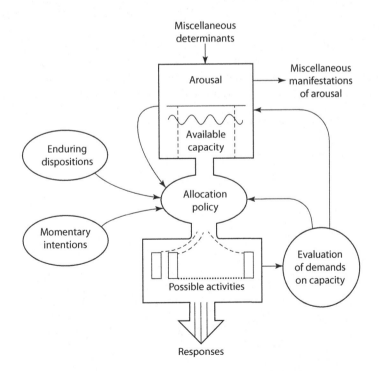

FIGURE 9.3 Unitary-resource model of attention.

is viewed as a limited-capacity resource that can be applied to a variety of processes and tasks. Executing several tasks simultaneously is not difficult unless the available capacity of attentional resources is exceeded. When capacity is exceeded, performance will suffer, and the information-processing system will need to devise a strategy for allocating the resources to different possible activities. This allocation strategy will depend on momentary intentions and evaluations of the demands being placed on these resources.

The unitary-resource model suggests that different tasks have different attentional requirements. Inspired by this idea, researchers began designing experiments that would allow them to measure these attentional requirements. Posner and Boies (1971) used a dual-task procedure in which a person is required to perform two tasks at once. They classified one task as primary and the other as secondary, and the person was made to understand that the primary task was supposed to be performed as well as possible. Under the hypothesis that attention is a single pool of processing resources, any available resources should be devoted to the primary task first. Any spare resources can then be devoted to the secondary task. If the attentional resources are depleted by the primary task, performance of the secondary task will suffer. Posner and Boies's procedure is sometimes called the *probe technique*, because the secondary task is usually a brief tone or visual stimulus that can be presented at any time during an extended primary task. Thus, the secondary stimulus "probes" the momentary attentional demands of the primary task. By looking at the responses to probes throughout the primary-task sequence, a profile of the attentional requirements of the primary task can be obtained.

For their primary task, Posner and Boies (1971) displayed a letter, followed by another letter 1 second later. Observers were to judge whether the pair were the same or different. The secondary task required observers to indicate when probe tones were presented by pressing a button. Reaction times to the probe were slowed only when the tone occurred late in the primary-task sequence (see Figure 9.4), leading Posner and Boies to conclude that it was the late processes of comparison and response selection that required attention. However, later studies showed small effects for tones presented early, suggesting that even the process of encoding the initial letter apparently requires a small amount of attentional resources (Johnson, Forester, Calderwood, & Weisgerber, 1983; Paap & Ogden, 1981).

These studies illustrate that dual-task procedures can provide sensitive measures of the momentary attentional demands on a person. Such procedures can be used to determine the difficulty of different tasks and task components, and therefore to predict when operator performance will suffer. For example, one study looked at ways to measure how much attention was

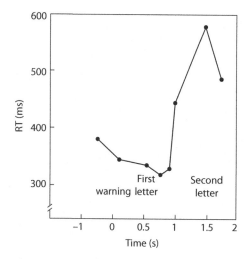

FIGURE 9.4 Probe reaction time (RT) at various times during a letter-matching task.

required to learn a manual motor task (Goh, Gordon, Sullivan, & Winstein, 2014). In addition to the motor task, people performed a task that required a vocal response to auditory stimuli. The reduction in reaction time with practice showed how attentional demands diminished while the manual motor task was learned. As we describe later, dual-task procedures like these have a long history of application in human factors "with the intention of finding out how much additional work the operator can undertake while still performing the primary task to meet system criteria" (Knowles, 1963, p. 156).

The ideas that we have been discussing have assumed that the amount of attentional resources that can be devoted to a task is the same regardless of when or how that task is performed. However, Kahneman (1973) suggested that available capacity may fluctuate with the level of arousal and the demands of the task. If a task is easy, the available attentional resources may be reduced. Young and Stanton (2002) made this aspect of resource theory, which they call *malleable attentional resources*, the basis for explaining why performance often suffers in situations of mental underload, or when the task is too easy. They used a dual-task procedure, where the primary task was simulated driving. The secondary task required the drivers to determine whether pairs of rotated geometrical figures in a corner of the driving display were the same or different. The driving task was performed under four levels of automation, in which some number of the subtasks were performed by the simulation (e.g., controlling velocity and steering). The number of correct same–different judgments increased with increasing automation, indicating reduced attentional demands of the primary driving task.

In addition to the accuracy of responses to the probe, Young and Stanton also measured how long drivers looked at the geometric figures. As automation increased, the ratio of the number of correct secondary-task responses to the amount of time spent gazing at the probe figures decreased. This finding indicates that drivers had to look longer at the figures to make their responses under conditions of low workload, which suggests that processing of those figures was less efficient. Young and Stanton suggested that available attentional capacity is reduced when the attentional demands of the primary driving task decrease. If this is true, then a potential hazard of automation (like driving with cruise control) is to reduce the alertness and attentional capacity of the driver, thus inadvertently reducing his or her performance.

Multiple-Resource Models

An alternative view that has been prominent, particularly in human factors, is *multiple-resource theory* (Navon & Gopher, 1979). Multiple-resource models propose that there is no single pool of attentional resource. Rather, several distinct cognitive subsystems each have their own limited pool of resources. Wickens and McCarley (2008) proposed a three-dimensional system of resources consisting of distinct stages of processing (encoding, central processing, and responding), information codes (verbal and spatial), input (visual and auditory), and output (manual and vocal) modalities, and visual channels (ambient and focal streams; see Figure 9.5). The model assumes that the greater the extent to which two tasks require separate pools of resources, the more efficiently they can be performed together. Changes in the difficulty of one task should not influence the performance of the other if the tasks draw on different resources.

Multiple-resource models were developed because the performance decrements observed for the performance of multiple tasks often depend on the stimulus modalities and the responses required for each task. For example, Wickens (1976) had observers perform a manual tracking task in which a moving cursor was to be kept aligned with a stationary target. At the same time, the observers performed another task involving either maintaining constant pressure on a stick or detecting auditory tones, to which they made vocal responses. Although they judged the auditory-detection task to be more difficult than the constant-pressure task, the observers performed better on the tracking task with the auditory-detection task than with the constant-pressure task.

The improvement in tracking performance with the auditory-detection task can be explained by the different kinds of resources needed for each task. Both the tracking task and the constant-pressure task require resources from the same output (manual) modality pool. The tracking task and

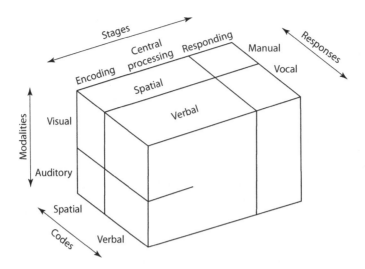

FIGURE 9.5 Multiple-resource model of attention.

the auditory-detection task require resources from different pools. The general principle captured by multiple-resource models is that performance of multiple tasks will be better if the task dimensions (stages of processing, codes and modalities) do not overlap and therefore do not pull from common resource pools. We will encounter other situations later (see Chapter 10) that reinforce this idea.

Wickens's (2002) characterization of the multiple-resource model has been extremely influential in human factors because of its "ability to predict *operationally meaningful* differences in performance in a multi-task setting" (p. 159). That is, it can predict how much two tasks will interfere with each other by looking at whether the tasks rely on the same or different resources. Even though the multiple-resource view is a useful way to evaluate the design of different task environments, multiple-resource models in general have not been widely accepted as a general theory of how attention works. This is because patterns of dual-task interference are much more complicated than we would expect from the simple concept of multiple resources (Navon & Miller, 1987).

EXECUTIVE CONTROL MODELS

The bottleneck and resource models just described explain decrements in multiple-task performance as consequences of a limited capacity for processing information. These models do not place much emphasis on executive control processes, which supervise how limited capacity is allocated to different tasks. However, voluntary control of capacity allocation, and how this control is exerted to accomplish specific task goals, is an important factor in human performance (Monsell & Driver, 2000).

One of the most prominent efforts to analyze performance in terms of executive control processes is called Executive Process Interactive Control (EPIC) theory (Meyer & Kieras, 1997a,b). This theory says that decrements in multiple-task performance are due to the strategies that people adopt to perform different tasks in different ways. One way in which this theory differs from the other models of attention we have discussed is that it assumes that there is no limitation in the capacity of central cognitive processes. EPIC takes into account the fact that, at a fundamental level, people's abilities to process information at peripheral perceptual-motor levels is limited (e.g., detailed visual information is limited to foveal vision, and an arm cannot move to the left and the right at the same time). At higher cognitive levels, EPIC accounts for decrements in multiple-task performance not through a limited-capacity bottleneck or shared resources but through flexible scheduling strategies used to satisfy task priorities and sensory limitations. Executive cognitive processes control these strategies to coordinate the performance of concurrent tasks.

EPIC computational models have been applied successfully to multiple-task performance, not only in the laboratory but also in real-world circumstances, including human–computer interaction and military aircraft operation (Kieras & Meyer, 2000). For example, one important application involves how two simultaneously presented spoken military commands are processed (Kieras et al., 2016). In agreement with findings suggesting that much of the limitation in the "cocktail party" phenomenon lies in perceptual processes involving segregation and streaming of sounds (Bronkhorst, 2015), and the assumption of EPIC's architecture of no capacity limits on cognitive processes, the researchers concluded that attention was being deployed in a way consistent with a "*very* late selection model" (p. 268). Performance in the task was limited by perceptual masking effects (see Chapter 5) and errors in assigning words to the auditory streams that provide the information required to implement a response strategy that results in appropriate choices.

SUMMARY

The early-selection filter theory explains the important fact that people have little awareness of or memory for stimulus events to which they are not attending. The early-selection attenuation model explains the fact that unattended stimuli of particularly high salience due to context or past experience may nevertheless enter awareness. The late-selection bottleneck model explains why major decrements in performance are often associated with processes that occur after perception and stimulus identification. The unitary-resource model accurately depicts how people can control how attention is divided across tasks, and the multiple-resource model explains why multiple-task performance is often worse when two tasks share the same sensory and motor modalities or processing codes than when they do not. Finally, the executive control processes theory emphasizes that, regardless of whether central processing limitations exist, an important part of multiple-task performance is strategic coordination of the tasks.

MODES OF ATTENTION

Attention can take many forms. Up to this point, we have discussed models of attention, which try to explain what it means to attend to something and the resources that power this attention. Attention can be focused on objects or tasks in different ways, and it is important to understand, in discussing the limits of human attention, what kind of attention, or *mode* of attention, is operating on those objects or tasks. In this section we make an important distinction between selective and divided attention, and we also discuss the twin issues of arousal and vigilance, which describe the levels of attention brought to bear on a task.

SELECTIVE ATTENTION

Selective attention is a component of many tasks. For example, when an operator reads an instruction manual, he needs to attend selectively to the written information in the manual and ignore the irrelevant auditory and visual information in his environment. The "cocktail party" phenomenon and Cherry's shadowing tasks are also examples of situations that require selective attention.

Questions about selective attention usually concern those characteristics of an environmental stimulus on which attention is focused and what characteristics of unattended stimuli disrupt the focus of attention. For example, when our operator is reading his instruction manual, what are the specific properties of the text that allow his attention to remain fixed on it? Furthermore, what kinds of environmental events could disrupt his reading? Many experiments have been performed to try to determine what holds attention and what draws it away. Most of these experiments have used either auditory or visual tasks.

Auditory Tasks

One task used to study selective attention is selective listening, in which a to-be-attended (target) auditory message is presented together with another (distractor) auditory signal. The distractor can interfere with the target message by masking it or by confusing the listener about which signal is the target, similarly to Kieras et al.'s (2016) model for processing of multiple commands that we described in the previous section.

We discussed already that selective listening is relatively easy when the target message is physically distinct from the distractor. Spatial separation of the target and distractor, induced by presenting the signals either from different loudspeakers or to separate ears through headphones, makes it easier to attend to the target message (Spieth, Curtis, & Webster, 1954; Treisman, 1964a). Similarly, selective listening is easier when the target and distractor are of different intensities (Egan, Carterette, & Thwing, 1954; Thompson et al., 2015) or from different frequency regions within the auditory spectrum (Egan et al., 1954; Woods et al., 2001). These findings are consistent with filter theory's emphasis on early selection of information to be attended based on gross physical characteristics.

However, it is not just the different physical characteristics of the target and distracting signals that influence performance. Meaning and syntax affect selective listening performance when both signals are speech messages. Listeners make fewer errors when the target and distractor are of different languages, when the target message is prose rather than random words, and when the target and distractor are distinctly different types of prose, for example, a novel and a technical report (Treisman, 1964a,b). Moreover, listeners may develop expectancies based on the context of each message, which can lead to misperception of words to make them consistent with the context (Marslen-Wilson, 1975).

We have already discussed how, when performing a selective listening task in which the distractor message is distinguished physically from the target message, say by spatial location, listeners cannot remember much about the distractor (e.g., Cherry, 1953; Cherry & Taylor, 1954). They can identify changes in basic acoustic features, such as that the voice switched from male to female in the middle of the message, but not particular words or phrases that occurred in the distractor. Cherry reported that only a third of his listeners noticed when the unattended message was switched to backward speech in the middle of the shadowing period. Wood and Cowan (1995a) confirmed this finding and found evidence that those who noticed the backward speech apparently did so because they diverted attentional resources to the distractor, as evidenced by disruption of shadowing several seconds after the distractor switched to backward speech.

Although listeners can only remember a little about the items in an auditory distractor message, the listener's ability to later recognize distracting information is better if the distracting information was presented visually instead of auditorily. Furthermore, recognition is better when the distractors are pictures or visually presented musical scores than when they are visually presented words, indicating that retention of the distractor information decreases as the content of the distractor message becomes more similar to the content of the target message (Allport, Antonis, & Reynolds, 1972).

Up to this point, we have discussed the factors that facilitate or inhibit a listener's ability to selectively attend to particular objects (messages). However, attention can be focused on particular features of a message. Scharf et al. (1987) showed that people can focus attention on a narrow band of the auditory spectrum. They required listeners to decide which of two time intervals contained a tone, which could vary in frequency. Events that occurred earlier in the experiment caused listeners to expect tones of a certain frequency. When the presented tone was near the expected frequency, it was detected well, but if it was significantly different from the expected frequency, it was not detected at all. Thus, under at least some conditions, focused attention alters sensitivity to specific auditory frequency bands.

Visual Tasks

Selective attention for visual stimuli has been studied by presenting several visual signals at once and requiring an observer to perform a task that depends on only one of them. Similarly to the messages presented in a shadowing task, the visual signal to be attended is called the target and all others are called distractors. As with auditory selective attention, the observer may show little awareness of events to which she is not attending (see Box 9.1).

BOX 9.1   CHANGE BLINDNESS

Change blindness is a remarkable phenomenon that has attracted the interest not only of people who study attention, but also of the popular media. Change blindness refers to a person's inability to detect gross or striking changes in a visual scene. A popular demonstration of change blindness asks observers to count the number of passes of a ball between players of a basketball game. While the game is in play, a person wearing a gorilla suit strolls through the players (Figure B9.1). Although the gorilla is directly in sight, and obviously and hilariously out of place, relatively few observers even notice that a gorilla has joined the game (see Simons & Ambinder, 2005, and Durlach, 2004, for reviews).

The most widely used procedure to study change blindness uses pictures (Rensink, O'Regan, & Clark, 2000). Two pictures that differ only in a single conspicuous element are shown one after the other, over and over again, with a blank screen intervening between pictures for a period of about 1/10 second. Changes between displays in the color of an object, its position, or even whether it is present in one version and absent in the other are difficult to detect. Some of these pictures might include a jet whose engine appears and disappears, a government building with a flagpole that moves from one side of the picture to the other, and a city street scene with a cab that changes from yellow to green. It may take many presentations of the pictures before the observer is able to identify what is changed in one display compared with the other.

Researchers who study change blindness are interested in why people are unaware of significant changes in displays and what conditions lead to this lack of awareness. We know, for example, that a change can be detected easily when the blank screen between the two displays is omitted. This may occur because the difference between the two pictures generates a "transient cue," a visual signal (such as apparent motion or an abrupt onset) that directs the observers' attention to the exact location of the change.

There are many real-world tasks that may be affected by change blindness. For example, operators of complex computer-based systems such as those in crisis response centers and air-traffic control centers must monitor multiple, multifaceted displays and perform appropriate control actions when needed (DiVita, Obermayer, Nugent, & Linville, 2004). Pilots in a flight-simulator study evidenced considerable blindness to changes of an aviation weather-report symbol from visual flight rules to instrument flight rules, with some kinds of symbols resulting in many more misses than others (Ahlstrom & Suss, 2015). Failure of operators to detect display changes signaling crucial events of this type can have serious consequences.

Change blindness occurs in a variety of situations in which the observer's attention is distracted or there is a brief break in the visibility of the information. O'Regan, Rensink, and Clark (1999) showed that presenting a "mudsplash" (a series of superimposed dots) on the screen at the time the change was made produced change blindness, even when the mudsplash was not in the area of the change. Change blindness also occurs if the change is

FIGURE B9.1   Change blindness in a pass-counting task.

timed to coincide with a blink (O'Regan, Deubel, Clark, & Rensink, 2000) or a saccadic eye movement (Grimes, 1996), which are essentially observer-induced blank periods. Levin and Simons (1997) demonstrated that the majority of people did not detect relatively salient changes between "cuts" in scenes from a video, including a change in the person who was the focus of the video.

One of the most striking demonstrations of change blindness was presented by Simons and Levin (1998). They had an experimenter stop a person on the street and ask for directions. As the person was providing directions, two other people carrying a solid door walked between the person and the experimenter. In a way that was carefully choreographed, the experimenter grabbed one end of the door and walked away while the person who had been carrying the door stayed behind. Only about 50% of the people providing directions noticed that they were now talking to a different person!

Many experiments on visual selective attention use letters as signals and require the observer to identify the letter that appears in a particular location. If distractors are at least 1° of visual angle away from a target (presented at a known location), they will produce little or no interference with the ability of the observer to identify the target (Eriksen & Eriksen, 1974). If the distractors are very close to the target, they will be identified along with the target, which can result in a decrement in task performance.

More generally, the required response to a target can be made more quickly when the distractors would require the same response as the target, but the response to a target is slowed when the distractors require a different response (e.g., Buetti, Lleras, & Moore, 2014). For example, an observer might be shown a letter triple, such as "X A X," and asked to identify the letter in the middle. If the letter in the middle is an A or a B, the observer is to press one key, but if it is an X or a Y, he is to press another key. The observer will have an easier time with displays like "B A B" or "Y X Y" than he will with displays like "X A X" or "B Y B." However, if the distance between the outer distracting letters and the central target increases, as in "X A X," he will not experience as much difficulty. Interference between the response to be given to the target and the response to be given to the distractors diminishes as the distance between the target and the distractors increases.

Results like this suggest that the focus of attention is a spotlight of varying width that can be directed to different locations in the visual field (Eriksen & St. James, 1986; Treisman, Sykes, & Gelade, 1977). Interference among visual stimuli occurs because the spotlight cannot always be made small enough to prevent distracting stimuli from being attended. In the case of stimuli like "X A X," the "X" distractors, which require a different response from the target "A," are included in the spotlight of attention and identified. The response to the target is inhibited by the competing response required for the distractors. If the Xs are separated from the A by a sufficient amount, they no longer fall within the spotlight and are not identified. Consequently, the response to X is never "activated" and cannot interfere with the response to A.

These studies suggest that the focus of attention has a lower limit: it can get smaller, but not too small. Another study showed that the focus of attention can be made larger. LaBerge (1983) had people perform different tasks with five-letter words. One task required the observer to determine whether the word was a proper noun, whereas the other task required the observer to determine whether the middle letter was from the set A–G or N–U. The word task required a larger focus of attention, at the level of the whole word, whereas the letter task required that the observer focus attention on the middle letter. During both tasks, on some "probe" trials no word was presented. Instead, a single letter or digit was presented in one of the five positions corresponding to where the letters of the word were presented, with # signs in the others. For example, rather than "HOUSE," the observer might instead see "#Z###" or "##7##." To this stimulus, the observer was required to quickly identify whether it contained a letter or a digit. If the observer was performing the word task (and her attention was focused on the whole word), where the letter or digit appeared on a probe trial did not influence how quickly it could be identified. However, if the observer was performing the letter task (and her attention was focused only on the middle letter), responses on probe trials were fastest when the letter or digit appeared in the middle position and became progressively slower as it moved farther away.

One way in which observers can selectively attend to different visual stimuli is by moving their eyes to different places in the visual field (e.g., Ranzini, Lisi, & Zorzi, 2016). Fixated objects will be seen clearly, whereas those in the visual periphery will not. However, the spotlight metaphor suggests that it should be possible to dissociate the focus of attention from the direction of gaze; that is, an observer should be able to selectively attend to a location in the visual field that is different from his fixation point. Such a process is referred to as *covert orienting*, as opposed to the overt orienting that occurs as a function of eye position.

Posner, Nissen, and Ogden (1978) showed that observers can use covert orienting to improve their performance in a simple visual task. Their task was to detect the onset of the letter X in a display. The X could appear 0.5° to either the left or the right of fixation. Prior to the presentation of the X, a cue was presented at the point of fixation. The cue was either a neutral plus sign or an arrow pointing to the left or right. The cue was intended to give the observer information about where the X was likely to appear. The X occurred on the side indicated by the arrow on 80% of the trials and on the opposite side on the remaining 20%. Reaction time to locate the X was fastest when the X appeared at the cued location and slowest when it occurred at the uncued location (see Figure 9.6). Observers apparently used the cue to shift the focus of attention from the fixation point to the most likely location of the X.

Experiments conducted after that of Posner et al. (1978) have tried to determine whether attention is moved gradually from the point of fixation to the cued position (like the movement of a spotlight over a surface) or whether the shift of attention occurs discretely. Some experiments have shown that the amount of time it takes to move from the fixation point to the position of the target stimulus is the same regardless of how far away the target position is. Therefore, we must conclude that attention "jumps" in a discrete way from one point to another (e.g., Yantis, 1988). We can see evidence for these discrete jumps in the activity of neurons in the frontal and parietal cortex (Buschman & Miller, 2010).

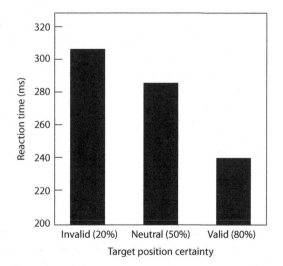

FIGURE 9.6 Reaction times as a function of target position certainty.

Posner et al.'s (1978) arrow cues, which point in one direction or another, are said to induce *endogenous orienting* of attention; that is, a shift of attention that is initiated voluntarily by the individual. Attention can also be drawn involuntarily to a location or object by the rapid onset or perceived motion of a stimulus (e.g., Bucker & Theeuwes, 2014), a type of shift that is called *exogenous orienting*. In other words, even when an observer does not move his eyes, his attention may shift reflexively and involuntarily to the location where he sees a stimulus appear or move unexpectedly.

Exogenous orienting of attention can both help and hinder the performance of a task. Consider a modification of the Posner et al. task whereby, rather than using arrow cues, the likely position of the target is cued exogenously by flashing a neutral stimulus at the target location. The abrupt onset (flash) of the neutral stimulus will draw attention to its location. If the time between this exogenous cue and the target is short (less than about 300 ms), responses to a target presented in that location will be made more quickly. However, if the time between the cue and the target is longer than 300 ms, responses to targets presented in uncued locations will actually be faster than those to targets presented in the cued locations (e.g., Palanica & Itier, 2015).

This phenomenon is called *inhibition of return* (Martín-Arévalo, Chica, & Lupiáñez, 2016). We may hypothesize that with longer cue-target delays, attention may shift (either voluntarily or involuntarily) to other locations in the visual field. Once attention has shifted away from an exogenously cued location, there is a tendency to avoid returning it to that same location. Although the purpose of this attentional mechanism is not yet well understood, it may make visual search of complex environments more efficient by preventing a person's attention from revisiting locations that have already been checked.

Switching and Controlling Attention

The distinction between endogenous and exogenous shifts of attention brings up questions about how attention is controlled. Many tasks, such as driving, require rapid switching of attention between sources of information. It turns out that people differ in their abilities to switch attention from one source of information to another. Kahneman and his colleagues (Gopher & Kahneman, 1971; Kahneman, Ben-Ishai, & Lotan, 1973) evaluated the attention-switching ability of fighter-pilot candidates and bus drivers using a dichotic listening task. The subjects were asked to shadow one of two messages presented to each ear. After they had selectively attended to information in one ear, a tone sounded in the other ear, indicating that the subject should shift his attention and shadow the message in the other ear. The number of errors made after the attention-switching signal was negatively correlated with the success of cadets in the Israeli Air Force flight school; successful cadets made fewer errors than unsuccessful cadets. The number of errors made was positively correlated with the accident rates of the bus drivers; drivers who made more errors had more accidents than drivers who made fewer errors. Similar results have been found for Royal Netherlands Navy air-traffic control applicants (Boer, Harsveld, & Hermans, 1997).

While we have concentrated our discussion on simple laboratory tasks that bear little resemblance to the complicated environments that people encounter in real-world settings, these studies indicate that attention shifting is a skill that affects the performance of complex navigation tasks outside of the laboratory. Furthermore, Parasuraman and Nestor (1991) note that attention-shifting skill deteriorates for older drivers with various types of age-related dementia. Because of these individual differences in attention-shifting ability, it may be appropriate to assess driving competence in part by evaluating attention.

DIVIDED ATTENTION

Whereas a selective-attention task requires a person to attend to only one of several possible sources of information, divided-attention tasks require a person to attend to several sources of information simultaneously. In many situations, people perform best when they must monitor only a single source of information, and they perform more and more poorly as the number of sources increases. This decrement in performance is usually measured as a decrease in accuracy of perception, slower response times, or higher thresholds for detection and identification of stimuli.

There are many applied settings in which operators must perform divided-attention tasks by monitoring several sources of input, each potentially carrying a target signal. Consider an environment in which an operator must monitor a large number of gauges, each providing information about some aspect of a complex system's performance. An operator may be required to detect one or more system malfunctions, which would appear as one or more gauges registering abnormal readings. Such environments are common in nuclear power plant control rooms, process control system interfaces, and aircraft cockpits.

How well an operator can monitor several sources of information depends on the task she is to perform. Suppose, for example, that if one or more of the gauges in an array registers an abnormal system condition, the operator's job is to shut down the system and inform her supervisor. In this case, the operator's ability to detect a target is only slightly degraded relative to when she must monitor only a single gauge (Duncan, 1980; Ostry, Moray, & Marks, 1976; Pohlman & Sorkin, 1976). This is because if more than one target occurs, the probability that the operator will detect at least one of them increases, although the probability that the operator will detect any particular target decreases. The likelihood of detecting a target from a single source diminishes further as the number of simultaneous targets from other sources increases.

Problems will arise when two or more targets must be identified separately. For example, the operator may need to shut a water intake valve in response to one abnormal reading but open a steam-pressure valve in response to another. In this scenario, the operator's ability to detect, identify, and respond to any particular target will be worse than when she is attending to only a single source of information.

Although some of the operator's difficulties in responding to multiple, simultaneous targets can be reduced with practice (Ostry et al., 1976), her performance will never be as good as when she is attending to only a single input source. In applied situations, if simultaneous targets are very likely to occur, and a failure to detect and respond to those targets may lead to system failure, then each source should be monitored by a separate operator.

For situations in which an operator must divide his attention between different tasks or sources of information, he may not need to give each task the same priority. For example, with the probe technique, which we described earlier, one of two tasks is designated as primary and the other as secondary. More generally, any combination of relative weightings can be given to the two tasks: For example, an operator might be instructed to pay twice as much attention to the primary task as the secondary task. That is, an operator can "trade off" his performance on one task to improve his performance on another task.

The tradeoff in dual-task performance can be described with a *performance operating characteristic* (POC) curve (Norman & Bobrow, 1976), sometimes called an *attention operating characteristic* (Alvarez, Horowitz, Arsenio, DiMase, & Wolfe, 2005), which is similar in certain respects to the ROC curve presented in Chapter 4. A hypothetical POC curve is shown in Figure 9.7. For two tasks, A and B, the abscissa represents performance on Task B, and the ordinate represents performance on Task A. In the POC, performance can be measured in any number of ways (speed, accuracy, etc.), as long as good performance is represented by high numbers on each axis. Baseline performance for each task when performed by itself is shown as a point on each axis. If the two tasks could be performed together as efficiently as when performed alone, performance would fall on the *independence point* P. This point shows performance when no attentional limitations arise from doing the two tasks together.

The box formed by drawing lines from point P to the axes defines the POC space. It represents all possible combinations of joint performance that could occur when the tasks are done simultaneously. The actual performance of the two tasks will fall along a curve within the space. *Performance efficiency*, the distance between the POC curve and the independence point, is an indicator of how

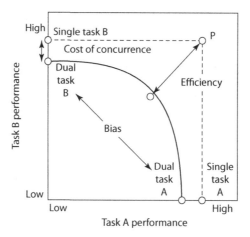

FIGURE 9.7   Performance operating characteristic.

efficiently the two tasks can be performed together. The closer the POC curve comes to P, the more efficient is performance. As with an ROC curve, the different points along the POC curve reflect only differences in bias induced by changing task priorities. The point on the positive diagonal reflects unbiased performance (equal attention given to both tasks), whereas the points toward the ordinate or the abscissa represent biases toward Task B or Task A, respectively. Finally, the *cost of concurrence* is shown by the difference between performance for one task alone and for dual-task performance in which all resources are devoted to that task.

A POC curve is obtained by testing people in single- and dual-task conditions and varying the relative emphases placed on the two tasks. Performance on a task in a dual-task scenario might approximate that when it is performed alone or might be substantially worse, depending on the conditions imposed in the dual-task scenario. POC analyses can be used to evaluate operator performance and task design in many complex systems in which operators must perform two or more tasks concurrently, such as monitoring radar or piloting aircraft.

To illustrate the use of POC curves, we will describe a study by Ponds, Brouwer, and van Wolffelaar (1988) that evaluated dual-task performance for young, middle-aged, and elderly people. One task involved simulated driving, whereas the other required counting a number of dots, which were presented at a location on the simulated windshield that did not occlude the visual information necessary for driving. Performance was normalized for each age group, so that the mean single-task performance for each group was given a score of 100%. This normalization makes it possible to evaluate age differences in dual-task performance independently from any overall differences that might be present across the groups.

POC curves were obtained for each age group by plotting the normalized performance scores obtained for dual-task performance under three different emphases on driving versus counting (see Figure 9.8). For the normalized curves, the independence point is (100, 100). The elderly show a deficit in divided attention, as evidenced by the POC curve for the older adults being further from the independence point than the POC curves for the middle-aged and young adults. This divided-attention deficit for the elderly corroborates Parasuraman and Nestor's (1991) work on attention shifting in elderly drivers and does not disappear with practice (McDowd, 1986).

AROUSAL AND VIGILANCE

A person's attentional ability is influenced by her level of arousal. Arousal level may influence the amount of attentional resources available to perform a task, as well as the policy by which attention is allocated to different tasks. This relation between attention and arousal underlies a widely cited law of performance, the *Yerkes–Dodson law* (Yerkes & Dodson, 1908). According to this law,

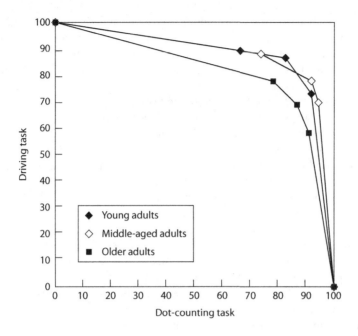

FIGURE 9.8   Normalized POC curves for older, middle-aged, and young adults in a divided-attention task.

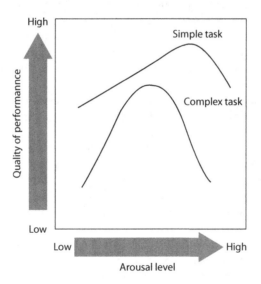

FIGURE 9.9   The Yerkes–Dodson law.

performance is an inverted U-shaped function of arousal level, with best performance occurring at a higher arousal for simple tasks than for complex tasks (see Figure 9.9).

It is not surprising that performance is poor at low arousal levels. Extremely low arousal may result in a person being unprepared to perform the task or failing to monitor performance and, as a result, failing to pay attention to changing task demands. Because the number of features to consider in difficult tasks typically is greater and the coordination of attention more crucial than in easy tasks, difficult tasks show a greater performance decrement at lower arousal levels than do simple tasks.

It is more surprising that performance tends to deteriorate at high arousal levels. Several factors contribute to this deterioration, but it is primarily due to a decrease in attentional control. At high arousal levels, a person's attention becomes more focused (either appropriately or inappropriately), and the range of cues he uses to guide his attention becomes more restricted (Easterbrook, 1959). Also, a person's ability to discriminate between relevant and irrelevant cues decreases. Thus, at high arousal levels, fewer and often less appropriate features of the situation control the allocation of attention. This theory suggests that performance will not decline at high levels of arousal if attention remains directed toward the task at hand (Näätänen, 1973).

The value of the Yerkes–Dodson law has been disputed (e.g., Hancock & Ganey, 2003; Hanoch & Vitouch, 2004), based in part on evidence that arousal depends on many different factors and consists of many different physiological responses. Therefore, it is not possible to attribute a benefit or decrement in performance to one general arousal level. However, as noted by Mendl (1999), "The law can be used descriptively as a shorthand way of summarising the observed relationship between a diverse range of apparently threatening or challenging stimuli and various measures of cognitive performance, without the implication that all relationships are mediated by a single stress or arousal mechanism" (p. 225). We devote the remainder of this section to two important effects of arousal on attention: perceptual narrowing and the vigilance decrement.

Perceptual narrowing refers to the restriction of attention that occurs under high arousal (Kahneman, 1973). Weltman and Egstrom (1966) used a dual task to examine perceptual narrowing in the performance of novice SCUBA divers. The primary task required the divers to add a centrally presented row of digits or to monitor a dial to detect a larger than normal deflection of the pointer, and the secondary task required them to detect a light presented in the periphery of the visual field. The level of arousal was manipulated by observing the divers in normal surroundings (low stress), in a tank (intermediate stress), and in the ocean (high stress). Performance on either primary task was unaffected by stress level, but as stress increased, it took the divers longer to detect the peripheral light. This finding suggests that the divers' attentional focus narrowed under increased stress. Janelle, Singer, and Williams (1999) replicated these results using college students performing a simulated driving task.

In contrast to perceptual narrowing, which occurs at high levels of arousal or stress, the vigilance decrement occurs under conditions in which arousal seems, at first, to be very low (Thomson, Besner, & Smilek, 2016). Before defining the vigilance decrement, we must define what we mean by vigilance. Many tasks involved in operating automated human–machine systems involve sustained attention, or vigilance. Consider the gauge-checker, discussed above, who must monitor many gauges simultaneously for evidence of system failure. If system failures are very infrequent, leaving the operator with almost nothing to do for very long periods of time, we say that he is performing a *vigilance task*. The defining characteristic of a vigilance task is that it requires detection of relatively infrequent signals that occur at unpredictable times.

Research on vigilance began in World War II, spurred by the problem that radar operators were failing to detect a significant number of submarine targets. As systems have become more automated, there are many more situations in which an operator's role is primarily one of passively monitoring displays for critical signals, so vigilance research is still important. Vigilance in part determines the reliability of human performance in such operations as industrial quality control, air-traffic control, jet and space flight, and the operation of agricultural machinery (Warm, Finomore, Vidulich, & Funke, 2015).

The *vigilance decrement* was first demonstrated in an experiment by Mackworth (1950). He devised an apparatus in which observers had to monitor movements of a pointer along the circumference of a blank-faced clock. Every second, the pointer would move 0.3 in. to a new position. Occasionally, it would take a jump of 0.6 in. Observers were required to execute a keypress response when a "target" movement of 0.6 in. was detected. The monitoring session lasted 2 hours. Mackworth found that the hit rate for detecting the target movement decreased over time, a finding that has since been replicated with many tasks. Figure 9.10 shows the vigilance decrement for three

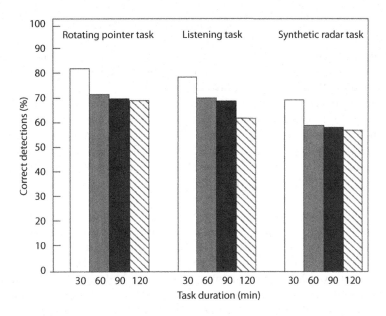

FIGURE 9.10 The vigilance decrement for three tasks.

tasks over a 2-hour period. The maximal decrement in accuracy occurs within the first 30 minutes. Not only does accuracy decrease, but other studies show that reaction times for hits (as well as for false alarms) become slower as the time spent at the task increases (Parasuraman & Davies, 1976).

Why does hit rate decrease? The decrease could reflect either a decrease in sensitivity to the signals or a shift to a more conservative criterion (requiring more evidence) for responding. To determine which of these is responsible for the decreased hit rate, we can perform a signal-detection analysis (see Chapter 4). When such analyses are performed on the vigilance decrement, for some situations there is an increase in the criterion (β), with sensitivity (d') remaining relatively constant, for detection performance early versus late in the task (e.g., Broadbent & Gregory, 1965; Murrell, 1975). The more frequently signals occur, the smaller the change in criterion. This suggests that the criterion can be maintained at a more optimal, lower level by using artificial signals to increase the frequency of events.

Initial applications of signal-detection theory suggested that a sensitivity decrement was rare in vigilance tasks, but subsequent research has suggested that sensitivity may decline across the vigil in many vigilance tasks (See, Howe, Warm, & Dember, 1995). Parasuraman and Davies (1976) and Parasuraman (1979) proposed that the sensitivity decrement was restricted primarily to tasks that require discrimination based on a standard held in memory, particularly if the event rate is high. An example of such a task would be trying to detect whether a light that comes on periodically is brighter than its usual intensity. However, a case has been made that, due to inappropriateness of the signal-detection sensitivity measures used to analyze vigilance task performance, shifts in response bias can inadvertently produce a decrease in the sensitivity measure (Thomson et al., 2016). Consequently, even under conditions of high memory demand and event rates, the vigilance decrement "may simply reflect a shift in response criterion rather than sensitivity" (Thomson et al., p. 70).

There are many factors that will influence the size of the vigilance decrement in sensitivity (Warm et al., 2015). See et al. (1995) concluded from an analysis of several studies that the vigilance decrement is different for discriminations based on sensory information (such as brightness detection) and those based on cognitive information (such as trying to detect a specific digit in a stream of digits). If the discrimination requires information from memory, there will usually be a larger vigilance decrement for cognitive discriminations than for sensory discriminations. However, the size of this difference will depend on event rate. At high event rates, there will be little difference in the size of the sensitivity decrement between sensory and cognitive discriminations. If the discrimination does not require information from memory, the vigilance decrement will be larger for sensory discriminations than for cognitive discriminations.

Performance in vigilance tasks will also be affected by other characteristics of the signal, as well as by the motivation of the observer. Stronger signals are easier to detect, and the vigilance decrement is not as pronounced (Baker, 1963; Wiener, 1964). Auditory signals are easier to detect than visual signals, and the vigilance decrement can be reduced by frequently alternating between auditory and visual modalities (e.g., every 5 minutes; Galinsky et al., 1990). The vigilance decrement can also be reduced by providing rest periods of 5–10 minutes or by financial incentives (Davies & Tune, 1969).

It might seem at first that arousal in vigilance tasks is affected by mental underload, with the vigilance decrement being a consequence of low arousal levels. However, there is now considerable evidence suggesting that performance of a vigilance task is quite effortful and that the vigilance decrement reflects a depletion of attentional resources rather than a decrease in arousal (Warm, Parasuraman, & Matthews, 2008). For example, Grier et al. (2003) had subjects perform two types of vigilance tasks, both of which produced a vigilance decrement. However, the subjects' assessments of mental workload (described in the next section) and stress showed elevated levels. Thus, contrary to what may seem to be the case, requiring someone to sustain attention for the detection of infrequently occurring events is actually very mentally demanding.

The primary applied message of the research on vigilance is that fairly substantial vigilance decrements can occur in a variety of situations. We can minimize these decrements by carefully selecting the stimulus types, the required discriminations, and the rate at which critical events occur. Also, we must keep in mind that vigilance tasks can be mentally demanding, and so it is important to provide observers with appropriate rest periods and performance incentives. Using appropriate workload assessment techniques, which we discuss in the next section, we may be able to modify the design of the vigilance task to reduce the mental demands on the operator.

MENTAL WORKLOAD ASSESSMENT

Models of attention have been profitably applied back to the solution of certain human factors problems. One area in which this application is evident is the measurement of mental workload (Wickens & Tsang, 2012). Workload refers to the total amount of work that a person or group of persons is to perform over a given period of time. Mental workload is the amount of mental work or effort necessary to perform a task in a given period of time. As task demands increase or the time allowed to perform a task decreases, mental workload increases. Young and Stanton (2006) defined mental workload as follows:

> The mental workload of a task represents the level of attentional resources required to meet both objective and subjective performance criteria, which may be mediated by task demands, external support, and past experience. (p. 507)

In work settings, similarly to the effect of arousal, performance may suffer if the mental workload is too high or too low. At the upper extreme, it is clear that performance will be poor if there are too many task demands. However, as we noted earlier, an undemanding task may also lead to a deterioration in performance by lowering an operator's level of alertness. Figure 9.11 illustrates the resulting inverted U-shaped function between mental workload and performance.

The purpose of mental workload assessment is to maintain the workload at a level that will allow acceptable performance of the operator's tasks. The workload imposed on an operator varies as a function of several factors. Most important are the tasks that the operator must perform.

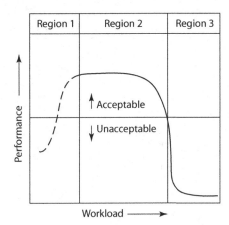

FIGURE 9.11 The hypothetical relation between workload and performance.

Workload will increase as required accuracy levels increase, as time demands become stricter, as the number of tasks to be performed increases, and so on. Workload will also be affected by aspects of the environment in which the tasks must be performed. For example, extreme heat or noise will increase the workload. Also, because the cognitive capacities and skills of individuals vary, the workload demands imposed by a given task may be excessive for some people but not for others.

The mental workload concept comes directly from the unitary-resource model of attention, in which the operator is believed to have a limited capacity for processing information (Kantowitz, 1987). This model lends itself nicely to the concept of spare capacity, or the amount of attentional resources available for use in additional tasks. However, most current workload techniques are more closely linked to the multiple-resource model, for which different task components are assumed to draw on resources from distinct pools of limited capacity. The primary benefit of the multiple-resources view is that it allows the human factors specialist to evaluate the extent to which specific processes are being overloaded.

There are many workload-assessment techniques, which differ in several ways (see Gawron, 2008, for summaries of many of the techniques). A useful taxonomy distinguishes between empirical and analytical techniques (Lysaght et al., 1989; see Figure 9.12). Empirical techniques are those that are used to measure and assess workload directly in an operational system or simulated environment, whereas analytical techniques are those used to predict workload demands early in the system development process. We will discuss each of these techniques, but we will pay most attention to the empirical techniques. It is an unfortunate consequence of the fact that many designers do not concern themselves with ergonomic issues until their systems are near completion that we have many more empirical than analytical techniques for assessing workload.

EMPIRICAL TECHNIQUES

There are four major empirical techniques. The first two are focused on performance measures of the primary task of interest or of a secondary task in a dual-task context. The last two include psychophysiological measures and subjective scales. A given situation may require using one or more of these techniques, and may preclude the use of other techniques. Table 9.1 outlines several criteria that can be applied to determine which workload-assessment technique is most appropriate for a given situation.

A technique should be sensitive to changes in the workload imposed by the primary task, particularly once overload levels are reached. It should also be diagnostic to the extent that the

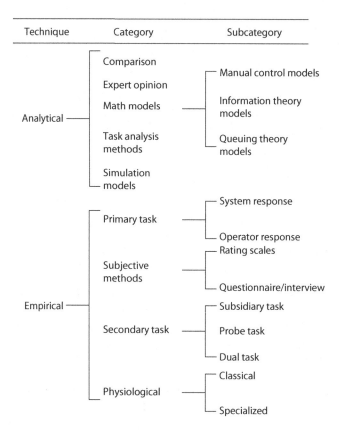

FIGURE 9.12   A taxonomy of mental workload techniques.

assessment can isolate particular processing resources being overloaded. On the basis of multiple-resource theory, this requires discriminating between capacities from the three dimensions of processing stages (perceptual-cognitive vs. motor), codes (spatial-manual vs. verbal-vocal), and modalities (auditory vs. visual). If the technique applied to a particular situation is unable to detect any change in mental workload or determine how mental capacity is being overloaded, it is, obviously, not very useful.

Imagine now a system that requires the operator to make a considerable number of (large or small) movements, such as the control of a robotic arm. You might decide to assess mental workload during the performance of different tasks with a technique that requires the operator to wear several physiological measuring devices, one in the form of a somewhat uncomfortable helmet, another strapped around his upper arm, and a third attached to the end of his finger. Each device has a few wires around which the operator must coordinate his movements, and some of them restrict his mobility.

These methods for assessing workload violate the remaining three criteria outlined in Table 9.1. We call such assessment techniques *intrusive*. Intrusive techniques will interfere with the operator's ability to perform his primary task, and any workload estimates obtained this way will be difficult to interpret. Any observed decrements in performance may be due to the measurement technique and not the task. The *implementation* of this kind of technique is also a problem, because such sophisticated measuring devices may be difficult to obtain or maintain. Techniques should be implemented that involve the fewest problems in doing so. Finally, imagine how dissatisfied the operator is going to be when the equipment he is wearing interferes with his ability to perform his duties, especially if he does not understand what purpose the equipment serves or

TABLE 9.1

Criteria for Selection of Workload Assessment Techniques

Criterion	Explanation
Sensitivity	Capability of a technique to discriminate significant variations in the workload imposed by a task or group of tasks.
Diagnosticity	Capability of a technique to discriminate the amount of workload imposed on different operator capacities or resources (e.g., perceptual vs. central processing vs. motor resources).
Intrusiveness	The tendency for a technique to cause degradations in ongoing primary-task performance.
Implementation requirements	Factors related to the ease of implementing a particular technique. Examples include instrumentation requirements and any operator training that might be required.
Operator acceptance	Degree of willingness on the part of operators to follow instructions and actually utilize a particular technique.

why he is being monitored. If the measurement technique is not *accepted* by the operators who are being evaluated, it will be very difficult to obtain meaningful workload measures. Not only will the operators be reluctant to perform at their best, but they may also actively sabotage the study (and the equipment).

It should be apparent from these considerations that selection of the workload measure that is appropriate for the specific use for which it is intended is a crucial part of workload evaluation.

Primary-Task Measures

Primary-task measures evaluate the mental workload requirements of a task by directly examining the performance of the operator or of the overall system. The assumption is that as task difficulty increases, additional processing resources will be required. Performance of the primary task deteriorates when the workload requirements exceed the capacity of the available resources. Some commonly used primary-task measures are glance duration and frequency (higher workload is associated with longer and/or more frequent glances) and number of control movements/unit time (e.g., number of brake actuations per minute).

It is important to use more than one primary-task measure of workload. Because different components of a task will require mental resources in different ways, a single performance measure may show no effect of workload, whereas other measures would. For example, in a study evaluating the impact of traffic-situation displays on pilot workload, Kreifeldt, Parkin, Rothschild, and Wempe (1976) obtained 16 measures of flight performance, including final airspeed error and final heading error, in a flight simulator. Some measures showed that the traffic-situation displays lowered workload demands, but other measures showed nothing. For example, the airspeed measure showed no improvement in flight performance with the display, while the heading error measure did. If only air speed were measured, then the display designer might conclude that the traffic-situation display did not reduce pilot workload. Using as many measures as we can, we can get a more accurate picture of workload.

Primary-task measures are good for discriminating overload from nonoverload conditions. However, they are not good for measuring differences in mental workload in conditions when performance shows no impairment. An alternative way of examining primary-task performance that sidesteps this problem is to examine the changes in strategies that operators employ as task demands are varied (Eggemeier, 1988). For example, our gauge-checker may, under low levels of workload, respond to system abnormalities by rote and without consulting an operating manual. However, under high levels of workload, she may rely on printed instructions such as those in an operating manual to recover the system. Any obvious strategy changes may be indicators of increased workload.

Primary-task measures also are not diagnostic of those mental resources that are being over-loaded. Furthermore, although they are usually nonintrusive, primary-task measures may require sophisticated instrumentation that renders them difficult to implement in many operational settings.

Secondary-Task Measures

Secondary-task measures are based on the logic of dual-task performance described earlier in the chapter. The operator is required to perform a secondary task in addition to the primary task of interest. Workload is assessed by the degree to which performance on either the primary or the secondary task deteriorates in the dual-task situation relative to when each task is performed alone. Thus, dual-task interference provides an index of the demands placed on the operator's attentional resources by the two tasks.

Secondary-task measures are more sensitive than primary-task measures. In nonoverload situations, where the primary task can be performed efficiently, the secondary-task measures can assess differences in spare capacity. Secondary-task measures also are diagnostic, in that specific sources of workload can be determined through the use of secondary tasks of different modalities. Possible drawbacks of secondary-task measures are that they can be intrusive and may introduce artificiality by altering the task environment. Also, the operator may need to practice the dual task considerably before his performance level stabilizes.

Operator workload can be assessed either by manipulating primary-task difficulty and observing variations in secondary-task performance, or by manipulating secondary-task difficulty and observing variations in primary-task performance. In the *loading task paradigm*, we tell operators to maintain performance on the secondary task even if primary-task performance suffers (Ogden, Levine, & Eisner, 1979). In this paradigm, performance deteriorates more rapidly on difficult than on easy primary tasks. For example, Dougherty, Emery, and Curtin (1964) examined the workload requirements of two displays for helicopter pilots: a (then) standard helicopter display and a pictorial display. The primary task was flying at a prescribed altitude, heading, course, and air speed. The secondary, or loading, task was reading displayed digits. Primary-task performance did not differ for the two display conditions when flying was performed alone or when the digits for the secondary task were presented at a slow rate. However, at fast rates of digit presentation, the pictorial display produced better flying performance than the standard display. Thus, the mental workload requirements apparently were lower with the pictorial display.

In the *subsidiary task paradigm*, we tell operators to maintain performance on the primary task at the expense of the secondary task. Differences in the difficulty of the primary task will then show up as decrements in performance of the secondary task. This paradigm is illustrated in a study by Bell (1978) that examined the effects of noise and heat stress. For the primary task, people had to keep a stylus on a moving target. The secondary task involved an auditory stream of numbers. The people pressed a telegraph key once if a number was less than the previous number and twice if the number was greater than the previous number. Secondary-task performance was degraded by both high noise levels and high temperature, although primary-task performance was unaffected.

The human factors specialist must decide which of several types of secondary tasks to use for measuring workload. The task should be one that draws on the processing resources required by the primary task. If it does not, the workload measure will be insensitive to the workload associated with that task. Moreover, several distinct secondary tasks can be selected to provide a profile of the various resource requirements of the primary task. Some commonly used secondary tasks are simple reaction time, which involves perceptual and response-execution resources; choice reaction time, which also imposes central processing and response-selection demands; monitoring for the occurrence of a stimulus, which emphasizes perceptual processes; and mental arithmetic, which requires central processing resources.

Verwey (2000) used two different secondary tasks to diagnose the workload imposed on drivers in different road situations (e.g., standing still at a traffic light, driving straight ahead, driving around a curve, etc.). While driving a designated route, drivers also performed one of two secondary tasks: saying "yes" when they detected a visual stimulus (a two-digit number) on a dashboard display or adding 12 to an auditorily presented number and speaking the answer. The visual detection task measures visual workload, whereas the addition task measures mental workload. Visual detection performance varied greatly among different road situations, suggesting that different roads produced low, intermediate, and high levels of workload. Auditory addition performance also varied across road situations, but to a lesser extent. So, although a large part of the effect of different road situations is on visual workload, there is some influence on mental workload.

One problem with the secondary-assessment procedure as we have described it so far is its artificiality. No one is forced to add 12 to random numbers while they are driving. To minimize the interfering effects of artificiality, an embedded secondary task can be used (Shingledecker, 1980). This is a task that is part of the normal operator duties but is of lower priority than the primary task. As one example, workload can be measured for pilots using radio communication activities as an embedded secondary task. Intrusiveness is minimized in this way, as is the need for special instrumentation, but the information about workload demands that can be obtained may be restricted.

Psychophysiological Measures

There are many popular psychophysiological indices of cognition, including measurement of EEGs, event-related potentials (ERPs), and functional neuroimaging. Some of these psychophysiological indices can be used to measure workload (Baldwin, 2003; Matthews, Reinerman-Jones, Barber, & Abich, 2015). Such measures avoid the intrusiveness of a secondary task, but they introduce a new problem of requiring sophisticated instrumentation. Moreover, the possibility exists that the equipment and procedures necessary to perform the measurements may be intrusive in a different way and interfere with primary-task performance, as we discussed with the operator of the robotic arm. The major benefit of psychophysiological measures is that they have the potential to provide online measurement of the dynamic changes in workload as an operator is engaged in a task.

Many kinds of psychophysiological measures have been used to measure workload, but they all generally fall into two classes: those that measure general arousal and those that measure brain activity. General arousal level is presumed to increase as mental workload increases, and indices of arousal thus provide single measures of workload. One such technique is *pupillometry*, or the measurement of pupil diameter (Sirois & Brisson, 2014). Pupil diameter provides an indicator of the amount of attentional resources that are expended to perform a task (Beatty, 1982; Kahneman, 1973). The greater the workload demands, the larger the pupil size. The changes that occur are small but reliable, and require a pupillometer to allow sufficiently sensitive measurements. While useful as a general measure of workload, pupil diameter cannot distinguish between the different resources that are being overloaded in the performance of a task.

A second psychophysiological measure of mental workload is heart rate. Increased heart rates are correlated with increased workloads (Wilson & O'Donnell, 1988). However, because heart rate is determined primarily by physical workload and arousal level, changes in heart rate do not always indicate changes in mental workload. A better measure seems to be heart rate variability, the extent to which heart rate changes over time, which is sensitive to overall demands of sustained attention (Luque-Casado, Perales, Cárdenas, & Sanabria, 2016).

The second category of measures estimate the brain activity associated with specific processes. The most reliable of these measures involves ERPs. Presentation of a stimulus causes a short-lived or transient electrical response from the brain, which arises as a series of voltage oscillations that originate in the cortex. These transient responses can be measured by electrodes attached to the scalp; many trials must be averaged to determine the waveform of the ERP for a particular situation. Components of the evoked response are either positive (P) or negative (N). They can also be identified in terms of their minimal latencies from the onset of the stimulus event (see Figure 9.13). The P300 (a positive component that occurs approximately 300 ms after the event onset) shows amplitude and latency effects that can be interpreted as reflecting workload. The latency of the P300

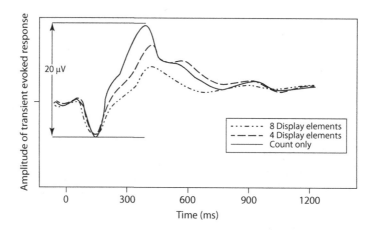

FIGURE 9.13   Amplitude of transient evoked response (P300) for tasks of different workloads.

peak is regarded as an index of stimulus-evaluation difficulty (Donchin, 1981). The amplitude of the P300 decreases as a stimulus is repeated but then increases again when an unexpected stimulus occurs (Duncan-Johnson & Donchin, 1977). Thus, the P300 seems to reflect the amount of cognitive processing performed on a stimulus.

The P300 is sensitive to the workload demands of real-world tasks. Kramer, Sirevaag, and Braune (1987) had student pilots fly a series of missions on a flight simulator. The flight was the primary task in a dual-task paradigm. The difficulty of the primary task was manipulated by varying wind conditions, turbulence, and the probability of a system failure. For the secondary task, the pilot pressed a button whenever one of two tones occurred. The P300 latency to the tones increased and the amplitude decreased with increasing difficulty of the mission, indicating that the tones were receiving less processing as the workload of the primary task increased.

The P300 is sensitive only to stimulus-evaluation processes and so can be used to assess workload associated with the detection of rare or novel stimulus events (Spencer, Dien, & Donchin, 1999). Other components of ERPs are more closely linked to early sensory processes and response-initiation processes and can be used to evaluate demands on these resources. For example, Handy, Soltani, and Mangun (2001) showed that high perceptual load can reduce the extent to which other visual stimuli are processed. High perceptual load not only reduced an observer's ability to detect a peripheral visual stimulus but also reduced the magnitude of the P100 ERP response to the stimulus, suggesting reduced processing of the stimulus in the primary visual cortex.

In sum, the P300 and other ERP measures are useful when we must assess workload in a way that does not disrupt performance of the primary task. However, recording ERPs requires sophisticated instrumentation and control procedures that may make these measures difficult to obtain. Moreover, Matthews et al. (2015) found that although several physiological measures showed sensitivity to workload variables, there was little correlation between the measures. Consequently, they advise that "Practitioners should exercise caution in using multiple metrics that may not correspond well" (p. 125).

Subjective Measures

Subjective assessment techniques evaluate workload by obtaining the operators' judgments about their tasks. Typically, we ask operators to rate overall mental workload or several components of workload. The strength of these techniques is that they are relatively easy to implement and tend to be easily accepted by operators. Given these virtues, it is not too surprising that subjective workload measures tend to be used extensively in the field. Indeed, Brookhuis and De Waard (2002) note, "In some areas such as traffic and transportation research, subjective measures and scales are rather common. It is hard to imagine research in the field without subjective measurement" (p. 1026).

Despite their usefulness, there are some limitations to subjective assessment techniques (Boff & Lincoln, 1988): (1) They may not be sensitive to aspects of the task environment that affect primary-task performance and, hence, it may be best to couple their use with primary-task measures; (2) operators may confuse perceived difficulty with perceived expenditure of effort; and (3) many factors that determine workload are inaccessible to conscious evaluation.

There are many subjective mental workload instruments, or standardized scales, in wide use. We will describe four of the most popular. The first, the modified Cooper–Harper scale, is appropriate when only an overall measure of workload is desired. The other two, the subjective workload index and the NASA Task Load Index, provide estimates of several distinct aspects of workload.

Cooper and Harper (1969) developed a rating scale to measure the mental workload involved in piloting aircraft with various handling characteristics. The scale has since been modified by Wierwille and Casali (1983) to be applicable to a variety of settings. Figure 9.14 shows how the scale involves traversal of a decision tree, yielding a rating between 1 (low workload) and 10 (high workload). The modified Cooper–Harper scale is sensitive to differences in workload and is consistent across tasks (Skipper, Rieger, & Wierwille, 1986).

The Subjective Workload Assessment Technique (SWAT) was designed initially for use with a variety of tasks and systems (Reid, Shingledecker, & Eggemeier, 1981). The procedure requires operators to judge which tasks have higher workload than others using a card-sorting procedure. Each card depicts a task that differs in three subcategories of workload (time load, mental effort load, and stress load), with three classifications for each (see Table 9.2). Time load refers to the extent to which a task must be performed within a limited amount of time and the extent to which multiple tasks must be performed at the same time. Mental effort load involves inherent attentional demands of tasks, such as attending to multiple sources of information and performing calculations.

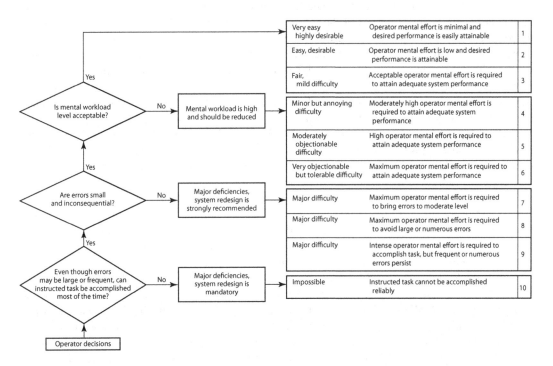

FIGURE 9.14 The modified Cooper–Harper scale.

TABLE 9.2

Three-Point Rating Scales for the Time, Mental Effort, and Stress Load Dimensions of the Subjective Workload Assessment Technique (SWAT)

Time Load	Mental Effort Load	Stress Load
1. Often have spare time. Interruptions or overlap among activities occur infrequently or not at all.	1. Very little conscious mental effort or concentration required. Activity is almost automatic, requiring little or no attention.	1. Little confusion, risk, frustration, or anxiety exists and can be easily accommodated.
2. Occasionally have spare time. Interruptions or overlap among activities occur frequently.	2. Moderate conscious mental effort or concentration required. Complexity of activity is moderately high due to uncertainty, unpredictability, or unfamiliarity. Considerable attention required.	2. Moderate stress due to confusion, frustration, or anxiety noticeably adds to workload. Significant compensation is required to maintain adequate performance.
3. Almost never have spare time. Interruptions or overlap among activities are very frequent, or occur all the time.	3. Extensive mental effort or concentration is necessary. Very complex activity requiring total attention.	3. High to very intense stress due to confusion, frustration. High to extreme determination and self-control required.

Stress load encompasses operator variables such as fatigue, level of training, and emotional state, which contribute to an operator's anxiety level.

Operators are asked to order all 27 possible combinations of the three descriptions according to their amount of workload. We then apply a process called *conjoint analysis* to the data to derive a scale of mental workload. Once the scale has been derived, we can estimate workload for various situations from simple ratings of the individual dimensions. The SWAT procedure is sensitive to workload increases induced by increases in task difficulty, as well as to those caused by sleep deprivation or increased time-on-task (Hankey & Dingus, 1990). However, the procedure is not very sensitive to low mental workloads, and the card-sorting pretask procedure is time-consuming. Consequently, Luximon and Goonetilleke (2001) developed a pairwise comparison version, in which operators choose which of two task descriptions has the higher workload. This version takes less time and yields a scale of high sensitivity.

Perhaps the most widely used subjective technique is the NASA-Task Load Index (NASA-TLX; Hart & Staveland, 1988). This index consists of six scales on which operators rate workload demands (see Table 9.3). The scales evaluate mental demand, physical demand, temporal demand, performance, effort, and frustration level. These scales were selected from a larger set on the basis of research showing that each makes a relatively unique contribution to the subjective impression of workload. An overall measure of workload can be obtained by assigning a weight to each scale according to its importance for the specific task, then calculating the mean of the weighted values of each scale. The NASA-TLX application can be downloaded from http://ece.eng.wayne.edu/~apandya/Software/NASA_TLX (Cao, Chintamani, Ellis, & Pandya, 2009).

One example of how the NASA-TLX can be used comes from studies of vigilance. Whereas previously it was thought that vigilance was relatively undemanding, observers rate the mental workload as high on the NASA-TLX, with mental demand and frustration being primary factors (Becker et al., 1991). Such results suggest that the vigilance decrement does not reflect simply a decrease in arousal and that vigilance performance requires considerable effort. Another example involves the use of NASA-TLX and SWAT measures by Airbus Industries (De Keyser & Javaux, 2000). Their goal was to demonstrate that their large aircraft could be flown safely by small, two-person crews.

TABLE 9.3
NASA-TLX Rating Scale Definitions

Title	Endpoints	Description
Mental demand	Low/high	How much mental and perceptual activity was required (e.g., thinking, deciding, calculating, remembering, looking, searching, etc.)? Was the task easy or demanding, simple or complex, exacting or forgiving?
Physical demand	Low/high	How much physical activity was required (e.g., pushing, pulling, turning, controlling, activating, etc.)? Was the task easy or demanding, slow or brisk, slack or strenuous, restful or laborious?
Temporal demand	Low/high	How much time pressure did you feel due to the rate or pace at which the tasks or task elements occurred? Was the pace slow and leisurely or rapid and frantic?
Performance	Low/high	How successful do you think you were in accomplishing the goals of the task set by the experimenter (or yourself)? How satisfied were you with your performance in accomplishing these goals?
Effort	Low/high	How hard did you have to work (mentally and physically) to accomplish your level of performance?
Frustration level	Low/high	How insecure, discouraged, irritated, stressed and annoyed versus secure, gratified, content, relaxed and complacent did you feel during the task?

The NASA-TLX and SWAT measures showed that the workload experienced by the two crew members was acceptable.

You may have noticed that although the NASA-TLX and SWAT measure different aspects of mental workload, they do not map very closely onto the multiple-resource model of attention. To address this issue, Boles, Bursk, Phillips, and Perdelwitz (2007) developed a subjective technique based on the Multiple Resources Questionnaire (MRQ). This questionnaire consists of 17 processes spanning the range of multiple resources (see Table 9.4). A crew member rates each process in a task on a scale of 0–100 (from no usage to extreme usage in the task; see Boles & Dillard, 2015). So, for example, a vocal process would be given a rating of 0 if no vocalizations were required for a task and close to 100 if there were extreme demands for vocalizations. The MRQ was better than the NASA-TLX at reflecting workload differences in multitask conditions for a vigilance task (Finomore, Shaw, Warm, Matthews, & Boles, 2013). Also, the MRQ's resource summaries conformed to the nature of the tested vigilance tasks, providing evidence that the MRQ has content validity.

Finally, there are at least two issues that limit the use of subjective measurements and their interpretation. First, the workload ratings obtained with them are sensitive only to the range of conditions to which the observers are exposed. Colle and Reid (1998) found that operators who experienced only a few levels of task difficulty rated their workloads as much higher than did operators who experienced a much broader range of difficulty levels. The authors recommend that experience with all possible difficulty levels for tasks be provided before ratings for particular task conditions are obtained. A second issue is that subjective estimates of mental workload can be different from psychophysiological or performance measures, so different that different conclusions might be reached about those situations that produce high versus low levels of workload.

ANALYTICAL TECHNIQUES

In contrast to empirical techniques, analytic techniques do not require the interaction of an operator with an operational system or simulator. Hence, they are used to estimate workload at early stages of system development. There are many analytic measurement techniques that rely on different

TABLE 9.4

Processes on the Multiple Resources Questionnaire for which a Task Is Rated

Auditory emotional process: Required judgments of emotion (e.g., tone of voice or musical mood) presented through the sense of hearing.

Auditory linguistic process: Required recognition of words, syllables, or other verbal parts of speech presented through the sense of hearing.

Facial figural process: Required recognition of faces, or of the emotions shown on faces, presented through the sense of vision.

Facial motive process: Required movement of your own face muscles, unconnected to speech or the expression of emotion.

Manual process: Required movement of the arms, hands, and/or fingers.

Short-term memory process: Required remembering of information for a period of time ranging from a couple of seconds to half a minute.

Spatial attentive process: Required focusing of attention on a location, using the sense of vision.

Spatial categorical process: Required judgment of simple left-versus-right or up-versus-down relationships, without consideration of precise location, using the sense of vision.

Spatial concentrative process: Required judgment of how tightly spaced are numerous visual objects or forms.

Spatial emergent process: Required "picking out" of a form or object from a highly cluttered or confusing background, using the sense of vision.

Spatial positional process: Required recognition of a precise location as differing from other locations, using the sense of vision.

Spatial quantitative process: Required judgment of numerical quantity based on a nonverbal, nondigital representation (e.g., bar graphs or small clusters of items), using the sense of vision.

Tactile figural process: Required recognition or judgment of shapes (figures), using the sense of touch.

Visual lexical process: Required recognition of words, letters, or digits, using the sense of vision.

Visual phonetic process: Required detailed analysis of the sound of words, letters, or digits, presented using the sense of vision.

Visual temporal process: Required judgment of time intervals, or of the timing of events, using the sense of vision.

Vocal process: Required use of your voice.

---

estimators of workload. Consequently, it is best to use a battery of techniques to assess the workload demands of any specific system. In the following sections, we discuss five categories of analytic techniques (Lysaght et al., 1989): comparison, expert opinion, mathematical models, task analysis methods, and simulation models.

Comparison

The comparison technique uses workload data from a predecessor system to estimate the workload for a system under development. One systematic use of the comparison technique was reported by Shaffer, Shafer, and Kutch (1986). They estimated the mission workload for a single-crewmember helicopter based on data from an empirical workload analysis of missions conducted with a two-crewmember helicopter. This technique is useful only if workload data from a predecessor system exist, which often is not the case.

Expert Opinion

One of the easiest and most extensively employed analytic techniques is expert opinion. Users and designers of systems similar to the one being developed are provided with a description of the proposed system and asked to predict workload, among other things. The opinions can be obtained informally or formally (and more formal methods are better). For example, SWAT has been modified for prospective evaluations from experts. The major modification is that the ratings are based on a description of the system and particular scenarios rather than on actual operation of the system. In the evaluation of pilot workload for military aircraft, prospective ratings using SWAT (and other methods) correlate highly with workload estimates made on the basis of performance (Eggleston & Quinn, 1984; Vidulich, Ward, & Schueren, 1991).

Mathematical Models

Many attempts have been made to develop mathematical models of mental workload. Models based on information theory (see Appendix II) were popular in the 1960s. One model by Senders (1964) assumed that an operator with limited attentional capacity samples information from a number of displays. The channel capacity for each display and the processing rate of the operator determined how often a display must be examined for the information in it to be communicated accurately. The amount of time that an operator devotes to any particular display could thus be used as a measure of visual workload.

In the 1970s, models based on manual control theory and queuing theory became popular. Manual control models apply to situations where continuous tasks, such as the tracking of a target, must be performed. These rely on minimization of error via various analytical and theoretical methods. Queuing models view the operator as a server that processes a variety of tasks. The number of times that the server is called upon provides a measure of workload. Although development of these mathematical models has continued, their use for workload estimation has diminished in recent years as computerized task analyses and simulations have been developed.

Task Analysis

As noted in earlier chapters, task analysis decomposes the overall system goal into segments and operator tasks, and ultimately into elemental task requirements. The analysis provides a time-based breakdown of demands on the operator. Consequently, most task-analytic measures of mental workload focus on estimation of time stress, which is the amount of mental resources required per unit time relative to those that are available. One exception is the McCracken–Aldrich technique (Aldrich & Szabo, 1986; McCracken & Aldrich, 1984), which distinguishes five workload dimensions: visual, auditory, kinesthetic, cognitive, and psychomotor. For each task element, ratings are made on a scale from 1 (low workload) to 7 (high workload) for each task dimension. Estimates of the workload on each dimension are made during half-second intervals by summing the workload estimates for all active task components. If the sum exceeds 7, then it is assumed that an overload exists for that component.

Simulation Models

A simulation model is probabilistic and, hence, will not yield the same result each time it is run. There are several simulation models that can be used to provide workload estimates. Most are variants of the Siegel and Wolf (1969) stochastic model discussed in Chapter 3. In that model, workload is indicated by a variable called "stress," which is affected by both the time to perform tasks and the quantity of tasks that must be performed. Stress is the sum of the average task execution times divided by the total time available. Several extensions of this technique have been developed that allow greater flexibility in the prediction of workload (Lysaght et al., 1989).

SUMMARY

Attention research exemplifies the ideal of a close relationship between basic and applied concerns in human factors. The resurgence of interest in attention arose from applied problems, but it has led to much basic, theoretical work on the nature of attentional control. This basic work, in turn, has led to better measures of the attentional requirements in applied settings.

Often, operators must perform tasks that require selectively attending to specific sources of information, distributing attention across multiple sources of information, or maintaining attention on a single display for long periods of time. We can apply what we know about how attention works to the design of systems for effective performance under these different situations. For example, we know that presentation of information in different modalities avoids decrements in performance due to competition for perceptual resources and improves memory for unattended information.

More generally, assessment of mental workload can help determine which tasks can be performed simultaneously with little or no decrement. Because mental workload varies as a function of the perceptual, cognitive, and motoric requirements imposed on an operator, the structure of a task and the environment in which it is performed can significantly affect workload and performance.

RECOMMENDED READINGS

Gopher, D., & Donchin, E. (1986). Workload: An examination of the concept. In K. R. Boff, L. Kaufman, & J. P. Thomas (Eds.), *Handbook of Perception and Human Performance. Vol. II: Cognitive Processes and Performance* (pp. 41-1–41-49). New York: Wiley.
Hancock, P. A., & Meshkati, N. (Eds.). (1988). *Human Mental Workload*. Amsterdam: North-Holland.
Johnson A., & Proctor, R. W. (2004). *Attention: Theory and Practice*. Thousand Oaks, CA: Sage.
Kramer, A. F., Wiegmann, D. A., & Kirlik, A. (Eds.) (2007). *Attention: From Theory to Practice*. New York: Oxford University Press.
Nobre, K., & Kastner, S. (Eds.) (2014). *Oxford Handbook of Attention*. Oxford, UK: Oxford University Press.
Parasuraman, R. (Ed.). (1998). *The Attentive Brain*. Cambridge, MA: MIT Press.
Pashler, H. E. (1998). *The Psychology of Attention*. Cambridge, MA: MIT Press.
Posner, M. I. (Ed.) (2004). *Cognitive Neuroscience of Attention*. New York: Guilford.
Styles, E. A. (2006). *The Psychology of Attention* (2nd ed.). Hove, UK: Psychology Press.
Wickens, C. D., & McCarley, J. S. (2008). *Applied Attention Theory*. Boca Raton, FL: CRC.

10 Retention and Comprehension of Information

Memory does not comprise a single unitary system, but rather an array of interacting systems, each capable of encoding or registering information, sorting it, and making it available by retrieval. Without this capability for information storage, we could not perceive adequately, learn from our past, understand the present, or plan for the future.

A. Baddeley
1999

INTRODUCTION

Human memory is intricate and diverse. Over a lifetime, you will learn vast amounts of information and retain it for various amounts of time. The important role played by memory in virtually all aspects of human life is evident when one considers the severe consequences of the memory deficits characteristic of Alzheimer's disease (Ryan, Rossor, & Fox, 2015). This disease is characterized in its late stages by its victims becoming lost in familiar environments and failing to recognize immediate family members. Memory is involved not only in identification and recognition of places and people, but also in remembering task goals and maintaining a "set," or appropriate readiness, to perform particular tasks. It is also involved in maintaining information in a readily available form for comprehending new information, solving problems, and retrieving facts and procedures that one has learned in the past.

It is an unappreciated fact that memories can be distorted and that a person's ability to retrieve his memories depends on many environmental and contextual conditions (Marsh & Roediger, 2013). This characteristic of memory can lead to many types of memory failures and errors. The human factors professional needs to know and appreciate that how well an operator learns and remembers plays an important role in his or her ability to perform within a human–machine system. In most circumstances, successful performance of the system depends on the operator's ability to recognize and retrieve information from memory. Human factors specialists can improve human performance by ensuring that environments and training materials will support learning, retention, and retrieval of important information.

There are a lot of ways that we can talk about memory, including the kind of information that is "stored" and the cognitive processes that allow storage and retrieval of information. One prevalent way of thinking about the kinds of information that can be stored in memory is to distinguish between semantic and episodic memory (Tulving, 1999): Semantic memory refers to a person's basic knowledge, such as the fact that a dog is a friendly animal with four legs that barks, whereas episodic memory refers to specific events (or episodes), such as that Fido bit the mail carrier this morning. Different processes might require us to talk about specific features of the environment that promote or interfere with the storage and retrieval of different kinds of information.

In the first part of the chapter, we focus primarily on episodic memory. A very popular way of thinking about episodic memory is called the *modal model* (Atkinson & Shiffrin, 1968; Thorn & Page, 2009; see Figure 10.1). In this model, when information is first presented, it is retained with almost perfect fidelity for no longer than a few seconds in the form of a sensory memory. Only some of this information is then encoded in a more durable form, called *short-term memory*. Short-term memories are retained for a period of around 10–20 s unless they are kept active through rehearsal,

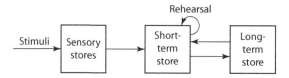

FIGURE 10.1   The three-store memory model.

or covert repetition. Finally, some of the information in short-term memory is transferred to long-term memory and retained for an indefinite duration.

The present chapter organizes our knowledge of human memory around the distinction between sensory, short-term, and long-term memories. Each type of memory has distinct properties that affect human performance in a wide variety of situations. We will describe these properties, as well as the important factors that affect the acquisition, retention, and retrieval of information of each. Although we know that the modal model of memory is not a completely accurate portrayal of the human memory system, it is useful for organizing our knowledge of human memory. In the last part of the chapter, we examine the role of memory in the comprehension and retention of written and spoken information.

SENSORY MEMORY

The sensory effects of stimuli persist for a short period of time after they have been removed from the environment. For example, when a letter is displayed briefly, its perceived duration exceeds its physical duration (e.g., Haber & Standing, 1970); that is, the display visibly persists. Researchers have shown that it is possible to retrieve information from the persisting representation of the display, in addition to the display itself. These and related findings have been taken as evidence for the existence of sensory memories that are thought to exist for each sensory modality.

VISUAL SENSORY MEMORY

Research on visual sensory memory was inspired by a memory limitation called the *span of apprehension*, known since the 1800s (Cattell, 1886). This span refers to the number of simultaneous, briefly displayed visual stimuli that can be recalled without error. For example, an array of letters may be presented to you briefly, and your task is to report as many of the letters as possible. This is a whole report task. If the array is small enough (four or five letters), then you can report all of the letters correctly. However, when the arrays are larger, you will report only a subset, usually around four or five letters—the same number of letters in a small display that you could report correctly. This is the span of apprehension.

Although large arrays of stimuli cannot be identified with complete accuracy, observers often claim that they can see the whole display at first but it "disappears" before all of the stimuli can be identified (Gill & Dallenbach, 1926). Ingenious experiments by Sperling (1960) and Averbach and Coriell (1961) established that observers could indeed see more letters than they could report. Instead of whole report, these researchers used a procedure known as *partial report*, in which only some of the letters are to be reported. Sperling showed people three rows of four letters very briefly. At varying times after the display, he then played a high-, medium-, or low-frequency tone as a cue to indicate whether they should report the top, middle, or bottom row. When the tone occurred immediately at the offset of the array, the cued row could be reported with almost perfect accuracy, regardless of which row was cued. Because people could not know in advance which row would be cued, this suggests that they could see all of the letters in the display. With whole report it appeared as though they could see only four or five letters, but with partial report it is apparent that they could see all of them. This difference is called the *partial report superiority effect*.

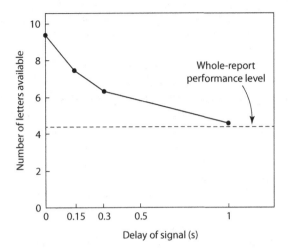

FIGURE 10.2 Partial report accuracy as a function of delay.

If the tone was delayed by as little as one-third of a second, report accuracy decreased to the four or five letters measured by the span of apprehension (see Figure 10.2). That is, report accuracy was at the level that would be expected if only four or five items were available. For delay times between the end of the letter array and one-third of a second later, partial report superiority could be seen to fade away. When a distracting array of random contours immediately followed the display, it interfered with sensory memory of the display, resulting in no partial report superiority even when the cue was not delayed.

These and other results obtained by Sperling (1960) led to the conclusion that visual stimuli persist visibly in a high-capacity sensory-memory store that decays within a second and is susceptible to disruption by subsequent visual stimulation. Sperling's research was extremely influential, resulting in many experiments exploring the properties of what came to be called *iconic memory* for visual stimuli and sensory memory more generally (see Cowan, 2008; Nairne & Neath, 2013). These experiments showed that *informational persistence* of the type evidenced by partial report superiority is distinct from *visible persistence* of the type demonstrated by tasks that require temporal integration of visual information (Coltheart, 1980; 2009; see Chapter 5). Completion of these tasks cannot be done unless the stimulus is "visible." For example, Haber and Standing (1970) asked people to estimate how long a briefly flashed array of letters was visible. They perceived that the array lasted longer than its actual duration. In another experiment, Eriksen and Collins (1967) asked people to report a three-letter nonsense syllable that could only be identified after integrating two successively presented random-dot patterns. They could accomplish this easily when the interval between the two patterns was less than 50 ms, but their report accuracy decreased drastically to about chance levels with an interval of a third of a second.

Although alternative measures of visible persistence correlate highly with each other, they do not correlate much with the information persistence measure of partial report performance (Loftus & Irwin, 1998). For example, the estimated duration of visible persistence is less than that of information persistence. Also, whereas visible persistence decreases as stimulus duration and luminance increase, partial report accuracy increases. Thus, it is generally accepted that the visible persistence that you can "see" is different from the informational persistence that results in the partial report superiority effect.

TACTILE AND AUDITORY SENSORY MEMORIES

Sensory stores with properties similar to those of iconic memory seem to exist for the other senses. In particular, sensory stores for touch and audition have been examined (Bliss, Crane, Mansfield, &

Townsend, 1966; Darwin, Turvey, & Crowder, 1972). In the same way that visual sensory memory can be disrupted by a distracting visual array, auditory sensory memory can be disrupted by a distracting auditory stimulus (e.g., Beaman & Morton, 2000).

The human factors specialist needs to remember that memory for auditory information can be disrupted by distracting auditory stimuli. This point is illustrated in a study by Schilling and Weaver (1983). Telephone operators working directory assistance at a local utility were told to say "Have a nice day" at the end of each transaction. Schilling and Weaver wondered whether this parting message could interfere with callers' memory for the telephone numbers. Subjects in their experiments were instructed to obtain and dial numbers under situations similar to those for real directory assistance clients. On each trial, the subject dialed 411, requested and received a prerecorded seven-digit phone number, and then attempted to dial the number. In one condition, the phrase "Have a nice day" immediately followed the prerecorded number, whereas in other conditions the subject heard either a tone or nothing after the phone number. Fewer phone numbers were dialed correctly in the "Have a nice day" condition than in the other two conditions. In the "Have a nice day" condition, subjects had the most difficulty remembering the last two digits of the number, the ones most likely to still be in auditory sensory memory. Thus, the telephone company's attempt to be polite may have actually interfered with the client's goal of remembering a phone number.

WHAT IS THE ROLE OF SENSORY MEMORY?

Our early understanding of sensory memory was that it served as temporary storage for sensory information to get ready for further processing. For instance, it was suggested that visual sensory memory creates a continuous perception of the world by integrating discrete visual images across saccadic eye movements (e.g., Breitmeyer, Kropfl, & Julesz, 1982). However, the duration of visible persistence is too short to serve this purpose, although it still sometimes may be important in integrating temporally separate events (Loftus & Irwin, 1998). Creating a continuous perception is more likely a role played by auditory sensory memory, because integration across short time periods is necessary for comprehension of complex stimuli such as speech and music (Crowder & Surprenant, 2000).

Another possibility is that persistence is simply a consequence of imperfect temporal resolution within the sensory systems and nothing more (Loftus & Irwin, 1998). No matter what the role of sensory memory in human information processing, the important point for the human factors specialist is that the effects of sensory stimulation will persist for a brief period of time after the stimulation is removed. These effects may serve as a substitute for the physical stimuli (Rensink, 2014) and might influence an operator's judgments about what she is perceiving.

SHORT-TERM MEMORY

You have probably at one time or another used a website—or even a telephone book!—to look up a phone number to call someone. After finding the number, you probably repeated the number to yourself until you dialed the number. If you were distracted, you probably forgot the number and had to look it up again. Experiences like these suggest that there is a short-term memory of limited capacity. Information in short-term memory must be "rehearsed" to be retained.

Short-term memory limits the performance of operators in a wide variety of situations. For example, an air-traffic controller must remember such things as the locations and headings of many different aircraft and the instructions given to each one (Garland, Stein, & Muller, 1999). Similarly, radio dispatchers for a taxi company must remember the taxis that are available and their locations. For tasks that rely on short-term memory, performance can be affected greatly by how information is presented and how the task is structured.

BASIC CHARACTERISTICS

Two studies that revealed a lot about how short-term memory works were conducted in different laboratories by Peterson and Peterson (1959) and Brown (1958). These researchers showed people three consonants (e.g., BZX) to remember on a trial. This triplet can be recalled easily not just seconds later but minutes later if the rememberer is not distracted. However, the people in this experiment were required to count backward by threes from a three-digit number until they were instructed to recall the letters. Brown and Peterson and Peterson assumed that the mental activity required to count backward would prevent rehearsal of the letters, causing them to be lost from memory. After 8 s of distraction, only about half of the letters could be recalled correctly, and after 18 s, very few could be recalled (see Figure 10.3). This suggests that without rehearsal, short-term memory for the letters lasted only a few seconds.

The rapid forgetting that occurs when rehearsal is prevented reflects two types of memory errors (Estes, 1972). A transposition or order error occurs when the correct items are recalled but in the wrong order (e.g., BZX could be recalled as BXZ). An intrusion or item error occurs when an item that was not in the list is recalled (e.g., BZX could be recalled as BGX). These two kinds of errors seem to be due to different kinds of processes (Nairne & Kelley, 2004). Order errors tend to occur more frequently than intrusion errors when the items must be remembered for only a short period of time. As time increases, item errors increase. This suggests that memory for the order in which items occurred is lost more quickly than memory for the items themselves. So, if an operator's task does not require remembering the exact order in which information occurred, task performance will not suffer as much from delays in responding.

An important constraint on the role of rehearsal was demonstrated by Keppel and Underwood (1962). They showed that recall is virtually perfect for a single set of items even after 18 s. However, after the third or fourth set of items is presented, recall deteriorates to pure guessing. So, short-term forgetting does not occur by "decay" alone. It seems as though sets of items presented at different times can interfere with each other. *Proactive interference* refers to memory for earlier presented items interfering with memory for later items. Proactive interference can be reduced, improving short-term memory, if items are presented only once every few minutes (Peterson & Gentile, 1963). Proactive interference also can be reduced by changing the semantic characteristics of later items from those of previous items (for example, changing word categories from fruits to flowers) or, to a lesser extent, by changing physical characteristics (for example, changing font type, size, or color;

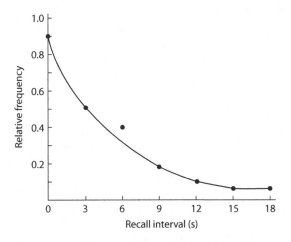

FIGURE 10.3 Recall performance as a function of a filled retention interval in a short-term memory task.

Wickens, 1972). In sum, the accuracy of short-term memory can be improved by increasing the intervals between successive messages and by making each message somehow distinctive.

The kind of information that is stored in short-term memory seems to have a strong acoustic component. That is, although semantic or visual information may be stored (Shulman, 1970; Tversky, 1969), much of the information is represented by how it sounds (Conrad, 1964). Conrad showed that intrusion errors were acoustically similar to the original items. For example, in an experiment similar to those of Brown (1958) and Peterson and Peterson (1959), the visual letter B was often remembered as the acoustically similar letter V, even though the two letters look nothing alike. This means that sets of items that are acoustically confusable (that is, sound alike) will produce more short-term retention errors than sets that are not.

You may already be familiar with a classic paper entitled *The Magical Number Seven, Plus or Minus Two: Some Limits on Our Capacity for Processing Information*, by George Miller (1956; see also Chapter 4). In this paper, Miller measured the capacity of short-term memory as seven plus or minus two *chunks*, or units, of information. If you try to remember isolated digits or letters, this capacity represents the number of them that you can recall correctly. However, if you group the items into larger chunks, recall can be vastly improved. For example, if a person attempts to remember the list CBSABCNBC as a string of separate letters, the string is nine chunks long, greater than the seven chunks that can be held easily in short-term memory. Because short-term memory is overloaded, recall of all nine letters may be difficult. However, if the person recognizes that the string can be coded as acronyms of three major television networks, CBS, ABC, and NBC, the string can be encoded as three chunks and will therefore be much easier to remember. More recent research confirms that short-term memory capacity is a function of chunks rather than the objective number of items to be remembered (Cowan, Chen, & Rouder, 2004).

Strings of digits are often used for such things as telephone numbers, bank accounts, customer identification, and so on. From a customer's perspective, these numbers are essentially random. It is very difficult to remember random strings of digits, so chunking is an important strategy that can be used to remember them. Some important chunking strategies involve the size of the chunk and the modality in which the information is presented. Wickelgren (1964) showed that lists of digits are easiest to remember if they are organized into groups of a maximum of four. Grouping provides a better benefit when digits are presented auditorily rather than visually, because people tend to chunk visual digits into pairs even when they are not grouped (Nordby, Raanaas, & Magnussen, 2002).

IMPROVING SHORT-TERM RETENTION

The limited capacity of short-term memory has implications for any situation that requires an operator to encode and retain information accurately for brief periods of time. Memory performance can be improved by using techniques that minimize activities intervening between presentation of the information and action on it, using sets of stimuli that are not acoustically confusable, increasing the interval between successive messages, making the to-be-remembered material distinct from preceding material, and grouping the information into chunks.

Several of these techniques were exploited in a study by Loftus, Dark, and Williams (1979) that examined communication errors between ground control and student pilots in a short-term memory task. Memory was tested for two types of messages: (1) a place for the pilot to contact plus a radio frequency (for example, "contact Seattle center on 1.829") and (2) a transponder code (for example, "squawk 4273," which means set the transponder code to 4273). The codes were presented in two-digit chunks (for example, "forty-two, seventy-three") or as unchunked single digits (for example, "four, two, seven, three"). In a low-memory-load condition only one of the two message types was presented, whereas in a high-load condition both were presented. After the message(s), the pilot had to read off a sequence of rapidly presented letters for varying amounts of time. When this task was completed, the pilot wrote down the original message(s) on a piece of paper.

Recall was worse in the high-load condition, which required more information to be retained. Moreover, recall of the radio frequency was better when the transponder code was chunked, suggesting that chunking made more short-term memory capacity available for other information. Loftus et al. (1979) concluded that as little information as possible should be conveyed to a pilot at one time. They also proposed that a delay of at least 10 s should intervene between successive messages, because they observed that on any trial, the longer it had been since the immediately preceding message had been presented, the more memory performance improved. Additionally, they showed that the response to a message should be made as quickly as possible to avoid error, and that alphanumeric strings should be chunked whenever possible.

Not only does the size of the chunks influence the accuracy of short-term memory, but so does the nature of the chunks, whether numbers or digits. Preczewski and Fisher (1990) examined the format of call signs used by the military in secured radio communications. The U.S. Army uses two-syllable codes of the sequence letter-digit-letter (LDL) followed by the sequence digit-digit (DD). These codes make radio communication very difficult, and they change at least once a day. Preczewski and Fisher compared the memorability of the current code format (LDL-DD) with that of three other formats: DD-LDL, DD-LLL, and LL-LLL. Operators were presented with a call sign, which they were to recall later. During a 10-s delay between presentation and recall, the operators read aloud strings of letters and digits. Performance was best when one syllable was composed only of digits and the other of letters (DD-LLL) and worst on the current code. Thus, mixing letters and digits within chunks seems to be harmful, whereas mixing them between chunks is beneficial.

Specific alphanumeric characters differ in their memorability. We have already seen that confusions can be reduced by using letters that do not sound similar to other letters (Conrad, 1964). Chapanis and Moulden (1990) investigated the memorability of individual digits, as well as of doublets and triplets, within eight-digit numbers. People viewed a number for 5 s, then immediately entered it on a numeric keyboard. Many errors were made, as would be expected, because the length of the number was greater than the normal memory span of seven. The digit that was remembered best was 0. The remaining digits, in order of their memorability, were 1, 7, 8, 2, 6, 5, 3, 9, and finally 4. Doublets were generally easier to recall if they contained a zero or if they contained the same digit twice. A similar pattern of results was found for triplets. Based on these findings, the authors provide tables that designers can use to construct numeric codes that are easy to remember.

MEMORY SEARCH

For many tasks, accurate performance requires not only that information be retained in short-term memory, but also that the information be acted on quickly. The time required for search and retrieval from short-term memory has been investigated extensively using a memory search task (Sternberg, 1966, 2016). In this task, which we introduced briefly in Chapter 4, observers are presented a set of one or more items (such as digits, letters, or words) to be held in short-term memory. Shortly thereafter, a single target item is presented. The observer is to indicate as quickly as possible whether or not the target was in the memorized set, usually by making one of two key presses indicating its presence or absence.

In Sternberg's (1966) study, the memory set was composed of one to six digits, followed by a single target digit. Reaction time increased as a linear function of the memory-set size (see Figure 10.4). The rate of increase was approximately 38 ms per item in the memory set. Sternberg interpreted these data as support for the idea that a rapid, serial scan was performed on the memory set. That is, when presented with the target, the observer compared the target with each item in the memory set, one at a time. If each comparison takes 38 ms, for example, the function relating response time to memory-set size will be linear with a slope of 38 ms and with an intercept equal to the time taken by all other processes not involved in the comparison, such as perceptual and response processes.

Although Sternberg's findings are consistent with those expected from a serial search of the memory set (see Sternberg, 2016), we can devise more complicated, non-serial processes (using, for

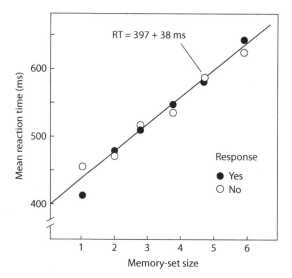

FIGURE 10.4 Memory search time as a function of set size and response.

example, variable comparison times) that will also predict linear reaction-time functions (Atkinson, Holmgren, & Juola, 1969; Townsend, 1974). Thus, Sternberg's findings are not conclusive evidence for serial comparison processes. Moreover, other experiments have shown findings, such as faster responses to repeated or the last items in the memory set (Baddeley & Ecob, 1973; Corballis, Kirby, & Miller, 1972), that are inconsistent with the basic assumptions of the serial search model.

Regardless of the exact nature of the search process, the slope of the function relating reaction time to memory-set size is an indicator of short-term memory capacity. Items such as digits, for which the memory span is large, show a slope that is considerably less than that for items such as nonsense syllables, for which the span is smaller (Cavanagh, 1972). The slope can be used as an indicator of the demands placed on short-term memory capacity. Moreover, this measure of memory capacity can be isolated from perceptual and motor factors that affect only the intercept. Consequently, the memory search task has been used as a secondary-task measure to assess mental workload (see Chapter 9).

The memory search task was used by Wickens, Hyman, Dellinger, Taylor, and Meador (1986) to study the mental workload demands imposed on instrument-rated pilots at various phases of flight. An instrument-rated pilot is one who is qualified to fly on instrument readings alone, without visual contact outside the cockpit window. Wickens et al. asked the pilots to fly an instrument-only holding pattern and instrument-only landing approach in a flight simulator while performing a memory search task. The intercepts of the search functions were greater during the approach phase than during the holding phase, but the slopes of the functions did not differ. Consequently, Wickens et al. (1986) concluded that the approach phase increases perceptual and response loads. Because there was no change in the slope of the search function, they concluded that short-term memory load was no greater during approach than during holding. They recommended that pilots should not be asked to perform tasks that require perceptual-motor processing while landing a plane.

MODELS OF SHORT-TERM, OR WORKING, MEMORY

Recent work on short-term memory has focused on its function, which seems to be primarily one of temporarily storing and manipulating information. To distinguish this approach from the study of short-term memory, the term *working memory* is often used. Working memory is involved in performing calculations necessary for mental arithmetic, comprehending the meaning of a sentence, elaborating the meaning of material, and so on. As Jonides, Lacey, and Nee (2005) indicate, "Without working memory, people would not be able to reason, solve problems, speak, and understand language" (p. 2).

Baddeley and Hitch's Working Memory Model

The most popular model of working memory yet developed is the one proposed by Baddeley and Hitch (1974), illustrated in Figure 10.5. The model has two storage systems, called the *phonological loop* and *visuospatial sketchpad*, and a control system, called the *central executive*. The phonological loop consists of a phonological store, in which the information is represented by phonological codes, and an articulatory rehearsal process that essentially involves saying the items over and over to yourself. Consistently with the findings from studies discussed earlier, memory traces in the phonological store are lost within a few seconds unless maintained by the articulatory rehearsal process. Evidence for this aspect of the model comes from a finding called the *word length effect*. In the word length effect, the number of words you can hold in short-term memory decreases as the number of syllables in those words increases. This might occur because it takes longer to rehearse words that have more syllables in them. The phonological loop might play a role in vocabulary acquisition, learning to read, and language comprehension.

The visuospatial sketchpad is a store for visuospatial information. Similarly to the phonological loop, it is limited to a capacity of only a few objects (Marois & Ivanoff, 2005). The sketchpad is presumed to be involved not only in the memory for visually presented objects but also in visual imagery. The primary role of the sketchpad is to hold and manipulate visuospatial representations, which are important for artistic and scientific creativity. The central executive is an attentional control system that supervises and coordinates the phonological loop and the visuospatial sketchpad. This emphasis on attentional control suggests that working memory is closely related to attention. In fact, some of the tasks the central executive performs include focusing and dividing attention, switching attention from one task to another, and coordinating working memory with long-term memory. Baddeley (2000, 2003) later proposed a fourth component to working memory, the *episodic buffer* (see Figure 10.6). This subsystem integrates information from the other working

FIGURE 10.5 Working-memory model.

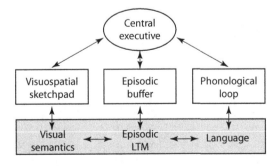

FIGURE 10.6 The revised working memory model.

memory subsystems and long-term memory into a common code. The central executive controls this subsystem, as it does the others, and the information in it plays a role in the formation of conscious experience.

This way of thinking about working memory implies that tasks should not interfere if they use different subsystems. For example, when people are asked to remember something (for example, a set of digits) to be recalled after another task is performed, the memory load (for example, the number of digits) often has little effect on the performance of that task. Baddeley (1986) varied the number of digits that people were to remember while they performed a reasoning or learning task that required central executive processes but not the phonological loop. Consistently with this theory of working memory, people were able to maintain a memory load of up to eight items without its interfering much with the performance of either task.

Another prediction from the model is that tasks sharing the same subsystem should interfere. Brooks (1968) reported evidence of such interference in the visuospatial sketchpad. Observers were asked to imagine a block letter, for example, the letter F (see Figure 10.7). They were told to mentally trace around the letter, starting from a designated corner and proceeding around the perimeter, responding to each successive corner with *yes* if it was either at the top or bottom of the figure, or *no* if it was in the middle. One group of observers responded vocally, and a second group tapped either of their index fingers. A third group had to point to a column on a sheet of paper that contained either Ys or Ns in a sequence of staggered pairs (see Figure 10.7). It took much longer to respond in the third group, presumably because the task required visually perceiving the locations of the Ys and Ns while at the same time visualizing the locations on the letter. These results are not due to the pointing responses being more difficult than the other kinds of responses, because pointing did not cause interference for a similar task that did not require the visuospatial sketchpad.

In considering why the working memory model has been popular for many years, one reason given by Baddeley (2016, p. 121) is "Because it has just four components, each of which is relatively easy to understand, it can be applied to practical issues …" Human factors specialists and designers need to be aware that different tasks may interfere with each other to the extent that they require common components in working memory. According to Fiore, Cuevas, and Salas (2003, p. 511),

An understanding of the relation between [working memory] and complex task performance is essential for the development of effective training and system design. Because many of today's tasks require one to monitor multiple system parameters, each potentially composed of input from differing modalities, and often require the integration of this information, we maintain that these differing task components uniquely impact systems of [working memory].

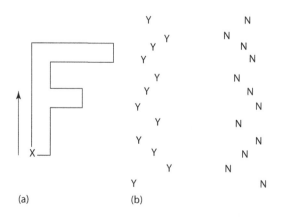

(a)   (b)

FIGURE 10.7   Block-letter stimulus and "yes"-"no" response display.

Cowan's Activation Model

Another influential memory model is that of Cowan (1997), illustrated in Figure 10.8. Cowan's model places even more emphasis than the working memory model on the relationship between attention and memory. In this model, the contents of the short-term store are activated long-term memories, and what we are consciously aware of at any moment (the objects or events that are in the focus of attention) is only a subset of the available information in the short-term store. Cowan's model includes a brief sensory store, which corresponds to the stimulus traces that produce sensory persistence. The other component of sensory memory, informational persistence, is part of the short-term store. Like the working memory model, Cowan's activation model includes a central executive that directs attention and controls voluntary processing.

IMAGERY

The nature of visual imagery has been the subject of many working memory experiments. We already reviewed in Chapter 4 some studies that showed how observers can mentally rotate objects into similar orientations to determine whether the objects are the same. Other researchers, most notably Kosslyn (1975; Kosslyn & Thompson, 2003), argue that imagery is very much like perception. Several of their studies suggested that images are mentally scanned in a manner similar to visually scanning a picture. For example, Kosslyn, Ball, and Reiser (1978) had observers memorize a map of a fictional island containing a number of objects (see Figure 10.9). The experimenter then read aloud the name of one of the objects on the island. The observer was to imagine the entire map but focus on the specific object. Five seconds later, the experimenter named a second object. The observer was then to mentally scan to the location of the second object and press a response key as soon as it was reached. The farther apart the two objects were, the longer it took to mentally scan from the first to the second object.

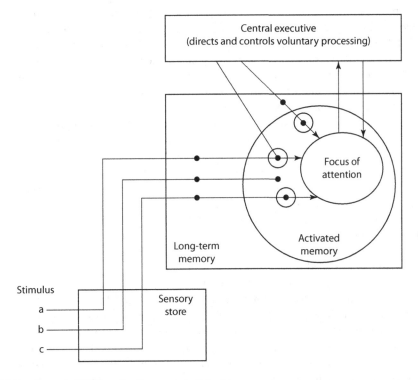

FIGURE 10.8 Cowan's (1997) activation model of short-term memory.

FIGURE 10.9   Map of a fictional island used by Kosslyn et al. (1978).

Kosslyn (1975) also provided evidence that it is more difficult to make judgments about components of small images than of large images. He asked observers to imagine a particular animal next to either a fly or an elephant. He argued that the animal would have a larger mental image next to the fly than next to the elephant. When the observer was asked to verify whether a certain property, such as fur, was part of the imagined animal, responses were faster if the animal was imagined as large (next to the fly) than if it was imagined as small (next to the elephant).

The concept of an imagery component to working memory has been extended to the notion of a *mental model*: a dynamic representation or simulation of the world (see Chapter 11; Johnson-Laird, 1983, 1989). Johnson-Laird has argued that mental models are a form of representation in working memory that provides the basis for many aspects of comprehension and reasoning (Johnson-Laird, Khemlani, & Goodwin, 2015). The key element of the mental model concept is that thinking about events involves mentally simulating different possible scenarios for that event. This means that if an operator has an accurate mental model of a task or system, she or he may be able to solve problems by visualizing a simulation of the task or of system performance. People with low working memory capacity are less likely to construct an accurate mental model from a verbal description, leading to more reasoning errors (Oberauer, Weidenfeld, & Hornig, 2006).

Fiore et al. (2003) emphasize that the connection between the concepts of working memory and mental models "can facilitate a deeper understanding of issues associated with cognitive engineering and decision making research" (p. 508). In this regard, Canas et al. (2003) confirmed that participants' mental models for a device controlling an electrical circuit relied heavily on visual working memory, by showing that the requirement to maintain a visuospatial memory load (the locations of dots) while making judgments about the circuit was highly interfering, whereas a requirement to maintain a verbal memory load (a set of letters) was not.

Our present understanding of short-term memory is considerably more detailed than it was in the 1960s, when research on short-term memory first began in earnest. Now, we know that short-term memory is not merely a repository for recent events but is closely related to attention and plays a crucial role in many aspects of cognition. Short-term memory is used to temporarily represent and retain information in forms that are useful for problem solving and comprehension. Careful consideration of the short-term and working memory demands imposed by different tasks will ensure better performance in many human factors contexts.

LONG-TERM MEMORY

Long-term memories persist from childhood throughout our lives. These memories are qualitatively different from the short-term memories that require continuous rehearsal if they are to persist. Cowan's (1997) model, which depicts short-term memory as information activated from long-term memory, makes it clear that long-term memory is involved in all aspects of information processing. An operator must often retrieve information from long-term memory to comprehend current system information and to determine what action is appropriate. Failure to remember such things as previous instructions may have catastrophic consequences.

The Japanese air raid at Pearl Harbor on December 7, 1941, one of the worst disasters in U.S. naval history, provides an example. Several factors contributed to the U.S. Navy's being unprepared for the attack. One important factor was that officers forgot how they were told to interpret events that could signal imminent attack (Janis & Mann, 1977). Five hours before the attack, two U.S. mine sweepers spotted a submarine that they presumed to be Japanese just outside of Pearl Harbor. The sighting was not reported, presumably because the officers had forgotten an explicit warning, given 2 months earlier, that a submarine sighting was extremely dangerous because it could signal the presence of a nearby aircraft carrier. Had the officers remembered the warning, naval forces could have been put on alert and been prepared for the attack that followed. If the information had been presented in a way that enhanced long-term retention of the warning, the ability of the officers to retrieve the information when required could have been improved.

In our discussion of long-term memory, we must distinguish between two tasks that are used to examine its nature: *recall* and *recognition* (Danckert & Craik, 2013). In recall tasks, people are presented with information that they have to retrieve later. Many of the experiments we discussed in the context of short-term memory used recall tasks. In recognition tasks, people are presented with a list of items to study. Then they are given a second list and required to indicate for each item whether it was in the original study list. So, whereas recall involves retrieving information from memory, often with no hints, for recognition the studied items are provided.

BASIC CHARACTERISTICS

Until the 1970s, most research on long-term memory was focused on episodic memory, and there is still a lot of interest in this topic. In contrast to the phonological codes of short-term memory, codes in long-term memory were assumed to reflect the meanings of items. For example, in a test of recognition memory, you are given a long list of words to remember; then you are tested a few minutes later with a second list. In that second list, words that were in the first list must be distinguished from those that were not. In such situations, when a word is falsely recognized, that word is often a synonym (that is, a word with a similar meaning) of a word in the original list (Grossman & Eagle, 1970). If the material to be remembered is a passage of text or a specific event, rather than a list of words, only the gist or meaning will be retained rather than any specific wordings used in the text or to describe the event.

Now, we understand that coding in long-term memory is flexible and not restricted to semantic codes. For example, there is some evidence that visual codes exist in long-term memory. In one experiment, people were asked to remember objects (like an umbrella) shown as line drawings. Later, they were shown other drawings of objects that they had seen and some new objects. For objects that had been seen, they responded more quickly when the drawing was identical to the original drawing than when it was not (Frost, 1972). When people are shown and fixate each of a series of objects in a natural scene, their memory is best for the two most recently fixated objects, indicating a role for visual short-term memory. Also, their memory is still well above chance for objects separated by many intervening items, indicating a visual long-term memory component (Hollingworth, 2004). Other evidence shows that concrete and imaginable words (like "umbrella") are remembered better than abstract words (like "honesty"), apparently because both semantic and visual information can be stored for concrete words but only semantic information for abstract words (Paivio, 1986).

The original modal model claimed that information was transferred from short-term memory into long-term memory through rehearsal (Atkinson & Shiffrin, 1968; Waugh & Norman, 1965). According to the model, the longer information is rehearsed in short-term memory, the greater the probability that it will be transferred into long-term memory. It is a well-established finding that long-term retention of information in fact increases with the number of times that the information is rehearsed (e.g., Hebb, 1961). However, how rehearsal is performed is much more important than how much is done.

We can distinguish between *maintenance rehearsal*, or rote rehearsal, which involves the covert repetition of material discussed in the previous section, and *elaborative rehearsal*, which involves relating material together in new ways and integrating the new information with information in long-term memory. Because long-term memory depends on connections between concepts (see below), elaborative rehearsal is much more important for long-term retention than is maintenance rehearsal (Rose, Buchsbaum, & Craik, 2014). Only elaborative rehearsal leads to better performance on recall tests, and although maintenance rehearsal improves performance on recognition tests, it does not do so as much as elaborative rehearsal does (e.g., Woodward, Bjork, & Jongeward, 1973). Even though elaborative rehearsal produces better recognition overall than maintenance rehearsal, it takes time to use the associations established during elaborative rehearsal. Consequently, if a person is pressured to make recognition decisions quickly, much of the benefit of elaborative rehearsal is lost (Benjamin & Bjork, 2000).

Questions about the capacity and duration of long-term memory are difficult to answer. There seem to be no limits on the capacity for acquiring, storing, and retrieving information (Magnussen et al., 2006). Psychologists have debated whether long-term memory is permanent (e.g., Loftus & Loftus, 1980). This is probably an unanswerable question. For many years, forgetting was assumed to reflect the loss of information from memory. Questions focused on whether the loss was due simply to time (decay theory) or to similar events that occurred either before (proactive) or after (retroactive) the events that were to be remembered (interference theory). The results of many experiments were consistent with predictions of interference theory (Postman & Underwood, 1973). However, forgetting in many circumstances is due not to information loss but to a failure to retrieve information that is still available in memory.

Tulving and Pearlstone (1966) had people learn lists of 48 words, 4 from each of 12 categories (e.g., flowers, foods, etc.). During learning, the appropriate category name (flowers) was presented with the words that were to be learned (tulip, daisy, etc.). Subsequently, people were asked to recall the words; half of them did so with the category names provided, and half without. The people provided with the category names recalled more words than those who were not provided with the names. Tulving and Pearlstone concluded that people who were not provided with the category names at recall must have had words available in long-term memory that were not accessible without the category cues.

The point of Tulving and Pearlstone's (1966) experiment is that effective retrieval cues enhance the accessibility of items in memory. This concept is sometimes called the *encoding specificity principle* (Tulving & Thomson, 1973): "Specific encoding operations performed on what is perceived determine what is stored, and what is stored determines what retrieval cues are effective in providing access to what is stored" (p. 369). In other words, a cue will be effective to the extent that it matches the encoding performed initially. Appropriate use of retrieval cues to reinstate context is a way to maximize the likelihood that an operator will remember information when it is needed at a later time. Reinstatement of context is particularly important for older adults, who have difficulty retrieving information (Craik & Bialystock, 2006).

PROCESSING STRATEGIES

The distinction between rote rehearsal and elaborative encoding we mentioned earlier is a difference in processing strategy. Different processing strategies can have drastic effect on long-term retention. Craik and Lockhart (1972) introduced the concept of *levels, or depth, of processing*. As Craik (2002) notes, "The concept of depth of processing is not hard to grasp—'deeper' refers to the analysis of meaning, inference, and implication, in contrast to 'shallow' analyses such as surface form, colour, loudness, and brightness" (p. 308). In the levels-of-processing framework, three basic assumptions about memory are required (Zechmeister & Nyberg, 1982). The first is that memories arise from a succession of analyses of increasing depth performed on stimuli. Second, the greater the depth of processing, the stronger the memory, and the better it will be retained. Third, memory improves only by increasing the depth of processing and not by repeating an analysis that has already been performed. The levels-of-processing view leads to the prediction that long-term retention will be a function of the depth of the processing performed during the initial presentation of items.

We examine the influence of levels of processing with "orienting tasks." An orienting task takes the place of an overt study session and familiarizes people with material that will later be tested. In other words, people perform a particular task on a set of items, unaware that they will receive a subsequent memory test on those items. For example, Hyde and Jenkins (1973) used five types of orienting tasks, two of which apparently required deep-level semantic processing of words' meanings (rating the pleasantness or unpleasantness of the words, estimating frequency of usage) and three that required shallow processing (checking words for the letters E and G, determining the part of speech for a word, and judging in which of two sentence frames a word fit best). Later recall for the words was better for those people who performed the deep tasks than for those who performed the shallow tasks. Moreover, recall performance on the deep tasks was equivalent to that of a group of people who received standard intentional memory instructions and studied the list without an orienting task. Thus, this study and others indicate that whether or not a person intends to remember the presented information is unimportant. What matters is that the information receives a deep level of processing.

Although the levels-of-processing idea seems to help explain how memory works in some circumstances, it has some limitations. First, the "depth" required for specific orienting tasks cannot be objectively measured. This means that we cannot be certain why recall is better in some conditions than in others. Our explanation can become circular: that is, this task led to better recall than another; therefore it involved a deeper level of processing. Second, another factor, elaboration, also influences retention. Elaboration refers to the number of details provided about material to be remembered. For example, Craik and Tulving (1975) showed that memory for words in complex sentences (e.g., "The great bird swooped down and carried off the struggling – – –") was better than that for words in simpler sentences (e.g., "He cooked the – – –"), where the word "chicken" could be used to complete either sentence.

Above all, it is the *distinctiveness* of encoding that is important for memory (Eysenck, 1979; Neath & Brown, 2007). Deeper and more elaborate processing improves retention by producing a representation of an item that is distinct from representations of other to-be-remembered items (Craik, 2002). We can distinguish between information based on the distinctive features of items (item information) and the common information shared by items (relational information; Einstein & Hunt, 1980). Whereas the quality of item information is important for recognition performance, relational information is important for recall. The type of information that is emphasized by particular materials and study strategies will in part determine how well the information is remembered.

Philp, Fields, and Roberts (1989) observed differences in recall and recognition performance for divers who performed memory tests while at surface pressure and during a "dive" in a hyperbaric chamber at a pressure equal to 36 m of seawater. The divers showed a 10% overall decrement in immediate recall of 15-word lists during the dive compared with performance at the surface, and a 50% decrement in a delayed recall test for words from all lists conducted 2 min after the immediate recall test was completed. However, they exhibited no such decrement on a subsequent old-new recognition test of the words. Thus, although the information apparently was intact in the long-term store, the divers had difficulty recalling it. This impairment of free recall suggests that, after people have been in a stressful environment, an accurate assessment of their memories may require that they be provided with cues during debriefing that encourage retrieval of relational information.

The relationship between the type of processing performed during study and the type of memory test is important. Morris, Bransford, and Franks (1977) had people perform either a shallow orienting task (deciding whether a word rhymed with another in a sentence frame) or a deep orienting task (deciding whether a word made sense in a particular sentence frame). The deep orienting task produced better performance on a recognition test for which the unstudied words were semantically similar to the studied words. However, when the new words sounded like the old words, the shallow orienting task led to better recognition. In other words, when the recognition test requires discriminations based on how words sound, it is better to study sound than meaning. This finding illustrates the principle of *transfer-appropriate processing*: the processing performed during study is effective for memory to the extent that the resulting knowledge is appropriate for the memory test. Lockhart (2002) has suggested that one reason why deep-level processing is usually beneficial to memory is that it increases the likelihood that the processing will turn out to be transfer appropriate to the retrieval contexts encountered later.

Long-term memory can also benefit from the use of strategies that organize the material in meaningful ways. Recall is better for lists of words that come from the same categories than from lists with no obvious category structure, and words from a given category tend to be recalled together even if they were not presented together (Bousfield, 1953). Moreover, studying lists of words grouped in their categories produces considerably better recall than does studying random-order lists of the same words (Bower, Clark, Lesgold, & Winzenz, 1969). Figure 10.10 shows a conceptual hierarchy for words belonging to the category "minerals." People studied either a random list of the words in the category or the conceptual hierarchy. More than twice as many words were recalled after studying the conceptual hierarchy. People apparently used the hierarchy to help them retrieve the studied minerals from long-term memory.

Even for words from different categories, the order in which the words are recalled often shows structure not provided in the study list. In one experiment, younger and older adults (average ages of 20 and 73 years, respectively) were asked to study a list of 20 words and then recall the words in any order (Kahana & Wingfield, 2000). Then they studied the same list of words in a different order, and then made another attempt at recall, until all 20 words were recalled correctly. Recall performance improved on average each time the list was studied, although more slowly for the older adults than

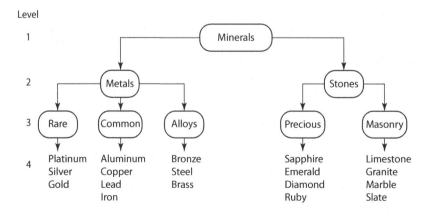

FIGURE 10.10 Conceptual hierarchy for minerals.

for the younger ones. Recalled words tended to be output in organized groups that were relatively constant from episode to episode, even though the order of words in the list was different each time it was studied. Although the older adults took longer to learn the lists, they showed the same level of organization as the younger adults when equated for degree of learning.

The effects of organizational strategies (or the lack of them) on memory performance are apparent not only in the laboratory but also in everyday life. Memory researchers, self-help gurus, and people who have to remember long lists of items have developed many organizational techniques for improving long-term memory. These techniques collectively are *mnemonics* (Barcroft, 2013; Worthen & Hunt, 2011). A mnemonic is an encoding strategy that organizes material in memorable ways. In addition to providing organization, mnemonic strategies make material more distinctive. This distinctiveness may arise in part from the generation of novel connections between to-be-remembered items induced by the mnemonic technique.

There are two classes of mnemonic techniques: visual and verbal. Visual mnemonics rely on imagery. For this technique, you need to imagine physical relationships between visual images and the items that you want to remember. An example is the method of loci, for which to-be-remembered items are imagined at various locations in a familiar environment, such as your home. When you need to recall the items, you mentally walk through your home to the different locations and "find" the items that you stored there. Visual mnemonics can significantly improve the learning of foreign language vocabulary (Raugh & Atkinson, 1975) and face–name associations (Geiselman, McCloskey, Mossler, & Zielan, 1984).

A verbal mnemonic relates items to be remembered to elements of well-known sentences or stories. Alternatively, the first letters of words to be remembered are combined into new words or phrases (acronyms) or used as first letters of new words in a meaningful sentence (acrostics). So, for example, one well-known verbal mnemonic used to remember taxonomic classification is the sentence "King Philip came over from good Spain," which represents kingdom, phylum, class, order, family, genus, and species. Both visual and verbal mnemonics can be effective, whether used by themselves or as components in more complex techniques (Cook, 1989).

Mnemonics can be particularly beneficial for elderly people, who may be at higher risk for memory failures (Poon, Walsh-Sweeney, & Fozard, 1980). However, those people who could most benefit from the use of mnemonics, such as the elderly, often forget to use them. Various sentence-based mnemonics have also been proposed to enable users to generate passwords that they will be able to remember but that will be difficult for a hacker to crack (Yang, Li, Chowdhury, Xiong, & Proctor, 2016). For example, think of a sentence that contains at least eight words, and then select a letter, number, or a special character to represent each word. The sentence might be, "I went to London four and a half years ago," and the resulting password could be iwtl4&ahya. A potential drawback of this approach is that it requires remembering not only the sentence but also the conversions used for each word in the sentence.

So far, we have discussed only internal aids to memory, that is, study and retrieval techniques. Many people also rely on external memory aids, including such things as tying a string around a finger, making notes for reminders, personal digital assistants (PDAs) such as smartphones, and so on. External memory aids are used most often to remember to do things rather than to remember information (Intons-Peterson & Fournier, 1986). One simple strategy explored by Sharps and Price-Sharps (1996) for aiding the memory of older adults was to put a colorful plastic plate in a prominent location in their homes. This plate was used as a place for objects that could be easily lost (keys, glasses, etc.) and for notes to remind the person of future activities. This simple strategy produced a 50% reduction in everyday memory errors for the older adults.

As electronic technology continues to advance, a greater array of commercial memory aids have become available. These range from alarm settings on digital watches to sophisticated smartphone applications. Such memory aids are particularly useful for older adults and patients with memory impairment (Armstrong, McPherson, & Nayar, 2012). For example, Kurlychek (1983) showed that a patient with early Alzheimer's disease could use an hourly alarm on his watch to consult a written schedule of activities and remind himself what he was supposed to be doing. Because an aid is intended to assist the performance of a particular task performed in a specific context, the effectiveness of an aid depends on the extent to which it corresponds with the situation for which it is being used (Herrmann & Petros, 1990). Inglis et al. (2004) evaluated electronic devices such as PDAs and concluded that the limitations of currently available technology can create difficulty in the use of such devices for memory-impaired and older users. With the widespread use of smartphones, interest has shifted to mobile apps for assisting older adults in remembering to perform tasks such as contacting family members and caregivers. Although several apps are available and progress is being made, "there have been very few successful mobile apps developed for assisting senior adults in their day-to-day activities" (Pang et al., 2015).

COMPREHENDING VERBAL AND NONVERBAL MATERIAL

One important role that long- and short-term memory plays is in understanding, or comprehending, language; that is, in reading and listening. Reading printed text can involve something as simple as reading the word "stop" on a sign or as complex as reading a technical manual providing instructions for operating a system. Similarly, spoken language can be a single utterance (the word "go") or a complex narrative. The comprehension and retention of a message are important for any situation involving verbal material. In this section, we examine the interplay between semantic memory (our storehouse of general knowledge) and the processes involved in the comprehension and retention of sentences, text, and structured discourse.

SEMANTIC MEMORY

To understand comprehension, we need to understand the processes involved in semantic memory (Tulving & Donaldson, 1972). Such processes include how knowledge is represented and how this knowledge is accessed (see McNamara, 2013, for a review). We know that people can judge faster whether an object is a member of a category when the category size is small rather than large. For example, the decision that a beagle is a dog can be made faster than the decision that a beagle is an animal. We call this the *category size effect*. Also, decisions about category membership can be made faster when the object is a typical member of the category than when it is less typical. For example, the decision that a canary is a bird can be made faster than the decision that a buzzard is a bird. We call this the *typicality effect*.

Two types of semantic memory models have been proposed to explain the category size and typicality effects. *Network models* assume that concepts are represented by distinct nodes that are interconnected within an organized network (see Figure 10.11). In a network model, the time to verify a sentence is a function of the distance between the concept nodes and the strengths of the connecting links between them. For example, Collins and Loftus (1975) proposed a spreading-activation model using two separate networks. In the conceptual network, a fragment of which is shown in Figure 10.11, concepts are organized according to semantic similarity. In the lexical network, the labels for each concept are organized according to auditory similarity. Retrieval in both networks occurs as the activation of one concept or name spreads through the network along connecting links, with nodes farther from the concept receiving weaker activation at longer delays than nodes that are close to the concept.

The second kind of model relies on *feature comparisons*. These models propose that concepts are stored as lists of semantic features, and verification occurs when the object and category in a statement share matching features. For example, in Smith, Shoben, and Rips's (1974) feature-comparison model, all features of the object and category are compared to yield an overall similarity value. When the similarity is very high or very low, a fast true or false response can be made. When the similarity is ambiguous, only the most important, defining features of the object and category are closely examined. The true or false response is then based on whether this subset of features matches or not.

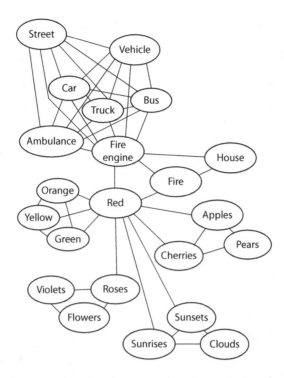

FIGURE 10.11 A representation of concept relatedness in a connected network of concepts.

Both network and feature models have provided good explanations of the category size and typicality effects. They have also been very useful in applied settings. Models like these have been used to structure knowledge data bases, such as those used for expert systems and decision-support systems (see Chapters 11 and 12).

Although network and feature models explain the category size and typicality effects in category verification, other important models can explain a wider range of data. Two such models include *distributed network models* and *high-dimensional spatial models*. Distributed network models are distinguished from the network models presented above in that concepts are not represented by distinct nodes but by patterns of activation across many nodes in the network. In distributed networks, knowledge is represented in the weights of the connections between nodes, which determine the patterns of activation induced by different stimuli. Distributed networks learn by applying rules to adjust the connection weights between nodes: The output of the network for a particular stimulus is compared with the desired output for that stimulus, and the weights are adjusted to reduce the mismatch between the actual and the desired output. Models of this type can explain many of the findings obtained from tasks relying on semantic memory (e.g., Kawamoto, Farrar, & Kello, 1994).

High-dimensional spatial models are similar to distributed network models in that concepts are represented as points in a multidimensional space. The most well-known model of this type uses Latent Semantic Analysis (LSA; Günther, Dudschig, & Kaup, 2015; Landauer, 1998; Landauer, McNamara, Dennis, & Kintsch, 2007). The basic idea behind LSA is that similarities between words (concepts) can be inferred from their co-occurrences in written text. When LSA is applied to a sample of text, it produces a semantic space in which different concepts cluster together. One analysis examined more than 90,000 words in over 37,000 different contexts representing material that an English reader might read from an early grade through the first year in college. The resulting LSA representation is a semantic network of a first-year college student.

It is surprising that a representation constructed solely on how words co-occur can be at all useful. However, LSA has yielded accurate simulations of many language phenomena and allowed automation of many tasks involving language. One of the more impressive applications of LSA has been the automatic grading of essays (Landauer, Laham, & Foltz, 2003). For this application, LSA is applied to a large sample of texts dealing with the topic of the essays. A representative sample of essays is then scored by human graders. Those graded essays and the essays to be graded automatically are represented as LSA vectors. The similarities between each ungraded essay and each previously graded essay are computed, and these similarities are used to assign grades to each essay. It has been demonstrated that the grades assigned automatically to the essays are as accurate as those assigned by a human. Landauer et al. suggest that one immediate application of this technology is in writing tutorial systems. Also, many standardized admissions tests, such as the Scholastic Aptitude Test (SAT) and Graduate Record Examination (GRE), now require an essay, and an automated system for grading those essays could greatly reduce the time necessary for grading and standardizing the results.

WRITTEN COMMUNICATION

Reading is a complex process that requires people to retrieve information from semantic memory and integrate information across sentences and passages in a text. Reading efficiency is affected by many factors. These include the purpose for reading, the nature of the materials, the educational background of the reader, and characteristics of the environment. In situations where the successful operation of a system depends on the operator's ability to read effectively, the human factors professional must consider these factors and how best to present written material. According to Wright (1988, p. 265), "As working life becomes more information intensive, so a better understanding of how to design and manage written information becomes more urgent." One study, looking at information in instruction booklets, recommended that written information be evaluated by way of usability testing prior to implementation (Brooke, Isherwood, Herbert, Raynor, & Knapp, 2012).

Many situations in which an operator must read information also require that she read it quickly. Reading speed is influenced by several factors, including the complexity of sentences and the goals of the reader. A sentence can be decomposed into some number of basic ideas, or propositions. The more propositions there are, the longer it takes to read the sentence, even if the number of words is held constant (Kintsch & Keenan, 1973). If the reader's goal is to remember a sentence word for word, it will take longer to read it than if the reader's goal is comprehension (Aaronson & Scarborough, 1976).

Reading speed is irrelevant if the information is comprehended poorly. Comprehension can be improved by making changes to the syntactic structure of the material. For example, comprehension is better when relative pronouns (e.g., *that*, *which*, *whom*) are used to signal the start of a phrase than when they are omitted (Fodor & Garrett, 1967). Consider the sentence, "The barge floated down the river sank." Such a sentence is often called a *garden path sentence*, because its construction leads the reader to interpret the word *floated* as the verb for the sentence, but this interpretation must be revised when the word *sank* is encountered. Changing the sentence to read "The barge that floated down the river sank" eliminates this ambiguity. However, not all garden path sentences can be remedied in this way. For instance, consider the sentence "The man who whistled tunes pianos." This sentence will be read more slowly and comprehended less well than the equivalent sentence "The whistling man tunes pianos," in which the word *tunes* is unambiguously a verb.

The effects that different syntactic structures have on reading comprehension have been used to construct theories of how people read and represent material (Frazier & Clifton, 1996). These theories focus on a variety of syntactic features, including word order and phrase structure rules (templates for possible configurations of words in phrases) and the case of a noun (which is related to the role it plays in the sentence structure; Clifton & Duffy, 2001). Sentences that contain nested clauses, such as "The man from whom the thief stole a watch called the police," are more difficult to comprehend than sentences that do not contain nested clauses (Schwartz, Sparkman, & Deese, 1970). This finding has been used to study the contribution of working memory limitations in reading comprehension.

Semantic structure is as important as syntactic structure. Some words seem to require more effort to understand than other words in some contexts. For example, the word *kicked* in the sentence "The man kicked the little bundle of fur for a long time to see if it was alive" takes longer to process than the word *examined* in the sentence "The man examined the little bundle of fur for a long time to see if it was alive" (Piñango, Zuriff, & Jackendoff, 1999). One explanation for this finding is that the verb *examine* implies an act that takes an extended period of time, consistently with the phrase *for a long time*. The verb *kick* must be changed from an act that is usually brief to an act that is repeated over an extended period of time to be consistent with the phrase *for a long time*. Changing the representation of the verb *kick* requires time and effort.

Another example demonstrating the importance of semantic structure can be seen in how events are ordered in a sentence. A sentence will be comprehended best when the order of events in the sentence follows the actual order of events (Clark & Clark, 1968). This has implications for the way that written instructions should be presented. Dixon (1982) presented people with multistep directions for operating an electronic device and measured the time it took for them to read each sentence. Reading times were shortest when the action to be performed (e.g., "turn the left knob") came before the specific details describing the desired outcome of the action (e.g., "the alpha meter should read 20"). This finding suggests that comprehension will be easiest when instructions are organized around the actions to be performed.

Complex communication involves integrating information across many sentences. Successful readers construct an abstract, rather than literal, representation of what is read. These readers make inferences about relations and events that are implied by the text but not directly stated. These inferences will later be remembered as part of the material that was read (Johnson, Bransford, & Solomon, 1973). Working memory plays an important role in comprehension because working memory allows new information to be interpreted in terms of and integrated with information and inferences already in memory. Poor readers differ from good readers primarily in the efficiency with which the integration of new propositions into the working representation can be performed (Petros, Bentz, Hammes, & Zehr, 1990).

Readers seem to form representations of material in the form of organized structures called *schemas* (Rumelhart & Norman, 1988; Thorndyke, 1984). Schemas are frameworks that organize our general knowledge about familiar objects, situations, events, actions, and sequences of events and actions. Most importantly, schemas cause a person to expect certain events to occur in certain contexts. These expectancies make interpretation of information easier and help the reader determine its relative importance (Brewer & Lichtenstein, 1981).

A famous experiment performed by Bransford and Johnson (1972) demonstrated the importance of an appropriate schema in text comprehension. Students read the passage shown in Table 10.1, then rated its comprehensibility and tried to recall the ideas contained in it. Students who were told prior to reading the passage that the topic was "washing clothes" rated comprehensibility higher and remembered many more of the details than did students who were not told the topic in advance.

Similarly, schemas for technical documents can aid in the comprehension and retention of technical material. Because failure to heed warnings may lead to injury, we should design instruction manual warnings around familiar schemas to enhance comprehensibility and memorability.

Young and Wogalter (1990) noted that comprehension and memory could be improved by increasing the likelihood that the warning is noticed in the first place, and then by providing visual information to accompany the verbal information. They conducted two experiments, one using instruction manuals for a gas-powered generator and the other instruction manuals for a natural-gas oven/range, in which the typeface of the warning messages was either plain or salient (larger type, orange shading) and the messages were either accompanied by pictorial icons or not (see Figure 10.12). Comprehension and memory of the warnings were significantly better when they were presented in

TABLE 10.1

Passage

The procedure is actually quite simple. First you arrange things into different groups. Of course, one pile may be sufficient depending on how much there is to do. If you have to go somewhere else due to lack of facilities, this is the next step; otherwise you are pretty well set. It is important not to overdo things. That is, it is better to do too few things at once than too many. In the short run this may not seem important, but complications can easily arise. A mistake can be expensive as well. At first the whole procedure will seem complicated. Soon, however, it will become just another facet of life. It is difficult to foresee any end to the necessity for this task in the immediate future, but then one can never tell. After this procedure is completed, one arranges the material into different groups again. Then they can be put into their more appropriate places. Eventually they will be used once more and the whole cycle will then have to be repeated. However, that is part of life.

salient type and included a pictorial icon. The authors suggest that this effectiveness of the salient print and icon combination may be due to better integration of the verbal and visual codes.

Spoken Communication

Teams of operators communicate with each other using speech to coordinate their performance. In many organizations, team leaders provide briefings to their team members at the beginning of the work shift. The points we just made about syntax, semantics, and schemas for reading

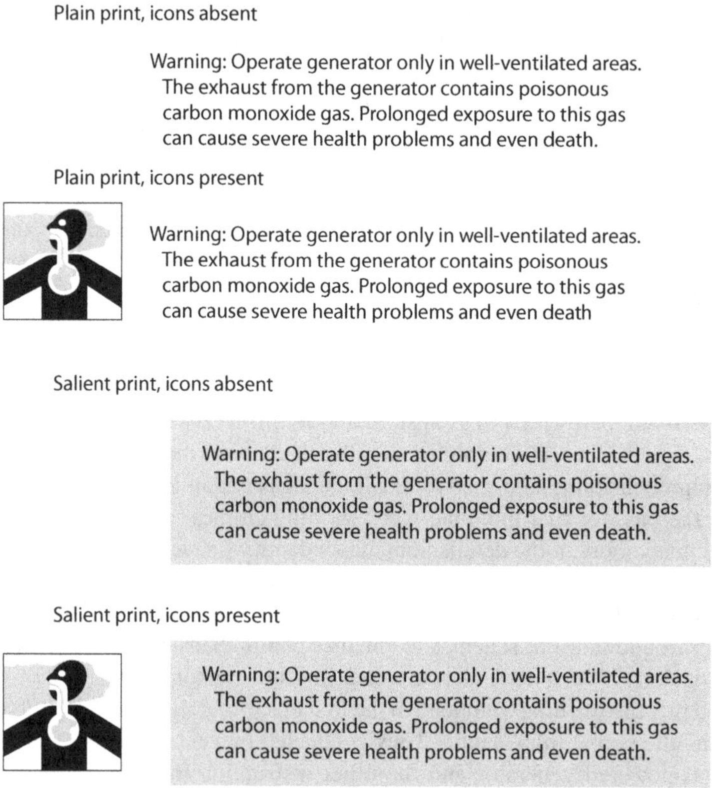

Plain print, icons absent

Warning: Operate generator only in well-ventilated areas. The exhaust from the generator contains poisonous carbon monoxide gas. Prolonged exposure to this gas can cause severe health problems and even death.

Plain print, icons present

Warning: Operate generator only in well-ventilated areas. The exhaust from the generator contains poisonous carbon monoxide gas. Prolonged exposure to this gas can cause severe health problems and even death

Salient print, icons absent

Warning: Operate generator only in well-ventilated areas. The exhaust from the generator contains poisonous carbon monoxide gas. Prolonged exposure to this gas can cause severe health problems and even death.

Salient print, icons present

Warning: Operate generator only in well-ventilated areas. The exhaust from the generator contains poisonous carbon monoxide gas. Prolonged exposure to this gas can cause severe health problems and even death.

Note: Shading represents orange highlighting.

FIGURE 10.12   Warnings differing in conspicuousness of print and the presence of an icon.

comprehension also apply to the comprehension of spoken language (Jones, Morris, & Quayle, 1987). Also important for comprehension of spoken language is "prosodic phrasing," or the grouping together of words by their tonal pitch, duration, and rhythm (Frazier, Carlson, & Clifton, 2006). Box 10.1 provides a vivid example of the importance of spoken and written communication in a real-life example.

Research on speech comprehension has been conducted by studying conversation, in which two or more people take turns conveying information to each other. When a person assumes the role of a listener, we assume that he or she tries to understand what the speaker wants to convey. Consequently, the listener assumes that what the speaker says is sensible and constructs an interpretation that hopefully is the one intended by the speaker.

Several components of a spoken utterance can be distinguished (Miller & Glucksberg, 1988). These include the utterance itself, the literal meaning of the utterance, and the meaning intended by the speaker. Comprehension requires more than just establishing the literal meaning; the speaker's intention also must be inferred. Many communication errors occur because a listener misinterprets a speaker's intentions.

The listener uses a number of rules to establish both the literal and the intended meaning of an utterance. According to Grice (1975), the most important rule is the *cooperative principle*: the assumption that the speaker is being cooperative and sincere, and is trying to further the purpose of the conversation. The cooperative principle specifies several conversational rules, or maxims. These maxims can be classified by conversational quantity, quality, relation, and manner (see Table 10.2). Overall, these maxims allow the listener to assume that the speaker is making relevant and unambiguous statements that are informative and truthful. If these maxims hold, then the listener's task of constructing an intended meaning from the speaker's literal meaning is made as easy as possible.

A speaker can improve a listener's comprehension by making direct references to information that was provided earlier in the conversation. Listeners seem to make use of a *given-new strategy* (Haviland & Clark, 1974). This strategy identifies two types of information in any utterance: given and new. The given information is assumed by the speaker to be already known by the listener, whereas the new information is to be added by the listener to the old. If the speaker can arrange sentences to distinguish between given and new information, the listener's task can be made easier.

The literal meaning of some utterances is not the same as the meaning intended by the speaker. This occurs for indirect requests, in which a request for some action on the listener's part is not stated directly. For example, the question "Do you know what time it is?" contains an implicit request for the listener to provide the speaker with the time of day. Much of spoken language is figurative and includes expressions that use irony, metaphor, and idiomatic phrases. An example of an idiomatic phrase that also uses metaphor is "making a mountain out of a molehill." In a case like this, the listener appears to construct both the literal and the nonliteral meaning of the utterance. The speed with which the listener comprehends figurative meanings of utterances is usually slower than the speed with which the listener comprehends their literal meaning (Miller & Glucksberg, 1988). One implication of this research is that figurative speech should be avoided in communication with operators unless the meaning is very clearly dictated by the context.

It will probably come as no surprise to learn that in group problem-solving situations, speech is the best way for group members to communicate. Chapanis, Parrish, Ochsman, and Weeks (1977) had two-member teams solve an equipment assembly problem using one of four modes of communication: typewriting, handwriting, speech alone, and a condition where communication was not restricted to any particular mode. The problem was solved approximately twice as fast in the speech alone and unrestricted conditions than in the typewriting and handwriting conditions. This finding held even though the communications in the speech and unrestricted conditions were much lengthier. One of the reasons for the better performance in these conditions was that the teams could engage in problem solving and communication activities simultaneously, whereas they could not do so in the typewriting and handwriting conditions.

BOX 10.1 TEXT COMPREHENSION AND
COMMUNICATION DURING AN AIR DISASTER

On the afternoon of July 19, 1989, United Airlines flight 232 took off from Denver. While en route to Chicago, the plane, a DC-10, experienced a catastrophic failure of its center engine. Debris from the engine collided with the tail of the aircraft, disabling not only the primary hydraulics system but the second and third backup systems as well. The flight control system was rendered inoperable, and the crew could only make right turns (and these only with great difficulty). The plane crash-landed in Sioux City, Iowa, using only the throttles, which operate much like the accelerator pedal in a car. The plane was torn apart in the landing, but 184 of the 292 passengers and crew survived.

There are two aspects of the crash of flight UA-232 that deserve consideration in our discussion of verbal and written communication. The first is the communications that the flight engineer had with the ground maintenance crew, and the second is the role of the flight manual during the emergency.

In the early stages of the disaster, the flight engineer, Dudley Dvorak, attempted to convey to the ground maintenance crew that all hydraulic systems were lost. We have to understand that although it was obvious to the crew on the flight deck that all three systems had failed, the ground crew persisted in interpreting Dvorak's messages from the perspective that a failure of three completely independent systems was impossible. In other words, the ground crew's schema did not allow for the possibility that all three systems could fail, which prevented them from understanding what had happened.

Dvorak's first contact with maintenance was to report that the number two engine was gone and that they had "lost all hydraulics." Maintenance responded to Dvorak by asking about the number two engine. Then maintenance attempted to confirm that (hydraulic) systems one and three were operating normally; Dvorak replied, "Negative. All hydraulics are lost. All hydraulic systems are lost." Even after this exchange, maintenance asked about hydraulic fluid levels. Dvorak reported back that there was no hydraulic fluid left. The maintenance crew then consulted the flight manual, and several more minutes of confused discussion took place between Dvorak and the maintenance crew before it was established that not a single one of the hydraulic systems was operational.

In Captain Haynes's words, "The hardest problem that Dudley had was convincing them that we didn't have any hydraulics. 'Oh, you lost number two,' 'No, we lost all three,' 'Oh you lost number three,' 'No, we've lost all of them,' 'Well, number one and two work,' 'No,' well we went on like this for quite a while" (http://yarchive.net/air/airliners/dc10_sioux_city.html).

The flight manual, to which the ground maintenance crew referred in an attempt to help recover the plane, is a book provided by the manufacturer of the aircraft. It contains descriptions of possible scenarios and the procedures appropriate for each. On most aircraft the flight engineer also has an abridged version of the flight manual, which can be consulted in the event of an unusual situation or an emergency. Before the ground maintenance crew began searching the manual (and logbooks and computer databases) for a solution to the hydraulics failure, Dvorak had consulted the manual for a solution to the engine failure. Emergency procedures in the manual are written in checklist form, without explanation. This reduces the demands on memory and problem-solving resources during an emergency. For example, the manual said that the first thing the flight crew should do is close the throttle to the failed engine. After that, the fuel is to be shut off. However, because of the hydraulics failure, these controls did not respond appropriately, signaling to the crew that something else had gone very badly wrong.

There were no procedures provided for total hydraulic failure in the flight manual, because the manufacturer believed that by engineering three such systems, each independent of the others, such a failure was impossible. Therefore, there was nothing for the ground mainte-nance crew to find in the manual. Because the maintenance crew kept making suggestions that the crew had already tried, as well as suggestions that required an operational hydraulics system, in frustration Captain Haynes instructed Dvorak to cease communications with main-tenance. In the final analysis, there really was nothing that maintenance could have done for the crew.

When Captain Haynes describes his experiences on flight UA-232, he is careful to empha-size that one of the factors that helped him and his crew to land the plane and save so many of the passengers was communication: the flow of information between the members of the crew, the communications between the Captain and the air traffic controllers at Sioux City, and the communications between the emergency teams on the ground in Sioux City. Verbal and written forms of communication were an integral part of the "success" of flight UA-232.

We should note that experts who investigated the crash of flight UA-232 deemed it impos-sible to successfully land a plane in a simulator under the conditions experienced by the crew. For this reason, Captain Haynes and his crew were commended for their extraordinary performance and credited with saving the lives of the 184 survivors. Captain Haynes attri-butes the crew's outstanding performance to a management technique called *crew resource management*, which encourages all members of a team to communicate with the team leader in solving problems.

SITUATIONAL AWARENESS

An important concept within the field of human factors that extends across issues of memory and comprehension is that of *situational awareness* (Endsley & Jones, 2012). Situation awareness is defined as "the perception of the elements in the environment ..., the comprehension of their mean-ing, and the projection of their status in the near future" (Endsley, 1988, p. 97). The term *awareness* emphasizes the importance of working memory, particularly the part that Baddeley (2000) calls the episodic buffer, which in part determines consciousness and awareness. The things of which

---

TABLE 10.2

Grice's Conversational Maxims

Maxims of Quantity

- Make your contribution as informative as is required.
- Do not make your contribution more informative than is required.

Maxims of Quality

- Try to make your contribution one that is true.
- Do not say what you believe to be false.
- Do not say that for which you lack adequate evidence.

Maxim of Relation

- Be relevant.

Maxims of Manner

- Be perspicuous.
- Avoid obscurity of expression.
- Avoid ambiguity.
- Be brief.
- Be orderly.

we are aware are affected by attentional factors (i.e., the central executive), which suggests that our situational awareness will be limited if our attention is directed to inappropriate elements of the environment. Endsley and Jones indicate that a person's failures of selective attention (such as those that might occur when talking on a mobile phone and driving) may restrict his or her situational awareness.

Loss of situational awareness is of particular concern when tasks are automated. Consider, for example, partially or fully autonomous cars in which the driver's responsibilities are reduced. Because automation of driving will only work during very predictable situations, such as highway driving, the driver must be prepared to take over when the situation changes unexpectedly and automation fails. Maintaining a level of awareness that permits the driver to re-engage attention when immediate action is needed is a topic of ongoing investigation in human factors (Sirkin, Martelaro, Johns, & Ju, 2017).

Because situational awareness depends on working memory, it will be limited by the same factors that limit working memory capacity and accuracy. Furthermore, a person's level of mental workload may also influence his or her level of situation awareness. If a person's mental workload is very high, his or her situation awareness may be poor. However, even when a person's mental workload is within acceptable limits, his or her situation awareness may still be poor. That is, comprehension of events in the environment may be limited even if the person's attentional and memory resources are not being overloaded.

Situation awareness is often measured by asking an operator to perform a primary task such as driving or flying in a simulated environment. We evaluate situation awareness by measuring the operator's immediate memory and understanding of the objects and events in the task, and his or her ability to predict the future behavior of the system. As with mental workload, there are both subjective and objective measures of situation awareness. Subjective measures require that the operator or an expert observer rate the operator's awareness for a specified scenario or period. Objective measures probe the operator for information about some aspect of the task.

Two accepted methods for obtaining objective measures (Vu & Chiappe, 2015) include Situation Awareness Global Assessment Technique (SAGAT; Endsley & Jones, 2012) and Situation Present Assessment Method (SPAM; Durso & Dattel, 2004). For SAGAT, the simulation is frozen and the operator is asked several questions; for example, the experimenter may stop a simulated driving task and ask the driver to recall where the other vehicles on the roadway were located, to predict what a vehicle at the side of the road will do, and so on. The accuracy of the responses to the questions is taken as an indicator of the extent of the operator's conscious awareness of the situation. For SPAM, on the other hand, questions are presented one at a time during the task, more like a dual-task procedure for measuring mental workload. To preclude interference from high momentary workload, the question is presented only after the operator has responded affirmatively in response to a prompt that she is ready to receive the question. SPAM allows use of response time as well as accuracy in the assessment of situational awareness within the ongoing dynamic task. In addition, the processes that the operator is using to maintain awareness, for example, off-loading subtasks to aids such as electronic navigation devices. The issues involved in measuring situation awareness are similar to those discussed for mental workload in Chapter 9—intrusiveness, ease of implementation, operator acceptance, and so forth—and consideration should be given to these issues when selecting a particular method for use.

SUMMARY

Successful performance of virtually any task depends on memory. If accurate information is not retrieved at the appropriate time during the performance of a task, errors may occur. Three general categories of memories can be distinguished on the basis of their durability. Sensory memories retain information in a modality-specific format for very brief periods of time. Short-term memories are retained in an active state and are used for reasoning and comprehension. Long-term memories are outside of awareness and not in a highly activated state but may be retained for extended periods of time. As our knowledge about memory has progressed, our view of memory has evolved into one of flexible, dynamic systems with multiple coding formats. The characteristics of different kinds of memory and the processes in which they are involved have predictable effects on human performance, which we need to remember while designing human–machine systems.

Memory is intimately involved in the comprehension and communication of information. Verbal materials and environmental events are identified by accessing knowledge in semantic memory. Comprehension of both written and spoken language is based on mental representations of the information being conveyed. These representations are constructed from the individual's perception of the material and the context in which it is perceived. To the extent that information is consistent with the observer's mental representation, language comprehension will be facilitated. Nonverbal events must also be comprehended, and the concept of situation awareness emphasizes the importance of accurate comprehension for the operation of complex systems. Memory and comprehension, along with perception and attention, play roles in situation awareness, which is assessed nowadays by human factors specialists in a variety of settings. They also play pivotal roles in thought and decision making, the topics of our next chapter.

RECOMMENDED READINGS

Baddeley, A. (1999). *Essentials of Human Memory*. Hove, UK: Psychology Press.

Baddeley, A., Eysenck, M. W., & Anderson, M. C. (2009). *Memory*. Hove, UK: Psychology Press.

Kahana, M. J. (2012). *Foundations of Human Memory*. New York: Oxford University Press.

Marsh, E., McDermott, K. B., & Roediger, H. L., III (Eds.) (2006). *Human Memory: Key Readings*. New York: Psychology Press.

Neath, I., and Surprenant, A. (2003). *Human Memory: An Introduction to Research, Data, and Theory* (2nd ed.). Belmont, CA: Wadsworth.

Radvansky, G. A. (2011). *Human Memory* (2nd ed.). New York: Routledge.

Rayner, K., Pollatsek, A., Ashby, J., & Clifton, C., Jr. (2012). *The Psychology of Reading* (2nd ed.). New York: Psychology Press.

Sirkin, D., Martelaro, N., Johns, M., & Ju. W. (2017). Toward Measurement of Situation Awareness in Autonomous Vehicles. Proceedings of CHI 2017 (pp. 405–415). New York: ACM.

Tulving, E., & Craik, F. I. M. (Eds.) (2000). *The Oxford Handbook of Memory*. Oxford, UK: Oxford University Press.

11 Solving Problems and Making Decisions

Like any goal-directed activity, thinking can be done well or badly. Thinking that is done well is thinking of the sort that achieves its goals.

J. Baron
2008

INTRODUCTION

Complicated problem-solving and decision-making processes are engaged for all sorts of human activities. You must make decisions about things as simple as what clothes to put on in the morning and as complex as how to raise your children. Your decisions can have long-lasting consequences. The CEO of a company could decide to expand based on an overestimate of the company's financial strength, which might result in bankruptcy. This in turn would result in the loss of many jobs and have devastating consequences for the local economy. Similarly, a government's decision to enter into war will result in loss of life, economic hardship, and aftereffects of varying types that carry far into the future. Scientists have tried to understand how human reasoning and decision making take place so that poor decisions can be prevented.

Consider the operator of a human–machine system. To operate the system effectively, he must comprehend system information and decide on appropriate actions. There are two ways that the operator can control the system (Bennett, Flach, Edman, Holt, & Lee, 2015). The first mode of operation will be used when the system is operating in a familiar and predictable way. Under these circumstances, the operator can control the system with very little effort, relying on well-practiced responses to the system's behavior (skill-based performance; see Chapter 3). The operator will face difficulty when system information indicates that an unusual condition has developed, requiring that the operator change to the second mode of operation. In this mode, the operator will need to make decisions based on his reasoning about the system state. This reasoning may involve recall of information from semantic or episodic memory (rule-based performance) or formulating a novel solution by integrating several different sources of information (knowledge-based performance).

For example, most of a pilot's efforts in flying an airplane involve monitoring the instruments in the cockpit. This does not require a great deal of mental effort. Only when the instruments indicate that a problem has occurred must the pilot engage in any effortful problem solving or decision making. When the pilot determines that an emergency has occurred, she must integrate information from the many visual and auditory displays in the cockpit, diagnose the nature of the emergency, and decide on the actions that she should take in response. Yet, as we have discussed in previous chapters, the pilot's capacity for processing such information is limited, and she may make a bad decision even though she is well-trained and well-intentioned.

This chapter examines how people reason about and choose between different actions. There are two ways to describe how people make decisions: *normative* and *descriptive*. A normative model specifies those choices that a rational person should make under ideal circumstances. However, as Johnson-Laird (1983, p. 133) observed, "Human reasoners often fail to be rational. Their limited working memories constrain their performance. They lack guidelines for systematic searches for counter-examples; they lack secure principles for deriving conclusions; they lack a logic." In other words, our decisions often deviate from those prescribed by normative models, primarily because of our limited capacity for processing information. Descriptive models of reasoning and decision making try to explain how people actually think. By understanding how and why people deviate from normative rationality, the human factors specialist can present information and design support systems that will help an operator devise optimal problem solutions.

PROBLEM SOLVING

In most problem-solving tasks, a person confronts a problem that has a clear goal. In the laboratory, problem solving is studied by presenting people with multistep tasks that take minutes or hours to perform. These tasks usually require a person to perform many different actions to attain a goal. One famous problem of this type is the Tower of Hanoi (see Figure 11.1), which is widely used to assess people's executive control functions (see Chapter 10; Welsh & Huizinga, 2005). The goal is to move all of the discs from peg A to peg C, under the restrictions that only one disc can be moved at a time and a larger disc cannot be put on top of a smaller disc. Problem solving in the Tower of Hanoi and similar tasks is studied by recording the problem solver's moves as well as his or her accuracy and time to solution.

Another way to study problem solving is to obtain verbal reports, sometimes called *protocols*, from the problem solver that describe the steps he took to solve the problem. Verbal protocols are especially useful for tasks in which intermediate steps to the solution are made mentally and are therefore not observable. Verbal protocol analysis has been used in applied settings, such as in the development of expert systems (see Chapter 12), as well as to understand problem-solving processes (Noyes, 2006). Protocols are assumed to reflect the information and hypotheses being attended in working memory, although in reality they are only reports of the thoughts that are occurring at the time (Ericsson & Simon, 1993).

Protocols are usually generated while the task is being performed, rather than after it is completed, because people may forget and fabricate information if the reports are obtained retrospectively (Russo, Johnson, & Stephens, 1989). When protocols are collected systematically, they can provide valuable information about the cognitive processes engaged for a particular task (Hughes & Parkes, 2003). However, if a person generates the protocol while performing the task, she may alter how she performs the task. This could occur because of competition between the resources required to generate the protocol and to solve the problem (Biehal & Chakravarti, 1989; Russo et al., 1989). We must also remember that the information supplied by a protocol is a function of the instructions that the problem solver has been given and the questions the investigator has asked (Hughes & Parkes, 2003). These will determine what information is reported, and how much, and poor instructions or bad questions will lead to useless protocols.

THE PROBLEM SPACE HYPOTHESIS

One way to think about problem solving is to imagine how objects are manipulated within an imaginary mental space. This space is constructed by the problem solver from his understanding of the problem, including those relevant facts and relationships that he thinks are important for the task. All problem solving takes place within this space. Objects are manipulated within the space according to the problem solver's knowledge of allowable actions defined by the rules of the problem. Finally, the problem solver has available a number of rules or strategies that can coordinate the overall problem-solving process.

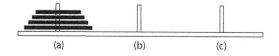

FIGURE 11.1   The Tower of Hanoi problem.

Newell and Simon (1972) proposed a framework for problem solving in which goals are achieved by movement through the problem space. Within this framework, different problem spaces are mental representations of different *task environments*. These problem spaces can be characterized by a set of states (positions in the problem space) and a set of operators that produce allowable changes between states (movement through the space). A problem is specified by its starting state and its desired ending, or goal, state. For the Tower of Hanoi problem, the starting state is the initial tower on peg A, the goal state is a tower on peg C, and the operators are the allowable movements of disks between the three pegs.

Two aspects of Newell and Simon's (1972) portrayal of problem solving are important: how the problem is represented and how the problem space is searched. First, because the problem space is only a mental representation of the task environment, it may differ from the task environment in certain important respects. With respect to product and system design, the editor of a special issue of a journal devoted to computational approaches for early stages of design noted, "The decisions we take in early design are often the most influential, imposing key constraints on our view of the problem space and thereby shaping later downstream decisions" (Nakakoji, 2005, p. 381). Similarly, advocates of support systems for developing groupware (computer software designed to facilitate interactions among group or team members) emphasize the need for developers and users to have a common understanding of the problem space (Lukosch & Schummer, 2006).

Second, searching the problem space requires consideration and evaluation of allowable moves between states. Some moves that are allowable within the task environment may not be included in the problem space, which means that they will not be considered. Furthermore, the limited capacity of working memory constrains the number of moves that can be considered simultaneously. This means that for complex problems, only a small part of the problem space can be held in memory and searched at any one time. Because only a limited number of moves can be examined, finding a problem solution quickly and effectively will require using strategies that direct search toward likely solution paths.

Consider the nine-dots problem in Figure 11.2. The goal is to connect the dots in the figure by drawing four straight lines without lifting the pencil from the paper. Many people find the nine-dot problem difficult. The reason for this is that the solution requires the lines to go outside of the boundaries of the square formed by the dots (see Figure 11.3). This allowable move is usually not included in the problem space, even though the problem description does not exclude such moves, possibly because Gestalt perceptual principles organize the dots into an object with boundaries and prior knowledge places inappropriate constraints on the representation (Kershaw & Ohlsson, 2004). When the move that allows the person to solve this problem is not included in the problem space, no amount of searching will allow the person to find it. As this example illustrates, an incomplete or inaccurate problem representation is a common source of problem-solving difficulty. Hence, one way to improve problem solving is for the person to spend more time constructing a mental

FIGURE 11.2 The nine-dots problem.

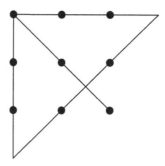

FIGURE 11.3   The solution to the nine-dots problem.

representation before seeking a solution (Rubinstein, 1986), and another is to provide hints that lead the person to change her representation of the problem space (Öllinger, Jones, & Knoblich, 2014).

Even when all allowable moves are contained in the problem space, a person needs a strategy to find a solution path through the problem space. A strategy will be most important when people are solving problems in unfamiliar domains, where their abilities to find a solution path is limited. Perhaps the weakest strategy, trial and error, consists of unsystematic or random selections of moves between states to attain the goal. Two stronger, more systematic strategies are forward chaining (working forward) and backward chaining (working backward). Forward chaining begins from the initial state. All possible actions are evaluated, the best one is selected and performed, and feedback tells the problem solver whether the action was a good one or a bad one. This process is repeated until a solution is achieved. Backward chaining begins from the goal state and attempts to construct a solution path to the initial state.

A third general strategy is called *operator subgoaling*. A person solving a problem selects a move (operator) without consideration of whether or not it is appropriate for the current state. If the move is inappropriate, the problem solver forms a subgoal in which he attempts to determine how to change the current state so that the desired move becomes appropriate.

These three strategies all incorporate heuristics to narrow the search for possible moves. You can think of heuristics as rules of thumb that increase the probability of finding a correct solution. Heuristics allow the problem solver to choose among several possible actions at any point in the problem space. For example, one heuristic is called *hill climbing*. In hill climbing, the problem solver evaluates whether or not the goal will be closer after making each possible move. The problem solver selects the move that brings him "higher" or closer to the goal state (the top of the hill). Because only the direction of each local move is considered, this heuristic works like climbing to the top of a hill while blindfolded. The problem solver may be left "stranded on a small knoll"; that is, every possible move may lead downhill although the goal state has not yet been reached. Consequently, the best solution may not be found.

Chronicle, MacGregor, and Ormerod (2004) proposed that a hill-climbing heuristic is one factor that underlies the difficulty that people have in solving the nine-dot problem. They argue that people evaluate potential moves against a criterion of satisfactory progress, which in the nine-dot problem is "that each line must cancel a number of dots given by the ratio of the number of dots remaining to lines available" (p. 15). Selecting moves that meet this criterion drives the problem solvers away from moves that lie on the correct solution path.

Means-end analysis is a heuristic that is similar to hill climbing in its focus on reducing the distance between the current location in the problem space and the goal state. The difference between means-end analysis and hill climbing is that in problem spaces appropriate for means-end analysis, the move needed to reach the goal can be seen, allowing an appropriate move to be selected to reduce the distance. Note that the heuristic described for the nine-dot problem above was called hill climbing rather than means-end analysis, because the criterion against which the problem solver was evaluating progress is inferred from the problem statement (dots must be cancelled) and not a known goal state.

Means-end analysis is a heuristic based on identifying the difference between the current state and the goal state and trying to reduce it. However, sometimes a solution path will require increasing the distance from the goal. Under means-end analysis, these kinds of actions are particularly difficult. For example, Atwood and Polson (1976) had people solve versions of water jug problems (in which water from a filled large jug must be distributed equally between it and another medium-sized jug, using a small jug). Their problem solvers had considerable difficulty with the problems for which finding a solution required them to move away from the known goal state of equal amounts of water in the two largest jugs.

The problem space hypothesis is particularly useful as a framework for artificial intelligence. This framework is embodied in the idea of a production system, which includes a data base, production rules that operate on the data base, and a control system that determines which rules to apply (Davis, 2001; Nilsson, 1998). One benefit of modeling human problem solving using production systems is that we can describe human performance with the same terminology used to describe machine performance. Consequently, insight into how human problem solving occurs can be used to advance artificial intelligence, and vice versa. This interaction lays the foundation for the design of cognitively engineered computer programs to assist human problem solving, called *expert systems*, which will be discussed in the next chapter.

ANALOGY

Analogy is another powerful heuristic in problem solving (Chan, Paletz, & Schunn, 2012; VanLehn, 1998). It involves a comparison between a novel problem and a similar, familiar problem for which the steps to a solution are known. An appropriate analogy can provide a structured representation for the novel problem, give an idea about the operations that will probably lead to a solution, and suggest potential mistakes. People tend to use analogies when the source and target problems have similar surface features (Bassok, 2003; Holland, Holyoak, Nisbett, & Thagard, 1986). A problem solver may attempt erroneously to apply an analogy when the surface characteristics are similar, even though the problems are structurally quite different and require different paths to solution. Conversely, analogical reasoning may not be used appropriately if the source and target problems have only structural similarity. Thus, the effective use of an analogy to solve a problem requires that the problem solver recognize structural similarity between the novel problem and the familiar analogous problem, and then apply the analogy correctly.

In general, people are adept at using analogies to solve problems, but they often fail to retrieve useful analogies from memory. Gick and Holyoak (1980, 1983) investigated the use of analogies in solving the "radiation problem" originated by Duncker (1945). The problem is stated as follows:

> Suppose you are a doctor faced with a patient who has a malignant tumor in his stomach. It is impossible to operate on the patient, but unless the tumor is destroyed the patient will die. There is a kind of ray that at a sufficiently high intensity can destroy the tumor. If the rays reach the tumor all at once at a sufficiently high intensity, the tumor will be destroyed. Unfortunately, at this intensity the healthy tissue that the rays pass through on the way to the tumor will also be destroyed. At lower intensities the rays are harmless to healthy tissue, but they will not affect the tumor either. What type of procedure might be used to destroy the tumor with the rays, and at the same time avoid destroying the healthy tissue? (Gick & Holyoak, 1983, p. 3)

In this problem, the actions that the problem solver might take are not well-defined. To arrive at a solution, she might use an analogy to transform the problem into a representation with clear actions. Before presenting the problem to be solved, Gick and Holyoak (1980) told the people in their study a military story in which a general divided his army to converge on a single location from different directions. Dividing the army is analogous to the solution of splitting the ray into several lower-intensity rays that converge on the tumor (see Figure 11.4). Approximately 75% of the people tested following the military story generated the analogous solution to the ray problem, but only if

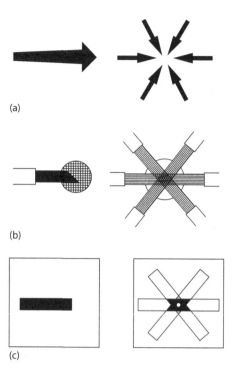

(a)

(b)

(c)

FIGURE 11.4 The visual analogs of the radiation problem used by (a) Gick and Holyoak (1983) and (b) by Beveridge and Parkins (1987).

they were told to use the analogy between the story and the problem. When they were not told that there was a relationship between the story and the problem, only 10% of them solved the problem. In short, the people had difficulty recognizing the analogy between the story and the problem, but they could easily apply the analogy when they were told that it was there.

When people read two different stories using the same convergence solution, they were more likely to solve the problem with the analogy than when they read only the military story. Gick and Holyoak (1983) attributed the ability to use the analogy after two stories to the problem solvers' acquisition of an abstract convergence schema. Specifically, when two stories with similar structure are presented, the problem solver generates an abstract schema of that structure. The problem is then interpreted within the context of that schema, and the analogy is immediately available. Requiring a person to generate an analogous problem after solving one seems to have a similar beneficial effect on the solution of another, related problem (Nikata & Shimada, 2005).

Given that the solution to the radiation problem has a spatial representation, we might predict that providing problem solvers with a visual aid should help them generate the problem solution. Gick and Holyoak (1983) used the diagram in Figure 11.4a for the radiation problem and found that it did not improve performance. However, Beveridge and Parkins (1987) noted that this diagram does not capture one of the essential features of the solution, which is that several relatively weak beams have a summative effect at the point of their intersection. When they showed people the diagram in Figure 11.4b or colored strips of plastic arranged to intersect as in Figure 11.4c, problem solvers' performance improved. Thus, to be useful, the visual aid must appropriately represent the important features of the task. Holyoak and Koh (1987) reached a similar conclusion about verbal analogies like the military story after they showed that problem solvers used them spontaneously more often when there were more salient structural and surface features shared by the two problems.

These findings suggest that to ensure the appropriate use of a problem-solving procedure, an operator should be trained using many different scenarios in which the procedure could be used.

Visual aids can be designed that explicitly depict the features important for solving a problem or for directing participants to attend to critical features (Grant & Spivey, 2003). By exploiting the variables that increase the probability that an analogy is recognized as relevant, the human factors specialist can optimize the likelihood that previously learned solutions will be applied to novel problems.

Although people have difficulty *retrieving* structurally similar, but superficially dissimilar, analogous problems to which they have previously been exposed, they seem to be much better at using structural similarities to *generate* analogies. Dunbar and Blanchette (2001) noted that scientists tended to use structural analogies when engaged in tasks like generating hypotheses. The results from their experiments suggest that even nonscientists can and do use superficially dissimilar sources for analogy if they are to freely generate possible analogies.

LOGIC AND REASONING

Recall again the concept of a problem space. We have presented problem solving as the discovery of ways to move through that space. Another way of thinking about problem solving is to consider how people use logic or reason to create new mental representations from old ones. Reasoning, which can be defined as the process of drawing conclusions (Leighton, 2004a), is a necessary part of all forms of cognition, including problem solving and decision making.

We can distinguish three types of reasoning: deductive, inductive, and abductive (Holyoak & Morrison, 2012). *Deduction* is reasoning in which a conclusion follows necessarily from general premises (assumptions) about the problem. *Induction* is reasoning in which a conclusion is drawn from particular conditions or facts relevant to a problem. *Abduction* is reasoning in which a novel hypothesis is generated to best explain a pattern of observations. People find all types of reasoning difficult, and they make systematic errors that can lead to incorrect conclusions. In the next two sections, we will discuss deductive and inductive reasoning, which have been studied extensively, and the ways that people can make mistakes when faced with different kinds of problems. We will then provide a briefer description of abduction.

DEDUCTION

Deduction depends on formal rules of logic. Formal logic involves arguments in the form of a list of premises and a conclusion. Consider the following statements:

1. Nobody in the class wanted to do the optional homework assignment.
2. Paul was a student in the class.
3. Therefore, Paul didn't want to do the optional homework assignment.

Statements 1 and 2 are premises, or assumptions, and statement 3 is a conclusion that is deduced from the premises. Together, these statements form a kind of "argument" called a *syllogism*. A syllogism is valid if the conclusion logically follows from the premises, as in this example, and invalid if it does not.

To the extent that any problem can be formulated as a syllogism, formal rules of logic can be applied to arrive at valid conclusions. We could characterize human reasoning for everyday problems as "optimal" if it really worked this way. However, it probably does not. Research on how people do reason deductively and the extent to which they use formal logic while reasoning has been performed using syllogisms (Evans, 2002; Rips, 2002). In particular, syllogisms are used to explore *conditional* and *categorical* reasoning.

Conditional Reasoning

To understand conditional reasoning, consider the statement, "If the system was shut down, then there was a system failure." This statement allows one to draw a conclusion (system failure) when given a condition of the system (being shut down). Deductive reasoning with conditional statements of this form is called conditional reasoning. More formally, we can write such statements as conditional syllogisms, which are of the form:

1. If the system was shut down, then there was a system failure.
2. The system was shut down.
3. Therefore, the system failed.

There are two rules of logic that allow us to come to a conclusion when given a syllogism of this form: affirmation (also called *modus ponens*) and denial (also called *modus tollens*). The syllogism just provided illustrates the rule of affirmation. Affirmation states that if *A implies B* (e.g., system shut down implies system failure) and *A is true* (system shut down), then *B* (system failed) must be true.

Now consider a different syllogism based on the same major premise:

1. If the system was shut down, then there was a system failure.
2. The system did not fail.
3. Therefore, the system was not shut down.

The rule of denial states that given the same major premise that *A implies B* and also that *B is false* (system did not fail), then *A* must also be false (the system was not shut down).

When we try to determine how people reason deductively, we present them with syllogisms and ask them to judge whether or not the conclusion of the syllogism is valid. People find some kinds of syllogisms easier than others. In particular, when a syllogism correctly makes use of the affirmation rule, people can easily distinguish between valid and invalid conclusions (Rips & Marcus, 1977). People begin to have problems, however, when the information provided by the premises is insufficient to draw a valid conclusion, or when drawing a valid conclusion requires use of the denial rule.

Consider for example, the premise, "If the red light appears, then the engine is overheating." Four syllogisms using this premise are shown in Table 11.1. The two valid syllogisms are shown in the top row of the table, using affirmation and denial. Whereas people have no problem judging the affirmation syllogism to be valid, they are less accurate at classifying the denial syllogism as valid. The table also shows two invalid syllogisms in the bottom row. For these syllogisms, the information provided by the second premise does not allow a person to draw any valid conclusion. The conclusions shown represent two common logical fallacies called the "affirmation of the consequent" and the "denial of the antecedent." In the original premise, the antecedent is the appearance of the red light, and the consequent is that the engine is overheating. To understand why these syllogisms are typically classified as fallacious, it is important to understand that nothing in the major premise

---

TABLE 11.1

Examples of Valid and Invalid Conditional Syllogisms

	Affirmation	Denial
Valid	1. If the red light appears, the engine is overheating. 2. The red light appeared. 3. The engine is overheating.	1. If the red light appears, the engine is overheating. 2. The engine is not overheating. 3. The red light did not appear.
	Affirmation of the consequent	**Denial of the antecedent**
Invalid	1. If the red light appears, the engine is overheating. 2. The engine is overheating. 3. The red light appeared.	1. If the red light appears, the engine is overheating. 2. The red light did not appear. 3. The engine is not overheating.

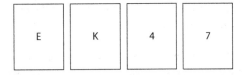

FIGURE 11.5 The four-card problem.

(statement 1) rules out the possibility that the engine could overheat without the light turning on. Both of the invalid conclusions are based on the unwarranted assumption that the conditional statement "if" implies "if and only if," that is, that the light always appears if the engine is overheating and never when it is not.

A famous experiment that explored how people engage affirmation and denial rules in reasoning was performed by Wason (1969). He showed people four cards, two with letters showing and two with digits showing, as shown in Figure 11.5a. He gave his subjects the following conditional statement:

If a card has a vowel on one side, then it has an even number on the other side.

The subject's task was to decide which cards would need to be turned over to determine whether the statement was true or false.

Most people turned over the E, demonstrating good use of the affirmation rule. But many people also turned over the 4, showing affirmation of the consequent. This is an error, because there is nothing in the statement that says that consonants could not also have an even number on the other side. According to the denial rule, the other card that must be turned over is the 7, since it must have a consonant on the other side. Very few people correctly turned over the 7.

It seems as though the problem solver's difficulty in applying the denial rule arises from an insufficient search of the problem space. Evans (1998) suggested that people are biased to select cards that match the conditions in the statement (i.e., the vowel and even number), regardless of whether or not they are relevant to the problem. Surprisingly, even students who had taken a course in formal logic did no better at this task than students who had not (Cheng, Holyoak, Nisbett, & Oliver, 1986), suggesting that this bias is a fundamental characteristic of human reasoning.

While people find the four-card problem difficult, they do very well with a completely equivalent problem as long as it is framed within a familiar context (Griggs & Cox, 1982; Johnson-Laird, Legrenzi, & Legrenzi, 1972). Consider the conditional statement "If a person is drinking beer, then the person must be over 20 years of age." If a police officer is attempting to determine whether a bar is in compliance with the minimum drinking age, students correctly indicated that the officer should examine the IDs of those people drinking beer and whether people under 20 were drinking beer.

If people were able to apply logical rules like affirmation and denial to conditional statements, we would not expect to see any difference in reasoning performance between the four-card problem and the minimum drinking age problem. The fact that people do better when problems are presented with familiar contexts suggests that people do not routinely use logical rules (Evans, 1989). Reasoning seems to be context-specific. For the drinking age problem, reasoning is accurate because people are good at using "permission schemas" (Cheng & Holyoak, 1985) to figure out what they are and are not allowed to do.

In the four-card problem, the tendency for people to turn over cards that match the conditions in the antecedent and the consequent can be viewed as a bias to look for confirming rather than disconfirming evidence. This bias affects reasoning performance in many other situations involved in verifying truth or falsity, such as medical diagnoses, troubleshooting, and fault diagnoses. Even highly trained scientists, when trying to confirm or disconfirm hypotheses through experimentation, may fall victim to their own confirmation biases.

One reason why confirmation bias is so strong is that people want to be able to retain their ideas of what is true and reject the ideas they wish to be false. One way to eliminate confirmation bias is to present premises that are personally distasteful to the problem solver, so that he will be motivated to reject them. Dawson, Gilovich, and Regan (2002) classified people according to their emotional reactivity. Each person took a test that classified him or her as having high or low emotional reactivity. Then they were told that people like themselves (high or low) tended to experience earlier death. This was not a belief that these people wanted to verify; they were highly motivated to disconfirm it. They were then shown four cards very similar to those used in the Wason four-card task, except that they were labeled with high and low emotional reactivity on one side and early and late death on the other. The labels that were exposed were high and low emotional reactivity and early and late death. The people were asked to test the early death hypothesis by turning over two cards. The two cards that correctly verify the hypothesis are the (confirming) card indicating the person's reactivity level and the (disconfirming) card indicating late death. The people told that they were at risk of an early death turned over the correct cards approximately five times more frequently than people who were told that they were not at risk. Dawson et al. found similar results by asking people to verify personally distasteful racial stereotypes.

Even when people intend to look for disconfirming evidence, this becomes hard to do when the task becomes more complex (Silverman, 1992). In such situations, reasoning can be improved through the use of computer-aided displays. There are several reasons why this may be true. First, the display can continuously remind the reasoner that disconfirming evidence is more important than confirming evidence, and second, the display can reduce some of the cognitive workload imposed by a complex task.

Rouse (1979) noted that maintenance trainees had difficulty diagnosing a fault within a network of interconnected units. This diagnosis required locating operational units in the network and tracing through their connections to potentially faulty units. However, the trainees tended to look for failures and to ignore information about which nodes had not failed. One condition in the experiment used a computer-aided display to help trainees keep track of tested nodes that had not failed. The trainees' fault diagnosis was better when the display was used than when it was not. Furthermore, people trained with the display performed better even after training was over and the display was no longer available.

Nickerson (2015, p. 1), in his book devoted to conditional reasoning, provides a clear statement that summarizes its importance:

> Conditional reasoning is reasoning about events or circumstances that are contingent on other events or circumstances. It is a type of reasoning in which we all engage constantly, and without the ability to do so, human beings would be very different creatures, and greatly impoverished cognitively.

Categorical Reasoning

Categorical syllogisms are different from conditional syllogisms in that they include quantifiers like *some*, *all*, *no*, and *some-not*. For example, a valid categorical syllogism is:

1. All pilots are human.
2. All humans drink water.
3. Therefore, all pilots drink water.

As with conditional syllogisms, judgments about the validity of a conclusion in a categorical syllogism can be influenced by the context of the syllogism, misinterpreting the premises, and confirmation bias.

Consider the syllogism:

1. Some pilots are men.
2. Some men drink beer.
3. Therefore, some pilots drink beer.

The conclusion that some pilots drink beer does not follow from the premises, although many people will judge it to be valid. To appreciate why this conclusion is invalid, consider the following very similar syllogism:

1. Some pilots are men.
2. Some men are older than 100 years.
3. Therefore, some pilots are older than 100 years.

There are no pilots older than 100 years. What has gone wrong here? If a logical rule is to be applied to a set of premises, the resulting conclusion should be valid regardless of the content of those premises. In both of these syllogisms, exactly the same logical rule was applied to reach a conclusion. Whereas it seems reasonable to conclude that some pilots drink beer, it does not seem at all reasonable to conclude that some pilots are older than 100 years. This means that the first conclusion, no matter how reasonable (or true), is not a valid deduction. The error made with syllogisms of this type is to assume that the subset of men who are pilots and the subset of men who drink beer (or who are older than 100 years) overlap, but none of the premises forces this to be true.

These types of errors have been explained in part by the atmosphere hypothesis (Woodworth & Sells, 1935). According to this hypothesis, the quantifiers in the premises set an "atmosphere," and people tend to accept conclusions consistent with that atmosphere (Leighton, 2004b). In the two syllogisms above, the presence of the quantifier *some* in the premises creates a bias to accept the conclusion as valid because it also uses the quantifier *some*.

Many errors on categorical syllogisms also may be a consequence of an inappropriate mental representation of one or more of the premises. For example, the premise *some men drink beer* could be incorrectly converted by a person to mean *all men drink beer*. The accuracy of syllogistic reasoning is also affected by how the premises are presented, and in particular, the ordering of the nouns in the premises. For example, the premises in the syllogism above present the nouns in the form pilots-men (premise 1) and men-beer (premise 2). When the premises are presented like this, people will be more likely to produce a conclusion of the form pilots-beer than of the form beer-pilots, regardless of whether it is valid (Morley, Evans, & Handley, 2004). We can refer to this form as "A-B, B-C," where A, B, and C refer to the nouns in the premises. For premises with the form B-A, C-B, people tend to judge conclusions of the form C-A as valid. The following syllogism illustrates an example of this form:

1. Some men are pilots.
2. Some beer drinkers are men.
3. Therefore, some beer drinkers are pilots.

Again, this is an invalid conclusion. One reason why people may endorse this conclusion is that they may change the order in which the premises are encoded, so that the second premise is represented first and the first premise is represented second (Johnson-Laird, 1983). Reordering the premises in this way allows a person to think about beer drinkers, men, and pilots as subsets of each other, and sets up an "easier" A-B, B-C representation.

Johnson-Laird (1983) proposes that reasoning occurs through the construction of a mental model of the relations described in the syllogism. For example, given the premises:

All the pilots are men.
All the men drink beer.

a mental tableau would be constructed. The first premise designates every pilot as a man but allows for some men who are not pilots. The tableau for this would be of the following type:

$$\begin{array}{l} \text{pilot} = \text{man} \\ \text{pilot} = \text{man} \\ \text{pilot} = \text{man} \\ \qquad (\text{man}) \\ \qquad (\text{man}) \end{array}$$

The parentheses indicate that men who are not pilots may or may not exist. The tableau can be expanded to accommodate the second premise for which all men drink beer but some beer drinkers may not be men. It leads to the following model:

$$\begin{array}{l} \text{pilot} = \text{man}\ \ = \text{beer drinker} \\ \text{pilot} = \text{man}\ \ = \text{beer drinker} \\ \text{pilot} = \text{man}\ \ = \text{beer drinker} \\ \qquad (\text{man}) = (\text{beer drinker}) \\ \qquad (\text{man}) = (\text{beer drinker}) \\ \qquad\qquad\qquad (\text{beer drinker}) \end{array}$$

When asked whether a conclusion such as *All the pilots are beer drinkers* is valid, the mental model is consulted to determine whether the conclusion is true. In this case, it is true.

Two factors can affect the difficulty of a syllogism, according to Johnson-Laird (1983). The first factor is the number of different mental models that are consistent with the premises. When trying to decide whether a conclusion is valid, a person must construct and consider all such models. This imposes a heavy load on working memory resources. The second factor is the order in which premises are presented and the ordering of nouns within the premises, as discussed above. The orderings dictate the ease with which the two premises can be related to form an integrated mental model. Again, it seems as though reasoning does not occur through the use of formal logical rules, but by cognitive processes that are subject to biases and working memory limitations.

INDUCTION AND CONCEPTS

Induction differs from deduction in that an inductive conclusion is not necessarily true if the premises are true, as is the case with valid deductions. Inductive reasoning is accomplished by drawing a general conclusion from particular conditions. We do this every day without using any formal rules of logic. For example, a student may arrive at the inductive conclusion that all midterm exams are held in middle week of the semester because all of hers have been held at this time. Although this conclusion may be generally true, the student may take a class next semester for which the midterm exam is given at some other time. Inductive reasoning involves processes like categorization, reasoning about rules and events, and problem solving (Holyoak & Nisbett, 1988).

Our understanding of how the world works grows by using induction (Holland et al., 1986). Induction modifies how we think about procedures, or ways to do things, and our conceptual understanding of the world, or how objects and concepts are related to each other. Concepts and procedures can be represented by interrelated clusters of rules (schemas). Rules and rule clusters operate as mental models that can simulate the effects of possible actions on different objects.

A concept is an abstraction of the rules and relationships that govern the behavior of certain objects. How concepts are learned from examples and used is a fundamental component of inductive reasoning. Concepts have at least two functions (Smith, 1989): minimizing the storage of information and providing analogies to past experience. Concepts minimize the amount of information stored in memory, because a general rule and the objects to which it applies can be represented more economically than specific relationships between all objects in a particular category. For example, the rule "has wings and flies" can be applied easily to most objects in the category "bird," whereas it seems wasteful to remember separately that "robins have wings and fly," "sparrows have wings and fly," "canaries have wings and fly," and so on.

Past experiences represented as concepts can be used as analogies for problem solving. Recall the student who has induced that all midterms occur in the middle week of the semester. If she habitually skips her classes, she may use this induction to attend classes during that middle week to avoid missing her midterms.

Induction cannot occur between just any conceptual categories using just any rules. We can conceive of induction taking place through the activation of conceptual categories and rules appropriate to those categories. Activated concepts are formulated into a mental model similar to the "problem space" we discussed earlier in this chapter. Induction will be limited by the information that a person is able to incorporate into his problem space and retain in working memory. A particular problem-solving context will activate only a limited number of categories of conceptual knowledge, and not all of the information that may be necessary for valid induction may be incorporated into the mental model. If the wrong categories are activated, any conclusions made within the context may not be accurate. Similarly, if important information is left out of the mental model, any inductive reasoning based on that mental representation will not be able to use that information, and again, the conclusions may not be accurate.

Mental models may be used to simulate possible outcomes of actions (Gentner & Stevens, 1983). That is, given a model that incorporates a particular conceptual category, induction may proceed by "running" the model through different possible configurations and "observing" the outcome. These models and a person's ability to use them in this way will depend on that person's experiences interacting with a system or other relevant experiences. As with any inductive reasoning, these simulations can result in an accurate conclusion, but nothing is guaranteed. The accuracy of a conclusion will depend on the accuracy of the mental model. The accuracy of the mental model is one factor that allows experts to reason better than novices in a particular domain (see Chapter 12).

McCloskey (1983) demonstrated how an incorrect mental model can lead to incorrect inferences. He examined naive theories of motion acquired from everyday interactions with the world. He asked his people to solve problems of the type shown in Figure 11.6. For the spiral tube problem, he told them to imagine a metal ball put into the spiral tube at the end marked by the arrow. For the ball and string problem, he told them to imagine the ball being swung at high speed above them. People then drew the path of the ball when exiting the tube in the first case and when the string broke in the second case. The correct path in each case is a straight line, but many people responded that the balls would continue in curved paths. This led McCloskey to propose that people used a "naive impetus theory" to induce the path of the balls: the movement of an object sets up an impetus for it to continue in the same path. This theory, when incorporated into a mental model, yields simulations that produce incorrect inductions.

One important issue in the development of conceptual categories is how particular objects are classified as belonging to particular categories. One idea is that an object is classified as belonging

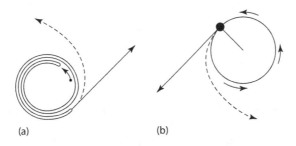

(a)   (b)

FIGURE 11.6   The spiral tube (a) and ball-and-string (b) problems with correct (solid lines) and incorrect (dashed lines) solutions.

to a particular category if and only if it contains the features that define the category (Smith & Medin, 1981). For example, "robin" may be classified as a "bird" because it has wings and flies. This idea, while important for the categorization of objects, does not explain how concepts are developed. Defining features do not exist for many categories. For example, there is no single feature that is shared by all instances of the concept *games*. Moreover, typicality effects, in which classification judgments can be made faster and more accurately if an object (robin) is typical of the category (bird) than if it is atypical (penguin; see Chapter 10), show that not all instances of a category are equal members of that category.

The effect of typicality results in fallacious reasoning (Tversky & Kahneman, 1983). For example, one experiment showed people personality profiles of different individuals, for example, "Linda," who was "deeply concerned with issues of discrimination and social justice and also participated in antinuclear demonstrations." They were then asked to judge how typical an example Linda was of the category "bankteller" or "feminist bankteller," or how probable it was that Linda was a bankteller or a feminist bankteller. Because feminist banktellers are a subset of the larger category "bankteller," people should estimate the probability that Linda is a bankteller as higher than the probability that she is a feminist bankteller. However, Linda was usually judged as more typical of a feminist bankteller, and people estimated the probability that she was a feminist bankteller as higher than the probability that she was just a bankteller.

This error is a *conjunction error* (Kahneman & Tversky, 1972; Shafir, Smith, & Osherson, 1990; Tversky & Kahneman, 1983), which arises from a *representativeness heuristic*. The representativeness heuristic is a rule of thumb that assigns objects to categories based on how typical they seem to be of those categories (see later discussion of decision-making heuristics in this chapter).

There are other ways that category membership can be determined. These include an assessment of how similar an object is to a category "prototype" (the ideal or most typical category member) and using other information to convert an inductive problem into one of deduction (Osherson, Smith, & Shafir, 1986). For example, when given a volume for an object and asked to decide whether it is a tennis ball or a teapot, objects that are closer in volume to the tennis ball than to the average teapot will be classified as a tennis ball (Rips, 1989). Apparently, the knowledge that tennis balls are of a fixed size is incorporated into the categorical judgment, changing the problem into one of deduction.

Abduction and Hypotheses

A third form of reasoning, introduced by Peirce (1940), is called *abduction* or *retroduction*. Abductive reasoning involves three interrelated elements (Holcomb, 1998; Proctor & Capaldi, 2006): explaining patterns of data; entertaining multiple hypotheses; and inference to the best explanation. With regard to explaining patterns of data, a person using abduction examines phenomena, observes patterns, and then develops a hypothesis that explains them. This form of reasoning is not deduction, because the hypothesis is not derived from the phenomena, nor is it an induction, because it is a generalization not about the properties shared by the phenomena but about their cause.

The latter two elements derive from the idea that people don't think about a single hypothesis in isolation. Thus, when reasoning abductively, people evaluate any given hypothesis relative to other hypotheses, with the goal of arriving at the best explanation. As we mentioned in Chapter 2, this form of reasoning is used widely in science (Haig, 2014). However, it also is used widely in other circumstances. For example, in medical diagnosis (Patel, Arocha, & Zhang, 2005) and judicial decision making (Ciampolini & Torroni, 2004), people will generate and consider alternative hypotheses, and their diagnosis or decision will be in favor of the hypothesis that provides the best explanation of the evidence. Likewise, when diagnosing a fault with a complex system, such as a chemical plant, operators will typically apply abductive reasoning in generating and evaluating different hypotheses (Lozinskii, 2000).

DECISION MAKING

The things that a person decides to do affect both the person (the decision maker) and the people around him. Also, the conditions under which the decision is made can influence what a person chooses to do. Decisions can be made under conditions of certainty, in which the consequences for each choice are known for sure, or under conditions of uncertainty, in which the consequences for each choice may be unknown. Gambling is an example of decision making under uncertainty. Most real-life decisions are made under uncertainty. If you decide to exceed the speed limit by a significant amount, one of several things could possibly happen: you might arrive at your destination early and save a lot of time, you could be stopped and cited for speeding and arrive late, or you could cause a serious traffic accident and never arrive at all. An example of an applied decision-making problem is the choice to include curtain air bags as standard equipment on a line of automobiles, given the estimated cost, prevailing market conditions, effectiveness of the air bags, and so on.

How do people choose what to do when they do not know what the consequences of their actions will be? The most interesting questions about how people make decisions are asked within this context. Decisions based on both certain and uncertain conditions are faced regularly by operators of human–machine systems as well as by human factors specialists. Therefore, it is important to understand the ways that decisions are made and the factors that influence them. There are two ways that we can talk about how people make decisions (Lehto, Nah, & Yi, 2012). Normative theories explain what people should do to make the best decisions possible. But people do not often make the best decisions, so descriptive theories explain how people really make decisions, including how people overcome cognitive limitations and how they are biased by decision-making contexts.

NORMATIVE THEORY

Normative theories of decision making concern how we should choose between possible actions under ideal conditions. Normative theories rely on the notion of *utility*, or how much particular choice outcomes are worth to the decision maker. Utility is a measure of the extent to which a particular outcome achieves the decision maker's goal. The decision maker should choose the action that provides the greatest total utility. If outcomes are uncertain, both the probabilities of the various outcomes and their utilities must be figured into the decision-making process.

How people incorporate utility into their decision-making processes has been studied using gambles. Table 11.2 shows two gambles. People in decision-making experiments are often given some amount of money to play with, and they can choose which gamble they would prefer to play. Of the choices in Table 11.2, which gamble should a decision maker choose? Is gamble A better than gamble B, or vice versa? *Expected-utility theory* provides an answer to these questions. For monetary gambles, assume that the utility of a dollar amount is equal to its value. An expected utility $E(u)$ of each gamble can be found by multiplying the probability of each possible outcome by its utility. That is,

$$E(u) = \sum_{i=1}^{n} p(i)u(i),$$

TABLE 11.2
Gambles with Different Expected Utilities

GAMBLE A		GAMBLE B	
OUTCOME	PROBABILITY	OUTCOME	PROBABILITY
WIN $10	.10	WIN $1	.90
LOSE $1	.90	LOSE $10	.10

---

where:

    $p(i)$ is the probability of the ith outcome, and

    $u(i)$ is the value of the ith outcome.

We can view an expected utility as the average amount that a decision maker would win for a particular gamble. In choosing between two gambles, the rational choice would be the one with the highest expected utility.

We can compute the expected utilities for the gamble in Table 11.2. For gamble A, the expected utility is

$$0.10(\$10) - .90(\$1) = \$1 - \$.90 = \$.10,$$

and for gamble B the expected utility is

$$.90(\$1) - .10(\$10) = \$.90 - \$1 = -\$.10.$$

Thus, gamble A has the highest expected utility and should be preferred to gamble B.

A rational decision maker makes choices in an attempt to achieve some goal. We have defined utility as the extent to which an outcome achieves this goal. Therefore, expected-utility theory provides a yardstick for rational action, because rational decisions must be consistent with those that yield the greatest utility.

Expected-utility theory forms the basis of a discipline called *behavioral economics*, which studies how people make economic choices. One reason why expected-utility theory has been so influential is the simple fact that rational choices must be based on numbers. This means that only a few fundamental rules of behavior, called *axioms* (Wright, 1984), can be used to deduce very complex decision-making behavior. One fundamental axiom is called *transitivity*. This means that if you prefer choice A to choice B, and you also prefer choice B to choice C, then when you are presented with options A and C, you should prefer A. Another is that of *dominance*: if, for all possible states of the world, choice A produces at least as desirable an outcome as choice B, then you should prefer choice A. Most importantly, preferences for different options should not be influenced by the way they are described or the context in which they are presented; only the expected utility should matter. As we shall see, these axioms do not always hold for real-life decisions, which is why psychologists have developed descriptive theories to explain human behavior.

DESCRIPTIVE THEORY

Do people perform optimally in the way proposed by expected-utility theory? The answer is no. People consistently violate the axioms of expected-utility theory and demonstrate what could be interpreted as irrational choice behavior. In this section, we will talk about the ways that people violate these axioms, the reasons for violating these axioms, and then, in the next section, ways to improve decision-making performance.

Transitivity and Framing

Consider the axiom of transitivity described above. If A is chosen over B, and B over C, A should be chosen over C. Yet violations of transitivity occur (Tversky, 1969), in part because small differences between alternatives are ignored in some situations but not in others. Consider the three health clubs in Table 11.3 (Kivetz & Simonson, 2000). Information about price, variety of exercise machines, and travel time is given if it is available. You can see that while you may prefer Health Club A to B on the basis of price, and B to C on the basis of travel time, you may nonetheless prefer Health Club C to A on the basis of variety. This violation of transitivity results from

TABLE 11.3

Health Clubs Described with Missing Information

	Health Club A	Health Club B	Health Club C
Annual membership fee (range: $200–$550)	$230/year	$420/year	(Information unavailable)
Variety of exercise machines (range: poor to excellent)	Average	(Information unavailable)	Very good
Driving time to health club (range: 5–30 minutes)	(Information unavailable)	6 minutes	18 minutes

a comparison between different features of each alternative and does not necessarily represent irrational behavior.

Another important violation of expected utility axioms results from *framing* (Levin et al., 2015). Choice behavior will change when the context of the decision changes, even when that context does not alter the expected utilities of the choices (Tversky & Kahneman, 1981). People can be manipulated to make different choices by restating identical problems to emphasize gains or losses. As one example, consider the following problem:

> Imagine that the U.S. is preparing for the outbreak of an unusual and virulent disease, which is expected to kill 600 people. Two alternative programs to combat the disease have been proposed. Assume that the consequences of the programs are as follows.

One description of the two programs, emphasizing lives saved, might look like this:

> If Program A is adopted, 200 people will be saved. If Program B is adopted there is a 1/3 probability that 600 people will be saved, and 2/3 probability that no people will be saved. Which of the two programs would you favor?

Another description, emphasizing lives lost, might look like this:

> If Program C is adopted, 400 people will die. If Program D is adopted, there is 1/3 probability that nobody will die, and 2/3 probability that 600 people will die. Which of the two programs would you favor?

Notice that the two descriptions are formally identical. For instance, in the first description 200 people are saved with the first program, which is the same as 400 people dying, as written in the second description. People who saw the first description usually chose Program A over Program B, whereas most people who saw the second version chose Program D over Program C. The first description provides a positive frame for "saving lives," whereas the second provides a negative frame for "lives that will be lost." This demonstrates that a person's decision may be affected greatly by the way in which important information is presented, primarily by influencing how people pay attention to various attributes of the choice.

Another axiom of expected-utility theory that is closely related to framing has to do with the stability of preference. If A is preferred to B in one situation, then A should be preferred to B in all other situations. However, it is easy to get people to reverse their preferences for A and B in different contexts. Lichtenstein and Slovic (1971) found that when choosing between a bet with a high probability of winning a modest amount of money and one with a low probability of winning a large amount of money, most people chose the high-probability bet. They were then asked to state their selling price for the gamble, where they would turn over the gamble to a buyer and allow that buyer to play the gamble. In this case, most people gave a higher selling price for the low-probability event than for the high-probability event.

This is a preference reversal, because the selling price indicates that the unchosen alternative has a higher value or utility than the chosen alternative. Tversky, Sattath, and Slovic (1988) concluded that such reversals occur because when the person must choose which gamble to play, he focuses his attention on probability, whereas when the person sets the selling price, he focuses his attention on the dollar amount. Again, the context in which the choice is framed has an effect on a person's preference, and a person will attend to different features of the choice in different contexts.

Bounded Rationality

Violations of transitivity and framing effects (among other findings) led Simon (1957) to introduce the concept of *bounded rationality*. This concept embodies the notion that a decision maker bases his or her decisions on a simplified model of the world. The decision maker

> behaves rationally with respect to this (simplified) model, and such behavior is not even approximately optimal with respect to the real world. To predict his behavior, we must understand the way in which this simplified model is constructed, and its construction will certainly be related to his psychological properties as a perceiving, thinking, and learning animal. (Simon, 1957, p. 198)

Bounded rationality recognizes that human decision makers have limitations on the amount of information that can be processed at any one time. For a difficult decision, it will not be possible to consider every feature of all of the alternatives. For example, when you go to choose a mobile phone plan, you cannot compare all plans on all possible features that differentiate them (Friesen & Earl, 2015). For a decision like this, you might think about only those features that you care about most (e.g., connection fees, price for on-network calls, data limits), and if you can come to a decision based only on these features, you will do so. This decision-making strategy is called *satisficing* (Simon, 1957). Whereas satisficing may not lead to the best decision every time, it will lead to pretty good decisions most of the time.

We defined heuristics earlier in this chapter as rules of thumb that allow people to reason in very complex situations. While heuristics will not always produce correct or optimal decisions, they help people bypass their cognitive and attentional limitations (Katsikopoulos & Gigerenzer, 2013). Satisficing takes place, then, through the use of heuristics.

Elimination by Aspects

One example of a satisficing heuristic applied to complex decisions is called *elimination by aspects* (Tversky, 1972). When people use this heuristic, they reduce the number of features they evaluate across their choices by focusing only on those that are personally most important. Beginning with the feature that you think is most important, you might evaluate all your choices and retain only those that seem attractive on the basis of this single feature. For you, price may be the most important aspect of a new car. You decide to eliminate all cars that cost more than $15,000. Size may be next important; so from the cars of $15,000 or less, you eliminate any compact cars. This elimination procedure continues through all of the personally important features until only a few alternatives remain that can be compared in more detail. Like many satisficing heuristics, although this procedure reduces the processing load, it can also lead to the elimination of the optimal choice.

Decision makers will often base their choice on a single dominant feature among alternatives and can be unwilling to consider other important attributes. In a study of proposals for coastline development in California, Gardiner and Edwards (1975) found that people could be grouped according to whether development or environmental concerns were most important to them. While members in the development group attended only to the development dimension across different development alternatives, members in the environmental group attended only to the environmental dimension. However, when people were forced to rate each proposal on both development and environmental dimensions, they gave some weight to both dimensions. This demonstrates that people can fairly evaluate alternatives on the basis of features that are not particularly salient to them if they are forced to do so.

The tendency to base decisions on only salient dimensions is even greater under stress. Stress increases the level of arousal, and, as discussed in Chapter 9, at high levels of arousal a person's attentional focus becomes narrowed and less controlled. Wright (1976) found evidence for both of these effects in the decisions of people who rated how likely they would be to purchase each of several automobiles. In one condition, Wright increased task-relevant time stress by reducing the time available for making the decision, and in another, he played an excerpt from a radio talk show as a distraction. Both manipulations caused the decision makers to focus more on the negative characteristics of the cars than they did in the baseline condition. This and other studies suggest that the tendency to narrow attention during decision making under stress can be minimized by eliminating unnecessary stressors and by structuring the decision process in such a way that the decision maker is forced to consider all the features important to making a good choice.

Availability

Another useful heuristic, which is used to estimate the probabilities or frequencies of events, is called *availability* (Kahneman, Slovic, & Tversky, 1982). Availability is the ease with which events can be retrieved from memory. More easily remembered events are judged as more likely than less memorable events. For example, if a person is asked to judge whether the letter *R* is more likely to occur in the first or third position of words in the English language, she will usually pick the first position as most likely. In reality, *R* occurs twice as often in the third position. Tversky and Kahneman (1973) argue that this happens because it is easier to retrieve words from memory on the basis of the first letter. Availability also biases people to overestimate the probability of dying from accidents relative to routine illnesses (Lichtenstein, Slovic, Fischoff, Layman, & Combs, 1978). Violent accidents such as plane crashes are much more available than most illnesses because they receive more media coverage, so their incidence tends to be overestimated.

Representativeness

The *representativeness* heuristic mentioned earlier in the chapter uses degree of resemblance between different events as an indication of how likely those events are to occur. More representative outcomes will be judged as more likely to occur than less representative ones. The following example from Kahneman and Tversky (1972) illustrates this point:

> All families of six children in a city were surveyed. In 72 families the *exact order* of births of boys (B) and girls (G) were GBGBBG.
>   What is your estimate of the number of families surveyed in which the *exact order* of births was BGBBBB?

Because there is a 50% chance of giving birth to a boy or a girl, the sequence BGBBBB has exactly the same probability of occurring as the sequence GBGBBG. Despite the fact that these two sequences are equally probable, the BGBBBB sequence is often judged to be less likely than the GBGBBG sequence. We can explain this mistake by noting that the sequence with five boys and one girl is less representative of the proportion of boys and girls in the population.

Representativeness is closely related to the gambler's fallacy, which is the belief that a continuing run of one of two or more possible events is increasingly likely to be followed by an occurrence of the other event. For example, suppose the births in the BGBBBB sequence above were presented sequentially to a person who made a probability judgment after each birth that the next birth would be a girl. The predicted probability that the subsequent birth would be a girl tends to become larger through the run of four boys, even though the probability is always 50%. The gambler's fallacy occurs because people fail to treat the occurrence of random events in a sequence as independent; that is, that having a boy does not change the future probability of having a girl.

Probability Estimation

People are very bad at making accurate probability estimates. For example, the gambler's fallacy is the failure to perceive independent events as independent. Shortly, we will see that people are also very bad at considering base rate information. Representativeness and anchoring (discussed below) are heuristics that permit people to make probability estimates for complex events. In particular, they allow people to make probability estimates for complex events composed of several simple events (such as the sequences of births presented above) without having to perform difficult mathematical calculations. As for all situations of satisficing, when a heuristic is used for probability judgments, such judgments will show systematic inaccuracies.

These inaccuracies can be demonstrated in real-life judgment situations (Fleming, 1970). Fleming asked people in his experiment to imagine themselves in a combat situation. He asked them to estimate the overall probability of an enemy attack on each of three ships, given the altitude, bearing, and type (e.g., size and armament) of an enemy plane. The person's goal was to protect the ship that was most likely to be attacked. Although each aspect of the plane was independent and of equal importance, people tended to add the different probabilities together rather than to multiply them, as was appropriate (see Chapter 4). Because of these mistakes, people underestimated the probability of very likely targets and overestimated the probability of unlikely targets. Decision makers apparently experience considerable difficulty in aggregating probabilities from multiple sources, which suggests that such estimates should be automated when possible.

When base rates or prior probabilities of events are known, the information from the current events must be integrated with the base rate information. In the previous example, if the prior probabilities of each of the three ships being attacked were not equal, then this information would need to be integrated with the altitude, bearing, and type information. Yet, in such situations, people do not typically consider base rates.

A famous example of a base rate problem is presented as an evaluation of the reliability of an eyewitness's testimony (Tversky & Kahneman, 1980). A witness, let's call him Mr. Foster, sees an accident late one night between a car and a taxi. In this part of town, 90% of the taxis are blue and 10% are green. Mr. Foster sees the taxi speed off without stopping. Because it was dark, Mr. Foster could not tell for sure whether the taxi was green or blue. He thinks it was green. To establish how well Mr. Foster can discriminate between blue and green taxis at night, the police showed him 50 green taxis and 50 blue taxis in a random order, all in similar lighting. Mr. Foster correctly identified the color of 80% of the green taxis and 80% of the blue taxis. Given Mr. Foster's identification performance, how likely is it that he correctly identified the color of the taxi involved in the accident?

Most people estimate Mr. Foster's testimony to have around an 80% probability of being accurate. However, this estimate "neglects" the base rate information provided early in the problem: only 10% of the taxis in that part of town were green to begin with. The true probability that Mr. Foster saw a green taxi is only about 31% when this information is considered.

We can demonstrate that people rely on the representativeness heuristic to solve some problems of this type. For example, Kahneman and Tversky (1973) gave people descriptions of several individuals supposedly drawn at random from a pool of 100 engineers and lawyers. One group was told that the pool contained 70 engineers and 30 lawyers, whereas the other group was told the reverse. These prior probabilities did not affect the judgments; the judgments were based only on how representative of an engineer or a lawyer a person seemed to be.

Decision makers can adjust their probability estimates when they are instructed to pay attention to base rate information, but their modified estimates are not adjusted enough. So, in the case of Mr. Foster, if a jurist were to be instructed to consider the fact that only 10% of the taxis are green, he might modify his estimate of Mr. Foster's accuracy from 80% down to 50%, but probably not all the way down to 31%. This tendency to be conservative in adjusting probability estimates can be linked to the use of an *anchoring* heuristic (Tversky & Kahneman, 1974). The evidence that Mr. Foster is

80% correct in judging blue from green taxis forms the basis of a preliminary judgment, or anchor. The base rate information is evaluated with respect to that anchor. The anchor exerts a disproportionate effect on the final judgment.

The importance of anchors was demonstrated by Lichtenstein et al. (1978), who had people estimate the frequencies of death in the United States for 40 causes. They were given an initial anchor of either "50,000 people die annually from motor vehicle accidents" or "1,000 deaths each year are caused by electrocution," and then they estimated the frequencies of death due to other causes. The frequency estimates for other causes were considerably higher with the anchor "50,000 deaths" than with the anchor "1,000 deaths."

In summary, when performing complex reasoning tasks, people often use heuristics that reduce their mental workload. These heuristics will produce accurate judgments in many cases, particularly when the reasoner knows something about the domain in question. The benefit of heuristics is that they render complex tasks workable by drawing on previous knowledge. The cost is that these heuristics are the source of many mistakes made by operators and decision makers.

IMPROVING DECISIONS

Individuals in an organization and operators of human–machine systems are often faced with complex decisions, sometimes under very stressful conditions. We have just discussed how people are forced to make less than optimal decisions because of their limited capacity for attending to and working with information. For this reason, one area in human factors has been concerned with the improvement of decision making through design. There are three ways in which we can improve the quality of decisions: designing education and training programs, improving the design of task environments, and developing decision aids (Evans, 1989).

TRAINING AND TASK ENVIRONMENT

We said earlier that people with formal training in logic make the same types of reasoning errors as people without such training. For example, Cheng et al.'s (1986) experiments found that people performed no better on Wason's four-card problem after a semester course in logic than before the course. What this means for us is that training focused on improving reasoning and decision making more generally is not going to be effective at improving reasoning and decision making for specific tasks. Rather, training should focus on improving performance in specific task environments, because most reasoning is based on context-specific knowledge.

One exception to this general rule involves probability estimation. Fong, Krantz, and Nisbett (1986) showed that people could be taught to estimate probabilities more accurately with training. Their task required their subjects to use a statistical rule called the *law of large numbers*. This law states that the more data we collect, the more accurate our statistical estimates of population characteristics will be. Some of Fong et al.'s subjects received brief training sessions on the law of large numbers, and then were given 18 test problems of the following type (Fong et al., 1986, p. 284):

> An auditor for the Internal Revenue Service wants to study the nature of arithmetic errors on income tax returns. She selects 4000 Social Security numbers by using random digits generated by an "Electronic Mastermind" calculator. And for each selected social security number she checks the 1978 Federal Income Tax return thoroughly for arithmetic errors. She finds errors on a large percentage of the tax returns, often 2 to 6 errors on a single return. Tabulating the effect of each error separately, she finds that there are virtually the same numbers of errors in favor of the taxpayer as in favor of the government. Her boss objects vigorously to her assertions, saying that it is fairly obvious that people will notice and correct errors in favor of the government, but will "overlook" errors in their own favor. Even if her figures are correct, he says, looking at a lot more returns will bear out his point.

The auditor's reasoning was based on the fact that she used random sampling, which should be unbiased, and a relatively large sample of income tax forms. Her boss's contrary stand is that the sample is not large enough to yield accurate estimates. The people who received training in the law of large numbers were much more likely to use statistical reasoning and to use it appropriately in their answers. In the above problem, for example, those who received training would be more likely to mention that the auditor's findings were based on random sampling of a large number of tax returns.

An important aspect of any training program or task environment is how information is presented to trainees or decision makers. We will talk more about training in general in the next chapter. For now, we wish to emphasize that if information is presented unclearly, too generally, or abstractly, people will be unable to perceive the relevance of the information to the task that they wish to perform, and they will be unable to apply their experience to solve novel problems.

We have already seen one important example of the effect that different ways of presenting information can have on decisions: the framing effect. Presenting a problem in one way may lead to a different decision than presenting it another way. For example, people have difficulty reasoning about negative information and do much better if the information is framed in such a way that important attributes are encoded positively rather than negatively in the mental representation (e.g., Griggs & Newstead, 1982).

Whereas framing can be used to draw a decision maker's attention to one or more features of a problem, many errors of inference and bias can be attributed to information being presented in such a way that it increases the decision makers' information-processing load (Evans, 1989). Unfortunately, this is very easy to do simply by presenting information in a complicated or unclear way. As an example, people who study consumer behavior are very concerned about how pricing information is presented for products on grocery store shelves. You are probably familiar with the little tags displaying unit price that appear under all of the products on a shelf in a U.S. grocery. These tags are supposed to provide the consumer with a unit price that allows him to make easy price comparisons across similar products. However, the units are often different on each tag, tags are not always aligned with the product that they identify, and often, tags across several meters of shelf space may need to be searched, memorized, and compared to determine the best price. Russo (1977) performed a simple experiment that compared the self-tag system with a simple list of unit prices for all products posted near the products. When the list was used, consumers reported that comparisons across brands were easier, and consumers purchased the less expensive brand more often.

Decision Aids

Decision-making performance can be improved by providing decision makers with aids that relieve some of the memory and information-processing demands of the task. There are many kinds of such aids, ranging from the very simple (like notes written on index cards) to the very complex (computer-based decision-support systems that use artificial intelligence). A decision aid may not even be an object. It could simply be a rule that one follows within a familiar, but uncertain, situation. For example, physicians often use something called the *Alvarado score* to diagnose acute appendicitis. Different symptoms, such as pain in the lower right abdomen, are assigned point values, and when the number of accumulated points becomes high enough, the physician will remove the patient's appendix.

Often the role of a decision aid is to force the decision maker to conform to the choices prescribed by normative theories. The Alvarado scale for appendicitis forces a physician to consider all relevant symptoms and weights them according to their diagnosticity, so it works a lot like an expected utility measure. One approach to complex decision making is *decision analysis*, a set of techniques for structuring complex problems and decomposing them into simpler components (Lehto et al., 2012). A decision analysis can be viewed as a decision aid in and of itself, or it can be performed for the purpose of better understanding a decision or constructing a new decision aid.

Structuring a problem for decision analysis usually involves the construction of a decision tree specifying all possible decisions and their associated outcomes. The probability and utility of each outcome are estimated. Then, the expected utility is computed for each possible decision and used to recommend an optimal choice (von Winterfeldt & Edwards, 1986). Decision analysis has been applied with success to problems like suicide prevention, landslide risks, and weather prediction (Edwards, 1998). This success is due in large part to adequate structuring of the complex problem.

Decision analysis must be used with care. Because probabilities and utilities are estimated by the decision analyst, biases can still arise when these quantities are inaccurately assessed. Furthermore, even during a decision analysis, it is possible that certain critical features of the decision-making problem will be overlooked. One of the more spectacular failures of decision analysis involved the decision to place the gas tank of the Ford Pinto (sold between 1971 and 1980) behind the rear axle (von Winterfeldt & Edwards, 1986). When the Pinto was hit from behind, there was a chance that the gas tank would rupture and explode. A decision analysis was performed in which the cost of relocating the tank in front of the axle ($11 per vehicle) was compared with the expected dollar value of lives saved ($200,000 per "soul") by tank relocation. The total cost of tank relocation was computed to be greater than the utilities associated with saving lives and avoiding injuries, so the gas tank was left where it was. Not considered in this analysis were the cost of punitive damages awarded in liability suits and the cost of the negative publicity resulting from publication of the analysis. The reputation of the Pinto never recovered, and it was discontinued in 1980.

One computer-based decision-analysis system is MAUD (Multi-Attribute Utility Decomposition; Humphreys & McFadden, 1980). MAUD contains no domain-specific knowledge but elicits information from the decision maker about the problem and the different alternatives available for solving the problem. Based on this input, it structures problems and recommends decisions using normative decision theory. Because of the way that MAUD asks questions of the decision maker, decision-maker bias is reduced.

In many disciplines, computer-based *decision-support systems* have been developed to aid complex decision-making processes (Marakas, 2003). The availability of mobile phones and table computers has allowed the opportunity for much more widespread use of decision-support systems than in the past (Gao, 2013). A decision-support system is used to guide operators through the decision-making process. A decision-support system has three major components: a *user interface*, a *control structure*, and a *fact base*. The interface solicits input from the user and presents information to the user that is relevant to the problem. Users may retrieve and filter data, request computer simulations or projections, and obtain recommended courses of action (Keen & Scott-Morton, 1978).

The control structure of a decision-support system consists of a data base management system and a model management system (Liebowitz, 1990). The data base management system is a set of programs for creating data files organized according to the needs of the user. The model management system is used to model the decision situation by drawing on information from the data base. Finally, the fact base of the decision-support system includes not only the data base but also the models that can be applied to the data.

A good decision-support system has a number of characteristics, and human factors engineering can make a positive contribution to its usability. Most important from the users' perspective is that it satisfy their needs. As Little, Manzanares, and Watson (2015, p. 273) note, "A decision support system that is mismatched with user needs benefits no one and can lead to poor decision making that results in unnecessary human and economic costs." In addition to usefulness, usability is a critical factor, which mainly involves the design of the interface. This interface should allow effective dialogue between the user and the computer. The design should consider how information is presented and elicited, providing flexibility in how data are analyzed and displayed. As we will see in later chapters, there is usually a tradeoff between flexibility and usability. Human factors engineers can help determine the level of flexibility appropriate for any particular application. It is important to recognize that decision-support systems do not replace decision makers but only provide them with important information in such a way that decision-making performance is improved. Even when a decision-support system is known to be effective, people's attitudes may prevent its widespread use, as we discuss in Box 11.1.

BOX 11.1 DIAGNOSTIC SUPPORT SYSTEMS

Computer-based Diagnostic Support Systems (DSSs) are used in a variety of situations, to aid medical diagnoses, for example of appendicitis and heart failure, and some DSSs are for very general use. Such aids can be extraordinarily effective. Some studies have shown that when physicians use a DSS, their diagnostic accuracy improves dramatically. Unfortunately, physicians demonstrate marked reluctance to use them.

One study examined physicians' ability to diagnose acute cardiac ischemia (ACI), which includes obstruction of the heart's arteries leading to chest pain, as well as full-blown "heart attacks," in which the obstruction is complete and the heart muscle is dying. Such diagnoses are extraordinarily expensive because of the procedures employed to protect the patient's life, but the cost of overlooking possible ACI is also very high, because the risk of the patient's death is very high. Because of this high risk of death, physicians tend to err on the side of caution by diagnosing ACI even when it is not present. That is, they make lots of false alarms.

There is a very accurate DSS that can be used to assist in the diagnosis of ACI, which takes into account the patient's actual risk of having ACI. For instance, a young, healthy woman who doesn't smoke but complains of chest pains is unlikely to be suffering from a heart attack, whereas an older, overweight man who smokes is far more likely to be suffering from a heart attack. When physicians used this DSS, their false-alarm rate dropped from 71% to 0%. However, when given the opportunity to use the DSS later, only 2.8% of physicians chose to do so, citing little perceived usefulness of the aid as the reason (Corey & Merenstein, 1987).

There are several reasons why physicians may be reluctant to use a DSS, but "usefulness" is surely not one of them. A more likely explanation is a physician's concern with how qualified he or she is perceived to be, both by patients and by colleagues. Even when told that such aids reduce errors, patients perceive physicians who use computer-aided DSSs to be less thorough, clever, and thoughtful than physicians who do not (Cruikshank, 1985).

Arkes, Shaffer, and Medow (2007) showed that this general finding persists even today, when, we might assume, patients are more accustomed to the presence of computer technology in medicine. Patients in their experiments read several scenarios in which physicians used either a computer-based DSS or no DSS at all. They rated physicians who used a computer-based DSS as less thorough, less professional, and having less diagnostic ability than physicians who used no aid at all. Furthermore, they also rated themselves as being less satisfied by the care they would receive from these physicians. These evaluations were mitigated somewhat, however, when they were also told that the DSS had been designed by the prestigious Mayo Clinic. Shaffer, Probst, Merkle, Arkes, and Medow (2013) found that seeking advice from another physician did not result in low ratings, suggesting that consultation of a nonhuman device specifically is the source of the negative evaluations.

These findings are troublesome for both patients and physicians. A physician trying to be as accurate as possible will have his or her best attempts at accuracy perceived negatively by his or her patients. There is also some evidence that these negative perceptions may extend also to the physician's colleagues. Such negative perceptions may lead to increased patient dissatisfaction, distrust, and, in the worst-case scenario, an increase in accusations of malpractice.

In sum, DSSs are a key component of modern medical practice. The system designer must be aware of the problems in their use. Some diagnostic systems, such as EEGs, contain within them DSSs that augment the system output with diagnostic guidelines. The challenge to the designer is to present this information in such a way that the physician is willing to use it. Patient acceptance is a more difficult problem, but one that would be easier to solve if physicians were more positive about the use of DSSs in their practice.

An alternative to support systems based on decision theory is *case-based aiding* (Lenz et al., 1998). Case-based aiding uses information about specific scenarios to support the decision maker. These computer-based systems try to provide the decision maker with appropriate analogies that are applicable to an ongoing problem. Kolodner (1991) argues that such an approach should be beneficial in many circumstances, because people reason through problems by using prior knowledge. A case-based support system, which stores and retrieves appropriate analogies, aids in decision making because people find it natural to reason using analogies but have difficulty retrieving them from memory.

As an example of how a case-based support system might be used, consider an architect who has been given the following problem:

> Design a geriatric hospital: The site is a 4-acre wooded sloping square; the hospital will serve 150 inpatients and 50 outpatients daily; offices for 40 doctors are needed. Both long-term and short-term facilities are needed. It should be more like a home than an institution, and it should allow easy visitation by family members. (Kolodner, 1991, p. 58)

The architect could highlight the key words in the problem specification, and the support system would retrieve cases of geriatric hospitals that are similar to the design criteria. The architect then could evaluate the successes and failures of one or more of those cases and adapt similar designs for his purpose.

A final example of decision support is a *recommendation system* (Stohr & Viswanathan, 1999). A recommendation system provides information about the relative advantages of alternative actions or products. Recommendation systems are used by online retailers to suggest books or recorded music that you might want to purchase, based on your previous purchasing patterns and those of other people. Web-based agents may make recommendations about various aspects of websites. An example is Privacy Bird®, which was developed in the first decade of the 21st century and is still available to download. It is a user agent that alerts users as to whether a website's privacy policy, posted in machine-readable form, is consistent with the user's preferences (Cranor, Guduru, & Arjula, 2006). A happy green bird indicates that the site's policy matches the user's preferences, an angry red bird indicates that it does not, and a yellow bird specifies that the site does not have a machine-readable privacy policy.

Recommendation systems are designed to provide users with information to assist in their decisions. In the case of Privacy Bird, this decision is whether to provide personal information to different companies. Most of you have experienced phishing attacks, in which you receive an apparently legitimate e-mail message directing you to a fraudulent website with the intent of getting you to enter personal information such as a credit card number. Recommendations to leave a suspected phishing site can be incorporated into the warnings, much as in Privacy Bird. However, for such warnings to be effective, various usability issues, such as the criteria for when to display the warning, what information should be displayed to users, how that information should be displayed, and how to encourage selection of the safe action, need to be taken into account when the warning/recommendation system is designed (Yang et al., 2017).

SUMMARY

Human problem solving, reasoning, and decision making are notoriously fallible. Even in very simple or straightforward cases, human performance deviates systematically from that defined as correct or optimal. However, these deviations do not imply that people are irrational. Rather, they reflect characteristics of the human information-processing system. Decision-making performance is constrained by a person's limited ability to attend to multiple sources of information, to retain information in working memory, and to retrieve information from long-term memory. Consequently, people use heuristics to solve problems and make decisions. These heuristics have the benefit of rendering complex situations manageable, but at the expense of increasing the likelihood of errors. In virtually all situations, a person's performance will depend on the accuracy of his or her mental representation or model of the problem. To the extent that this representation is inappropriate, errors will occur.

Human factors engineering has focused on training programs, methods for presenting information, and the design of decision-support systems to improve reasoning and decision-making performance. We have discussed how training in statistics and domain-specific problem solving can, to a limited extent, improve performance. However, it is easy to present information in ways that mislead or bias the decision maker. Computer-based decision-support systems circumvent many of these problems. Recommendation systems provide users with suggestions about actions or products, usually during their interactions with the World Wide Web. However, many decision-support systems are intended for use by experts in a particular field and cannot be used to help untrained individuals perform like experts. The knowledge possessed by an expert in a domain differs substantially from that of a novice. These differences and how expertise is incorporated into expert systems are topics of the next chapter.

RECOMMENDED READINGS

Baron, J. (2008). *Thinking and Deciding* (4th ed.). New York: Cambridge University Press.

Davidson, J. E., & Sternberg, R. J. (Eds.) (2003). *The Psychology of Problem Solving*. Cambridge, UK: Cambridge University Press.

Gigerenzer, G., Hertwig, R., & Pachur, T. (Eds.) (2015). *Heuristics: The Foundations of Adaptive Behavior*. New York: Oxford University Press.

Hastie, R., & Dawes, R. M. (2010). *Rational Choice in an Uncertain World: The Psychology of Judgment and Decision Making* (2nd ed.). Thousand Oaks, CA: Sage.

Holyoak, K. J., & Morrison, R. G. (Eds.) (2012). *The Oxford Handbook of Thinking and Reasoning*. New York: Oxford University Press.

Kahneman, D., Slovic, P., & Tversky, A. (Eds.) (1983). *Judgment under Uncertainty: Heuristics and Biases*. New York: Cambridge University Press.

Kahneman, D., & Tversky, A. (Eds.) (2000). *Choices, Values, and Frames*. New York: Cambridge University Press.

Kern, G., & Wu, G. (Eds.) (2015). *The Wiley Blackwell Handbook of Judgment and Decision Making* (2 vols.). Chichester, UK: John Wiley.

Leighton, J. P., & Sternberg, R. J. (Eds.) (2004). *The Nature of Reasoning*. Cambridge, UK: Cambridge University Press.

12 Experts and Expert Systems

The study of expertise covers remarkably diverse domains, such as sports, chess, music, medicine, and the arts and sciences, and examines the entire range of mastery from beginners to world-class performer.... Very high levels of achievement in virtually all domains are mediated by mechanisms acquired during an extended period of training and development.

K. A. Ericsson
2005

INTRODUCTION

In the previous chapters in this section we have discussed the processes involved in attention, memory, and thought, with a special emphasis on how people are limited in their ability to process information. In Chapter 9, we discussed the limited capacity that people have for attending to multiple sources of information. In Chapter 10, we emphasized that similar capacity limitations influence our ability to retain and perform computations in working memory and retrieve information from long-term memory. In Chapter 11, we showed that a person's ability to perform abstract reasoning is also limited, and because of these limitations, human reasoning relies heavily on simplifying heuristics and past experience.

Despite their information-processing limitations, people can develop expertise and become highly skilled in specific domains. We say that someone is an expert in a domain when they have achieved elite, peak, or exceptionally high levels of performance (Bourne, Kole, & Healy, 2014). An expert in a domain solves problems much faster, more accurately, and more consistently than does a novice. The question of how experts differ from novices, and therefore how novices can be trained and supported to perform like experts, is a central concern for human engineering. As we will see in this chapter, differences in performance between novices and experts arise to a large extent from the expert's specialized knowledge acquired from years of practice (Ericsson, 2006a), although other traits and genetic factors also play a role (Ullén, Hambrick, & Mosing, 2016). Experts see problems differently from novices and use different strategies to obtain solutions.

The present chapter focuses on how people acquire specialized knowledge and how this knowledge affects their information processing and performance. We will examine the way that speed and accuracy of performance vary as a function of how a task is learned and practiced. In order to explain and understand the effects of training, it is useful to consider several different perspectives of skill acquisition. To help us understand what expertise really is, we can compare expert and novice performance on a task. These comparisons can reveal why experts are able to think more efficiently and how novices may best be trained.

Large, complex systems often require expertise to operate and to troubleshoot problems. However, experts are usually in high demand and may not be available when the need arises. Consequently, expert systems have been designed to help novices perform those tasks usually performed by experts. These computer-based systems are designed using our understanding of the knowledge and reasoning processes that experts use in problem solving. This understanding, of course, derives from research on the skilled performance of experts. Eberts, Lang, and Gabel (1987) noted, "To design more effective expert systems, we must understand the cognitive abilities and functioning of human experts" (p. 985). In this chapter, we will also describe the characteristics of expert systems and the crucial roles that human factors specialists can play in the development, implementation, and evaluation of such systems.

ACQUISITION OF COGNITIVE SKILL

How skill is acquired has been a topic of interest since the earliest research on human performance (e.g., see Bryan & Harter's, 1899, study of telegraph skill in Chapter 1). However, much of this research has focused on the development of perceptual-motor skills (see Chapter 14). Today's technologically specialized jobs require cognitive expertise rather than perceptual-motor expertise, although both are required to some extent in expert performance of essentially any task (Suss & Ward, 2015). Consequently, we are now more interested in how cognitive skills are acquired, and research is now focused on cognitive differences between experts and novices. This research has improved our understanding of how changes in cognitive processing occur as knowledge and skills in a specialized domain are acquired (e.g., Anderson, 1983; Healy & Bourne, 2012).

A person is skilled in a particular domain when her performance is relatively precise and effortless. Cognitive tasks can be as simple as pressing a key when a stimulus event occurs or as complex as air-traffic control, and therefore there may be many or only few components of a task at which a person can become skilled (Johnson & Proctor, 2017). For these components, it is important that we distinguish between different task requirements and identify the kinds of information necessary to complete each one.

One dichotomy we can make is between convergent and divergent tasks. Tasks are said to be convergent if there is only one acceptable, predetermined response, whereas divergent tasks require novel responses. Another dichotomy is between algorithmic and non-algorithmic tasks. Algorithmic tasks can be performed in a sequence of steps that infallibly lead to a correct response, and so no deeper understanding of the task requirements is necessary. Tasks that are not algorithmic require an understanding of the principles that underlie the problem to be solved. Furthermore, task-performance skills can require deductive reasoning or inductive reasoning, and they can be performed in closed (predictable) or open (unpredictable) environments. Finally, there are highly specialized cognitive skills such as chess playing, or nearly universal skills such as reading. Given these distinctions, it should be clear that we must always qualify general principles of skill acquisition according to the particular task requirements.

POWER LAW OF PRACTICE

It is universally appreciated that practice has beneficial effects on performance. In particular, people get faster and more accurate the longer they practice a task. Across a wide variety of perceptual, cognitive, and motor tasks, speed-up in performance for a group of people is characterized by a power function (Newell & Rosenbloom, 1981). The simple form of the function is

$$T = BN^{-\alpha}, \tag{12.1}$$

where:

T represents the time to perform the task,

N represents the amount of practice or the number of times the task has been performed, and

B and α are positive constants.

This function is the *power law of practice*. One characteristic of the power law is that the more people practice a task, the less effect a fixed amount of additional practice will have. Early on, when they have little experience with the task, performance time will decrease rapidly, but later on, improvements will not be so great. The rapidity with which improvement decreases is a function of the rate parameter α. For power law curves, the amount of improvement on each trial is a constant proportion of what remains to be learned.

Neves and Anderson (1981) asked people to practice 100 geometry-like proofs. It took people 25 minutes on average to solve the first proof, and they got faster and faster with each proof. People took only 3 minutes to complete the last proof. Figure 12.1a shows the solution time as a function of the number of proofs completed, plotted on a linear scale. Two characteristics of the power law are apparent. First, the benefits of practice continue indefinitely: People were still speeding up even on the last proof. Second, the greatest benefits occur early in practice: Initial speed-ups are quite large, whereas later ones are small. Figure 12.1b shows the same data plotted on a log-log scale. The function should be linear if it follows the power law, because

$$\ln(T) = \ln(BN^{-\alpha}) = \ln(B) - \alpha\ln(N).$$

As shown by the straight line fit to the data, the effect of practice on the geometry-like proofs closely approximates a power function.

One problem with Equation 12.1 is that as N grows very large (i.e., practice is very extensive), the time to complete a task should approach zero. This makes no sense, because task performance, even at the most expert level, should take at least some time. For this reason, we might rewrite the power law with a nonzero asymptote. Also, we might want to take into account any previous experience a person has with the task. The generalized form of the power law that incorporates these factors is

$$T = A + B(N + E)^{-\alpha},$$

where:

A is the fastest possible performance, and

E is the amount of practice that a person brings to the task, that is, the amount of previous experience.

This generalized function still yields a straight line with slope $-\alpha$ when plotted on a log-log scale.

The family of generalized power functions characterizes performance across a wide range of tasks. For example, one study found that the time spent at particular e-commerce websites (e.g., Amazon) decreased with each successive visit following power functions (Johnson, Bellman, & Lohse, 2003). The researchers concluded that people can quickly learn to navigate a site, with the speed-up being faster for a well-designed website than for a poorly designed site. Thus, initial differences in usability may become even more pronounced after a few visits to the sites, leading users to prefer the better-designed website even more than they did originally.

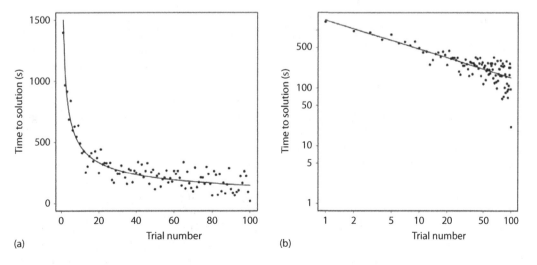

FIGURE 12.1 The power law of practice, as illustrated by the time to generate geometry-like proofs, on (a) a linear scale and (b) a log-log scale.

An additional point to note is that the power law also describes increasing productivity in a production process that occurs as a result of a group of operators' increasing experience with the system (Lane, 1987; Nanda & Adler, 1977). Such manufacturing process functions predict how quickly products will be produced, but they do not necessarily predict how quickly individual operators will perform. More generally, research has shown that individual improvements in performance may be described better by at least two power functions (Donner & Hardy, 2015), reflecting a shift in processes or strategies operating at different phases of skill acquisition, as discussed in the next section.

TAXONOMIES OF SKILL

The power law of practice suggests that improvement occurs continuously across time, but the fact that at least two functions seem to be needed to characterize an individual's improvement implies that there are qualitative changes in performance as well. By this we mean that people seem to do things in completely different ways depending on their skill level. As expertise is acquired, people transition from one way of performing a task to another. Several taxonomies have been developed to capture these differences in performance. Two complementary and influential taxonomies are Fitts's phases of skill acquisition and Rasmussen's levels of behavior.

Phases of Skill Acquisition

Fitts (1964; Fitts & Posner, 1967) distinguished between three phases of skill acquisition, which are, from least skilled to most skilled, *cognitive*, *associative*, and *autonomous*. Performance in the initial cognitive phase is determined by how well instructions and demonstrations are given. Fitts used the term *cognitive* to refer to the fact that the novice learner is still trying to understand the task, and therefore must attend to cues and events that will not require attention in later phases. During the associative phase, the task components that have been learned in the cognitive phase begin to be related to each other. This is accomplished by combining these components into single procedures, much like the subroutines of a computer program. The final autonomous phase is characterized by the automatization of these procedures, which makes them less subject to cognitive control.

Automatic processes are those that do not require limited-capacity attentional resources for their performance. There are four general characteristics of automatic processes (Schneider & Chein, 2003; Schneider & Fisk, 1983; Shiffrin, 1988). They (1) occur without conscious intention during the performance of the task; (2) can be performed simultaneously with other attention-demanding tasks; (3) require little effort and may occur without awareness; and (4) are not affected much by high workload or stressors.

It is easy to demonstrate that with increasing practice, task performance appears to shift from being effortful and attention-demanding to requiring little effort and attention (Kristofferson, 1972; Schneider & Shiffrin, 1977). Most of these demonstrations use a simple task, like visual or memory search, in which a person is asked to determine whether an item (such as a letter or digit) is present in a visual display or a set of items memorized previously. As the number of items to be searched increases (usually by adding distracting items), response time increases, reflecting greater cognitive demands. However, with increased practice, response times become more independent of the number of items to be searched, as long as the items are consistently "mapped" to the same stimulus category, target or distractor, throughout the task. That is, if the digit "8" is a target to be searched for, it will never appear among the items to be searched when any other target is presented.

After task procedures have become automatic, it may be very difficult to perform the task in any other way, even after the task requirements change. Shiffrin and Schneider (1977) asked people to practice a memory search task with a consistent mapping for 2100 trials. Then the mapping was reversed so that former distractors became targets, and vice versa. Nine hundred trials of practice with the reversed task were required before people reached the level of accuracy they demonstrated on the original task without any practice at all (see Figure 12.2). Accuracy on the reversed task

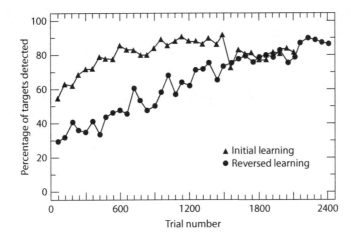

FIGURE 12.2   Learning with initial and reversed consistent mappings.

remained poorer than on the original task until 1500 trials after the reversal. Shiffrin and Schneider argued that the automatic procedures for target identification developed during the initial training apparently continued to "fire" when their stimulus conditions were present, even though the task requirements had changed.

Skill-Rule-Knowledge Framework

Whereas Fitts's taxonomy focuses on different phases of skill acquisition, Rasmussen's (1986) taxonomy, introduced in Chapter 3, focuses on three levels of behavioral control that inter-act to determine performance in specific situations. These levels (skill-, rule-, and knowledge-based) correspond approximately to Fitts's phases of skill acquisition, except that Rasmussen acknowledges that even a skilled performer will revert to earlier levels of control in certain circumstances.

Skill-based behavior involves relatively simple relations between stimuli and responses. Task performance is determined by automatic, highly integrated actions that do not require conscious control. The performance of routine activities in familiar circumstances would fall within this category, as would the memory search performance described above. For some skills, such as simple assembly and repetitive drill operations, highly integrated, automatic routines are desirable to maximize performance efficiency (Singleton, 1978). However, many skills require not only fast and accurate performance but also a high degree of flexibility. Flexibility arises from an ability to organize many elemental skilled routines in different ways to accomplish different, sometimes novel, goals.

Rule-based behavior is controlled by rules or procedures learned from previous experi-ences or through instructions. This level of behavioral control arises when automatic perfor-mance is not possible, such as when the performer experiences a deviation from the planned or expected task conditions. Rule-based performance is goal-oriented and typically under conscious control.

Knowledge-based behavior is used in situations for which no known rules are applicable. People may engage in knowledge-based behavior after an attempt to find a rule-based solution to a problem has failed. Knowledge-based behavior relies on a conceptual model of the domain or system of interest. A person must formulate a concrete goal and then develop a useful plan. Knowledge-based behavior involves problem solving, reasoning, and decision making of the type we described in Chapter 11. Consequently, performance depends on the adequacy of the mental model of the per-former and can be influenced by the many heuristics that people use to solve problems and make decisions.

According to Reason (1990; 2013), distinct types of failures can be attributed to each different performance level. For skill-based performance, most errors involve misdirected attention. Failures due to inattention often occur when there is intent to deviate from a normal course of action but "automatic" habits intrude. Conversely, errors of over-attention occur when the performer inappropriately diverts his or her attention to some component of an automatized sequence; that is, the performer "thinks too hard" about what he or she is trying to accomplish. At the rule-based level, failures can result from either the misapplication of good rules or the application of bad rules. At the knowledge-based level, errors arise primarily from fallibilities of the strategies and heuristics that problem solvers use to address their limited capacities for reasoning and representing the problem.

In sum, most skills require not only that routine procedures become automatized, but also that enough appropriate knowledge is learned for efficient rule-based and knowledge-based reasoning.

THEORIES OF SKILL ACQUISITION

Theories of skill acquisition and skilled performance are of value for several reasons. First, they help us understand why people do better in some situations than in others. Second, they provide us with a foundation for designing new experiments that may potentially lead to greater understanding of skill and expertise. With this greater understanding, the human factors engineer can help design and implement training programs to optimize skill acquisition. These theories are formalized in models of skill acquisition.

There are two major types of models of skill acquisition: *production system models* and *connectionist models* (Ohlsson, 2008). Production system models view skill acquisition as similar to problem solving and describe how production rules change and how people use them differently across different phases of practice. Connectionist models are based on networks of connected units, like the network memory models in Chapter 10. These units will be activated to greater or lesser degrees depending on task demands and the strength of their connection to other units. The result is a pattern of activation levels across the units of the network, which determines performance. Like learning and memory in other contexts, skill acquisition arises from changes in the connections within the network.

A Production System Model

Anderson's ACT-R (Adaptive Control of Thought–Rational; Anderson et al., 2004) cognitive architecture distinguishes three phases of skill acquisition similar to those proposed by Fitts (1964). The model relies on a procedural memory that contains the productions used to perform tasks, a declarative memory that contains facts in a semantic network, and a working memory that is used to link declarative and procedural knowledge. The first phase of skill acquisition is called the *declarative stage*, because it relies on declarative knowledge. In this stage, performance depends on general problem-solving productions that use weak heuristic methods of the type we described in Chapter 11. The person learning a task encodes the facts necessary to perform that task in declarative memory. The learner must retain these facts in working memory, perhaps by rehearsal, to be useful for the general productions.

In the second, associative phase, the learner gradually detects and eliminates errors. He begins to develop domain-specific productions that no longer require declarative memory for their operation. The process that leads to the acquisition of domain-specific productions, called *knowledge compilation*, has two subprocesses: composition and proceduralization. The composition subprocess collapses several productions into a single, new production that produces the same outcome. The proceduralization subprocess removes from the productions those conditions that require declarative knowledge. Composition and proceduralization together can be referred to as *production compilation*.

The domain-specific productions acquired in the associative phase become increasingly specific and automatic in the third, procedural phase as performance becomes highly skilled. These productions are further tuned through subprocesses of generalization (development of more widely applicable productions), discrimination (narrowing the conditions in which a production is used to only those situations for which the production is successful), and strengthening through repeated application.

Imagine an air-traffic controller who must learn to direct her attention to the bottom left of a display screen to read a list of the planes that can be landed (Hold Level 1; Taatgen & Lee, 2003). Table 12.1 shows three general rules needed initially to interpret when/how to do this. These rules are (1) retrieve an instruction, then (2) move attention, and finally (3) move the eyes to the appropriate location on the display. You can see that each of these rules is composed of an "if statement," which describes the conditions under which the rule should be applied, and a "then statement," which lists the actions to be taken.

As the air-traffic controller becomes more skilled, production compilation combines these general procedures with the specific declarative instructions for the air-traffic controller task, producing the new set of rules in Table 12.2. You can see that the new rules are combinations of pairs of the original rules. At the highest level of skill, these combination rules will be compiled with the

---

TABLE 12.1

Rules for Learning to Direct Attention to the Bottom Left of the Display Screen to Read a List of the Planes That Can Be Landed

Retrieve instruction

IF	You have to do a certain task
THEN	Send a retrieval request to declarative memory for the next instruction for this task

Move attention

IF	You have to do a task AND an instruction has been retrieved to move attention to a certain place
THEN	Send a retrieval request to declarative memory for the location of this place

Move to location

IF	You have to do a task AND a location has been retrieved from declarative memory
THEN	Issue a motor command to the visual system to move the eyes to that location

---

TABLE 12.2

Set of Rules Developed from Production Compilation

Instruction and attention

IF	You have to land a plane
THEN	Send a retrieval request to declarative memory for the location of Hold Level 1

Attention and location

IF	You have to do a task AND an instruction has been retrieved to move attention to Hold Level 1
THEN	Issue a motor command to the visual system to move the eyes to the bottom left of the screen

remaining rule from the original set ("move to location" with "instruction & attention" or "retrieve instruction" with "attention & location"). This results in the following single, task-specific rule:

All three:

IF:　　　You have to land a plane,
THEN:　Issue a motor command to the visual system to move the eyes to the bottom left of the screen.

The process of production compilation produces a single, compact production rule that can be executed much more quickly than the original rules.

A Connectionist Model

One early connectionist model of skill acquisition was proposed by Gluck and Bower (1988). Their model described the performance of students learning to make medical diagnoses based on descriptions of patients' symptoms. The different symptoms had different probabilities of occurring with each disease, so it was impossible for the students to be 100% accurate. Students made a diagnosis for each patient, and each diagnosis was followed by feedback about the accuracy of the diagnosis.

The different symptoms were represented in the model by activations across a network of input units (see Figure 12.3). These activations are weighted and summed at an output unit. The activation of the output unit reflects the extent to which one disease is favored over the other. This activation is used to classify the disease, and the feedback about the accuracy of the diagnosis is used to modify the weights. Modifying the weights gives the network the ability to detect correlations between symptoms and diseases, and to use these correlations to arrive at a diagnosis.

Some models incorporate properties of both production system and connectionist models (Schneider & Chein, 2003). These more complex models are implemented with connectionist components, such as a data matrix of input, internal operation, and output modules, a control system with multiple processors that receive input and transmit output. Schneider and Chein's model has automatic and controlled processing modes, and can explain many findings on controlled processing, automaticity, and improvements with performance as skill is acquired.

TRANSFER OF LEARNING

A significant issue in human factors is the extent to which the benefits of practice at one task or in one domain can "transfer" to related tasks and domains. By *transfer*, we mean the extent to which a person will be able to perform a new task because of his or her practice with a related task. Transfer has been studied in both basic and applied research (Cormier & Hagman, 1987; Healy & Wohldmann, 2012).

Views of Transfer

There are two extreme points of view regarding transfer (Cox, 1997). At one end of the continuum is the idea that expertise acquired in any domain should improve task performance in any other domain. This is the doctrine of formal discipline, originated by John Locke (Dewey, 1916). This doctrine attributes expertise in any area to general skills that are required for the performance of a

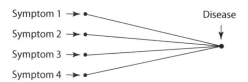

FIGURE 12.3　Network model.

broad range of tasks. From a production system perspective, extended practice at solving problems within a specific area allows the learner to acquire procedures related to reasoning and problem solving. These general procedures can then be used for novel problems in other areas.

At the other end of the continuum is Thorndike's (1906) theory of identical elements. This theory states that transfer should occur only to the extent that two tasks share common stimulus-response elements. Practice at solving problems within a specific area should benefit problem-solving performance within a different area if the two areas share common elements. Thus, the extent to which transfer will occur will depend on the characteristics of the practiced and novel tasks, and could be very limited or nonexistent.

Results from experiments investigating the extent of transfer between different tasks indicate that neither of these extreme views is correct. The evidence for transfer of general problem-solving skills (the doctrine of formal discipline) has been primarily negative. For example, students in one study received several weeks of training in solving algebraic word problems using a general problem-solving procedure intended to teach heuristics that could be applied to a variety of problems. These students did no better at solving new problems on a subsequent test than students who had not received training, leading the authors to conclude, "The results of this study suggest that formal instruction in a heuristic model suggesting *general* components of the problem-solving process is not effective in promoting increased problem-solving ability" (Post & Brennan, 1976, p. 64).

The lack of evidence for transfer of general skills may be due to the fact that the training regimens used in these and similar experiments are focused on those generally applicable weak methods (see Chapter 11) that are already highly practiced for most adults (Singley & Anderson, 1989). Other evidence indicates that transfer is not as specific as envisioned by Thorndike. Studies using tasks such as those interpreted in terms of a permission schema (see Chapter 11) show that transfer can occur when the stimulus and response elements are not identical (Cheng, Holyoak, Nisbett, & Oliver, 1986). However, skill acquisition seems to occur more along the lines of Thorndike's identical-elements view rather than the formal discipline view.

An alternative proposal made by Singley and Anderson (1989) relates the identical-elements view to mental representations. Focusing on the ACT architecture's distinction between declarative and procedural phases of performance, they proposed that the specific productions developed with practice are the elements of cognitive skill. Transfer will occur to the extent that the productions acquired to perform one task overlap with those for a second task. In other words, the specific stimulus and response elements do not have to be identical for transfer to occur; rather, the acquired productions must be appropriate for the second task.

This point is emphasized by an experiment Singley and Anderson (1989) conducted on learning calculus problems. Students who were unfamiliar with freshman calculus learned to translate word problems into equations and select operations to perform on those equations. These operations included differentiation and integration, among seven possible operations. When problems stated as applications in economics were changed to applications in geometry, there was total transfer of the acquired skill of translating the problem into equations. They also observed transfer in operator selection from problems that required integration to ones that required differentiation, but only for the operations that were shared between the two problem types. In short, transfer occurred only when the productions required for integrating and differentiating economics and geometry problems were similar.

Part-Whole Transfer

The operator of a human–machine system typically has to perform a complex task composed of many subtasks. The issue of *part-whole transfer* involves the question of whether the performance of the whole task can be learned by learning how to perform the subtasks. Training the subtasks is called *part training*, while training the whole task is called *whole training*. From a practical standpoint, there are many reasons why part training might be preferable to whole training. For example, (1) whole-task simulators are typically more complex and expensive than part-task simulators; (2)

as a consequence of lack of salience or emphasis, subtasks critical to successful performance of the whole task may receive relatively little practice in the whole-task situation; (3) experienced operators could be trained more efficiently on only the new subtasks required for a new machine or task; and (4) relatively simple training devices could be used to maintain essential skills.

There are three ways that tasks can be broken into subtasks (Wightman & Lintern, 1985). Segmentation can be used for tasks that are composed of successive subtasks. The subtasks can be performed in isolation or in groups and then recombined into the whole task. Fractionation is similar to segmentation, but applies to tasks in which two or more subtasks are performed simultaneously. This procedure involves separate performance for each of the subtasks before combining them. Finally, simplification is a procedure used to make a difficult task easier by simplifying some aspect of the task. It is more applicable to tasks for which there are no clear subtasks.

When the use of the part method seems appropriate, it is important to plan how the components will be reassembled into the whole task once they have been individually mastered. There are three schedules for part-task training: pure-part, repetitive-part, and progressive-part (Wightman & Lintern, 1985). With a pure-part schedule, all parts are practiced in isolation before being combined in the whole task. With a repetitive-part schedule, subtasks are presented in a predetermined sequence and progressively combined with other components as they are mastered. A progressive-part schedule is similar to the repetitive-part schedule, but each part receives practice in isolation before being added to the preceding subtasks. In certain circumstances, the whole task may be presented initially to identify any subtasks that may be especially difficult. These subtasks are then practiced using the part method.

No single method of training is best for all situations. Part-task training is most beneficial for complex tasks composed of a series of subtasks of relatively long duration, but it can be detrimental if subtasks overlap or have to be, at least in part, performed at the same time (Wickens, Hutchins, Carolan, & Cumming, 2013). The reason for this seems to be due to the training method not permitting development of a time-sharing skill. Wickens et al. concluded that for tasks in which time sharing is crucial, a training schedule in which the whole task is performed but with varying emphases on its component subtasks is most beneficial.

EXPERT PERFORMANCE

The research that we have discussed up to this point has focused on skill acquisition in laboratory tasks. These artificial, oversimplified tasks are easily mastered, but they bear little resemblance to most real-world tasks. To say that someone is an expert after performing a few sessions of a laboratory task stretches the definition of the word "expert." An expert is someone who has acquired special knowledge of a domain (like an entomologist or physician) or a set of complex perceptual-motor skills (like a concert pianist or an Olympic athlete). Ten years of intensive practice and training is typically required before a person's abilities reach an expert level in these real-world domains (Ericsson, Krampe, & Tesch-Romer, 1993).

The benefit of laboratory studies is that they give us the ability to see how simple skill acquisition occurs under controlled conditions (see, for example, Proctor & Vu, 2006a). However, because the acquisition of true expertise cannot be studied in the laboratory, studies of expertise focus on how experts think and behave differently from novices. Such studies have enhanced our understanding of cognitive skill and provided a foundation for the development of expert systems.

Distinctions between Experts and Novices

There is no argument that experts are able to do things that novices cannot (Glaser & Chi, 1988). Table 12.3 summarizes some characteristics of experts' performance. These characteristics reflect the expert's possession of a readily accessible, organized body of facts and procedures that can be applied to problems in his or her domain. That is, the special abilities

TABLE 12.3

Characteristics of Experts' Performance

1. Experts excel mainly in their domains.
2. Experts perceive large meaningful patterns in their domain.
3. Experts are fast; they are faster than novices at performing the skills of their domain, and they quickly solve problems with little error.
4. Experts have superior short-term and long-term memory for material in their domain.
5. Experts see and represent a problem in their domain at a deeper (more principled) level than novices; novices tend to represent a problem at a superficial level.
6. Experts spend a great deal of time analyzing a problem qualitatively.
7. Experts have more accurate self-monitoring skills.
8. Experts are good at selecting the most appropriate strategies to use in a situation.

of experts arise mainly from the substantial amount of knowledge that they have about a particular domain and not from a more efficient general ability to think. For example, expert taxi drivers can generate many more secondary, lesser-known routes through a city than can novice taxi drivers (Chase, 1983). As another example, although both chemists and physicists are presumed to be of equal scientific sophistication, chemists solve physics problems much like novices do (Voss & Post, 1988).

Chess is a domain that lends itself well to the study of expertise (Gobet & Charness, 2007). Experts are ranked and designated by an explicit scoring system that is monitored by several national and international organizations. Chess is also a task that can be easily brought into the laboratory. Therefore, although we cannot observe the development of expertise over time, we can observe differences between experts and novices under controlled conditions. Some of the most influential research on expertise has compared the performance of chess masters with that of less skilled players (Chase & Simon, 1973; de Groot, 1966).

One famous experiment presented chess masters and novices with chess board configurations to remember (de Groot, 1966). The pieces on the board were placed either randomly or in positions that would arise during play. When the configuration was consistent with actual play, the masters demonstrated that they could remember the positions of more than 20 pieces, even when the board had been shown for only 5 seconds. In contrast, novices could remember the positions of only five pieces. However, when the configuration was random, both masters and novices could remember only five pieces.

Chase and Simon (1973) later examined how the masters "chunked" the pieces on the board together. They found that the chunks were built around strategic relations between the pieces. Chess masters can recognize approximately 50,000 board patterns (Simon & Gilmartin, 1973). We can hypothesize that each of these patterns has associated with it automatized procedures composed of all the moves that could be made in response to that pattern. Because chess masters have already learned the legitimate board configurations that could be presented and their associated procedures, they can access the configurations effortlessly and hold them in working memory, whereas novices cannot.

What we learned from the studies investigating chess mastery is that experts have elaborate mental representations that maintain the information and associations between objects and procedures. Other studies have demonstrated that an expert's mental representation of their domain can be used as a scaffold to remember other kinds of information. Chase and Ericsson (1981) examined skilled memory in more detail for one man (S.F.), a long-distance runner. He practiced a simple memory task for over 250 hours during the course of 2 years. This task, a digit span task, required that he remember randomly generated lists of digits and recall the digits in order. In the beginning, his digit span was seven digits, about what we would expect given normal limitations of working memory (see Chapter 10). However, by the end of the 2-year period, his span was approximately 80 random digits.

How did S.F. accomplish this more than 10-fold increase in memory performance? Verbal protocols and performance analyses indicated that he did so by using mnemonics. S.F. began, as most people would, by coding each digit phonemically. However, on day 5, he started using a mnemonic of running times, exploiting his mental representation of long-distance running. S.F. first used three-digit codes, switching in later sessions to four-digit running times and decimal times. Much later in practice, he developed additional mnemonics for digit groups that could not be converted into running times.

Based on the performance of S.F. and other people, Chase and Ericsson (1981) developed a model of skilled memory. According to this model, increases in a person's memory span beyond that which we consider normal reflect not an increase in the capacity of short-term memory, but more efficient use of long-term memory. The model attributes five characteristics to skilled memory: (1) to-be-remembered information is encoded efficiently using existing conceptual knowledge, like S.F.'s knowledge of running times; (2) the stored information is rapidly accessed with retrieval cues; (3) the to-be-remembered information is stored in long-term memory; (4) the speed of encoding can be constantly improved; and (5) the acquired memory skill is specific to the stimulus domain that was practiced: in S.F.'s case, strings of digits.

Ericsson and Polson (1988) used this model as a framework to investigate the memory skills of a headwaiter at a restaurant who could remember complete dinner orders from more than 20 people at several tables simultaneously. Unlike S.F., this headwaiter did not rely on his expertise in some other domain, but only on his expertise as a waiter. However, like S.F., he used a highly organized mnemonic scheme. He organized his orders into groups of four and represented them in a two-dimensional matrix for the dimensions of location and course (e.g., entrée). In addition, he used imagery to relate each person's face to her or his order and other special encoding schemes for different courses.

The headwaiter's memory showed all the characteristics predicted by the skilled memory model, with the exception that his skills transferred to other stimulus materials. Ericsson and Polson (1988) attribute the relatively broad generality of the headwaiter's memory skills to the wide range of situations that he had to remember. Consequently, he had developed not only considerable flexibility in encoding dinners, but also a more general understanding of his own memory structure, of long-term memory properties, and of broadly applicable "metacognitive" strategies.

We discussed the fact that experts have different mental representations for remembering information. Another way that experts seem to differ from novices is in the quality of their mental models. Recall that a mental model allows a person to simulate the outcome of different actions on the basis of a mental representation. Because experts have better mental representations and, therefore, better mental models, their performance is better. In one experiment, Hanisch, Kramer, and Hulin (1991) evaluated the mental models of novice users of a business phone system. They asked the users to rate the similarity between each pair of nine standard features of the phone. They then compared these ratings with those of system trainers, who were highly knowledgeable about the system features. The users' mental models were quite different from those of the trainers. The trainers' mental models corresponded closely to documentation about the system features, but the novice users' mental models contained many deficiencies and inaccuracies. Hanisch et al. proposed that good training programs should highlight and explain clusters of features of the phone system in a way that is similar to how the trainers cluster the features of the phone system.

Up to this point, we have mostly discussed why experts are more accurate than novices. Experts also differ from novices in terms of how long they take to perform a task. Although experts are faster overall than novices in task performance, they take longer to analyze a problem qualitatively before attempting a solution. Experts may engage in this lengthy, qualitative analysis to construct a mental model incorporating relationships between elements in the problem situation. The extra time they spend on qualitative analysis may also allow them to add constraints to the problem and so reduce the scope of the problem. These analyses, while time-consuming, allow the expert to generate solutions efficiently.

One reason why experts spend more time analyzing a problem is that they are better able to recognize the conceptual structure of the problem and its correspondence to related problems. Experts are also better able to determine when they have made an error, failed to comprehend material, or need to check a solution. They can more accurately judge the difficulty of problems and the time it will take to solve them. This allows experts to allocate time among problems more efficiently. Chi, Feltovich, and Glaser (1981) found that physicists sorted physics problems into categories according to the physical principles on which they were based, whereas novices were more likely to sort the problems in terms of the similarities among the literal objects (balls, cannon, etc.) described in the problems. However, there are both good and poor experts (Dror, 2016): The best experts are unbiased by irrelevant contextual information, as were the physicists in Chi et al.'s study, and reliably reach the same conclusion from the same relevant information.

NATURALISTIC DECISION MAKING

In Chapter 11, we introduced the topic of decision making under uncertainty. Most of the research we described examined choices made by novices in relatively artificial problems with no real consequences. However, most decisions in everyday life are made under complex conditions, often with time pressure, that are familiar and meaningful to the individuals making the decisions. Consequently, reliance on expert knowledge seems to play a much larger role in natural settings than in most laboratory studies of decision making. Beginning in 1989, a naturalistic approach to decision making was developed, which emphasizes how experts make decisions in the field (Gore, Flin, Stanton, & Wong, 2015; Lipshitz, Klein, Orasanu, & Salas, 2001).

Klein (1989) conducted many field studies in which he observed how fireground commanders (leaders of teams of firefighters), platoon leaders, and design engineers made decisions. The following is one example of the thoughts and actions of a decision maker in the field:

> The head of a rescue unit arrived at the scene of a car crash. The victim had smashed into a concrete post supporting an overpass, and was trapped unconscious inside his car. In inspecting the car to see if any doors would open (none would), the decision maker noted that all of the roof posts were severed. He wondered what would happen if his crew slid the roof off and lifted the victim out, rather than having to waste time prying open a door. He reported to us that he imagined the rescue. He imagined how the victim would be supported, lifted, and turned. He imagined how the victim's neck and back would be protected. He said that he ran his imagining through at least twice before ordering the rescue, which was successful. (Klein, 1989, pp. 58–59)

This example illustrates how expert decision makers tend to be concerned with evaluating the situation and considering alternative courses of action. Klein concluded that mental simulation ("[running] his imagining through") is a major component of the expert's decisions. These mental simulations allow the expert to quickly evaluate possible consequences of alternative courses of action.

Most explanations of naturalistic decision making emphasize the importance of recognition-primed decisions (Klein, 1989; Lipshitz et al., 2001). A skilled decision maker must first recognize the conditions of a particular situation in making their judgments. The decision maker will recognize many situations as typical cases for which certain actions are appropriate. In such situations, the decision maker knows what course of action to take, because he or she has had to deal with very similar conditions in the past. However, many situations will not be recognized, and for these the decision maker may adopt a strategy of mental simulation to clarify the conditions of the situation and the appropriate actions to take.

These decision strategies rely heavily on expertise. As Meso, Troutt, and Rudnicka (2002) note, "Real life decision making requires expertise in the problem domain in which the problem being solved belongs" (p. 65). For example, police officers who were experts in firearms were shown to use similar processes as those who were novices, but the experts were able to use their experiential knowledge to perform much better (Bolton & Cole, 2016). Their knowledge allowed them to accurately categorize incidents, recognize irregularities, adapt rapidly to a dynamically changing environment, and use their training automatically—freeing up cognitive resources for mental simulation of the immediate situation. According to the recognition-primed decision model, expert decision makers should be trained by improving their recognition and mental simulation skills in a variety of contexts within their domain of expertise (Ross, Lussier, & Klein, 2005).

EXPERT SYSTEMS

Our comparisons between experts and novices demonstrated that many of the differences between them arise from the large amount of domain-specific knowledge that the experts possess. We mentioned earlier that experts are not always available for consultation when a problem arises and that they can also be very expensive. This has led to the development of artificial systems, called *expert systems*, designed to help nonexperts solve problems (Buchanan, Davis, & Feigenbaum, 2007). Expert systems, also known as *knowledge-based systems*, have been developed for problems as diverse as lighting energy management in school facilities (Fonseca, Bisen, Midkiff, & Moynihan, 2006), selection of software design patterns (Moynihan, Suki, & Fonseca, 2006), optimization of sites for waste incinerators (Wey, 2005), and financial performance assessment of healthcare systems (Muriana, Piazza, & Vizzini, 2016).

Unlike decision-support systems, which are intended to provide information to assist experts, expert systems are designed to replace the experts (Liebowitz, 1990). More specifically:

> An expert system is a program that relies on a body of knowledge to perform a somewhat difficult task usually performed by only a human expert. The principal power of an expert system is derived from the knowledge the system embodies rather than from search algorithms and specific reasoning methods. An expert system successfully deals with problems for which clear algorithmic solutions do not exist. (Parsaye & Chignell, 1987, p. 1)

Most expert systems are not simply data bases filled with facts that an expert knows; they also incorporate information-processing strategies and heuristics intended to mimic the way an expert thinks and reasons. This design feature is called *cognitive emulation* (Slatter, 1987). That is, the expert system is intended to mimic the thoughts and actions of the decision maker in all respects (Giarratano & Riley, 2004).

A host of human factors issues are involved in designing an effective expert system. An expert system may contain technically accurate information but still be difficult to use and fail to enhance the user's performance. A good expert system will be technically accurate (and make appropriate recommendations) and adhere to good human engineering principles (Madni, 1988; Preece, 1990; Wheeler, 1989). In this section, we review characteristics of expert systems, with a special emphasis on how the contribution of human factors is important.

CHARACTERISTICS OF EXPERT SYSTEMS

Expert systems have a modular structure (Gallant, 1988). The system modules include a knowledge base that represents the domain-specific knowledge on which decisions are based, an inference engine that controls the system, and an interactive user interface through which the system and the user communicate (Laita et al., 2007).

Knowledge Base

Knowledge can be represented in an expert system in many different ways (Buchanan et al., 2007; Ramsey & Schultz, 1989; Tseng, Law, & Cerva, 1992). Each choice of representation might correspond to alternative ways of representing knowledge in models of human information processing (see Chapters 10 and 11). Three such choices are production rules, semantic networks, and structured objects. Recall from Chapter 10 that production rule representations specify that if some condition is true, then some action is to be performed. A semantic network system is a connected group of node and link elements. Each node represents a fact, and each link a relation between facts. Structured objects represent facts in abstract schemas called *frames*. A frame is a data structure that contains general information about one kind of stereotyped event and includes nonspecific facts about, and actions performed in, that event. Frames are linked together into collections called *frame systems*.

Why a designer would select one kind of representation over another depends on the following three considerations:

1. *Expressive power*: Can experts communicate their knowledge effectively to the system?
2. *Understandability*: Can experts understand what the system knows?
3. *Accessibility*: Can the system use the information it has been given? (Tseng et al., 1992, p. 185)

None of the representations we have described satisfy these or any other criteria perfectly, so the best representation to use will depend on the purpose of a particular expert system. Production systems are convenient for representing procedural knowledge, since they are in the form of actions to be taken when conditions are satisfied (see examples earlier in this chapter). They also are easy to modify and to understand. Semantic networks are handy for representing declarative knowledge, such as the properties of an object. Frames and scripts are useful representations in situations where consistent, stereotypical patterns of behavior are required to achieve system goals. Sometimes an expert system will use more than one knowledge representation (like the ship design system we describe later on).

Inference Engine

The inference engine module plays the role of thinking and reasoning in the expert system. The inference engine searches the knowledge base and generates and evaluates hypotheses, often using forward or backward chaining (see Chapter 11; Liebowitz, 1990). The type of inference engine used is often closely linked to the type of knowledge representation used. For example, for case-based reasoning systems (see Chapter 11; Prentzas & Hatzilygeroudis, 2016), the knowledge base is previously solved cases; the inference engine matches a new problem against these cases, selecting for consideration the ones that provide the best matches.

Because many decisions are made under situations of uncertainty, the inference engine, together with the fact base, must be able to represent those uncertainties and generate appropriate prescriptions for action when they are present (Hamburger & Booker, 1989). One function of the inference engine is to make computations of utility that account for preferences of outcomes, costs of different actions, and so on, and base final recommendations on that utility. One way to do this is to incorporate a "belief network" into the system. A belief network represents interdependencies among different facts and outcomes, so that each fact is treated not as a single, independent unit but as a group of units that systematically interact.

User Interface

The user interface must support three modes of interaction between a user and the expert system. These modes are (1) obtaining solutions to problems, (2) adding to the system's knowledge base, and (3) examining the reasoning process of the system (Liebowitz, 1990). The creation of a useful dialogue structure requires the designer to understand what information is needed by the user, and how and when to display it. The dialogue structure should be such that the information requested by the computer is clear to the user and the user entry tasks are of minimal complexity.

An important component of an expert system's user interface is an explanation facility (Buchanan et al., 2007). The explanation facility outlines the system's reasoning processes to the user when he or she requests this information. By examining the reasoning process, the user can evaluate whether the system's diagnoses and recommendations are appropriate. Often, mistakes made while inputting information to the system can be detected using this part of the interface.

Human Factors Issues

The development of an expert system usually involves several people (Parsaye & Chignell, 1987). A domain expert provides the knowledge that is collected in the knowledge base. An expert-system developer, or "knowledge engineer," designs the system and its interface as well as programs for accessing and manipulating the knowledge base. Users are typically involved from the very beginning of the process, especially while designing and evaluating different user interfaces, to ensure that the final product will be usable by the people for whom it is intended.

Several human factors concerns must be addressed during the construction of an expert system (Chignell & Peterson, 1988; Nelson, 1988): (1) selecting the task or problem to be modeled; (2) determining how to represent knowledge; (3) determining how the interface is to be designed; (4) validating the final product and evaluating user performance.

Selecting the Task

Although it seems as though selecting the task would be the easiest part of developing an expert system, it is not. Many problems that might be submitted to an expert system are very difficult to solve. An expert-system designer, when faced with an intractable problem, must determine how to break that problem up into parts and represent it in the system. Expert consultants and the system designer must agree on how the task in question is best performed. Once the design team agrees on the parts of the task, work can begin on how the system represents those tasks. Clearly, easily structured tasks that rely on a focused area of knowledge are the best candidates for representation in expert systems. Furthermore, to the extent that a task can be represented as a deductive problem, it will be easy to implement as a set of rules to be followed. Inductive problems are more difficult to implement (Wheeler, 1989).

Representation of Knowledge

The representation of knowledge and the accompanying inference engine must reflect the expert's knowledge structure (Arevalillo-Herráez, Arnau, & Marco-Giménez, 2013). One way to ensure this is to acquire the knowledge and inference rules for the system from an expert (see Box 12.1). Most often, the knowledge is extracted through interviews, questionnaires, and verbal protocols collected while the expert performs the tasks to be modeled. As with any naturalistic study, certain factors will determine how useful data collected with these instruments will be. These factors include (1) whether or not the knowledge engineer and subject matter expert share a common frame of reference that allows them to communicate effectively, (2) making sure that the instruments used to elicit information from the expert are compatible with the expert's mental model of the problem, and (3) detecting and compensating for biases and exaggerations in the expert's responses (Madni, 1988). These factors are important because much of an expert's skill is highly automatized, so verbal reports may not produce the most important information for system design.

User-friendly expert system shells are available for developing specific expert systems. The shells include domain-independent reasoning mechanisms and various knowledge representation modes for building knowledge bases. One widely used expert system shell is CLIPS (C Language Integrated Production System; Giarratano & Riley, 2004; Hung, Lin, & Chang, 2015). CLIPS was

BOX 12.1   KNOWLEDGE ELICITATION

If an expert system is to incorporate expert knowledge, we must first extract that knowledge from experts in the domain. Without an accurate representation of the information and the strategies used by the experts to solve specific problems, the expert system will not be able to carry out its tasks appropriately. There are two fundamental questions that define the core problem of knowledge elicitation (Shadbolt & Burton, 1995): "How do we get experts to tell us, or else show us, what they know that enables them to be experts at what they do?" and "How do we determine what constitutes the expert's problem-solving competence?" These questions are difficult to answer because (1) much of an expert's knowledge is tacit; that is, it is not verbalizable; (2) an expert often solves a problem quickly and accurately with little apparent intermediate reasoning; (3) the expert may react defensively to attempts at eliciting her or his knowledge; and (4) the knowledge elicitation process itself may induce biases in the extracted information (Chervinskaya & Wasserman, 2000).

We can use a variety of techniques effectively to elicit knowledge from experts. These include, for example, verbal protocol analysis, in which experts describe the hypotheses they are considering, the strategies they are using, and so on, as they perform tasks in their domains, and concept sorting, in which experts sort cards with various concepts into related piles. To illustrate, these two methods were used to elicit from information security experts their knowledge about risks in use of applications on mobile devices, from which three main dimensions were identified: personal information privacy, monetary risk, and device stability/instability (Jorgensen et al., 2015). Because different experts within a domain may have dissimilar knowledge representations, it is often a good strategy to elicit knowledge from more than one expert, as was done in that study.

Knowledge elicitation is also important for purposes other than building an expert system (Hoffman, 2008). For example, Peterson, Stine, and Darken (2005) describe knowledge elicitation and representation for military ground navigators, with the goal of using this information in design of training applications. Knowledge elicitation is also a necessary step in determining what content to include in an e-commerce website and how to manage that content (Proctor, Vu, Najjar, Vaughan, & Salvendy, 2003). Knowledge elicited from experts can reveal what information needs to be available to the user, how that information should be structured to allow easy access, and the strategies that different users may employ to search and retrieve information. When we elicit knowledge for usability concerns, as in the case of website design, we must obtain that knowledge not only from experts but also from a broad range of users who will be accessing and employing the information. Obtaining information from users is often conducted in the context of trying to understand their knowledge, capabilities, needs, and preferences, rather than of obtaining specifications for design.

Some knowledge elicitation techniques, such as verbal protocol analysis, are aimed at eliciting knowledge from experts, but others (such as questionnaires and focus groups) are targeted more toward novices and end users. Also, some methods are based on observations of behavior, whereas others are based on self-reports. Table B12.1 summarizes many of these methods along with their strengths and weaknesses. As a rule, we recommend that several methods be used to increase the quantity and quality of relevant information that is elicited

TABLE B12.1
Knowledge Elicitation Methods and their Major Advantages and Disadvantages

Method	Brief Description	Major Advantages	Major Disadvantages
Interviews	Interviewer asks the expert or end-user questions relating to a specific topic.	• Most well-known method for eliciting knowledge • Qualitative data	• Time-consuming • Expensive
Verbal protocol analysis	Experts report thought processes involved in performing a task or solving a problem.	• Qualitative data • Document thought processes related to performance	• Time-consuming • Hard to analyze
Group task analysis	A group of experts describes and discusses processes pertaining to a specific topic.	• Obtain different viewpoints • Document thought processes and information related to performance	• No research validating this method
Narratives, scenarios, and critical incident reports	The expert or end-user constructs stories to account for a set of observations.	• Provide insight to reasoning processes and implicit knowledge • Good for ill-defined problems	• Reliance on self-reports
Questionnaires	User groups report information or preferences relating to a topic.	• Quantitative data • Easy to code	• Low return rate • Response may not correspond with actual behavior
Focus groups	A group of users discusses different issues regarding the features of a system.	• Allows exchange of ideas • Good for generating lists of functions and features for products	• An individual may dominate the discussion • Not good for discovering specific problems
Wants and needs analysis	User groups/experts brainstorm about content they want and need in a system.	• Exchange of ideas • Determine areas of focus • Prioritized list of functions and features	• What users say they want and need may not be realistic
Observation and contextual inquiries	Observe as users interact with a product in a natural environment.	• Studied in natural environment • Qualitative and quantitative data	• Time-consuming • Dependent on detailed notes of the observer
Ethnographic studies	User culture and work environment are observed.	• Studied in natural environment • Good for discovering new products	• Time-consuming • Hard to generalize results to other product designs
User diary	Users record and evaluate actions over a period of time.	• Real-time tracking • Qualitative data	• Can be invasive or difficult to implement • May be delay in entries by users
Concept sorting	Users/experts establish relations among a fixed set of concepts.	• Determine relations among components • Helps structure information	• Grouping may not be optimal • Resulting structure may be too elaborate
Log files	Users' behaviors are logged to understand the users' interactions with the system.	• Uses actual recorded behaviors • Can collect data from a range of users	• Irrelevant or wrong information may be recorded • Data do not reflect cognitive processes

developed by NASA in 1985, and version 6.30 is available as public domain software (www.clip-srules.net/). This expert system shell allows knowledge to be represented as production rules, which specify the actions to be performed when certain conditions are met. It also supports the representation of knowledge as an object-oriented hierarchical data base, which allows systems to be represented as interconnected, modular components.

One benefit of expert system shells is that they allow an opportunity for the domain expert to be directly involved in the development of the expert system, rather than just serving as a source of knowledge. This may allow more information about the expert's knowledge to be incorporated into the system, because the input is more direct. Naruo, Lehto, and Salvendy (1990) describe a case study in which this method was used to design an expert system for diagnosing malfunctions of a machine for mounting chips on an integrated circuit board. A detailed knowledge elicitation process was used to organize the machine designer's knowledge. This process took several weeks, but the implementation as a rule-based expert system using the shell then took only about a week. On-site evaluation showed that the expert system successfully diagnosed 92% of the malfunctions of the chip-mounting machine.

An alternative to basing the knowledge of an expert system on an expert's verbal protocols is to have the system develop the knowledge from the actions taken by the expert in a range of different situations. The connectionist approach to modeling is particularly suited to this approach, because connectionist, neural network systems acquire knowledge through experience (Gallant, 1988). The system is presented with coded input and output, corresponding to the environmental stimuli and the action taken by the expert, for a series of specific problems. A learning algorithm adjusts the weights of the connections between nodes to closely match the behavior being simulated. Unlike the previous approaches, the connectionist approach does not rely on formal rules or an inference engine but only on how often the expert performs certain tasks and how frequently different environmental situations occur.

Hunt (1989) describes the results of several experiments that illustrate the potential for the connectionist approach. In one, students were instructed to imagine that they were learning how to troubleshoot an internal combustion power plant. Readings of instruments, such as coolant temperature and fuel consumption, were displayed as in Figure 12.4. From the configuration of readings, one of four malfunctions (radiator, air filter, generator, gasket) had to be diagnosed. In a series of such problems, the students' performance began at chance level (25% correct) and increased to 75% by the end of the experiment.

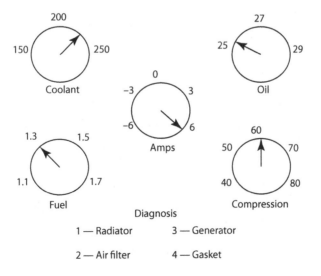

FIGURE 12.4 A typical display shown in the diagnostic task.

Hunt (1989) developed a connectionist model for each student based on their responses. The models accurately approximated the performance of the individual students, with the mean classification accuracy being 72%. In contrast, a rule-based system for which the knowledge was acquired by interviews averaged only 55% correct. These results suggest that more objective methods of knowledge elicitation, coupled with a system that learns from the expert's actions, may allow the development of better expert systems.

Interface Design

We have already discussed the importance of good interface design. If the interface is poorly designed, any benefit provided by the expert system may be lost. Errors may be made, and, ultimately, the users may stop using the system. The human factors specialist can provide guidelines for how the expert system and the user should interact. These guidelines will determine the way that information is presented to the user to optimize efficiency. Of particular concern, as we discussed earlier, is the presentation of the system's reasoning in such a way that the user will be able to understand it.

Two modes of interaction are commonly used for expert system interfaces (Hanne & Hoepelman, 1990). Natural language interfaces present information in the user's natural language and are the most common. Such interfaces are usable by almost everyone, but they can lead the user to overestimate what the system "understands." Because the system seems to talk to the user, the user may tend to anthropomorphize the system.

An alternative to the natural language interface is a graphical presentation of the working environment. Graphical user interfaces are effective for communicating such things as system change over time and paths to solution. In many situations, a combination of language and graphical dialogue will be most effective. A good design strategy is to have users evaluate prototypes of the intended interface from early in the development process, so that interface decisions are not made only after the rest of the expert system is developed.

Validating the System

Even a perfectly designed expert system must be validated. There may be errors in the knowledge base or faulty rules in the inference engine that can lead to incorrect recommendations by the system. Incorrect recommendations can be hazardous, because there will be many people who tend to accept the system's advice without question. Dijkstra (1999) had people read three criminal law cases and the defense attorney's arguments, which were always correct. After reading the materials for each case, the people consulted an expert system that always gave incorrect advice, the basis of which could be examined by using three explanation functions. Seventy-nine percent of the decisions made by the people were in agreement with the incorrect advice provided by the expert system instead of the correct advice provided by the attorney, and slightly more than half of the people agreed with the expert system for all three cases. In contrast, only 28% of decisions made by persons who judged the criminal cases without the advice of either the attorney or the expert system were incorrect. Several measures indicated that those people who always agreed with the expert system did not put in the effort to study the advice of the expert system but simply trusted it.

A system can be tested by simulating its performance with historical data and having experts assess its recommendations. Because the system ultimately will be used in a work environment by an operator, it is also important to test the performance of the operators. It can be difficult to modify an imperfect expert system once it is installed in the field, so tests of system performance and knowledge validity need to be performed prior to installation. These tests may be accomplished by establishing simulated conditions in a laboratory environment and evaluating operator performance with and without the expert system.

Unfortunately, expert-system designers often neglect the important step of evaluating the operator's performance, and this can have negative consequences. Nelson and Blackman (1987) evaluated two variations of a prototype expert system developed for operators of nuclear reactor facilities. Both systems used response trees to help operators monitor critical safety functions and to identify a particular problem-solving route when a safety function became endangered. One system required the operator to provide input about failed components when they were discovered and to request a recommendation when one was desired. The other system automatically registered any failures that occurred, checked to determine whether a new problem-solving recommendation was necessary, and displayed this new recommendation without prompting.

Neither system significantly improved performance over that of operators who had no expert system at all. Even the automated system, which was much more usable than the operator-controlled system, did not improve performance. For the operator-controlled system, it was easy to enter incorrect information, which resulted in erroneous and confusing recommendations. Therefore, we cannot assume that an expert system will always improve operator performance, even when that system is easy to use.

An important part of the validation process is to assess how acceptable a system is to its users. The introduction of new technology in the workplace always has the potential to generate suspicion, resentment, and resistance of the users for many reasons. The expert-system designer can minimize acceptance problems by involving users in all phases of the development process. He must also develop training programs to ensure that operators understand how the expert system is to be integrated into their daily tasks. He will also need to develop maintenance procedures that will ensure the reliability of the system. Finally, he should evaluate possible extensions of the system into areas for which it was not initially designed.

EXAMPLE SYSTEMS

As noted earlier, expert systems have been used successfully in a variety of domains, including the diagnosis of device and system malfunctions (Buchanan et al., 2007). One of the earliest expert systems, MYCIN, was used for the diagnosis and treatment of infectious diseases. Digital Equipment Corporation used the XCON system successfully to configure a computer hardware/software system specific to the customer's needs. Telephone companies have used the ACE system to identify faults in phone lines and cables that may need preventative maintenance. Although expert systems have their limitations, their uses and sophistication can be expected to continue to expand in the future. We describe in detail below two expert systems, one for diagnosing faults in the shapes of steel plates and the other for ship design.

DESPLATE

A system called DESPLATE (Diagnostic Expert System for Steel Plates) was developed to diagnose faults in shapes of rolled steel plates (Ng, Cung, & Chicharo, 1990). Slabs of reheated steel are rolled into plates of specified thickness and shape. The final products are to be rectangular, but perfect rectangular shapes rarely occur. Figure 12.5 illustrates five examples of faulty shapes. Some plates may be sufficiently deviant that they must be cut into smaller dimensions, which is a costly and time-consuming process. Therefore, DESPLATE is designed to locate the cause of particular faulty shapes and recommend adjustments to correct the problem.

DESPLATE uses a mixture of forward and backward chaining to reach a conclusion. The user is prompted for a set of facts observed prior to or during the session in which faulty plates were produced. From this set of data, DESPLATE forward chains until the solution space is sufficiently narrow. If a cause can be assumed, backward chaining is then used to prove this cause; otherwise, forward chaining continues.

DESPLATE searches a knowledge base that is arranged hierarchically. The entire knowledge base is organized according to the time required to test for a fault and the frequency of that fault. Observations or tests that are easily performed have priority over those that are more difficult, and faults that rarely occur are only tested when everything else has failed. Information is presented in order of these priorities. There are three kinds of information in the knowledge base: (1) the observations, or symptoms, that are used to identify different types of faults; (2) the tests used to diagnose faults; and (3) the faults themselves, which are hierarchically arranged according to their nature.

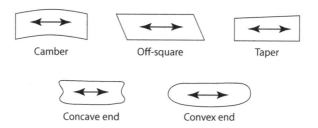

FIGURE 12.5 Examples of faulty shapes.

DESPLATE was installed in 1987 at the plate mill of the BHP Steel International Group, Slab and Plate Products Division, Port Kembla, Australia. It produced satisfactory solutions and recommendations for three faulty shapes: camber, off-square, and taper.

ALDES

ALDES (Accommodation Layout Design Expert System) was developed to provide expert assistance in the ship design process (Helvacioglu & Insel, 2005). Task modules were developed to provide expertise about three tasks involved in ship design: (1) generating a general arrangement plan; (2) determining the minimum number of crew members required; and (3) generating layouts of decks for the accommodations area. To develop ALDES, a visual programming interface shell was paired with a CLIPS expert system shell as the inference engine. The interface shell provides functions for data input from the user and outputs results to the user, to maintain a database of objects during the design process; to visually depict the layout; and to perform some fundamental calculations (e.g., container capacity in the hull).

Knowledge in ALDES was acquired from interviews with ship designers, investigation of national and international regulations, examination of social rules and accommodation in ships, and data bases from ships of the same general type. The ship was represented as a hierarchical data base of objects, and the procedural knowledge acquired from experts was represented as production rules. Reasoning in ALDES proceeds through refinement and adaptation of an initial prototype of the ship. A prototype is selected and then decomposed into its main components. Each main component is decomposed into subcomponents, and so on, with the decomposition continuing until a stage is reached at which the design description can be generated using deductive logic.

SUMMARY

Skill acquisition takes place in an orderly way across a range of cognitive tasks. Early in training, task performance depends on generic, weak problem-solving methods. With practice, a person acquires domain-specific knowledge and skills that can be brought to bear on the task at hand. It is this knowledge that defines an expert. The domain-specific knowledge possessed by experts, and how that knowledge is organized, allows them to perceive, remember, and solve problems better in that domain than nonexperts can. Expert behavior can be characterized as skill-based. When the expert encounters an unfamiliar problem to which a novel solution is required, he or she engages a range of general problem-solving strategies and mental models.

Expert systems are knowledge-based computer programs designed to emulate an expert. An expert system has three basic components: a knowledge base, an inference engine, and a user interface. The potential benefits of expert systems are limited by human performance issues. Human factors specialists can assist in design decisions by providing input about tasks that can be successfully modeled, the most appropriate methods for extracting knowledge from domain experts, the best way to represent this knowledge in the knowledge base, the design of an effective dialogue structure for the user interface, evaluations of the performance of the expert system, and the integration of the system into the work environment.

RECOMMENDED READINGS

Berry, D., & Hart, A. (Eds.) (1990). *Expert Systems: Human Issues*. Cambridge, MA: MIT Press.

Bourne, L. E. Jr, & Healy, A. F. (2012). *Train Your Mind for Peak Performance: A Science-Based Approach for Achieving your Goals*. Washington, DC: American Psychological Association.

Chi, M. T. H., Glaser, R., & Farr, M. J. (Eds.) (1988). *The Nature of Expertise*. Hillsdale, NJ: Erlbaum.

Ericsson, K. A. (Ed.) (1996). *The Road to Excellence: The Acquisition of Expert Performance in the Arts and Sciences, Sports, and Games*. Hillsdale, NJ: Erlbaum.

Ericsson, K. A., Charness, N., Feltovich, P., & Hoffman, R. R.. (Eds.) (2007). *Cambridge Handbook of Expertise and Expert Performance*. Cambridge, UK: Cambridge University Press.

Giarratano, J. C., & Riley, G. D. (2004). *Expert Systems: Principles and Programming* (4th ed.). Boston, MA: Course Technology.

Gluck, M. A., & Bower, G. H. (1988). Evaluating an adaptive network model of human learning. *Journal of Memory and Language*, 27, 166–195.

Healy, A. F., & Bourne, L. E.., Jr. (Eds.) (2012). *Training Cognition: Optimizing Efficiency, Durability, and Generalizability*. New York: Psychology Press.

Johnson, A., & Proctor, R. W. (2017). *Skill Acquisition and Training: Achieving Expertise in Simple and Complex Tasks*. New York: Routledge.

Reason, J. (2013). *A Life in Error: From Little Slips to Big Disasters*. Burlington, VT: Ashgate.

Sternberg, R. J., & Grigorenko, E. L. (Eds.) (2003). *The Psychology of Abilities, Competencies, and Expertise*. Cambridge: Cambridge University Press.

Part IV

Action Factors and Their Applications

13 Response Selection and Principles of Compatibility

It's the le .... It's the right one.

David McClelland

*copilot of a British Midland flight
that crashed in January, 1989, before
the crew turned off the wrong engine.*

INTRODUCTION

Human–machine interaction requires that the operator perceives information, cognitively processes that information, and ultimately selects and executes an action. Even if perception and cognition are accomplished flawlessly, an operator may still take an inappropriate or inaccurate action. The quotation with which this chapter begins illustrates an occasion on which a flight crew turned off the wrong engine, resulting in the crash of a commercial aircraft. This kind of error is called a *response-selection error*. Response-selection errors cannot be entirely avoided, but proper design can increase the speed and accuracy with which operators can select responses.

In the laboratory, response selection is studied using reaction-time tasks. As we described earlier, such tasks require an observer to be instructed to make a rapid response to a stimulus. We can distinguish between three basic processes that intervene between the onset of a stimulus and the completion of the response initiated by the stimulus (see Figure 13.1): stimulus identification, response selection, and response execution. How well stimulus identification can be performed is a function of stimulus properties such as brightness, contrast, and so forth. How well response execution can be performed is a function of response properties such as the complexity and accuracy of movements that must be made. The focus of this chapter, response selection, is on how quickly and accurately people can determine which response they are to make to a stimulus. How well response selection is performed is influenced primarily by the relationships between the members of the stimulus set and the responses assigned to each.

If a person is expected to operate a machine effectively, the interface through which he or she controls the machine must be designed to optimize the efficiency with which displayed information can be transformed into the required controlling responses. Understanding the processes involved in response selection is critical to our understanding of action more generally. This chapter will discuss those factors that influence the time taken to choose between different responses and how experiments are designed to evaluate alternative interface designs.

SIMPLE REACTIONS

Some tasks require an operator to react to a signal with a single, predetermined response (Teichner, 1962). For example, if an alarm sounds, the operator may have to press a button to shut off a piece of equipment as quickly as possible. Situations in which a single response must be made to a stimulus event are called *simple reaction tasks*. Response-selection processes are presumed to play a minimal role in simple reaction tasks, because only one response is to be made to any event: there are no choices among responses to be made (Miller, Beutinger, & Ulrich, 2009). However, even for a simple reaction, a decision still must be made about the presence or absence of the stimulus itself (Rizzolatti, Bertoloni, & Buchtel, 1979).

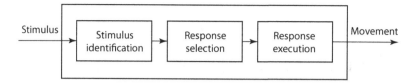

FIGURE 13.1 The three-stage model for reaction-time tasks.

It helps to understand what is happening in a simple reaction task by considering a model of the response process in which evidence about the presence of the stimulus accumulates over time and is stored somewhere in the brain. The observer can execute a response as soon as he or she has obtained enough evidence that a stimulus is present (e.g., Diederich & Busemeyer, 2003; Miller and Schwarz, 2006). In some situations, the observer will need a lot of evidence before deciding to respond. For example, if the observer's task is to shut down a large machine and interrupt a production line, she may need to feel very sure that the signal has occurred before responding. In other situations, the observer may not need as much evidence. So, even though there is no response selection to be made in a simple reaction task, the reaction time will still be influenced by the amount of evidence that the response requires.

Reaction time in a simple reaction task will be more strongly affected by stimulus factors. As the stimulus is made more salient, reaction time will decrease. There are several ways that a stimulus can be made more salient. For example, as the intensity, size, or duration of a visual auditory or tactile stimulus is increased, reaction time to its onset will decrease (Miller & Ulrich, 2003; Schlittenlacher & Ellermeier, 2015; Teichner & Krebs, 1972). The fastest that a simple reaction can be is approximately 150 ms for visual stimuli and 120 ms for auditory and tactual stimuli (Boff & Lincoln, 1988). Referring again to a model in which evidence accumulates over time, stimulus factors such as intensity affect primarily the rate at which information about the presence of the stimulus accumulates. Consistently with such a model, reaction time is shorter when two or more redundant signals (e.g., an auditory and a visual stimulus) are presented simultaneously (Miller et al., 2009), allowing a more rapid accumulation of information.

Another major factor influencing simple reaction time is whether or not the observer is prepared to respond to a stimulus. When an observer is unprepared, his response times will be longer than when he is prepared. In some situations (as in most laboratory experiments), an observer can be prepared for, or warned about, an imminent signal. A warning signal can increase an operator's state of readiness, lowering the amount of evidence necessary for responding. However, preparation to respond is attention-demanding, and when an observer must perform another task simultaneously, simple reactions will be slowed significantly (Henderson and Dittich, 1998).

A task that is closely related to the simple reaction task is a "go/no go" task (see our discussion of Donders in Chapter 1). In this task, a person responds only when a "target" stimulus occurs but does not respond to any other stimulus. The person can make two kinds of errors in performing this task: omissions (failing to respond to a target stimulus) or commissions (responding to the other stimulus; see Chapter 3). Omission errors are sometimes attributed to failures of attention, whereas commission errors are thought to be due to failures of response inhibition (Bezdjian, Baker, Lozano, & Raine, 2009).

CHOICE REACTIONS

Usually, an operator will need to choose one of several possible alternatives when an action is required. Situations in which one of several possible responses could be made to a stimulus are called *choice reaction tasks*. In most choice reaction tasks, many responses could be made to many different stimuli. For example, several distinct auditory alarm signals, each with its own assigned response, may be used in a process control room. When an alarm sounds, the control room operators must identify it and choose the appropriate course of action.

Response selection has been investigated most thoroughly for tasks in which one of two or more responses is to be made to one of two or more possible stimuli. Most of the variables that affect simple reaction time, such as stimulus intensity and state of readiness, also influence choice reaction time. However, other factors are also important, including the relative emphasis on speed or accuracy, warning interval, amount of uncertainty, compatibility of stimuli and responses, and practice.

SPEED–ACCURACY TRADEOFF

In a simple reaction task, an inappropriate response cannot be selected, because there is only one response; errors are made by either not responding or responding at the wrong time. However, in choice reaction tasks, an inappropriate response can be made. The time to make a response in a choice reaction task depends on how accurate the choice must be. For example, if accuracy is of no concern, you could make any response you wanted whenever you detected the onset of a stimulus (extreme speed emphasis in Figure 13.2). You would be simply guessing at the appropriate response, which you could then make very quickly, but your accuracy would be no better than chance. Alternatively, you could wait until you were sure about the identity of the stimulus and its associated response (extreme accuracy emphasis in Figure 13.2). Your responses would be much slower, but your accuracy would be nearly perfect.

Between these two extremes, you could choose any level of speed, resulting in some level of accuracy (or vice versa). The function relating speed and accuracy shown in Figure 13.2 is called the *speed–accuracy tradeoff* (Osman et al., 2000). The speed–accuracy tradeoff function shows different combinations of speed and accuracy that can be obtained for a single choice situation, much as a receiver operating characteristic (ROC) curve shows different hit and false-alarm rates for a given sensitivity (see Chapter 4). As with the ROC curve, the selection of a point on the speed–accuracy function is affected by such things as instructions and payoffs. A person's speed–accuracy criterion can be manipulated experimentally by imposing different response deadlines or presenting a signal to respond at varying delays after stimulus onset (Vuckovic, Kwantes, Humphreys, & Neal, 2014).

One of the easiest ways to account for the speed–accuracy tradeoff is through information accumulation models similar to the one we presented for simple reaction time tasks (Ulrich, Schröter, Leuthold, & Birngruber, 2015). Assume that evidence accumulates and is stored for each response separately, and a response is selected when enough evidence for it has accumulated. Responses made on the basis of less information will be faster but less accurate. Moreover, just as the criterion in signal-detection theory reflects an observer's bias to respond "yes" or "no,"

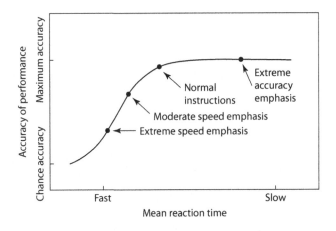

FIGURE 13.2 A theoretical speed–accuracy tradeoff function.

the threshold amounts of evidence across the alternative responses reflect an observer's bias. The lower a threshold is in relation to other thresholds, the greater is the bias toward that response. For human factors, the primary implication of the speed–accuracy tradeoff is that variables may affect either the efficiency of information accumulation or the threshold amount of evidence required for each response.

Alcohol has a marked effect on a person's speed–accuracy tradeoff (Rundell & Williams, 1979). In a study investigating effects of alcohol, people performed a two-choice task in which they were to make left or right keypresses to low- or high-frequency tones. Asking them to respond at three different speeds (slow, medium, and fast) generated a picture of each person's speed–accuracy trad-eoff function. People who had drunk alcohol prior to performing the task made their responses just as quickly as people who had not, but they had higher rates of error. The shape of the speed–accu-racy tradeoff functions suggested that alcohol impaired how well information accumulated. These results suggest that an alcohol-impaired person will have to respond more slowly to avoid making mistakes.

TEMPORAL UNCERTAINTY

As we described for simple reactions, knowledge about when a stimulus is going to occur affects the speed of responding in choice reactions. If a person knows that a stimulus will be occurring at a particular time, he can prepare for it. Warrick, Kibler, and Topmiller (1965) had secretaries respond to the sound of a buzzer by reaching to and pressing a button located to the left of their typewrit-ers. Half the secretaries were given warning that the buzzer was going to sound, and the other half were not given warning. The buzzer was sounded once or twice a week for 6 months. Their reac-tion times decreased over the 6-month period, but secretaries who were not warned were always approximately 150 ms slower to respond than secretaries who were warned.

The effects of preparation are examined in the laboratory by varying the time between a warn-ing signal and the stimulus to which an observer is to respond. We can plot a "preparation func-tion" for such a task by showing the response times as a function of the time between the warning signal and the stimulus. The data shown in Figure 13.3 come from a study by Posner, Klein, Summers, and Buggie (1973). They asked observers to respond with a left or a right key to an X that occurred to the left or right of a vertical line. On each trial, there was either no warning or a brief warning tone, followed by the X after a variable delay. The left panel of the figure shows that

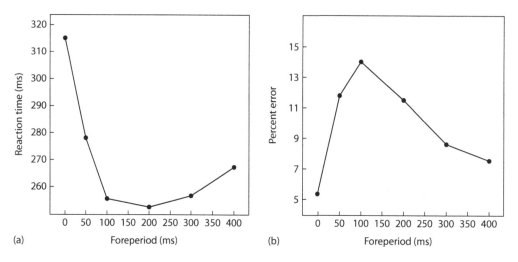

FIGURE 13.3   Reaction times (a) and percentages of error (b), as a function of foreperiod, in a two-choice task.

reaction times were a U-shaped function of warning interval, with the fastest responses occurring when the warning tone preceded the stimulus by 200 ms. This function, in which reaction time decreases to a minimum at a warning interval of 200–500 ms and then increases as the interval becomes longer, is fairly typical. Both psychophysiological and behavioral evidence suggests that the U-shaped preparation function is *not* due to processes involved in executing the response (in the motor stage; Bausenhart, Rolke, Hackley, & Ulrich, 2006; Müller-Gethmann, Ulrich, & Rinkenauer, 2003).

However, note that, as the right panel illustrates, the percentages of errors in this study were the inverse of the reaction time function. Errors increased and then decreased as the warning interval increased. Slower response times were associated with higher accuracy, demonstrating a speed–accuracy tradeoff. In terms of information accumulation models of the type described earlier, this outcome implies that being prepared to respond results in a lower threshold for responding but not in improved efficiency of information accumulation.

Warning signals are sometimes referred to as *alerting signals* in operational environments (e.g., Gupta, Bisanz, & Singh, 2002). The effectiveness of such signals, or how well they improve responses to an event, will depend on how much time the operator has to respond. If the response must be made very rapidly, processing an alerting signal will use up some of the time that could have been used for processing and responding to the event. Simpson and Williams (1980) had airline pilots respond to a synthesized speech message while flying a simulator. Reaction time measured from the onset of the message was faster when an alerting tone occurred 1 s prior to the message than when one did not. However, when the additional time for the alerting tone was taken into account, overall system response time was actually longer (see Figure 13.4). Consequently, Simpson and Williams concluded that an alerting tone probably should not be used with synthesized speech displays in the cockpit environment.

STIMULUS-RESPONSE UNCERTAINTY

Our discussion of stimulus-response uncertainty necessarily involved discussion of information theory, which expresses the amount of information (H) in a set of stimuli or responses as a function of the number of possible alternatives and their probabilities. Information theory is described in more detail in Appendix II. It became popular in psychology in the 1950s because choice reaction time was shown to be a linear function of the amount of information transmitted, a relation that has come to be known as Hick's law, or the *Hick–Hyman law*, after the researchers who first discovered it (see Proctor & Schneider, 2018, for a detailed description of the impact of the law on research and interface design). To understand the Hick–Hyman law, it is important to understand

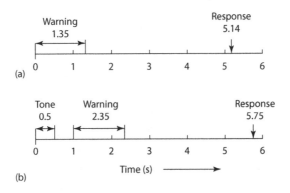

FIGURE 13.4 Mean times to respond to a warning message alone (a) or preceded by an alerting tone (b).

that "information" as it is computed in information theory is a measure of uncertainty: Reaction time increases as uncertainty increases.

Hick (1952) used a display of 10 small lamps arranged in an irregular circle and 10 corresponding response keys on which a person's fingers were placed. When one of the lamps came on, the corresponding keypress response to that stimulus was to be made. In different sets of trials, the number of possible stimuli (and responses) was varied from 2 to 10. Reaction time was a linear function of the amount of information in the stimulus set (see Figure 13.5).

In Hick's (1952) experiment, the stimuli were equally likely, and performance was measured for sets of trials in which no errors were made. If no errors are made, information is perfectly transmitted, and the amount of information transmitted is equal to the stimulus information. If accuracy is not perfect, the amount of information transmitted depends on the stimulus-response frequencies (see Chapter 4). Therefore, in a second experiment, Hick encouraged people to respond faster but less accurately. The decrease in reaction time was consistent with the reduction in information transmitted.

Information (uncertainty) also decreases when some stimuli are made more probable than others. Hyman (1953) showed that when uncertainty was reduced in this way, or by introducing sequential dependencies across trials (which altered the probabilities on a trial-to-trial basis), average reaction times were still a linear function of the amount of information transmitted.

The Hick–Hyman law is written as

$$\text{Reaction time} = a + b\Big[T(S,R)\Big],$$

where:

a is a constant reflecting sensory and motor factors,
b is the time to transmit one bit of information, and
$T(S,R)$ is the information transmitted between stimulus (S) and response (R) (see Figure 13.5).

The Hick–Hyman law fits data from many other choice reaction tasks, but the rate of information transmission ($1/b$) varies greatly as a function of the specific task. The general point of the Hick–Hyman law is that in most situations, an operator's reaction time will increase by a constant amount for each doubling of the number of distinct possible signals and associated responses. The fewer the alternatives, the faster the operator can respond.

The studies by Hick (1952) and Hyman (1953) led to a resurgence of interest in choice reaction time, which had been studied extensively in the latter part of the 19th century and the early part of the 20th century. Much of the research following Hick and Hyman placed qualifications on

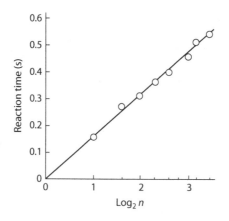

FIGURE 13.5   Reaction times plotted as a function of number of alternatives on a $\log_2$ scale.

the Hick–Hyman law. For instance, Leonard (1958) showed that although the Hick–Hyman law describes the effect of the information on reaction time, a model based on binary decisions of the type implied by information theory cannot explain how people select responses in choice-reaction tasks.

Usher, Olami, and McClelland (2002; see also Usher & McClelland, 2001) provided evidence that the Hick–Hyman law is a consequence of subjects attempting to maintain a constant level of accuracy for all stimulus-response set sizes. They assumed that people used the information accumulation mechanism discussed above to select one of N possible responses: the response for which evidence reaches threshold first is selected. Lower thresholds result in faster responses than higher thresholds, because a low threshold is reached sooner after stimulus onset. However, alternatives with lower thresholds will be more easily reached regardless of what kind of information is coming in, so these alternatives will be more likely to be selected by mistake, resulting in a higher error rate.

When the number of response alternatives increases, another accumulator mechanism must be added to those already operating. Each additional alternative therefore adds another opportunity for an incorrect response to be selected. To prevent the error rate from increasing to potentially very high rates with many response alternatives, all the response thresholds must be adjusted upward. Usher et al. (2002) showed that the probability of an error will not increase if the increase in the threshold as N increases is logarithmic. This logarithmic increase in thresholds results in a logarithmic increase in reaction time. The Hick–Hyman law is therefore a byproduct of a person's attempts not to make too many mistakes as the number of possible responses increases.

Although reaction time typically increases as a function of the number of stimulus-response alternatives, the slope of the Hick–Hyman function is not constant and can be reduced to approximately zero with sufficient practice (Mowbray, 1960; Mowbray & Rhoades, 1959; Seibel, 1963). Moreover, for highly compatible stimulus-response mappings, the slope approximates zero even without practice. Leonard (1959) demonstrated this using stimuli that were vibrations to the tip of a finger and responses that were depressions of the stimulated fingers. With these tactile stimuli, there was no systematic increase in reaction time as the number of choices was increased from two to eight. We see the same lack of effect of stimulus-response uncertainty on reaction time for responses requiring eye movements to the locations of visual stimuli (Kveraga, Boucher, & Hughes, 2002) or aimed movements to a target (Wright, Marino, Chubb, & Rose, 2011), and for responses that require saying the names of the digits or letters (Berryhill, Kveraga, Webb, & Hughes, 2005). Thus, the number of alternatives has little effect for highly compatible or highly practiced stimulus-response relations—topics to which we turn next.

PRINCIPLES OF COMPATIBILITY

STIMULUS-RESPONSE COMPATIBILITY

About the same time as the work of Hick and Hyman, Fitts and Seeger (1953) reported another classic choice reaction time study. They used three different stimulus and response sets, all with eight alternatives. Thus, the sets were equivalent in terms of the amount of stimulus and response information they contained. The stimulus sets differed in the way that the information was signaled (see Figure 13.6). For set A, any one of eight lights could come on; for stimulus sets B and C, any of four lights alone or any four pairs of these lights could occur. The three response sets corresponded conceptually to the displays. People were required to move a single stylus to a target location for sets A and B, and two styli to locations for set C. For response set A, there were eight locations in a circular configuration; for response set B, there were also eight locations, but responses were made by moving from the start point along one of four pathways, which then branched into Ts; for response set C, there were up-down locations for the left hand and left-right locations for the right hand, with the eight responses signaled by combinations of the two hand movements.

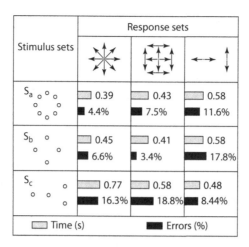

FIGURE 13.6 Stimulus sets, response sets, and data (reaction time and error rates).

Responses in this study were faster and more accurate for the pairings of stimulus sets and response sets that corresponded naturally than for those that did not. Fitts and Seeger called this phenomenon *stimulus-response (S-R) compatibility* and attributed it to cognitive representations or codes based on the spatial locations of the stimulus and response sets.

In a second classic study, Fitts and Deininger (1954) manipulated the mapping of stimuli to responses within a single stimulus and response set (the circular sets from the previous study). The operator's task was to move the stylus to an assigned response location when one of the stimuli was lit. There were three stimulus-response assignments: direct, mirrored, and random. With the direct assignment, each stimulus location was assigned to the corresponding response location. With the mirrored assignment, the left-side stimulus locations were assigned to their right-side counterparts of the response set, and vice versa. Finally, with the random assignment, no systematic relation existed between the stimuli and their assigned responses. Responses were faster and more accurate with the direct assignment than with the mirrored assignment. Even more strikingly, the reaction times and error rates for the random assignment were over twice those for the mirrored assignment.

Morin and Grant (1955) further explored the effects of relative compatibility by having people respond to eight lights arranged in a row with combinations of keypresses. Responses were fastest when there was a direct correspondence between stimulus and response locations (see Figure 13.7). Responding was also fast when the stimulus and response locations were perfect mirror images of each other (i.e. the rightmost response would be made if the leftmost stimulus occurred, and so on), as in Fitts and Deininger's (1954) study. When the stimulus-response assignments were quantified by a correlation coefficient, they found that reaction time increased as the correlation approached zero. These findings indicate that people can translate quickly between a stimulus and response when a simple rule (e.g., respond at the location opposite to the stimulus location) describes the relation (Duncan, 1977). It has been suggested that the correlation coefficient might provide a metric for compatibility in real-world situations for which several stimuli and responses are involved (Kantowitz, Triggs, & Barnes, 1990).

Since these early experiments on S-R compatibility, many basic and applied studies of compatibility effects have been completed (see Proctor & Vu, 2016). These compatibility studies have implications for the design of displays and control panels. The most important of these implications is the fact that machine operation will be easiest when the assignment of controls to display elements is spatially compatible. A classic applied illustration of the compatibility principle comes from studies of four-burner ranges conducted by Chapanis and Lindenbaum (1959) and Shinar and Acton (1978). A common design is to have two back burners located directly

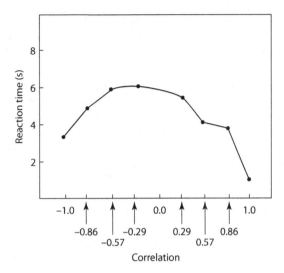

FIGURE 13.7 Reaction time as a function of the correlation between stimulus and response locations.

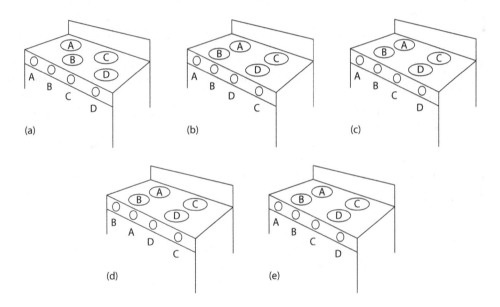

FIGURE 13.8 Control–burner arrangements of a stove.

behind two front burners with the controls arranged linearly across the front of the range (see Figure 13.8b–e). For this design, there is no obvious relation between controls and burners, so a cook may be confused about which control operates a particular burner. However, if the burners are staggered in a sequential left-to-right order (see Figure 13.8a), each control location corresponds directly to a burner location, and confusion about the relation between controls and burners is eliminated.

Relative Location Coding

A widely used procedure for studying S-R compatibility is a two-choice task in which visual stimuli are presented to the left or right of a central fixation point. In the compatible condition, the observer is to respond to the left stimulus with a left keypress and to the right stimulus with a right keypress (see Figure 13.9a). In the incompatible condition, the assignment of stimulus locations to response keys is reversed (see Figure 13.9b). Responses in the two-choice task are faster and more accurate when they are compatible with the stimulus locations (Proctor & Vu, 2016).

S-R compatibility effects also occur when stimulus location is irrelevant for determining the correct response (Hommel, 2011; Proctor, 2011). For example, if an operator is to respond to the color of an indicator light, the location of the light will have an effect on the operator's response. More concretely, suppose you are to make one response to a red light and another to a green light, and that these lights can occur to the left or right of a central point. If you are to respond to the red light by pressing a key on the right and to the green light by pressing a key on the left, your responses will be fastest when the red light occurs to the right or the green light to the left. The influence of the spatial correspondence between the locations of the lights and the responses is called the *Simon effect* (Simon, 1990). Note that the Simon effect is a special case of S-R compatibility that arises when the location of the stimulus is irrelevant to the response that is to be made. We will discuss the Simon effect again later in this chapter.

When a person's left and right hands operate the left and right response keys (as shown in panels a and b of Figure 13.9), the distinction between left and right response locations is redundant with the distinction between left and right hands. This means that we do not know whether S-R compatibility is due to the locations of the responses or to the hands used to make those responses. To determine which is more important, the left and right hands can be crossed, so that the left hand is at the right response location and the right hand is at the left location (see Figure 13.9c,d). With this arrangement, responses will still be faster when there is a direct correspondence between the stimulus and response locations (e.g., Brebner, Shephard, & Cairney, 1972; Roswarski & Proctor, 2000; Wallace, 1971). This means that response location is more important than the hand used to execute the response. This benefit of correspondence between stimulus and response positions occurs even when a person's hands are not crossed but hold sticks that are crossed to press the opposite-side keys (Riggio, Gawryszewski, & Umiltà, 1986).

Spatial S-R compatibility effects will be observed in many situations (Heister, Schroeder-Heister, & Ehrenstein, 1990; Reeve & Proctor, 1984), even when "left" and "right" stimulus or response locations are not defined relative to the locations on a person's body (Nicoletti, Anzola, Luppino, Rizzolatti, & Umiltà, 1982; Umiltà & Liotti, 1987). In other words, it is not the absolute physical locations of the stimuli

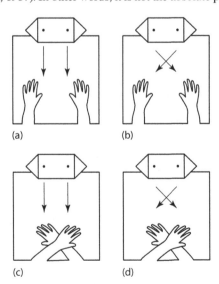

(a) (b)

(c) (d)

FIGURE 13.9 Compatible (a and c) and incompatible (b and d) stimulus-response assignments in two-choice tasks, with the hands uncrossed (a and b) and crossed (c and d).

and responses that determine the degree of compatibility, but their locations relative to each other. These findings and those above suggest that stimuli and the locations at which an action is effected are mentally coded categorically (e.g., left or right) by their relative positions (Umiltà & Nicoletti, 1990).

S-R compatibility effects also occur when stimuli and/or responses are not located in physical space. For example, a person will respond faster to the words "left" and "right" when they are assigned to left and right keypresses or "left" and "right" vocal responses (Proctor, Wang, & Vu, 2002). Similar results are obtained with left- and right-pointing arrows presented in a fixed location (Miles & Proctor, 2012) and even when the stimuli vary along a vertical dimension (up or down positions) and the responses along a horizontal dimension (left or right keypresses or movements; Cho & Proctor, 2003). Moreover, compatibility effects arise for a variety of nonspatial stimulus and response dimensions, such as numerosity (one or two brief stimulus occurrences mapped to responses for which a single key is tapped once or twice; Miller, Atkins, & Van Ness, 2005). These results imply that S-R compatibility effects occur whenever there is similarity in the cognitive representations of the stimulus and response sets.

Theoretical Interpretations

Most explanations of S-R compatibility rely on the idea that stimuli and responses are coded in terms of perceptual and conceptual features. S-R compatibility arises to the extent that the feature dimensions are similar between the stimulus and response sets (called *dimensional overlap*; Kornblum, 1991). Compatibility effects occur whenever S-R dimensions overlap conceptually. For example, both stimuli and responses could be defined by the concepts of left and right, and a compatibility effect would occur even if the stimuli were the words "left" and "right" and the responses were left and right keypresses. However, compatibility effects are strongest when the dimensions are also physically similar. For example, changing the response to the words "left" and "right" from keypresses to the spoken words "left" and "right" would give a larger compatibility effect (Proctor & Wang, 1997).

Kornblum and Lee (1995) described another way in which stimuli and responses can be similar, which they called *structural similarity*. Structural similarity can be illustrated in the following example. The letters A, B, C, and D and the numbers 1, 2, 3, and 4 have a common natural order that does not depend on conceptual or physical similarity. The letters A, B, C, and D might be used as stimuli and assigned arbitrarily to the spoken responses "1," "2," "3," and "4." Responses will be fastest and most accurate when the assignment is consistent with the natural order: A to "1," B to "2," etc. Structural similarity may be the reason why a variety of compatibility effects arise in tasks for which people have to make binary decisions and the stimulus and response dimensions are unrelated (as, for example, with up and down stimulus locations mapped to right and left responses; see Proctor & Xiong, 2015).

"single-route" models of S-R compatibility focus on the computations that a person makes in selecting a response when stimulus and response dimensions overlap. In these models, response selection is an intentional process of selecting the instructed response. The most well-defined of these models can be used to predict a person's performance in applied settings.

Rosenbloom (1986; Rosenbloom & Newell, 1987) developed a model of S-R compatibility that attributes compatibility effects to the number of transformations that have to be performed to select the response. Their model "is based on the supposition that people perform reaction-time tasks by executing *algorithms* (or programs) developed by them for the task" (Rosenbloom, 1986, p. 156). This kind of model is commonly referred to as a GOMS model (e.g., Kieras, 2004), where GOMS stands for Goals, Operators, Methods, and Selection rules (see Chapter 19 and Box 3 in Chapter 3). When this kind of model is used to explain compatibility effects, a researcher must first perform a task analysis to determine the algorithms that can be used to perform the task. Tasks with low S-R compatibility will require the use of algorithms with more steps than will tasks with high compatibility. Following the task analysis, the researcher must estimate the time required for each operation in the algorithm. Based on these estimates, the researcher can predict how large the compatibility effects on response time will be. Box 13.1 describes an application of this approach to predicting

BOX 13.1 COMPATIBILITY EFFECTS IN
HUMAN–COMPUTER INTERACTION

Issues of stimulus-response compatibility arise frequently in human–computer interaction (John, Rosenbloom, & Newell, 1985). There are obvious spatial relationships that affect performance (e.g., alternative arrays of cursor movement keys) but also nonspatial relationships. For example, many applications use command name abbreviations, like "DEL" or "INS," to indicate actions like "delete" or "insert." John et al. examined how well people could interpret abbreviations like these in a simple experiment.

In the experiment, a spelled-out command (like "delete") appeared on the screen, and the person was asked to type its abbreviation as quickly as possible. John et al. (1985) looked at two methods of abbreviation: *vowel deletion*, in which the vowels were deleted to yield the abbreviation (e.g., *dlt* as the abbreviation for *delete*), and *special character*, in which the abbreviation was the first letter of the command preceded by an arbitrary special character (e.g., */d* as the abbreviation for *delete*). Each person also performed in a no-abbreviation condition (e.g., type the word *delete* in response to the command *delete*) and a nonsense abbreviation condition (e.g., type an assigned three-letter meaningless nonsense syllable to the command *delete*). Each person studied and practiced the abbreviations before they were tested. Their typing times were recorded as initial response times (the time to press the first key) and execution times (the time between the first and the last keypress).

Initial response times were shortest in the no-abbreviation condition (842 ms), followed in increasing order by the vowel-deletion (1091 ms), nonsense abbreviation (1490 ms), and special character (1823 ms) conditions. Execution times were shortest in the special character condition (369 ms), followed by the nonsense abbreviations (866 ms), no-abbreviation (1314 ms), and vowel-deletion (1394 ms) conditions

By analyzing the tasks and observing how people performed them, John et al. (1985) described the processing steps, or sequence of operators, from stimulus presentation to completion of response execution. They suggested that people used four kinds of operators: perceptual (for word identification); mapping (or cognitive; for figuring out an abbreviation); retrieval (for retrieving information from memory); and motor (for typing responses). Each task requires a different combination of these operators (although all of them use the perceptual operator). John et al. estimated that each mapping operator took 60 ms to complete, each retrieval operator took 1200 ms to complete, and each motor operator took 120 ms to complete.

While John et al.'s (1985) data tell us that the special character mapping is more complex than the vowel-deletion mapping, their examination of the tasks and the steps people took to complete them tells us more precisely how much more complex character mapping is and explains changes in both the initial response time and the execution time.

John et al. explained performance algorithmically; that is, they specified the number and sequence of operators a person had to complete to do the task. John and Newell (1987) used the same approach to evaluate two additional abbreviation rules: a minimum-to-distinguish rule, in which abbreviations were determined by the minimum number of letters to distinguish the commands in a set (e.g., *def* for *define* and *del* for *delete*), and a two-letter truncation with exceptions (e.g., *de* for *define* and *dl* for *delete*). The minimum-to-distinguish rule is commonly used in many applications that allow keyboard shortcuts to augment or replace "point-and-click" command selection. While the algorithms for these tasks were different from those in the earlier experiment, they used the values of the mapping, motor, and retrieval operators from John and Newell's (1985) earlier study to generate initial response and execution time predictions for the new tasks. The predictions derived from the algorithms matched people's response times very well, even though they were based entirely on the estimates from the earlier experiment.

How important is this kind of work when we have to make decisions about the design of computer applications? John and Newell (1990) applied the results of these experiments to some transcription typing tasks and several stimulus-response compatibility tasks. They showed that, even for tasks that were not very similar to those examined in the original experiments, they could develop appropriate algorithms using the four operators and accurately predict response times to within about 20% of actual performance. Tools like these can therefore give the designer a good feel for what kinds of application "widgets" will be easiest to interpret and what commands will be easiest to type, although final design decisions must depend on careful testing. Design tools such as these will be discussed more in Chapter 19.

the extent of compatibility for alternative mappings of abbreviations to command names, a problem in human–computer interaction.

Although single-route models can account for many compatibility effects, they do not provide an account of phenomena like the Simon effect, for which a stimulus dimension is defined as irrelevant for the task. Consequently, the most successful models of S-R compatibility are "dual-route" models. These models include not only an intentional process but also an automatic response-selection mechanism.

Kornblum, Hasbroucq, and Osman (1990) proposed a dual-route dimensional overlap model in which a stimulus automatically activates the most compatible response, regardless of whether or not that response is the correct one. The correct response is identified by way of the intentional response-selection route. If the automatically activated response is not the same as the one identified by the intentional route, it must be inhibited before the correct response can be programmed and executed. Inhibition of the automatically activated response when it conflicts with the correct response explains the Simon effect. Compatibility effects for relevant stimulus dimensions arise in the model from both response inhibition and the time required by the intentional response-selection route.

Hommel, Müsseler, Aschersleben, & Prinz (2001) developed the *Theory of Event Coding* to explain perception–action relationships more generally, including S-R compatibility effects. This theory assumes that codes for stimuli and responses share the same cognitive system that subserves perception, attention, and action. The theory emphasizes structures called *event codes* or *event files* (Hommel, 2004; Hommel et al., 2001). A file is a temporary, linked collection of features that define an event (Kahneman, Treisman, & Gibbs, 1992). Hommel et al. proposed that a stimulus and its associated response are coded as linked and integrated features in an event file. The features that are bound into the event file will be less available for other perceptions and actions.

The theory of event coding emphasizes that actions are coded in terms of their effects. Kunde (2001) performed an experiment in which subjects responded to one of four colored stimuli with a keypress made with two fingers on each hand. Pressing a key lit up one box in a row of four on the lower part of the display. The box that lit up is called the *effect of the response*. In one condition the location of the box corresponded to that of the key, whereas in another it did not. Responses were faster with the corresponding response–effect mapping than with the noncorresponding mapping, even though the effect did not occur until after the key was pressed. If no effects were coded for the actions, the location of the box should not have influenced response times; the fact that it did provides strong support for the theory. We will see more evidence for the role of effects in the control–display population stereotypes we discuss later, in which people anticipate a particular outcome when they operate a control.

S-C-R Compatibility

Most research on S-R compatibility has focused on situations in which simple rules relate stimuli to responses (Kantowitz et al., 1990). However, by attributing compatibility effects to the cognitive codes used to represent the stimulus and response sets, the implication is that central cognitive processes must be responsible for the effects. Because the role of cognitive mediation in response selection will be larger for more complex tasks that do not involve simple rules or response tendencies, Wickens, Sandry, and Vidulich (1983) have used the term *S-C-R compatibility* to emphasize the central processes. The mediating processes (C) reflect the operator's mental model of the task. Compatibility will be observed to the extent that stimuli and responses correspond with the features of the mental model.

Wickens et al. (1983) structured their theory of S-C-R compatibility around the multiple-resources view of attention and, hence, stressed the importance of the cognitive codes (verbal and visual) used to represent the task. They proposed that codes must be matched with input and output modes for S-C-R compatibility to be maximized. Wickens et al. provided evidence that tasks represented by a verbal code are most compatible with speech stimuli and responses, whereas tasks represented by a spatial code are most compatible with visual stimuli and manual responses. Robinson and Eberts (1987) obtained evidence consistent with the coding relations proposed by S-C-R compatibility in a simulated cockpit environment. Either a synthesized speech display or a picture display was used to communicate emergency information. As predicted by the S-C-R compatibility hypothesis, when responses were made manually, they were faster for the picture display than for the speech display.

Greenwald (1970) proposed that the highest compatibility between stimuli and responses occurs when they have high ideomotor compatibility. Ideomotor feedback refers to the sensations resulting from an action. Stimuli and responses have high ideomotor compatibility when the modality of the stimulus is the same as the ideomotor feedback from the response. Ideomotor compatibility is high, for example, when spoken responses (e.g., saying "A") are required to auditory letters (e.g., "A" played over headphones). Greenwald conducted an experiment in which he presented letters either auditorily and visually and asked people to respond with the name of the letter either vocally or in writing. He found that response times were fastest when auditory letters were paired with naming responses or visual letters were paired with written responses. When the assignment of response modality to stimulus modality was reversed, response times were slower.

The concept of S-C-R compatibility was expanded by Eberts and Posey (1990) to emphasize the role that mental models play in explaining compatibility effects. Specifically, they propose that a good mental model that accurately represents the conceptual relations of the task should enable better and more efficient performance than a poor mental model. Eberts and Schneider (1985) performed several experiments with a difficult control task and showed that people perform better when their training incorporates an appropriate mental model. From experiments such as these and John and Newell's (1990) work on compatibility effects in human–computer interaction, described in Box 13.1, we know that compatibility effects influence performance in a broad range of tasks that are more complex than those performed in the laboratory.

Practice and Response Selection

Like any other human activity, performance on choice reaction tasks improves with practice. However, as noted in Chapter 12, benefits will be small when the mapping of stimuli to responses varies. For example, the late psychologist Richard Gregory noted,

> There's the famous case about a mine trolley that had a brake pedal and accelerator—so the accelerator was on the right and the brake pedal on the left for one direction of travel—but reversed on the way back—as one sat on a different seat and used the same pedals. Believe it or not, they had an awful lot of crashes. (p. 38, quoted in Reynolds & Tansey, 2003)

The improvement in performance that occurs when the S-R mapping does not vary on a regular basis is characterized by the power law described in Chapter 12 (also see Newell & Rosenbloom, 1981). This means that although performance continues to improve indefinitely, the additional benefit for a constant amount of practice becomes less and less as people continue to perform the task. Moreover, because practice effects are larger for tasks that have more stimulus-response alternatives, the slope of the Hick–Hyman law function relating reaction time to number of alternatives becomes progressively smaller as people become more practiced (Teichner & Krebs, 1974).

A classic study that illustrates the extent to which human performance can improve with practice was conducted by Crossman (1959). He examined the time required for people to make cigars on a hand-operated machine. Operators were tested whose experience levels ranged from novice to 6 years on the job. Speed of performance increased as a power function of experience up to 4 years, at which point the operators worked faster than the machine, and so the machine processing time determined performance time. In other words, the machine reached its limits before the operators reached theirs.

Seibel (1963) performed a 1023-choice reaction task in which a subset of 10 horizontal lights were illuminated. In response to each presentation of lights, Seibel (who was his own subject) pressed each of the response keys indicated by the lights as a chord. Initially, his reaction times averaged over 1 s, but they decreased to 450 ms after 70,000 trials of practice. As can be seen in Figure 13.10, performance improved continuously over the course of the study.

There are several explanations of how response-selection processes change as a person becomes skilled at a task. The power law seems to show a gradual, continuous change from unpracticed to practiced states, leading some people to treat practice effects as quantitative changes over time. For example, using the idea of production systems that we discussed in the last chapter, the change might arise from increasing the number of procedures that are incorporated into a single chunk (e.g., Rosenbloom, 1986). Alternatively, a qualitative change over time may occur, such that the procedures used to do the task at the end of practice are not the same as those that were used at the beginning of practice (e.g., Teichner & Krebs, 1974).

Not only do responses get faster, but S-R compatibility effects also decrease with practice. However, they never completely go away (Dutta & Proctor, 1992). For example, Fitts and Seeger (1953) found that responses for an incompatible display–control arrangement were still considerably slower than those for a compatible arrangement after 26 sessions of 16 trials for each arrangement. Such findings are consistent with proposals by Eberts and Posey (1990) and Gopher, Karis, and

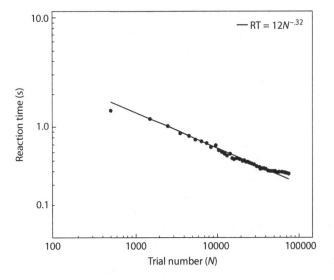

FIGURE 13.10 Reaction time as a function of practice in a 1023-choice task.

Koenig (1985) that mental representations continue to play an important role in translating between stimuli and responses even for well-practiced performers. Although we still do not know why S-R compatibility effects persist, the point to remember is that incompatible display–control arrangement can result in decrements in performance that cannot be remedied entirely by practice.

IRRELEVANT STIMULI

The previous section discussed compatibility effects on performance due to features of the stimulus and response sets that were relevant to the task. Earlier, we also discussed the Simon effect, which is similar to compatibility effects except that the factors resulting in a Simon effect are irrelevant to the task.

Simon's (1990) original experiments used auditory stimuli: People made a left or right response as determined by the high or low pitch of a tone that occurred in the left or right ear. Simon initially proposed that the effect reflects an innate tendency to orient, or respond, toward the auditory stimulus (Simon, 1969). When this response tendency conflicts with the correct response, for example, when a stimulus detected on the left requires a response to the right, the tendency to respond to the left must be inhibited before the correct response can be made. Other explanations of the Simon effect maintain this response competition idea and link it to automatic activation of the spatial response code corresponding to the stimulus location, which produces interference when the wrong codes are activated (Umiltà & Nicoletti, 1990; see our earlier discussion of Kornblum et al.'s, 1990, model).

Why is stimulus location processed even though it is defined as irrelevant to the task? The answer seems to be that the act of discriminating left and right response alternatives activates a similar coding of stimulus locations (Ansorge & Wühr, 2004), causing stimulus location to be added into the information used during the decision process (Yamaguchi & Proctor, 2012). The additional information tends to activate the corresponding response location code. These kinds of activations between codes for stimuli and responses have been attributed to highly overlearned, and possibly even innate, associations between the stimulus and response locations (e.g., Barber & O'Leary, 1997).

Like other S-R compatibility effects, the Simon effect persists even after considerable amounts of practice. However, the Simon effect can be reversed by giving people prior practice in performing a two-choice task in which the response locations are mapped incompatibly to the stimulus locations (Proctor & Lu, 1999; Tagliabue, Zorzi, Umiltà, & Bassignani, 2000). Thus, an operator's experience in specific contexts may act to reverse the normal advantage for spatial correspondence that a designer might expect to find.

An important consideration is how location of stimuli and responses is represented or coded relative to a person's body or relative to the machine she is operating. For instance, a stimulus location can be coded as left or right of person's body midline, or above or below a line on a computer screen. Factors like the relative salience of different reference frames and instructions emphasizing one frame over another can determine which reference frame will affect performance the most.

For example, while driving, turning the steering wheel clockwise will usually result in the car turning to the right. However, when the wheel is held at the bottom, a clockwise turn results in leftward movement of the hands, which is opposite to the right turn of the wheel. If we look at how people code their responses in this situation (when instructions do not indicate one reference frame or the other), approximately half of them will be using a wheel-based reference frame and the other half a hand-based reference frame (e.g., Guiard, 1983; Proctor, Wang, & Pick, 2004). However, if we instruct people to turn the wheel to the left for one stimulus and to the right for another, everyone reverts to a wheel-based reference frame. For situations in which stimulus and response locations can be coded with respect to multiple reference frames, we can predict what the most compatible mapping will be only if we know what frame will dominate coding.

A closely related phenomenon to the Simon effect is the *Stroop effect* (Stroop, 1935/1992). People performing the Stroop task name the ink colors of words that spell out color names. For example, the word "green" could be printed in red ink, and a person's task is to say "red." The Stroop effect occurs when the word and the ink color conflict. People find it very difficult to say "red" to the word "green" printed in red ink. Stroop interference occurs in many tasks, not all of which involve colors (MacLeod, 1991). The difference between the Stroop and Simon effects is that the Stroop effect seems to arise from conflicting stimulus dimensions, whereas the Simon effect seems to arise from conflicting response dimensions. Explanations of the Stroop effect have tended to focus on response competition, much like explanations of the Simon effect (e.g., De Houwer, 2003).

One last related phenomenon is called the *Eriksen flanker effect* (Eriksen & Eriksen, 1974), introduced in Chapter 9. People are asked to identify a target stimulus (usually a letter) presented at the point of fixation. On most trials, the target is surrounded on each side (flanked) by irrelevant letters. For example, a person may be asked to press a left key in response to the letter H and a right key in response to the letter S. On a trial, the person may see the letter string "XHX" or "SHS." Responses to strings of the form "SHS" are slower and less accurate than to strings of the form "XHX." The explanation for this effect again involves response competition (e.g., Sanders & Lamers, 2002): The letter S activates the right response, and this activation must decrease or be inhibited before the left response can be made.

All three of these effects, the Simon, Stroop, and Eriksen flanker effects, illustrate that irrelevant stimulus attributes can interfere with performance when they conflict with the stimulus and response attributes relevant to the task at hand. They reflect people's limited ability to selectively attend to relevant task dimensions and to ignore irrelevant ones. The design of display panels and other interface devices should minimize the possibility for interference and conflict between sources of information.

DUAL-TASK AND SEQUENTIAL PERFORMANCE

In most real-world situations, people are required to perform several tasks or task components at once. For example, navigating a vehicle will require the pilot to perform several manual control actions, often at the same time, and also to negotiate obstacles in his or her environment, some of which may appear or disappear unexpectedly. In this situation, several stimuli may require responses in rapid succession. In these more complex tasks, we need to consider how well a person can select and coordinate multiple responses.

PSYCHOLOGICAL REFRACTORY PERIOD EFFECT

How people coordinate several responses to several stimuli at the same time has been studied using a simple laboratory task called the *dual-task paradigm*. In this paradigm, two stimuli occur in rapid succession, one after the other. Each stimulus requires a different response, usually something simple like a keypress. In this situation, reaction time to the second stimulus becomes longer the closer in time it appears after the first stimulus. As the time between the stimuli becomes longer, reaction time to the second stimulus speeds up, until it almost reaches that obtained when the stimulus is presented alone. This phenomenon was discovered by Telford (1931), who named it the *psychological refractory period* effect.

Most explanations of the psychological refractory period (PRP) effect have attributed it to a central response-selection bottleneck (Pashler, 1994; Welford, 1952; see Figure 13.11). According to this account, response selection for the second stimulus cannot begin until that for the first task is completed. When the interval between onsets of the two stimuli is short, the response to the first stimulus is being selected and prepared while the second stimulus is being identified. If the duration between the stimuli is short enough, response selection and preparation for the second stimulus may have to wait until the first response is prepared. The response-selection bottleneck model predicts that there should be a linear decrease in reaction time to the second stimulus as the interval between stimulus onsets increases. When the interval duration is long enough that the two response selection

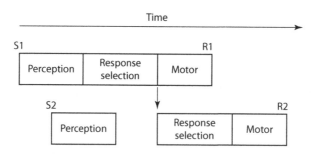

FIGURE 13.11   The sequence of stages in response-selection postponement.

processes no longer overlap, no further decrease in response time to the second stimulus should occur. This is approximately what we see in these kinds of tasks.

The response-selection bottleneck model also predicts that increasing the difficulty of stimulus identification for the second task (by making the stimulus smaller or more difficult to see) should not have the same effect on response time that increasing the difficulty of response selection will have (Schweickert, 1983). Because there is no bottleneck for stimulus identification, making the second stimulus more difficult to identify should not produce any change in the response time for the second task as long as the increase in stimulus-identification time is no greater than the waiting time at the bottleneck. In contrast, because variables that affect response-selection difficulty for the second task have their influence after the bottleneck, the extent of the refractory effect will not depend on the interval between the two stimulus onsets.

Several studies have confirmed this basic prediction. In one study, people were asked to identify a high- or low-frequency tone by pressing one of two keys with fingers on their left hand (Pashler & Johnston, 1989). After the tone, they were shown a letter (A, B, or C, visually), which was to be classified by pressing a key with one of three fingers on the right hand. This is a standard dual-task paradigm. The experimenters carefully manipulated stimulus identification and response selection difficulty together with the length of the interval between the tone and the letter. The delay between the tone and the letter was either short or long. Stimulus identification was made difficult by reducing the contrast of the letter (gray on a dark background) on half of the trials and by increasing the contrast (white on a dark background) on the other half. Response selection was made easy or difficult by exploiting the fact that it is easier to repeat a response on successive trials than to switch to a new one. On a percentage of the trials, the letter (and therefore the response) was a repetition of the one from the previous trial, and on the remainder, the letter was different from the previous trial.

Figure 13.12 shows that responses to the letter were slower at the short intervals than at the longest one, indicating a PRP effect. More important, as predicted by the response-selection bottleneck model, the PRP effect was larger when letter identification was easy than when it was more difficult (see Figure 13.12a), but the effect size did not vary as a function of whether the letter was repeated from the previous trial (see Figure 13.12b).

Although we can find a lot of data that are consistent with the response-selection bottleneck model, we can also find data that are inconsistent. For example, according to the model, response selection for the second task cannot begin until that for the first is completed. This means that performance of the first task should not be affected by variables related to response selection for the second task. But, several experiments demonstrate such *backward crosstalk* effects (Hommel, 1998; Lien & Proctor, 2000; Ko & Miller, 2014; Logan & Schulkind, 2000). In one, Hommel asked people first to respond to a red or green colored rectangle with a left or right keypress and second to respond to the letter H or S by saying "green" or "red." When the rectangle and the letter appeared close together in time, people made faster keypresses when the letter response was consistent with the rectangle color than if it was not.

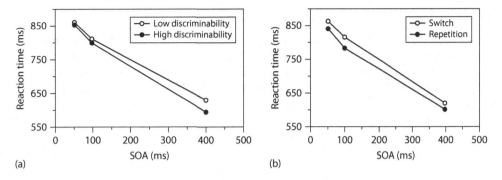

FIGURE 13.12 Mean reaction times for task 2 as a function of interval between stimulus onsets (stimulus onset asynchrony; SOA) and (a) difficulty of stimulus identification and (b) difficulty of response selection.

Backward crosstalk effects have led to two alternative models for explaining the response-selection bottleneck. One assumes that response selection uses a limited capacity central resource that can be partially allocated to each task (Navon & Miller, 2002; Tombu & Jolicœur, 2005). The other model assumes that the central resource for response selection is of unlimited capacity, and the bottleneck is strategically created to ensure that the response for the first task is made before that for the second task (Meyer & Kieras, 1997a,b; see Chapter 9).

These two models imply that there should be conditions under which no PRP effect will be obtained. Indeed, when people practice making their responses in any order, the PRP can disappear (Schumacher et al., 2001). Greenwald and Shulman (1973) provided some evidence to suggest that even without any practice, the PRP effect can be eliminated if the two tasks are ideomotor compatible and processing is thus relatively automatic (see earlier). Although the PRP effect is reduced greatly when individuals are practiced at responding in any order or both tasks are ideomotor compatible, it has been difficult to determine whether the bottleneck is indeed being bypassed or eliminated, as Schumacher et al. and Greenwald and Shulman suggest (see, for example, Lien, Proctor, & Allen, 2002; Ruthruff, Johnston, & Van Selst, 2001, for opposing arguments and evidence). The primary message for human factors specialists is that response selection will be slowed when two or more tasks must be performed close together in time, but this slowing can be reduced under certain circumstances.

STIMULUS AND RESPONSE REPETITION

As we just mentioned, reaction times will be faster on a trial in which the stimulus and response are the same as on the preceding trial. This repetition effect will be greatest when the stimulus for the next trial occurs very quickly after the response. The magnitude of the repetition effect is influenced by several other factors as well (Kornblum, 1973): it will get bigger for larger numbers of stimulus-response alternatives than for fewer, and smaller for stimulus-response alternatives with high compatibility. When responses are not repeated, stimulus-response compatibility effects are larger than when they are. In other words, response repetition is most beneficial when response selection is difficult.

Pashler and Baylis (1991) presented evidence that the interaction between repetition and ease of response selection occurs because repetition enables the person to bypass the normal process of response selection when the stimulus-response link already is in an active state. Their results showed a benefit of repetition only when both the stimulus and the response were repeated: No benefit of response repetition alone was observed. In the most extreme case, when the assignment of responses to stimuli was changed on every trial, neither stimulus repetition nor response repetition produced any speed-up in response time. One way to explain this is that people have expectancies about the next stimulus and its associated response that can either help or hinder the response-selection process.

PREFERENCES FOR CONTROLLING ACTIONS

All of the research discussed in this chapter has examined situations in which a person must choose the right response for different stimuli. These situations resemble the interactions an operator might have with display and control panels, where particular buttons must be pushed or switches flipped in response to displayed information. More realistically, there may be many possible ways to manipulate a control device; for example, a knob could be rotated clockwise or counterclockwise. An operator will need to choose an action from these response alternatives that will achieve a specific goal; for example, to increase volume. We know much less about how selection of action is controlled in these circumstances, but research on grip patterns and display–control relationships provides some insight.

GRIP PATTERNS

Grip patterns are the limb movements and finger placements that people use to grasp and manipulate an object (see Chapter 14). Grip patterns are affected by at least two factors (Rosenbaum, Cohen, Meulenbroek, & Vaughan, 2006). The first involves the properties of the object for which a person is reaching, such as size, shape, texture, and distance. For example, when a person reaches out to grasp an object, the grip aperture (the extent to which your hand is open) is directly related to the size of the object (Cuijpers, Smeets, & Brenner, 2004; Jeannerod, 1981). The larger the object is, the larger a person's grip aperture will be. However, the grip aperture is not affected much by the distance of the object from the person's hand.

The second factor affecting grip patterns is the intended use of the object. Figure 13.13a shows an experimental apparatus used by Rosenbaum, Marchak, Barnes, Vaughan, Slotta, and Jorgensen (1990). People in this experiment were asked to pick up a horizontal bar and place either its left end or right end on a platform to the left or right of the bar. Everyone who used their right hand used an overhand grip to bring the white (right) end of the bar to either target disc (Figure 13.13b) and an underhand grip to bring the black (left) end to either disc (Figure 13.13c).

Rosenbaum et al. (1990) emphasized the importance of minimizing movement effort in the context of constraints imposed by joint angles and object locations. People apparently selected their initial grip to minimize the awkwardness of the final position they anticipated. A similar study

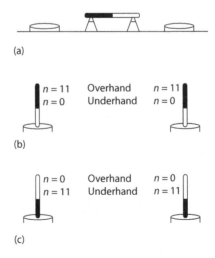

FIGURE 13.13 Apparatus used by Rosenbaum et al. (1990) (a) with the number of right-handed subjects who used the overhand or underhand grip to bring (b) the white (right) end of the bar to both target discs and (c) the black (left) end of the bar to both target discs.

confirmed this: When people had to grasp two bars, one in each hand, and place them in specific target locations, their grips maximized the comfort of their postures at the end of the movement even if the grips had to be different for the two hands (Weigelt, Kunde, & Prinz, 2006). In other words, the goal of achieving final comfortable postures for both hands determines how people carry out the task.

POPULATION STEREOTYPES

Another situation in which selection among controlling actions has been studied involves display–control relationships. In a typical task, people either are asked to demonstrate the most natural relation between a control and a visual indicator (e.g., a needle) or are required to use a control to align the indicator with a particular dial setting. For many types of displays and controls, certain display–control relationships are preferred over others (Loveless, 1962). In the simplest case, consider a horizontal display whose settings are controlled by movements of a parallel, horizontal control stick (see Figure 13.14). It should be clear from our earlier discussion of S-R compatibility that rightward movement of the control stick should cause rightward movement of the indicator, and so on, rather than having rightward responses associated with leftward movements of the indicator. Because most people would intuitively make this association between the stick and the indicator, the association is called a *population stereotype*.

More interesting is the fact that population stereotypes are found when there is no direct relation between the display and the control. Often, the settings of linear displays are controlled by rotary knobs, as in some car radios. For such situations, four principles act to determine the preferred relationship:

1. Clockwise-to-right or -up principle: A clockwise turn of the control is expected to move a pointer to the right for a horizontal display or upward for a vertical display.
2. Warrick's principle: When the control is at one side of the display (see Figure 13.15), the pointer should move in the same direction as the side of the control nearest the display.
3 Clockwise-to-increase principle: Clockwise rotation is expected to correspond with an increased reading on the display scale.
4. Scale-side principle: The indicator is expected to move in the same direction as the side of the control that is next to the display's scale.

As with Gestalt organization, it is possible to vary the extent to which a particular display–control relationship is consistent with these principles. Hoffman (1990, 1997) evaluated the relative contributions of each of these principles for horizontal displays. Groups of engineers and psychologists

FIGURE 13.14 Horizontal display and control arrangements, with preferred movements indicated.

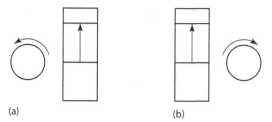

(a)   (b)

FIGURE 13.15 Illustration of Warrick's principle. The stereotypic control movement to produce an upward movement of the display indicator is counterclockwise for (a) and clockwise for (b).

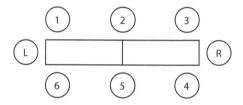

FIGURE 13.16 Locations of the knobs tested by Hoffman (1990).

indicated their preferred direction of movement for 64 display–control arrangements composed of 8 control locations (see Figure 13.16), 2 directions of scale increase (left, right), 2 types of indicator (a neutral line or a directional arrow), and 2 scale sides (top, bottom). For these situations, the dominant principles were the clockwise-to-right and Warrick's principles, with the preferred direction of motion predicted well by a weighted sum of the strengths of the individual principles. For engineers, Warrick's principle was most important, whereas for psychologists, the clockwise-to-right principle was. Hoffman attributed this difference to the engineer's knowledge of the mechanical linkage between control and pointer, with which Warrick's principle is consistent. It may be speculated that the engineers' mental models of the control–display relationship incorporate this knowledge. More generally, the difference between engineers and psychologists illustrates that the population of interest must be considered when evaluating display–control relations.

Another factor that influences expected display–control relations is the orientation of the control operator. Worringham and Beringer (1989) had people guide a cursor to one of 16 target locations with a joystick. The joystick was always operated with the right hand, but the positioning of the arm, hand, and trunk was varied across 11 experimental conditions (see Figure 13.17). With this procedure, the effects of three types of compatibility could be distinguished. Visual field compatibility is display movement that mirrors the control movement while the person looks at the control. Control–display compatibility is defined in terms of the actual direction of movement of the control relative to the display; visual–trunk compatibility arises when the control movement is in the same direction as the display movement relative to the operator's trunk. Visual field compatibility is the most important determinant of a person's expectations, regardless of the person's physical orientation.

The implication of visual field compatibility being the most important influence on expectations can be understood by considering the condition in which the person's head is turned to view a moving cursor located on the left side of the body and the control is located on the opposite side, to the right of the trunk. This arrangement has display–control compatibility when the joystick moves forward to move the cursor forward and backward to move it backward, but visual field compatibility when the joystick moves backward to move the cursor forward and forward to move it backward. Yet, the mapping with visual–motor compatibility yielded better performance than the one with display–control compatibility.

Visual field compatibility applies not only to the operation of horizontally moving controls but also to the operation of vertically moving controls and rotary controls (Hoffmann and Chan, 2013). This relation is called the *Worringham and Beringer principle*. In Hoffman and Chan's words,

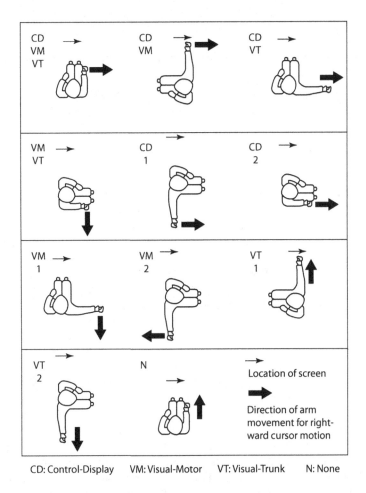

CD: Control-Display   VM: Visual-Motor   VT: Visual-Trunk   N: None

FIGURE 13.17 Relationship between direction of arm movement and cursor movement, and positions of hand, trunk, and arm, for Worringham and Beringer's (1989) experimental conditions.

> For design purposes, the Worringham and Beringer principle for the relationship between display and control movements, when the operator is moving and the display may not be in the same plane as the control, is the most powerful design principle available for stereotype strength prediction. Designers, where possible, should use this principle in design. (p. 1623)

Up to this point, we have only discussed population stereotypes for two-dimensional movements (left-right, up-down). There are also population stereotypes for complex, three-dimensional displays. One study demonstrated how, for three-dimensional displays, pushing motions were preferred over pulling motions for rightward shifts, backward shifts, and clockwise rotations in the frontal plane (Kaminaka and Egli, 1984). Pulling motions were preferred for upward shifts and rotations toward the operator.

Stereotypic responses also have been demonstrated for controls that are not associated with displays. Hotta, Takahashi, Takahashi, and Kogi (1981) conducted a survey of preferred direction of motion for controls used regularly in daily life. People were shown cubes, each of which had a rotary lever, a slide lever, or a pushbutton on the front, top, bottom, or left or right sides. Given the task of turning a doorknob, turning on water, gas, or electricity, or producing a more generic "output increase" (increasing volume, temperature, speed, etc.), people selected different stereotyped movements shown in Table 13.1. Preferred directions depend on the purpose of a control and the plane in which it is located.

TABLE 13.1

Common Direction-of-Motion Stereotypes in Relation to the Control Purpose and the Control Plane

Purpose	Plane	RotaryKnob	Rotary Lever	Button	Slide Lever		Two Buttons	
Door	F	Clockwise	Counterclockwise	Pull		Right		
Water/gas	F	Clockwise	Counterclockwise		Downward	Right		
	T		Counterclockwise		Backward	Right		
	B	Clockwise	Clockwise	Pull	Backward	Right		
	R	Clockwise	Clockwise	Pull	Downward	Backward		
	L	Counterclockwise	Counterclockwise	Pull	Downward	Backward		
Electricity	F	Clockwise	Counterclockwise	Push	Downward	Right	Upward	Right
	T	Clockwise	Counterclockwise	Push	Backward	Right	Backward	Right
	B	Clockwise	Clockwise		Backward	Right	Backward	Right
	R	Clockwise	Clockwise	Push	Downward	Backward	Upward	Backward
	L		Counterclockwise	Push	Downward	Backward	Upward	Backward
Increase	F	Clockwise	Counterclockwise		Upward	Right		
	T	Clockwise	Counterclockwise		Forward	Right		
	B	Clockwise	Clockwise	Pull	Forward	Right		
	R	Clockwise	Clockwise	Pull		Forward		
	L		Counterclockwise			Backward		

A person's ability to manipulate controls may be degraded under normal operating conditions when display–control relations are incompatible or inconsistent with population stereotypes. Errors can be minimized by assigning control functions to be consistent with the stereotypes. Table 13.2 summarizes some of the recommended relations between control actions and functions. In emergencies, control actions are more automatic and less deliberate, and stereotypic response tendencies tend to emerge. Loveless (1962) describes a case where the ram of a heavy hydraulic press was raised by pushing a lever down. When an emergency occurred that required the ram to be raised, the ram operator mistakenly made the more stereotypic response of pulling the lever up, which caused the ram to move down and destroy the press. The moral of this story is that it always is best to use display–control relationships that are highly consistent with population stereotypes.

TABLE 13.2

Recommended Control Movements

Control Function	Response Outcome
On	Up, right, forward, pull
Off	Down, left, rearward, push
Right	Clockwise, right
Left	Counterclockwise, left
Up	Up, rearward
Down	Down, forward
Retract	Rearward, pull, counterclockwise, up
Extend	Forward, push, clockwise, down
Increase	Right, up, forward
Decrease	Left, down, rearward

SUMMARY

Response selection is a critical aspect of human performance. The operator of a human–machine system is faced with displays of information that indicate specific actions that he or she needs to take. In many systems, the time with which these response-selection decisions are made is crucial, as is their accuracy. The relative speed and accuracy of responding in a particular situation will be influenced by the threshold used to evaluate the accumulating information. With a high threshold, responses will be slow but accurate, whereas with a low threshold, they will be fast but inaccurate.

The efficiency of response selection is affected by many factors. These include the number of possible stimuli, the number of possible responses, the interrelationships between stimuli and responses, and the amount of practice a person has had in making responses. Moreover, many limitations in the performance of simultaneous multiple tasks can be traced to the response-selection stage. Probably the most important factor in response-selection efficiency is the compatibility of stimuli and responses. Principles of compatibility can be applied to ensure that the easiest or most natural control actions are required in response to displayed information.

When manipulating objects in the environment, the operator has a range of alternative actions for accomplishing a goal. Information about the objects to be grasped and the resulting postures for the limbs is involved in the selection of any particular action. In the next chapter, we will examine the way that actions are controlled.

RECOMMENDED READINGS

Hommel, B., Brown, B. R. E., & Nattkemper, D. (2016). *Human action control: From intentions to movements*. Switzerland: Springer.

Hommel, B., & Prinz, W. (Eds.) (1997). *Theoretical issues in stimulus-response compatibility*. Amsterdam: North-Holland.

Newell, A., & Rosenbloom, P. S. (1981). *Mechanisms of skill acquisition and the law of practice*. In J. R. Anderson (Eds.), *Cognitive skills and their acquisition* (pp. 1–55). Hillsdale, NJ: Lawrence Erlbaum.

Pachella, R. G. (1974). *The interpretation of reaction time in information-processing research*. In B. H. Kantowitz (Ed.), *Human information processing: Tutorials in performance and cognition* (pp. 41–82). Hillsdale, NJ: Lawrence Erlbaum.

Proctor, R. W., & Vu, K.-P. L. (2006). *Stimulus-response compatibility principles: Data, theory, and application*. Boca Raton, FL: CRC Press.

Sanders, A. F. (1998). *Elements of human performance: Reaction processes and attention in human skill*. Mahwah, NJ: Lawrence Erlbaum.

14 Control of Movement and Learning of Motor Skill

Rather than viewing perceptual-motor behavior as a series of motor responses made to reach some goal, it is possible, and I believe considerably more profitable, to view such behavior as an information-processing activity guided by some general plan or program.

P. M. Fitts
1964

INTRODUCTION

Any interaction between a person and a machine or the natural environment ultimately requires the person to execute a motor response—to move his or her body. This movement can be as simple as pushing a button or as complex as the coordinated actions required to perform heart surgery or to operate heavy machinery. In all cases, the person must not only perceive information correctly, make suitable decisions, and select appropriate responses, but also successfully carry out the intended actions. Often, the limiting factor in performance will be the speed and precision with which these actions can be executed. Because motor control is a major component of many tasks, the human factors specialist must understand the ways that simple and complex movements of various types are planned and executed.

Every movement requires the cooperation of different muscle groups and the neural mechanisms that control them. The limbs that will execute the action (the effectors) must be selected and prepared, sequences of movements must be timed and coordinated, and the final movements must be executed with the right force and speed to accomplish the goal. The role of the nervous system is to activate the proper muscle groups in a precise order and use feedback from the various senses to coordinate and modify ongoing movements, maintain posture, and plan future actions.

Think for a moment about the skills required to ride a bicycle. Maintaining balance is an important part of the process. Your legs must pedal with force sufficient to attain the speed you want. You need to be able to steer the bicycle with your arms, and brake (with your hands or legs) when necessary. The coordinated performance of all of these actions requires constant monitoring of proprioceptive, visual, and vestibular feedback. When you first tried to ride a bicycle, it probably seemed as if it was impossible to do all of these things at the same time. Yet, most people learn to ride a bicycle very quickly. In this chapter, we will discuss the principles underlying the control of movement and how complex motor skills are learned. Most research in these areas is conducted from one of two perspectives: cognitive science and human information processing, or ecological psychology and dynamical systems (Rosenbaum, Augustyn, Cohen, & Jax, 2006). Our emphasis in this chapter will be on the cognitive science and human information processing approach, but you should keep in mind that the two approaches are complementary rather than antagonistic (Anson, Elliott, & Davids, 2005).

PHYSIOLOGICAL FOUNDATIONS OF MOVEMENT

In Chapters 5 and 7, when we introduced the visual and auditory perceptual systems, we provided a brief description of the sensory structures that underlie perception. We will do the same thing in this chapter for the motor system. We will first discuss how the human body is engineered for movement and those structures in the nervous system that are involved in movement.

THE MUSCULOSKELETAL SYSTEM

There are 200 bones in the adult human skeleton that provide support for the body. The bones are joined by connective tissues called ligaments, and similar tissues, called tendons, connect muscles to the bones. Movement is accomplished by muscular contractions acting on the bones. All movement takes place through the changes that muscles make on the joint angles.

Movements at some joints occur only in two dimensions and are said to involve one degree of freedom (movement in a single plane). The movement of the forearm in relation to the upper arm at the elbow is of this type. Other movements can occur in more dimensions and involve multiple degrees of freedom. For example, movement of the upper arm at the shoulder involves three degrees of freedom (left-right, up-down, and rotation). Because most limb movements involve motion at multiple joints, there are many different combinations of joint motions, and corresponding trajectories, that could move a limb from one location to another. Yet, somehow, the motor system constrains the degrees of freedom to arrive at a single, smoothly executed trajectory. The question of how this is accomplished is called the *degrees of freedom* problem (Rosenbaum, 2010). Some of the constraints used by the motor system are biomechanical, such as the range of motion of a particular joint or the stiffness of a particular muscle, whereas other constraints are imposed by cognitive processes that coordinate action.

Muscles are arranged in groups with opposite actions. One group (the agonist) engages in flexion, and its opposing group (the antagonist) engages in extension. Movement at the joints is controlled by the coupled actions of the agonists and the antagonists. A muscle contracts when it receives a signal from motor neurons. We will now discuss various mechanisms involved in the control of movement.

CONTROL OF MOVEMENT

The nervous system controls movements through a hierarchy of mechanisms. At the lowest levels, the elastic properties of the muscles themselves and motor neurons control muscle contraction. At higher levels, the central nervous system controls movement through the spinal cord. We will examine these different levels of control, beginning with the physical properties of muscles and bones, and motor neurons, and ending with the higher-level organization and execution of action mediated by the brain.

Mass-Spring Property and Motor Unit

A convenient way to describe the behavior of the muscle is as a *mass-spring system* (Bernstein, 1967; de Lussanet, Smeets, & Brenner, 2002). The stretchy muscle can be thought of as a spring attached to the bone. Every spring has an equilibrium point, or resting length, the length to which it stretches when a mass is attached to it. When the spring is moved from its equilibrium point and then released, it will return to its resting length. Leg stiffness, a predictor of athletic performance and injury risk, can be modeled effectively with a simple mass-spring system (Brauner, Sterzing, Wulf, & Horstmann, 2014).

The resting amount of flexion and extension of the agonist and antagonist muscle groups at each joint, and hence the resting joint angle, is determined by the muscles' equilibrium points. When some external force changes the joint angle, the muscles in one group will be stretched, and the muscles in the other will be compressed. When the force is removed, the muscles will return to their equilibrium points, and the joint angle will return to its original position. This is the most basic level of movement, in that it can occur in the absence of any neural signal. However, more complex movements (controlled by higher levels of the nervous system) may also rely on the mass-spring properties of muscles. In some situations, it may be useful to talk about movement, such as elbow flexions, as being accomplished by a systematic change in the stiffness, and hence the equilibrium point, of each muscle. The elastic properties of the muscles cause the position of the limb to change in accord with the changing equilibrium points.

Change in muscle stiffness, caused by contraction of the muscle, is a result of signals from motor neurons. Each muscle is composed of muscle fibers, which are innervated by hundreds of motor neurons. An individual motor neuron innervates many muscle fibers spread throughout a muscle; the neuron and fibers together are called a *motor unit*. When the motor neuron "fires," all of the muscle fibers within the unit will be affected. The motor unit is considered to be the smallest unit of motor control. The contraction of a single muscle will be determined by which and how many motor units are activated at the same time (e.g., Gielen, van Bolhuis, & Vrijenhoek, 1998).

Although motor unit activity may seem far removed from ergonomic concerns, a study by Zennaro et al. (2004) illustrates that this is not the case. They measured surface electromyograms (EMGs; an overall measure of muscle activity) and the intramuscular motor unit activity of a shoulder muscle while participants tapped a key for 5 minutes on a desktop. The desktop was either an appropriate height (where the upper arm could hang relaxed, with the lower arm parallel to the desktop) or 5 cm above that height. EMG activity was greater for the too-high desktop than for the one at the appropriate height. This greater EMG was due mainly to individual motor units being active for longer periods of time, rather than to an increased number of active motor units. This study demonstrates that ergonomic issues can be evaluated at the motor unit level and that incorrectly adjusted office equipment can lead to prolonged activity of the motor units, which in turn may result in musculoskeletal disorders (see Chapter 16).

Spinal Control

The spinal cord controls certain actions by way of *spinal reflexes* (Abernethy et al., 2013; Bonnet, Decety, Jeannerod, & Requin, 1997). Such reflexes begin with stimulation of the sensory receptors that provide information about limb position (proprioception). Proprioceptive receptors are located within the muscles, tendons, joints, and skin. Their signals are sent to the spinal cord, where a motor signal is quickly evoked and sent to the appropriate muscles. Spinal reflexes allow movements to be made within milliseconds of the initiating stimulus. For example, when a sensory neuron receives a painful signal, the limb withdrawal reflex will cause the appropriate muscle to contract, removing the limb from the source of the pain. This happens very quickly because the signal does not have to travel all the way to the brain and then back again.

Spinal control of movement is not limited to the initiation of reflex responses. Both spinal reflexes and the higher central nervous system contribute to maintenance of posture (Tanaka, 2015). Gait and other movement patterns, though initiated by the brain, are controlled by the spinal cord once they are initiated (Grillner, 1975; Pearson & Gordon, 2000). There is even some evidence that the spinal cord can perform complex controlling operations (Schmidt & Lee, 2011). The sophisticated information-processing capabilities of the spinal cord free the brain so that it can engage in other activities.

Control by the Brain

After the spinal cord, the next three structures critical to motor control are the brainstem, cerebellum, and basal ganglia (see Figure 14.1). The brainstem controls movements of the head and face, respiration and heart rate, and in part controls movement of the eyes.

The cerebellum is involved in several aspects of motor control (Rosenbaum, 2010). These include the maintenance of muscle tone and balance, the coordination and timing of rapid action sequences, and the planning and execution of movement. The cerebellum may help plan and initiate movements, but without input from sensory feedback (Bastian, 2006).

The basal ganglia are low-level brain structures that are also involved in the planning of movements. Evidence suggests that the basal ganglia form an action-selection circuit that helps choose between actions that use common motor pathways (Humphries, Stewart, & Gurney, 2006). The

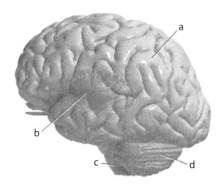

FIGURE 14.1   The cerebrum (a), basal ganglia (b), brain stem (c), and cerebellum (d).

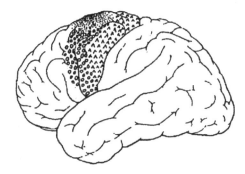

FIGURE 14.2   The motor (^), premotor (°), and supplementary motor (·) cortices.

basal ganglia also control the size or amplitude of movements and integrate perceptual and motor information. They control slow, smooth movements, such as those required for postural adjustments and those requiring the continual application of force. Motor learning requires both the cerebellum and the basal ganglia (Doya, 2000).

After the brainstem, cerebellum, and basal ganglia, the highest levels of motor control are located in the cortex. The motor, premotor, and supplementary motor cortices are located in adjacent regions in the rear part of the frontal lobes (see Figure 14.2). The relationships between these different motor areas are very complicated and not yet well understood (Pockett, 2006), although we are learning more about them every day (Sosnik et al., 2014). All the areas communicate with each other to greater or lesser degrees.

We know that the motor cortex is structured as a topographic map that takes the form of a "homunculus," or little man (see Figure 14.3). It is involved in the initiation of voluntary movements. The premotor cortex controls movement of the trunk and shoulders. It is also implicated in the integration of visual and motor information, and we believe that it plays a role in turning the body to prepare for forthcoming movements. The supplementary motor cortex is involved in the planning and execution of skilled movement sequences. It is different from the premotor cortex in that its activity does not seem to depend on perceptual information.

CONTROL OF ACTION

Motor performance is controlled by cognitive processes. Our understanding of these processes is based primarily on measures of human performance under a variety of conditions. In particular, we want to understand how movements are selected and controlled, and how perception and action are related (Rosenbaum, 2005).

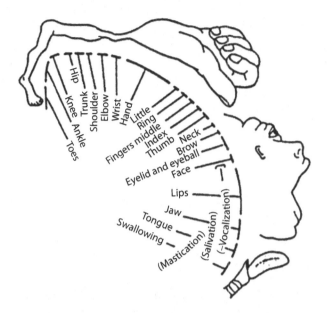

FIGURE 14.3 The homunculus on the motor cortex.

Movements are made to accomplish a task. Different kinds of tasks will require different kinds of movements, which in turn will require different cognitive processes. Some tasks, such as throwing a ball or pressing a key, are discrete, because the action has a distinct beginning and ending. In contrast, tasks such as steering a car are continuous. Some tasks, such as those that might be performed on an assembly line, involve a series of discrete actions and fall in between. To understand the processes involved in motor control, we need to understand the demands placed on the cognitive system, which may not be the same across these different kinds of tasks.

We also need to distinguish between open and closed motor skills (Poulton, 1957). Open skills are performed in dynamic environments, and the speed and timing of the movements tend to be determined by events occurring in the environment. Closed skills are performed in static environments and are self-paced. Football and basketball are sports that require mainly open skills, whereas gymnastics and track and field require primarily closed skills. Because open skills are paced by the environment, they require rapid adaptations to the environment, which closed skills do not. When athletes are asked to perform tasks unrelated to their sport, athletes in open-skills sports are more influenced by environmental factors than are those in closed-skills sports (Liu, 2003).

The type of task, whether discrete or continuous, and the type of skill, whether open or closed, will interact to determine how movements will be controlled. Movement control is referred to as *open-* or *closed-loop*, terms that refer to the extent to which feedback is used in the performance of the action (Heath, Rival, Westwood, & Neely, 2005). Do not confuse the kind of control with open and closed skills, because open skills actually use closed- more than open-loop control, and closed skills use open- more than closed-loop control.

Closed-Loop Control

Recall from Chapter 3 that closed-loop systems are characterized by the use of negative feedback to regulate their outputs. To apply this idea to motor control, we assume that a person knows what movement needs to be made and generates a mental representation of the desired movement (see Figure 14.4). During the action, sensory feedback produced as the movement is executed is compared with this representation. A person will perceive as error any difference between the actual and desired positions of her limb. The person will then use this error to select corrective movements

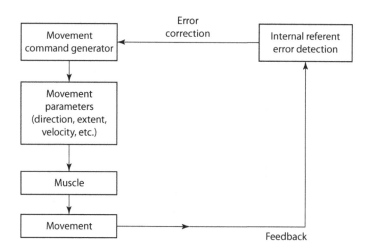

FIGURE 14.4   A closed-loop system for motor control.

to bring her actual limb position closer to the desired one. She continuously makes comparisons between her actual and desired positions until the difference between them is minimized.

Closed-loop control depends on the sensory feedback produced as a person acts on the environment. What sources of feedback are available? As Smetacek and Mechsner (2004) state, "Our daily doings are coordinated and run by a trinity of independent sensory systems: proprioception, vision, and the vestibular organs of the inner ear" (p. 21). Earlier in this chapter we mentioned the role of proprioception in reflexes, but Smetacek and Mechsner emphasize that "all purposeful movements, both conscious and unconscious, are controlled by proprioception" (p. 21). As a person moves his hand toward an object, visual feedback provides information about the locations of both his hand and the object that can be used to take corrective actions. The vestibular sense provides feedback about his posture and head orientation. Feedback is also provided by touch (Niederberger & Gerber, 1999) and audition (Winstein & Schmidt, 1989). For example, a person walking down carpeted stairs can determine when the uncarpeted floor has been reached by the feedback conveyed by changes in both touch and sound.

OPEN-LOOP CONTROL

In contrast to closed-loop control, open-loop control does not depend on feedback (see Figure 14.5). Movements subject to open-loop control are either too rapid to allow modification from feedback or

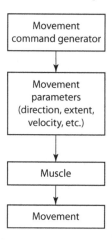

FIGURE 14.5   An open-loop system for motor control.

overlearned. Open-loop control is achieved by developing a sequence of movement instructions generated by a general mental representation called a *motor program*. The selection of a motor program and generation of a movement sequence occur prior to the initiation of the movement.

The idea of a motor program was first formalized by Keele (1968) and later developed and extended by Schmidt (1975). A motor program is an abstract plan for controlling a specific class of movements. To execute a particular movement within a class, parameters such as the muscles to be used, their order, and the force, duration, and timing of their contractions must be specified. In the rest of this section, we present certain implications of the motor program idea and describe the invariant characteristics of a motor program, its modular, hierarchical structure, and the role of feedback.

Implications

The motor program concept has several implications. First, because the motor program contains a general template, or schema, of the desired movements, coordinated movement should be possible when feedback is not available. Studies have demonstrated that monkeys and humans who cannot process proprioceptive feedback (because of surgical intervention or disease) can still perform some skilled actions, such as grasping, walking, and running (Bizzi & Abend, 1983; Rothwell et al., 1982; Taub & Berman, 1968). These findings are consistent with the motor program concept.

The motor program concept also implies that rapid movements can be made accurately even when the transmission and processing of sensory feedback take more time to accomplish than the movement itself. These kinds of movements are made during keyboarding, in which keystrokes are made very rapidly (Lashley, 1951; see Box 14.1). Based on the speed of the keystrokes, it makes sense to assume that the keystroke movements are programmed in advance. Another implication of the motor program concept is that it should take longer to program more complex movements. It is true that response times to initiate a movement are longer for more complex movements (Henry & Rogers, 1960; Klapp, 1977). Thus, the speed with which a person can react to a stimulus is directly related to the complexity of the movements that must follow.

Invariant Characteristics

Motor programs have invariant characteristics that specify critical structural aspects of the class of movements that a particular program controls. The invariant characteristics are independent of the particular muscles used to execute the movement. These characteristics may include the order of movement components, the relative amount of time that each component takes, and the way in which force is distributed to each component.

Evidence for invariant characteristics can be seen in writing samples produced by different muscle groups (Merton, 1972; see Figure 14.6). For example, you can see similarities between a writing sample produced by arm and shoulder muscles (as on a blackboard) and a writing sample produced by hand and finger muscles (as on a piece of paper), as long as both samples are produced with the dominant hand. Writing produced with the nondominant hand is also similar to writing produced with the dominant hand, suggesting that many aspects of the motor program for writing are independent of the effectors used for execution (Lindemann & Wright, 1998).

Modular Organization

Considerable evidence suggests that motor programs are composed of several modules that reflect the invariant characteristics of a movement. For example, timing of movement seems to involve an independent module or component of control (Keele, Cohen, & Ivry, 1990). The evidence for this comes from studies in which people are asked to produce timed movements with different parts of their bodies. Not everyone can time their movements accurately, and these differences across people show up as differences in the variance of movement times. People who are less accurate have larger movement time variances. Moreover, the extent of variation for a person will be similar regardless of the effector he or she uses to make the movement.

BOX 14.1   KEYBOARD ENTRY AND TYPING

Typing is the most common way that people interact with computers and other machines. Consequently, it is a topic of considerable interest to the field of human factors and human–computer interaction. Moreover, it is easy to find typists at every level of skill, so typing can be easily studied in the laboratory (Salthouse, 1986).

The typewriter was first marketed in 1874 (Cooper, 1983). Early typists used a "hunt-and-peck" method with only two fingers from each hand, and they had to look at the keyboard at all times. In 1888, Frank McGurrin demonstrated that touch typing was much faster by competing with another typist who was skilled in the hunt-and-peck method. His well-publicized victory in that contest led to the gradual adoption of the touch-typing method over the next decade.

Salthouse (1984) proposed that to type requires the typist to perform four processes. First, the typist must read the text and convert it into "chunks." The typist then decomposes these chunks into strings of characters to be typed. He or she must then convert the characters into movement specifications (motor programs) and implement these (ballistic) movements. Therefore, we can see that skilled typing requires perceptual, cognitive, and motor processes.

Skilled typists are fast. On average, a professional typist can type 60 words per minute, or approximately 5 keystrokes per second (a median interstroke interval of 200 ms). World champion typists can type as fast as 200 words per minute, with a median interstroke interval of 60 ms. These speeds are well below the minimum time for a choice reaction, but skilled typists make choice reactions at about the same speed as everyone else. For example, Salthouse (1984) showed that typists who could type from printed copy with an average of 177 ms between successive keystrokes needed about 560 ms between keystrokes when performing a serial, two-alternative choice reaction task in which the response to one stimulus triggered the onset of the next. Such results suggest that typists do not prepare to type each letter one at a time, but instead, prepare chunks of letters and their keystrokes together.

Typists seem to encode and prepare whole words for typing rather than individual letters. The word, therefore, seems to be the smallest unit upon which a chunk is based. In support of this point, typists cannot type as well when the text is changed from words to chunks of random letters (e.g., Shaffer & Hardwick, 1968; West & Sabban, 1982). Also, typists can type strings of random words just as quickly as meaningful text. Because the semantic and syntactic context provided by meaningful text fails to facilitate typing, we can conclude that typists perform no cognitive processing any more complicated than recognition of the words. For skilled typists, each word may specify a motor program that controls the execution of the component keystrokes (Rumelhart & Norman, 1982).

Both fast and slow typists make mistakes. Consistently with the conclusion that typing uses representations at the level of words, most errors are misspelled words. Misspellings can arise in one of four ways: letter substitutions (*work* for *word*), letter intrusions (*worrd* for *word*), letter omissions (*wrd* for *word*), and letter transpositions (*wrod* for *word*). All of these errors seem to be based in the movement-related translation and execution processes (Salthouse, 1986).

Substitution and intrusion errors usually arise when the incorrect key is adjacent to the correct keys, which suggests that the source of these errors is faulty movement specifications or mispositioning of the hands. When a typist makes an omission error (*wrd* for *word*), the time between the keystroke preceding the omission (w) and the following keystroke (r) is approximately twice the normal keystroke time. This suggests that the typist attempted to strike the missing letter (o) but did not press the key hard enough. Transposition errors usually happen when the adjacent letters are typed with fingers on each hand, so we might assume that they arise from errors in timing the two keystrokes.

Other than becoming faster overall, how else does typing performance change as skill is acquired? Certain kinds of movements speed up, but not all. For example, typing digraphs (pairs of letters) either with two hands or with two fingers on the same hand gets much faster (Gentner, 1983). For novices, typing digraphs with two fingers is more difficult (slower) than typing one-finger doubles, in which the same key is pressed twice in a row with the same finger. However, for skilled typists, this pattern is reversed. This is because typing skill involves coordinating rapid *parallel* movements of the fingers (e.g., Rumelhart & Norman, 1982). Because movements involving two fingers can be prepared in parallel, skilled typists can type digraphs with two fingers much faster than digraphs with the same finger.

Skilled typists also type digraphs that use two fingers on different hands much faster than digraphs using two fingers on the same hand. This difference can be explained in terms of biomechanical constraints on the movements. Specifically, the simultaneous coordination of fore-aft and lateral finger movements on the same hand is difficult (Schmuckler & Bosman, 1997).

For many different kinds of digraphs, there are differences in the speed with which they can be typed that cannot be accounted for by physical difficulty alone (Gentner, Larochelle, & Grudin, 1988). Difficulty is determined by digraph frequency, word frequency, and syllable boundaries with words. The combined effect of these factors is similar in magnitude to those based on physical constraints.

The most important changes that occur during the acquisition of typing skill are more efficient translation of characters to movements, and more efficient execution and coordination of the movements of successive keystrokes. These improvements in response selection and control are accompanied by perceptual changes that increase the span for encoding the written material. As Ericsson (2006) concluded, "In sum, the superior speed of reactions by expert performers, such as typists and athletes, appears to depend primarily on cognitive representations mediating skilled *anticipation*" (p. 697).

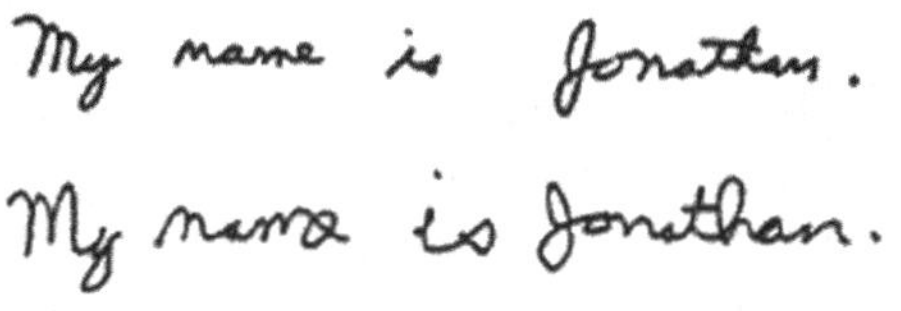

FIGURE 14.6 Writing samples produced using the hand (on a piece of paper (top)) and the arm (at a blackboard (bottom)). We thank Bob Hines for contributing these samples.

This correlation between variances across parts of the body occurs mostly in tasks for which movement timing is especially salient, such as tapping in beat to a metronome (Ivry, Spencer, Zelaznik, & Diedrichsen, 2002; Zelaznik et al., 2005). Other tasks may also put a strong premium on movement time, but the movement trajectory (or some other feature of the movement) may be more salient. For example, some researchers have asked people to draw circles at a timed rate (e.g., Ivry et al., 2002). The variability of movement time in these tasks is not related to the variability of movement time for tapping tasks. Thus, when movement timing is not as salient as the trajectory of the movement, the timing of the movements seems to arise from the trajectory control process itself.

Hierarchical Arrangement

Timing and trajectory control are examples of high-level modules of a motor program. Motor programs are hierarchical, in that high-level modules pass control to lower-level modules. Evidence for hierarchical control comes from studies of tapping. For example, one study asked people to tap their fingers in response to sequences of six numbers (Povel & Collard, 1982). The sequence 123321 was used to indicate the order for three fingers, labeled 1, 2, and 3, respectively. Regardless of the fingers

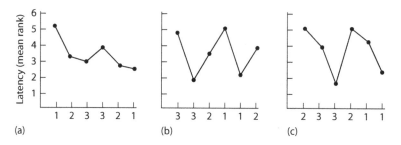

FIGURE 14.7 Mean latency of intertap intervals for the sequences 123321 (a), 332112 (b), and 233211 (c).

used, the first and fourth response latencies were longer than the others (see Figure 14.7). This finding suggests that a hierarchical program was executed to perform the sequence, with the top level passing control to the first response subgroup (123) and then to the second response subgroup (321). The longer response latencies for the first and fourth taps reflect the time required to shift control to the next subgroup.

Role of Feedback

Although the motor program idea allows movement to be executed without feedback, feedback is still assumed to contribute to the control of action in several ways. First, information fed back through the senses specifies the position of the body part that is to be moved and where it is to be moved. Without this information, it would not be possible to select the appropriate parameters for the motor program. Second, for slow movements, feedback can be used to correct the parameters of an ongoing program, or a new program can be selected, as appropriate. Third, rapid corrections of movements based on proprioceptive and visual feedback can occur in less than 100 ms in some situations (Saunders & Knill, 2004).

AIMED MOVEMENTS

Aimed movements are those that require an arm or some other part of the body to be moved to a target location. The movement made by an operator to bring a finger to a pushbutton is an example, as is moving the foot from the accelerator pedal to the brake pedal in an automobile. The speed and accuracy with which such movements can be made are influenced by many factors, including the effector used, the distance of the movement, and the presence or absence of visual feedback. To ensure that an operator's movements will be made within the necessary speed and accuracy limits, a designer must consider the way in which aimed movements are controlled.

Aimed movements were first studied by Woodworth (1899), who was interested in the amount of time needed to use visual feedback. To investigate this issue, Woodworth used a task in which people repeatedly drew lines on a roll of paper moving through a vertical slot in a table top. The lines were to be of a specified length, with the rate of movement set by a metronome that beat from 20 to 200 times each minute. One complete movement cycle (up and down) was to be made for each beat. The role of visual feedback was evaluated by having people perform with their eyes open or closed. Movement accuracy was similar for the two conditions at rates of 140 cycles/min or greater; thus, visual feedback had little or no effect on performance. However, at rates of less than 140 cycles/min or less, the eyes-open condition yielded better performance. This result suggests that the minimum time required to process visual feedback is longer than $(60 \text{ s}/140 \text{ cycles}) \times 1000 \text{ ms/s} = 428.6 \text{ ms/cycle}$ (although modern estimates of this time are much shorter; see below).

To explain these and other results, Woodworth (1899) proposed that rapid aimed movements have two phases, which he called *initial adjustment* and *current control*, and which correspond to open- and closed-loop modes, respectively. The initial phase transports the body part toward the target location. The second phase uses sensory feedback to correct errors in accuracy and home in on the target. Elliott, Helsen, and Chua (2001) noted, in a review of the impact of Woodworth's research on the study of goal-directed arm movements, "his theoretical contribution to the understanding of speed-accuracy relations and the control of rapid goal-directed movements that has stood the test of time" (p. 354).

Not much research was conducted during the first half of the 20th century to follow up Woodworth's (1899) work. However, since the 1950s, aimed movements have been examined extensively. In addition to the time-matching task used by Woodworth, a time-minimization task has been widely used (Meyer, Smith, Kornblum, Abrams, & Wright, 1990), in which people are to minimize their movement times while approximating a specified target accuracy value. For both tasks, single movements as well as repetitive movements have been examined.

Fitts's Law

Fitts (1954) established a fundamental relation between aimed movement time and the variables of distance and precision that has come to be known as *Fitts's law*. He examined performance in a repetitive tapping task, in which a person was required to move a stylus back and forth between two target locations as quickly as possible. As the distance between the targets increased, movement time increased. Conversely, as the widths of the targets increased, movement time decreased. Because the width of the targets dictates the precision of the movements, this relation can be viewed as a speed–accuracy tradeoff (see Chapter 13).

From these relations, Fitts (1954) defined the *index of difficulty* (*I*) for an aimed movement as

$$I = \log_2(2D / W),$$

where:

    D   is the center-to-center distance between the targets, and
    W   is the width of the targets.

Fitts found this index to be related to movement time (MT) by the linear equation

$$\text{MT} = a + bI,$$

where *a* and *b* are constants (see Figure 14.8). According to Fitts's law, when the distance required for a movement is doubled, the movement time will not change if the target width is also doubled.

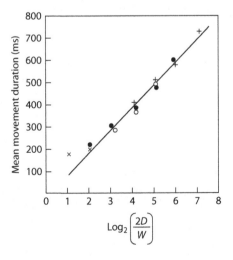

FIGURE 14.8 Movement time as a function of the index of difficulty.

Fitts's law applies across a wide range of tasks. Fitts (1954) obtained similar results with tasks that required washers to be placed onto pegs or pins into holes. Other researchers have found the law to hold in tasks as diverse as angular positioning of the wrist (Crossman & Goodeve, 1963/1983), arm extension (Kerr & Langolf, 1977), positioning a cursor with a joystick or a head-movement controller (Jagacinski & Monk, 1985), working with tweezers under a microscope (Langolf & Hancock, 1975), and making aimed movements underwater (Kerr, 1973). Fitts's law is also important in the study of human–computer interaction, where many tasks require moving a cursor to a target location (Seow, 2005), and for interacting with a mobile device such as a tablet or a smartphone (Alexander, Schlick, Sievert, & Leyk, 2008).

In 2004, a special issue of the *International Journal of Human-Computer Studies* was devoted to the 50th anniversary of Fitts's original study. In that issue, the editors emphasized, "Fitts' law has ... [made] it possible to predict reliably the minimum time for a person in a pointing task to reach a specified target" (Guiard & Beaudouin-Lafon, 2004, p. 747). Newell (1990) characterized the generality of Fitts's law as follows: "Fitts' law is extremely robust .... Indeed, it is so neat, it is surprising that it doesn't show up in every first-year textbook as a paradigm example of a psychological quantitative law" (p. 3).

There have been many explanations of Fitts's law since 1954. The most widely accepted explanation is the *optimized initial impulse model* (Meyer, Abrams, Kornblum, Wright, & Smith, 1988). An aimed movement toward a specified target location is presumed to involve a primary submovement and an optional secondary submovement that is made if the initial submovement is "off target." The submovements are programmed to minimize the average time for the total movement.

The idea that aimed movements are controlled by executing a single corrective submovement has become a central idea in motor control (Hoffmann, 2016). The evidence supporting this idea includes the facts that movements rarely show more than two submovements and that the time to execute these submovements is constrained by Fitts's law. It is interesting to note that the optimized impulse model is consistent with Woodworth's (1899) original proposal that movements are controlled in two phases.

Application

Fitts's law is used to evaluate the efficiency of different movements in a wide variety of real-world situations. Efficiency is measured as the slope (b) of the function relating movement time to the index of difficulty. This measure of efficiency can be used to evaluate different workspace designs. For example, Wiker, Langolf, and Chaffin (1989) noted that many manual assembly tasks require people to use hand tools raised above their shoulders. Wiker and his colleagues examined people's ability to make repetitive movements of a stylus to a hole with their hands raised to different positions (-15 to $+60°$ relative to shoulder level; see Figure 14.9). Movement times were longer (by 20%) when people performed the task at the highest position (compared with the lowest). Wiker and colleagues attributed the longer movement times to the increased tension in the muscles needed to raise the hand. They recommended that sustained manual activity be restricted to below shoulder level when possible.

Another example involves assistive technology devices such as chin, head, and mouth sticks, which are used by people with limited mobility to press keys on a computer keyboard. Andres and Hartung (1989) asked people to tap between targets of varying width and separation with a chin stick. Fitts's law still held, but the mean information transmission rate was 7 bits/s. This value is considerably less than what we usually see for hand or foot movements. While part of the reason for the lower transmission rate is due to how neck and shoulder muscles are controlled, part of the problem is in the design of the stick.

Baird, Hoffmann, and Drury (2002) asked people to perform aimed movements with hand-held pointers of different lengths. The longer the pointer, the longer the movement time was for the same index of difficulty. The ends of longer pointers "wiggle" more, because the pointer amplifies small

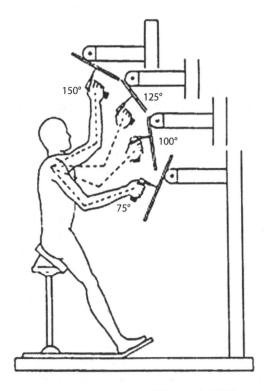

FIGURE 14.9   Task postures tested in an experiment by Wiker et al. (1989).

muscle tremors. This wiggling makes it more difficult for people to bring the end of the pointer to a target. This finding suggests that, for assistive technology devices like sticks and for other hand-held tools like screwdrivers and soldering irons, the shorter the length of the tool, the easier it will be to bring the tip of the tool to the object of interest. When the tool cannot be shortened, then the size of the object (the target area) must be increased to compensate.

Visual Feedback

Another issue in the control of aimed movements is the role of visual feedback. Remember that Woodworth (1899) estimated the time to process visual feedback to be 450 ms, because people benefited from having their eyes open only for slow movements. This estimate seems quite long, because we know that people can make accurate choices based on visual information in half that time.

Since Woodworth's (1899) original experiment, this question has been revisited several times (Keele & Posner, 1968; Zelaznik, Hawkins, & Kisselburgh, 1983). We now know that the time to process visual feedback depends on the type of task to be performed and whether or not people know in advance that feedback will be available. For example, Zelaznik and his colleagues asked people to make timed, aimed movements toward a target, and on some trials they turned the lights off at the beginning of the movement. When people knew that the lights were going to be turned off, they were able to accurately aim their movements with as little as 100 ms preparation time. In a similar study, when the target of an aimed movement jumped to a different position during the movement, people were able to resolve the mismatch between the movement and the new perceived target position within 100 ms (Dimitriou, Wolpert, & Franklin, 2013). The fact that the time for visual processing in aimed movements can be quite short means that even very rapid movements may be more accurate if visual feedback is available.

Bimanual Control

The experiments investigating manual control that we have discussed up to this point have allowed people to make only single aimed hand movements. Natural body movements usually involve coordination of several limbs. Some tasks, such as light assembly, require people to make two different aimed movements at once, one with each hand. It is not hard to think of situations in which these two different movements would have different indices of difficulty. However, Fitts's law does not apply to each limb separately.

Kelso, Southard, and Goodman (1979) asked people to perform two movements simultaneously, for which the indices of difficulty differed, one with each hand. The right hand moved to a close, large target (a movement with a low index of difficulty) and the left hand to a distant, small target (a movement with a high index of difficulty). If the two hands moved independently, the right-hand movement should have taken less time than the left-hand movement, because the index of difficulty was lower. Instead, people moved their hands at different speeds, so that both lifted off and reached the targets simultaneously. People also accelerated and decelerated the two movements at the same times. Overall, the time to make both movements was approximately equal to the time to make the more difficult movement when it was performed alone. Thus, the easier movement was coupled to the harder movement.

We can explain why movements are coupled like this by referring back to the motor program. If the program is responsible for executing classes of similar movements, then both movements may require the same program (Schmidt et al., 1979). Movement characteristics like velocity and distance to be traveled can be determined separately for each limb. If similar hand movements are controlled by the same motor program, then we would expect that a hurdle placed in the path of one limb would produce a change in the movement paths for both limbs, and this is what we see (see Figure 14.10; Kelso, Putnam, & Goodman, 1983). Also, with practice, people can learn to move two limbs independently. Schmidt et al. (1998) proposed that, in this case, the motor program is modified so that it allows for different trajectories for the limbs involved in the coordinated movement.

These experiments show that the index of difficulty does not have to be equal for different controls (e.g., pushbuttons) when designing a control panel. When these controls must be activated simultaneously, the operator will be able to coordinate his or her movements without too much difficulty. However, although bimanual movements seem to be controlled by a single motor program, they cannot be completely synchronized. Warrick and Turner (1963) had people hold down keys with their left and right index fingers, which were to be released simultaneously at any time after a "ready" signal was presented. The average difference between the release times of the two fingers was zero, so people were able to release their fingers virtually simultaneously. However, there was a considerable amount of variability in the difference between the release times, so that one finger could precede the other by as much as 20 ms. For more complex tasks, Warrick and Turner worried that these differences could be even larger, which could create potential problems for the operation of equipment or machinery requiring near-simultaneous activation of left- and right-hand controls.

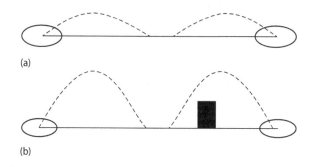

(a)

(b)

FIGURE 14.10 Bimanual movement without (a) and with (b) a hurdle.

GRASPING AND INTERCEPTING OBJECTS

Grasping is a fundamental component of actions as diverse as picking up a cup of coffee, grabbing a hammer, opening a door, or flipping a light switch. Movements that culminate in grasping an object can be broken into two components: a transport phase (reaching) and a grip formation phase (grasping). The transport component is very similar to the aimed movements we have been discussing and involves moving the hand to the object. The grasp component involves positioning the fingers for grabbing the object. The fingers of the hand gradually open to a maximum aperture and then close until the grip conforms to the object size (see Figure 14.11).

Researchers have been interested in the factors that influence the transport and grasp components, and how transport affects grasp, and vice versa. Most experiments have looked at movements when the object to be grasped changes size or location (e.g., Castiello, Bennett, & Stelmach, 1993; Paulignan, MacKenzie, Marteniuk, & Jeannerod, 1990). When the object is moved farther away, the transport component takes longer, but the grip component does not change. Regardless of the transport duration, the grip aperture reaches a maximum within the first 60%–70% of the movement (Jeannerod, 1981, 1984). However, the transport and grip components are not independent of each other. When an object unexpectedly changes shape or location after the reach begins, modifications of both the transport and grip components often occur together, suggesting that the components are coupled to each other (Rosenbaum, Meulenbroek, Vaughan, & Jansen, 2001; Schmidt & Lee, 2011).

While most experiments on reach have used stationary objects, people often need to grasp objects that are moving. For a moving object, not only do people need to reach toward the object in the same way as they would if it were stationary, but they also need to time their movements so that they grasp the object at an appropriate point in its trajectory. This may require movements not only of the arm and hand but also of the entire body. For example, an outfielder can catch a fly ball only after he or she estimates the ball's flight path (e.g., Whiting, 1969). The player must calculate the best location at which to intercept the ball, execute the movements to reach that location, and then execute the appropriately timed grasping movements to make the catch.

A variable important for determining where a moving object can be intercepted is based on how quickly the object's retinal image is growing. The inverse of the speed of growth, called *tau*, determines the time to contact with the object (Lee, 1976). If the image is growing very quickly, tau will be small, and "time-to-contact" will be short. If the image is growing very slowly, tau will be large, and time-to-contact will be long. When people are asked to estimate time to contact, tau is important, but so are other factors such as pictorial depth cues (Hecht & Salvesbergh, 2004; Lee, 1976).

OTHER ASPECTS OF MOTOR CONTROL

Human movements can be very simple, but are usually amazingly intricate and complex. We cannot do justice to all of the many aspects of motor control that are relevant to issues in human factors. However, before we leave this topic, there are three important aspects of motor control that we need to discuss briefly: posture, locomotion, and eye and head movements.

FIGURE 14.11   The grasping phase of movement.

Posture

Posture and balance control takes place mostly in the spinal cord and is maintained through closed-loop control. Adjustments to posture are made on the basis of information provided by the proprioceptive, vestibular, and visual senses. Some of the parameters that are controlled by the feedback loop include force, velocity, and distance of corrective movements. It is interesting that these parameters will eventually be modified under low-gravity conditions. After extended space flight, astronauts often have difficulty maintaining posture and balance (Cohen et al., 2012), apparently because their spinal neurons adapt to low gravity and are slow to re-adapt to normal gravity (Lackner, 1990).

Locomotion

Most people spend a great deal of time walking, or locomoting, through their environment. Locomotion occurs in a four-phase step cycle, shown in Figure 14.12. Like posture and balance, the step cycle is controlled by the spinal cord. Like posture and balance, the step cycle can be modified by information from the proprioceptive, vestibular, and visual systems. Visual feedback serves two important purposes in locomotion (Corlett, 1992). People use visual cues to plan routes from their current position to a desired location, and they use these cues during locomotion to initiate each step, which is then executed ballistically (or without modification from visual feedback; Matthis, Barton, & Fajen, 2015).

Eye and Head Movements

We have emphasized the importance of visual feedback in maintaining posture and balance and in locomotion. Delivering this information to motor control centers requires frequent and extensive eye and head movements to ensure a complete picture of the environment. For example, consider the eye movements a person must make to track a moving target. His smooth pursuit eye movements must match the velocity of the eye with that of the target, and if the target moves a significant distance, his head must move as well.

Eye movements are coordinated with head movements through the vestibulo-ocular reflex, which is triggered by rotation of the head or body while looking at an object. The eye will move in the direction opposite to the head, and so compensate for any change in the visual image caused by the head movement. This compensation is not very accurate when the head is turning very quickly (Pulaski, Zee, & Robinson, 1981). Furthermore, when a person fixates on a target image that rotates with the head, like the images in a helmet-mounted display, the vestibulo-ocular reflex is suppressed. Because the reflex is suppressed, the person will no longer be able to track objects in the environment well.

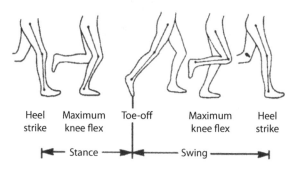

FIGURE 14.12 The step cycle.

MOTOR LEARNING

Part of understanding how actions are controlled involves understanding how people learn to make complex movements. Some of the questions addressed by researchers in this area include how movements are represented and retained in memory, what role feedback plays in the acquisition of motor skill, what kinds of practice and feedback result in optimal learning, and how motor skill relates to other skills. The answers to these questions have implications for the structuring of training programs and design of equipment for use in the workplace (Druckman & Swets, 1988; Druckman & Bjork, 1991; Schmidt & Bjork, 1992).

A lot of contemporary research in motor learning has been inspired by Schmidt's (1975) *schema theory*. At the heart of schema theory is the concept of the motor program. Recall that a motor program is an abstract plan for controlling a specific class of movements. Accurate performance requires not only that the appropriate motor program be selected, but that parameters such as force and timing be specified correctly. Two kinds of motor schemas act to determine these parameter values: recall and recognition schemas.

When a movement is to be made, the initial conditions (where a person is now) and outcome goal (where a person wants to be) are used by a recall schema to select the response parameters for a motor program. A recognition schema specifies the expected sensory consequences of the movement. The recall and recognition schemas are used in different ways for fast and slow movements. For fast movements, the recall schema both initiates and controls the movement. Then, after the movement is completed, perceived consequences can be compared against those expected from the recognition schema. Any mismatch between the two is used as the basis for correcting the recall schema. The recall schema initiates slow movements as well. However, comparison between sensory feedback and the sensation predicted by the recognition schema can occur during the movement, and correction to the movement parameters can be made as soon as an error is detected.

The schema theory assigns a prominent role to sensory feedback, incorporates motor programs, has two memory components (one involved in movement initiation and the other in evaluating feedback), considers feedback to be important during learning, and provides a way for error to be detected during or after execution of the movement. Though we know now that schema theory is not completely accurate (Shea & Wulf, 2005), these general features are central to contemporary views of motor-skill acquisition.

CONDITIONS AND SCHEDULES OF PRACTICE AND TRAINING

The term *practice* refers to repeated execution of a task with the goal of attaining mastery of that task. How a person practices a motor skill determines how quickly he or she will attain mastery, how long he or she will remember the skill, and the extent to which the skill will result in improved performance for other tasks. There are many ways that specific motor skills can be taught. Different methods of practice can be viewed as different kinds of training programs. Not all training programs are equally effective. Consequently, a lot of effort has been devoted to investigating different training programs for different kinds of motor skills. In particular, most training programs are designed with the goal of optimizing performance with a minimal amount of training time.

Training programs often are evaluated by the amount of practice required to attain a criterion level of performance and not how long that level can be retained. The level of performance that a person reaches during training can be influenced by many factors and does not always indicate the amount that the person has learned. Therefore, it is important to distinguish between variables that affect learning and cause a relatively permanent change in behavior and those that only temporarily affect performance. For example, a person's performance may deteriorate near the end of a long, difficult practice session due to fatigue, but that person's performance may be much improved once he or she is no longer fatigued.

The extent of learning can be demonstrated by measuring performance after a delay following training, a procedure referred to as a *retention test*. More effective training programs will result in better performance after such a delay. Another way of measuring learning is to look at a person's ability to perform new tasks that are related to, but different from, the tasks learned in the training program, a procedure referred to as a *transfer task*. In the rest of this section, we will focus on how different practice conditions contribute to retention and transfer.

Amount of Practice

Usually, retention will increase with more practice. Even after a performer has attained an acceptable level of skill, if she continues to practice (or "overlearns" the skill), she may retain the skill better. One experiment examining this issue required soldiers to disassemble and assemble an M60 machine gun (Schendel & Hagman, 1982). Three groups of soldiers had to disassemble and assemble the gun until they had made no errors. One group, the control, received no further training and was retested 8 weeks later. The two other groups received overtraining; soldiers in these groups performed additional assemblies equal to the number of assemblies they had performed before the first errorless execution. For one group, the overtraining was performed on the same day as the initial training, whereas for the other, it was performed 4 weeks later, halfway between the initial training and the retention test. Both experimental groups had greater retention than the control group. Although overtraining may seem excessive, it is an effective way to ensure that skills will be remembered.

Another experiment asked college basketball players to perform free throws (a distance of 15 ft) and shots at six other distances (three closer and three farther; Keetch, Schmidt, Lee, & Young, 2005). The players made 81% of their shots from the free throw line, a percentage that was much higher than their performance at any other distance. One explanation for this is that, because basketball players have more practice at free throws than at any other kind of shot, free throws are overlearned. Note that it is difficult to explain this result with schema theory, because the movements are all from the same class, differing only in their distance (Keetch et al., 2005).

Fatigue and Practice

Extended physical activity results in fatigue. Consequently, we must ask whether or not practice is as effective when the learner is fatigued as when he is rested. Fatigue has a detrimental effect on the speed with which a person can acquire a new motor skill, at least for some laboratory tasks (Pack, Cotton, & Biasiotto, 1974; Williams & Singer, 1975). However, skills acquired under conditions of fatigue are retained almost as well as those acquired under conditions of non-fatigue (Cotten, Thomas, Spieth, & Biasiotto, 1972; Heitman, Stockton, & Lambert, 1987).

As one example, one study asked people to perform a task that required them to rotate a handle in a clockwise direction and then rotate it in a counterclockwise direction, and finally to knock down a wooden barrier (Godwin & Schmidt, 1971). On the first day, people in a fatigued group had to turn a stiff crank for 2 minutes before each performance of the task. People in a non-fatigued group only tapped their fingers. The fatigued group performed the task more slowly than the non-fatigued group (see Figure 14.13). However, after a three-day rest, there was only a small difference in performance between the two groups. This difference was eliminated by the fourth retention trial, indicating that fatigue had little effect on learning.

The effects of fatigue on learning are greatest at highest levels of fatigue. One study asked people to perform a ladder-climbing task that required them to maintain their balance (Pack et al., 1974). Three levels of fatigue were induced in three groups of people by having them perform strenuous exercise that maintained their heart rates at 120, 150, or 180 beats/s between trials of the task. People who maintained their heart rates at the two highest levels did not perform as well on retention trials as people who maintained their heart rates at the lowest level, who showed no difference from people who did not perform any strenuous exercise.

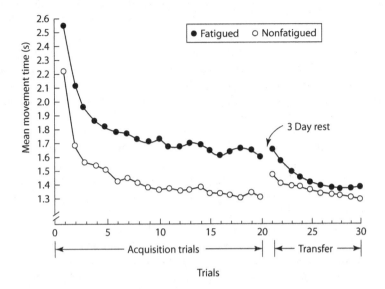

FIGURE 14.13   Effect of fatigue on initial performance and learning.

Distribution of Practice

Distribution of practice refers to the influence that scheduling of practice sessions and work periods has on the acquisition of motor skill. There are two kinds of practice sessions: massed or distributed. With *massed practice*, a person practices the same task repeatedly for an extended period of time, whereas with *distributed practice*, the person rests occasionally between trials. When we compare performance for massed versus distributed practice, we will look at conditions in which the number of practice trials is the same whether massed or distributed. Under distributed practice, the trials are performed over a longer period of time than under massed practice, where the trials are performed all at once.

Massed practice can result in a much slower rate of skill acquisition than distributed practice. Lorge (1930) asked his subjects to trace a star by watching their movements in a mirror. People who used distributed practice performed the tracing task better than people who used massed practice. Then, Lorge shifted some people from distributed to massed practice, and their performance dropped to the level of the people who had used massed practice all along. This suggests that the difference in performance under massed and distributed practice is only temporary.

The extent to which the type of practice schedule influences retention is unclear. Several reviews have suggested that massed versus distributed practice will influence mainly tasks that require continuous movements. These reviews showed that massed practice hurts retention for these kinds of tasks, and that distributed practice improves retention (Donovan & Radosevich, 1999; Lee & Genovese, 1988). Distributed practice of continuous tasks may be more beneficial when the sessions are separated by a day rather than shorter periods within the same day (Shea, Lai, Black, & Park, 2000).

A few studies have examined distribution of practice effect using tasks with discrete movements. One asked people to pick a dowel up out of a hole, flip the ends of the dowel, and put it back in the hole (Carron, 1969). In contrast to what happens in continuous tasks, massed practice produced slightly better retention than distributed practice. This result was replicated when a stylus was to be moved from one metal plate to another in 500 ms (Lee & Genovese, 1989). However, Dail and Christina (2004) found that people acquired and retained golf-putting ability better under distributed practice. This suggests that there is no simple relation between task type (discrete or continuous) and whether massed or distributed practice produces better learning. Some evidence suggests that training schedules incorporating both massed and distributed practice may be best for some perceptual-motor tasks (Paik & Ritter, 2016).

Variability of Practice

Variability of practice refers to the extent that the movements required for each practice trial differ. When a person executes the same movement on each trial, there is little variability, but when he or she performs different movements on each trial, there is greater variability. According to schema theory, variable practice should lead to better performance than practice with only a single movement. This is because variable practice will produce a more detailed recall schema that can be used when a new variation of the movement is encountered.

Variability of practice has its greatest influence on transfer tasks, when people are asked to apply their skills to a task for which they haven't practiced. People who receive variable practice perform better on transfer trials than those who do not. For example, one study asked people to perform a two-part timed movement in which they knocked down two barriers (Lee, Magill, & Weeks, 1985). They were told how long each part of the movement should take. One group of people practiced the movements under four different time requirements (the random practice group), and another group practiced the movements under a single time requirement (the constant practice group). Each group performed exactly the same number of trials and then was shifted to a new, unfamiliar time requirement. People in the random practice group were better able to meet the new time requirement than people in the constant practice group (see Figure 14.14). A third group of people received variable practice, but time requirements were blocked so that people performed all of the trials with one set of time requirements before moving on to the next block of trials (the blocked practice group). The blocked practice group showed no better performance on the transfer trials than the constant practice group (see Figure 14.14).

Schema theory can explain why variable practice is helpful mainly for random practice if we consider the concept of *contextual interference* introduced by Battig (1979) to explain verbal learning performance (Sherwood & Lee, 2003). This concept refers to disruption of short-term memory, and thus performance, as a consequence of practicing multiple-task variations within the same practice session. Contextual interference will be least when successive movements are identical and greatest when they are different. When contextual interference is minimized, a movement can be performed with relative ease, even retained well, but the resulting schema may not be very detailed or accurate. The schema may contain only the requirements for a single movement because the movement parameters needed for earlier blocks of trials have been

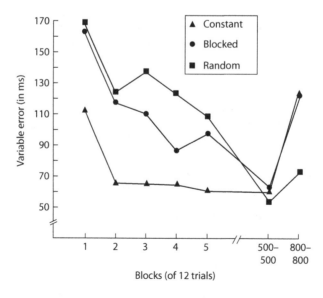

FIGURE 14.14 Accuracy of movement time for constant, blocked, and random practice.

"written over." Conversely, when contextual interference is maximized by using random practice, movements are more difficult to learn, but the resulting schema contains the parameters for all of the movements required.

The more elaborate and flexible schema arising from random practice has its greatest benefits in performance of transfer trials. These benefits imply a range of practical considerations, such as dealing with problems that arise with aging. Older adults show the same improvements in performance on transfer trials with random practice schedules as do younger adults, suggesting that appropriate design of training schedules is an important consideration for offsetting declining motor skills in older adult populations (Pauwels, Vancleef, Swinnen, & Beets, 2015).

Most accounts of the benefits of random variable practice on motor skill acquisition emphasize how cognitive factors contribute to learning. However, motivational factors are also important (Holladay & Quinones, 2003; Wulf & Lewthwaite, 2016). Consider the concept of self-efficacy, which refers to judgments about one's capabilities for performing various tasks. Random variable practice conditions, which lead to better learning, also lead to higher generality of self-efficacy across a range of transfer conditions. The implication is that a person may be more highly motivated to try to do a transfer task well if their assessment is that they are capable of performing the task than if it is that they are not.

Traditionally, repetitive drill-type training has been used to teach motor skills. For example, the Web Institute for Teachers' (2002) instructions for teaching keyboarding skills to elementary-school students state, "Teachers must provide repetitive drill for developing skill." Although repetitive training works, the studies we have reviewed in this section suggest that learning and retention of the skills can be improved significantly by varying the routine on a trial-to-trial basis.

Mental Practice

Mental practice is the term used to describe mentally imaging the execution of a desired action for performing a task. Athletes routinely engage in mental practice before performing a difficult routine or motor sequence. Despite how common this practice is, there is some question about the extent to which it actually improves performance or facilitates the acquisition of a skill.

To answer this question, experimenters compare the performance of groups of people who have acquired a skill with and without the assistance of mental practice. For example, one group of people would physically practice a task, a second group would mentally practice the task for the same amount of time, a third group would perform both physical and mental practice, and a control group would receive no practice at all (Druckman & Swets, 1988). Mental practice usually results in better performance than no practice at all (e.g., Driskell, Copper, & Moran, 1994; Feltz & Landers, 1983; Wulf, Horstmann, & Choi, 1995). However, for transfer tasks, where we would expect that learning one skill would make it hard for people to perform a new skill, mental practice does not seem to hurt performance as much as physical practice (Shanks & Cameron, 2000). This means that while mental practice can be helpful, it is not the same as performing physical practice.

One benefit of mental practice is to allow rehearsal of the cognitive components of the practiced task (Sackett, 1934). For example, a tennis player must not only execute her backhand flawlessly, but she must also be able to anticipate her opponent's return and position herself appropriately. This means that mental practice should be more effective for motor tasks that have a large cognitive component (e.g., card sorting) than for those that do not (e.g., repetitive tapping).

A study of sequence learning confirmed this hypothesis (Wohldmann, Healy, & Bourne, 2007). People in this study either practiced rapidly typing four-digit strings or only imagined typing the same strings instead of actually typing them. On a later test, those who had "practiced" using mental imagery showed the same improvement in performance as those who physically practiced the task. The lack of difference between the mental practice and physical practice conditions, as well as other results in the study, suggests that the benefits of practice were on cognitive representations and not the physical effectors.

Another prediction from the hypothesis that mental practice benefits a cognitive component is that the extent of mental practice should be independent of the movements required by the task: If two tasks share the same cognitive component, but require very different movements, the benefit of mental practice should be the same for the two tasks. One motor task with a large cognitive component is reading sentences in a foreign language. MacKay (1981) asked bilingual people to read sentences in German and English as rapidly as they could. Silent reading (mental practice) of the sentences in one language not only decreased reading time; it also resulted in faster reading of those same sentences translated into the other language. Mental practice resulted in a greater decrease in reading time for the translated sentences than did physical practice: The benefits of mental practice were not dependent on the different patterns of muscular activity required to read the sentences in German or English.

Acquisition of motor skill is usually best when performers combine mental practice with physical practice (Allami, Paulignan, Brovelli, & Boussaoud, 2008; Druckman & Swets, 1988). It may be that the combination of mental and physical practice results in more detailed and accurate motor programs. An optimal training routine will use some combination of both mental and physical practice. Apart from improved acquisition, there are other benefits to mental practice, including no need for equipment, no physical fatigue, and no danger.

TRAINING WITH SIMULATORS

We have talked about transfer and the extent to which practice of one set of movements improves the performance of a novel set of movements. How well practice transfers is of particular concern in the design and use of military and industrial simulator-training devices (Baudhuin, 1987; Rogers, Boquet, Howell, & DeJohn, 2010), such as those used to train pilots. Simulators are used for situations in which it is not feasible to have operators train in the real system. For example, student pilots should not train in real, fully loaded Boeing 767s, but they can operate a simulator.

The goal of using a simulator for training is to ensure the greatest possible amount of transfer to the operational system that is being simulated at the lowest possible cost. If training on a simulator transfers to the operational system, then money that would have been spent can be saved for the operation of the system itself. Moreover, the risk of physical harm and damage to the real system can be minimized. A "crash" in a flight simulator causes no real harm. For these reasons (and because of their increasingly wide availability), simulators are used extensively for research in air-traffic management (Vu, Kiken, Chiappe, Strybel, & Battiste, 2013), driving (Rendon-Velez et al., 2016), construction equipment operation (So, Proctor, Dunston, & Wang, 2013), and laparoscopic surgery (Luursema, Verwey, & Burie, 2012).

A major issue in simulator design involves the fidelity of the simulation to the real system. Designers often assume that physical similarity is important, and that the physical characteristics of the simulator should closely resemble those of the real system. This is clearly the case for flight simulators used for commercial aviation, which attempt to duplicate the natural cockpit environment closely (Lee, 2005). The cockpit of a full-motion flight simulator is an exact replica of that of the simulated aircraft. High-fidelity, wrap-around visual displays are used, auditory cues are provided, and changes in forces on the controls that would occur in real flight are simulated. The cockpit is mounted on a platform that moves in three dimensions, simulating the forces on the vestibular system that arise in flight. The end result is an experience that closely approximates that of an actual flight.

However, high fidelity is not necessary for effective simulator training. Practice on low-fidelity simulators such as desktop flight simulators and flat-screen construction equipment simulators results in positive transfer to real environments (Rogers et al., 2010; So et al., 2016). The extent to which simulated practice transfers to real systems depends on the extent to which the procedures to be executed are the same in the simulated and operational environments, even if the specific stimulus and response elements of the tasks are not identical.

Given the cost of high-fidelity simulation, along with the technological limitations that prevent perfect resemblance between the simulated and operational environments in many situations, training programs must emphasize functional equivalence over realism: the equivalence between the tasks that the operator will be required to perform in the simulation and in the real systems (Baudhuin, 1987). Functional equivalence, not realism, will determine how well practice will transfer.

With the widespread availability of relatively low-cost computers with powerful image-generation systems, it has become easy to develop inexpensive simulators based on personal computers. For example, X-Plane is a powerful flight simulator that will run on a personal computer. It can be used as a desktop simulator on any computer powerful enough to support it, and even to provide displays for an unmanned aerial vehicle (UAV) simulator (Garcia & Barnes, 2010). Lower-fidelity simulators like these provide only a restricted view on the computer monitor and omit many of the sensory cues available in a high-fidelity, full-motion simulator. However, they are sufficient for teaching basic perceptual-motor control skills, spatial orienting skills, and how to read flight instruments (Bradley & Abelson, 1995).

Virtual environment systems, based on virtual reality generators, construct a simulated environment in which a person is completely immersed. Within the virtual environment, a person interacts with a system in much the same way as with a "hard" simulator. Virtual environments can be used in place of more expensive training (Lathan et al., 2002). Because these environments do not have many of the constraints of physical simulators, they have the potential to ensure good skill transfer to the operational environment very cheaply.

FEEDBACK AND SKILL ACQUISITION

Many sources of sensory feedback are available both during a person's movement (concurrent feedback) and upon its completion (terminal feedback). These sources of information are intrinsic: They come from the performer. Intrinsic feedback is not only important in control of movements, as described earlier in the chapter, but it also provides a basis for the learning of motor skill (e.g., Anderson, Magill, Sekiya, & Ryan, 2005). However, motor learning solely on the basis of intrinsic feedback can be very slow, and so most training programs use some form of augmented feedback that comes from a trainer or other source. This feedback usually takes the form of *knowledge of results* (KR), *knowledge of performance* (KP), or *observational learning*.

KNOWLEDGE OF RESULTS

KR refers to feedback about a performer's degree of success in achieving a desired goal. Such feedback can be provided by an instructor or by an automated device. For example, a flight instructor may tell a student whether or not the goal of a particular maneuver was accomplished, or a flight simulator may indicate whether a landing was accomplished safely. KR reliably improves both the initial performance of a motor-learning task and its subsequent retention (Newell, 1976; Salmoni, Schmidt, & Walter, 1984). However, there are many ways that KR can be presented, and some forms of presentation are better than others. Research has focused on the effects of the precision, frequency, delay, and control of KR.

Precision of KR

There are two kinds of KR. Qualitative KR provides general information about the quality of performance (e.g., correct/incorrect), and quantitative KR specifies the direction and magnitude of error. Quantitative KR is therefore more precise than qualitative KR. Typically, quantitative KR produces better performance during acquisition than qualitative KR (Salmoni et al., 1984). Quantitative KR also leads to better retention, even under conditions where people receiving quantitative and qualitative KR perform equally well during acquisition (Magill & Wood, 1986; Reeve, Dornier, & Weeks, 1990).

Frequency of KR

Schema theory suggests that KR should be most beneficial when it is given after every trial, and that the benefits of KR will decrease as the percentage of trials on which KR is given decreases. This is generally true if performance is measured by the rate at which a skill is acquired (Salmoni et al., 1984), but more KR results in poorer retention. This suggests that less KR may produce better learning.

One experiment asked people to learn a pattern of lever movements (Winstein & Schmidt, 1990). This pattern consisted of four movements that were to be produced in 800 ms. KR was given on a computer that showed the actual movement together with the goal movement. One group of people received KR after each trial, while the other group received KR on only half of the trials. While both groups learned the task equally well, the group that received KR less often retained the task best. Another study showed similar results when people learned a pronunciation task (Steinhauer & Grayhack, 2000). However, some very complex tasks, such as a slalom-type task performed on a ski simulator, are not retained as well with low-frequency KR. This suggests that if a motor task is very complex, then more frequent KR will be beneficial (Wulf, Shea, & Matschiner, 1998).

The fact that retention is often better when KR is not provided on every trial suggests that it might be most effective to provide summary KR only after sets of trials are completed. Several experiments demonstrated that, in fact, this is true (Lavery, 1962; Vieira et al., 2014). One of these experiments (Schmidt et al., 1989) had people learn a timed lever-movement task like the one we just described (Winstein & Schmidt, 1990). They provided summary KR after sets of 1, 5, 10, or 15 trials to 4 groups of people. Everyone's performance improved during the acquisition phase of training, but people who received KR less frequently did not perform as well as those who received it more frequently (see Figure 14.15). However, people's performance on a delayed retention test showed an inverse relation between the length of the set and accuracy. Performance was best when summary KR was given every 15 trials and worst when KR was provided every trial.

Delay of KR

After a movement has been executed or a trial completed, there is some delay before KR is presented. This interval is called *KR delay*. KR delay is only important when it is very short (Salmoni et al., 1984). When KR is provided immediately, it interferes with learning the task (Swinnen, Schmidt, Nicholson, & Shapiro, 1990). In one experiment, people performed a timed movement. One group of people received KR immediately, and the other half got KR after a brief delay. After a delay, people who had received immediate KR showed poorer retention than those who had received delayed KR. The researchers hypothesized that the time after a trial was important for evaluating

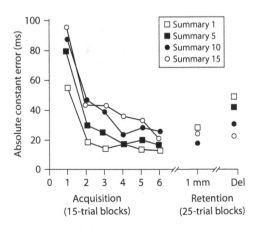

FIGURE 14.15 Effects of summary knowledge of results.

intrinsic performance feedback, and that this evaluation helps people to detect their own mistakes. Immediate presentation of KR may interfere with this process.

In another experiment, Swinnen (1990) asked people to perform an attention-demanding secondary task during the KR delay. These people showed poor retention for the primary motor task, demonstrating that the secondary task interfered with learning. However, when the secondary task was performed after KR and before the next trial, retention was much better. In another study, people were asked to perform an extra movement during KR delay, and this extra movement affected retention only when it had to be remembered along with the primary task (Marteniuk, 1986). When people were asked to solve a number problem during KR delay, their retention performance was equally poor. All of these results demonstrate that part of learning a movement involves processing information about that movement after it has been attempted, and that any higher-level cognitive activity performed during the KR delay that is unrelated to the movement will interfere with retention.

KR and Self-Control

Allowing people to choose when they receive KR can also improve learning. One study paired people together to learn a task and allowed one person of each pair to choose when they would receive KR. The other person did not get to choose, but received KR at the same time as her partner. The person who chose when to receive KR learned better than the person who didn't, even though they both received KR on exactly the same schedule (Sanli, Patterson, Bray, & Lee, 2013). One explanation for the benefit of self-control attributes better learning to improved motivation and perceived self-autonomy.

However, another study found that when people were required to perform a mentally demanding task during the interval after motor execution but prior to KR, the benefit for self-control of the KR schedule was eliminated (Carter & Ste-Marie, 2017). This finding suggests instead that the information-processing activity in which a person typically engages immediately after motor execution (e.g., evaluating how accurately the task was performed) is what determines the benefit, and not just a person's self-control over whether feedback is provided.

Role of KR

Clearly, KR is important for the acquisition of movement skill. There are three major roles that KR may play (Salmoni et al., 1984). KR may improve motivation, one of the explanations of the effect of self-control, which in turn may result in greater exertion or effort when KR is present than when it is not. KR may help the formation of associations in memory. This is especially important for schema theory, in which KR helps form associations between stimulus and response features to create recall and recognition schemas. Finally, KR may provide guidance and help direct performance during acquisition (Anderson et al., 2005). When KR is provided for every trial, this can allow accurate performance without requiring the deeper processing necessary for learning to occur.

In summary, you should remember the following three points. First, a person learning a motor skill must actively process the information that provided by KR if it is to be of any benefit. Second, KR will be most effective if it is precise, controlled by the learner, and presented when the learner is not required to process other information at the same time. Third, when KR is presented too frequently, the learner may fail to process intrinsic information about his or her performance and instead rely only on the guidance provided by KR.

KNOWLEDGE OF PERFORMANCE

KR provides information about the outcome of a movement, but KP provides information about the performance of the movement, such as how the movement was controlled and coordinated (Nunes et al., 2014). We have good reason to believe that KP should be more effective than KR (Newell, Sparrow, & Quinn, 1985; Newell & Walter, 1981), but we must distinguish between kinematic KP, which describes some aspect of the motions involved in an action, and kinetic KP, which describes the forces that produce those motions. Studies of the effectiveness of KP show people kinematic and kinetic feedback about their movements, often with the kinematic and/or kinetic information for a successful movement presented for comparison.

Kinematic KP includes information about the spatial position, velocity, and acceleration of the limbs. An old, classic study of kinematic KP watched workers who operated a machine used to cut discs from tungsten rods (Lindahl, 1945). The machine was operated through a coordinated pattern of hand and foot movements. The foot movements were particularly important in determining the workers' cutting efficiency and the quality of the discs. Lindahl recorded the feet of the most skilled workers (see Figure 14.16) and used these records to train new workers. This kinematic KP not only resulted in new workers learning the task faster and performing better (see Figure 14.16), but it also improved the performance of more experienced workers.

You may already have noticed that the kinematic KP in Lindahl's experiment was actually just a kind of quantitative KR, because the goal of the movement was the same as the movement itself. Also, Lindahl did not examine the operator's retention of the skill, but only acquisition performance. If our goal is to determine whether kinematic KP is better than KR, or in what circumstances KP or KR should be used, then we need to separate the movement goal from the movement and look at what happens with retention and/or transfer (Schmidt & Young, 1991). One experiment that did this is shown in Figure 14.17. People were shown a sequence of illuminated light-emitting diodes that looked like a moving ball. They were asked to manipulate a horizontally mounted lever in a way similar to a tennis racquet. When the light began to move, the performer made a backswing with the lever, bringing it anywhere to the left of the lights. Then the performer attempted to "bat the ball" by swinging the lever forward to intercept the moving light. In a task like this, KP is easily separated from KR. KR tells the performer whether or not the ball was successfully hit, whereas KP

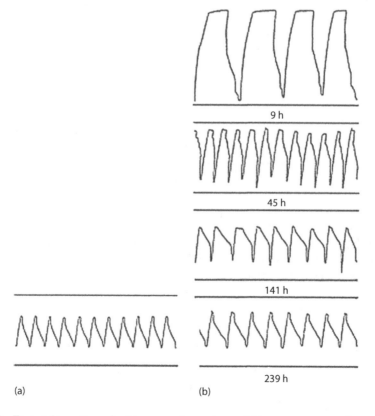

(a)　　　　　　　　　　　　　　　　(b)

FIGURE 14.16 Foot action patterns for (a) an expert operator and (b) a new operator after various amounts of practice with kinematic KP.

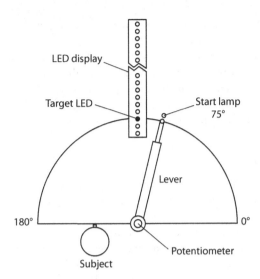

FIGURE 14.17 Schematic diagram of the coincident-timing apparatus used by Schmidt and Young.

gives information about the swing. Note that KP is important in this task, because the apparatus did not constrain the performer's movement much: the performer was required to select the position of the backswing, and the force and timing of the forward swing.

The best batting accuracy was achieved when the performer moved the lever back to approximately 165°. People were given only KR about the accuracy of their swing or the KR plus KP about position of their backswing. The people who received KR and KP performed better on acquisition trials and also on a retention test. Therefore, we can conclude that kinematic KP provides a benefit to motor learning over and above that provided by KR alone.

The benefits of kinetic KP have been examined for skills that require control of force and movement durations. For example, some isometric tasks (which require changes in muscle force without limb movement) should benefit greatly from kinetic KP, because accurate performance of such tasks is completely determined by kinetic variables. An early experiment demonstrating this was conducted on new U.S. Army recruits learning to shoot a rifle (English, 1942). Good marksmanship requires that the soldier squeeze the stock of the rifle at the same time as the trigger. Because this technique is difficult to learn, English implemented a new training program that provided kinetic KP. The stock of a rifle was hollowed out, and a fluid-filled bulb was placed within it. The bulb was attached to a fluid-filled tube that displayed the amount of force applied to the stock. The soldier could compare the level of the liquid in the tube when he shot the rifle with the level produced by an expert marksman. This method was remarkably effective, with even soldiers "given up as hopeless" achieving minimum standards quickly.

Some more recent experiments looked at the role of kinetic KP for tasks that required not only the application of a specific amount of force but also a gradual change in the application of force over time (Newell & Carlton, 1987; Newell, Sparrow, & Quinn, 1985). People performing one task produced a maximum force of 30 N against an immovable handle, whereas people in the other produced a specific force–time curve. Both groups received kinetic KP in the form of their force–time curves. KP improved both initial performance and retention for the task with the force–time criterion but not for the task with the peak force criterion. Further investigation into the way that this kind of KP should be presented evaluated the efficiency of performance when the actual and desired force–time curves were superimposed. Superimposing the curves was only beneficial when the desired force–time curve was asymmetric, or of a shape that was unfamiliar to the performers.

One popular way to present KP is by showing the performer a video replay of his movements (Newell & Walter, 1981; Rothstein & Arnold, 1976). However, in many situations this kind of KP can provide too much information, and can therefore confuse the performer rather than clarify what he needs to do to improve his performance. Video replay is most effective when the performer is told to pay attention only to the specific aspects of his performance that are important for learning.

The general principle emerging from this work is that the success of any type of augmented feedback will depend on the extent to which it provides relevant information in a form that is useful for improving performance. This means that before deciding what types of KR or KP to provide, a trainer must first analyze in detail the requirements of the task.

OBSERVATIONAL LEARNING

Sometimes people learn how to perform a motor task by observing someone else (a model) performing it. This is called *observational learning*. Explanations for how people learn through observation can be based on Bandura's (1986) social cognitive theory. According to this theory, the observer forms a cognitive representation of the task to be learned by attending to the salient features of the model's performance. This representation can then guide the production of the action when the observer is asked to perform it. The representation also provides a referent against which feedback from the observer's own performance can be compared. Not too surprisingly, many of the same variables that affect motor learning when a person performs a task, such as frequency of KR, have similar effects on motor learning when a person watches a model perform the task (Badets & Blandin, 2004).

A movement sequence can be learned partly through observation, but not entirely (Adams, 1984). This is because important task factors (such as static force and muscle tension, and any unseen components of the movement) can only be learned by performing the task, and the inaccuracies or ambiguities in the cognitive representation of the movement cannot be resolved until the task is performed. Consequently, it is often helpful to combine observational learning with physical practice (Shea, Wright, Wulf, & Whitacre, 2000).

To the extent that a trainer can provide information about relevant task factors and resolve ambiguities during observational learning, a learner's performance might benefit. One experiment explored this idea and demonstrated how observational learning can be improved by showing the observer the parts of a movement that are otherwise unobservable to a learner (Carroll & Bandura, 1982). This experiment asked learners to manipulate a paddle device in a complicated way (see Figure 14.14–14.18). A demonstration video recording was made of a model performing the components of the paddle-manipulation task, with the recorded image being of the back right of the model's body (as in the figure) such that the orientation of the model's arm and hand corresponded exactly to the observer's arm and hand. Each learner saw the demonstration video of the modeled pattern six times, and after each demonstration, he or she had to execute the movement pattern. The experimenters presented simultaneous video of the learner executing the movement during his or her performance (visual feedback) on none of the trials, the first three trials only, the second three trials only, or all six trials. After the six acquisition trials, the learner had to execute the movement pattern three additional times without the demonstration video or visual feedback. After each set of three trials, the experimenters measured the accuracy of the learner's cognitive representation of the task by asking him or her to put in order nine photographs representing the components of the action sequence.

Visual feedback during only the first three acquisition trials did not help performance, but visual feedback during only the second three trials was as helpful as visual feedback on all six trials. The accuracy of ordering the photographs also was higher after the second and third sets of trials than after the first set. The authors interpreted their findings as indicating that, as implied by social cognitive theory, an accurate cognitive representation of the observed behaviors must be established before visual feedback of one's own behavior can be beneficial.

Carroll and Bandura (1985, 1987) showed that the video augmentation was an effective training tool, but only when it was provided simultaneously with the learner's movements. When it

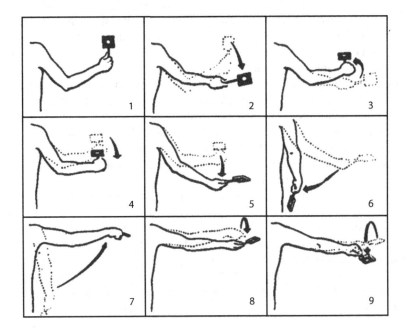

FIGURE 14.18 Response components of the action pattern investigated by Carroll and Bandura (1982), with the components numbered in the order in which they were enacted.

was delayed by about a minute, performance was no better than when the video was not provided. The improvement in learning obtained by having the learner match the actions of a model was equivalent to that obtained by the video augmentation, but this did not depend on whether the model's action was presented simultaneously or later. Further investigation also showed that the more frequently the learner is shown the movements of the model, the better his performance will be (Carroll & Bandura, 1990).

Finally, it should be noted that observational learning may occur in either a more "bottom-up" or "top-down" fashion. People in one study practiced a motor sequence with a computer mouse (Roberts, Bennett, Elliott, & Hayes, 2015). The sequence trajectory either mimicked natural, biological motion or was an artificial trajectory that moved at a constant velocity. When the learners were led to think that the biological motion was human-generated, the learning was bottom-up: that is, it was primarily in the sensorimotor system and relatively automatic. But when they were led to think that the movement was computer-generated, both the biological and artificial motion trajectories were learned in a more top-down manner: that is, the learning required effortful cognitive processing. These results imply that a person's understanding about agency in reproducing a motor pattern influences how they learn from observation.

To summarize, observational learning can be an effective training tool, but only to the extent that it promotes the learner's development of an accurate cognitive representation of the task. The extent to which observational learning is useful for the acquisition of complex movement skills, relative to other KP methods, is still an open question. It may be that while observation is effective for learning the coarse aspects of tasks, such as the order and extent of different movements in a sequence, it will not be useful for learning exact details (Newell, Morris, & Scully, 1985).

SUMMARY

Understanding how people execute movements and control their actions is a fundamental part of understanding human factors. We have presented several important ideas in this chapter. First, control of action is hierarchical. The motor cortex receives proprioceptive feedback and delivers signals for control and correction of movement. These signals travel through the spinal cord, which alone can control movement to some degree. Our current understanding of higher-level motor control is that the brain develops plans for the execution of complex actions, whereas the spinal cord is involved in control of the fine adjustments.

The brain's action plans, called motor programs, are hierarchical and modular, just like the organization of the nervous system. Complex actions involving more than one muscle group can be controlled by a single motor program. Motor programs rely on sensory feedback to determine the appropriate parameters, such as force and distance, for a particular movement, and the way that sensory feedback is used depends on whether an action requires open- or closed-loop control. Sensory feedback can be used to modify slower closed-loop actions as they are being executed, but it plays a smaller role in actions that are executed very quickly.

Perhaps the most interesting questions about how people control their movements are directed toward understanding how highly skilled behavior is learned. The ease with which motor skills are acquired varies greatly with the kind of training program used. High levels of practice variability will lead to better performance, retention, and transfer of similar types of movements. Learning and performance also will benefit from augmented feedback. Providing knowledge of both results and performance can improve learning when the information provided is chosen appropriately. The human factors specialist has an opportunity to provide input on optimal training programs that can speed an operator's progress through the phases of skill acquisition.

RECOMMENDED READINGS

Enoka, R. M. (2015). *Neuromechanics of Human Movement* (5th ed.). Champaign, IL: Human Kinetics.

Jeannerod, M. (Ed.) (1990). *Attention and Performance XIII: Motor Representation and Control*. Hillsdale, NJ: Erlbaum.

Jeannerod, M. (1990). *The Neural and Behavioural Organization of Goal-Directed Movements*. New York: Oxford University Press.

Magill, R. A., and Anderson, D. I. (2014). *Motor Learning: Concepts and Applications* (10th ed.). New York: McGraw-Hill.

Rosenbaum, D. A. (2010). *Human Motor Control* (2nd ed.). San Diego, CA: Academic.

Schmidt, R. A., and Lee, T. D. (2011). *Motor Control and Learning* (5th ed.). Champaign, IL: Human Kinetics.

15 Controls and Controlling Actions

How do the operators avoid the occasional mistake, confusion, or accidental bumping against the wrong control? Or misaim? They don't. Fortunately, airplanes and power plants are pretty robust. A few errors every hour is not important—usually.

D. A. Norman
2013

INTRODUCTION

Machines communicate information about their status to their operators through displays. Operators communicate how they want the machine's status to change by manipulating controls. There are many kinds of physical devices available for use as controls, including pushbuttons, toggle switches, joysticks, knobs, and touchscreens. They can be operated using the hands, feet, and, in some cases, eye and head movements. Sound-sensitive controls that respond to human speech are used for machines that restrict the operator's ability to divert her gaze to a control panel. In cars, for example, interactive voice navigation systems can respond to voice commands and give driving directions without the driver ever having to look away from the road (Mehler et al., 2016).

Different kinds of controls require different types of actions. This means that a control that works well in one situation may not necessarily be the best in another situation. Moreover, rapid changes in technology ensure that there will always be new problems to overcome in control design. For example, hand-held devices such as smartphones pose unique problems for data entry (and display) because they are so small. Users may benefit from novel controls (such as pressure-sensitive controls, devices, and even clothing; Paepcke et al., 2004; Porta, 2007; Zhou & Lukowicz, 2017). The job of the human factors engineer is to use ergonomic data to determine the particular controls and layout of the panel that will optimize both operator and system performance.

Effective controls have at least three characteristics. First, they are easily operated by their users. Second, their sizes and shapes are determined by biomechanical and anthropometric factors (see Chapter 16) as well as by population stereotypes for the mapping of the control settings to system states (see Chapter 13). Third, they are appropriate for the controlling action they were designed to facilitate, they can accommodate the muscle force required to move them, and they can respond with the necessary speed and accuracy (Bullinger, Kern, & Braun, 1997).

Usually, several controls are arranged together on a panel. A good control panel will ensure that the operator can easily determine the identities and functions of each control. Also, the operator must be able to reach all of the controls and apply the forces necessary for their operation. In this chapter, we will discuss the features of controls and control panels that make them usable and how human factors engineers can contribute to their design.

CONTROL FEATURES

Controls can differ in many ways. Some require considerable force to operate (perhaps to avoid accidental activation), whereas others require little force. Some are pushed, some pulled, and some turned. They can move in one, two, or three dimensions. The operator may use his hands, his feet, or even his head, mouth, or eyes to manipulate them. Their surfaces may be rough or smooth.

The system may respond to different controlling actions in many different ways. It may respond immediately when the operator moves a control, or it may take a long time. The resulting system change may be very large or very small. Applying the brake pedal in a moving car produces a large system change immediately, whereas turning up the heater produces a small system change very slowly. In this section, we discuss those features of controls that we need to consider when deciding which control to use for a particular application.

BASIC DIMENSIONS

The easiest way to classify controls is into whether they are discrete or continuous. *Discrete controls* can be set to one of a fixed number of states. For example, a light switch has two settings, one for "light off" and another for "light on." Some discrete controls have several states. A stereo amplifier may use a discrete control to select the listening device (compact disc, radio tuner, audiotape, or video source). A car's gear shift is also a discrete control.

Continuous controls (sometimes called *analog controls*) can be set to any value along a continuum of states. A light "dimmer switch" is the continuous analog to a light switch. Analog radio tuners use a continuous tuning control to select radio frequencies. A car's steering wheel is also a continuous control. Discrete controls are used when there are a small number of discrete system states or when accuracy in selecting an exact state is important. Continuous controls are used when there is a continuum of system states or a large number of discrete control states. Continuous controls also are used for cursor control during interactions with visual computer displays.

We can also classify controls as linear or rotary. A light switch is not only discrete but also linear, because its movement is along a single axis. Stereo equalizers often use continuous linear controls to select the output level of different frequency bands. The input knob for the amplifier and the dimmer switch described above could be either linear or rotary. A car's steering wheel is rotary.

Figure 15.1 gives more general examples of linear and rotary controls. This figure classifies the controls by the kinds of movements a user must make to manipulate them. Controls with swiveling motion, for example, require movements around one or more axes of rotation, usually located relative to the point at which the control is connected to the controlling mechanisms. Linear motion controls generally encompass the linear controls we have discussed. Turning motion controls include rotary controls, but also rollerballs (which are sometimes used as positioning devices).

Controls are either unidimensional or multidimensional. Both the light and the dimmer switch are unidimensional, because they adjust the single dimension of lighting level. In contrast, a joystick or computer mouse is two-dimensional, because it controls position in two-dimensional space. Controls that determine the position of a system in unidimensional or multidimensional space often use a unidimensional or multidimensional representation that corresponds to the desired changes in the system. For example, moving a computer mouse up and to the left will (usually) result in a cursor moving up and to the left on a computer monitor. These kinds of controls, movable and responsive

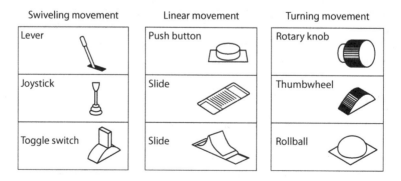

FIGURE 15.1 Examples of control types, categorized according to the path of movement.

TABLE 15.1
Uses for Discrete Controls

Type	Uses
Linear	
Pushbutton	Where a control or an array of controls is needed for momentary contact or for activating a locking circuit
Legend	Where an integral legend is required for pushbutton applications
Slide	Where two or more positions are required
Toggle	Where two positions are required or space limitations are severe
	Three-position toggles used only as spring-loaded center-off type or where rotary or legend controls are not feasible
Rocker	In place of toggles where toggles may cause snagging problems or where scarcity of panel space precludes separate labeling of switch positions
	Three-position rockers used only as spring-loaded center-off type or where rotary or legend controls are not feasible
Push-pull	Where two positions are required and such configuration is expected (e.g., auto headlights, etc.) or where panel space is scarce and related functions can be combined (e.g., ON-OFF/volume control)
	Three-position push-pulls used only where inadvertent positioning is not critical
Rotary	
Selector	Where three or more positions are required
	In two-position applications where swift visual identification is more important than positioning speed
Key operated	In two-position applications to prevent unauthorized operation
Thumbwheel	Where a compact digital control-input device with readout is required

to displacement are called *isotonic controls*. In some cases, similar results can be obtained using *isometric controls*, which are fixed and responsive to force. Some positioning devices, such as the pointing stick used for cursor positioning on some laptop computers, are isometric.

Controls also vary in mass, shape, range of motion, and resistance to movement. The usefulness of a particular control, and the relative ease with which a person can operate it, is a function of these and other factors. For example, performance using a joystick may be a function of the size of the joystick handle (because this will affect the way that the operator grips the joystick), its physical range of motion, and whether joystick movement is accomplished primarily by the wrist and hand or by the arm (Huysmans, de Looze, Hoozemans, van der Beek, & van Dieen, 2006). Tables 15.1 and 15.2 summarize the uses of the various types of discrete and continuous controls, respectively. These guidelines were compiled by Boff and Lincoln (1988), adapted from U.S. military design criteria. They can be used to help determine which type of control may be best for a particular application.

CONTROL RESISTANCE

Any control will have at least some resistance to movement and require some force to move it. The designer of an interface can modify (to some degree) the type and amount of resistance in each control. Changes in resistance will affect not only the amount of force required to move the control, but also the feel of the control, the speed and accuracy with which it can be operated, and how smoothly continuous control movements can be made.

Types of Resistance

There are four distinct kinds of resistance: elastic, frictional, viscous, and inertial (Adams, 2006; Bahrick, 1957). Each kind has different effects that a designer must consider when designing a control. Spring-loaded controls have *elastic resistance*. Elastic resistance increases as the control is moved farther from its neutral position. The direct relationship between the control's resistance and

TABLE 15.2

Uses for Continuous Controls

Type	Uses
Linear	
Lever	When large amounts of force or displacement are involved or when multidimensional control movements are required
Isotonic (displacement) joystick	When precise or continuous control in two or more related dimensions is required
	When positioning accuracy is more important than positioning speed
	Data pickoff from CRT or free-drawn graphics
Isometric (force) joystick	When a return to center after each entry or readout is required, operator feedback is primarily visual from system response rather than kinesthetic from the stick, and there is minimal delay and tight coupling between control and input and system response
Track ball	Data pickoff from CRT; when there may be cumulative travel in a given direction; zero-order control only
Mouse	Data pickoff or entry of coordinate values on a CRT; zero-order control only
Light pen	Track-oriented readout device; data pickoff, data entry on CRT
Rotary	
Continuous rotary	When low forces and precision are required
Ganged	Used in limited applications where scarce panel space precludes the use of single continuous rotary controls
Thumbwheel	Used as an alternative to continuous rotary controls where a compact control device is required

its position gives the operator proprioceptive feedback about how far the control has been moved. Many people thought for a long time that this property of elastic resistance improved an operator's performance with the control, at least when the position of the control was directly related to the position of the machine or display element that it controlled (like a computer mouse). However, there is not much scientific evidence to support this idea (Anderson, 1999). It turns out that discrete aiming movements are equally accurate when they are made with or without elastic resistance, even when the person making the movements has had extended practice with knowledge of results. Moreover, when people were tested on their ability to use the control after practice, people who used controls with elastic resistance had poorer movement accuracy. Muscle fatigue may be partially responsible for this poorer performance: Controls that require the fingers to use a pinch grip, effortful movements similar to those required to use a pencil or to turn a key in a lock, show no performance benefits from elastic resistance (Han, Waddington, Anson, & Adams, 2013).

Controls with elastic resistance will return to their neutral position when released. Some controls with elastic resistance, called *deadman switches*, exploit this property. If something happens to the operator, deadman switches ensure that the machine will not continue to operate without anyone in control. For example, federal standards in the U.S. require that all walk-behind self-propelled lawnmowers have a spring-loaded control on the handle bar that is gripped (depressed) by the operator while mowing. If the operator releases his or her grip on the control, an automatic brake must stop the blade within 3 s. This switch prevents the mower from traveling without someone guiding it and also prevents people from attempting to clear jammed grass clippings from the blade while it is still turning.

Friction is the second type of resistance. There are two kinds of friction. A control with static friction has the most friction at rest, and this friction decreases once motion begins. Sliding friction, in contrast, arises when the control moves. The amount of sliding friction is not influenced by the velocity or position of the control. Frictional resistance of either type does not usually make the control easier to use, and it can interfere with how well a user can control a device. One study showed that reducing static friction helped people make the movements of a prosthetic arm much more smoothly and precisely (Farell, Weir, Heckathorne, & Childress, 2005). High-friction controls are good for use as on-off switches, because the friction reduces the likelihood of accidental operation of the control.

One source of friction that is not part of the control itself occurs between the control (or tool) and the operator's hand (contact friction). Movement accuracy will be improved if there is at least some contact friction. Contact friction is important for safe controlling movements of controls with low static friction. Without contact friction, the operator's hand could slip, and system failure or injuries could occur as a result (Seo, Armstrong, & Drinkaus, 2009).

Viscous resistance, or viscous damping, is an increasing function of the speed with which a control is moved. Imagine stirring a thick liquid like molasses with a spoon. The faster you move the spoon, the greater the resistance you will encounter. Because viscous resistance is a direct function of control velocity, it provides important proprioceptive feedback about how quickly the control is moving. Viscous resistance also helps make controlling movements smooth, because the control will not respond to abrupt changes in velocity. Viscous resistance has been used to mimic the effect of gravity on movements. A prototype space suit to be worn on space flights uses measurement devices that keep track of the wearer's inertia. These devices then control gyroscopes located at different places on the suit corresponding to parts of the wearer's body segments to provide viscous resistance against whatever direction is specified as "down" (Duda et al., 2015). This kind of suit might be used to counter the deleterious side effects of low gravity, such as reduced bone density and muscle atrophy.

Inertial resistance, that is, resistance to change of state of motion, is a function of control acceleration. For these controls, it may be difficult to move them from their resting position, but they are easier to move once they get going. Once the control begins moving, inertia helps keep it moving and prevents it from stopping. This means that a lot of force may be required both to start and to stop the control.

Revolving doors typically have high inertial resistance. Because of their mass, you have to push on them fairly hard to get them to start turning, but then they tend to keep turning even after you stop pushing. For controls that need to be moved in a particular direction to a particular target location or setting, inertial resistance creates a tendency for the operator to overshoot the target setting.

Performance and Resistance

How much resistance should a control have, and how does resistance influence a person's ability to use that control? Knowles and Sheridan (1966) conducted a study to answer these questions. They were particularly interested in frictional and inertial resistance in rotary controls. First, they used psychophysical methods (see Chapter 4) to measure how sensitive people were to changes in friction and inertia. For both friction and inertia, the just-noticeable-difference in resistance was between 10% and 20% of the resistance of a "standard." This means, for example, that if an operator is accustomed to a control with a certain level of friction or inertia, she will not notice any difference in that control until the resistance increases by 10%–20%. After Knowles and Sheridan determined the just-noticeable-difference for resistance, they asked operators to rate their preference for controls of different weights under different levels of frictional, viscous, and inertial resistance. Operators consistently preferred lighter controls over heavier controls and viscous resistance over frictional resistance.

People in this experiment also preferred controls with some inertial resistance over those with none. But when people need to use a control for a continuous task, like steering, inertia can make performance of the task harder. Howland and Noble (1953) asked people to use a rotary control to move a cursor to follow a target moving in a sine-wave pattern. The control was loaded with all combinations of elastic, viscous, and inertial resistance, and people's ability to keep the cursor on the target was measured for each combination. Performance was worst for all controls with inertial resistance (see the combinations with the letter J in the last position in Figure 15.2). The best performance was obtained with elastic resistance alone (the KOO combination). Because the task was spatial in nature, the control with elastic resistance probably provided important proprioceptive feedback about control location.

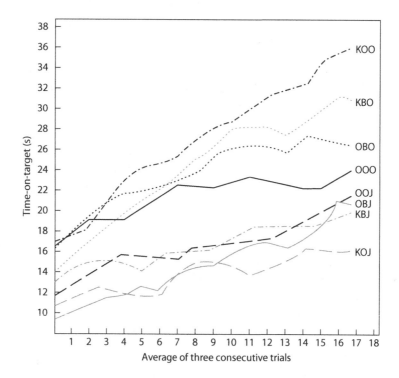

FIGURE 15.2 Mean time-on-target for the eight experimental conditions, as a function of practice, examined by Howland & Noble (1953). For the labels, the letter K in the first column indicates elastic resistance, the letter B in the second column indicates viscous resistance, and the letter J in the third column indicates inertial resistance. An O in any column indicates the absence of the corresponding type of resistance.

Control designers need to remember that, based on these findings, different kinds of resistance can produce interactive effects when combined within a single control. For example, while elastic resistance improved performance relative to a control with no resistance (KOO vs. OOO in Figure 15.2), combining elastic resistance with inertial resistance made performance much worse (OOJ vs. KOJ). These effects are not predictable by looking at performance with controls with only a single type of resistance.

MANIPULATION–OUTCOME RELATIONS

A person manipulates a control with the intent of producing a response in the system being controlled. With a continuous control, the speed and accuracy with which a person can manipulate the control will be a function of several factors. We have discussed some effects of different kinds of control resistance. Other factors include deadspace and backlash, the control–display ratio, and control order. The influence of these factors on continuous control is often investigated by looking at people's performance in *tracking tasks*.

Tracking Tasks

Formally, a tracking task has a path, or target track, and a device, or control object, used to follow the path (Jagacinski & Flach, 2003). Driving is a tracking task, where the roadway is the path and the car is the device that must follow the roadway. The task used by Howland & Noble (1953), described above, is an example of a laboratory tracking task.

There are two kinds of tracking tasks, or tracking modes: pursuit and compensatory (see Figure 15.3; Hammerton, 1989; Jagacinski & Flach, 2003). A pursuit tracking task has a track and control object that are simultaneously visible. For example, one pursuit task shows a moving dot or

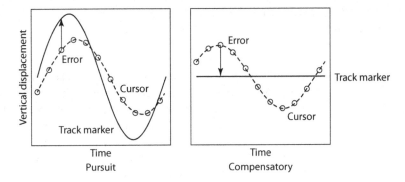

FIGURE 15.3 Pursuit and compensatory tracking displays.

track marker on a computer display. The person operates a joystick that controls a cursor, which is to be kept on top of the track marker. Any discrepancy between the position of the cursor and the location of the track marker can be easily identified as "error." In contrast, a compensatory tracking task has the same task requirements as a pursuit tracking task, but the person is only shown the difference between the track location and the cursor position. A person might see only a single dot on a screen moving around a "zero" point. All actions in a compensatory tracking task are made to reduce the error to nothing and keep the dot fixed at zero.

Performance in tracking tasks can be measured in several ways. One simple measure is the total or percentage of tracking time spent on target (see Figure 15.3). The larger this time (or percentage) is, the better the tracking performance. Other, more widely used measures examine location error. In particular, root mean square error (Jagacinski & Flach, 2003) is computed from the difference between the track location and the cursor position at fixed moments in time throughout the task (e.g., every 300 ms). These differences (errors) are squared and averaged, and the root mean square error is the square root of the result. Other error statistics can be calculated that may be more appropriate in certain situations (Buck et al., 2000). People do better with a pursuit display than with a compensatory display because the relation between the display and the actions needed to correct tracking error is more compatible (Chernikoff, Birmingham, & Taylor, 1956).

Deadspace and Backlash

The *deadspace* for a control is the extent to which it can be moved around its neutral position without affecting the system. For example, the keys on a computer keyboard have a little bit of "play" in them, so that you can push them down a bit without anything happening as a result. *Backlash* is deadspace that is present at any control position. For example, a car steering wheel, after having been turned clockwise, must be turned some distance in the counterclockwise direction before the wheels will start to turn back.

To understand backlash, imagine placing a hollow cylinder like a toilet paper tube over a joystick control (Poulton, 1974). If you move the cylinder to the left, the joystick will not start moving until the cylinder comes into contact with it on the right side. If you then want to change direction to the right, the joystick will not start moving back to the right until the cylinder comes into contact with it on the left side.

Deadspace is sometimes used for computer-control devices like spring-centered joysticks or force sticks so that users can more easily find the neutral or null position (Jagacinski & Flach, 2003). However, while a little bit of deadspace might be useful, too much deadspace or backlash will decrease the accuracy of a person's control actions, particularly with sensitive control systems (Rockway & Franks, 1959; Rogers, 1970). For example, Rockway (1957) systematically varied deadspace and Rockway and Franks (1959) systematically varied backlash for a tracking task. Tracking performance worsened as both increased. We described a study earlier that looked at users' ability to control a prosthetic limb (Farell et al., 2005). Backlash had negative consequences in this study as well. Eliminating backlash eliminated jerkiness in the limb.

Control–Display Ratio

We often measure a control's sensitivity by the *control–display ratio*, which is the ratio of the magnitude of control adjustment to the magnitude of the change in a display indicator (see Figure 15.4). When we are interested in the relation between the control movement and the response of the system, we use instead the term *control–response ratio*. We can talk more generally about a control's *gain*, which is a term used to describe how responsive the control is: A control with a low control–display ratio has high gain, whereas a control with a high control–display ratio has low gain.

Control movement and system response can be described either in terms of linear distance or in terms of radial angle or revolutions. Linear distance is used for lever-like controls, whereas radial angle or revolutions are used for wheels, knobs, and cranks. How control–display ratios are computed depends on how the control moves. For a linear lever paired with a linear display, the control–display ratio (C/D) is the linear displacement of the lever (C) divided by the corresponding displacement of the display element (D). For a joystick control, the displacement (C) is

$$C = \left(\frac{a}{360}\right) \times 2\pi L,$$

where:
    a is angular movement in degrees, and
    L is length of the joystick.

For a rotary control paired with a linear display, the control–display ratio is the reciprocal of the display movement (D) produced by one revolution of the control: $1/D$ revolutions per unit of distance.

When we are interested in how quickly people can position a control, we have to distinguish between *travel time* and *fine adjust time*. The travel time is how long it takes to move the control into the vicinity of the desired setting, whereas the *fine adjust time* is how long it takes to set the control precisely where it needs to be. Figure 15.5 shows how the control–display ratio affects travel time and fine adjust time. A low control–display ratio minimizes travel time, whereas a high ratio minimizes fine adjust time. To understand this, imagine trying to tune in a radio station with

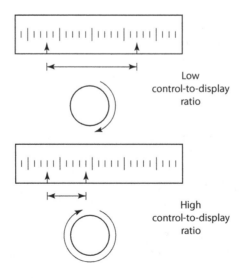

FIGURE 15.4   Illustrations of low and high control–display ratios.

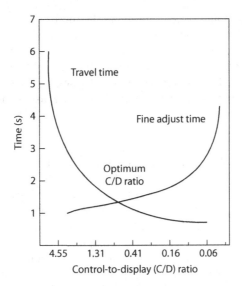

FIGURE 15.5 Travel time and fine adjust time as a function of control–display ratio.

a highly sensitive tuning knob. You will be able to move the indicator quickly from one end of the dial to the other, but it will be hard for you to home in on the station you want when you get there. The reverse relation holds with a control of low sensitivity.

The optimum control–display ratio will be a value that is not too high and not too low. An intermediate ratio will allow relatively fast travel time coupled with relatively fast fine adjust time. The optimum control–display ratio will also depend on the type of control, the size of the display, the tolerance of the system to adjustment errors, and the lag between the control movement and the corresponding display or system change. All controls have a limited range of travel, and this limits how sensitive the control can be (Hammerton, 1989). For example, the maximum displacement of a joystick is restricted to about 40° from vertical, so a high control–display ratio (i.e., low sensitivity) is not possible. When the control's range is limited, the optimum control–display ratio is usually the one with the lowest possible sensitivity. Alternatively, we could design two controls for these kinds of adjustments: a control with high gain for coarse adjustments and a control with low gain for fine adjustments, but there will be a cost (in time) for having to switch between two controls.

Many variables will influence the optimum control–display ratio, or gain, for a control. For example, the optimal gain for a joystick control will be less if it has a short handle than if it has a long one, because the short handle requires less hand movement to cover the same angular-displacement range (Huysmans et al., 2006). This influence of many variables means that, for any design problem, a control will need to be tested in realistic settings by a representative sample of its potential users. Ellis et al. (2004) examined surgeons' abilities to make aimed movements with a surgical robot of the type used in robot assisted surgery. The surgeons watched the robot's movements on a video display that allowed several different levels of zoom, or magnification of the region around the robotic end effector and the target location, by varying the settings of a surgical microscope. The control–response ratio (i.e., the ratio between control movement and robot movement) was varied between 7.6/1 and 2.8/1. Movement time was shortest when gain was set at an intermediate level, with the optimum gain increasing as optical zoom decreased. This result suggests that, with lower magnification, travel time is relatively more important than total movement time to the target.

The control–display ratio alone does not completely characterize the performance of a control (Arnaut & Greenstein, 1990). A control and its output have four components: display and control amplitude (or movement distance) and display and control target width (the range of positions that will be on target). As you now know, gain is defined by the ratio of the display and control amplitudes. However, Arnaut and Greenstein found that performance in cursor positioning with a mouse was also affected by display target width, as would be expected from Fitts's law (see Chapter 14). Thompson, Slocum, and Bohan (2004) also demonstrated this by showing that, just as movement times for longer distances are reduced by high gain controls, movement times to larger targets are also reduced by high gain controls. However, when gain was very high, movement times were longer than those predicted by Fitts's law because of increases in fine adjust time for smaller targets.

Control System Order

All of our discussion of control–display ratios has so far focused on tasks for which control position determined the position of the display element. However, control position may not always relate directly to display (or system) position. *Control system order* describes how changes in a display or other system response occur as derivatives of control position with respect to time. Up to now, we have only discussed zero-order controls, where there is a direct relation between the displacement of the control and the position of the display element. Computer mice and radio tuning knobs are zero-order controls. First-order controls determine velocity. The (misnamed) accelerator pedal in a car is a first-order control: If you hold it at a particular position, the car will move at a constant velocity on a flat road. Some joysticks can operate like a first-order control: When they are at a fixed position, the display cursor will move in a particular direction at a constant velocity.

A second-order control determines the acceleration of the control object or system output. A vehicle steering wheel is usually described as a second-order control, because the amount the wheel is turned in a particular direction determines a change in rate of velocity in that direction. Also, some processes in nuclear reactor control rooms and chemical plant use second-order controls. Even higher-order control is found in other complex systems. For example, the steering mechanism of a ship or submarine is best characterized as third order, with considerable lag between the controlling action and the system response, whereas flying an aircraft uses third-order longitudinal and fourth-order lateral controls (Roscoe, Eisele, & Bergman, 1980).

People have an easier time using lower-order controls than higher-order controls. To understand this, consider a joystick and the actions required to move a cursor from the center of the computer screen to a target location on the left and then back to center. A zero-order joystick is moved left to the position that corresponds to the target location and then held in that position. To return to center, the joystick is moved back to the neutral position. For a first-order joystick, at least two movements are required for each segment of the task. Positioning the joystick to the left imparts leftward velocity to the cursor. To stop the cursor at the target location (i.e., to impart zero velocity), the joystick must be returned to the neutral setting. These actions then must be repeated in the opposite direction to return the cursor to center. For a second-order joystick, a minimum of three movements are required for each segment. Deflection of the joystick to the left produces a constant rate of acceleration toward the target location. To decelerate, the joystick must be moved to the corresponding position on the right, and then returned to center just as the cursor reaches the target. With complex systems with third or fourth control orders, the relation between control actions and the changes of the system can become even more complex and obscure. You should now understand why one general human factors principle is to use the lowest possible order of control.

Order of control also plays a role in determining optimal gain. Kantowitz and Elvers (1988) asked people to perform a tracking task with zero-order positional and first-order velocity controls using an isometric joystick. Consistently with our principle that lower-order controls are easier to use than higher-order controls, their performance was considerably better with the zero-order control. However, high gain improved their performance with the first-order control but hurt it with the zero-order control.

Anzai (1984) used a simulated ship to examine how people learn to manage a complex system with a high order of control. He asked people to steer a ship through a series of gates as quickly as possible, as if they were steering a large tanker into a narrow harbor. The steering system used a second-order control. There are several factors that make this task very difficult. First, to effectively use a second-order control, the pilot has to remember a sequence of past control actions to determine the most appropriate control actions to make in the future. Second, large ships have a great deal of inertia and a long delay between when particular control actions are taken and when the ship responds to them. Finally, the direction that the ship must go (as seen as a change in its angle) and the angle that the steering wheel must be turned to accomplish a change in direction are very different.

Anzai (1984) collected verbal protocols from his subjects, in which they tried to tell him how they steered the ship. These protocols showed that novice pilots spent most of their time trying to figure out how control actions changed the ship's trajectory. The pilots devoted most of their attention to the immediate effects of control actions rather than to predicting the future or to selecting strategies. More experienced pilots focused instead on making predictions and using strategies designed to attain more distant goals. Anzai described the process of skill acquisition for these pilots as one in which general heuristics such as "The ship is going straight, but the next gate is at the right, so I turn the control dial to the right" are refined over time. The pilots monitor the course of the ship with the goal of detecting errors, and the actions associated with these errors are incorporated into more sophisticated heuristics and strategies. In short, learning how to control a complex system involves extensive use of cognitive strategies that ultimately lead to an accurate mental model of control manipulation–outcome relationships.

One way to improve performance with higher-order control systems is to use augmented displays that provide the pilot with visual feedback in a form consistent with the display–control relationship (Hammerton, 1989). For example, a rate-augmented display would show not only the current state of the system but also the rate at which it is changing. So, the pilot of an airplane may see information about altitude and rate of change in altitude. The rate information is useful during approach and leveling out at a desired altitude.

Another kind of augmented display is a predictor display. This display shows both the current status of the system and how it is likely to change in the immediate future. The future state of the system is predicted from its current state, velocity, acceleration, and so on. How far in the future the display can show will be a function of the speed at which the system responds. This display presents what the state of the system will be if the pilot keeps the control at its current setting. Figure 15.6 shows a predictor display that is used to assist a submarine pilot to level off the ship at an intended depth (Kelley, 1962).

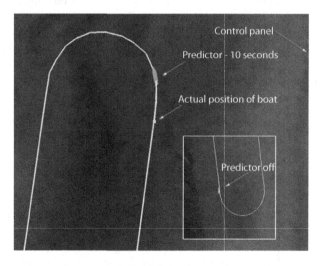

FIGURE 15.6 A predictive display to assist a helmsman in controlling a 40 ft survey water vessel, with predictor on and off (inset).

A similar type of predictor display for high-speed rail service was shown to be effective in simulation studies (Askey & Sheridan, 1996). The display includes a preview of the route ahead, including signal locations, station locations, speed limits, and so on. It also shows the operator the predicted speed over the next 20 s and curves for where the train will be if normal braking or emergency braking is applied. A more sophisticated version of the display includes an advisory function that provides the optimal speed profile for getting the train to the station on time while adhering to speed limits and minimizing energy consumption. The predictor display reduced the deviation of station stops from 12.7 m to less than 1 m. Reaction time for the operator to return the throttle to neutral at onset of a signal decreased from 8.6 to 1.8 s.

Augmented displays may also help the operator acquire an appropriate mental model of the system. Eberts and Schneider (1985) had people learn to perform a second-order tracking task, some with an augmented display that showed the required control positions for a given acceleration (see Figure 15.7) and some without. The display showed the operator the relation between the position of the joystick and the acceleration of the cursor. Those who used the augmented display during training were able to perform the tracking task accurately even after the display was removed. The augmented display not only aided performance but also gave the operators the information necessary to learn the second-order relationship.

Many systems are highly automated, so that the operator's role is supervisory control. Process control systems or the sort found in chemical plants and manufacturing require this kind of control. The operator monitors the performance of semiautomated subsystems and executes controlling actions when he or she detects a problem. This kind of monitoring requires complex problem solving, and so the cognitive processes of the operator are even more important than in difficult manual control tasks. Good performance will depend heavily on how accurate the operator's mental model is. However, it is important to realize that manual control remains important as well. As Wieringa and van Wijk (1997) emphasize, "The reader should realize that supervisory control tasks still include manual actions. In fact, these actions, such as pressing a button or moving a switch, play a very important role in supervisory control because they are the realization of the human-machine communication" (p. 251).

Moray (1987) noted that because the operator has had experience with only a limited range of states of the system, the operator's mental model will be a reduced version of the complete model that characterizes the system. This reduced mental model will consist of subsystems in which some states are combined and labeled idiosyncratically. The reduced mental model has the virtues of being simpler and less capacity demanding than the complete version. However, it does not allow the true state of the system to be deduced unambiguously.

The reduced model will be sufficient during normal operating conditions. Moray (1987) suggests that decision aids and displays for normal operation will be most useful if they are based on analyses of the subsystems contained in the operators' mental models. However, solutions to emergencies typically require consideration of the complete system model. Support systems for

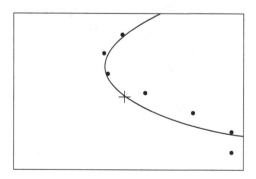

FIGURE 15.7 A parabola display indicating the required positions for a given acceleration.

emergencies should assist the operator in "breaking the cognitive set" imposed by the restricted mental model and encourage the operator to think in terms of the complete model. Decision-support tools based on expert knowledge provide a promising means for assisting both the normal and emergency procedures in supervisory control (Bugarski, Bačkalić, & Kuzmanov, 2013; Wieringa & van Wijk, 1997).

CONTROL PANELS

In most situations, a control is not placed in isolation but is included among other controls on a panel. There may also be a visual display that indicates the state of the system, and in the case of vehicles, shows a view of the outside world as well. The operator must be able to reach, identify, select, and operate the appropriate control in response to displayed information, while at the same time continuing to monitor visual input and avoiding the operation of inappropriate controls. We need to be aware of the factors that influence operators' performance in such situations and take these factors into consideration when we design control panels.

CODING OF CONTROLS

When a control panel contains more than one control, the operator must be able to identify each of the controls on the panel. He will not be able to make the appropriate control action if he selects the wrong control, and the change in the system that results from manipulating the wrong control may cause the system to fail. Above all, operators must be able to identify the appropriate control quickly and accurately. To minimize confusions in identification, controls must be coded so that they can be easily distinguished and recognized. Good control coding increases the chance that operators can locate controls on a panel rapidly and accurately.

Several coding methods can be used in a particular application. Which is most appropriate will depend on (1) the demands on the operator, (2) the coding methods already being used, (3) the illumination level, (4) the speed and accuracy required for control identification, (5) the available space, and (6) the number of controls that must be coded (Hunt, 1953).

Location Coding

In most applications, controls are distinguished by their locations, like the accelerator and brake pedals in a car. The brake is always to the left of the accelerator. Location coding is only effective when the distance between the controls is far enough that the operator can reliably distinguish between the different control locations.

How far is far enough? There are no consistent guidelines to answer this question. One laboratory study suggested that in some circumstances, the final positions of continuous movements can be reliably discriminated if the positions are as little as 1.25 cm apart (Magill & Parks, 1983). However, for most situations, controls must be separated considerably further apart than this.

Many of the airplane accidents in World War II were caused by pilots who failed to discriminate between flaps and landing-gear controls (Fitts & Jones, 1947). The two controls were situated in close proximity, with no other distinguishing features beside their different locations. Consequently, when the flaps were to be adjusted during landing, the landing gear was often raised instead. Despite the fact that it has been known for many years that this problem can be alleviated by using other forms of control coding, the National Transportation Safety Board reported that control misidentification, due to the exclusive use of location coding, was the major source of accidents in one popular small aircraft (Norman, 2013).

Localizing controls along a vertical dimension results in more accurate identification than localizing them along a horizontal dimension. Fitts and Crannell (1953) had blindfolded individuals reach to and activate one of nine toggle switches arrayed in a vertical column or horizontal row. People were more accurate with the vertical arrangement than with the horizontal arrangement.

FIGURE 15.8 Chalkboard keg tap handles for labeling. From www.homewetbar.com/whats-tap-chalkboard-tap-handle-engraveable-p-4416.html.

They made fewer errors when vertically arranged controls were separated by 6.3 cm or more and when horizontally arranged controls were separated by 10.2 cm or more.

Because the operator's representation of control location is not very precise, location coding alone is not sufficient for most applications. Control panel designers will usually augment location coding with some other form of coding. In fact, when other forms of coding are not provided by system designers, operators often institute their own augmented coding systems (see Figure 15.8).

Labels

Labels are demarcated by different materials, paint, or other surfaces that contrast with the control panel. They are placed very near or on controls and provide information in written or symbolic form about the use of that control. There are a number of factors involved in the design of labels, and one important consideration is how well they can be interpreted. The use of labels assumes that all control operators will be able to interpret them correctly. Uninterpretable labels can result in poor system performance or even failure. One study examined the labels on mechanical ventilating systems in low-carbon housing developments. Such developments usually rely on heating and air conditioning systems that use centralized resources and energy storage that may be unfamiliar to most residents. The residents of the housing units in the study found the labels on the control panels for the ventilation systems to be confusing, leading to an inability to adjust the controls (Stevenson, Carmona-Andreu, & Hancock, 2013).

Either alphanumeric or symbolic labels can be used to identify controls. However, using only labels to indicate control functions is not a good idea. In addition to concerns about label interpretability, if only labels are used, the operator must always be able to see the labels, which will not be possible in dim lighting or if the operator is unable to look at the control. Also, when a large number of similar controls are distinguished only by labels, people's responses tend to be slow and inaccurate. Finally, labels require space on the control panel, and not all panels will have enough room for them.

Designers of control panels use the following general principles when they label controls (Chapanis & Kinkade, 1972):

1. Locate labels systematically relative to the controls.
2. Make labels brief, without using technical terms.

3. Avoid using abstract symbols that may require special training, and use common symbols in a conventional manner.
4. Use standard, easily readable fonts for alphanumeric characters.
5. Position labels so that they can be referred to while the operator engages the control.

You may not realize it, but a TV remote control is actually a control panel. Because digital TV provides much more functionality than standard analog TV, the digital TV remote control is an example of a control panel where labeling is important. The larger number of controls on the digital remote and the limited size of the remote mean that labels must be carefully designed. However, many users find these labels hard to understand.

The process by which designers select labels can be complex, and may make use of more than one research method. One study used focus groups, questionnaires, and behavioral experiments (Lessiter, Freeman, Davis, & Dumbreck, 2003) to design a remote control. The focus groups were small groups of British citizens who generated several possible labels for different functions on the remote, for example, "system Set-up Menu." These possibilities were then screened by a panel of digital television experts. These experts eliminated suggestions that were inappropriate for one reason or another, such as being too long to place on a button.

The remaining suggestions were mailed out as a questionnaire to a larger sample of the British public. The questionnaire asked respondents to rank their preferences for the labels nominated for each function. The designers then selected a small number of labels and functions to test in controlled experiments, in which people were asked to use the controls with different labels. The designers found that people's speed and accuracy of button-label identification were usually highly correlated with the preferences shown in the questionnaire data, although there were some exceptions. The results of this study suggest that people's preferences can be used as a basis for selecting control labels, and controlled testing that measures objective performance can also give important information about how intuitive users will find different labels.

Color Coding

Another way to code visible controls is to color them differently. Recall from Chapter 8 that an operator's capacity to make absolute judgments is limited to about five categories along a single dimension. This means that, for most situations, no more than five different colors should be used to distinguish between controls. When the controls are close enough to allow side-by-side comparisons of color, the number of colors can be greater than five. The primary disadvantage of color coding is that perceived color will vary as a function of the illumination in the workspace. If this is a serious concern, designers can conduct psychophysical experiments to determine how discriminable different colors are under the illumination conditions in the workspace. The designers also must remember that a significant portion (up to 10%) of people are color blind, which means that any attempt to use color coding should be paired with some other type of coding (labels, shapes, etc.).

Color coding can improve performance when the displayed signals are also colored. One study used a 2×2 display of four stimulus lights paired with a 2×2 panel of four toggle switches (Poock, 1969). In the control condition, all of the lights were red and the switches white (no color coding), and in the experimental conditions, the lights were of four different colors and the assigned responses were the same colors (color coding). When the toggle switch was in the same relative position as the light to which it was assigned, people performed equally well in the control and color-coded conditions. However, when the assignment was spatially incompatible (i.e., if the top left toggle switch was to be activated for the bottom right light), people responded much faster in the color-coded conditions than in the control conditions. Thus, when display and control elements are mismatched spatially, a color match allows people to rapidly determine which control is assigned to which display element.

Shape Coding

Shape coding of controls is particularly useful, especially under conditions in which vision is unreliable. While shape can provide a visual feature for distinguishing controls, it is a tactual feature useful when the viewing conditions are poor or the operator's gaze needs to be directed elsewhere. People can accurately distinguish a large number of shapes (between 8 and 10, if they are selected carefully) through touch. The principal drawback of shape coding is that it may make it harder to manipulate a control and to monitor its setting.

An early study on shape coding was performed by Jenkins (1946). People in this study were blindfolded and then asked to identify the shape of a knob. There were a total of 25 different shapes. Jenkins analyzed the mistakes that people made—how different shapes were confused with each other—and identified two groups of knobs for which the within-group confusions were minimal (see Figure 15.9). Since this early study, other studies have been performed, resulting in more sets of shapes that are difficult to confuse (Hunt, 1953).

Size Coding

Size coding is another useful scheme to use when vision is restricted. However, people cannot accurately discriminate between as many sizes as they can shapes. This means that two controls coded by size will need to be very different in size to be distinguishable, and this may make one of the controls difficult to operate. Very large or very small controls may be difficult to grasp and manipulate. Hence, size coding is best used in conjunction with other coding methods.

Controls coded by size may vary not only in the diameter of the control knobs but also in their thicknesses. As a general guideline, knobs that differ by 1.27 cm in diameter or 0.93 cm in thickness will not be confused (Bradley, 1967). Designers can, by combining changes in diameter with changes in thickness, create larger sets of easily discriminable controls.

Size coding is important for "ganged" controls, where two (or more) knobs are mounted concentrically on the same shaft to save space. Your car radio's volume and tone controls may very well be ganged. In this kind of situation, the different knobs must be of different sizes so that they can be both discriminated from each other and operated separately.

Texture Coding

Surface texture is another dimension on which controls can vary. In Bradley's (1967) original study investigating the confusability of knobs, he also investigated people's ability to tactually discriminate the knobs shown in Figure 15.10. He found that people could reliably identify three classes of textures: smooth, fluted, and knurled. That is, people never confused the smooth knob with any other knob, and rarely confused the fluted knobs with the knurled knobs. However, people could not distinguish between different types of knurled or fluted knobs. Bradley proposed that these three texture classes can be used for the coding of controls. When a designer couples size coding by diameter and thickness with texture, he can construct a large set of tactually identifiable knobs.

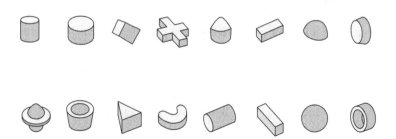

FIGURE 15.9 Two groups of eight knobs for which within-group confusions are minimal.

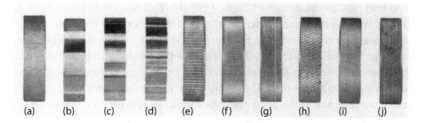

FIGURE 15.10   Examples of smooth (a), fluted (b–d), and knurled (e–j) knobs.

Other Codes

In addition to the common coding dimensions of labels, color, shape, size, and texture, coding can be based on the type of control operation. For example, a rotary control for amplitude is not likely to be confused with a pushbutton "on-off" switch. However, coding by mode of operation will not be an effective way to prevent selection errors in most situations, because the operator has to select the control before operating it. An alternative is redundant coding, which we discussed briefly above. Redundant codes use two or more dimensions. If a designer combines some type of visual coding in the form of labels or colors together with some type of tactual coding, she can ensure that the operator can use more than one sensory modality to identify a control. Given these alternative coding schemes, designers can choose from a wide array of optimal coding systems for any system environment.

CONTROL ARRANGEMENTS

When we discussed visual displays in Chapter 8, we stressed the importance of functional grouping of displays. Functional grouping is also an effective way to organize control panels (Proctor & Vu, 2006b). It is particularly beneficial when the display groups match the control groups. Grouping of controls is accomplished by placing related controls close together on the panel or by designing them to be of similar size and shape.

In Chapter 13, we talked about how people have stereotypical preferences relating the spatial location of visual stimuli to the location of responses. In addition to these sorts of stereotypes, people exhibit population stereotypes about the locations of controls. One study evaluated the position controls on tractors (Casey & Kiso, 1990). There are three critical controls (throttle, range shift, and remote hydraulic) on a tractor, and the researchers obtained subjective ratings of the positions of these controls from many tractor drivers for many different tractors. The locations of these controls across the different tractors were highly variable and had a strong influence on the users' ratings. The best-preferred locations for each of the controls depended on their functions. Investigations of other kinds of machines show a similar dependence between control function and location preference (i.e., heavy mining vehicles; Hubbard, Naqvi, & Capra, 2001).

After a designer decides where controls should be located on a panel, he must then make sure that the controls can be reached and that the controls that will be used most frequently are most easily accessible. Consider, for example, the problems posed in the design of the "cockpit" of a standard automobile. The driver of a car may have to look away from the road to operate the correct control. The farther a control is from the driver's normal line of sight, the longer the driver will need to look away from the road, and the larger and less well-controlled her steering movements will be (Dukic, Hanson, & Falkmer, 2006). This is dangerous, and the time required to locate and operate such a control should be as short as possible (Abendroth & Landau, 2006). In response to this concern, designs incorporating steering-wheel-mounted switches are now quite common (Mossey, 2013).

As discussed in the next chapter, anthropometric data (physical characteristics of the population of users) are used to design workspaces to accommodate most people. Similar concerns and strategies come into play in the design of control panels. To ensure that 95% of the users will be able to reach the controls on a panel, a designer can establish a *reach envelope* from the 5th percentile of the population's reach distance. Figure 15.11 shows two such reach envelopes for seated male operators. The immediate reach envelope specifies the region within which people can reach controls without bending, whereas the maximum reach envelope specifies the region within which people can reach controls with bending. Designers try to locate frequently used controls within the immediate reach envelope and rarely used controls within the maximum reach boundary.

Control panel designs can be evaluated using an index that incorporates the distance of each control and the frequency of its use (Banks & Boone, 1981). This *index of accessibility* (IA) is based on the operator's immediate reach envelope, the frequencies with which individual controls are used, and the relative physical positions of the controls with respect to the operator.

A limitation of the accessibility index is that it does not take into account grouping by function, sequence, or spacing. Many variables go into the computation of the IA. If a designer wanted to design a control panel so that the IA was as large as possible, she might be faced with a problem that was very difficult to solve. The way that we handle problems like these is to employ an optimization algorithm, usually implemented in a computer program. Based on the factors that the designer identifies as most important, an appropriate mathematical index (like the IA) is optimized. For example, one study designed a control panel with the goal of minimizing the distance moved by the operator (Freund & Sadosky, 1967). The distance computation incorporated the frequency of use for each control, the distances of the controls from a central point, and the distance between controls.

The index to be optimized can incorporate a great many design concerns (Holman, Carnahan, & Bulfin, 2003). For instance, to optimize (minimize) movement distance, measuring the distance of each control from a single central point may not be appropriate if an operator is going to be manipulating some controls with the left hand and some with the right hand. An expanded index can use two origin points, one on the left and one on the right. It can also take into account the sequence of control use, how controls are clustered (e.g., the numeric keypad on a computer keyboard), how controls are aligned to the left or right side of the control panel, and whether the control must be operated by the left hand, the right hand, or both.

A model called the *index of functionality*, developed specifically for agricultural tractors (Drakopoulos & Mann, 2008), takes into account both the physical characteristics and frequency of use of the controls in the workstation, and four attributes of the workstation design (placement of

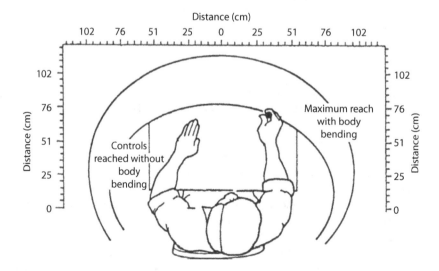

FIGURE 15.11 Immediate and maximum reach envelopes, based on the 5th percentile for adult males.

controls, suitability of controls, functional reach, and labeling of controls). The researchers showed that the index of functionality has increased greatly over the years, demonstrating ergonomic improvements in tractor design.

PREVENTING ACCIDENTAL OPERATION

If a person inadvertently operates a wrong control on a panel, the system being controlled may fail. When the consequences of accidental activation of a control are serious, designers take steps to ensure that it is very difficult and unlikely for the control to be unintentionally activated. However, making a control hard to operate by accident will also make it hard to operate on purpose.

There are several ways designers can minimize the likelihood of a control's accidental activation (Chapanis & Kinkade, 1972; NASA, 1995). Perhaps the easiest way is to make sure the control is far enough away that the operator is unlikely to come into accidental contact with it. We could also increase the control resistance, which in turn will increase the force required for the operator to activate the control. We might recess the control into the control panel or place it behind a barrier. If the control is not used frequently, we can place a protective cover over it, or lock the control into position. These options do not only make the control harder to operate; they create a sequence of several actions that the operator must take to activate the control (e.g., lift the cover, unlock the control, and then move the control). If the operation requires more than a single step, the operator must go through a sequence not only of actions, but also of decisions to make each action. Each point in the activation process then provides an opportunity for the operator to reconsider her choice to activate the control.

SPECIFIC CONTROLS

Up to this point, we have discussed general principles about controls that can be applied to design decisions. However, unique controls have features that make them more or less suited to different applications. Consequently, we will talk in greater detail about several types of controls and their applications.

HAND-OPERATED CONTROLS

Most controls are hand operated. As we indicated at the beginning of the chapter, such controls come in various shapes and sizes. Among the most widely used are pushbuttons, toggle switches, rotary selector switches, and rotary knobs. Table 15.3 summarizes the characteristics of each of these four types of controls.

Pushbutton and Toggle Switches

We see pushbutton controls on a range of devices, from telephones to calculators to industrial control panels. They are used for stopping and starting machines and engaging and disengaging particular operating modes. For example, the interface for an automobile audio system may use a pushbutton to move between Bluetooth, AM radio, and FM radio modes. As with all controls, designers must design and place pushbuttons so that they can be reached and pressed. When pushbutton telephones began to replace rotary telephones in the 1960s, the layout and pushbutton properties were based on detailed human factors experiments conducted to evaluate button arrangement, the force required for depression, the amount of movement of the buttons, and feedback concerning the press (Deininger, 1960; Lutz & Chapanis, 1955).

The factors that determine pushbutton usability include resistance, displacement, diameter, and separation between pushbuttons. Table 15.4 shows the recommended physical dimensions of a pushbutton depending on its function and whether the button is to be activated by a finger or a thumb (Chengalur et al., 2004; Moore, 1975). Optimal pushbutton resistance depends on factors such as

TABLE 15.3
Comparison of Common Control Types

Characteristic	Push Button	Toggle Switch	Rotary Selector Switch	Continuous Knob
Time required to make control setting	Very quick	Very quick	Medium to quick	—
Recommended number of control positions (settings)	2	2–3	3–24	—
Likelihood of accidental activation	Medium	Medium	Low	Medium
Effectiveness of coding	Fair	Fair	Good	Good
Effectiveness of visually identifying control position	Poor	Good	Fair	Fair
Effectiveness of check reading to determine control position when part of a group of like controls	Poor	Good	Good	Good

TABLE 15.4
Recommended Minimum (Min), Maximum (Max), and Preferred Physical Dimensions of Pushbuttons for Operations by Finger or Thumb

	Diameter (mm)	Displacement (mm)		Resistance (g)		Control Separation (mm)	
	Min	Min	Max	Min	Max	Min	Preferred
Type of operations fingertip							
One finger—randomly	13	3	6	283	1133	13	50
One finger—sequentially	13	3	6	283	1133	6	13
Different fingers—randomly or sequentially	13	3	6	140	560	6	13
Thumb (or palm)	19	3	38	283	2272	25	150
Applications							
Heavy industrial pushbutton	19	6	38	283	2272	25	50
Car dashboard switch	13	6	13	283	1133	13	25
Calculating machine keys	13	3		100	200	3	
Typewriter	13	0.75	4.75	26	152	6	6

the extent to which accidental activation must be avoided and the strength of the user population. Older adults, for example, are not as strong as younger adults and are more likely to have a physical disability such as arthritis or hand tremor. Therefore, they may have difficulty operating pushbuttons that require a great deal of force or travel distance (how far the button moves), particularly if they have to press on the button for any significant length of time (like a fast-forward or reverse button on a remote control; Rahman, Sprigle, & Sharit, 1998). Apart from concerns about resistance, displacement, and so forth, we should design pushbuttons so that they provide the user with some visual or auditory feedback that the button has been activated, such as the audible clicks of keys on most computer keyboards. Simple feedback like this can increase the usability of the pushbutton for many applications (Ivergaard, 2006).

We presented Fitts's law in Chapter 14. Recall that Fitts's law states that movement time is a linear function of the index of difficulty, which in turn is a function of the size of a target (button) and its distance. If we want different pushbuttons to be equally easy to operate, we will need to increase the size of the button with the distance of the button from the operator. The separation between buttons is also important (Bradley & Wallis, 1958). If we hold the distance between button centers constant, decreasing the size of the buttons will reduce operator error without affecting response time. If we hold the distance between the edges of the buttons constant, both accuracy and speed of responding increases with the diameter of the buttons. This means that as we put buttons closer together, performance will suffer. This decrease in performance will be even more dramatic if visual feedback is not available, such as when the operator must perform in poor lighting. If the operator is expected to perform with limited vision, we will need to design control panels with widely spaced pushbuttons.

We can improve an operator's ability to identify different pushbuttons using any of the coding methods we described above. Most likely we will use some form of button labeling. In situations where the buttons are close together, we can place the identifying labels directly on the buttons to avoid confusion about which label goes with which button. However, we may want to consider placing labels above the buttons, so that the operator can see the label at the same time as she presses the button. When the control panel is spacious enough, this is a viable option. In some situations, we can use software to display programmable "virtual" labels on a screen, so that a single pushbutton can serve different functions in different applications. We will discuss this option in more detail later.

We might also consider tactile coding. To be effective, the shapes and textures used for such codes must be distinguishable by touch using the tip of the index or middle finger, which is used to activate most pushbutton controls. In simple applications, we will probably want to do some testing to find out which shapes are most easily discriminable for our application, and which shape we should associate with which function. As an example, we will discuss an experiment conducted by Moore (1974).

The application that Moore was interested in was complex mail sorting equipment used by the Post Office in the U.K. This application needed six pushbuttons to perform the control functions Start, Stop, Slow, Delayed Stop, Inch, and Reverse. He designed 25 shapes on 2-cm diameter buttons (see Figure 15.12). To determine how confusable the different buttons were, he asked people to identify each shape by reaching through a hole in a curtain and touching each button with the top of their forefinger. As a result of people's ability to identify the buttons, he determined that shapes 1, 4, 21, 22, 23, and 24 (see Figure 15.12) were seldom confused and provided an easily discriminable set.

After Moore (1974) determined the set of six most easily discriminable controls, he was still faced with the problem of which control to assign to which function. Accordingly, he conducted another experiment in which blindfolded people ranked each of the six buttons in terms of their suitability for each control function. He used these rankings to determine the final assignments. In today's terms, what Moore did was to determine the stereotypic function for each pushbutton shape so that their shapes could be assigned consistently. For larger or more complex problems, we can use optimization algorithms to select codes for sets of any size (Theise, 1989).

Many of these same issues will arise if we are interested in toggle switches. However, toggle switches also have a direction, which may create design problems. As a rule of thumb, we will want to place toggle switches so that they can be flipped without moving the hand much. People will be able to respond most quickly when the toggle switches are arranged horizontally and switched up and down (Bradley & Wallis, 1960). If we decide to place toggle switches vertically, then the switches should move to the left and right. An operator will be less likely to accidentally throw a toggle switch than he will be to accidentally push a button, particularly when the buttons and switches are very close together. These facts lead us to the recommendation that for panels in which the spacing between centers is restricted to less than 2.54 cm, we should use toggle switches instead of pushbuttons if our goal is to minimize accidental activation of the control. However, even though errors will be less likely for toggle switches, people will still need more time to operate the switches as the density of controls increases (i.e., separation decreases; Siegel, Schultz, & Lanterman, 1963).

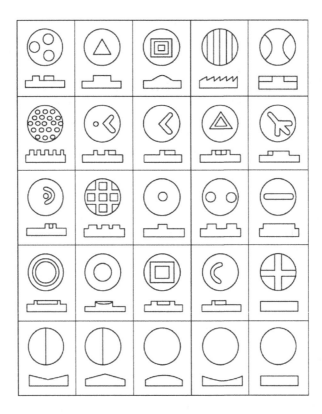

FIGURE 15.12 Pushbutton shapes examined by Moore (1974).

Rotary Selector Switches and Knobs

Rotary selector switches can accommodate up to 24 discriminable settings (Ivergaard, 2006). The primary drawback of these switches is that they cannot be operated as quickly as toggle switches or pushbuttons. Figure 15.13 shows recommended dimensions for rotary switches. In addition to the physical dimensions of the switch, we need to worry about the pointer and the scale. The pointer should be easy to see and mounted close to the scale. The beginning and end of the scale ranges should have stops, so that the switch can't move past the scale limits, and the switch should "click" into each setting, so that there is no confusion about which setting has been selected.

A switch that "clicks" into position is a discrete control. Rotary switches and knobs can also be continuous. The volume control on a stereo amplifier is usually a continuous rotary knob. Many of the same considerations that apply to other controls apply to continuous rotary knobs. For instance, spacing between knobs, knob diameter, and knob configuration will determine the usability of the control (Bradley, 1969b). People do best when knobs are separated by at least 2.54 cm (measured between knob edges). People also make fewer mistakes with controls of smaller diameters, all other factors (like the distance between knob centers) being held constant. However, if the distance between knob edges is constant, people get better as the knobs get bigger.

This means that if we have the space on the control panel, we should use larger knobs spaced farther apart. If space is limited, we need to make sure that the knobs are at least 2.54 cm apart and small. Other considerations include the arrangement of knobs on the panel and the coding schemes we use to identify different knobs. People make fewer contact errors with knobs arranged

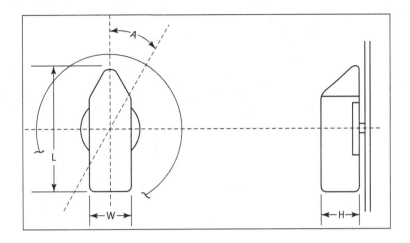

Dimensions			
Length L	Width W	Depth H	Resistance
Minimum 25 mm (1 in.)		16 mm (0.625 in.)	113 mN-m (1 in.-lb)
Maximum 100 mm (4 in.)	25 mm (7 in.)	75 mm (3 in.)	678 mN-m (6 in.-lb)

Displacement A		Separation	
Regular*	Large**	One-hand random	Two-hand operation
Minimum 15°	30°	25 mm (1 in.)	75 mm (3 in.)
Maximum 40°	90°		125 mm (5 in.)
Preferred —	—	50 mm (2 in.)	125 mm (5 in.)

* For facilitating performance.
** When special engineering requirements demand large separation.

FIGURE 15.13   Recommended dimensions for rotary switches.

vertically: horizontal arrays result in more accidental bumping and brushing of the knobs with arms, elbows, and so forth. If we decide to use size coding, the difference in sizes between different knobs must be at least 1.27 cm in diameter or 0.95 cm in thickness to minimize confusion (Bradley, 1967).

Controls like switches, buttons, and knobs connect to equipment behind the control panel. One way to reduce the amount of space needed for these connections is to use concentrically mounted rotary knobs. Such controls are useful when the control functions are related, when the controls must be operated in sequence, when some knobs of necessity must be large, and if inadvertent activation of one of the ganged knobs is not critical. Figure 15.14 outlines some general recommendations for ganged knobs (Bradley, 1969a). Consider a knob that is approximately 4 cm in diameter. If that knob is ganged with a smaller knob (on the top), the smaller knob should be no greater than 1.5 cm in diameter. If that knob is ganged with a larger knob (on the bottom), the larger knob should be no smaller than 7 cm in diameter. These size differences will make sure that the operator will be able to identify each knob and operate them easily. Unfortunately, because we need these large differences in size to ensure that the ganged knobs are usable, we won't reduce the amount of space for the controls on the front of the panel.

Some people will have trouble using rotary controls. For example, people with arthritis or muscular dystrophy may not be able to use some kinds of rotary controls. Arthritis reduces the amount of torque (turning force) that a person can apply to a knob (Metz, Isle, Denno, & Li, 1990). Women

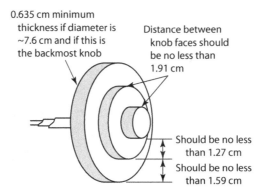

FIGURE 15.14 Recommended dimensions for concentric controls.

can, on average, apply only 50% of the torque that men can (Matsouka et al., 2006). These and other population differences will influence how we decide the best design for a control panel.

Multifunction Controls

A complex system will have many more control functions than a simpler system. The number of controls that we can put on a panel is limited, as is the number of controls that a person can operate at one time. Like ganging rotary knobs, we can recover some space and decrease the number of controls by using multifunction controls (Wierwille, 1984). For example, computer joysticks also have pushbuttons located close to the grip.

Military aircraft have used controls like this for many years. The F-18 fighter jet, one of the first aircraft to make extensive use of multifunction displays/controls, was designed with two multifunction controls, one for the pilot's left hand and one for the right (Wierwille, 1984). These controls are shown in Figure 15.15. The main purpose of the left-hand control is to determine engine thrust by sliding controls forward/backward for each engine. The right-hand control is a two-dimensional joystick that manages pitch and roll. In addition, each control includes a variety of auxiliary controls to be operated by the thumb and fingers. These multifunction controls were designed so that the pilot could operate the individual controls without having to look at them. Coupled with a head-up display, this arrangement allows the pilot to fly the jet, fire its weapons, and so on, without ever moving his gaze from targets in the airspace.

The cockpits of most modern commercial aircraft are highly automated flight management systems. They are called "glass cockpits" in reference to the electronic visual display units used on the flight deck. The pilot interacts with the flight management by way of a multifunction control display unit (MCDU; see Figure 15.16). This interface, situated between the pilot and the co-pilot, is used for communication of the flight information between the flight crew and the aircraft (Kaber, Riley, & Tan, 2002). The MCDU consists of a video display terminal and a keypad that contains an alphabetic keyboard and a number pad for entry of flight information. It also contains mode keys, which determine what information is presented on the screen, and function keys for certain critical functions. The main use of the MCDU by the crew is to pre-program the whole flight, from takeoff through landing. As well, it is used to revise a flight plan. Because the MCDU is the primary means of communication between cockpit crew and the flight management system, maximizing usability of the MCDU is a priority.

One major drawback of an MCDU is the potential confusion that might arise in trying to remember which functions go with which keys—these assignments are sometimes called "keybindings." To avoid this confusion, Kaber et al. (2002) recommend that a function key should execute only a single function. When a designer must, from necessity, assign multiple functions to a single key, the unit should display the keybindings for each operating mode. These displays are sometimes called "soft" keys.

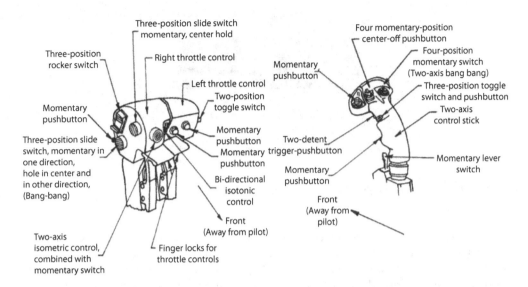

FIGURE 15.15   Left- and right-hand multifunctional controls for the F-18 aircraft.

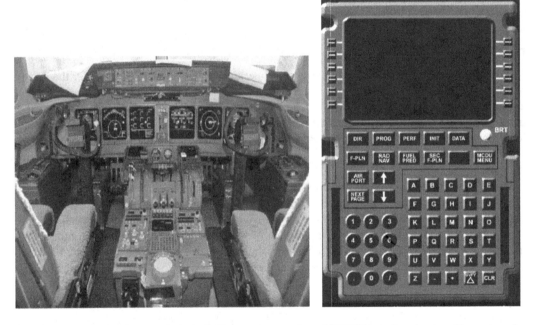

FIGURE 15.16   An MD-11 cockpit, including center console and MCDU interface (center of image), and a detailed depiction of the MCDU.

Complexity of control is a problem not only for airplane pilots but also for automobile drivers. Multifunction interfaces can also be used in cars to reduce this complexity. For example, many cars have several electronic devices, such as global positioning system (GPS) navigation systems, audio systems with radios and CD players, DVD players, cellular phones, environmental control systems, and so on. Bhise (2006) designed a multifunction interface for displaying and retrieving information such as music files and phone numbers from a microprocessor while driving. The interface is similar to that of a typical automobile radio but has "soft" buttons that can be bound to several functions (see Figure 15.17).

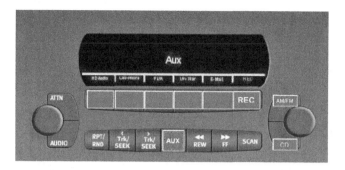

FIGURE 15.17 "Soft" button interface.

The interface has a modifiable center screen, below which are located six buttons. The buttons are labeled in the lower part of the screen, allowing the keybindings for each button to be changed as the driver progresses through a menu of control operations. Although this is an effective way to increase the number of virtual pushbuttons while maintaining the same number of actual ones, the menu structure can quickly become complex enough to distract the driver from the primary task of driving while searching for information. A range of issues arise in the use of soft controls for computer interfaces, as Box 15.1 describes.

FOOT-OPERATED CONTROLS

Although we tend not to think much about foot-operated controls when we consider the design of control panels, foot-operated controls are very common. They are used in automobiles and airplanes. They are used to power vehicles, such as bicycles, and to operate some musical instruments (e.g., pianos) and electrical machinery (e.g., sewing machines). As a general rule, foot controls can be used in situations where a person needs to use his hands for other tasks, or if the control requires more force to operate than a person can easily apply with his hands or arms.

If we decide that a foot-control is most appropriate for a particular design problem, we need to worry first about how quickly and accurately an operator can activate it. Remember that Fitts's law dictates how quickly a person can move to a target of fixed size and distance. We can apply Fitts's law to foot controls in the same way as we evaluated pushbutton controls. One interesting difference, though, is that a person's shoe size will need to be incorporated into the effective width of the target (the foot-operated switch; Drury, 1975). A modified index of difficulty (*I*) is

$$I = \log_2\left(\frac{D}{W + S} + 0.5\right),$$

where:
    D = target size,
    W = target width, and
    S = shoe sole width.

As long as the separation between pedals is set at a minimum safe separation of 130 mm (which is the 99th percentile shoe size), pedal width will not influence movement time. With practice, people can move to a foot-operated switch in any direction (front, back, left, right)

BOX 15.1   COMPUTER INTERFACE CONTROLS

Think about the interface provided on your favorite computer. You interact with graphical objects on the screen using a mouse, keyboard, touchpad, or touchscreen, and these objects direct the computer to perform different functions. When we refer to a computer interface control, we are referring to "an object that represents an interface to the underlying software" (Blankenship, 2003).

The controls for a computer interface differ from the kinds of controls you might find in your car or in an airplane cockpit. First, many computer interface controls are indirect or generated by software: to operate them, you have to perceive and identify graphical objects or read words on the display. The effect of clicking on a mouse or tapping on a screen depends on over which item in a menu or which button on the screen your finger or the cursor is located. Second, you need to make both discrete controlling actions (as when you tap or click on an icon or menu item) and continuous movements (like positioning the cursor with the mouse or swiping your fingers over the screen).

Direct computer controls are hardware, like the keyboard and mouse. Most computer keyboards consist of a typewriter keyboard, a number pad, and a collection of "function keys" that can be programmed to execute a variety of commands. Even though the typewriter has number keys arrayed above the upper row of letter keys, these keys are not arranged well for numeric entry; hence the addition of the number pad.

When computers shifted from text-only, command line interfaces to graphical interfaces, the computer mouse was developed to allow easy positioning of the cursor on the screen. There are other such positioning devices, like trackballs, light pens, tablets, and joysticks, but the mouse has the advantage of being shaped approximately to a user's hand and easily controlled by the larger muscles of the arm and shoulder, allowing simultaneous positioning of the fingers over keys to click when the cursor is positioned over a particular icon or location. Most computer interfaces rely heavily on pointing, clicking, scrolling, and other actions that can be performed easily with a mouse.

Smaller laptop computers, for which space is limited, use the standard typewriter keyboard but different kinds of controls. A number pad may not be included. If it is, it may take the form of a subset of the typewriter keys arranged roughly like a standard number pad and accessible by toggling a function key. Many of the keys, particularly function keys, may be smaller than those on a full-size keyboard. The mouse may be replaced with a small pointing stick, usually a force transducer, located in the middle of the keyboard, or a touch-sensitive pad located below the bottom row of keys. Most people find these cursor controls more difficult to use than the computer mouse. The difference between full-sized and laptop keyboards shows how functionality and space/size trade off against each other.

The number and kind of on-screen controls we can implement in software are limited only by the designer's imagination. Consider, for example, the kinds of controls you might find on a typical Web page. If you want to purchase something from an online store, the page might be the page for a particular item in the store's catalog; say, a digital camera. The page will display information about the camera you have selected. It might provide controls that let you look at pictures of the camera from different angles. It might have controls that let you select optional features of the camera, like color or memory size. It might give links to pages for similar cameras or for accessories, like bags, lenses, and memory cards, that you could purchase at the same time. The page will also contain navigation tools to take you to other pages where you can order the camera, get help, shop for other items, read consumer reviews of the product, and so forth.

All of these page elements, which you can click on and modify, are controls. Sometimes, you will see these controls called "widgets." We can classify the different controls by what they do (Blankenship, 2003):

1. *Information display controls*: These controls are to inform the visitor to the page. Examples include a progress indicator (which shows the status of a process that must be completed before the user can proceed) and a site map.
2. *Function controls*: These kinds of controls send a command to the host system. Examples include buttons that a visitor can click to specify preferences, and links, which take the visitor to other pages.
3. *Input controls*: These controls allow the visitor to enter information on the page. Textboxes permit the visitor to enter extended text, such as written comments or opinions. Others may permit the visitor to choose from a limited number of possible inputs. Examples include radio buttons, listboxes, and checkboxes, each of which allows visitors to select one or more options from a fixed list.
4. *Navigation controls*: These controls will transport the visitor to information located somewhere else. Links are the primary example of this kind of control, but another example is breadcrumbs. Breadcrumbs are often located at the top of a page, and they show the visitor their location in the site's hierarchy. If you are buying a digital camera, for example, the breadcrumbs might appear like this: "Electronics > Digital > Cameras > Less than $200." Each breadcrumb in the sequence may also be a link that can take the visitor to that page in the hierarchy.
5. *Containment controls*: These controls group other controls within a restricted area. The simplest example of a containment control is a group, which is simply a visual element like a box or colored area on which a set of controls is displayed. Tabstrips, which appear like the jutting tabs on file folders (see Figure B15.1), act as a series of dividers, allowing access to several Web pages in the same region on one page.
6. *Separation controls*: These controls consist of a line that separates two or more controls. These controls are usually inactive elements that appear on pages or in toolbars. For example, a toolbar separator is displayed between controls grouped on a toolbar.
7. *Layout controls*: These are controls that put other controls into rows or columns. Checkboxes and radio buttons may be organized by checkbox and radio button groups.
8. *Technical controls*: These are nonvisual controls that typically are not noticed by the visitor. Examples include drag source and drop target, which oversee information about mouse positions and clicks. These controls allow drag and drop actions, such as moving a file into a text box.

An important part of Web-page design involves deciding what kinds of controls to use, and this decision will depend on the purpose of the control and how it behaves. Consider, for example, the tabstrip control in Figure B15.1. Here are some of the questions that we need to ask ourselves if we choose to use a tabstrip control (Blankenship, 2003):

If the tabstrip control has many tabs, should a scrolling or paging mechanism appear? If so, when should it appear, and how should it be implemented?

Consider the contents of each tab. How are these aligned or arranged? Should the alignment be the same or different for various kinds of content (e.g., an image, a form, a table)?

Within each control, there is some "padding" that spaces out the items contained in the control. Is this padding always the same, or can it change depending on the kind of content?

Can different areas of the tabstrip have different background colors that signal the kind of contents they contain?

Is it appropriate to use icons in the tabs? If so, are there any restrictions on how we use them?

How should error and messages handling operate within the control?

How will disabled tabs be indicated visually? Will they be grayed out, differently colored, or invisible, for example?

After we answer all of these kinds of questions, we must decide on a design and develop a prototype page. We then need to test the page on many different computers, operating systems, and Web browsers to make sure it works well for everyone who might visit the page.

It is easy to design bad Web pages. Hypertext markup language (HTML), which is the language of the Web, is not difficult to learn, and so almost anyone with some programming experience can design his or her own site. These designers are not necessarily interested in good design, but only in getting their pages on the Web. You have probably visited some of these pages yourself. Remember how hard it was to read the page where the designer decided that the way to emphasize her message was to make the text blink on and off?

Using visual elements and controls well is not as easy as learning to use HTML. Even very sophisticated programmers and designers have produced useless Web pages. You can learn a lot by examining them. A collection of them is housed at Vincent Flanders' *Web Pages That Suck* (www.webpagesthatsuck.com), together with advice on good design and how to repair bad design

FIGURE B15.1 Screen of file manager with tabstrips across the top and file locations on the left.

as quickly as they can for a hand-operated switch (Kroemer, 1971). The time to actuate a foot pedal once it is reached will be a function of factors such as the amount of control resistance (Southall, 1985).

Most research on foot controls has investigated how well people use the brake and accelerator pedals in automobiles. In most cars, the accelerator is mounted lower than the brake. However, people are measurably slower to brake with this configuration than when the pedals are set at the same height (Davies & Watt, 1969; Snyder, 1976). People can brake even faster when the brake pedal is placed 2.5–5.1 cm lower than the accelerator (Glass & Suggs, 1977). The reason for these differences in braking time arises because a driver must use the brake at those times when she is pressing on the accelerator. The driver must therefore move her foot up a greater distance when the brake is higher than the accelerator. Lowering the brake reduces the movement distance.

Incidents of unintended acceleration are often reported in which an automobile appears to have accelerated out of control, even though the operator claims to have had his or her foot on the brake. These incidents are not due to mechanical failure: such episodes have been reported for a wide variety of automatic transmission designs, and testing has shown no mechanical defects in the cars involved. Instead, unintended acceleration is caused by foot placement errors (see Brick, 2001). That is, the driver inadvertently depresses the accelerator when intending to depress the brake. It is easy to get people to make mistakes like this in the laboratory. The Texas Transportation Institute reported that differences in the way brakes and accelerator pedals are positioned in different car models are sufficiently large that people are more likely to make a pedal error when they drive an unfamiliar car. The variability in nerve impulses in muscles and the spinal cord may also contribute to pedal errors (Schmidt, 1989), and this kind of variability would also explain why drivers are unaware of the error. The drivers moved their foot in exactly the same way they always do to brake, but the slight random difference in the movements resulted in pressing on the accelerator. Because the drivers are unaware of the error in their foot placements, they cannot stop the car.

We talked earlier about the kinds of feedback that different control resistances can provide and how this feedback can actually improve performance. Similar feedback systems can be designed for brakes and accelerators. For example, the active accelerator pedal provides feedback about the car's speed relative to the speed limit (Várhelyi, Hjälmdahl, Hydén, & Draskóczy, 2004). It uses a GPS receiver that identifies the position of the vehicle and digital maps identifying the speed limits on each road. When the car reaches the speed limit, the accelerator pedal applies a counterforce that makes the pedal harder to press. When people use an active accelerator pedal, their compliance with the speed limit is better, and their speed variability is lower. Furthermore, vehicle emissions are reduced.

SPECIALIZED CONTROLS

For certain applications, we will need controls that people can use without moving their arms and legs. In some systems people will need to be using their arms and legs for other things. These controls can also be used by people with limited mobility. The controls that we will discuss are automated speech controls, gaze- and head-movement controls, gesture-based controls, and teleoperators.

Speech Controls

Speech-activated, or vocal, controls can be found on computers, smartphones, navigation, and other systems (Simpson, McCauley, Roland, Ruth, & Williges, 1985). These controls rely on voice-recognition software. Speech controls can be used (1) for controlling computing devices, allowing a person's hands to do other things, (2) for dictation, where people can perform text entry much faster than if text were typed, and (3) by people who have physical disabilities (Noyes, Haigh, & Starr, 1989). Given the physical limitations of older adults, it is perhaps not too surprising to find that older adults are more accepting of speech-activated controls than are younger adults (Stephens, Carswell, & Dallaire, 2000).

Speech controls require the user's vocalizations to be captured by a microphone, which may be very small or hidden within the device to be controlled. The speech signal is converted into a digital signal. Speech-recognition software processes the digital signal, using an algorithm to recognize the words or phrases. There are two kinds of speech-recognition systems: speaker-dependent and speaker-independent (Entwistle, 2003). Speaker-dependent systems work best for only a single person, who must train the system with examples of his own speech. Speaker-independent systems are designed to recognize speech spoken by virtually anyone. Speaker-independent recognition systems work best with a small vocabulary and a relatively homogeneous population of speakers.

Another distinction between speech-recognition systems is whether they process isolated words, connected words, or continuous speech. Isolated-word systems respond to individual words and require the speaker to pause for at least 100 ms between words. Connected-word systems do not require artificial pauses, but the speaker can use no inflection, as if she were reading the words from a list. Continuous-speech systems are intended for use with natural speech. The complexity of the systems increases as the speech that they can recognize becomes more natural. A major problem for early continuous-speech systems was that there are no clear separations between the beginning of one word or sentence and the next. This increases the likelihood that the system will not recognize the speech.

Modern speech-recognition systems must process continuous-speech signals with high talker variability and in conditions of background noise (Li et al., 2014). An effective system not only has high recognition accuracy, but it must also be fast. High-performance systems make use of intensive computational models implemented on devices of considerable processing power. Recognition accuracy has been the major limitation of any speech-recognition system. Conditions that produce variability in speech patterns, such as environmental stressors, will reduce recognition accuracy, as will increases in vocabulary, and even whether the speech is conversational or read from a script (Fiscus, Ajot, & Garofolo, 2007). For example, one study found that speech-recognition accuracy decreased from 78% when a person was rested to 60% immediately after she had been engaged in hard exertion (Entwistle, 2003). The successful implementation of a speech system requires that it be integrated with other equipment. Speech controls are likely to improve performance only for complex tasks with high demands on visual and manual performance, such as driving.

Speech-based controls can be used in air battle management. Vidulich, Nelson, and Bolia (2006) examined how 12 Airborne Warning and Control System (AWACS) operators from the U.S. Air Force interacted in a simulated environment. The AWACS aircraft monitors all aircraft flights in a conflict area and directs missions within the area. Many tasks are performed by many personnel, including a weapons team composed of weapons directors. A weapons director directs the movements of various air vehicles from a console. The director's task is complex, composed of many subtasks, and imposes a high mental workload. Vidulich et al. conducted simulated battle exercises that allowed the use of speech controls for some tasks. The weapons directors were able to perform some tasks faster with speech controls than with the standard controls. Having speech controls available enabled them to time-share more efficiently between subtasks, and the participants elected to use the speech controls when they were given a choice between the speech and standard controls.

Gaze-Based and Head-Movement Controls

Another way to control a system is through eye and head movements. We can mount equipment to monitor eye and head movements on a user's head, and then the direction of the user's gaze will activate controls. Head-movement controls require mounting a stick or pointing device to the user's head, which can be used to tap.

People using a gaze-based control will select an item (e.g., an icon) by looking at it on the screen for a criterion amount of "dwell time" (Calhoun, 2006). For instance, a person may wish to open a document file on a computer. She would look at an icon ("Open New File") and maintain her gaze long enough that the control activates or acquires the icon. In some situations, the user might then press a single button to trigger the control action (opening a new file). Gaze-based control can allow faster object selection and cursor positioning than manual control.

Head-movement controls allow people with limited mobility to use a computer or other similar device (LoPresti, Brienza, & Angelo, 2002). In Chapter 14, we discussed the fact that Fitts's law applies to tapping that is controlled by the head. It also describes movement times for head movements that control a cursor (Radwin, Vanderheiden, & Li, 1990; Spitz, 1990). The slope of the function relating movement time to the index of difficulty is substantially greater with head movements than it is with either a manually operated mouse or a digitizing tablet. This means that increasing the index of difficulty will lead to much greater slowing of head movement times. Because pointing with head movements is not as efficient as with manual input devices, head-movement control should be restricted to situations in which movement time is not a factor or in which the operator has restricted mobility.

People with restricted mobility may also not be able to move their heads, or they may not be able to control their head movements well. These problems can be solved by using different kinds of controls with different kinds of resistance. One study asked people with multiple sclerosis to perform icon acquisition tasks with head-movement controls (LoPresti et al., 2002). They explored several kinds of controls designed to compensate for head-movement limitations. Their results showed that, compared with their performance with a standard head control interface, their subjects were more accurate using an interface with increased sensitivity. They also found that first-order controls (in which the user's head movements controlled cursor velocity instead of cursor position) improved the subjects' aim.

Gesture-Based Control

A gesture-based control uses dynamic hand or body movements (Rautaray & Agrawal, 2015). There are a variety of these kinds of devices, but they can be classified into two categories. Contact-based devices require people to use an interface device such as a touch screen, data glove, or accelerometer, as in the Nintendo Wii Remote. Consider, for example, how people use their tablet computer touch screens. Most gestures permit the use of a single finger, and people can adjust how their gestures are interpreted. Default tablet settings usually have the display move in the direction of a swiping motion; "zooming in" is by an "expanding out" gesture and "zooming out" with a "pinching in" gesture (Rakubutu, Gelderblom, & Cohen, 2014). People tend to use single-finger gestures more on smartphones than on tablet computers with larger screens, but most devices use the same or similar gesture controls (Billinghurst & Vu, 2015).

The gestures interpreted by a touchscreen are two-dimensional (2-D) gestures. With depth sensors and camera systems, we can also design three-dimensional (3-D) gesture controls. 3-D gesture controls are important for human–computer interaction (HCI). 3-D gestures may allow a more intuitive interactive experience with the computer or other device than 2-D gesture input, because they permit the user to interact with the computer with no hardware interface (touchscreen, mouse, or keyboard). However, implementing 3-D requires determining how different gestures should be mapped to different commands. A suggested 3-D gesture set for common HCI tasks (such as moving an icon from one location to another on the screen) was developed by Pereira, Wachs, Park, and Rempel (2015) after they studied different hand posture risks and user preferences. They were able to determine some stereotypic mapping between gestures and commands, but the amount of agreement across users varied with the particular command, with more stereotypical commands being executed more quickly. Research evaluating gesture lexicons is very active because of the potential for 3-D gesture interfaces to revolutionize HCI (e.g., Cheng, Yang, & Liu, 2016).

Teleoperation

Teleoperators are robots that perform as remote extensions of the operator's arms, hands, legs, and feet (Johnsen & Corliss, 1971; Kheddar, Chellali, & Coiffet, 2002). Teleoperators are the devices that allow people to do things like pick up samples of the lunar surface while on earth or manipulate radioactive compounds. Teleoperation can be used to guide micro-instruments in surgery and control micro- and nano-scale devices (Kheddar et al., 2002). Remotely controlled robots, like the one that the Dallas, Texas, police department used on July 8, 2016, to deliver a bomb that killed a sniper who was shooting police officers, are teleoperators.

Many of the human factors issues we have confronted for other design problems must also be confronted for the design of teleoperators. We have to decide which control-oriented tasks to assign to the operator and which to the teleoperator (Sheridan, 2016). There are problems of spatial correspondence between the positions of the controls and those parts of the teleoperator that they control. Vision and other sensory feedback must be incorporated into the control. We also have to decide, for the teleoperator itself, what kinds of controls to use.

How the teleoperator moves is determined by many of the same factors that constrain human movements. Among other things, Fitts's law applies to the relation between speed and accuracy of aimed teleoperator arm movements, although the slope is much greater than for human arm movements (Draper, Handel, & Hood, 1990). The operator's performance will be best when the dynamics of the teleoperator movements, that is, its motions, are consistent with the operator's own movements (Wallace & Carlson, 1992).

The Internet makes possible the remote control of a teleoperator from anywhere around the world. However, the usual delays that Internet users experience will cause similar delays between the teleoperator and the operator, and these delays will degrade performance. Sheik-Nainar, Kaber, and Chow (2005) evaluated the effect of network delays on the control of a "telerover," a robotic vehicle on wheels. They found that the deterioration in control performance caused by network delays was reduced by the use of a system gain adaptation algorithm. This algorithm automatically adjusts the gain (sensitivity) of the teleoperation system controller, reducing it when an increase in network delay is detected. The idea is that reducing the sensitivity will lessen the impact of any control adjustments made by the operator during the delay period, reducing navigation errors and collisions with objects. In fact, the algorithm did reduce the deterioration in user performance and provided an enhanced "telepresence" experience, that is, experience of control of the robot.

As with many computerized systems, the direct control of robots by humans, as in teleoperation, is diminishing. Instead, much of the control is managed by software, which may require occasional reprogramming by a human supervisor, in which case the machine is called a *telerobot* (Sheridan, 2016).

SUMMARY

People communicate with machines by operating controls. Controls come in a variety of types, shapes, and sizes. Their mechanical properties produce different "feels" that can be exploited to optimize a person's performance in a variety of applications. Population stereotypes can also be exploited to ensure that the movements associated with a control function are the most natural ones associated with the operators' movements. An operator's performance will vary as a function of the relation between control displacement and system response.

Often, we encounter many controls arranged on a single control panel. Designing control panels well avoids confusion about which control to operate and about the relation between display elements and controls. Control codes can be used to aid identification, and frequently used controls should be readily accessible. We should design panels so that controls critical to the integrity of the system will not be accidentally activated.

This chapter concludes our discussion of the ways that operators control the movement not only of themselves but of the objects and machines in the environment around them. We have described the relationship between operators and machines from an information-processing perspective. The operator and machine form a closed-loop system in which information is passed back and forth through the human–machine interface. You need to recognize, however, that the human–machine system does not operate in isolation but in the context of the surrounding environment. The last part of the book will discuss how the environment affects the performance of the operator, which in turn determines the performance of the entire system.

RECOMMENDED READINGS

Adams, S. K. (2006). Input devices and controls: Manual, foot, and computer. In W. Karwowski (Ed.), *International Encyclopedia of Ergonomics and Human Factors* (2nd ed.; 1: pp. 1419–1439). Boca Raton, FL: CRC Press.

Bullinger, H.-J., Kern, P., & Braun, M. (1997). Controls. In G. Salvendy (Ed.), *Handbook of Human Factors* (2nd ed.; pp. 697–728). New York: Wiley.

Chapanis, A., & Kinkade, R. G. (1972) Design of controls. In H. P. Van Cott & R. G. Kinkade (Eds.), *Human Engineering Guide to Equipment Design* (345–379). Washington, DC: U. S. Superintendent of Documents.

Chengalur, S. N., Rodgers, S. H., & Bernard, T. E. (2004). *Kodak's Ergonomic Design for People at Work*. Hoboken, NJ: Wiley.

Ivergard, T. (2006). Manual control devices. In W. Karwowski (Ed.), *International Encyclopedia of Ergonomics and Human Factors* (2nd ed.; 1: pp. 1457–1462). Boca Raton, FL: CRC Press.

Part V

Environmental Factors and Their Applications

16 Anthropometrics and Workspace Design

Anthropometry is a major component of the total systems point of view that is a hallmark of good human factors or ergonomics practice.

John A. Roebuck, Jr.
1995

INTRODUCTION

The measurement of human physical characteristics is called *anthropometrics*, and *engineering anthropometry* refers to the design of equipment, tasks, and workspaces so that they are compatible with the physical characteristics of the people who will be using them (Kroemer, Kroemer, & Kroemer-Elbert, 2010). The reach envelope discussed in Chapter 15 is an example of how human factors specialists use anthropometric data. Designing the envelope around the 5th percentile for reach distance ensures that 95% of potential users can reach the controls within the envelope.

Good workspace design depends on far more than just making sure that users can reach all the controls or objects in the workspace. In addition, we must consider the motions of the joints of the body and the range of those motions. *Biomechanics* is the field of study concerned with how the body moves (Peterson & Bronzino, 2014). Human factors specialists routinely apply biomechanical data to equipment design so that equipment and tasks will accommodate the biomechanics of the user population.

A workspace or workstation is any area in which a person works for an extended period of time (Grobelny & Karwowski, 2006). Workspaces are desks, control panels, computer workstations, assembly line stations, truck cabs, and so on. Working in a poorly designed workspace for long periods of time can be physically and psychologically damaging to a worker and may harm the worker's ability to operate equipment. We have discussed some components of workspace design, such as the display of information and the organization of control panels, in previous chapters. However, the entire ensemble of equipment that makes up the workspace must be designed and arranged to be compatible with the operator's physical capabilities (see Figure 16.1).

In this chapter we summarize the important principles of engineering anthropometry and biomechanics. When these principles are violated, operators may receive painful and incapacitating injuries. Anthropometrics and biomechanics play an important role in tool design and manual materials handling. Because tool usage and manual materials handling are involved in many jobs and are the sources of many work-related injuries, we will evaluate the factors that influence the efficiency and safety of tools and materials handling. We will also consider the practical aspects of how anthropometric and biomechanical factors are incorporated into workspace design.

ENGINEERING ANTHROPOMETRY

Anthropometrics refers to measurements of the dimensions of the human body. When we measure a particular body dimension, such as reach distance, we will do so for as many individuals within a population as we can. It is important that the sample be randomly selected from the target population, since our goal is to get as accurate a picture as possible of the distribution of the measurements

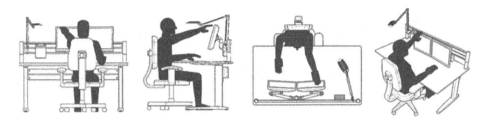

FIGURE 16.1 Workspace designed to be compatible with a person's physical capabilities.

of interest. All the measurements that we could make (height, weight, reach distance, leg length, etc.) taken together describe the anthropometric characteristics of that population. It is almost always the case that all such measures are approximately normally distributed. Consequently, published tables of anthropometric data include measures of central tendency (mean or median) and variability (standard deviation) and also sometimes quantiles, published in tabular form so that design engineers can use them.

The most commonly used anthropometric percentile ranks are the 5th, 50th, and 95th percentiles, below which 5%, 50%, and 95% of the population fall. For example, Table 16.1 gives these quantiles for the anthropometric characteristics of females and males in the U.S. The purpose of the quantiles is to provide a minimum, average, and maximum value of each measurement. These data can be used to establish design criteria for equipment and to provide criteria for evaluating existing equipment. They can also be used to select operators to fit the workspace dimensions (Kroemer, 1983a). For example, the Apollo Command Module, used for manned space missions to the moon from 1968 to 1972, was designed to accommodate up to the 90th percentile for standing height, so astronaut recruits could not exceed 1.83 m (approximately 6 ft.).

The quantiles in a table of anthropometric data are used by design engineers to ensure that equipment will be usable by almost all members of a population. For instance, problems of "clearance," which include head room, knee room, elbow room, and access to passageways and equipment, require the engineer to design for the largest or tallest individuals in the user population. Most commonly, the 95th percentile values for height or breadth measurements will be used to ensure adequate clearance. For problems of reach, which involve such concerns as the locations of controls, the designer should be concerned with the smallest individuals in the user population, or the 5th percentile. If an object is intended to be out of reach, such as a control that should not be unintentionally activated, then this criterion is reversed.

Other design issues focus on the average person (the 50th percentile). For example, work surfaces should not be placed at a height ideal for either the tallest or the shortest individuals in a population, but instead somewhere in the middle. This means, however, that for half the population the work surface will be too high, and for the other half it will be too low. This problem can be solved by incorporating adjustable chairs and work surfaces, so that each person can adjust the workspace to his or her specifications.

It is important to exercise care when designing for the minimum, maximum, or average. Robinette and Hudson (2006) caution, "since as early as 1952 … we have known that anthropometric averages are not acceptable for many applications" (p. 322) and "Designing for the 5th percentile female to the 95th percentile male can lead to poor and unsafe designs" (p. 322). One reason for these cautions is that when multiple dimensions are involved, some people will be large on some dimensions but small on others. If the design is based on given percentile values for the single dimensions independently, the percentage of people who will be able to use the equipment comfortably may be much lower than the designer might intend.

TABLE 16.1

U.S. Civilian Body Dimensions. Female/Male. In Centimeters for Ages 20–60 Years.

	Percentiles			
	5th	50th	95th	Standard Deviation
Heights				
Stature (height)[f]	149.5/161.8	160.5/173.6	171.3/184.4	6.6/6.9
Eye height[f]	136.3/151.1	148.9/162.4	159.3/172.7	6.4/6.6
Shoulder (acromion) height[f]	121.1/132.3	131.1/142.8	141.9/152.4	6.1/6.1
Elbow height[f]	93.6/100.0	101.2/109.9	108.8/119.0	4.6/5.8
Knuckle height[f]	64.3/69.8	70.2/75.4	75.9/80.4	3.5/3.2
Height, sitting[s]	78.6/84.2	85.0/90.6	90.7/96.7	3.5/3.7
Eye height, sitting[s]	67.5/72.6	73.3/78.6	78.5/84.4	3.3/3.6
Shoulder height, sitting[s]	49.2/52.7	55.7/59.4	61.7/65.8	3.8/4.0
Elbow rest height, sitting[s]	18.1/19.0	23.3/24.3	28.1/29.4	29/3.0
Knee height, sitting[f]	45.2/49.3	49.8/54.3	54.5/59.3	2.7/2.9
Popliteal height, sitting[f]	35.5/39.2	39.8/44.2	44.3/48.8	2.6/2.8
Thigh clearance height[f]	10.6/11.4	13.7/14.4	17.5/17.7	1.8/1.7
Depths				
Chest depth	21.4/21.4	24.2/24.2	29.7/27.6	2.5/1.9
Elbow-fingertip distance	38.5/44.1	42.1/47.9	56.0/51.4	2.2/2.2
Buttock-knee distance, sitting	51.8/54.0	56.9/59.4	62.5/64.2	3.1/3.0
Buttock-popliteal distance, sitting	43.0/44.2	48.1/49.5	53.5/54.8	3.1/3.0
Forward reach, functional	64.0/76.3	71.0/82.5	79.0/88.3	4.5/50
Breadths				
Elbow to elbow breadth	31.5/35.0	38.4/41.7	49.1/50.6	5.4/4.6
Hip breadth, sitting	31.2/30.8	36.4/35.4	43.7/40.6	3.7/2.8
Head dimensions				
Head breadth	13.6/14.4	14.54/15.42	15.5/16.4	0.57/0.59
Head circumference	52.3/53.8	54.9/56.8	57.7/59.3	1.63/1.68
Interpupillary distance	5.1/5.5	5.83/6.20	6.5/6.8	0.44/0.39
Foot dimensions				
Foot length	22.3/24.8	24.1/26.9	26.2/29.0	1.19/1.28
Foot breadth	8.1/9.0	8.84/9.79	9.7/10.7	0.50/0.53
Lateral malleolus height	5.8/6.2	6.78/7.03	7.8/8.0	0.59/0.54
Hand dimensions				
Hand length	16.4/17.6	17.95/19.05	19.8/20.6	1.04/0.93
Breadth, metacarpal	7.0/8.2	7.66/8.88	8.4/9.8	0.41/0.47
Circumference, metacarpal	16.9/19.9	18.36/21.55	19.9/23.5	0.69/1.09
Thickness, meta III	2.5/2.4	2.77/2.76	3.1/3.1	0.18/0.21
Digit 1: Breadth of interphalangeal	1.7/2.1	1.98/2.29	2.1/2.5	0.12/0.21
Crotch-tip length	4.7/5.1	5.36/5.88	6.1/6.6	0.44/0.45
Digit 2: Breadth of distal joint	1.4/1.7	1.55/1.85	1.7/2.0	0.10/0.12
Crotch-tip length	6.1/6.8	6.88/7.52	7.8/8.2	0.52/0.46
Digit 3: Breadth of distal joint	1.4/1.7	1.53/1.85	1.7/2.0	0.09/0.12
Crotch-tip length	7.0/7.8	7.77/8.53	8.7/9.5	0.51/0.51
Digit 4: Breadth of distal joint	1.3/1.6	1.42/1.70	1.6/1.9	0.09/0.11
Crotch-tip length	6.5/7.4	7.29/7.99	8.2/8.9	0.53/0.47
Digit 5: Breadth of distal joint	1.2/1.4	1.32/1.57	1.5/1.8	0.09/0.12
Crotch-tip length	4.8/5.4	5.44/6.08	6.2/6.99	0.44/0.47
Weight (in kg)	46.2/56.2	61.1/74.0	89.9/97.1	13.8/12.6

[f] Above floor.

[s] Above seat.

ANTHROPOMETRIC MEASUREMENT

In traditional anthropometry, static (or structural) measurements are obtained while a person holds different postures. For example, a person might be measured while standing and sitting upright (Roebuck, 1995). Static measures like standing height and sitting height form the core of an anthropometric database. However, dynamic (or functional) anthropometric measurements, which incorporate biomechanical constraints, are also important when our goal is to determine whether an operator can execute a particular task. The reach envelope is an example of a functional measurement, because a person's maximum reach distance will vary with different postures, different grasps, and different tasks. Workspace dimensions are usually determined with functional anthropometric measures rather than static anthropometric measures.

Anthropometric measurements are made with mechanical instruments such as measuring tapes, calipers, and scales. When describing a particular measurement, we use the following definitions (Kroemer, Kroemer, & Kroemer-Elbert, 2010, pp. 322–325):

> *Height* is a straight-line, point-to-point vertical measurement.
> *Breadth* is a straight-line, point-to-point horizontal measurement running across the body or a segment.
> *Depth* is a straight-line, point-to-point horizontal measurement running from the front to the back of the body.
> *Distance* is a straight-line, point-to-point measurement between landmarks on the body.
> *Curvature* is a point-to-point measurement following a contour; this measurement is neither closed nor usually circular.
> *Circumference* is a closed measurement that follows a body contour; hence, this measurement usually is not circular.
> *Reach* is a point-to-point measurement following the long axis of an arm or leg.

Anthropometric measures are described in terms of the position of the body, the part of the body being measured, and the direction of the dimension being measured. A "map" of the body incorporating these terms is shown in Figure 16.2, and the postures a person assumes while being measured are shown in Figure 16.3.

The three-dimensional (3-D) planes that pass through the body are the transverse, sagittal, and coronal planes. The sagittal plane cuts longitudinally and separates the left half of the body from the right half. Transverse planes cut horizontally and separate top from bottom. Coronal planes also cut longitudinally and separate front from back. Directional terms are used in opposite pairs and are specific to the plane of measurement. A body part above a transverse plane is superior, and one below it is inferior. A body part to the left or right of the sagittal plane is lateral, while one close to it (to the center of the body being measured) is medial. A body part in front of a coronal plane is anterior, and one behind it is posterior. Finally, a body part that is far from the trunk is distal, whereas one that is close to the trunk is proximal.

Modern anthropometric measurements no longer depend entirely on calipers and tape measures. 3-D body-scanning technologies can provide very accurate measurements of the body surface (Bubb, 2004). The scanners typically use optical techniques, although the specific technologies vary (Bragança, Arezes, & Carvalho, 2015). The person being measured assumes a specified stance in the scanner, usually wearing only form-fitting shorts and (for women) a halter top, and an image of the entire body is captured. Scanning technologies supply more complete, precise, and reproducible measurements than traditional methods. But, because they capture shapes and not direct measurements, they produce massive amounts of data that can be difficult to summarize and require software to extract specific measurements (Seidl & Bubb, 2006). Markers can be attached to the body at landmark locations commonly used for manual measurements to allow these specific measurements to be calculated directly by the software.

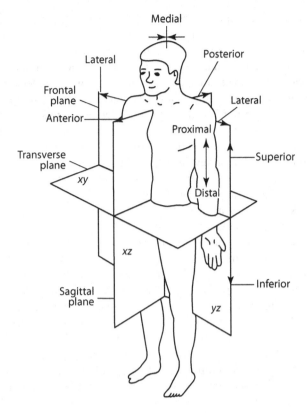

FIGURE 16.2   Descriptive terms and measuring planes used in anthropometry.

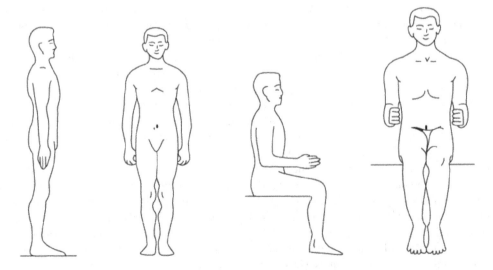

FIGURE 16.3   Postures assumed by the subject for anthropometric measurements.

The Civilian American and European Surface Anthropometry Resource (CAESAR) project was the first survey to use 3-D scanners to provide 3-D body measurements (Robinette & Daanen, 2003). This project, which was a collaborative effort between government agencies and private industries, collected anthropometric data on over 6000 U.S., Canadian, and European civilians. Each individual was scanned and measured, and also measured with calipers and tape measures, in standing and seated postures. The resulting database contains detailed anthropometric data for men and women ranging from ages 18 to 65, with people of different weights and socioeconomic status, and from different ethnic groups and geographic regions, represented in the sample.

The CAESAR study was just the beginning of the new era of anthropometric measurement. Another large-scale anthropometric survey, Size NorthAmerica, is an anthropometric measurement survey of more than 17,000 U.S. and Canadian citizens from 6 to 65 years of age. The purpose of this survey, taking place in the U.S. and Canada from March 2016 to September 2017 at the time the revision of this book was occurring, is to document changes in body measurements with aging (Seidl, Trieb, Wirsching, Smythe, & Guenzel, 2016).

Whereas anthropometric measurements are made on a stationary body, biomechanical measurements are made while the body is in motion. Biomechanical measurements are more complex than anthropometric measurements. Not all biomechanical measurements can be made directly from a person's body or movements, but some can. For example, the force that a person generates, which is of particular interest in engineering anthropometry, can be measured directly. Static strength, which is the maximum force that a person can exert isometrically in a single effort, is relatively easy to measure (Kroemer, 2006b). Static measurements provide good indicators of the exertion possible for slow movements but not for fast movements such as hammering. Dynamic strength, which is the amount of force that a person can exert throughout the range of motion, is more complicated to measure (Kroemer et al., 1997). Consequently, in many situations where we want to estimate the range of motion and stress on a particular body part, we have to rely on models of the musculoskeletal system (e.g., Sesto, Radwin, & Richard, 2005). These models take anthropometric data and biomechanical measures as their inputs and compute projected tolerances as their outputs. Differences in output for different designs are used to determine which designs are best.

Sources of Anthropometric Data

Designers do not have to collect their own data before selecting design parameters. There are several sources of anthropometric data that designers can consult. Among these are the National Aeronautics and Space Agency (NASA) *Anthropometric Source Book* (1978) and the *Human Systems IAC Anthropometric Data Sets*, which include the results of many separate surveys, as well as the aforementioned *CAESAR 3-D Anthropometric Database*.

As with the use of any tabled data source, a designer must make sure that the data are appropriate for his or her application. If the designer is constructing workspaces to be used by a particular population, then only data about that population will be useful. For example, people from Asian countries are significantly shorter on average than people from the U.S. and Europe (Li, Hwang, & Wang, 1990; Seidl & Bubb, 2006). If a designer decides on workspace parameters for the Japanese market based on the anthropometric characteristics of the U.S. population, the final design will not be attractive, appropriate, or usable by Japanese.

Data for civilian populations are somewhat limited, because most published data were obtained from military populations (Van Cott, 1980). The CAESAR anthropometry project has remedied this limitation to some extent. Military and civilian populations are similar in terms of head, hand, and foot size, but differ on most other dimensions (Kroemer et al., 1997). For example, for most measures of girth the military population will be smaller, because it mainly consists of people under age 40 and excludes individuals who are either very small or very large (Chengaluer et al., 2004).

It is possible to include individuals who do not appear in military populations by shifting the military median and the quantiles of the measurement distribution. Adjustment of a military population is based on the assumption that the measurements are normally distributed but requires information about civilian means and variances so that the measurements can be appropriately rescaled. Such adjustments are common for estimating measurements for civilian male populations from military male measurements, where civilian men are matched to the military data by height and weight (McConville, Robinette, & Churchill, 1981). However, this strategy does not work for female populations: even when civilian women are matched to military women by height and weight, the population differences are still substantial.

Anthropometric data obtained from a population at large also cannot always be applied to specific subpopulations. For example, in the U.S., the median weight for farm equipment operators is about 14% greater than the weight of the general population (Casey, 1989). As a consequence, if tractor seats are designed to satisfy the anthropometric criteria for the general population, they will be too small for most farm equipment operators. Anthropometric data for U.S. farm workers in three postures—standing, seated in a tractor seat, and bent forward (as in a rollover accident)—obtained using 3-D full body scans and traditional measures are provided by Hsiao et al. (2005). According to these data, the current standards for vertical clearance in tractor cabs are too short, but also smaller workers are poorly accommodated by current designs. Hsiao et al. developed 3-D tractor driver models to assist designers in determining where to place controls to best accommodate the drivers.

Kroemer (2006a) has written a book that focuses on designing for special populations, such as expectant mothers, older adults, children, and people with disabilities. As an example, the girth of a pregnant woman is much larger, and her shape very different, than that of a nonpregnant woman. This greater girth can cause problems with steering wheel clearance and proper seatbelt positioning in a typical automobile (Ascar & Weekes, 2005). Culver and Viano (1990) collected anthropometric data for U.S. women in different stages of pregnancy, and Yamana et al. (1984) and Ascar and Weekes (2005) did the same for Japanese and U.K. women, respectively. These data can be used to design automobile interiors and restraints to accommodate the rapidly changing girth of a pregnant woman and maximize her safety and that of her fetus. Fryar, Gu, and Ogden (2012) have collected another anthropometric dataset that contains measurements for children and adults in the U.S.

Unfortunately, there are not enough anthropometric data available for people over age 65 (Kroemer, 1997). Most studies of older adults have been conducted with healthy white males (Kelly & Kroemer, 1990). Consequently, we do not have good data for elderly females or for populations suffering from the common diseases of old age, such as arthritis and osteoporosis. These conditions diminish a person's functional capabilities and mobility, but such restrictions are not represented in anthropometric data. More generally, anthropometric characteristics for older adults change with increasing age (Shatenstein, Kergoat, & Nadon, 2001), and the elderly population is very heterogeneous. Thus, it is a mistake to treat older adults as a single, homogeneous group.

Similarly, although anthropometric measurements for developmentally and physically challenged people differ from those of the general population, relatively few data are available for designing workspaces and tools that are ergonomically acceptable for these special populations. For example, Goswami (1997) said, "Despite scattered attempts in some specific areas, the data about the physical dimensions of the physically disabled is not sufficient" (p. 339). He concluded that this is because of wide variation in the nature of the disabilities. A study that collected anthropometric data for use in designing seats for people with cerebral palsy illustrated that even people with the same disability will show differences in posture, muscle development, and bone structure (Hobson & Molenbroek, 1990). For anthropometric data from these populations to be useful, such differences must be noted and taken into account.

BIOMECHANICAL FACTORS

Good workspace design depends on more than accurate anthropometric measures. Most people spend their entire work day within a workspace, moving within the environment and using equipment for up to 8 hours or more. People perform many actions, some repetitively and some infrequently. Consequently, biomechanical constraints are major factors in the design and evaluation of tasks and workspaces. The application of biomechanics to workspace design is called *occupational biomechanics*, which can be defined as "the study of the physical interaction of workers with their tools, machines, and materials so as to enhance the worker's performance while minimizing the risk of musculoskeletal disorders" (Chaffin, Andersson, & Martin, 2006, p. 2). By considering these biomechanical factors, as well as anthropometric factors, we can eliminate conditions that promote injury and discomfort quite early in the design of a workspace. It is always more expensive to modify a poorly designed workspace after it has been implemented than it is to design it correctly the first time.

Tichauer (1978) defined *work tolerance* as "a state in which the individual worker performs at economically acceptable rates, while enjoying high levels of emotional and physiological well-being" (p. 32). This definition emphasizes the desire for a worker to be both productive and healthy. Three categories of biomechanical factors contribute to work tolerance (see Table 16.2).

The first category deals with posture. Good posture minimizes skeletal and muscular stress, and can be encouraged by designing the workspace so that a person can keep his or her elbows close to his or her body and minimize his or her head movements. This in turn helps ensure that the forces (moments) acting on the spine are small and stresses are minimized. Because men and women have different bodies, postural concerns will be different for men and women. For example, differences between the center of mass in men and women can result in a 15% increase in lifting stress for a woman over that experienced by a man lifting the same object.

The second category deals with the engineering considerations involved in the design of the system interface. Improperly designed or misused equipment can result in compression ischemia, or obstruction of the blood flow. Exposure to vibrations, discussed in Chapter 17, can cause tissue damage and psychological stress. A worker's chair must provide proper support, especially if it is used for long periods of time. Repetitive tasks can concentrate stress on particular tissues, which may in turn result in chronic inflammation and permanent injury. Specialized equipment, such as tools that allow the wrist to be kept straight, can be used to prevent injuries.

The third category deals with kinesiological factors, or the type and range of movements that are performed. Long forward reaches produce stress on the spinal column and so should be avoided. Such reaches may also result in "muscular insufficiency," which is a decrease in the range of a person's movement due to overextended (or completely contracted) muscles. It can be prevented by designing the workspace so that people can manipulate controls, tools, and other objects within the extreme limits of muscular contraction. A person's movement trajectories should be curved rather than straight, because they are easier to make and learn, and they are less tiring.

It is sometimes important to consider a worker's clothing. A person's movements can be obstructed by protective clothing such as gloves and chemical suits. People required to wear such clothing will have a limited range of movement, and these limitations will help determine the design parameters of a workspace. The muscle groups used in different tasks will also be a factor in workspace design. Because antagonist muscles are smaller than agonist muscles, and smaller muscles

---

TABLE 16.2

Factors to Maximize Biomechanical Work Tolerance

Postural		Engineering		Kinesiological	
P1	Keep elbows down.	E1	Avoid compression ischemia.	K1	Keep forward reaches short.
PI	Minimize moments on spine.	E2	Avoid critical vibrations.	K2	Avoid muscular insufficiency.
P3	Consider sex differences.	E3	Individualize chair design.	K3	Avoid straight-line motions.
P4	Optimize skeletal configuration.	E4	Avoid stress concentration.	K4	Consider working gloves.
PS	Avoid head movement.	ES	Keep wrist straight.	K5	Avoid antagonist fatigue.

fatigue more quickly than larger muscles, the engineer must design tasks to prevent fatigue of the smallest muscles involved in performing the task.

CUMULATIVE TRAUMA DISORDERS

When certain types of manual actions are performed repetitively, they lead to *cumulative trauma disorders*. Such disorders are a collection of "syndromes characterized by discomfort, impairment, disability or persistent pain in joints, muscles, tendons and other soft tissues, with or without physical manifestations" (Kroemer, 1989, p. 274). Cumulative trauma disorders are associated with many work activities, including manual assembly, packing, keyboarding, and mousing, as well as with leisure activities such as sports and playing video games. The disorders arise from repeated physical stress at a person's joints, which in turn causes damage to the tissues and/or to nerve fibers. They can cause a worker extreme pain and physical impairment, as well as reduced productivity, and they are very expensive in terms of medical costs and disability compensation.

In the U.S., the incidence of cumulative trauma disorders rose from 3.6 per 10,000 workers in 1982 to 23.8 and 27 per 10,000 workers in 2001 and 2014, respectively (Brenner, Fairris, & Ruser, 2004; Bureau of Labor Statistics, 2015). This increase may be due to changes that have occurred in the workplace in the past decades. For example, many industries now use "just-in-time" inventory systems, in which materials are scheduled to arrive at precisely the moment in the production process when they are needed. These systems are very popular because they decrease storage costs, improve production quality, and reduce labor, among other things. However, while this practice may improve productivity, it decreases the control that workers have over the timing and pacing of their work (Brenner et al., 2004). This lack of control, in addition to other pressures to increase productivity, may result in more cumulative trauma disorders (Houvet & Obert, 2015).

The symptoms of cumulative trauma disorders include pain, swelling, weakness, and numbness in the affected region. The onset of the symptoms usually occurs in three stages (Chatterjee, 1987). In the first stage, a person may experience pain and weakness during work, but these symptoms subside after resting. During the second stage, the symptoms may persist even after resting, and the person's ability to perform repetitive work decreases. Finally, in the third stage, a person's pain is continuous. His sleep may be disrupted, and he may experience difficulty in performing a range of tasks. Each of the first two stages may last for weeks or months, while the third stage may last for years. Early (first-stage) detection of these disorders is important; they are usually completely reversible at this stage if the source of the physical stress is removed or brought within acceptable limits.

Cumulative trauma disorders can occur at any joint and its surrounding anatomy. However, most occur in the shoulders, arms, and hands, with 60% of all cases involving the wrist and hand. Table 16.3 lists several of the disorders for the hand and wrist and their associated risk factors. The most widely publicized of these disorders is carpal tunnel syndrome (Ledford, 2014), for which the symptoms are tingling and numbness in the thumb, index, and middle fingers (see Box 16.1).

As described in Chapter 15, tendons are the fleshy bands that connect muscle to bone. Most tendons are protected by a sheath that contains lubricating fluid. Injury and overuse can result in tendonitis, tenosynovitis, and ganglion cysts. Tendonitis is the inflammation of a tendon that is tensed or moved repetitively. Tenosynovitis is the inflammation of both the tendon and its sheath, and ganglionic cysts, swellings of the sheath with excess fluid, are visible through the skin. Disorders of tendons and nerve entrapment are not restricted to the wrist and hand. Such disorders also occur in the elbow and arm, the knees and ankles, and the neck and shoulder. Some shoulder and elbow tendons do not have sheaths, and tendonitis in these areas can progress to calcification of the tendon.

The risk for suffering a cumulative trauma disorder depends on several factors, including ergonomic deficiencies of the job or workspace, management practices, and factors specific to each individual (You et al., 2004). We have already talked a great deal about the importance of proper design. A tool, workstation, or job should be designed to fit a person's physical capabilities by incorporating anthropometric and biomechanical limits into the design. Furthermore, tasks should not

TABLE 16.3

Some Reported Occupational Risk Factors for Cumulative Trauma Disorders of the Upper Extremity

Disorder	Reported Occupational Risk Factors
Carpal tunnel syndrome	1. Accustomed and unaccustomed repetitive work with the hands 2. Work that involves repeated wrist flexion or extreme extension, particularly in combination with forceful pinching 3. Repeated forces on the base of the palm and wrist
Tenosynovitis and peritendonitis crepitans of the abductor and extensor pollicus tendons of the radial styloid (DeQuervain's disease)	1. More than 2,000 manipulations per hour 2. Performance of unaccustomed work 3. Single or repetitive local strain 4. Direct local blunt trauma 5. Simple repetitive movement that is forceful and fast 6. Repeated radial deviation of the wrist, particularly in combination with forceful exertions of the thumb 7. Repeated ulnar deviation of the wrist, particularly in combination with forceful exertions of the thumb
Tenosynovitis of finger flexor tendons	Exertions with a flexed wrist
Tenosynovitis of finger extensor tendons	Ulnar deviation of the wrist outward rotation
Epicondylitis	Radial deviation of the wrist with inward wrist rotation
Ganglionic cysts	1. Sudden or hard, unaccustomed use of tendon or joint 2. Repeated manipulations with extended wrist 3. Repeated twisting of the wrist
Neuritis in the fingers	Contact with hand tools over a nerve in the palm of sides or fingers

require repetitive movements, prolonged exertion of force involving more than 30% of a person's muscle strength, awkward or extreme postures, or the maintenance of postures for long time periods (Kroemer, 1989).

We also mentioned management practices, such as the just-in-time inventory system. These practices can also affect the incidence of cumulative trauma disorders. Managers must be willing to analyze tasks and jobs for their potential to produce disorders, and to redesign them to minimize the risk. Moreover, workers and medical personnel must be made aware of the early symptoms associated with the disorders so that diagnosis and treatment can be made at an early, easily reversible stage in the disease. People who have been diagnosed with cumulative trauma disorders will need to be reassigned to a different job with different postural and movement characteristics, and policies and procedures for such reassignments must be implemented. If managers are unwilling to assign workers to different jobs and/or redesign the tasks and workspaces that created the trauma in the first place, the workers' problems will return or worsen when they go back to their old job.

There are many individual risk factors associated with the disorders (e.g., Gell et al., 2005). More women than men develop the disorders, and the incidence increases with age. Some hobbies (woodworking, piano playing, etc.) can increase a person's chances of getting a cumulative trauma disorder. Family history, pregnancy, and nutritional habits can also affect risk. Diseases that reduce circulation, as well as past injuries and other traumatic conditions, increase the risk. People who are physically fit have a lower incidence of cumulative trauma disorders than people who are not.

BOX 16.1 CARPAL TUNNEL SYNDROME

Carpal tunnel syndrome is a cumulative trauma/repetitive stress disorder that arises when large forces within the wrist produce inflammation and swelling of the ligaments and tendons in the small passageway at the bottom of the hand through which the median nerve passes (see Figure B16.1). This passage is called the *carpal tunnel*. The swelling puts pressure on the median nerve, which delivers neural signals to and from the thumb and the index and middle fingers. Chronic pressure damages the median nerve.

Early symptoms of carpal tunnel syndrome include pain, numbness, or tingling of the fingers, often at night. At more advanced stages, a patient may experience atrophy of the muscles and significant reduction in dexterity of the fingers and grasping strength. Similar symptoms occur for the ring and little fingers when the ulnar nerve is entrapped in the Guyon tunnel. However, Guyon tunnel syndrome is less common and less disabling than carpal tunnel syndrome.

Carpal tunnel syndrome arises during repetitive or forceful exertions of the wrist (Dillon & Sanders, 2006). Activities that require repetitive or forceful grips and grasping, deviation of the wrist in the ulnar-radial plane to angle the hand outward from the arm, resting the forearm on a hard surface or edge, and repetitive flexion of the elbow can all contribute to carpal tunnel syndrome. Carpal tunnel syndrome is associated with surgery, dentistry, manual assembly, carpentry, piano playing, and other occupations that require extensive and repetitive tool use.

Because of widespread media coverage, most people believe that carpal tunnel syndrome arises from the use of computer keyboards for typing and data entry. Although typing is by no means the only cause of carpal tunnel syndrome, one study attributed 21% of the cases reported in a year to typing or data entry (Szabo, 1998). One reason why the use of a computer keyboard can cause carpal tunnel syndrome is that the standard computer keyboard requires a user to rotate his hands outward from the wrists, creating pressure on the median nerve (Amell & Kumar, 1999). A typist must maintain his hands in this awkward posture for extended periods of time, as a skilled typist makes as many as 100,000 keystrokes per day (Adams, 2006).

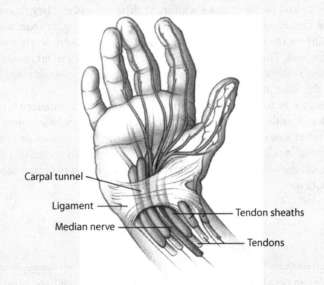

Carpal tunnel

Ligament

Median nerve

Tendon sheaths

Tendons

FIGURE B16.1   The carpal tunnel and associated ligaments, tendons, and nerve.

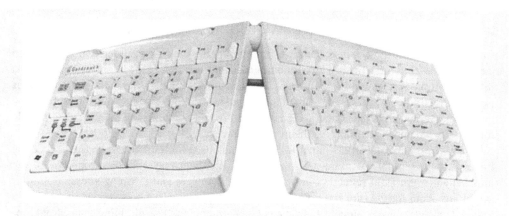

FIGURE B16.2   Split keyboard.

The split keyboard (see Figure B16.2) may provide a solution to this problem (Marklin & Simoneau, 2006). These keyboards are split in the middle, and each half is rotated so that the user's wrists can be straight, reducing the pressure on the median nerve. When the opening angle between the two halves is approximately 25° (i.e., each half is angled at 12.5° from horizontal), the user's wrist has a neutral deviation posture (Marklin, Simoneau, & Monroe, 1999). This more neutral posture might reduce the incidence of carpal tunnel syndrome among typists.

However, the extent to which split keyboards reduce the incidence of carpal tunnel syndrome is questionable. Measurements of pressure in the carpal tunnel are not smaller when the user's wrist posture is neutral than when it is angled for typing on a standard keyboard. Also, the split keyboards eliminate only ulnar deviation in the ulnar-radial plane and not extension of the wrist in the flexion/extension plane, which also contributes to development of carpal tunnel syndrome.

Extended use of a computer mouse has effects similar to those observed for extended keyboard use. Keir, Bach, and Rempel (1999) measured carpal tunnel pressure while people performed dragging and pointing tasks with three different mice. They found high pressure within the carpal tunnel when people performed these tasks (higher than when their hands were simply resting on the mouse). For many people, these pressure levels were high enough to affect nerve function. This finding suggests that computer tasks should be designed to avoid prolonged dragging with the mouse, and also that computer users should periodically perform other tasks with the hand that operates the mouse.

A perhaps better way to reduce carpal tunnel syndrome in computer users is to use different methods for inputting data: for example, speech. This can be accomplished by using voice-enhanced interfaces (Zhang & Luximon, 2006), which use voice-recognition software to allow speech input in addition to keyboard and mouse input. By increasing the use of voice commands as input, the overall physical load on the hand and wrist associated with manual entry will be reduced.

HAND TOOLS

Baber (2006) defined tools as "objects external to the user that support engagement with objects in the world" (p. 8). We can divide hand tools into two categories: manual and power. The forces necessary to operate manual tools are provided entirely by human muscles, whereas those for operating power tools come in part from external sources. Regardless of what type of tool it is, the purpose of a tool is to facilitate the performance of tasks that would be difficult or impossible to perform without it, such as removing a screw, tightening a lug nut, or cutting a piece of sheet metal. Power hand tools have the additional benefit of replacing or augmenting a user's physical strength with a different primary energy source, thereby reducing the amount of physical energy expended by the user and increasing the amount of force that he or she can generate.

An efficient tool must satisfy several requirements (Anonymous, 2000; Drillis, 1963). It must (1) effectively perform the function for which it is intended, (2) be proportioned to fit the user's body, (3) be adjusted to the strength and work capacity of the user, (4) minimize fatigue, (5) be adapted to the user's sensory capacities, and (6) be inexpensive and easy to maintain.

Hand tools are simple devices, and so there is a tendency to underestimate the importance of human factors considerations in their design. However, as you can see from the requirements listed above, there are many ergonomic guidelines that constrain the making of a good tool. The proportion of industrial injuries attributable to hand-tool use is approximately 9% (Cacha, 1999). Many of these problems can be traced to inadequate tool design, and include

1. Pinching, crushing, and amputation of the fingertips or of entire fingers;
2. Entry of foreign objects into the eyes, with possible loss of vision;
3. Straining or "tearing" of muscle tendons, causing acute and chronic pain with reduced function;
4. Inflammation of the wrist/hand tendon sheaths and nerves, making finger and wrist motion very painful and limited;
5. Back pains, with resulting difficulty of torso motion and lifting;
6. Muscle fatigue, causing decreased capability for performing manual work;
7. Mental fatigue, producing slow and error-prone work; and
8. Prolonged operator learning times (Greenberg & Chaffin, 1978, p. 7).

Many of the above requirements for an efficient tool are anthropometric or biomechanical considerations. The incidence of musculoskeletal disorders caused by failing to satisfy these requirements can be reduced by the design and use of "ergonomic" tools. As with most human factors implementations, the process should start with a task analysis of how a specific tool is being used, and how often, as well as the risks involved in its use (Armstrong, 2010). A change in design or tool is warranted if the tool is used frequently and the associated risks are significant.

The incidence of mental fatigue and slow learning, problems 7 and 8 in the list, illustrates that there are also significant cognitive factors to be considered in tool design (Baber, 2006). Part of the reason for this might be that people have mental representations for the shapes of tools and how they are to be used. These mental representations will determine, for example, how a person holds and manipulates a tool, and the movement sequences that he or she attempts to perform with it. A better understanding of these representations could lead to tool designs that improve learning and decrease cognitive load.

DESIGN PRINCIPLES FOR HAND TOOLS

Hand tools are designed with the primary goal of maximizing the forces a person can produce with the tool while minimizing the physical stress to which the person's body is subjected. This goal is particularly important for tasks that are performed for extended periods of time and with a tool that requires considerable force to use (Sperling et al., 1993). The following principles are important for obtaining this goal.

Bend the handle, not the wrist

As we have already discussed, when a person's wrist is bent, the amount of stress on the supporting tissues and the median nerve increases significantly. Consequently, one step toward minimizing cumulative trauma disorders is to reshape tools so that bending of the wrist is avoided. For example, Figure 16.4 shows a bent-handled soldering iron that can be used while keeping the wrist straight, and also a more standard straight-handled iron that must be used with the wrist bent.

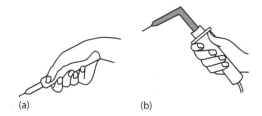

(a)　　　　　　(b)

FIGURE 16.4   Straight-handled (a) and bent-handled (b) soldering irons held in the posture for soldering on a horizontal surface. The forearm is in a more natural posture with the bent-handled iron.

A study examined the angle at which handles are attached to American-style woks, large, heavy cooking pans used for stir-frying (Lim, Liu, Wang, & Joines, 2011). Cooking with a wok can involves repetitive wrist motions to fling the heavy wok to stir the food. The wok can have a "negative" handle that slopes downward from the side of the pan or a "positive" one that slopes upward. Although previous research had suggested that a negative handle improved cooking performance while keeping the exertion of the cook at an acceptable minimum, the researchers demonstrated that the negative handle produced greater wrist bending and muscular effort, leading to a recommendation against using woks with the negative handle design.

When using a tool requires bending the wrist, the user will often compensate by raising his or her arm (abduction). The greater a user's abduction, the more quickly he or she will become fatigued (Chaffin, 1973). This means that tools with angled handles will not only reduce a user's risk of a cumulative trauma disorder, but may also minimize the user's fatigue due to abduction.

A study compared two groups of trainees on an electronics assembly line using either bent-handled or straight-handled pliers (see Figure 16.5; Tichauer, 1978). Incredibly, after 12 weeks on the job, 60% of the trainees using the straight-handled pliers had developed some type of wrist-related cumulative trauma disorder, while only 10% of those using the bent-handled pliers did.

Another study looked at the effects of handle angle for two hammering tasks (Schoenmarklin & Marras, 1989a,b). People hammered on either a horizontal surface (a bench) or a vertical surface (a wall) using a hammer with a handle that was angled 0°, 20°, or 40° from its center of mass (see Figure 16.6). The amount of wrist flexion at impact was lowest for the 40° handle and greatest for the 0° handle.

People could hammer very accurately on the horizontal surface no matter what the handle angle was. On the vertical surface, accuracy was not as good (but still pretty good) for the angled handles. While there were no benefits (or costs) in terms of fatigue or discomfort with the angled handles, everyone preferred hammering horizontally to hammering vertically. From these findings, the researchers recommend the use of bent-handled hammers, because their use does not significantly impair performance or increase fatigue or discomfort, but does reduce the extent to which the user must bend his or her wrist.

Allow an Optimal Grip

There are two kinds of grips that a user can apply to a tool: a power grip and a precision grip. In a power grip, the four fingers of a person's hand wrap around the tool grip while his or her thumb reaches around the other side and touches the index finger. Hammers, saws, and crowbars require a power grip. The power grip allows force to be applied parallel to the forearm (as in sawing), at an angle (as in hammering), and about the forearm (as in turning a screwdriver).

In a precision grip, the person's thumb opposes his or her fingertips. Pencils, forks, and soldering irons require a precision grip. Precision grips allow much finer control over the movements of a tool than power grips. There are two varieties of precision grips: external and internal. An internal

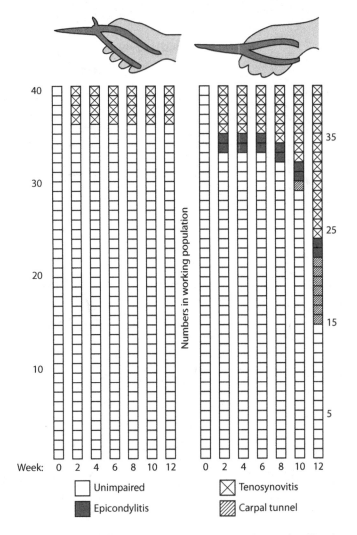

FIGURE 16.5 Percentages of electronics trainees showing cumulative trauma disorders after using either bent or straight pliers for 0–12 weeks.

precision grip places the handle of the tool against the user's palm and is used for such things as table knives. An external precision grip rests the tool handle on the webbing between a user's thumb and forefinger and is used for things such as pencils.

Tools can be designed so that their handles encourage and accommodate hand grips appropriate for the task they are to help perform (Cal/OSHA Consultation Service & NIOSH, 2004). Some important parameters include the tool handle diameter, length, and surface material. Each of these parameters will depend on the task that the tool is designed for and how it is to be used. For example, single-handled tools like hammers will require a larger handle diameter than double-handled tools like pliers (Cal/OSHA Consultation Service & NIOSH, 2004; see Table 16.4 for recommended handle diameters as a function of grip and handle type). A tool that requires a power grip must have a handle long enough for all four fingers. Similarly, a tool manipulated by an external precision grip must be long enough that the handle can rest against the base of the user's thumb, which provides support. Compressible grip surfaces with nonslip surfaces are better than hard, slippery unyielding surfaces, because they prevent tissue compression and loss of local blood circulation (Konz,

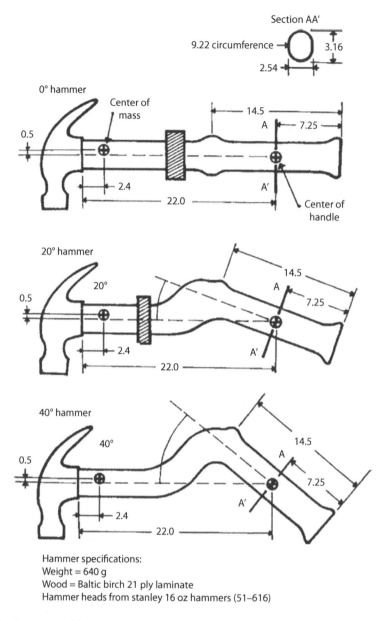

FIGURE 16.6 Hammers with handles angled 0°, 20° and 40°.

TABLE 16.4

Recommended Handle Diameters as a Function of Grip and Handle Type

	Single Handle	Double Handle
Power grip	3.8–5.1 cm	5.1 cm (closed)–8.9 cm (open)
Precision grip	0.6–1.3 cm	2.5 cm (closed)–7.6 cm (open)

1974). In general, tool handles should have smooth surfaces and should not conduct either heat or electricity.

Use Compressible Grip Surfaces

A good grip surface will provide good contact between the user's hand and the tool, and at the same time avoid pinch points and tissue compression. Nonslip materials that compress slightly, such as foam rubber, will allow the pressure on the handle to be more evenly distributed across the user's hand. Moreover, they also will resist vibrations and extreme temperatures, both of which can create problems for tool users.

An Example Design Problem

Food scoops are used by workers in the food service industry for long periods, often several different times during a day. The amount of force required to scoop the food and transfer it to a plate or bowl ranges from very little (as for macaroni and cheese) to very high (as for ice cream). Most food scoops were not designed with ergonomic considerations in mind, having straight handles and grips that are not compressible (see Figure 16.7, top). Based on ergonomics principles, Williams (2003) proposed a redesigned food scoop that allowed a straight wrist to be maintained and had an optimal lift angle, handle size, and handle composition. The resulting food scoop had the handle angled 70° and the scoop head angled 35° relative to the shaft of the scoop (see Figure 16.7, bottom). The handle of the scoop was wider than usual to allow optimal grip diameter and covered with a compressible material.

While we might believe (for good reasons) that Williams's (2003) design would be better than the old design, to our knowledge it has not yet been tested. Before switching to a new tool, the new tool must be evaluated for effectiveness (how accurately users can apply it) and its potential to increase fatigue or cumulative trauma disorders. In some situations, a new tool might require some retraining of its users. So, while the redesigned scoop might seem attractive, you should remember that ergonomic designs should always be tested and evaluated before replacing old, effective (but perhaps not optimal) designs.

MANUAL OR POWER TOOLS

Power tools can generate more power than human muscles and hence perform tasks that could not be accomplished otherwise. Moreover, power tools can perform tasks faster, and they reduce user

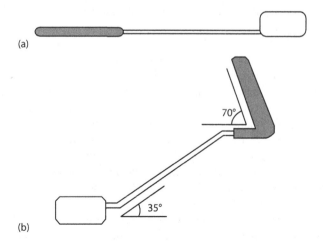

FIGURE 16.7 Standard and redesigned food scoops.

fatigue. In most cases, it makes sense to consider the use of power tools over manual tools (Konz, 1974).

Using power tools may minimize the potential for cumulative trauma disorders associated with repetitive movements. For example, the twisting or ratcheting motion required to operate a manual screwdriver is largely eliminated by a power screwdriver. Although this may not be of much consequence if a screwdriver is used only occasionally, it may be an important consideration for workers who must use a screwdriver repetitively. For instance, if a worker must tighten 1000 screws a day with a ratcheting screwdriver, he or she may perform 5000 effortful movements. With a power screwdriver, that number of movements is reduced to 1000 (less effortful) movements (Armstrong, Ulin, & Ways, 1990). Effort is reduced with the power screwdriver because the load on the user's forearm muscles is reduced (Cederqvist, Lindberg, Magnussen, & Ortengren, 1990).

However, we cannot always assume that a power tool will automatically reduce fatigue or the risk of cumulative trauma disorders. Controlling the tool may require the user to grip the tool much harder than would be required for a manual tool. Also, the user may need to adopt a different or more stressful posture, which may negate any benefit of reducing the number of repetitive movements. For example, an electric screwdriver may require the operator to exert more force on the screw against the workpiece, possibly using two arms rather than one.

Another drawback to the use of power tools is their vibration. Vibration can cause trauma disorders, and we will discuss the effects of vibration and the family of trauma disorders related to vibration in Chapter 17. Power tools also have several unique hazards, including the risk of electric shock, unintended activation, and severe injury, which must be controlled (Cacha, 1999). Electric shock can be prevented with proper grounding and insulation. Safeguards such as safety and "deadman" switches can prevent unintended activation. Moving parts that could cause injury, such as circular saw blades, must be shielded.

ADDITIONAL PRINCIPLES

There are several additional principles that apply to hand-tool design: the role of special-purpose tools, the handedness of the user, and limitations created by the muscle groups that will move the tool (Konz, 1974).

Special-purpose tools are, in most circumstances, better to use than general-purpose tools. Because general-purpose tools are useful for a wide variety of tasks, they are cheaper to provide than special-purpose tools. However, using a general-purpose tool instead of a special-purpose tool may mean that it will take longer to do most tasks: special-purpose tools are more appropriate for performing the specific tasks for which they were developed than general-purpose tools are. If the tool is used often, those delays will, in the long run, be more expensive than the early savings gained by not purchasing the special-purpose tool.

It should not matter whether a user is left- or right-handed. However, some tools cannot be used by left-handed people. Paper scissors, for example, usually have a wider aperture on the bottom half of the handle, for two or more fingers on the right hand, and a smaller aperture on the top half of the handle, for the right thumb. The scissors cannot be operated easily by either hand, because the position of the blades matters. To operate the scissors with the left hand, they must be turned upside-down, which puts the smaller aperture on the bottom handle and the larger one on the top. This means that the left-handed user will have less control over the blades and significantly more discomfort from an ill-fitting grip than a right-handed user. Effective, well-designed tools are operable by either hand for two reasons: first, so that left-handed people can use them as effectively as can right-handed people; and second, because most tool work produces muscle fatigue, and the ability to switch hands can relieve this fatigue.

Finally, a tool should exploit the strength of the most appropriate muscle group. For example, forearm muscles are stronger than finger muscles, and so, when a job requires a lot of force, a tool should be operated by the forearm rather than the hand. Also, the hand can exert more force by squeezing than by opening, so tools such as heavy scissors should normally be held open by a spring, so that excessive force is not required to open them.

MANUAL MATERIALS HANDLING

"Materials handling" is what people do when they move things around. For example, in a factory, workers routinely load and unload boxes of equipment or products and move materials from one location to another, either by hand or using mechanical devices such as trucks and forklifts. This kind of work has a high risk of physical injury, often from acute trauma. As an example, consider the story of a supervisor in the bakery and delicatessen shop of a supermarket (Showalter, 2006). She demonstrated to new employees how they should drain grease into a 20-gallon container and then picked up the bucket. At this point, in her words, "It was bones cracking. [The employees] heard it too, which made it worse."

A survey conducted in Great Britain in 2001–2002 estimated that manual materials handling accounted for approximately 38% of all injuries that caused 3 or more days of missed work (HSE, 2004). This risk is even higher for jobs that require heavy lifting, such as baggage handling and nursing (for which people are the heavy loads that may need to be lifted and moved), where more than 50% of injuries sustained on the job are due to handling and lifting (Pheasant & Haslegrave, 2006). Excessive load and awkward posture are the two most critical causes of manual materials handling injuries (Al Amin, Nuradilah, Isa, Nor Akramin, & Febrian, 2013).

There are two important components to reducing the number of injuries due to materials handling. First, workers must be taught appropriate handling methods, such as how to lift, lower, and carry objects that may be heavy or of irregular shape. Second, the work system must be structured with the goal of preventing injury.

Risk factors that affect the likelihood of manual handling injuries include the following (Chaffin et al., 2006): worker characteristics, material-container characteristics, task-workplace characteristics, and work practices. Worker characteristics include an individual's level of physical fitness and how well he or she is trained. While some worker characteristics are amenable to change, some are not. The characteristics of the materials and containers being handled include such things as their weight, shape, and dimensions. While some aspects of the materials can be changed, some are determined by the product that is being manufactured and the equipment used to produce it. Some task characteristics, such as the pace of work, the design of the workspace, and the tools and equipment to be used, can also be altered to reduce the risk of injury, but again, there may be some factors that are determined by the nature of the job. Airline baggage handling, for example, will always have a rapid pace that is determined by flight schedules. Work practices, such as administrative safety policies and incentives, scheduling of work shifts, and management styles are perhaps the easiest practices to change to reduce the risk of injury.

In this section, we will discuss the tasks that a worker performs in manual materials handling (lifting and lowering, pushing and pulling, and carrying) and the factors associated with each task that influence the likelihood of physical injury. These factors include the direction, speed, and frequency of the movement. The container or load being handled affects the risk of injury directly through its bulk, shape, and weight, and indirectly through the limitations that it imposes on the way in which it can be held and carried (Drury & Coury, 1982).

Lifting and Lowering

There are three static force components that are important to lifting and lowering (Davis & Marras, 2005; Tichauer, 1978). These components are called *moments*, which are measurements of force made within a system that rotates around one or more axes (like the human body). Sometimes we refer instead to *torque* to describe these forces. The moments are named according to the plane of anatomy in which they are measured (see Figure 16.2). A particular moment will tend to produce rotational movement in its plane. For lifting and lowering, the three moments are the sagittal, lateral (coronal), and torsional (transverse).

The sagittal moment is a measure of the forces acting downward in the sagittal plane. The magnitude of a sagittal moment depends on the weight of a person's body, the height of his work surface, his position (seated or standing), and so on. Sagittal moments produce forward and backward movements. The lateral moment is a measure of the forces acting downward in the lateral plane. Lateral moments arise when a person's weight is shifted from one foot to another. Torsional moments are measured in the transverse plane and arise when a person twists at the waist. All of these moments, sagittal, lateral, and torsional, stress the spine and other joints of the body. Minimizing such stresses reduces the risk of injury.

You may recall from physics class that the force of an object is equal to its mass times its acceleration. This means that as the weight of a lifted object increases, the forces pulling on the person's spine increase. Other factors that influence the amount of stress to the spine include how easily the object can be grabbed, how asymmetric the weight distribution is in the object, and how unusual its shape is. Spinal stress also increases when starting and ending heights for the lift are too low or high, for asymmetric lifts in which the person's body is twisted or bent, and with increasing distance, speed, and frequency of lifts.

For example, Davis and Marras (2005) had people lift a box and move it from one shelf to another. They found that the heights of the first and second shelves and the asymmetry of the person's posture during the lift determined the degree to which the person's spine was compressed. The height of the second shelf was the most significant factor in determining the magnitude of sagittal moments and the shearing forces on the spine. Also, when the two shelves had widely varying heights, stress on the spine was increased. The largest moments were measured when the box was lifted from knee height to shoulder height. These results emphasize that it is important to evaluate both the starting and ending position for lifted objects and their relation to each other when trying to decide if a lifting task is safe.

Guidelines for manual lifting were first established by the National Institute for Occupational Safety and Health (NIOSH) and published in the *Work Practices Guide for Manual Lifting* (NIOSH, 1981). These guidelines distinguish between jobs that require infrequent lifts, frequent lifts for less than an hour, and frequent lifts for an entire workday. The guide includes a "lifting equation" that incorporates several of the factors listed above. This equation determines the maximum recommended weight for two-handed symmetrical lifts, called the *action limit*. The action limit is the upper bound for lifting conditions requiring intervention or special equipment. Three times the action limit is the *maximum permissible limit*, a higher bound not to be exceeded under any circumstances.

In 1991, the NIOSH equation was revised to accommodate a broader range of lifting conditions (Waters, Putz-Anderson, & Garg, 1994; Waters, Putz-Anderson, Garg, & Fine, 1993). The revised weight limit (RWL) specifies the weight of a load that a healthy individual could lift for up to 8 hours per day without an increased risk of lower back pain. The equation is:

$$RWL = LC \times HM \times VM \times DM \times AM \times FM \times CM,$$

where LC is the load constant, 23 kg (51 lb), which is the maximum weight of a load that can be lifted safely under the best lifting conditions. The other terms reduce the recommended weight when conditions are not optimal. HM is the horizontal multiplier, which depends on the horizontal distance of a person's hands measured from a point midway between her ankles. VM is the vertical multiplier, which depends on the change in the vertical distance of the person's hands above the ground at the origin and destination of the lift. DM is the distance multiplier, which depends on the carrying distance of the lift. AM is the asymmetric multiplier, which is based on the angle of the object in front of the lifter, or how far to the person's side the object is. FM is the frequency multiplier, which is based on the average number of lifts per minute over a 15 min period. CM is the coupling multiplier, which depends on how well the hand and object are coupled, or how easy the object is to grab.

The RWL is used to provide a lifting index (LI), which estimates the amount of stress associated with lifting a load (L) of a given weight:

$$LI = \frac{L}{RWL}.$$

The LI should never exceed 1, and the lower it is, the lower is the risk of lower back injury. The RWL and LI equations are only used to evaluate two-handed manual lifting tasks performed while standing. They do not take into account unstable loads, slips, or falls that might result from inadequate worker/floor coupling, or other task factors that can influence the likelihood of injury. Applications that rapidly calculate the NIOSH lift index are available for iPhone and Android devices.

The NIOSH guidelines assume that the physical stresses that a person endures are constant over time. That is, they are static. Mirka and Marras (1990) argued that measurements based on static forces are misleading and cannot completely determine safe limits for lifting tasks. They measured velocity and acceleration of lifters' bodies during low- and high-velocity lifts, lifts that can eventually lead to damage of the spinal discs. For symmetric and asymmetric lifting postures, Mirka and Marras found that trunk acceleration peaked at a much higher level for fast lifts than for slow lifts. However, for slow lifts, trunk acceleration peaked more often, resulting in forces on the spine that were sometimes greater overall than for the fast lifts. Most guidelines emphasize the importance of smooth, controlled lifting motions in minimizing (static) forces on the spine. However, the assumption of constant forces throughout a lift may result in an underestimate of the load on the spine during slow lifts. These data suggest that many other variables, such as external forces and the cumulative effects of trauma, need to be considered before determining the best conditions for lifting.

A device called a *lumbar motion monitor* (LMM) was developed by Marras et al. (1993). The LMM tracks the motion of a person's trunk during the performance of various tasks. The LMM determines when in the task a person's trunk changes position, velocity, and acceleration in 3-D space. It can be used to quantify the risk for low back disorder using a combination of trunk kinematics and workplace measures.

For example, Ferguson, Marras, and Burr (2004) used the LMM to compute risk estimates for workers who had recently experienced low back injury and workers who had no signs of back injury. There were no differences in the way the two groups moved their trunks. Computed risks based on the LMM measurements were not different for the two groups, either. These findings suggest that risk is determined primarily by job design.

One job that involves lifting heavy weights during high levels of activity (and therefore great stresses on the spine) is garbage collecting. Kemper et al. (1990) reported the results from an extensive study on garbage collectors conducted in the Netherlands. They compared the workers' ability to handle (carry, lift, and throw) trash cans and plastic bags. They found that, compared with cans, workers could carry more weight with bags and could throw bags with more force. Garbage collectors gathered 70% more garbage with bags than with cans. As a result, the city of Haarlem replaced garbage cans with plastic garbage bags and changed from twice- to once-a-week collection. The study also found that, even though the garbage collectors were performing more efficiently, they still were working above the acceptable limits of workload tolerance and, therefore, were at higher risk for back injuries.

We mentioned already that one way to reduce back injuries is to screen workers according to their lifting capabilities; only those people whose health and physical strength were sufficient to handle the materials safely would be assigned to the job. Such screening can be accomplished in three different ways: isometric strength testing, which measures the amount of force a person can exert against a stationary object; isokinetic testing, which measures a person's strength while he or she is moving at a constant speed; and isoinertial testing, which measures the maximum mass that a person can lift. Isometric and isokinetic strength testing do not require much movement from the person being tested. Isoinertial testing measures force dynamically, as the person is engaged in the act of lifting. Just as Marras and Mirka discouraged the use of static strength measures, Kroemer (1983b) concluded that such measures are often inappropriate for screening a person's handling capability. He demonstrated that the dynamic measures provided by isoinertial testing methods were more reliable than static test measures.

There are many situations in which the load to be lifted is more than one worker can handle. In such situations, workers use mechanical aids. For example, an overhead lifting device is commonly used to lift and move residents in extended care nursing facilities. This device reduces the injury risks for healthcare staff, who otherwise would be doing the lifting manually (Engst, Chokkar, Miller, Tate, & Yassi, 2005). Alternatively, lifting may be performed by a team of two or more people. Team lifting reduces the biomechanical stress that each person must endure relative to the stress endured if each person were to attempt the lift alone. However, the total lifting capacity of a team is less than the sum of the capacities of the individual team members (Barrett & Dennis, 2005). For instance, team lifting may reduce the ability of each team member to grasp the object or may restrict the range of motion of the team members. Many factors can influence the level of biomechanical stress a person will experience during team lifting.

CARRYING AND PUSH/PULLING

Package delivery, mail delivery, warehouse loading operations, and many other jobs require carrying materials in addition to lifting. Carrying requires a worker to exert the force needed to lift the object but also to maintain this force until the destination is reached. This means that the maximum weight that a person can carry will be less than the weight that he or she can lift.

Carrying tasks will require the use of either both hands or only one hand. Carrying a suitcase, for example, is a one-handed task. A person cannot generate as much force in a one-handed carrying task as she can in a two-handed carrying task. One-handed carrying also means that a person must endure the harmful stresses associated with lifting asymmetric loads. Figure 16.8 shows how the maximum recommended weight changes as a function of carrying distance and posture. As distance increases, the maximum weight decreases.

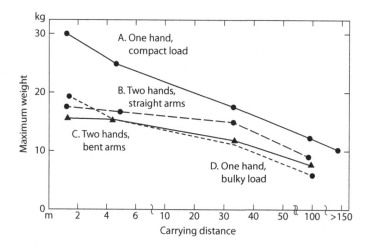

FIGURE 16.8 Maximum load weight as a function of carrying distance and carrying posture.

Some industrial tasks, such as tool operation, require people to push or pull. Some jobs are designed so that heavy loads are moved by a cart rather than being lifted and carried, which substitutes lifting and carrying operations with pushing and pulling operations. Usually, people can move heavier objects by pushing and pulling than they can by lifting and carrying.

The force that a person can apply as a push or a pull varies as a function of the person's weight, the height and angle at which force is applied, the distance of the object from the person, the degree of friction between the person's shoes and the floor, and how long the force is applied (Chengalur, Rodgers, & Bernard, 2004). To determine acceptable parameters for pushing and pulling tasks, we can consult a data bank collected by Snook and Ciriello (1991), which is generally considered to provide the best source of information for these kinds of tasks. It specifies the maximum acceptable push and pull forces for men and women as a function of task frequency, distance, height, and duration (see also Ciriello, 2004).

Chaffin, Andres, and Garg (1983) recorded the postures assumed by men and women of varying sizes when they were asked to exert push and pull forces against an isometric strength-testing fixture with handles. When these people kept their feet side-by-side, there was no difference between pushing and pulling strength for either men or women. When they staggered their feet, putting one in front of the other, their strength was much greater, because they could lean into and away from the apparatus more. Also, men had greater pushing than pulling strength. When the handle was raised, everyone's strength decreased.

This study investigated the factors influencing strength for horizontal pushes and pulls. Similar factors affect a person's ability to push or pull vertically and to apply force laterally. Vertical pushing and pulling strength will be determined by the height at which the force must be exerted. If the point at which force must be applied is too high or too low relative to the person's body, strength will be reduced. Much depends on the extent to which a person can use his or her leg muscles to assist with the push or pull, as we saw with horizontal pushes and pulls made with staggered feet. A person can generate much less force when sitting than when standing. Similarly, a person can generate much smaller lateral forces (approximately half) than those generated for horizontal pushes and pulls, because the leg muscles can't assist as much in lateral applications.

When we attempt to reduce the risk of injury by redesigning jobs that require lifting, carrying, and pushing, we have to take into account the entire system in which the tasks are performed. Consider, for example, a study of dairy truck drivers that was done by Nygard and Ilmarinen (1990). Dairy truck drivers must load and unload large volumes of dairy products daily. In Finland, a rolling delivery system was instituted that allowed the drivers to move the products from the dairy to the truck and from the truck to the store on transport dollies. The purpose of the new system was to reduce the physical workload imposed on the drivers. With this method, the amount the drivers had to carry decreased, but they had to do more pushing.

It was disappointing, then, that measures of physiological strain showed only a slight improvement with the rolling delivery system. This improvement was evident at the dairy but not at the stores. The reason for this was that the dairies could accommodate the dollies, but many of the stores could not. Some of the stores had no platform for loading and unloading, and the driver had to negotiate steps or stairs. For the system to have been effective, the authors noted, ergonomic improvements would have to have been made throughout the entire delivery system.

WORKSPACE DESIGN

We have now discussed several elements of workspace design. Bringing all these elements together, we see that the design of a workspace will depend on the hardware that is to be operated, where displays and controls are located, the worker's posture, the computer software the worker will use (if any), the physical environment, and how work is organized and scheduled (see Figure 16.9).

A good workspace design helps ensure that most people can do their work safely and efficiently. Good workspaces minimize extraneous motions, make it easy for workers to reach controls and

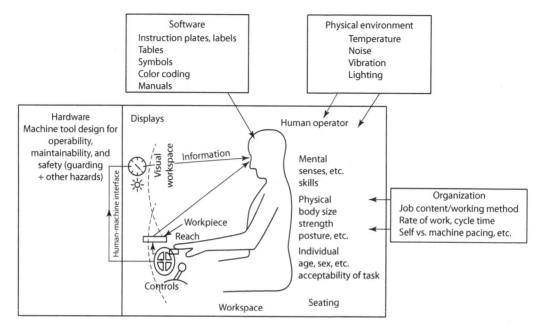

FIGURE 16.9   Ergonomic considerations in workspace design.

other equipment, and eliminate biomechanical stresses that can cause fatigue and injury (Chaffin, 1997). Toward these ends, we must consider (1) whether the worker will be sitting or standing, (2) the layout and height of the work surface, (3) chair design and height, and (4) the location of visual displays.

WORKING POSITION

A worker can use a workspace while standing, sitting, or both (Chengalur et al., 2004; Kroemer & Grandjean, 1997). Usually, workspaces are designed for a single posture. There may not be too much debate about which type of workspace is most appropriate for a particular job. Most modern jobs are done by people seated in vehicles or in an office (Robinette & Daanen, 2003). In those situations where posture is an option, we need to consider several factors.

Workers who stand have more mobility. If a job requires a worker to move around frequently, standing workstations make more sense than sitting workstations. Similarly, if a job requires the worker to exert large forces (e.g., to handle heavy objects or press downward as in packaging), a standing workstation will be more appropriate, because more force can be exerted from a standing position. A standing workstation may also be appropriate when the location of the workstation has limited space so that, for example, there is no knee clearance.

The optimal height of a standing work surface will depend on the type of task being performed. For tasks like writing and light assembly, the optimal working height is 107 cm (see Figure 16.10). This allows the details of the visual work to be seen while minimizing neck discomfort and allowing the arms to be stabilized, but at the expense of discomfort for the shoulders (Marras, 2006). For tasks requiring large downward or sideward forces, or other heavy work, the optimal height is lower (91 cm), because a person can generate more force at lower heights. In both cases, the height of the objects to be handled will in part determine the height of the work surface. A height-adjustable work surface lets different users select the heights that are best for them and also allows the surface to be adjusted for different kinds of jobs.

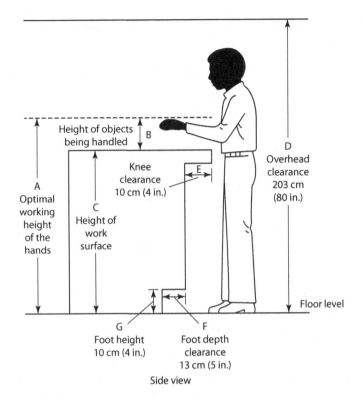

FIGURE 16.10   Recommended standing workspace dimensions.

Activities that require fine manual control are best performed at sitting workstations. Such work is performed best when the person doing it can keep his or her body as free from movement as possible. Sitting is one way to increase stability, as is the person's ability to rest his or her forearms against the work surface. Close visual work that does not require strong exertion or reaching is also best performed at sitting workstations.

As with standing workspaces, the optimal working height of a sitting workstation will depend on the tasks that the worker will perform. For most tasks, such as writing, the height of the work surface should be the same as the person's elbow height when seated. Higher surfaces may be used for fine detail work, which requires more stability and discrimination of detail For example, the table height for sewing machines should be a least 5 cm above elbow height (Delleman & Dul, 1990). Sitting work surfaces should also be adjustable, from 5 cm above the 5th percentile seated elbow height of the user population to 15 cm above the 95th percentile elbow height.

Some workstations can serve dual purposes for both sitting and standing work. Sit/stand workspaces are useful when the person needs to preserve his or her mobility and to perform a range of different tasks, some that are performed best while seated and some while standing. The design issues that arise for sit/stand workspaces can be complicated. One problem that the designer must solve is how the workspace will serve both sitting and standing postures. What should the work surface height be in this situation? When the work surface height is fixed, the designer might put the work surface at the lowest possible level for standing tasks and provide a high seat for sitting.

Adjustable height workstations might provide a solution for some issues that arise for sit/stand workspaces. Some offices provide adjustable height workstations for their workers who would otherwise be in a seated position all day. The motivation for providing such workstations comes from the idea that allowing work to be performed while standing might prevent some adverse health out-comes associated with sitting. Some researchers have recommended that the amount of time spent seated should approximately equal that spent standing, and the worker should alternate between seated and standing postures every 15 minutes (Callaghan, Carvalho, Gallagher, Karakolis, & Nelson-Wong, 2015).

SEATING

A person's chair plays a big role in how comfortable he or she is and, hence, how well he or she performs. The main function of a chair is to provide support and reduce the stress on the sitter's spine and joints. As with any other equipment, a chair can have good or poor ergonomic design (Colombini, Occhipinti, Molteni, & Grieco, 2006). Table 16.5 lists some of the factors that must be considered for a work seat to function adequately (Corlett, 2005). The optimal seat design will depend on the tasks workers are to perform and the characteristics of those workers. For example, a backrest can assist in the performance of a task that requires pushing forces. Similarly, to prevent leg discomfort, no less than two-thirds of the worker's weight should be supported by the seat and backrest, rather than the feet.

When a person sits, the weight of his body is transferred primarily to the seat of the chair (see Figure 16.11). Some part of his weight also may be transferred to the backrests and armrests, as well as to the floor and the work surface. The distribution of force onto the chair and over the parts of the person's body depends on that person's posture, which in turn is directly influenced by the chair design and the task that is performed (Andersson, 1986; Chaffin et al., 2006). Sitting increases pressure on the lumbar discs of the spine (see Figure 16.12), which in turn can restrict the flow of spinal fluid (Serber, 1990). Good posture can minimize this pressure. If a seat induces poor posture, these factors can lead to chronic back pain, herniated discs, and pinched nerves.

Good posture is encouraged by appropriate seat height and support. When the seat height is correct, the sitter's spine will be straight and, simultaneously, his or her thighs should exert very little pressure on the seat. Too much pressure on the backs of the thighs can interfere with blood circulation to the legs ("compression ischemia"), which in turn may result in pain and swelling. The sitter's body weight should be supported by his or her feet and buttocks, and a footstool may be necessary to take pressure off the sitter's thighs.

TABLE 16.5
Factors to Consider for Adequate Work Seat Design

The task	The sitter
Seeing	Support weight
Reaching	Resist accelerations
The seat	Under-thigh clearance
Seat height	Trunk-thigh angle
Seat shape	Leg loading
Backrest shape	Spinal loading
Stability	Neck/arms loading
Lumbar support	Abdominal discomforts
Adjustment range	Stability
Ingress/egress	Postural changes
	Long-term use
	Acceptability
	Comfort

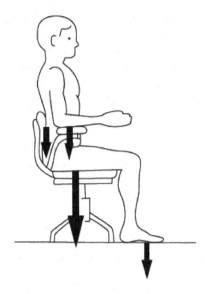

FIGURE 16.11 Transfer of body weight when sitting.

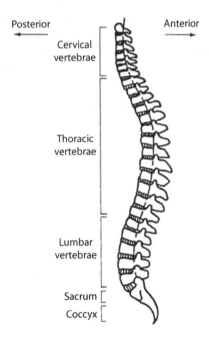

FIGURE 16.12 The human spinal column.

Appropriate seat height will be determined by a person's knee height. Even people of the same stature may require very different seat heights. In most workspaces, chairs with adjustable seat heights are most appropriate, especially when workstations are shared by more than one person. As long as the seat height can be adjusted without sacrificing stability in the chair, adjustability allows workers to select the heights that are right for them. It also allows the worker to change her position relative to the work surface height when she has to perform different tasks.

The shape of the chair, as well as its height, can be changed to reduce pressure on the sitter's body. For instance, tilting the chair back toward the sitter can increase the load on the chair back and reduce pressure on the spinal discs (Andersson, 1986; Kroemer & Grandjean, 1997). A lumbar support will also reduce pressure on the lumbar spine. Armrests, when appropriate, can help support the weight of the sitter's arms.

As we have mentioned already, selecting a chair that is appropriate for the demands of the task to be performed can decrease the pressures on a person's body and make her more comfortable. These benefits translate directly into improvements in task performance (Eklund & Corlett, 1986). An experiment evaluated three chair designs for use with three types of tasks (see Figure 16.13). One chair had a high backrest, another had a low backrest that provided lumbar support, and the final chair was a sit-stand seat. People performed each of three tasks. In the forward-pushing task (c), a handle was gripped with both hands and pushed hard in a forward direction. In the sideways-viewing task (b), people viewed a television set placed 90° to the left. Finally, an assembly task (a), which involved screwing nuts on bolts, was performed with restricted knee space. People performed the forward-pushing task best when their chair had a high backrest, the sideways-viewing task best when their chair had a low backrest, and the assembly task best when their chair had a sit-stand seat.

A work seat for industrial sewing-machine operators was designed and evaluated by Yu and Keyserling (1989). The new seat allowed workers to maintain a low sit/stand posture, which they preferred. They redesigned the chair back to give more thoracic and lumbar support and the seat pan to give more pelvic and thigh support. The workers reported significantly less discomfort with the new chair, and 41 of 50 workers preferred it over their old chair.

So we see that something as simple as seat redesign may be a first step toward enhancing the work environment and increasing work tolerance. Conversely, poorly designed seats can lead to pain in both the upper and lower parts of the body (Hunting & Grandjean, 1976). However, there are no nerve endings in the lumbar discs, the part of the spine most affected by a badly designed chair. This means that people cannot always feel the difference between a good chair and a bad chair. According to Helander (2003),

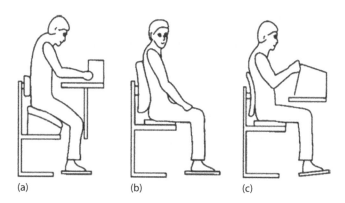

(a)   (b)   (c)

FIGURE 16.13 Postures for the assembly task (left), the sideways-viewing task (middle), and the pushing task (right).

Unless there are no obvious violations of biomechanics design rules, chairs users will not complain about discomfort. Important ergonomic design features include a rounded front edge of the seat pan, so that the blood circulation in the legs is not cut off, a back rest angle adjustable to about 120 degrees, a cushioned seat pan and back rest, and support for the legs. Most chairs presently have these features. Users are, however not particularly sensitive to minor design changes in ergonomic variables; they can simply not be perceived. (p. 1316)

Consequently, Helander recommends that more emphasis in chair design be placed on "comfort factors" such as aesthetics and plushness.

A chair designer will also face challenges from the needs of special populations. For example, older adults may have certain physical limitations that require special design considerations. In particular, the ease with which a sitter can get in or out of the chair is an often overlooked factor that is important for the elderly (Corlett, 2005). Ingress/egress is particularly a problem with car seats, and older adults may have difficulty getting in and out of their vehicles (Namamoto, Atsumi, Kodera, & Kanamori, 2003).

POSITIONING OF VISUAL DISPLAYS

Regardless of whether a worker sits or stands, any visual displays required for her job must be positioned to be viewed easily without imposing excessive stress on her musculoskeletal system (Straker, 2006). Before deciding where such displays should be positioned, the workspace designer needs to know the worker's *line of sight* and *field of view* (Kroemer & Grandjean, 1997; Rühmann, 1984). The line of sight is the direction in which a person's eye is fixated. The field of view is the region of the workspace that a person can "see" effectively at a particular line of sight.

The horizontal line of sight is the direction of a person's eyes when his head and eyes are straight (see Figure 16.14). This position is not the most comfortable, or necessarily the most effective for a

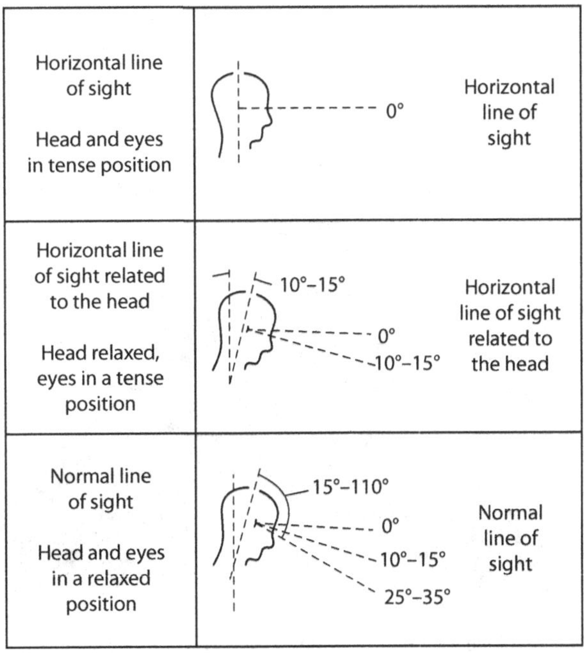

FIGURE 16.14   Lines of sight.

particular task. A relaxed, comfortable head posture, with the eyes straight ahead, has the person's head inclined forward approximately 10°–15° from vertical. So, the line of sight relative to the head is 10°–15° below the horizontal line of sight. Finally, the normal line of sight, for which the person's eyes are also relaxed, is 25°–30° below horizontal. If visual displays are positioned so that the worker can maintain a normal line of sight most of the time, fatigue (of both the eyes and the neck) will be minimized.

The maximum direct field of view is that part of the workspace for which visual receptors (on the retina) can be stimulated for a fixed line of sight. For binocular conditions, when the person is using both eyes, this region extends 45° above and below the line of sight and 95° to either side. The maximum direct field of view is not the same as the functional field of view, which is usually smaller and decreases when viewing the workspace with only one eye, when the person must distinguish between different colors, and so on. The functional field can be larger than the direct field of view when the person moves his eyes, and larger still when he can move his head.

To determine the best location for different displays, the designer must be able to rank them by their priority, or how important they are for the performance of the task. High-priority displays are usually positioned within the direct field of view along the normal line of sight. Medium-priority displays are positioned so that they can be seen when the person moves only his eyes, and low-priority displays can be placed outside the field of view (so that the person must move his head and/or rotate his trunk to see them).

Positioning of Controls and Objects

We introduced the two-dimensional (2-D) reach envelope in Chapter 15 when we discussed the positioning of controls. We can extend the reach envelope into a 3-D surface that partly surrounds the body (see Figure 16.15). This figure shows the range of a *normal reach surface* as all the locations

FIGURE 16.15 3-D reach envelope. (From https://reducedeffort.wordpress.com/2010/09/27/do-you-like-wasting-time/.)

in a workspace that a person can reach without leaning or stretching. All frequently used controls or objects should be located within the normal reach surface. Objects that are used only occasionally can be located outside of this area but still within the *maximum reach surface*, which is similar to the normal reach surface but at the maximum distance that a person can reach by stretching and leaning.

The shape of the 3-D reach envelope depends on whether the workspace is sitting or standing. It will also change with other factors that influence the operator's mobility, such as table height, whether one or two arms must be used, clothing, and physical restraints. The extent of a person's forward reach will be impaired by increased table height (see Figure 16.16) and when he performs tasks that require the use of both arms. Similarly, bulky protective clothing and safety restraints (see Figure 16.17) restrict movement and so decrease the range of the 3-D reach envelope.

STEPS IN WORKSPACE DESIGN

Although anthropometric and biomechanical data provide the foundation for effective workspace design, the design process itself involves many steps of development and evaluation. This process relies on converting tabled data into concrete drawings, scale models, mockups, and prototypes. Roebuck (1995; pp. 104–116) lists these steps as follows:

1. *Establish requirements*. Determine the goals of the system and other relevant requirements.
2. *Define and describe the population*. Appropriate use of anthropometric data involves definition of the user population and corresponding anthropometric values. If available tables do not fit the population or measurements of interest, new measurements may be necessary.

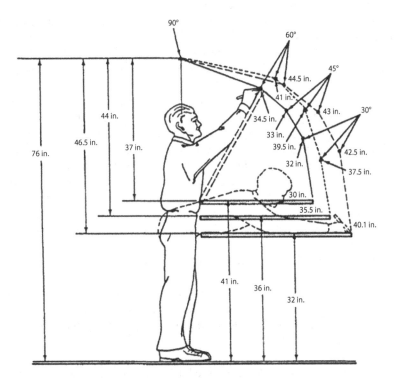

FIGURE 16.16   Maximum reach for a 5th-percentile operator at drafting tables of different lengths.

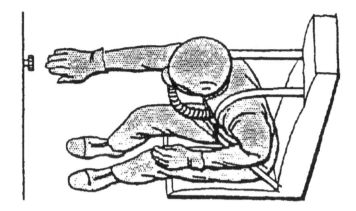

FIGURE 16.17 Maximum reach for a 5th-percentile operator wearing protective clothing and safety restraints.

3. *Select design limits*. A design criterion population is specified by determining the range of percentiles for which the workspace is designed.

4. *Prepare engineering drawings of basic body dimensions*. The appropriate anthropometric data are used to design an "individual" who corresponds to the desired percentile.

5. *Prepare drafting aids*. Traditionally, physical overlays of the individual have been prepared at this step, sometimes along with 2-D drafting mannequins. Computerized 3-D digital human models can now assist in this step.

6. *Prepare workspace layout*. In this step, the designer uses the anthropometric data to construct the functional layout to accommodate individuals at the intended percentile. Computer modeling can be invaluable at this step.

7. *Mathematical analysis*. Mathematical calculations of geometric interrelationships between the person and the workspace are computed.

8. *Develop a small-scale physical model*. Scale models are constructed to verify the requirements being developed in the earlier steps and to determine whether obvious design flaws exist.

9. *Prepare functional test requirements*. Explicit experimental test or evaluation requirements are developed to verify that the system-design criteria have been met.

10. *Prepare mockups and prototypes*. Full size mockups are built for use in evaluating the design adequacy with real users.

11. *Prepare reach and clearance envelopes*. People are used in the mockup, positioned as they would be in actual operation, to determine the functional envelopes. The geometry of these envelopes can be easily incorporated into a 3-D computer model, which greatly simplifies the procedure.

12. *Prepare special measuring devices*. New measuring devices may need to be built to evaluate the workspace.

13. *Test the mockup and prototype*. Batteries of tests are conducted with appropriate test subjects. They may be used to revise the workspace design.

14. *Prepare design letters, memoranda, standards, and specifications*. Based on the results of the workspace evaluation, documented recommendations for design are made. The expected consequences and costs of following or not following the recommendations must be conveyed clearly.

Some variation of these steps should be followed in the design and evaluation of any workspace. The issues we have discussed in this chapter are summarized as the general design principles shown

TABLE 16.6
The General Principles for Workspace Design

1) Avoid static work.

2) Avoid extreme position of the joints.

3) Avoid overloading the muscular system.

4) Aim at best mechanical advantage.

5) Avoid unnatural postures.

6) Maintain a proper sitting position.

7) Permit change of posture.

8) Allow the small operator to reach and the large operator to fit.

9) Train the operator to use the physical facility.

10) Match the job demands and operator capacity.

11) Allow the operator to maintain an upright and forward-facing posture during work.

12) Where vision is a requirement of the task, permit the necessary work points to be adequately visible with the head and trunk upright or with just the head inclined slightly forward.

13) Avoid work performed at or above the level of the heart.

in Table 16.6. If a designer adheres to these principles and follows these procedures, she will maximize the likelihood that the final workspace will be suited to both the task and the user.

SUMMARY

Anthropometric and biomechanical measures provide information about the dimensions and physical properties of the human body that must be accommodated when designing equipment and workspaces for human use. A good design will allow most of the intended user population to operate the equipment and manipulate objects effectively. Poor designs can result in cumulative trauma disorders, which arise when a person must make movements that are repetitive and/or that require considerable force. Tools and equipment can be designed to minimize such disorders by preventing extreme joint angles and/or the generation of large forces.

Because the handling of heavy or bulky objects can cause acute injuries, we have to be careful to structure tasks that involve lifting and carrying so that the risk of injuries is minimized. Workspace design also contributes to the risk of injury. We need to design workspaces so that psychological, biomechanical, and anthropometric factors act to maximize comfort and usability. Within the physical constraints of the environment, necessary equipment must be placed within reach of the user so that he or she will be able to operate it effectively. Displays must be positioned so that they are easily visible, repetitive movements that stress the musculoskeletal system need to be avoided, and seats should provide appropriate support and permit the user to maintain good posture.

RECOMMENDED READINGS

Cacha, C. A. (1999). *Ergonomics and Safety in Hand Tool Design*. Boca Raton, FL: Lewis.

Martin, B. J., Nussbaum, M. A., & Andersson, G. B. J. (2016). *Chaffin's Occupational Biomechanics* (5th ed.). Hoboken, NJ: John Wiley.

Chengalur, S. N., Rodgers, S. H., & Bernard, T. E. (2004). *Kodak's Ergonomic Design for People at Work*. Hoboken, NJ: John Wiley.

Corlett, E. N., & Clark, T. S. (1995) *The Ergonomics of Workspaces and Machines: A Design Manual* (2nd ed.). Boca Raton, FL: CRC.

Department of Health and Human Services (2007). *Ergonomic Guidelines for Manual Material Handling*. Cincinnati, OH: National Institution for Occupational Safety and Health.

Karwowski, W., Wogalter, M. S., & Dempsey, P. G. (Eds.) (1997). *Ergonomics and Musculoskeletal Disorders: Research on Manual Materials Handling, 1983–1996*. Santa Monica, CA: Human Factors and Ergonomics Society.

Pheasant, S., & Haslegrave, C. M. (2006.) *Bodyspace: Anthropometry, Ergonomics, and the Design of Work*. Boca Raton, FL: CRC.

Roebuck, J. A., Jr. (1995). *Anthropometric Methods: Designing to Fit the Human Body*. Santa Monica, CA: Human Factors and Ergonomics Society.

17 Environmental Ergonomics

There are numerous factors that can make up a working environment. These include noise, vibration, light, heat and cold, particulates in the air, gases, air pressures, gravity, etc. The applied ergonomist must consider how these factors, in the integrated environment, will affect the human occupants.

K. C. Parsons
2000

INTRODUCTION

So far, we have talked about those issues in workspace design, controls, and tools that have an obvious and direct influence on human performance. But a person's performance is also influenced by the physical environment in which he or she must carry out a task. Anyone who has tried to mow a lawn in the heat of a summer afternoon or balance a checkbook while a baby is crying can appreciate the influence of these sometimes subtle environmental variables. Often, the effect that the environment will have is not obvious during the design of a workspace or task. Some physical factors will become evident only when the workspace is implemented within the larger work environment.

The study of human factors issues with respect to the physical environment is called *environmental ergonomics*. According to Hedge (2006), "Environmental ergonomics studies our physiological and behavioral reactions to the ambient environment, and the design of effective barriers that allow us to survive in otherwise inhospitable settings" (p. 1770). By anticipating possible problem areas, such as glare on a visual display screen, human factors experts seek to design tasks and workspaces so that the consequences of noxious environmental variables are minimal. However, despite all attempts to reduce the impact of harmful environmental factors, some issues may arise only through the synergy of the workplace. Action often must be taken by the designer "on the spot" to remedy problems as they are detected. In addition to environmental ergonomics' focus on the person, "green ergonomics" attempts to remedy environmental problems while simultaneously considering concerns about sustainability of the physical environment (Pilczuk & Barefield, 2014).

In this chapter, we will examine four powerful environmental factors: lighting, noise, vibration, and climate. We encounter these factors within larger environments, such as offices, buildings, and other contained environments. We must also recognize that these factors can be sources of psychological stress, and so they may have harmful physiological and psychological consequences.

LIGHTING

Lighting affects how well people can perform tasks by how it restricts visual perception. However, poor lighting may also be responsible for certain health problems and adverse effects on mood. Human factors experts are most often concerned with the conditions that promote good interior lighting, which is essential in the home and at work. There are some situations where we also must consider lighting conditions outdoors, such as along roadways, for fields and stadiums where sports are played, and so on. In this chapter, we will focus mainly on interior lighting problems.

Lighting considerations are determined by four major human factors issues (Megaw & Bellamy, 1983): (1) how important light levels are to a person's ability to perform the task; (2) the speed and accuracy with which a person must perform the task; (3) the person's comfort; and (4) the person's subjective impressions of the quality of the lighting. As in all design problems, different lighting solutions will be more or less expensive. The human factors expert will need to balance costs against outcomes. "Good" lighting solutions will provide the best visual conditions for the lowest cost.

Our discussion of lighting will cover four topics. First, we will describe how light is measured. Second, we will discuss the characteristics of different kinds of artificial lights. Third, we will talk about how lighting can influence a person's ability to perform certain tasks. Finally, we will expand on the relationship between lighting and performance in a discussion of the effects of glare, which is the reflection of light from surfaces in the work environment.

LIGHT MEASUREMENT

An evaluation of lighting conditions in a home or work environment must begin with the measurement of effective light intensity, or photometry (Kitsinelis, 2015). However, we must distinguish between light that is reflected from and light that is generated by a surface on which the measurement is made. *Illuminance* is the amount of light falling on a surface, and *luminance* is the amount of the light generated by a surface (either a light source or a reflection). Both luminance and illuminance are determined by *luminous flux*, which is measured in units called *lumens*. Lumens represent the amount of visible light in a light source and thus the power of the light source corrected for the spectral sensitivity of the visual system. Illuminance is the amount of luminous flux per unit area (one square meter), and luminance is the luminous flux emitted from a light source in a given direction. The luminance of a reflection is a function of both the illuminance and the reflectance of a surface.

Both luminance and illuminance are measured with a device called a *photometer*. The photometer measures light in the same way that the human visual system does in daylight viewing conditions: each wavelength coming into the photometer is weighted by the corresponding threshold on the spectral sensitivity curve. For measures of luminance, a lens with a small aperture is connected to the photometer. The lens is focused on the surface of interest from any distance. If the light energy within the focused region is not uniform, the photometer integrates over the focused area to give an average luminance. The photometer gives the measure of luminance in candelas per square meter (cd/m^2). (A *candela* is a fixed amount of luminous flux within a fixed cone of measurement.)

For measures of illuminance, an illuminance probe is connected to the photometer and placed on the illuminated surface to be measured, or a special illumination meter can be used. Unlike luminance, the amount of illuminance will vary with the distance of the surface from the light source. The photometer or illumination meter gives the measure of illuminance in lumens per square meter (lm/m^2) or lux (lx).

We talked about how important contrast is in Chapter 6. Recall that contrast is the difference in luminance between two areas in the visual field. Contrast can also be measured with the help of a photometer (Kitsinelis, 2015). The contrast (C) between the luminance (L_o) of an object and that of its background (L_b) is often defined as:

$$C = \frac{L_o - L_b}{L_b}.$$

One way of thinking about the difference between luminance and illuminance is that while luminance measures the amount of light coming from a surface, illumination measures the amount of light falling on it. Designers of workspace and workspace lighting are most often concerned with illumination, because it is a measure of the effective amount of light energy for a particular work surface or area. The illuminance for an office should be between 300 and 500 lux at work surfaces, with lower lighting levels needed in homes.

LIGHT SOURCES

Different kinds of light sources will have different illuminances and different costs. One important factor the designer must consider with different light sources is that the accuracy of color perception (which we will call *color rendering*) will also depend on the light source. Optimal lighting solutions will provide a quality of light that is appropriate for the tasks performed in the environment while minimizing the expense of the lighting system. This means that there will never be a "one-size-fits-all" solution for different work environments. Each environment will potentially require different kinds of lighting.

Daylighting

The most basic distinction we can make is between natural lighting (sunlight) and artificial lighting. Sunlight contains energy across the entire spectrum of visible wavelengths, but with relatively more energy devoted to the long (red) wavelengths. This fact accounts for sunlight's yellow color. Windows and skylights provide natural lighting in building interiors. Sunlight is inexpensive and allows good color rendering. However, it is not very reliable. Illumination levels will vary as a function of such factors as time of day and year, and the weather. The distribution of natural light cannot be easily controlled. Some workspaces may be easy to position near a window or skylight, but others may not.

Using natural lighting for building interiors is sometimes called *daylighting*. While natural light through windows and skylights is inexpensive, the heating and cooling costs of daylighting are not negligible. The ratio of skylight to floor area that minimizes annual total building energy use in commercial buildings is estimated to be 0.2 (Nemri & Kwartri, 2006). This means that a 10×10 ft. room (100 ft^2) should use a skylight no larger than 4×5 ft. (20 ft^2). While this size of skylight may be appropriate in some circumstances, it may be too small in others, and the problem of light distribution within the room may remain.

There are several architectural options for daylighting. Roof monitors are boxes placed on building roofs through which daylight can enter, and a series of diffusers and mirrors then distribute the light through the building. Light shelves are reflective, horizontal shelves placed above window exteriors that can "catch" and distribute light more evenly through the interior space. Tubular skylights, like roof monitors, use roof-mounted light collectors. The light is then directed down a tube and through a diffusor lens that distributes the light evenly across an interior space.

Most office buildings were constructed without plans for daylighting, and so the light distribution through windows and skylights is not always uniform throughout the interior. One way to remedy this problem is the PSALI (Permanent Supplementary Artificial Lighting Installation) approach. This design approach first analyzes the availability of natural light throughout the interior, with the goal of relying on it as much as possible. Then, artificial lighting is added to supplement the natural light, creating a uniform light distribution over all areas (Hopkinson & Longmore, 1959). With the PSALI approach, more light fixtures would be installed to illuminate desks and areas in the interior that are located farthest from windows, and fewer in the areas that are closest to windows.

Artificial Lighting

Artificial lighting systems are discriminated first and foremost by the kinds of bulbs that the light fixtures use. The two most common types of artificial light bulbs are incandescent and fluorescent (Mumford, 2002).

Incandescent light bulbs were the primary source of home lighting for many years. Their use has been restricted in some countries and they are being phased out in others, including the U.S., because of their energy inefficiency. That is, the number of lumens these bulbs can produce per watt is very low; as much as 95% of the energy that goes into powering the bulb is converted into heat. Because incandescent bulbs that meet more stringent energy efficiency standards are still being manufactured and sold, and because people tend to prefer the qualities of incandescent lighting to those of the alternatives, design issues surrounding incandescent lighting are still relevant in modern home and work environments.

Incandescent light is produced by current flowing through a tungsten filament inside a glass bulb that contains an inert gas, which becomes heated and glows. Tungsten halogen lamps are also incandescent. They differ from ordinary light bulbs in that a halogen gas instead of an inert gas is used in the bulb. Tungsten lamps are long-lasting sources of bright light that are used for vehicle headlights and floodlights (Kitsinelis, 2015).

Incandescent lamps have several benefits, including that their initial cost is low, they generate light energy across the visual spectrum, and they provide a full output of light immediately upon being turned on (Wolska, 2006b). Consequently, despite its low efficiency, incandescent lighting has been useful for residences, where only a small number of lamps are needed, but not for businesses and other larger organizations that may have much greater lighting demands.

Fluorescent lamps are gas discharge bulbs. An electric current is alternated through an inert gas, producing invisible ultraviolet light, which excites phosphors coating the inside of the bulb. Although the light appears steady, it actually flickers at a high frequency (equivalent to the frequency of the alternating current). Fluorescent lamps can require as little as 25% of the power required to light an incandescent lamp, and they have a longer life. Because of their relative efficiency, fluorescent lamps are used for lighting in schools, offices, and industrial buildings, and bulb-like compact fluorescent lamps are now replacing incandescent bulbs for home use.

One disadvantage of fluorescent lamps is that their light output decreases over the lamp's lifespan (Wolska, 2006b). Also, the spectral distribution for the common cool-white fluorescent lamp is not much like that of daylight, which results in poor color rendering. Mayr, Köpper, and Buchner (2014) showed that people can't discriminate between colors as well under compact fluorescent lighting as they can under halogen lamps. An alternative to cool-white fluorescent bulbs are more expensive full-spectrum or high color rendition bulbs, which use a mix of different phosphors to more closely match the spectrum of natural light. Fluorescent lighting technology has continuously improved, leading Mumford to state as far back as 2002 that, "Compared with the fluorescent lighting available only 20 years ago, there are fluorescent products available today which reduce office and video terminal fatigue, are economical in use and can help some of the readers with special needs" (p. 3). You should be aware that there are more speculative claims about the benefits of full-spectrum fluorescent lighting for a person's performance of perceptual and cognitive tasks, and even a person's physical and psychological health! However, there is little scientific evidence for these purported benefits (McColl & Veitch, 2001; Veitch & McColl, 2001).

Although they are more efficient than incandescent lamps, fluorescent lamps are still not very efficient. There are other gas discharge lamps, including induction, metal halide, and sodium lamps, that are far more efficient. However, these other lamps are much more expensive, and their color rendering capabilities may be much poorer because of peaks in their spectral frequency distributions. Figure 17.1 shows the spectra for four types of lamps: incandescent, fluorescent, metal halide, and sodium. The incandescent and fluorescent spectra are much smoother than those of the metal halide and sodium lamps. The extreme peaks in the metal halide and sodium spectra mean that these lamps provide large amounts of light from only a few wavelengths. These few wavelengths can "wash" the work area with the hues of those wavelengths, making accurate color rendering difficult or impossible. Low-pressure sodium lamps, for example, are used almost exclusively for street lights. They generate yellow/orange light and provide no color rendering (Wolska, 2006b).

The last kind of lighting we will discuss is called *solid-state lighting*, and it uses arrays of light emitting diodes (LEDs). LED bulbs can fit into the sockets of light fixtures manufactured for incandescent bulbs. LEDs are particularly useful for applications where small size and long lifetimes are important, such as for color indicator lamps. For example, they are now widely used for traffic lights, toll booth lane indicators, light rail signals, vehicle tail lights, and airport runway lighting (Boyce, 2014). While high-brightness LEDs can be used for a variety of lighting applications, they have a number of drawbacks (Žukauskas et al., 2002). In addition to the fact that they can be very expensive, like fluorescent bulbs, color rendering can be poor, and there is some evidence that blue and cool-white LEDs can cause glare, damage the retina, and interfere with normal sleep cycles (Algvere, Marshall, & Seregard, 2006; American Medical Association, 2016).

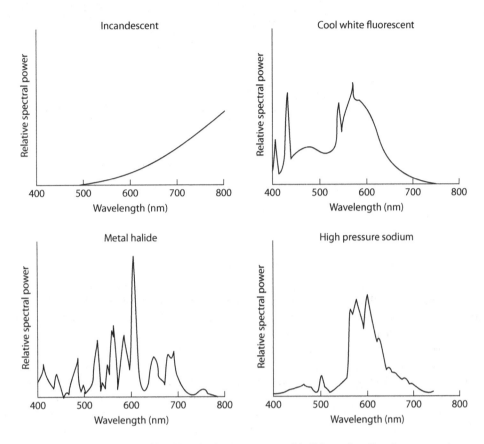

FIGURE 17.1 Light spectra for incandescent, fluorescent, metal halide, and sodium lamps.

Apart from the kind of bulb (incandescent, fluorescent, gas discharge, or solid-state) that a light source uses, the next common distinction between light sources is whether the lighting is direct or indirect (Wolska, 2006b). Direct lighting falls on a surface directly from the light source. In contrast, indirect lighting has been reflected from other surfaces, often the ceiling, before falling on the work surface. Technically, if 90% or more of the light from a source is directed toward the work surface (downward), then the lighting is classified as direct. If 90% of the light is directed away from the work surface (upward), then the lighting is classified as indirect. Lighting is called semi-direct or semi-indirect when 60%–90% of the light is directed toward or away from the work surface, respectively. Direct lighting often results in glare, and indirect lighting is not very effective for work requiring close visual inspection.

ILLUMINATION AND PERFORMANCE

We have emphasized throughout this book that the tasks that people perform have perceptual, cognitive, and motor information-processing components. We have also discussed how tasks can be classified according to the extent to which each of these components is required for a task's completion. Illumination will have the greatest effect on those tasks that depend on the visual perception component.

We can characterize the visual difficulty of a task by the size of the smallest critical details or items with which a person must work and the contrast of those details with the background. Often, simply increasing illumination does not make a visually difficult task easy. Figure 17.2 shows the

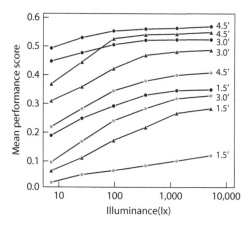

FIGURE 17.2 The effects of size (in minutes of visual angle), illuminance, and contrast on performance. Black circles = high contrast; triangles = medium contrast; gray circles = low contrast.

effects of object size, illuminance, and contrast on performance (Weston, 1945). While performance improves as size, illumination, and contrast increase, performance with small, low-contrast objects is always much poorer than that for larger, higher-contrast objects.

Many kinds of field studies are conducted to directly evaluate performance under different levels of illumination and types of lighting. For example, one study measured productivity under changes of lighting in a leather factory over a 4-year period (Stenzel, 1962). The workers' tasks involved cutting shapes from skins to make leather goods such as purses and wallets. For the first 2 years of the study, work was performed in daylight with additional fluorescent fixtures, which provided an overall illuminance of 350 lx on the factory floor. Before the third year, the daylight was removed, and fluorescent lighting was installed that provided a uniform 1000-lx illumination. As shown in Figure 17.3, workers' performance clearly improved after the installation of the lighting.

Can we conclude that the increase in illumination level caused the increase in productivity? Unfortunately not, although the findings suggest that this was the case. The change in performance could be due to other factors, such as the increased uniformity of illumination, color modifications, or unrelated variables (such as pay raises or different work schedules) that may have been altered at about the same time as the change in lighting.

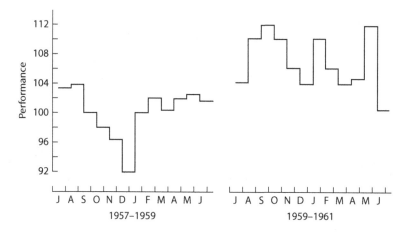

FIGURE 17.3 Normalized performance in a leather factory as a function of lighting conditions (left panel is old lighting, right panel is new lighting) and month (from July through June).

The well-known Hawthorne lighting experiments demonstrated how hard it is to control extraneous variables in the work environment. Three experiments were conducted from 1924 to 1927 at the Hawthorne Plant of the Western Electric Company to assess the effects of lighting on productivity (Gillespie, 1991). The impetus for the experiments came from the electrical industry, which claimed that good lighting would increase productivity significantly. The workers were informed about the nature of the study in order to obtain their cooperation.

In the initial experiment, three test groups of workers involved in the assembly of telephone parts performed under higher than normal lighting levels, while a control group performed under normal lighting. Production increased dramatically in the three test groups in comparison to their level of productivity before the experiment, but it increased a similar amount for the control group. Also, within each experimental group, there was no correlation between productivity and the lighting levels under which each group performed. The researchers concluded that the increased productivity was due to an increase in management's involvement with the workers, which was required for measuring lighting levels and productivity, rather than to the lighting level itself. That is, because the workers either knew they were being more closely watched by their managers or were enjoying the increased attention being given to them by their managers, they worked harder than they had before the lighting experiment began.

The researchers in charge of the Hawthorne study conducted more experiments, in which they made explicit attempts to control the effect of management attention. Even in these experiments, the lighting level had little effect on performance, except at very low illumination levels. One explanation for this is that the workers may have expended more effort than usual under conditions of low illumination to compensate for any increased difficulty.

There are several other mechanisms by which new lighting may affect performance in the work environment (Juslén & Tenner, 2005). These include visual comfort, visual ambience, interpersonal relationships, and job satisfaction, as well as biological effects such as the timing of circadian rhythms and alertness (Boyce, 2014; van Bommel, 2006; see also Chapter 18). One study showed that the productivity of assembly workers increased when their work stations were equipped with a controllable task-lighting system, which allowed the workers to adjust the lighting to their preferred intensity levels (Juslén, Wouters, & Tenner, 2007). The increase in productivity could have been due to improved visual perception or to some other psychological or biological mechanism associated with the controllable lighting system. Juslén et al. argue: "Seeing lighting change as a process with several mechanisms, which are partly 'light related mechanisms' and partly general mechanisms, will help designers and managers to estimate whether a lighting change is worth the investment" (p. 853).

While it is difficult to draw conclusions about the relationship between lighting and task performance in field studies, we can be much more confident about what we observe when we move the task environment into the laboratory. This also requires that the researchers design a simulation of the real-world task that preserves its critical elements while eliminating those that would make a causal relationship difficult to establish. Stenzel and Sommer (1969) performed such a laboratory experiment, in which people either sorted screws or crocheted stoles. They varied illuminance during task performance from 100 to 1700 lx. The number of errors that were observed decreased with increasing illuminance for crocheting but not for sorting. For the sorting task, errors decreased as illuminance increased to 700 lx, but then increased as illuminance rose to 1700 lx. Therefore, the effect of increasing illumination depended on the specific task that was performed.

Illumination and contrast are particularly critical in designing workplaces appropriate for older adults, because visual acuity declines rapidly with age. One study asked young (18–22 years) and older (49–62 years) adults to proofread paragraphs for misspelled words (Smith & Rea, 1978). The researchers presented paragraphs that were of good-, fair-, or poor-quality text on white or blue paper, and varied illumination from 10 to 4900 lx. The readers' performance increased as the copy quality increased and also as the illumination increased. However, the young adults showed very little improvement with increased illumination, whereas the older adults showed marked improvement

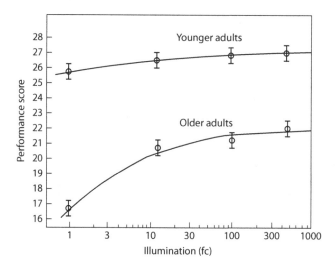

FIGURE 17.4   Proofreading performance by younger and older adults as a function of illumination level.

(see Figure 17.4). Therefore, illumination and print quality are more important for older than for younger people. This fact has been confirmed in other studies showing that a higher level of illumination is more critical for older adults who are visually impaired through age-related macular degeneration than for those who have normal vision for their age (Fosse & Valberg, 2004).

Performance is not the only issue involved in designing a lighting scheme. Visual "comfort" is also important. A visual environment is comfortable when workers in that environment can complete perceptual tasks with little effort or distraction, and without stress; that is, when sources of visual discomfort are absent. Visual discomfort can occur when the visual task is difficult (e.g., having to resolve fine details; driving under foggy conditions), irrelevant objects provide distraction by attracting the worker's attention away from the task, and lighting conditions produce confusable reflections on the objects in the workspace (Boyce, 2014).

An important factor in predicting visual comfort is the ratio of the luminance of an object or task area being viewed to the luminance of its surroundings. Visual comfort can be maintained as long as the luminance ratio does not exceed 5:1 (Cushman & Crist, 1987). However, comfort and performance may not be significantly affected even when the luminance ratio is as large as 110:1. Cushman, Dunn, and Peters (1984) had people make prints of photograph negatives under luminance ratios ranging from 3.4:1 to 110:1. As the luminance ratio decreased, the printing rate declined slightly, but so did the error rate. The study participants reported less ocular discomfort and overall fatigue when they were allowed to adjust the luminance ratio.

Much of what makes one lighting scheme preferable to another is subjective: that is, it cannot be objectively measured by a designer. Some environmental qualities, such as clarity, pleasantness, spaciousness, and how relaxing a space is, are not functions of luminance flux. Flynn (1977) published the results from studies in which he asked people to give subjective ratings of these and other qualities under different types of lighting. The lighting schemes used in these studies varied along several dimensions. An overhead versus peripheral dimension determined whether the lights were mounted on the ceiling or on the wall. A nonuniform versus uniform dimension described the distribution of light in the room as a function of the location of objects and surfaces in the office. Lighting was also adjusted to be either bright or dim, and either warm or cool. Table 17.1 shows how the values of different lighting dimensions evoke positive qualities of, for example, spaciousness and privacy. Some of these qualities will be more important than others, depending on the task.

Hedge, Sims, and Becker (1995) conducted a field study investigating productivity and comfort with two different lighting systems installed in a large, windowless office building. These were

TABLE 17.1
Lighting Reinforcement of Subjective Effects

Subjective Impression	Reinforcing Lighting Modes
Visual clarity	Bright, uniform lighting mode
	Some peripheral emphasis, such as with high-reflectance walls or wall lighting
Spaciousness	Uniform, peripheral (wall) lighting
	Brightness is a reinforcing factor, but not a decisive one
Relaxation	Nonuniform lighting mode
	Peripheral (wall) emphasis, rather than overhead lighting
Privacy or intimacy	Nonuniform lighting mode
	Tendency toward low light intensities in the immediate locale of the user, with higher brightnesses remote from the user
	Peripheral (wall) emphasis is a reinforcing factor, but not a decisive one
Pleasantness and preference	Nonuniform lighting mode
	Peripheral (wall) emphasis

lensed-indirect uplighting (LIL) and direct parabolic lighting (DPL). The LIL used fixtures suspended from the ceiling, which projected light upward to be reflected from the walls and ceiling. The DPL used fixtures recessed into the ceiling and shielded by parabolic louvers. Office workers responded to a questionnaire that asked them about their satisfaction with the lighting system installed in their offices. The DPL system generated significantly more complaints than the LIL system about problems like glare and harshness, and workers estimated up to four times more productivity loss because of such lighting problems. Workers in the DPL group also reported three to four times more productivity loss due to visual health problems, such as focusing problems, watery eyes, or tiredness.

GLARE

Glare is high-intensity light that can cause discomfort and interfere with the perception of objects of lower intensity. There are different kinds of glare: direct and reflected. Light sources within the visual field, such as windows and light fixtures, can produce direct glare. Reflected glare is produced by objects and surfaces that reflect light. Reflected glare can be avoided by locating light sources and work surfaces so that light sources are not in an "offending zone." The offending zone is where light from the source will reflect from the work surface into the eyes (see Figure 17.5).

There are different kinds of reflected glare. Specular reflection produces images of objects in the room on the viewing surface. Veiling reflection results in a complete reduction of contrast over parts of the viewed surface. Both direct and reflected glare can be particularly debilitating for workstations with visual display units (VDUs).

Glare also can be classified according to its severity. *Disability glare* reduces the contrast ratio of display characters by increasing the luminance of both the display background and the characters. This reduces the detectability, legibility, and readability of the display characters. It usually results from a light source that is located close to the line of sight. *Discomfort glare*, which may or may not be accompanied by disability glare, will cause the worker discomfort when the work surface is viewed for a period of time.

Discomfort from glare increases as the luminance and number of glare sources increase (Wolska, 2006a). However, because discomfort is a subjective event, it will be affected by many factors other than light intensity. For instance, a person will report greater discomfort produced by glare when

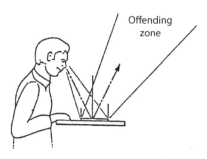

FIGURE 17.5 The offending zone for glare.

she is performing a visually demanding task than when the task is not as visually demanding. Her prior experience with the task may be important as well.

For an example illustrating the role of prior experience, consider the fact that automobiles in Europe have low-intensity amber (filtered) headlights. U.S. automobiles, in comparison, have very bright white (unfiltered) headlights. Sivak, Olson, and Zeltner (1989) reasoned that European drivers, because of their experience with low-intensity amber headlights, may be more subject to discomfort glare than U.S. drivers when driving on U.S. roadways. This was the case; West German drivers rated filtered and unfiltered headlights of different brightnesses higher in discomfort than did U.S. drivers. The drivers' past experience helped determine the degree of discomfort.

There are several measures of visual discomfort. One that can be used to assess the potential for direct discomfort glare is the visual comfort probability (VCP) method (Guth, 1963). These measures take into account the direction, luminance, and solid angle of the glare source, as well as the background luminance. The VCP method relies on calculation of a glare sensation index (M):

$$M = \frac{L_S Q}{2PF^{0.44}},$$

where:
L_S is the luminance of the glare source,
P is an index of the position of the glare source from the line of sight,
F is the luminance of the entire field of view including the glare source;

and
$$Q = 20.4\omega_S + 1.52\omega_S^{0.2} - 0.075,$$

where ω_S is the (solid) visual angle of the glare source. Sometimes, there may be more than one glare source affecting a single location. In this case, the glare sensation M_i for each source ($i = 1, 2, 3, ..., n$) at that location can be calculated and the results compounded into the single discomfort glare rating (DGR) by the formula

$$DGR = \left[\sum_{i=1}^{n}(M_i)\right]^a,$$

where n is the number of glare sources

and
$$a = n^{-0.0914}.$$

The *VCP* is defined as the percentage of people who would find the level of direct glare in the environment acceptable. The *DGR* can be converted into *VCP* using the formula

$$VCP = 279 - 110(\log_{10} DGR)$$

for the range of primary concern (*VCP* from 20% to 80%). It is generally agreed that direct glare will not be a problem for a lighting application if the *VCP* is 70 or higher.

We can reduce glare in many ways. Window luminance can be controlled with blinds or shades. Similarly, shades and baffles on light fixtures can reduce the amount of light coming directly from the fixtures. We can position VDUs or other displays so that bright sources of light are not in the field of view and reflections are not seen on the screen. Some displays allow the user to avoid glare by tilting or swiveling the screen. Anti-glare devices, such as screen filters, can be used for VDUs, but they reduce contrast and so may degrade visibility. Liquid crystal displays (LCDs) can replace older cathode ray tube (CRT)-type displays and are less susceptible to glare.

Glare is a significant factor in night driving. Direct glare from the headlights of oncoming vehicles and indirect glare from the headlights of trailing vehicles can produce discomfort and impair a driver's vision. A study examined how well drivers performed with direct glare by mounting a simulated headlight source on the hood of an instrumented vehicle (Theeuwes, Alferdinck, & Perel, 2002). In direct glare conditions, when the simulated headlight was turned on, drivers drove more slowly and were less likely to detect pedestrians at the side of the road. Older drivers were more adversely affected than younger ones. Another study showed that older adults reported more discomfort from the same levels of glare in driver-side mirrors than did younger adults. When the flat mirrors were replaced with curved ones, both older and younger adults suffered less discomfort (Lockhart, Atsumi, Ghosh, Mekaroonreung, & Spaulding, 2006).

Another study examined how well truck drivers performed with indirect glare reflected in the side-mounted rearview mirrors of a truck simulator (Ranney, Simmons, & Masalonis, 2000). Drivers were asked to detect stationary pedestrians and to determine the location of a target X presented on vehicles in the truck's mirrors. The researchers created glare by directing beams of light into the side mirrors, which were set for either no glare reduction or high glare reduction (which reduced the reflectiveness by 80%). When the mirrors did not reduce glare, truckers could not detect targets in the mirror well, and they had poorer control of the truck: their lane variability increased, speed on curves slowed, and steering variability increased. However, glare-reducing mirrors did not improve the truckers' performance much, either in target detection or in vehicular control. Nonetheless, the drivers indicated that they preferred having the glare-reducing mirrors.

NOISE

Noise is undesirable background sound that is irrelevant to the task that someone is trying to accomplish (School, 2006). It is present to some extent in any work environment, as well as in almost all other settings. Noise can be generated by office equipment, machinery, conversation, and ventilation systems, as well as by traffic and miscellaneous events such as doors slamming. A high noise level can be uncomfortable; it can also reduce performance, and workers may experience permanent hearing loss. Human factors experts can help design and modify work environments to reduce the deleterious effects of noise by determining what noise is tolerable and, as with lighting, establish suitable aesthetic criteria for the well-being of the people who work in the environment.

In this section, we will first discuss how noise levels are measured, and then how noise can affect a person's performance. Then we will discuss how noise causes hearing loss, and the effects of hearing loss on performance. Finally, we will discuss strategies for reducing noise in the workplace.

NOISE MEASUREMENT

Remember that an auditory stimulus (a tone or a sound) can be broken into its component frequencies just as a light source can be broken into its component wavelengths. Each frequency in a sound will have an amplitude that describes how much of that frequency contributes to the sound as a

FIGURE 17.6   Model CEL-354 Sound-Level Meter.

whole. When we measure the intensity (amplitude) of a noise, we have to worry about these different frequencies, because people are better at hearing some frequencies than others.

A sound-level meter (see Figure 17.6) will give a single measure of sound amplitude averaged over the auditory spectrum. Just as the photometer is calibrated for human sensitivity to light of different wavelengths, the sound-level meter is calibrated according to human sensitivity to tones of different frequencies. However, remember also that *relative* sensitivity (the loudness a person perceives for tones of different frequencies) is a function of the amplitude of a tone. This means that the sound-level meter will need to be calibrated differently to measure noise at different intensity levels.

A sound-level meter often has three calibration scales, one appropriate for low intensities (the A scale), one for intermediate intensities (the B scale), and one for high intensities (the C scale), although the B scale is omitted from some meters. Figure 17.7 shows how the scales differ, and that the difference is primarily in how the meter weights frequencies below 500 Hz. If we measure the same sound twice, using the A and the C scales, the difference between the two measurements gives

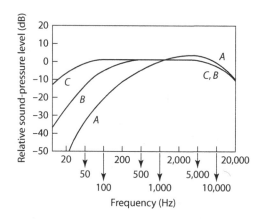

FIGURE 17.7   The spectral weightings for the A, B, and C scales on a sound-level meter.

an indication of the intensity of low-frequency components in the sound. If the two measures are very similar, then the sound energy contains components that are mostly above 500 Hz, whereas if the C measure is much higher than the A measure, then a substantial portion of the sound energy is below 500 Hz. Some sophisticated meters include band-pass filters that let us measure sound energy within specified frequency regions.

In most environments, noise levels will not be constant but will fluctuate, either quite rapidly or more gradually over time. Most sound-level meters can accommodate changes like this because they are equipped with "slow" and "fast" settings that differ in the length of the time interval over which the noise is averaged. The meter will average the noise for a longer period of time on the slow setting (1 s) than the fast setting (125 ms). If the noise level changes rapidly, the slow setting will show less fluctuation on the meter. Some meters have "hold" buttons to use with the fast setting that will display maximum and/or minimum intensity levels.

We might also be concerned about a person's total noise exposure across the course of a day. We can get a cumulative measure of a worker's total exposure with a device called an *audio-dosimeter* (Casali, 2006), which is worn by the worker for an entire day. These meters are small and inexpensive, but their measurements can be inaccurate. This is because the meter will register high noise levels that arise because the noise source is close to the microphone. Although some of these sounds may be of concern, others, such as the worker's own voice, may not be.

NOISE LEVEL AND PERFORMANCE

A person's performance may suffer in many ways if he is forced to work in a very noisy environment. We discussed in Chapter 8 how noise can mask both speech and nonspeech sounds. Masking will interfere with a person's attempts to communicate with other people and to perceive auditory displays. When a person shouts to try to overcome a high background noise level, his speech patterns will change, and these changes will also impair communication. Even when a worker is not trying to communicate with anyone else, other people's conversation in the background can prevent him from concentrating on reading or listening to other verbal material.

Noise can evoke highly emotional responses. Anyone who has been exposed to the sound of fingernails being scraped over a blackboard can appreciate how compelling some sounds can be. The startle reaction, for example, is something that everyone experiences when they hear a sudden loud noise. It consists of muscle contractions and changes in heart and respiration rate, and is usually followed by an increase in arousal. Fortunately, such reactions are usually very brief, and their intensity tends to diminish with repeated exposure.

Sonic booms are examples of unpleasant noise that evoke strong emotional responses. A sonic boom occurs when an aircraft travels faster than the speed of sound. The booms occur unexpectedly, have rapid onset, and are loud enough to shake buildings and startle people. One of the most notorious studies of the effects of sonic booms on people was conducted in 1964 by the U.S. Federal Aviation Administration (FAA; Borsky, 1965). From February 3 through July 30 of that year, residents of Oklahoma City, which during the latter part included one of the authors (RWP), were subjected to eight sonic booms per day to assess the possible effects of supersonic transport flights on residents' attitudes. Interviews were conducted with nearly 3000 persons at 11, 19, and 25 weeks after the beginning of the testing period, and complaints filed by all residents were recorded. As Gordon Bain, then Deputy Administrator for Supersonic Transport Development of the FAA, commented, "The Oklahoma City sonic boom study … was the first major effort anywhere in the world to determine the nature of public reaction to sonic boom at specified, measured levels over a reasonably extended period of time" (Borsky, 1965, p. ix).

Almost all of the respondents reported that the booms rattled and shook their houses, and the booms broke many windows in the city's largest buildings. Otherwise, physical damage was minimal. Some 35% of the interviewees reported having startle and fear responses to the sonic booms, and 10%–15% reported interference with communication, rest, and sleep. Only 37% indicated annoyance with the booms during the first interview period, but by the last period, more than half (56%) did. This suggests that sonic booms may in fact become more annoying with prolonged exposure. But, because the intensity of the booms was increased after each interview, the increased annoyance could have been due to the increase in boom intensity.

At the last interview, approximately 75% of the residents indicated that they did not find the eight booms per day too bothersome, but 25% said that they did. Moreover, 3% of the entire population, or about 15,000 people, were sufficiently bothered to file a formal complaint or lawsuit. This number is most likely an underestimate, as the report notes that one reason for the low complaint level was that "there was widespread ignorance about where to complain" (Borsky, 1965, p. 2).

Not all emotional responses are necessarily detrimental to performance. Background noise that increases arousal will produce better performance on a vigilance task, in which performance tends to decline as arousal decreases (see Chapter 9; McGrath, 1963). However, this is not true for all vigilance tasks. Some researchers have found that vigilance performance can sometimes be worse with noise (e.g., Becker, Warm, Dember, & Hancock, 1995).

Noise levels as low as 80 dB (about as noisy as a vacuum cleaner) can have a detrimental effect on performance. People may have trouble with the following activities if they try to do them in a noisy environment: (1) tasks of extended duration, if the background noise is continuous; (2) tasks that require a steady gaze or fixed posture, which can be disrupted if a person is startled by sudden noise; (3) unimportant or infrequent tasks; (4) tasks that require comprehension of verbal material; and (5) open tasks, in which a rapid change of response may be required (Jones & Broadbent, 1987).

A comprehensive evaluation of the noise levels in an environment can, therefore, be complicated. This is because the acceptability of different noise levels depends on the task to be performed, and the way that noise levels are measured depends on the background intensity and frequencies of other noise in the environment. These background noises, produced by mechanical systems such as air conditioners, can also generate intense low-frequency sound waves that vibrate floors and walls. These vibrations produce rattles, and even audible noise, called rumble.

There are several established methods for rating noise and assessing its acceptability (Broner, 2005), and each method is based on "noise criterion" curves like the ones shown in Figure 17.8. A noise criterion specifies the maximum intensity level for noise of different frequencies in different environments that will not interfere with speech or be otherwise disturbing.

The specific noise criterion curves shown in Figure 17.8 were developed by Beranek (1989). They are called the *Balanced Noise Criterion* (NCB) curves, intended to be applicable to vehicles and buildings. Noise frequencies in a task environment are measured in octave bands, which are ranges of frequencies from one half to double the reference or "center" frequency. So, for example, the frequencies measured for a 500-Hz octave band center frequency range from 250 to 1000 Hz. Each NCB corresponds to a different kind of environment, with louder environments allowing higher intensities before exceeding the NCB values. Similarly, lower noise frequencies can have higher intensities before exceeding any NCB value. The A and B regions to the upper left of the figure indicate those combinations of intensity and frequency that produce clearly and moderately noticeable vibrations, respectively.

To use the curves, we must first decide what the appropriate NCB level is for the environment in question. We do this by consulting tables such as Table 17.2. For example, the environment might be a telemarketing office, and the task people are expected to perform is talking with potential clients on the telephone. We might decide that this not quiet but not loud environment, with many people talking on the telephone, rates an NCB level of 35 ("moderately good listening conditions" in Table 17.2). We will then measure the sound levels in decibels for each octave frequency band. If the average of the noise levels in the four bands most important for speech (the 500-, 1000-, 2000-, and 4000-Hz bands) exceeds the value of the chosen NCB (in this case, 35), then the environment is too noisy, and we will need to take measures to reduce the noise. We evaluate rumble similarly by considering only the sound-pressure levels in the octave bands of 31.5–500 Hz, and vibration by determining whether the levels in any of the three lowest frequency octave bands fall in the A or B regions.

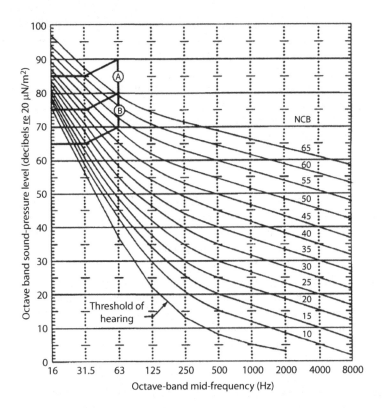

FIGURE 17.8   Balanced noise criterion curves.

TABLE 17.2
Recommended NCB Curves and Sound-Pressure Levels for Several Categories of Activity

Acoustical Requirements	NCB Curve[a]	Approximate[b] LA (dBA)
Listening to faint music or distant microphone pickup used	10–20	21–30
Excellent listening conditions	Not to exceed 20	Not to exceed 30
Close microphone pickup only	Not to exceed 25	Not to exceed 34
Good listening conditions	Not to exceed 35	Not to exceed 42
Sleeping, resting, and relaxing	25–40	34–47
Conversing or listening to radio and TV	30–40	38–47
Moderately good listening conditions	35–45	42–52
Fair listening conditions	40–50	47–56
Moderately fair listening conditions	45–55	52–61
Just acceptable speech and telephone communication	50–60	56–66
Speech not required but no risk of hearing damage	60–75	66–80

[a]   NCB curves are used in many installations for establishing noise spectra.

[b]   These levels (LA) are to be used only for approximate estimates, since the overall sound-pressure level does not give an indication of the spectrum.

The NCB curves are widely used in the U.S., but a similar set of curves, called the *Noise Rating curves*, proposed by Kosten and Van Os (1962), are used more extensively in Europe. The Noise Rating curves weight the intensity of each octave frequency band to correct for the sensitivity of the auditory system. This system, therefore, gives higher weight to higher frequencies. The Noise Rating system also allows "correction" in the curves depending on the duration (e.g., noise is present

5% of the time) and quality of the noise (e.g., intermittent vs. continuous). There are other noise assessment methods that focus on other dimensions of the acoustic environment. Room Criterion curves, for example, optimize sound quality (Blazier, 1981, 1997).

Which noise assessment method you use will depend on the environment you are evaluating and the priorities of the people working in that environment. An evaluation of the noise levels of a concert hall, for example, will not require you to resolve the same kinds of issues as an evaluation of a factory floor. Whichever method you choose, the first step toward optimal acoustic design of a workplace is the measurement and evaluation of noise levels.

Hearing Loss

Many popular musicians, including Sting, Pete Townshend, Jeff Beck, Eric Clapton, and will.I.am, have severely impaired hearing and tinnitus (discussed below) from their many years of standing on stage in front of their amplifiers. For these musicians, the constant exposure to very high noise levels resulted in a permanent decrease in their auditory sensitivity. Such decreases are called *threshold shifts* (Haslegrave, 2005). Short-term exposure to high noise levels can cause temporary threshold shifts in anyone. A temporary threshold shift is defined as an elevation in a person's auditory threshold measured 2 minutes after exposure. The magnitude of the temporary threshold shift is a function of the noise level and frequency, and the length of exposure time (see Figure 17.9).

Human factors engineers need to determine whether the noise exposures in the environments they are studying are large enough that short-term hearing impairment is possible. For example, military planes do not usually have insulated cockpits, and so engine and wind noise requires pilots to wear ear protection. Nonetheless, the U.S. Air Force routinely loses many pilots a year because of permanent hearing loss. With earplugs or other suitable ear protection, pilots can avoid permanent damage, but may still experience a temporary threshold shift (Kuronen, Sorri, Paakkonen, & Muhli, 2003). These shifts, which occur for pilots flying many different kinds of aircraft, are small enough that the pilots are not at high risk for permanent threshold shifts.

A permanent threshold shift is an irreversible increase in the auditory threshold; that is, permanent damage. The magnitude of a permanent threshold shift will depend on the years of exposure

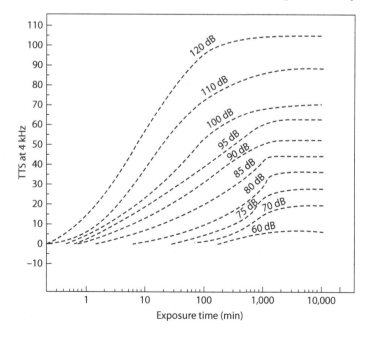

FIGURE 17.9 Temporary threshold shift as a function of noise level, frequency, and exposure time.

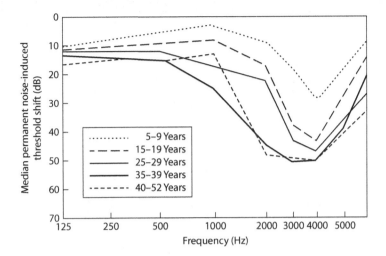

FIGURE 17.10 Permanent threshold shift as a function of exposure duration and noise frequency.

and the frequencies in the noise. A person's degree of hearing loss is quantified by the magnitude of the threshold shift, with up to 40 dB impairment considered "mild," between 41 and 55 dB "moderate," between 56 and 70 dB "moderately severe," between 71 and 90 dB "severe," and 90 dB or greater "profound." Usually, hearing loss due to long-term noise exposure is concentrated on frequencies around 4000 Hz. Figure 17.10 shows hearing losses in workers in a jute-weaving factory. Workers who had been in the factory the longest (some over 50 years) showed moderate damage for sounds varying from 500 to 6000 Hz, with the most severe losses for frequencies around 4000 Hz.

The relationship between noise intensity and frequencies is shown in Figure 17.11. This figure shows permanent threshold shifts for workers exposed to different noise levels and different frequencies for 8-hour shifts over 10 years. The accumulated effect of exposure to 80 dB noise for 10 years of 8-hour shifts is negligible (Passchier-Vermeer, 1974), but these effects increase dramatically for noise levels of 85 dB and above. Figure 17.12 shows the maximum amount of time that a worker can be exposed to potentially damaging noise levels without producing a permanent threshold shift. This maximum exposure duration decreases with increasing decibels.

In Chapter 7, we discussed how delicate the anatomy of the inner ear is. We also discussed how sound energy is nothing more or less than changes in air pressure that push against these delicate structures. Sudden loud sounds deliver extremely high pressures to the inner ear, like poking the

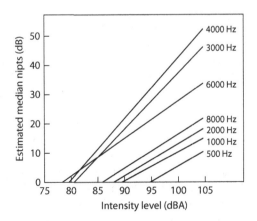

FIGURE 17.11 Accumulated effects of exposure to noise (noise-induced permanent threshold shifts) of different intensity levels for 10 years.

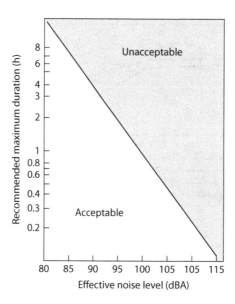

FIGURE 17.12   The maximum acceptable exposure duration to high-level noise.

mechanism with a pencil. Therefore, these kinds of noises may result in acoustic trauma and permanent damage to the structures of the inner ear.

Consider the effect of sounds like gunshots and claxons, for which the onset of the sound is rapid or stepped. Combat soldiers, for example, are frequently exposed to noises like these when they shoot their rifles or when they are in close proximity to bomb blasts. One study reported that of 29% of soldiers exposed to acute acoustic trauma while in the military were still experiencing tinnitus (ringing in the ears) when discharged (Mrena, Savolainen, Kuokkanen, & Ylikoski, 2002). Moreover, of this group, more than 60% reported still experiencing tinnitus 10 years later. A 2005 report from the U.S. Institute of Medicine indicated that 62% of soldiers treated for blast injuries also experience acute acoustic trauma. This report also estimated the number of veterans with permanent damage at over 25%. The consequences of acute acoustic trauma can be severe: Disability payments to U.S. veterans with hearing loss total approximately $1 billion dollars annually, and soldiers experiencing tinnitus describe it as a source of difficulties in their lives (Schutte, 2006).

NOISE REDUCTION

Because the effects of noise can have such profound physical consequences, reduction of noise is a fundamental concern in human factors engineering. Machinery and equipment designers should work to minimize the noise output of their products. After all engineering efforts have been exhausted, workers can be protected from noisy equipment by baffles, which provide a physical sound-absorbing barrier between the worker and the source of the noise, and by ear protection devices.

Ear protection devices are the simplest resource available for noise control. These devices fall into two categories: earplugs and earmuffs. Several types of earplugs and earmuffs are readily available over the counter, and the degree of sound reduction that they are supposed to provide is usually clearly marked on the packaging. However, the level of protection they provide is frequently less than the manufacturers' ratings (Casali, 2006; Park & Casali, 1991). One solution to this problem is to use earmuffs, which are usually more expensive than earplugs but more effective. Another solution is to use custom-molded earplugs, which, because they can only be inserted correctly, provide better fit and protection than standard earplugs (Bjorn, 2004).

The reason why standard earplugs are not effective as they should be is probably that users do not know how to fit the earplugs properly (Park & Casali, 1991). Noise reduction ratings for three types of earplugs and a popular earmuff with and without user training are shown in Figure 17.13, together with the manufacturers' ratings. The earmuff provided more protection for untrained users, probably because the earmuff is easy to fit. For users trained on its fit, however, a malleable foam plug, the earplug provided maximum noise reduction. Note, though, that for both trained and untrained users, the noise reduction ratings were uniformly less than those claimed by the manufacturers.

Although earmuffs are generally effective at attenuating sound, they are more effective for frequencies above 2000 Hz than for those below (Zannin & Gerges, 2006). A benefit of earmuffs is that, when combined with headphones, they can provide both a source of protection against external noise and a means to deliver acoustic information in a noisy environment. Custom-molded earplugs with built-in electronics are also available, but they can be very expensive. Noise-cancelling headphones are equipped with active noise control (Casali, Robinson, Dabney, & Gauger, 2004). For such headphones, a microphone senses the frequency and amplitude of the sound inside the headphones and then produces an inverted signal, 180° out of phase with the sensed one, which cancels out the energy of the sound wave. Active noise reduction works best for repetitive noise and for low-frequency noise below 1000 Hz. Noise-reduction headphones have been shown to reduce miscommunication errors in pilots (Jang, Molesworth, Burgess, & Estival, 2014).

There are two issues that we need to keep in mind when trying to minimize people's exposure to noise, regardless of the type of ear protection we eventually decide is most appropriate. First, because of the discrepancy between manufacturers' noise reduction ratings and the attenuation any ear protection device actually provides, we must allow a large safety margin between the actual noise level and the noise reduction rating of the device. Second, the user needs to be trained to use the ear protection device and made to understand why ear protection is important (Behar, Chasi, & Cheesman, 2000; Park & Casali, 1991). Many companies and agencies (included the U.S. Armed Forces) now have hearing conservation programs, in which the use of ear protection devices is only part of an overall training program that emphasizes compliance. In companies that do not have comprehensive programs, 40%–70% of workers typically will not use the devices at all.

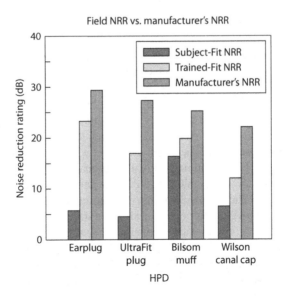

FIGURE 17.13   Noise reduction ratings for four hearing protection devices (HPDs) for trained and untrained subjects and as provided by the manufacturer.

VIBRATION

The term *vibration* refers to any oscillatory motion around a central point and is usually described in three dimensions. As with sound waves, vibration can be characterized in terms of amplitude and frequency. Often, the same mechanisms that produce noise also produce mechanical vibration. If this vibration is in a piece of equipment or machinery, the operator will also experience vibration. For example, offshore oil rigs vibrate a lot and are noisy, and so all the workers stationed on a rig are exposed to quite a bit of noise and vibration (Health & Safety Executive, 2002).

Vibration is measured with an accelerometer, which can be attached to the vibration source or to a bony spot on a person's body. This device measures displacement acceleration in one or more dimensions. The most common descriptive measure of vibration is the root mean square (RMS) value:

$$RMS = \sqrt{\frac{1}{T}\int_0^T x^2(t)\, dt},$$

where $x(t)$ is displacement along a particular dimension (usually specified as X, Y, or Z in three-dimensional space) as a function of time. RMS is, roughly, the square root of the average squared displacement for a fixed interval of time T.

An accelerometer should be as small as possible and should be sensitive to the ranges of acceleration and frequencies expected from the vibration source.

Operators who work with powered equipment, such as heavy vehicles or hydraulic devices, experience body vibrations for extended periods of time. As with any repetitive motion, this extended exposure can be detrimental to the health of the operator. The presence of vibration also has the more immediate effect of degrading an operator's performance by interfering with her motor control. When we are evaluating vibration, we make a distinction between whole-body vibration and segmental, or hand-transmitted, vibration applied to particular body parts (Griffin, 2006; Wasserman, 2006). We will talk about each kind of vibration separately.

WHOLE-BODY VIBRATION

Whole-body vibration is transmitted to an operator through supports such as floors, seats, and backrests. Vibration discomfort will increase as the amplitude of the vibration increases. Most people rate RMS magnitudes of 1 m/s² or larger as uncomfortable (Griffin, 2006). However, RMS is not the only important factor in determining whether a vibration is uncomfortable. Discomfort will increase with increasing exposure times, so even minor vibration may become intolerable if a person has to endure it long enough. The frequency of the vibration is also important. Every object, even the human body, has a resonant frequency. If vibration is transmitted to an object at a frequency near the object's resonant frequency, then the object will vibrate with amplitude higher than that of the vibration source. For the human body, the resonant frequency is approximately 5 Hz. This means that frequencies in the neighborhood of 5 Hz can have an even more damaging effect on a person's body than frequencies outside of this neighborhood.

As we mentioned already, vibration occurs in all three dimensions. How can we assess vibration and predict whether or not it will cause discomfort? One study looked at assessment of vibration both in the laboratory and in the field (Mistrot et al., 1990). In the lab, they exposed people to vibration in either one or two axes of motion and asked them to rate their discomfort. In the field, several professional truck drivers drove a truck with different loads, at different speeds, over good or poor sections of road. They measured vibration in all three directions by putting accelerometers in the driver's chair, and the drivers estimated their degree of discomfort. The researchers compared the drivers' judgments of discomfort with those from the people in the lab, and concluded that discomfort is best predicted using the RMS of the displacement in each axis. That is, there is no single direction of vibration that will produce discomfort.

Whole-body vibration interferes with vision and manual control. It also can have health effects, particularly lower back pain and damage to the lumbar region of the spine, for individuals who are exposed to whole-body vibration on a regular basis. This would include truck drivers, helicopter pilots (Smith, 2006), and operators of heavy construction equipment (Kittusamy & Buchholz, 2004). Helicopter pilots experience a lot of vibration through their seats (Bongers et al., 1990). Relative to other nonflying U.S. Air Force officers, helicopter pilots are more prone to both transient and chronic back pain. The pilots' degree of pain was related to the amount of vibration they experienced and their age. The older the pilot, and the more vibration to which they were exposed, the higher was the prevalence of chronic back pain.

One way to reduce whole-body vibration for vehicle operators is by redesigning the operator's seat. Designs that minimize the contact between the operator and the seat will reduce whole-body vibration. One study showed that a new car seat that tilted the back of the seat down to minimize seat contact, and included a padded protruding cushion for increased lumbar support, decreased the amplitude of whole-body vibration by about 30% (Makhsous, Hendrix, Crowther, Nam, & Lin, 2005). An alternative to reducing contact with the seat is to construct seats with suspension systems that counteract unwanted vibration from the road, much as in noise-reduction headphones. These kinds of seats reduce driver fatigue and should reduce the risk of musculoskeletal disorders for drivers of trucks, vans, buses, and tractors (Wang, Davies, Du, & Johnson, 2016).

SEGMENTAL VIBRATION

Segmental vibration to the hand and arm occurs while using power tools. We will talk mostly about hand-arm vibration, but in some situations a human factors expert may need to deal with head-shoulder and head-eye vibrations. Recall that the resonant frequency of the human body is around 5 Hz. However, the human arm has very little resonance. This means that most vibrations are absorbed into the hand and transmitted up the arm. As vibration frequency increases, less of the vibration is transmitted up the arm (Reynolds & Angevine, 1977). For vibrations of approximately 100 Hz, the entire vibration is absorbed by the hand.

Segmental vibration can cause a person to misperceive the movements and location of the vibrated segment, which in turn can lead to inaccurate aimed movements and possibly accidents (Goodwin, McCloskey, & Matthews, 1972). Vibration applied directly to the tendon of one of the major elbow muscles, either the biceps or the triceps, will induce a reflexive movement of the arm. If you ask a person to match the movement of the vibrated arm by moving his other arm, the mismatch between what the person feels and what is actually happening will be apparent. The person will move his arm too much and in the wrong direction—the direction that his arm would have moved if the vibrated muscle were being stretched. Vibration feels like muscle stretch, which results in misperception of the movement at the elbow. We can do this for other joints as well, which means that vibration could severely hamper motor performance, particularly when a person can't see or doesn't look at the vibrated segment to verify how it is moving.

Long-term use of vibrating power tools can result in a cumulative trauma injury called *vibration-induced white finger*, *Raynaud's disease*, or *hand-arm vibration syndrome* (HAVS; Griffin & Bovenzi, 2002; Poredos, 2016). HAVS arises from structural damage to blood vessels and nerves and is characterized by extreme numbness and intermittent tingling in the affected hand or finger. In later stages of the disease, the finger alternates between blanching (or whiteness) and cyanosis (or blueness), which is symptomatic of interruptions in the blood supply to the finger. The interruption may ultimately lead to gangrene, requiring amputation of the affected finger (or fingers).

The amount of exposure that a person can tolerate is a decreasing function of the intensity of the vibration. Figure 17.14 shows how many people will develop blanching symptoms

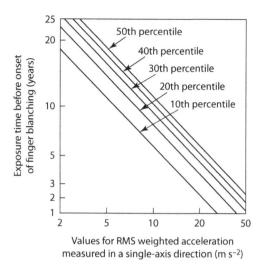

FIGURE 17.14 Years of exposure before developing symptoms of advanced stages of vibration-induced white finger, as a function of the intensity of vibration.

after exposure to different magnitudes of vibrations (RMS on the abscissa) for extended periods of time (years on the ordinate). Most hand-arm vibration data come from men. Women show greater sensitivity to and discomfort from vibration, which means that we need to be careful to consider possible gender differences when assessing segmental vibration (Neely & Burström, 2006).

HAVS can be aggravated by many factors. For example, a person who uses a tight grip and works in the cold, which causes the arteries to constrict, may develop HAVS more quickly. Also, some vibration frequencies are more problematic than others. In particular, exposure to vibrations between 40 and 125 Hz increases the likelihood that a person will develop HAVS (Kroemer, 1989).

THERMAL COMFORT AND AIR QUALITY

The climate of a working environment usually refers to the temperature and relative humidity of people's surroundings. There are some workplaces where it is easy to maintain a normal temperature and humidity. However, there are other workplaces where this is not possible. A frozen-food warehouse cannot be kept comfortably warm; a tent in the desert cannot be kept comfortably cool. Extremes of temperature and humidity can severely restrict a person's capabilities, diminishing his or her stamina, motor function, and overall performance.

To evaluate the climate in a workspace, we often refer to a *comfort zone* (Fanger, 1977). A comfort zone is a range of temperature and humidity that people will find acceptable given restrictions imposed by the tasks they are trying to perform, their clothing, air movement, and so forth. Figure 17.15 plots the comfort zone for moderate air speeds (0.2 m/s), light work, and light clothing. The zone is shown as a dashed rectangle in the center of the temperature–humidity range. For this zone, dry-bulb temperature (that measured by a typical thermometer) varies from 19°C to 26°C (66°F–79°F), and relative humidity varies from 20% to 85% at the extremes (Eastman Kodak Company, 1983). Under the specific conditions being depicted, people should be comfortable as long as the combination of temperature and humidity remains inside of the comfort zone.

A person's impression of comfort will be influenced by several factors. Heavy work will shift the comfort zone to lower temperatures. Workers do not often perform heavy work continuously throughout a shift, and not all workers on a shift will be performing heavy work. This means that

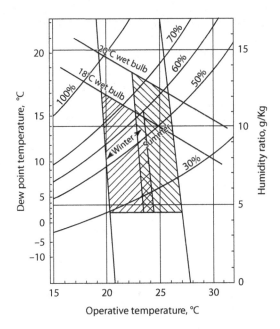

FIGURE 17.15 The comfort zones for winter and summer.

the temperature in the work area must be a compromise between the comfort zone for sedentary work and that for heavy work. We can solve this problem by providing those workers who perform less strenuous tasks with warmer clothing, such as sweaters or jackets. High air velocities will reduce the insulating ability of clothing and will require that we set the temperature higher. In the range of comfortable temperatures, relative humidity has only a minor influence on thermal sensation.

Fanger's (1977) comfort zone concept is the basis for attempts to develop active customized thermal comfort controls (Andreoni, Piccini, & Maggi, 2006). Customized control uses sets of sensors and transducers to measure the temperature and relative humidity in a room and on the body of a person who is working in the room. The climate control system takes these measurements and computes an estimate of thermal comfort based on the comfort zone, and responds quickly to changes in climate conditions and activity levels to maintain the person's comfort.

We can also determine discomfort zones. Discomfort will arise when a person's body's thermal regulatory system is strained beyond its normal bounds. This will happen for some combinations of temperature, humidity, and workload. For example, when you get very hot, you may sweat a lot. Sweating is often uncomfortable. Furthermore, your tools and controls might get slippery, and when your clothes get wet it may be harder to move. Your mental acuity may suffer (Parsons, 2000), and your dexterity as well (Ramsey, 1995). People are less able to perform tracking and vigilance tasks in temperatures of 30°C–33°C (86°F–91°F) or higher.

Hot environments require managers to implement certain work practices to prevent heatstroke or hyperthermia. Workers must be provided with ample water and a cool area in which to rest. They must also be trained to recognize the symptoms of hyperthermia, and they must be given enough time to adjust to the heat when they first arrive on the job, or when they return after vacation or a leave of absence.

A person's performance will deteriorate in the cold, too. Manual dexterity will deteriorate as a consequence of physiological reactions such as stiffening joints (Marrao, Tikuisis, Keefe, Gil, & Giesbrecht, 2005; Parsons, 2000). Also, cold environments require additional clothing, which will restrict a person's range and speed of movement. We may need to restructure some tasks to accommodate the workers' decreased mobility. We also need to be aware that the work environment may be hazardous for someone wearing bulky clothing: some open machinery may allow fabric to become entangled. Drafts will increase the workers' discomfort, and so we need to make every effort to eliminate them and to provide a source of radiant heat. An increased workload will also make the cold environment more tolerable. As with excessively hot environments, workers must be trained to recognize the symptoms of hypothermia and frostbite.

Both extreme cold and extreme heat can have deleterious effects on a person's ability to perform complex tasks (Daanen, Vliert, & Huang, 2003; Pilcher, Nadler, & Busch, 2002). Cognitive task performance can be impaired up to 14% for temperatures less than 10°C (50°F) or greater than 32°C (90°F). Similarly, driving performance decreased by 13%–16% in hot and cold temperatures. These findings imply that environmental and human factors engineers need to make the greatest effort possible to maintain moderate temperatures in the workplace, because this has direct implications for safety.

Apart from temperature and humidity, we often need to be concerned about indoor air quality. Usually, we will focus our attention on the presence of gaseous and particulate pollutants in the air, concentrations of which can build up to be many times higher indoors than outdoors.

Pollutants can be classified into three categories (ASHRAE, 1985):

1. Solid particulates, such as dust, pollen, mold, fumes, and smoke
2. Liquid particulates in the form of mists or fogs
3. Nonparticulate gases

To evaluate the air quality of an environment, we measure each of these categories of pollutants. If we find high levels of any pollutant, we have to determine its source. Some common sources of indoor pollutants include living organisms (pets, rodent and insect pests, bacteria, and mold), tobacco smoke, building materials and furnishings, central heating and cooling systems, chemicals used for cleaning, copy machines, and pesticides (U.S. Environmental Protection Agency, 2017).

Pollutants are spread from their sources by way of air movement, which is wind in the outdoor environment and the ventilation within an indoor environment. Because ventilation systems bring in air from the outside, the source of an indoor pollutant can be from either inside or outside a building. Poor ventilation can create conditions in which molds and fungi flourish (Peterman, Jalongo, & Lin, 2002). Molds can cause allergies and (depending on the type of mold) can be toxic. Molds often grow in damp areas, such as ceilings. Musty smells may give away the presence of mold. Air conditioning cooling towers can also harbor mold and bacteria, such as the bacteria responsible for Legionnaire's Disease, and spread those bacteria through the ventilation system.

Poor air quality can also have a negative effect on performance. In one study, the level of air pollution was manipulated by introducing or removing a pollution source, an old carpet (Wargocki, Wyon, Bake, Clausen, & Fanger, 1999). The carpet had previously been removed from an office building for having an unpleasant odor and irritating employees' eyes and throats. Participants were exposed to the pollution source or its absence for 265 minutes, unaware of the condition, since the carpet was placed behind a partition. During this period, the participants performed tasks simulating office work and filled out assessments of perceived air quality. Headaches were reported as greater when the pollutant was present, and the air quality was rated as worse than without it. The participants typed 6.5% more slowly when the pollutant was present than when it was not, consistently with reported lower levels of effort. Thus, the discomfort caused by poor air quality can have a negative impact on performance and productivity.

When many occupants of a building experience recurring respiratory symptoms, headaches, and eye irritation, they are said to suffer from *nonspecific building-related symptoms*, or *sick building syndrome* (Norbäck, 2009; Runeson & Norbäck, 2005). The syndrome is controversial, because it is a function of several medical, psychological, and social factors (Thörn, 2006), but it can have a significant negative impact on the people in the affected building (Söderholm, Öhman, Stenberg,

& Nordin, 2016). Sick building syndrome is blamed on the tightly sealed buildings that were constructed beginning in the late 1970s to conserve energy. These buildings had minimal ventilation from the outside, resulting in buildups of pollutants within the building. Perhaps as many as 30% of the buildings worldwide that were built or remodeled during this period could be diagnosed with sick building syndrome (World Health Organization, 1984). Sick building syndrome is corrected by improving indoor air quality.

We can improve air quality by one of two methods. We can use devices like high-energy particulate absorbing (HEPA) filters in air purifiers. These filters remove over 99% of particles of 0.3 micrometers (microns) diameter, and larger and smaller particles are filtered even better. (For reference, the HIV virus is 0.1 micron in diameter). Alternatively, the contaminated air can be diluted with outdoor air by increasing ventilation rates (Cunningham, 1990), assuming that the outdoor air is not also polluted. All ventilation systems work by bringing outdoor air inside the building, but different buildings will require higher or lower rates of air circulation. In the U.S., state building codes will state the amount of outdoor air required for specific applications, such as the combustion of wood, dry cleaning, painting, hospitals, and so forth.

STRESS

Stress is a physiological and psychological response to unpleasant or unusual conditions, called *stressors* (Sonnentag & Frese, 2003). These conditions may be imposed by the physical environment, the task performed, one's personality and social interactions, and other stressful situations at home and at work. Although specific stressors, such as temperature extremes, produce specific physiological responses in a person, they all cause the same nonspecific demand on the body to adapt itself. This demand for adaptation is stress. Acute stress associated with immediate events can be intense and affect performance; chronic stress over a period of time can have harmful physical, as well as psychological, effects on a person.

GENERAL ADAPTATION SYNDROME AND STRESSORS

In 1936, Hans Selye first characterized stress as a physiological response. He noticed that rats injected with different toxic drugs exhibited many of the same symptoms even though the drugs were different. He also discovered a characteristic pattern of tissue changes in the adrenal and thymus glands, and in the lining of the stomach wall taken from sick rats. Sick rats, or stressed rats, had swollen adrenal glands, atrophied thymus glands, and stomach ulcers. These three symptoms are the *General Adaptation Syndrome* (Selye, 1973). This syndrome is characterized by stages of physiological responses of increasing intensity.

The first stage in the syndrome is the alarm reaction, which is the body's initial response to a change in its state. It is characterized by discharge of adrenaline into the bloodstream. If the stressor inducing the alarm reaction is not so strong that the animal dies, then the body enters the second, "resistance," stage. In this stage, adrenaline is no longer secreted, and the body acts to adapt to the presence of the stressor. As exposure to the stressor continues, the body enters the final "exhaustion" stage, in which its resources are depleted and tissue begins to break down.

Stress is a function not just of physiological factors, but of psychological factors as well. Probably the most important factor is how a person appraises or construes his or her situation (Lazarus & Folkman, 1984). The person appraises the harm that has already occurred, the threat of harm that may take place in the future, and the available resources for dealing with the stressor. On the basis of this appraisal, the person will decide whether the stressful environment is merely unpleasant or intolerable, and then how he or she will react to the stressor. The appraisal can be affected by such things as the degree of control that the individual has over the situation and his or her understanding of why situations are as they stand.

Extremely high stress can severely impair a person's ability to make decisions, particularly if the person feels that he is under time pressure. In such situations, he may react in a way that is called *hypervigilance* (Janis & Mann, 1977). Hypervigilance is a panic state in which his memory span is reduced and his thinking becomes overly simplistic. He may search frantically and haphazardly for a solution to a problem and fail to consider all of the possible solutions. In an attempt to beat a decision deadline, he may make a hasty, impulsive decision that has some promise for immediate relief but also has negative longer-term consequences.

Hypervigilance may contribute to some of the errors that people make in emergency situations. Emergencies are characterized by acute stress, which is induced by a sudden potentially life-threatening situation for which a solution must be found quickly. Hypervigilance has been linked to incidents of unintended acceleration: The panic induced when a person's car is rocketing out of control reduces her ability to detect that her foot is on the accelerator instead of the brake pedal.

There are three classes of stressors: physical stressors, social stressors, and drugs (Hockey, 1986). We can also distinguish between external versus internal sources of stress and transient versus sustained stress. External stressors arise from changes in the environment, such as heat, lighting level, or noise, whereas internal stressors arise from the natural dynamics of a person's body. Transient stressors are temporary, whereas sustained stressors are of longer duration.

Figure 17.16 shows the relationship between different stressors and a person's internal cognitive states. The person is designated by the larger broken box, and his internal cognitive states by the smaller broken box within it. Drugs and physical and social factors provide external stress, whereas cyclical changes (such as a woman's menstrual cycle) and fatigue provide internal stress. Physical stress is caused by annoying and uncomfortable environmental conditions of the type discussed in this chapter. Physical stressors directly influence the stress state of the individual, although their effects are mediated to some extent by the person's cognitive appraisal of the situation. Physical stressors also can produce fatigue.

The influence of social stressors, such as anxiety about evaluations of one's performance and incentives, is mediated by cognitive appraisal. In response to stress, a person may try to regulate the state of her body by taking antianxiety drugs. Thus, drugs have their primary effect on her stress state. They also can influence the person's level of fatigue, which in turn affects her stress state.

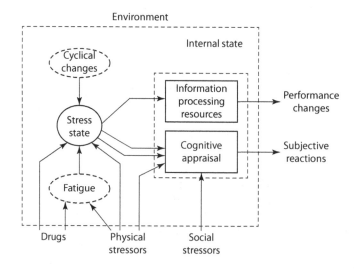

FIGURE 17.16 The relation between stressors and internal states.

TABLE 17.3

Classes of Stressors

Class of Stressor	Examples	Mode of Effect	Interacting Variable
Physical	Heat-cold, noise-vibration, lighting conditions, atmospheric conditions	Direct effect on central nervous system via changes in sensory receptors	Individual differences, task, possibility of control, other stressors
Social	Anxiety, incentives	Cognitive mediation	Individual differences, type of task, presence of other stressors
Drug	Medical (tranquilizers), social (caffeine, nicotine, alcohol)	Direct effect on central nervous system	Individual differences, task, other stressors
Fatigue	Boredom, fatigue, sleep deprivation	Both direct physiological and cognitive mediation	Individual differences, type of task, time of day, other stressors
Cyclical	Sleep–wake cycle, body-temperature rhythm, other physiological rhythms; usually studied are disruptions of the rhythms by shift work or transzonal flight	Some are dependent on environmental changes: others seem internally driven	Individual differences, task, form of the disruption

Fatigue is the wide range of situations in which a person feels tired. It can be caused by excessive physical and mental workloads and loss or disruption of sleep. Fatigue results in feelings of not only tiredness but also boredom. Cyclical stressors are those involving natural, physiological rhythms. These stressors are usually studied by investigating performance when rhythms are disrupted, for example, by shift work or jet flight. High fatigue and disruptions of circadian rhythms (see Chapter 18) increase stress.

Table 17.3 gives a summary of the different classes of stressors and the locus of their effects. There are large differences in the extent that different people are susceptible to stress. The same stressor applied to two different people may have different effects. Moreover, the effect of a given stressor may vary depending on what a person is trying to do. The level of stress induced by a particular variable (e.g., cold temperature) may not be as great with an undemanding task as with a demanding task. Note also that the effect of a particular stressor on the stress state may be larger when other stressors are present.

OCCUPATIONAL STRESS

The term *occupational stress* specifically refers to stress associated with a person's job (Gwóźdz, 2006). Healthcare workers are particularly susceptible to this kind of stress. For example, Marine, Ruotsalainen, Serra, and Verbeek (2006) noted:

> Healthcare workers suffer from work-related or occupational stress often resulting from high expectations coupled with insufficient time, skills and/or social support at work. This can lead to severe distress, burnout or physical illness, and finally to a decrease in quality of life and service provision. The costs of stress and burnout are high due to increased absenteeism and turnover. (p. 2)

This quote demonstrates how stress in the work environment can arise from the physical and social environment, organizational factors, and a person's tasks, but, in addition, a person's personality attributes and skills are important factors (Smith, 1987).

Environmental sources are, as we have discussed, the climate, lighting, and so on in the work-place. Physical environmental stressors are more of a factor for manual laborers than for office workers and managers. Organizational factors involve job involvement and organizational support. For example, an autocratic supervisory style can lead to a person's job dissatisfaction and, hence, increased stress. Lack of performance feedback or continually negative feedback can also be stress-ful. Workers in an organization that allows employees to participate in decisions that impact on their jobs will experience less stress than workers in an organization that does not. Opportunities for career development also serve to lessen occupational stress.

Job-task factors influencing stress include high mental and physical workload, shift work, dead-lines, and conflicting job demands. To some extent, an individual must be "matched" to certain jobs (Edwards, Caplan, & van Harrison, 1998). Training must be appropriate, the job must be acceptable to the individual, and the individual must have the physical and mental capabilities necessary to perform the job. The degree to which a worker is not well matched to his job in training, desire, and capability in part determines the level of stress that he will experience.

There are several types of intervention that can help relieve occupational stress (Kivimäki & Lindström, 2006). Those that focus on the individual include stress management training, in which the person learns stress reduction and coping strategies such as muscle relaxation. Cognitive-behavioral interventions have the goal of changing a person's appraisal of the situation. In the case of a single traumatic event, like the accidental death of a co-worker, debriefing programs conducted within a day or two after the event can minimize stress. For the workforce as a whole, organizations may also provide employee assistance programs, promote healthy work organizations, and institute ergonomic improvements. Job redesign and organizational change, topics of the next chapter, can also be effective tools for reducing occupational stress.

Some work environments, such as a space station (see Box 17.1) or an Antarctic research station, are "contained": a person can't leave the work environment because it is the only environment that supports life. Forced containment restricts the actions that a person can take to reduce stress, and this restriction introduces stressors of its own. These include (1) the surrounding hostile environ-ment, (2) a limited supply of life-supporting resources, (3) cramped living spaces and enforced intimacy, (4) the absence of friends and family, (5) few recreational activities, (6) an artificial atmo-sphere, and (7) an inability to leave the contained environment (Blair, 1991).

The stress of a contained environment manifests itself in several ways. A person may experi-ence increased appetite and weight gain, as eating becomes very important as entertainment. Her decreased activity and the loss of light and dark cycles disrupt her sleep patterns. Because of her enforced proximity with other people in the environment, her sleep/wake cycle can be very disrup-tive to others with different sleep/wake cycles. Anxiety and depression are common, and her sense of time may be distorted.

Because these stressors cannot be removed from a contained environment, the best way to con-trol stress in such environments is through careful screening of the applicants. People should be selected who adapt well and are not unduly affected by the stressors induced by the contained envi-ronment. Blair (1991) describes a good candidate as one whose predominant interest is in work and who is comfortable with, but has no great need for, socializing. He describes the best candidates for work in these environments as "often not very interesting people."

SUMMARY

Human performance, health, and safety are not determined solely by the design of displays, con-trols, and the immediate work station. Additionally, the larger environment in which a person lives and works makes a difference between tolerable and intolerable working conditions. The goal of environmental ergonomics is to ensure that engineers appropriately consider the physical envi-ronment when designing workspaces. Some critical factors include appropriate illumination, noise

BOX 17.1   SPACE ENVIRONMENT

As technology has progressed, humans have moved into more and more hostile environments. People now take jet flights across continents, live underwater in submarines for extended durations, and travel in space. All such exotic settings require contained environments in which the atmosphere, lighting, and heat are provided artificially. They also possess unique properties enforced by the external environment that must be accommodated in designing for the human.

Most notable is the extension of human life to outer space. As Harrison (2001) notes, "Spacefaring is a partnership involving technology and people" (p. xi). People who go into space are entirely dependent on technology for their survival. They also must adapt to new features of the extraterrestrial environment they inhabit.

The first spaceflight by Yuri Gagarin in 1961 was less than a day long. Since such early brief excursions into space, the durations of space voyages have increased greatly (Grigoriev & Potapov, 2013). The NASA space shuttle program requires astronauts to live outside the earth's atmosphere for several weeks at a time, and occupants of the International Space Station stay for about 6 months (Mount, 2006). The current record length of time in space for an individual is 438 days on the Russian Mir space station, but the crews of planned missions to the planet Mars will be in the relatively cramped quarters of their spacecraft for several years. It is not sufficient just to keep the astronauts alive during their time in space; they also must function well for the entire flight.

Spaceflight for humans involves a number of unique physical, psychological, and cultural factors that could create problems. Here, we will focus only on the physical environment factors (Mount, 2006; Woolford & Mount, 2006). A breathable atmosphere must be provided to support life. Resources, such as food and water, must also be supplied. On brief trips, all of the air and water that are needed can be supplied, but on longer ones they must be recycled to reduce the needs for replacement.

Noise is potentially a problem, because space vehicles contain hardware systems necessary for life-support and other functions that may create high noise levels. Lighting is a concern as well, with glare produced by the sun being particularly problematic for reading displays while in earth orbit. For missions to Mars, dust from the planet's environment poses a potential problem that needs to be addressed.

Of course, the most obvious difference in the environment for space flight and on earth is the absence of normal gravity. During launch and reentry, the crew face periods of hypergravity from acceleration and deceleration lasting up to 17 minutes (Harrison, 2001). In space, however, they experience microgravity for long periods of time. This lack of gravity and experience of weightlessness adds a new perspective to design of the workplace. For example, in the absence of gravity, the human body increases in height by about 3%, and the natural body posture changes to become more flexed (see Figure B17.1; Louviere & Jackson, 1982; Woolford & Mount, 2006). Due to the reduced gravity, the legs and back muscles lose about 10%–20% of their strength within a few days (Fitts, Riley, & Widrick, 2000; Jahweed, 1994). A person's vestibular sense will need to adapt to the lack of gravity over a period of about 3 days, during which motion sickness and disturbances of movement may occur (Shelhamer, 2015).

Cognitive tasks typically show little impairment during spaceflight (e.g., Manzey & Lorenz, 1998). However, impairments of manual tracking performance do occur (Heuer, Manzey, Lorenz, & Sangals, 2003). There has been debate as to whether these tracking impairments are due to specific effects of microgravity on motor control. Heuer et al. reported that a cosmonaut who showed the tracking impairment also showed similar changes in rapid aiming movements, suggesting that the tracking impairments are direct consequences of the impact of microgravity on motor control.

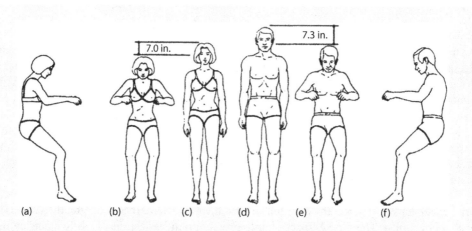

FIGURE B17.1 Differences between gravity-present (c and d) and gravity-absent (a, b, e, and f) neutral body positions.

It may be that, in the future, entire colonies of people will be living and working in space. The range of people that must be accommodated and the tasks that they perform will be much more varied. Issues in the design of work and living spaces thus will become more prominent, with the specific characteristics of the extraterrestrial environment taken into account. The best place for studying the effects of isolated groups of people working for extended durations under harsh conditions is thought to be the winter research stations in Antarctica (Harrison, Clearwater, & McKay, 1991). With the exception of the absence of gravity, these stations exist in a hostile environment and have most of the characteristics that would be associated with space colonies. Harrison (2001) points out that missions to Mars, in particular, will benefit from polar human factors research, in which the mission conditions and crew size and composition could be matched to those anticipated for the actual mission to Mars itself.

levels within tolerable ranges, protection against extreme noise and vibration, task-appropriate climate, and high air quality.

Inadequate environmental conditions are major contributors to stress. Stress is also produced by a variety of other factors, including the social environment, task demands, and long-term confinement. High levels of stress can result in illness and poor performance. By selecting candidates using appropriate screening methods and designing the environment and tasks to minimize stress, we can keep stress within acceptable limits.

RECOMMENDED READINGS

Behar, A., Chasin, M., & Cheesman, M. (2000). *Noise Control: A Primer*. San Diego, CA: Singular.
Boyce, P. R. (2014). *Human Factors in Lighting* (3rd ed.). Boca Raton, FL: CRC.
Harrison, A. A. (2001). *Spacefaring: The Human Dimension*. Berkeley, CA: University of California Press.
Kroemer, K. H. D., & Kroemer, A. D. (2001). *Office Ergonomics*. London: Taylor & Francis.
Mansfield, N. J. (2005). *Human Response to Vibration*. Boca Raton, FL: CRC.
Oborne, D. J., & Gruneberg, M. M. (Eds.) (1983). *The Physical Environment at Work*. New York: Wiley.

18 Human Resource Management and Macroergonomics

Work systems are becoming more complex, creating numerous challenges for those involved in making decisions that affect the design of work systems ... Human factors and ergonomics (HFE) has responded to this challenge by embracing and adapting models and concepts that incorporate the organizational and sociotechnical context of work such as macroergonomics.

P. Carayon et al.
2015

INTRODUCTION

In addition to the influence of the physical design of workspaces and the larger work environment, a person's performance and well-being are influenced by many social and organizational factors. Job satisfaction is a function of the tasks that a person must perform, work schedules, and whether the person's skills are adequate for the job. The degree of participation that the workers have in policy decisions, the avenues of communication available within an organization, and the social interactions experienced every day with coworkers and supervisors also play a major role in determining job satisfaction. Job satisfaction in turn affects a person's physical and psychological health, as well as her or his level of productivity.

Productivity bears directly on an organization's "bottom line" (profits), as well as on other measures of organizational success. The recognition of the fact that organizational performance is determined by the productivity of individual employees has drawn the human factors expert into areas traditionally left to management. The problems associated with job and organizational design, employee selection and evaluation, and management issues form the basis of the field of industrial/organizational psychology and human resource management. However, such organizational design and management problems are also of concern in human factors, because an employee must perform within the context of specific job expectations and the organizational structure.

The unique viewpoint that human factors specialists bring to job and organizational design issues is the systems perspective. An organization is a "sociotechnical system" that transforms inputs into outputs (e.g., Clegg, 2000; Emery & Trist, 1960). A sociotechnical system has technical components (the technical subsystem) and social components (the personnel subsystem) and operates within a larger environment (Kleiner, 2006, 2008). The technical and personnel subsystems must function effectively in an integrated manner, within the demands imposed by the external environment, if the sociotechnical system is to function effectively as a whole.

Macroergonomics is a term that describes the approach to the "human–organization–environment–machine interface," beginning with the organization and then working down to issues of environment and workspace design for the individual worker in the context of the organization's goals (Hendrick, 1991). Macroergonomics therefore focuses on organizational issues and their relation to human–machine interface issues (Kleiner, 2004). The goal of the human factors expert is to optimize human and system performance within the overall sociotechnical system, which in this case is the organization or business. Some issues that the expert will consider include job design, personnel selection, training, work schedules, and the influence of organizational structure on decisions. The focus of macroergonomics on the relation between human factors and the organization as a system provides a perspective for human factors specialists that can allow them to make significant contributions to the workplace (Kleiner, 2008).

In the present chapter, we examine the social and organizational factors that affect the performance of the organizational system and the individual employee within it. We begin with the individual employee, discussing job-related factors such as employee recruitment and job design. We then move to employee interactions and some of the ways that social psychology can be exploited to the benefit of the organizational group. We conclude with issues pertaining to the organizational structure, such as how groups interact, employee participation in decision making, and the process of organizational change.

THE INDIVIDUAL EMPLOYEE

We have already discussed many factors that influence worker performance and satisfaction. These involved the physical aspects of the human–machine interface and the surrounding environment, as well as task demands. The core of human factors involves analysis of the psychological and physical requirements for performing specific tasks. Because a job usually involves the performance of many tasks, the human factors expert may be called upon to perform a job analysis.

Job Analysis and Design

The broad range of activities demanded by most jobs usually means that employees must possess a similarly broad range of skills. A job analysis is a well-defined and rigorous procedure that provides information about the tasks and requirements of a job, and the skills required for a person to perform the job (Brannick & Levine, 2007).

The job analysis is used to describe, classify, and design jobs. As Sanchez and Levine (2012) emphasize, "Job analysis constitutes the preceding step of every application of psychology to human resources (HR) management including, but not limited to, the development of selection, training, performance evaluation, job design, deployment, and compensation system" (p. 398). Job analysis provides the basis for choosing appropriate selection criteria for prospective employees, determining the amount and type of training that is required for employees to perform the job satisfactorily, and evaluating employee performance. It also is used to determine whether jobs are well-designed ergonomically; that is, whether they are safe and can be performed efficiently. If we determine that a job is not well-designed, the job analysis can be used as a basis to redesign the job. A job analysis, then, can have a profound effect on an employee's activities.

Any job analysis uses a systematic procedure to decompose a job into components and then describe those components. There are two types of job analysis: work-oriented and worker-oriented (Shore, Sheng, Cortina, & Yankelevich Garza, 2015). A work-oriented job analysis will provide a job description that lists the tasks that must be performed, the responsibilities that a worker in that position holds, and the conditions under which the tasks and responsibilities are carried out (Brannick & Levine, 2007). Techniques that concentrate on the individual elements of specific jobs include, for example, Functional Job Analysis (Fine, 1974), which focuses on the functions performed on the job. Such an analysis might produce functions like supervising people, analyzing data, and driving a vehicle.

A worker-oriented job analysis focuses instead on the knowledge, skills, and abilities required of a person to perform the job. The Position Analysis Questionnaire (PAQ; McCormick, Jeanneret, & Mecham, 1972), described below, which measures the psychological characteristics of the worker and the environment, is an example of a person-oriented analysis. We might also use a hybrid approach in which work- and worker-oriented analyses are combined (Brannick & Levine, 2007). Any job analysis provides as an end result a job specification or description, which spells out the characteristics of the job and those required of a person holding that job.

TABLE 18.1

Information about a Job that Can Be Gathered in Job Analysis

1. Job Content
   - What the person with the job does (tasks, procedures, responsibilities)
   - Machines, tools, equipment, and materials used during the performance of the job
   - Additional tasks that might be performed
   - Expectations of the person performing the job (products made, word standards)
   - Training or educational requirements

2. Job Context
   - Working conditions (risks and hazards, the physical environment)
   - Physical and mental demands
   - Work schedule
   - Incentives
   - How the job is positioned in the management hierarchy of the company

3. Job Requirements
   - Knowledge and information necessary for the person to meet expectations
   - Specific skills (e.g., computer programming, data analysis, communication)
   - Ability and aptitude (e.g., mathematical, problem solving, reasoning)
   - Educational qualifications
   - Personality characteristics (e.g., willingness to embrace change, eager to learn)

Table 18.1 provides an outline of the information that might result from a job analysis. This information might come from several sources (Jewell, 1998). We could start with the records provided by supervisory evaluations, company files, and the U.S. Department of Labor Employment and Training Administration's Dictionary of Occupational Titles. Most of it will come from the people actually performing the job. We can use interviews and questionnaires to determine not only the actual tasks involved in the job we are analyzing, but also the employees' perceptions about the task requirements and the necessary skills. We can also do field studies and watch people on the job. Finally, we can ask other people who know the job, such as supervisors, managers, and outside experts, to contribute their knowledge to the job analysis data base.

Clearly, the most valuable source of information is the job incumbent, the person who is doing the job now. If we decide to interview the incumbents, we will get a lot of data from unstructured questions that will be difficult to organize and analyze. An alternative to interviews is a structured questionnaire. This is a set of standard questions that can be used to elicit the same information that we might obtain in an open interview.

One popular structured questionnaire used to conduct job analyses is the previously mentioned PAQ (García-Izquierdo, Vilela, & Moscoso, 2015; McCormick et al., 1972). The PAQ contains approximately 200 questions covering 6 major subdivisions of a job (see Table 18.2). These subdivisions are (1) information input (how the person gets information needed to do the job); (2) mediation processes (the cognitive tasks the person performs); (3) work output (the physical demands of the job); (4) interpersonal activities; (5) work situation and job context (the physical and social environment); and (6) all other miscellaneous aspects (shifts, clothing, pay, etc.). Each major subdivision has associated with it a number of job elements. For each job element, the PAQ has an appropriate response scale. So, for example, the interviewer may ask the incumbent to rate the "extent of use" of keyboard devices on a scale from 0 (not applicable) to 5 (constant use).

Although the PAQ has many positive features, there are jobs for which structured questionnaires such as the PAQ may not be appropriate. For instance, the PAQ is not appropriate for workers with poor reading skills, because it requires a high level of reading ability to understand and respond to

TABLE 18.2
Position Analysis Questionnaire (PAQ)

Information Input (35)
  Sources of job information (20): Use of written materials
  Discrimination and perceptual activities (15): Estimating speed of moving objects

Mediation Processes (14)
  Decision making and reasoning (2): Reasoning in problem solving
  Information processing (6): Encoding/decoding
  Use of stored information (6): Using mathematics

Work Output (50)
  Use of physical devices (29): Use of keyboard devices
  Integrative manual activities (8): Handling objects/materials
  General body activities (7): Climbing
  Manipulation/coordination activities (6): Hand-arm manipulation

Interpersonal Activities (36)
  Communications (10): Instructing
  Interpersonal relationships (3): Serving/catering
  Personal contact (15): Personal contact with public customers
  Supervision and coordination (8): Level of supervision received

Work Situation and Job Context (18)
  Physical working conditions (12): low temperature
  Psychological and sociological aspects (6): Civic obligations

Miscellaneous Aspects (36)
  Work schedule, method of pay, and apparel (21): Irregular hours
  Job demands (12): Specified (controlled) work pace
  Responsibility (3): Responsibility for safety of others

Note: Numbers in parentheses refer to the number of items on the questionnaire dealing with the topic.

the questionnaire items. Also, for some jobs, the PAQ may yield a large number of "not applicable" responses. In these circumstances, we may have to consider an open interview, an alternative questionnaire, or even a different source of information. Regardless of how we collect the information, when we compile the data, we must be able to write job descriptions and specifications that capture the features of the jobs.

You might suspect that proactive job analysis, that is, reliance on job analysis in advance of any problems, correlates positively with organizational performance. This seems to be the case (Siddique, 2004). One study of 148 companies in the United Arab Emirates showed that those organizations that performed proactive job analyses performed better than those that did not, particularly if human resource management was generally prominent in the company. Although these data are only correlational, they suggest that job analysis may be important to an organization's well-being.

A task analysis performed proactively can provide a basis for job design, whereas one performed retroactively can provide a basis for job redesign. Job design (Oldham & Fried, 2016) will require us to make decisions about the tasks that will be performed by the workers and the way in which these tasks are to be grouped together and assigned to individual jobs (Davis & Wacker, 1987). A properly designed job benefits performance, safety, mental health, and physical health in multiple ways (Morgeson, Medsker, & Campion, 2012).

There are many approaches that we could take to job design. We may want to emphasize the most efficient work methods, the workers' psychological and motivational needs, their information-processing abilities, and human physiology. In addition, we may want to design jobs from a team perspective, examining teams of workers and their social and organizational needs. Two popular approaches to job design are sociotechnical theory and the jobs characteristics approach (Holman, Clegg, & Waterson, 2002).

Sociotechnical theory is based on the sociotechnical systems concept we described earlier. In this approach, we recognize that desirable job characteristics include sensible qualitative and quantitative demands on the worker, an opportunity for learning, some area over which the worker makes decisions, and social support and credit (e.g., Holman et al., 2002). The jobs characteristics approach more specifically attributes job satisfaction and performance to five job characteristics: autonomy (the extent to which a worker can make her own job-relevant decisions), feedback (information about how well she is performing her job), skill variety (the range of skills she uses in her job), task identity (the degree to which a job requires completion of an entire, identifiable portion of work), and task significance (e.g., Hackman & Oldham, 1976). The extent to which a worker perceives that she has opportunities to use her skills (and not just the extent to which she actually uses those skills) also correlates with job satisfaction (Morrison, Cordery, Girardi, & Payne, 2005).

Both the sociotechnical and the job characteristics approach can contribute to our efforts to design and redesign a job. In the best of all worlds, we should try to design a job well with respect to any approach (Morgeson et al., 2012). However, sometimes alternative approaches may lead to conflicting recommendations. For example, under the sociotechnical approach, the requirement for sensible demands may preclude the autonomy emphasized under the job characteristics approach. When such conflicts arise, they should be resolved by considering which alternative would be best overall for the person doing the job and what we need that person to accomplish in the job.

Personnel Selection

Few brain surgeons possess the technical skills to fly a Boeing 787, although their level of training is extensive. No matter how smart and well-trained a brain surgeon is, it would not make much sense to hire her to fly a commercial airliner.

An employer's primary goal is to select individuals whose training is appropriate for the work that a job requires. How does an employer decide the minimum requirements for employment and select the most highly qualified personnel from a pool of applicants? Many of these decisions are made on the basis of a job analysis, which is the first step in personnel selection (Shore et al., 2015). We often use the job specification from a job analysis to develop employment criteria, training programs, and employee evaluations.

To fill a position, an employer must generate a suitable pool of applicants and narrow this pool down to the most qualified individuals. The job description developed from the job analysis can be the basis of a job advertisement and other recruiting efforts. It is often difficult to write a job description that targets the right applicants, and to make that pool of applicants aware of the position.

Recruiting is done either internally or externally. Internal recruiting generates a pool of applicants already employed by the organization. Internal recruiting has a number of advantages. Not only are the difficulties involved in recruiting solved, but internal recruiting is inexpensive, and the opportunity for job advancement provides a significant psychological benefit to employees. External recruiting comprises those activities directed toward the employment of persons not already associated with the organization. External recruiting usually produces a larger pool of applicants, which allows the employer more selectivity.

Selecting applicants from the pool is a screening process. The employer's goal is to determine who is most likely to be successful at the job, or, in other words, who best matches the job performance criteria defined by the job analysis. Almost everyone has applied for a job, and so you are already familiar with the most common screening device: the application form. This form elicits important information, such as an applicant's educational background, which the potential employer uses to quickly sort through the applicant pool. After the employer has examined the application forms, he will usually call potential employees in for interviews.

The unstructured personal interview is probably the most widely used screening device (Sackett, 2000). Despite its wide use, the unstructured interview is neither a reliable nor a valid predictor of future job performance. As with any other subjective measure, the biases of the interviewer affect his evaluation of the applicant. We can reduce the effects of biases and increase reliability and validity by using standardized procedures such as tests (van der Zee, Bakker, & Bakker, 2002). However, using unstructured interview information in addition to standardized test information can increase interviewers' overconfidence in their decisions and reduce the validity of their decisions (Kausel, Culbertson, & Madrid, 2016).

Standardized tests used for employment screening often measure cognitive and physical abilities and personality. Other tests might provide a prospective employer with a work sample. For example, if you apply for a cashier's position in a store, you will probably be tested on your arithmetic skills. Of all of these tests, it turns out that, for people with no experience, "the most valid predictor of future performance is general cognitive ability" (Hedge & Borman, 2006, p. 463).

No matter what screening method an employer uses, it is considered a "test" by U.S. law. This is because the applicants' responses to several critical items will determine whether the employer will consider them further. Under U.S. law, application questions and tests that are used as screening criteria must be valid indicators of future job performance. In the U.S., employers are bound by Title VII of the Civil Rights Act of 1964, which was revised in 1991 (Barrett, 2000). Title VII prohibits unfair hiring practices. An unfair hiring practice is one where screening of potential employees occurs on the basis of race, color, gender, religion, or national origin. Title VII was necessary to prevent employers from administering difficult, unnecessary exams to Black applicants for the purpose of removing them from the applicant pool, as was also done sometimes when Black citizens attempted to register to vote.

The Civil Rights Act has been expanded over the years. The Age Discrimination in Employment Act of 1967 made it illegal in the U.S. to discriminate on the basis of age, and the Americans with Disabilities Act of 1990, amended in 2008, did likewise for mental and physical disabilities, and pregnancy.

The Equal Employment Opportunity Commission (EEOC) established the first guidelines for fair hiring practices in 1966. Even with the Civil Rights Act in place, the EEOC was not legally empowered to enforce Title VII until 1972. The EEOC can now bring suit against employers for violating EEO laws. Discrimination laws similar to those in the U.S. exist for many high-income countries in the world.

A selection procedure "discriminates" or has adverse impact if it violates the "four-fifths rule." This rule is a general guideline which states that any selection procedure resulting in a hiring rate (of some percentage) for a majority group, such as white males, must result in a hiring rate no less than four-fifths of that percentage for an underrepresented minority group. If a selection procedure has adverse impact, the employer is required to show that the procedure is valid, that is, that it has "job relatedness." Furthermore, the law requires an affirmative action plan for larger organizations that outlines not only the organization's hiring plans but also its plans for recruiting underrepresented minorities.

Training

Rarely do new employees come to work completely trained. Usually, their employer will provide them with a training program that will ensure that the employees will become equipped with all the necessary skills to perform their jobs. How training is designed and delivered is of great importance and great concern in an organization.

We must also keep in mind that there is an alternative to designing and providing training programs. We could, instead, hire highly skilled applicants. However, highly skilled applicants require higher salaries and so increase cost. Furthermore, it may be difficult to find applicants with the specific skills required for the job we need done, and even if found, those applicants will often need at least some training to become familiar with all aspects of the specific job they are to perform. So, training programs will always play an important role in organizations.

The more effective training is, the better will be the workforce and the products the organization sells (Salas, Wilson, Priest, & Guthrie, 2006). The systems approach provides a strong framework for instruction and training (Goldstein & Ford, 2002). Using a systems approach to instructional design, we will rely on a needs assessment. This needs assessment will specify the potential learners, the necessary prerequisite knowledge or skills, and the instructional objectives. It will help us design training programs to achieve those objectives, provide criteria against which the effectiveness of the training can be evaluated, and allow us to assess the appropriateness of the training (Johnson & Proctor, 2017).

Many factors can influence the effectiveness of a training program. These factors include the variability of the conditions under which training occurs, the schedule with which training is administered, and the feedback that is provided, as discussed in Chapters 12 and 15. An additional factor is the skill of the trainer. How well does this person teach, motivate, and persuade the employee, and convince the employee that he or she will be successful at the job?

Training can occur on-the-job, on-site, or off-site. Each option has benefits and drawbacks, and we will discuss each in turn.

On-the-job training

Employees who receive on-the-job training are immediately productive, and no special training facilities are needed. However, mistakes made on the job by a trainee may have serious consequences, such as damage to equipment needed for production or personal injury.

Because on-the-job training is often informal, we might worry that the employee is not learning correct and safe procedures. There are a number of analyses we might use to evaluate an employee's learning progress, even in informal settings (Rothwell, 1999). Structured analyses similar to those performed for job analysis, job design, and training needs assessments can be performed for on-the-job training. The results from these analyses will determine whether or not the training program is effective and, if not, highlight changes to the program that we need to make to ensure that trainees are learning appropriate procedures. Overall, evidence indicates that on-the-job training is beneficial for both the workers and their employers, although which of several underlying processes contribute to that benefit remains unclear (De Grip & Sauermann, 2013).

Closely related to on-the-job training is job rotation, in which an employee moves from one work station or job to another on a periodic basis. Some organizations use job rotation to teach employees a range of tasks. This strategy helps ensure that the organization has a pool of employees who are broadly trained, and so operations will not be disrupted if a worker is absent or suddenly quits. In peacetime, the U.S. military employs a form of job rotation (rotational and operational reassignments) in which soldiers and civilian employees receive a new assignment every two or three years. According to the U.S. Department of Defense, this strategy ensures a flexible and quickly deployable source of "manpower" (Wolfowitz, 2005).

Job rotation can develop a flexible workforce whose members are familiar with many of the jobs critical to the functioning of the organization. Job rotation may also provide a way for organizations to learn about employees' productivities on different jobs and so help determine which employees best match which jobs (Ortega, 2001). For physically demanding jobs, it may also be an effective ergonomic intervention to reduce musculoskeletal injuries (Leider, Boschman, Frings-Dresen, & van der Molen, 2015). However, job rotation may be inappropriate when a high degree of skill is necessary to perform a specific job proficiently.

On-site and off-site training

On-site training occurs somewhere at the job site. The training area may be a room reserved for training purposes, or an entire facility constructed for training purposes, or it might be as informal as the employee lunch room.

On-site training is more controlled and systematic than on-the-job training. However, even though the new employee is drawing a salary, he or she is not immediately contributing to productivity. There may be an additional cost associated with designing and furnishing the training facility.

Off-site training takes place away from the job site. In some organizations, a technical school or university may be contracted to conduct the training sessions. For example, "continuing education" classes, which may be workshops lasting a day or two or classes that meet regularly for several months, give employees the opportunity to learn skills to improve their job performance and earn raises and promotions. Some professional and licensing organizations (such as individual state medical licensing boards and the American Medical Association) require professionals to routinely complete continuing education courses throughout their careers to maintain their license to practice their discipline.

Some employees may need more extended training. In this case, which may arise for people wanting to change their jobs, many companies encourage employees (by covering tuition or promising job advancement) to return to school to become certified or to receive advanced degrees. Distance education, for which employees take courses toward advanced degrees while remaining on-site or at home (Moore & Kearsley, 2012), is a more and more popular way to continue employee education, especially with the growing popularity of web-based courses.

PERFORMANCE APPRAISAL

Employee performance in any organization is evaluated both informally and formally. Informal evaluations occur continuously, forming the basis of an employee's acceptance by his or her peers, the manager's impressions of the employee, and the employee's own perception of "belonging to" the organization. Performance appraisal is the process of formally evaluating individual employee performance, and it is usually conducted in a structured and systematic way (Denisi & Smith, 2014). Performance appraisals provide feedback to both the employee and management. They give an employee the developmental information he needs to improve his performance, and they give management the evaluative information necessary for administrative decisions, such as promotions and raises in salary (Boswell & Boudreau, 2002).

A performance appraisal can have several positive outcomes. These include motivating the employee to perform better, providing a better understanding of the employee, clarifying job and performance expectancies, and enabling a fair determination of rewards (Mohrman, Resnick-West, & Lawler, 1989). However, if an employee perceives that the appraisal is unfair or manipulated by a manager to punish the employee, the employee may become less motivated and satisfied, or may even quit (Poon, 2004). Also, relationships among employees may deteriorate, and, if the problems in the evaluation process are widespread, some employees may even file lawsuits. Because performance appraisal can have devastating effects on an employee's position within an organization, and because an organization needs to have the best workforce possible, managers need to make every effort to ensure that performance appraisals are done well. However, developing an effective performance appraisal procedure is not easy.

An effective performance appraisal begins with an understanding of what it is that must be evaluated. Recall that a manager will have an impression of an employee before the formal performance appraisal, based on his or her daily interactions with the employee. This impression (or bias; see below) may be colored by factors that are not relevant to the employee's performance, such as the employee's appearance or personality quirks, which may or may not be relevant to the job. Managers must be very careful not to let these biases impinge on the formal performance appraisal, and to do their best to evaluate employees according to the factors that are relevant for job performance.

A company may choose to evaluate its employees on the results of job performance, such as the number of sales a person accomplished or the number of accidents in which he or she was involved (Spector, 2017). Evaluation on the basis of results sounds attractive, because managers can measure performance objectively, and the effects of any personal bias on the appraisal process will be minimized. However, objective measures may have serious limitations. Usually, objective measures like "number of sales" emphasize quantity of performance, rather than quality, which is more difficult to measure. Consider, for example, two employees whose jobs are to enter records into a data base. One employee completes a great many record entries, but these entries are all very easy. The second employee completes fewer than half of the entries completed by the first employee, but these entries were based on very complicated records that required much more problem-solving skill. Which is the more productive employee?

The results of an employee's performance may be subject to many factors outside of the employee's control. For example, a salesperson's poor sales record could be due to an impoverished local economy. For the data-entry jobs described above, there may not have been as many records this year as there were in previous years. This means that, while it may be appropriate to include some objective measures of performance results in an employee's appraisal, a fair appraisal will also need to use subjective measures of the employee's performance.

A good performance appraisal begins with the job description. An accurate job description gives a list of critical behaviors and/or responsibilities that can be used to determine whether an employee is functioning well. In other words, the description distinguishes employee behaviors that are relevant to the job from ones that are not. It also helps identify areas where performance might be improved or where additional training may be needed.

Once we have established the basis on which an employee is to be evaluated, we must decide who is to appraise the employee's performance and how the evaluation is to be made. Both decisions will depend on the organizational design and the purpose of the evaluation (Mohrman et al., 1989). Appraisers can be immediate supervisors, higher-level managers, the appraisees themselves, subordinates, and/or independent observers. Each appraiser brings a unique perspective that will emphasize certain types of performance information over other types. The choice of appraisal method depends on the reason for the evaluation. If our purpose is to select from a group of employees only a few to receive incentives or to be laid off, then we will use a comparative evaluation scheme in which employees are rank ordered. Usually, however, evaluations are individual, and our goal is to provide feedback that the employee can use to improve his or her performance.

There are several standardized rating scales that we can use for individual performance evaluations (Landy & Farr, 1980; Spector, 2017). The two most popular are the behaviorally anchored rating scale (BARS) and the behavior observation scale (BOS). A BARS performance assessment will typically use several individual scales. A specific "anchor" behavior is associated with each point on a BARS scale to allow the ratings to be "anchored" in behavior. Figure 18.1 shows an example BARS item for "cooperative behavior." We would rank an employee from 1 to 5 based on her behavior when asked to assist other employees. Associated with each point is a specific behavior, such as "This employee can be expected to help others if asked by supervisor, and it does not mean working overtime," which rates 3 points on the scale.

The BOS differs from the BARS in that the rater must give the frequency (percentage of time) with which the employee engages in different behaviors, and points are determined by how frequently each behavior occurred. Both the BARS and the BOS represent significant improvements over evaluation methods that have been used in the past, in large part due to their dependence on job descriptions and their decreased susceptibility to appraiser bias.

It is the specter of appraisal error, particularly rater bias, that haunts the process of performance appraisal. As we hinted above, an appraiser who brings his or her bias into the appraisal process risks evaluating an employee unfairly on irrelevant factors. This can potentially alienate

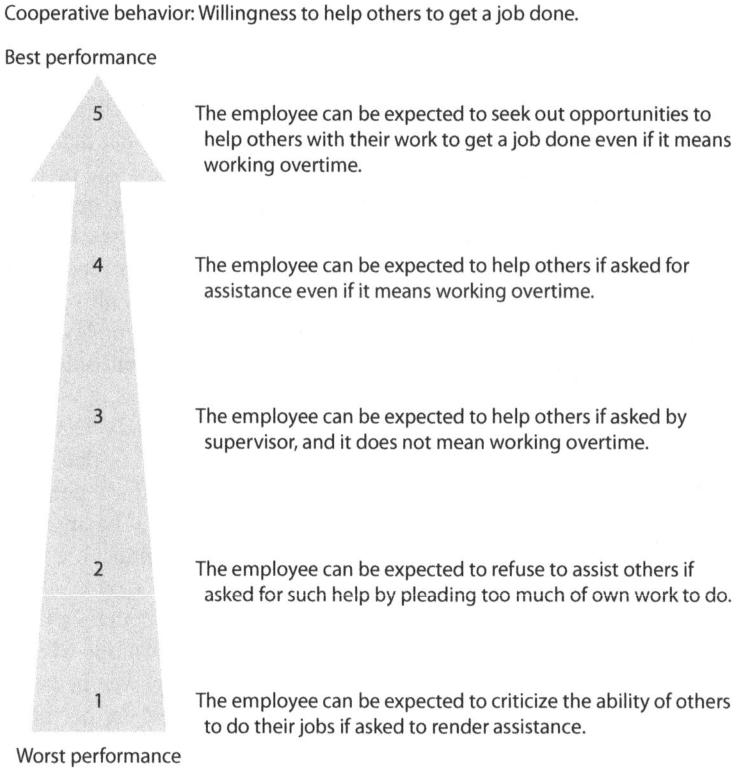

Cooperative behavior: Willingness to help others to get a job done.

Best performance

5   The employee can be expected to seek out opportunities to help others with their work to get a job done even if it means working overtime.

4   The employee can be expected to help others if asked for assistance even if it means working overtime.

3   The employee can be expected to help others if asked by supervisor, and it does not mean working overtime.

2   The employee can be expected to refuse to assist others if asked for such help by pleading too much of own work to do.

1   The employee can be expected to criticize the ability of others to do their jobs if asked to render assistance.

Worst performance

FIGURE 18.1 An item from a behaviorally anchored rating scale.

that employee and other employees who perceive the process as unfair. While rating scales like the BOS and BARS may reduce this kind of error, these rating instruments may also be biased, or they may play on the appraiser's biases. For example, the words "average" and "satisfactory" may mean different things to different appraisers. If a scale asks an appraiser whether some aspect of a person's performance has been below average, average, or above average, this ambiguity will result in very different ratings from different appraisers. A scale may also fail to accommodate all aspects of performance relevant to a job. For instance, a social worker who interacts with children may need to be nurturing and compassionate, but there may not be a scale for "nurturing behaviors" on the rating instrument. Conversely, a scale may be contaminated because it forces the evaluation of irrelevant aspects of performance. The social worker may be forced to work odd hours away from the office, but the scale may ask the appraiser to rate his punctuality or how frequently he interacts with other people in the office. Finally, a scale may be invalid in the same way that any standardized test can be invalid: it may not measure what it was intended to measure.

To some extent, appraisal bias is unavoidable. The appraiser brings a host of misperceptions, biases, and prejudices to the performance appraisal setting. She does not do this intentionally, but because she is subject to the same information-processing limitations that arise in every other situation requiring cognition and decision making. Biases have the effect of reducing the load on the appraiser's memory. Categorizing an employee as "bad" means that the appraiser does not have to struggle to evaluate each behavior that employee makes; they are all "bad." This kind of bias, called *halo error*, is the most pervasive. Halo error occurs when an appraiser evaluates mediocre (neither bad nor good) behavior by "bad" employees as bad, but that same behavior by "good" employees as good. Although appraiser bias cannot be completely eliminated, we can reduce it considerably by extensively training appraisers (Landy & Farr, 1980).

The social and organizational contexts in which the appraisal takes place can also contribute to bias and error (Breuer, Nieken, & Sliwka, 2013; Levy & Williams, 2004). The organizational climate and culture can affect not only the performance appraisal, but also how the appraisal process is conducted and the structure of the appraisal itself. For example, we have already discussed how important it is to evaluate an employee only on the job-relevant aspects of his or her performance. But a loose or informal management structure in an organization may lead to different managers expecting different things from the same employee, resulting in difficulties in determining how best to gather and organize appraisal data, and deciding what feedback to emphasize and what to downplay or ignore.

CIRCADIAN RHYTHMS AND WORK SCHEDULES

In 2004, Lewis Wolpert said, "It is no coincidence that most major human disasters, nuclear accidents like Chernobyl, shipwrecks or train crashes, occur in the middle of the night." When we work is an important factor in how effectively we work. An important component of job design is the development of effective work schedules. How well people work with different schedules will depend in part on their biological rhythms.

Circadian rhythms

Biological rhythms are the natural oscillations of the human body. In particular, circadian rhythms are oscillations with periods of approximately 24 hours (Refinetti, 2016). We can trace a person's circadian rhythms by examining his sleep/wake cycles and body temperature. Some circadian rhythms are endogenous, which means that they are internally driven, and others are exogenous, which means that they are externally driven.

We can track a person's endogenous rhythms from his or her body temperature, which is easy to measure and reliable. A person's body temperature is highest in the late evening. It will then decrease steadily until early morning, when it begins to rise throughout the day (see Figure 18.2). This cycle persists even when the person is in an environment without time cues, such as the day/night light cycle (in a polar research station, for example), though it will drift toward a slightly longer period of 24.1 hours (Mistlberger, 2003). Evidence suggests that the systems in the brain that drive the circadian rhythm in body temperature are used by other timing systems throughout the body (Buhr, Yoo, & Takahashi, 2010).

Performance on many tasks, such as those requiring manual dexterity (e.g., dealing cards), or inspection and monitoring, is positively correlated with the cyclical change in a person's body

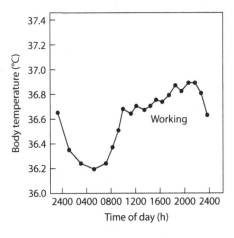

FIGURE 18.2 Body temperature throughout the day.

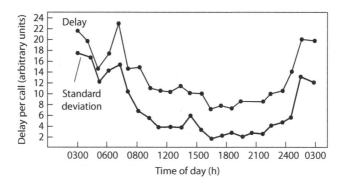

FIGURE 18.3 Delay to make a telephone connection, as a function of time of day.

temperature (Mallis & De Roshia, 2005). Performance levels have a very similar pattern that lags slightly behind the temperature cycle (Colquhoun, 1971). For example, Browne (1949) investigated the performance of switchboard operators over the course of 24 hours and found that it paralleled the circadian temperature rhythm (see Figure 18.3): Operators' performance improved during the day and then decreased at night.

Long-term memory performance also follows the circadian temperature cycle in a similar manner (Hasher, Goldstein, & May, 2005). However, short-term memory performance follows an opposite cycle: People do better early in the day, with performance decreasing to evening and then increasing during the night (Folkard & Monk, 1980; Waterhouse et al., 2001).

A person's performance is only partly a function of endogenous circadian rhythms. The performance of more complicated tasks that involve reasoning or decision making does not seem to follow the rhythm of body temperature (Folkard, 1975). When people are asked to complete tests of logical reasoning, they will get faster and faster from morning until early evening, when they will start to slow down. There is a period where both speed and body temperature increase together, but speed starts decreasing much sooner than body temperature. Regardless of a person's speed, accuracy does not seem to depend on the time of day.

Many of us have experienced jet lag. Jet lag occurs after long-distance flights that cross multiple time zones. Symptoms include difficulty sleeping and decreased performance (Waterhouse, Reilly, Atkinson, & Edwards, 2007). Similar symptoms arise when people try to sleep at an unusual time because of overwork or unexpected demands that kept them from sleeping at their usual time. For instance, if you had to work all night to finish a term paper, you may have tried to sleep during the day after you turned it in. Even though you might have been exhausted, you may have had trouble getting to sleep. It is your circadian rhythms that make it difficult to sleep during the day when your normal nighttime sleep pattern is disrupted.

A person accustomed to sleeping at night has endogenous rhythms that promote alertness during the day rather than drowsiness, and so these rhythms make sleep during the day difficult. Moreover, exogenous factors such as sunlight and increased noise can contribute to daytime sleep disturbance. Even when these factors are eliminated (by using blinds, earplugs, and so forth), people who try to sleep in the morning after their usual nighttime sleep is delayed will sleep for only 60% as long as they would have at night (Åkerstedt & Gillberg, 1982). Good sleep will happen only when it is aligned with the sleep phase of the circadian wake/sleep cycle.

When your standard day/night, wake/sleep cycle is disrupted for an extended period of time, your circadian rhythms will start to adjust. This is because external stimuli, such as the light cycle, entrain your internal circadian oscillations. Eventually you will adapt to the change in your work schedule or new time zone, but this change will take time. On average, it will take about one day per hour difference for your sleep patterns to adjust to a new schedule, but the effects of the change on your work performance may last much longer than that (see below). It will also take at least a week for your circadian temperature rhythm to adjust to an 8-hour phase delay of the type that occurs when a person starts working at night.

Work schedules

The standard work schedule in the U.S. is 5 days a week for 40 hours, and most people work during the day. Some people work more than this and some work less, and some people work strange hours. People who work more than 40 hours a week are on overtime schedules. An overtime schedule may have a person working more than 8 hours a day, or more than 5 days a week. In both cases, people experience fatigue, which can result in decreased productivity relative to that for a 40-hour week. In addition, when a person works extra hours on a given day, these hours typically will be during a phase of his circadian cycle during which his alertness and performance are depressed. This means that there is a tradeoff between a person's total productivity on an overtime schedule and the quality of the work he does during overtime hours. Although the absolute amount of productivity may increase with an overtime schedule, this occurs at a cost that may not justify the overtime.

Many people who work nights and evenings are shift workers. Shift workers are employed by organizations that must be in operation for longer than 8 hours per day. Many manufacturers operate continuously (24 hours a day), as do hospitals, law enforcement agencies, custodial services, and so on. Approximately 20% of the U.S. workforce is on shift work (Monk, 2003). Organizations in continuous operation usually employ at least three shifts of workers: morning, evening, and night shifts. Compared with the morning shift, accident rates are considerably higher for the evening shift, and even higher for the night shift (Caruso, 2012; Folkard & Lombardi, 2006).

Most people have a preference for one shift over another, but not everyone prefers, say, morning shifts over evening or night shifts. You probably already have a good idea of whether you are a "morning person" or not. These preferences determine not only the kind of shift you might prefer to work, but also how well you can do different tasks (Horne, Brass, & Pettitt, 1980). One experiment asked people classified as morning types and evening types to perform a vigilance task. The results are shown in Figure 18.4, together with their body temperatures throughout the day (from 7:00 a.m. to 12:00 a.m. or midnight). While the number of detected targets steadily increased throughout the day for evening types, it decreased throughout the day for morning types. The morning types had peaks in their endogenous circadian rhythms (measured by body temperature) earlier in the day than the evening types did.

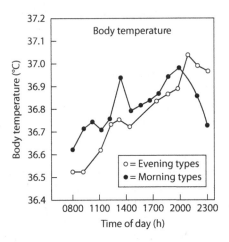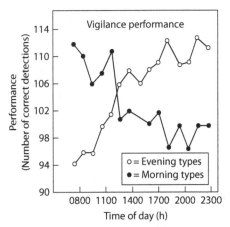

FIGURE 18.4 Body temperature and performance, as a function of the time of day, for subjects preferring morning or evening work.

Morning types may be less suited for shiftwork than evening types, because their rhythms do not adjust as easily as those of evening types (Dahlgren, 1988). Young adults show a stronger preference for evening activities, whereas older adults show a strong preference for morning activities (Hasher et al., 2005). This means that if you call yourself a night person now, you can expect to find yourself preferring earlier and earlier activities as you get older. This shift toward being a morning type with age may account in part for the fact that older adults have more difficulty adapting to shift work (Monk, 2003).

Given that it is difficult to adapt to a new schedules, many people generally dislike rotating shift schedules. In a rotating schedule, a person's days off are followed by a change to a different shift. The problem is that it often takes longer to adjust to a shift change than the time that a person is assigned to that shift (Hughes & Folkard, 1976). One study observed six people working in an Antarctic research station, who were asked to perform various simple tasks after different changes in their schedules. Even 10 days after an 8-hour change in the schedule, their performance rhythms had not yet adapted to match their pre-change rhythms. It may take as long as 21 days after changing to the new shift to completely adapt (Colquhoun, Blake, & Edwards, 1968).

There are a lot of reasons why adaptation to a new schedule takes a long time. Probably the most important reason is that the cycle of daylight and darkness only aligns with the day shift. Consequently, most people work the day shift, which means that the social environment is also structured for the day shift. However, if a person works a fixed schedule, even if it is the evening or night shift, her circadian system will usually adapt completely for that schedule.

Organizations in the U.S. that use shift rotations usually cycle the schedules on a weekly basis. The employees work for 4–7 days on one shift, then switch to another. If we consider how long it takes a person's circadian rhythms to adapt to a shift change, this is the worst possible rotation cycle. Weekly changes mean that a person's endogenous rhythms will always be in the process of adapting, and she will be working with chronic sleep deprivation. We could either rotate her shifts more quickly (1–2 days per shift) or more slowly (3–4 weeks per shift; Knauth & Hornberger, 2003). European organizations favor rapid rotation schedules (Monk, 2003). Rapid rotation maintains the worker's normal circadian cycle. She will experience some sleep deprivation, but not very much, because she can maintain her customary sleep schedule when she is on the day or evening shift, and on her days off.

In contrast, slower rotation schedules of 3–4 weeks allow a person to adapt completely to a new shift before it changes again. While the person spends a week out of each period adapting to the new schedule, there will be 2 or 3 weeks when she is working at her best regardless of the time of day. Slower rotations also tend to prevent any extreme sleep deprivation.

There are two other kinds of schedules we must present: flextime and compressed schedules. Each of these alternatives has its own costs and benefits. In organizations that need continuous operation, flextime and compressed schedules may work well, especially for a workforce that values flexibility.

We have already discussed how different people prefer different work schedules. For this reason, flextime is a popular alternative to the traditional shift schedule (Thompson, Payne, & Taylor, 2015). Flextime schedules allow a significant amount of variability in a person's working hours (Baltes et al., 1999) and sometimes the workplace. An employee is required to be on the job for some predetermined amount of time each day, for example, 8 hours, and must be on the job during some pre-designated interval, for example, 10:00 a.m.–2:00 p.m. This interval is called *core time*. The employee has control over all work time outside of core time.

Flextime has the benefit of allowing the employee to coordinate her or his personal needs with work responsibilities. This flexibility in scheduling may in turn reduce stress. For example, a study of commuters working in Atlanta, Georgia, found that those working with a flextime schedule reported less driver stress and time pressure from the commute (Lucas & Heady, 2002). As the employee is allowed to structure the work schedule so that he/she feels best, the employee's productivity may also increase.

Flextime is used primarily in organizations that are not involved in manufacturing. It is more difficult to allow flexibility like this when the operation of assembly lines and other continuous processes is at stake (Baltes et al., 1999). Because of the difficulties involved in coordinating flextime with continuous manufacturing processes, manufacturing organizations use compressed schedules most frequently as an alternative to traditional shift scheduling.

Compressed schedules require an employee to work 4 days a week for 10 hours a day (Baltes et al., 1999). The U.S. government provides an alternative compressed schedule in which, over a 2-week period, employees work eight 9-hour days and one 8-hour day, and then they have one "regular day" off. Many working couples take advantage of compressed schedules to reduce the amount of time their children need to spend in daycare. Employees who must commute long distances also benefit from the reduction in the time spent driving. Despite these and other potential advantages, fatigue may be a problem for employees working longer days. That is, a person's productivity or overall performance may be better on a 5-day rather than a 4-day schedule. It is also difficult to synchronize a compressed schedule with rotating shifts, unless the employee is cautious to maintain the same sleep schedule for days off as for days on.

The benefits of flextime and compressed schedules compared with traditional 5-day/8-hour work schedules can be examined scientifically (Baltes et al., 1999). Under flextime schedules, employees were absent fewer days and had higher productivity and higher reported satisfaction, both with their jobs and with their work schedule. Under compressed schedules, employees also reported higher satisfaction with their jobs and work schedules, but they were absent the same number of days and no more productive than they were on a traditional schedule.

Shift work is not easy. People doing shift work often deal with other problems that can affect their job performance (Monk, 1989; 2003). Employees on a fixed night shift are much more likely to try to work a second job, possibly out of financial need, than employees on the day shift. Employees working two jobs will be more tired and stressed, and their performance will suffer as a result. Shift schedules also introduce domestic and social problems. Employees may go for days without seeing their spouses and children. They may feel isolated from their families and community. If these domestic and social factors are not addressed by the employer, they can have a significant adverse impact on productivity.

One way an employer can combat the problems that arise from shift work is to provide adequate employee education and counseling (Monk, 1989; 2003). Employees can be taught how they can facilitate the adaptation of their circadian rhythms with their work schedules. For example, an employee who is working nights on a fixed or slowly rotating schedule needs to be able to identify and strengthen those habits that enable rapid change to and maintenance of a nocturnal orientation. Employees on the night shift also need to learn good "sleep hygiene": they should maintain a regular schedule of sleep, eating, and physical and social activity, but sleep during the day and be active at night, just as if they were at work.

Employees who learn good sleep hygiene will maximize the amount of sleep that they get, and a training program can emphasize the little tricks that people do who successfully sleep during the day. These tricks include installing heavy, light-blocking shades and curtains to block out sunlight and unplugging the telephone. It is also important that the employee avoids caffeine in the hours before bedtime, which can be difficult when the rest of the world is sitting down to their first cups of coffee. The employee's entire family has to be trained in the same way, so that they don't inadvertently sabotage the employee's efforts to sleep during the day.

The employee's family needs more training than just in sleep hygiene, however. Because of the domestic and social issues arising with shift work, the employee's family should be included in any broad training or counseling program. If the entire family is aware of potential domestic difficulties that accompany shiftwork and possible ways to cope with them, the employer can maximize the employee's chances for a healthy, productive, and satisfying work experience.

INTERACTIONS AMONG EMPLOYEES

In most organizations, employees must interact with at least a few other employees on a daily basis. Sometimes these interactions are minimal, but sometimes the employee may rely on dozens of other people every day.

The type of relationship two people share is often reflected in the distance that they preserve between them. The way that people manage the space around them is called *proxemics* (Hall, 1959; Harrigan, 2005). The study of proxemics emphasizes how people use the spaces around them and their distances from other people to convey social messages. As robots have become more common, interest has developed in human–robot proxemics, or the impact of physical and psychological distance in human–robot interactions (Mumm & Mutlu, 2011; Walters et al., 2009). Proxemics is important to human factors experts because a person's proximity to other people (or robots) will affect his or her levels of stress and aggression, and so also his or her performance. Some environmental design recommendations are based on the considerations of personal space, territoriality, and privacy (Oliver, 2002).

Personal Space

Personal space is an area surrounding a person's body that, when entered by another, gives rise to strong emotions (Sommer, 2002). The size of the personal space varies as a function of the type of social interaction and the nature of the relationship between the people involved. There are four levels of personal space, each having a near and a far phase: intimate distance, personal distance, social/consultative distance, and public distance (Hall, 1966; Harrigan, 2005).

Intimate distance varies from 0 to 45 cm around a person's body. The near phase of intimate distance is very close, from 0 to 15 cm, and usually involves body contact between the two people. The far phase is from 15 to 45 cm and is used by close friends. Personal distance varies from 45 to 120 cm—within arms' length. The near phase is from 45 to 75 cm and is the distance at which good friends converse. The far phase varies from 75 to 120 cm and is used for interactions between friends and acquaintances.

Social/consultative distance varies between 1.2 and 3.5 m. At this distance, no one expects to be touched. Business transactions or interactions between unacquainted people occur in the near phase, from 1.2 to 2.0 m. In the far phase, from 2.0 to 3.5 m, there is no sense of friendship, and interactions are more formal. Public distance is greater than 3.5 m of separation. This distance is characteristic of public speakers and their audience. People must raise their voices to communicate. The near phase, from 3.5 to 7.0 m, would be used perhaps by an instructor in the classroom, whereas the far phase, beyond 7.0 m, would be used by important public figures giving a speech.

When the near boundary of a person's space is violated by someone who is excluded from that space (by the nature of the relationship), the person usually experiences arousal and discomfort. The distance at which a person first experiences anxiety varies as a function of the nature of the interaction. It also is affected by cultural, psychological, and physical factors (Moser & Uzell, 2003). For example, some evidence suggests that older people with reduced mobility tend to have larger personal spaces (Webb & Weber, 2003).

Personal space can also be used as a cue about the nature of the relationship, both by the people interacting and by people watching the interaction. If Person A is unsure whether his acquaintance with Person B has passed into the friendship stage, he can use the distance between himself and Person B to help make this decision. Similarly, if Person A and Person B are almost touching, a third person watching them can interpret their relationship as an intimate one.

The personal distances maintained by members of a group will influence how well group members perform tasks. People will perform better when their distance from other group members is appropriate for the job the group is to accomplish. For example, if the group members are competing with each other, individuals perform better when the distance between them is greater than

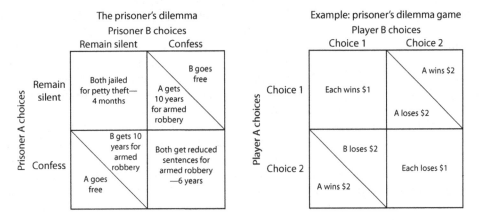

FIGURE 18.5   The prisoner's dilemma.

the personal distance (Seta, Paulus, & Schkade, 1976). Similarly, if the task requires cooperation, people do their tasks better when they are seated closer together, at the personal distance.

A game called the *Prisoner's Dilemma* (see Figure 18.5) is often used to study how people cooperate and compete. Two players are "prisoners" accused of a crime (robbery). Each player must decide whether he or she will confess to the police (implicating the other player) or remain silent, without knowing what the other player has decided to do. However, the best choice depends on what the other player has decided to do. If both players choose to confess, they both get reduced but significant sentences (6 years for armed robbery). If both players choose to remain silent, they both get minimal sentences (4 months for petty theft). However, if one person chooses to confess while the other person remains silent, the confessor will go free whereas the silent person will get a maximum sentence (10 years). The game can be easily translated into a monetary game, in which patterns of competition and cooperation result in monetary gains and losses for the two players.

We can use the Prisoner's Dilemma to explore how competitive and cooperative behavior evolves as a function of proximity and eye contact between the players (Gardin, Kaplan, Firestone, & Cowan, 1973). Cooperative behavior occurs more frequently when players are seated close together (side-by-side versus across a table). However, when the players can also see each other's eyes, more cooperative behavior actually occurs when seated across a table. Thus, when interpersonal separation exceeds the personal distance, we can still ensure a degree of cooperation by making sure they can maintain eye contact.

TERRITORIALITY

Territoriality refers to the behavior patterns that people exhibit when they are occupying and controlling a defined physical space, such as their homes or offices (Moser & Uzzell, 2003). We can extend the definition of territoriality beyond physical spaces to ideas and objects. Territorial behavior involves personalization or marking of property, the habitual occupation of a space, and in some circumstances, the defense of the space or objects. People also defend ideas with patents and copyright protections.

Territories are primary, secondary, or public, depending on the levels of privacy and accessibility allowed by each (Altman & Chemers, 1980). Primary territories are places like your home. You own and control it permanently. This place is central to your daily life. Secondary territories are more shared than primary territories, but you can control other people's access to them, at least to some extent. Your office or desk at work could be an example of either a primary or a secondary territory. Public territories are open to everyone, although some people may lose their access to them because of their inappropriate behavior or because they are being discriminated against. Public territories are characterized by rapid turnover of the people who use them.

A person might infringe on your territory by invading, violating, or contaminating it (Lyman & Scott, 1967). Invasion occurs when she enters your territory for the purpose of controlling it. Violation, which may be deliberate or accidental, occurs when she enters your territory only temporarily. Contamination occurs when she enters your area temporarily and leaves something unpleasant behind. Intruders differ in their styles of approach. They can use either an avoidant or an offensive style (Sommer, 1967). An avoidant style is deferential and nonconfrontational, whereas an offensive style is confrontational and direct.

You may defend your territory in two ways: prevention and reaction (Knapp, 1978). Prevention defenses, such as marking your property with your name, take place before any violation occurs. Reactions are defenses you make after an infringement and are usually physical. For example, the posting of a "no trespassing" sign is preventative, whereas ordering an intruder to leave your land at gunpoint is reactive. How intense your reaction will be depends on the territory that was violated. You will feel worst for infringements of your primary territory and least bad for infringements of public territory. When someone infringes on public territory, your response will probably be abandonment—you will leave and go somewhere else.

Your primary territory is important because it is where you feel safest and in control. As a designer, you can exploit this fact by creating workspaces that foster the perception of primary territory, and so create areas where people feel comfortable. Architectural features that demarcate distinct territories for individuals and groups can be built into homes, workplaces, and public places (Davis & Altman, 1976; Lennard & Lennard, 1977). Something as simple as allowing people to personalize their workspace can encourage self- and group-identities.

CROWDING AND PRIVACY

Personal space and territoriality can be viewed as ways that people achieve some degree of privacy. When someone violates your territory, this can be a source of considerable stress. Similarly, crowding, which can occur in institutions such as prisons (Lawrence & Andrews, 2004) and psychiatric wards (Kumar & Ng, 2001), as well as many other environments, can have a profound effect on your behavior. This happens because crowding leads to limitations on your territory and continuous, unavoidable violations of what little territory you have. We can link crime, poverty, and other societal ills to crowding.

Crowding is an experience that is associated with the density of people within a given area, similarly to the way the perception of color is associated with the wavelength of light. Whereas density is a measure of the number of people in an area, crowding is the perception of that density. Your perception of crowding is also based on your personality, characteristics of the physical and social settings, and your skills for coping with high density. The same density will lead different people to different perceptions of crowding in different settings. For example, culture has a strong influence on the perception of crowding. While no one likes being crowded, and there are in fact no consistent cultural differences in how well people tolerate being crowded, Vietnamese and Mexican Americans tend to perceive less crowding at the same densities than do African and Anglo-Americans (Evans, Lepore, & Allen, 2000).

There are two types of density: social and spatial (Gifford, 2014). When new people join a group, social density increases. When a group of people moves into a smaller space, spatial density increases. However, density is not the same thing as proximity, which is the distance between people. Crowding is directly related to the number of people in an area and inversely related to their distance (Knowles, 1983). When people are asked to rate their impressions of crowding of each photograph, relative to another, of groups that varied in size and distance, the resulting ratio scale values of crowding increased with the number of people in a slide and decreased with their distance (see Figure 18.6). These relationships can be quantified with a proximity index:

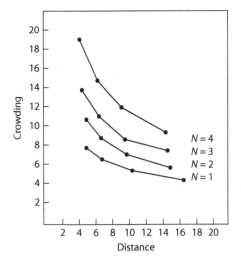

FIGURE 18.6 Estimates of crowding as a function of number of people (*N*) and their distance.

$$E_i = k\sqrt{\frac{N}{D}},$$

where:

E_i is the total energy of interaction at point *i*, or impression of crowding,

D is the distance of each person from point *i*, and

N is the size of the group.

In other words, crowding increased as the square root of group size and decreased as a square root of distance.

Crowding produces high levels of stress and arousal. We can measure these responses using blood pressure, the galvanic skin response, and sweating. As stress and arousal increase, these physiological measures increase. Experiments on crowding show that the level of stress that a person experiences in high density situations is a function of the size of her personal space (Aiello, DeRisis, Epstein, & Karlin, 1977). People who prefer large separations between themselves and others are more susceptible to stress in high density situations than those who prefer small separations (see Figure 18.7). This means that individual factors are important in determining whether high density will produce stress.

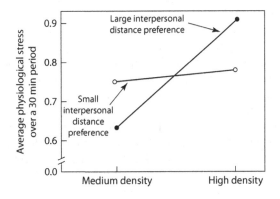

FIGURE 18.7 Physiological stress as a function of preferred distance and crowding.

The Yerkes–Dodson law, discussed in Chapter 9, describes a person's performance as an inverted U-shaped function of his or her arousal. To the extent that crowding influences arousal, it will impair performance. This effect is greatest for complex tasks (Baum & Paulus, 1987), because the performance of complex tasks suffers more than the performance of simpler tasks at high levels of arousal. One study asked people outside a supermarket during crowded and uncrowded times to complete a shopping list comprised of both physical and mental tasks (Bruins & Barber, 2000). Under crowded conditions, people were less able to perform the mental tasks than the physical tasks, probably because the mental tasks were more difficult. Another study found that when emergency departments in hospitals were crowded, there was an increase in inpatient deaths and small increases in costs and stay lengths for admitted patients (Sun et al., 2013).

People's behavior in response to crowding can be classified as either withdrawal or aggression. If the option is available, people will try to avoid crowding by escaping from crowded areas. When escape is impossible, a person may withdraw by "tuning out" others or attacking those perceived as responsible for their stress. In these circumstances, aggression is seen as a means to establish control. When a person believes she has no control over her environment, she may stop trying to cope or to improve her situation, choosing instead to passively accept the conditions. This reaction is called *learned helplessness*.

Many explanations of human responses to crowding emphasize how people feel as though they have lost control in high density situations (Baron & Rodin, 1978). We can distinguish four types of control that a person loses: decision, outcome, onset, and offset. Decision control is the person's ability to choose his or her own goals, and outcome control is the extent to which the attainment of these goals can be determined by the person's actions. Onset control is the extent to which exposure to the crowded situation is determined by the individual, and offset control is the person's ability to remove himself or herself from the crowded situation.

Social survey studies confirm that crowding directly influences a person's perception of control. Pandey (1999) had residents from high and low density areas of a large city fill out questionnaires asking them about crowding, perceived control, and health. The higher the reports of crowding, the lower control people perceived over their surroundings, and the higher were the rates of reported illness.

OFFICE SPACE AND ARRANGEMENT

Psychosocial factors, such as territoriality and crowding, are prominent in the workplace. We can anticipate problems that might arise from these factors by taking a macroergonomics approach to the analysis and design of offices (Robertson, 2006). Through appropriate design, we can take advantage of the benefits of these factors, avoid their negative consequences, and increase the quality of life at work (Vischer & Wifi, 2017).

We must first systematically evaluate a room or office design from the perspective of those people who will be using it (Harrigan, 1987). We will need to answer questions about the purpose of the room or building, the characteristics of the operations that will take place there, and the nature and frequency of information exchange between people and groups. Who will be using the facility, and what are their characteristics? How many people must it accommodate, and what circulation patterns will facilitate their movements through the space?

After acquiring information about the purpose of the structure, the tasks to be performed, and the users, we will use this information to determine design criteria and objectives. We will do space planning to determine the spaces needed, their size, and how they are arranged. Within the rooms and offices of the building, we must choose what furniture and equipment to provide, as well as the utilities that maintain an acceptable ambient environment. Depending on the size of the workplace, this kind of project will involve not only a human factors engineer but also a team of managers, engineers, human resources managers, designers, architects, and workers.

The office is a workplace where we can apply concepts of social interaction to design problems. The human factors evaluation begins with consideration of the office's purpose, the workers and other users, and the tasks to be performed in the office. Our goal in designing the office workplace is to make these tasks, and any related activities, as easy to complete as possible.

Facilitating the activities and tasks performed by the office workers is just one dimension of office design, which we refer to as *instrumentality* (Vilnai-Yavetz, Rafaelli, & Yaacov, 2005). There are two additional dimensions of importance: aesthetics and symbolism. Aesthetics refers to the perception of the office as pleasant or unpleasant. Symbolism is the dimension that refers to status and self-representation. A well-designed office should afford instrumentality, be aesthetically pleasing, and allow appropriate symbolic expression.

There are two kinds of offices: traditional and open. Traditional offices have fixed (floor-to-ceiling) walls and typically hold only a small number of workers. Such offices provide privacy and relatively low noise levels. Open offices have no floor-to-ceiling walls and may hold a very large number of workers. These offices do not provide much privacy, and the noise levels can be quite high.

Traditional offices

The primary human factors consideration in the design of traditional offices is the selection and placement of furnishings. One of the earliest studies of office design was published in 1966 by Propst, who reported the results of several years' investigation into the design and arrangement of office equipment. He obtained information from experts in several disciplines, studied the office patterns of workers that were considered exceptional, experimented with prototype offices, and tested several different office environments. As a result of his investigation, he emphasized the need for flexibility, while pointing out that most office plans of that time relied on oversimplified and restrictive concepts. He also argued that an office needs to be organized around an active individual rather than the stereotypic sedentary desk worker. The furniture and layouts that Propst designed have come to be known collectively as the *action office* (see Figure 18.8).

Propst claimed that the action office would not improve creativity and decision making, but rather, would facilitate fact-gathering and information-processing activities, and so make performance

FIGURE 18.8 The action office.

FIGURE 18.9   Bullpen office.

more efficient. Regardless of Propst's claims, formal evaluations of worker perceptions of the action office showed that the workers greatly preferred the action office design (Fucigna, 1967). After being switched from a standard office to the action office, workers felt that they were better organized and more efficient, that they could make more use of information, and that they were less likely to forget important things. Thus, the action office served the needs of its occupants better than the standard office.

Open-plan offices

An alternative to the traditional office is the open-plan office. The open office is intended to facilitate communication among workers and to provide more flexible use of space. However, this is at a cost of increased disturbances and distractions (Kim & de Dear, 2013). There are three kinds of open offices: bullpens, landscaped offices, and nonterritorial offices.

The bullpen office is the oldest open-plan office design. It has many desks arranged in rows and columns. Figure 18.9 shows an example of one of several such offices used at the 2004 Olympic Games in Athens. This arrangement allows a large number of people to occupy a limited space, while still allowing traffic flow and maintenance. However, most workers find a bullpen-style office dehumanizing. Employees of a Canadian company were shifted from traditional offices to open bullpens housing up to nine people (Brennan et al., 2002). The employees expressed deep dissatisfaction that did not abate with time. The employees felt that the lack of privacy actually decreased communication, rather than facilitating it.

A landscaped design does not arrange desks in rows, but groups desks and private offices according to their functions and the interactions of the employees (see Figure 18.10). The landscaped design uses movable barriers to provide greater privacy than in the bullpen design.

One study investigated the efficiency of the landscaped-style office by surveying employees who worked in a rectilinear, bullpen office about the productivity, group interaction, aesthetics, and environmental description of the office (Brookes & Kaplan, 1972). The office was then redesigned using a landscape plan, in which private offices and linear flow between desks were eliminated. Nine months after the office redesign, they surveyed the employees again with the

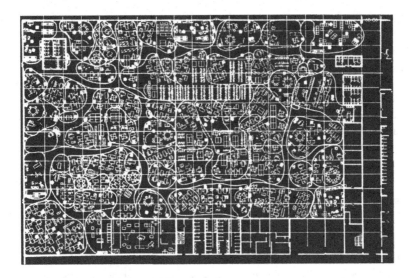

FIGURE 18.10 Blueprint of a landscaped office.

same questionnaire. Although the employees agreed that the new office looked much better than the old one, they did not judge it to work better. There was no increase in productivity, and the employees disliked the noise, lack of privacy, and visual distractions associated with the landscape plan.

This study illustrates that a major problem in any open-plan office is the presence of visual and auditory distractions. Table 18.3 gives several ways to control the influence of these distractions. We can easily prevent visual distractions by using barriers. Auditory distractions pose a more serious problem. Noise in open offices comes primarily from two sources: building services, such as the air conditioning system, and human activities (Tang, 1997). Clerical workers exposed to 3 hours of low-level noise typical of open offices showed physiological and behavioral effects indicative of increased stress levels (Evans & Johnson, 2000). Some research suggests that when

TABLE 18.3
Reducing Visual and Auditory Distractions in the Workplace

Install partitions and barriers between workstations that are at least 1.5 m high and 2.4 m wide and made of sound-absorbing materials to block both visual and auditory distractions. Glass should be avoided because of glare, acoustic transmission, and transparency.

Orient workstations "face-to-back" to reduce directional noise from conversation.

Use sound-absorbing materials with high noise-reduction coefficients for ceilings and walls.

Install noisy equipment such as copiers and paper shredders in a closed room exclusively dedicated to their use. Limit use of such equipment to certain times of the day.

Reduce the number of overhead lighting fixtures, which can increase glare and transmit noise, and increase the number of task lighting fixtures.

Install sound-proofed windows, or angle windows outward at the top to reflect sound toward a sound-absorbent ceiling.

Install sound-absorbing components such as heating and air conditioning diffusers, carpet, acoustic panels, and other work surfaces with high noise-reduction coefficients.

Use an acoustic noise-masking system that uses speakers to broadcast white noise throughout the work space, or provide portable noise-masking devices for each workstation.

open-plan offices fail, as in the study of Brookes and Kaplan (1972), it is because of noise reflected from hard ceilings (Turley, 1978). While there are some remedies for these kinds of problems, such as sound-absorbent material on ceilings and sound masking devices like white noise generators, we must consider potential problems associated with noise in the early design phases for open office spaces.

After noise issues, thermal discomfort is the next most common complaint of open-office occupants (Maula et al., 2016). Because of the larger open space, an individual cannot control the temperature at a specific workstation, with the consequence that some office occupants will be too hot or cold at essentially any office temperature. Windows in the office allow daylighting (which reduces energy costs) and a view of the outside environment, but this can produce not only temperature fluctuations throughout the office but also variations in lighting levels and glare, leading to significant visual discomfort (Konis, 2013). One solution may involve computerized smart systems to control the lighting and temperature levels across locations to maintain comfortable working conditions for people at all workspaces in the office (Konstantzos, Tzempelikos, & Chan, 2015).

Another style of open office is the nonterritorial office. In this type of office, employees are not assigned their own spaces. All work is performed at benches, tables, or desks. An employee may decide to work anywhere that suits him, but may also need to reserve the workspace in advance. One study looked at the effects of this layout on performance and communication within a product-engineering department (Allen & Gerstberger, 1973). Product engineers completed a questionnaire before and after the removal of office walls and permanently assigned workstations. The communication rate among department members increased over 50% in the nonterritorial office. Although performance levels did not change, the engineers preferred the nonterritorial office over the traditional office arrangement.

This preference for the nonterritorial office seems contradictory to the implications of territoriality research discussed earlier. It may arise from the collaborative nature of the research group, and we might not see it in settings where a high degree of interaction between employees was not required (Elsbach, 2003). Employees of a high-tech corporation that had implemented a new, nonterritorial work arrangement in most of its offices did not respond in the same way that the research group did. These employees felt that their workplace identities were threatened because they were not able to personalize an office work area.

The value of systematically considering the users in office design is illustrated by a case study reported by Dumesnil (1987). A small office housing two separate work activities, commercial designing and political consulting, needed to be redesigned. Four people worked in the office (three in commercial design and one in consulting), which was small and had ground and main levels (see Figure 18.11, left panel). The options for remodeling were restricted by a limited budget, the fact that no architectural changes could be made, and the desire to use most of the existing office furniture.

Dumesnil (1987) used unobtrusive observation and focused interviewing techniques at various times throughout the work schedule to determine social-behavioral problems with the existing office. These included (1) territorial confusion, (2) lack of privacy, leading to many nonwork-related distractions and difficulty in protecting the privacy of communications, (3) lack of definition of public and private territory, and (4) lack of personal space to maintain the appropriate interpersonal distances. She solved these problems by moving the political consultant from the reception desk on the main level to the ground level, installing tall modular cabinets as barriers between workstations, designating a new waiting area, and placing the reception area between it and the workstations. The location of the reception area provided a visual cue that kept visitors from entering the work areas. She also created separate, distinct conference areas for the designers and the political consultant. These changes eliminated conflicts over space, resulted in more task-oriented verbal interactions, and most importantly, increased productivity and user satisfaction.

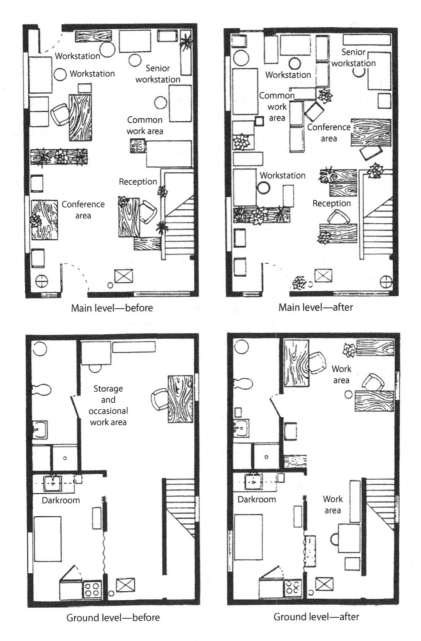

FIGURE 18.11 Office arrangements for the main and ground levels, before and after redesign.

INTERACTIONS BETWEEN ORGANIZATIONAL GROUPS

Employee performance is influenced not only by the design of jobs and the office environment, but also by the organizational environment. Such things as management style, benefit packages, and the opportunity for advancement all impact on an employee's feelings of well-being, company loyalty, and willingness to perform. A healthy organization possesses the following four attributes (Dettinger & Smith, 2006): (1) a clearly articulated set of goals that it strives to attain; (2) a culture of respect; (3) flexibility and the ability to respond to changing situations in an agile and efficient manner; (4) timely and capable decision making. Perhaps of primary importance is the way that the managers in the organization choose to communicate with the employees, and how employees communicate among themselves.

COMMUNICATION IN ORGANIZATIONS

Organizational communication refers to the transmission of information between two or more individuals or groups. Aside from compensation, communication probably has the greatest impact on job satisfaction. It is fundamental to all organizational operations. It is how organizations and teams within them achieve their goals (see Box 18.1). The transmission of information can use formal or informal channels, and the information may or may not be work-related.

To understand how information flows in an organization, it is important to understand the organizational hierarchy (see Figure 18.12). The topmost level of the hierarchy is the president or CEO of the company. The hierarchy proceeds through the levels of management down to the rank-and-file employees. Communication can go in three directions in this hierarchy: up, down, or horizontally. Upward communications are from subordinates to superiors and are used to inform or persuade. For example, a subordinate may say, "I have completed my assignment," or "I think we need more people on this project." Downward communications take place from superiors to subordinates and are also used to inform, as well as to command. For example, a superior may say, "Jane is in charge of this project," or "We've met our production goals for this month." Horizontal communication occurs across a single level in the hierarchy and is a means to influence coworkers and integrate information. For example, one employee might say to another, "I have finished Part A of the project, and I am waiting for you to finish Part B so that I can start on Part C."

Informal communications circumvent the organization's official communication protocols. This information often comes through the "grapevine," which may take the form of water-cooler or lunchtime conversations (e.g., Sutton & Porter, 1968). While manifestations and uses of the grapevine are classes of phenomena that receive a great deal of attention from social psychologists, our concern is with the formal flow of information through an organization; that is, the communication network.

Our focus on formal communications should not be interpreted to mean that informal communications are unimportant. Often people classify informal communications as merely rumor and gossip. Keep in mind that rumor and gossip are not necessarily harmful (Michelson & Mouly, 2004) and that informal communication is important for an organization to function effectively (von Bismarck, Bungard, Maslo, & Held, 2000). Consequently, we should probably devote more effort toward developing tools to facilitate informal communication.

The formal communication network is either centralized or decentralized. A centralized network is one where information comes from a single source to subgroups in the hierarchy and little communication occurs across those subgroups. A decentralized network is one with no single information source; subgroups communicate between each other and superiors. Centralized networks are effective when employees' tasks are simple and well-defined, whereas decentralized networks are more effective when employees must communicate with each other to solve problems (Jewell, 1998).

Generally speaking, knowledge is power. How information moves through a network conveys social messages. That is, individuals who are selected to receive information can be viewed as more important than those who are not selected. A superior may choose to "reward" particular subordinates by making them the recipients of information. Job satisfaction is directly related to the amount of information that an employee is receiving relative to that which he or she believes is needed to perform the job (O'Reilly, 1980). Thus, information dissemination should be done with care.

As information flows through a communication network, that information changes. These changes are a function of the direction the information flows through the network (Nichols, 1962). Information that flows downward from superiors to subordinates is often subjected to filtering, in which parts of the message are lost. In contrast, information that flows upward through the network is often subjected to distortions, in which parts of the message are changed. Similar distortions occur for information that flows horizontally through the network, often in the form of exaggeration. A message will not be transmitted through the organizational hierarchy with perfect fidelity.

BOX 18.1   TEAM PERFORMANCE AND GROUPWARE

You have probably worked or played on a team before. Formally, a team is "a distinguishable set of two or more people who interact, dynamically, interdependently, and adaptively toward a common and valued goal/objective/mission, who have each been assigned specific roles or functions to perform, and who have a limited life-span of membership" (Salas, Dickinson, Converse, & Tannenbaum, 1992, p. 4). Teamwork is important, because many people working together can accomplish more (and get it done better) than one person acting alone.

Many organizations rely on teams of skilled people to solve problems. These organizations realize that complex problems are more likely to be solved, and solved in new, creative ways, by bringing together different people with a wide range of knowledge and skills.

Teamwork is not just important for creative problem solving. Some complex systems, such as nuclear power plants and commercial aircraft, require teams to operate them. Some jobs, such as paving a road or performing a heart transplant, require many people performing different tasks to finish the job.

How do effective teams work to get things done? How do teams "think"? We know quite a bit now about team performance (Bowers, Jentsch, & Morgan, 2006) and the factors that influence it. We also know a bit about "team cognition" (Salas, Fiore, & Letsky, 2012), or how teams understand and think about the problems on which they are working. Some factors that influence team performance include the size of the team, the differences among the team membership, and the leadership structure within the team.

Regardless of the job, as the number of team members increases, communication and coordination between the members will become more difficult. The different attitudes and skill sets that team members bring to the group will also affect team performance: Homogeneous teams will be able to function together more effectively, but heterogeneity may promote creativity. Authority also influences team performance: Teams with structured hierarchies tend to perform better under time pressure than less formally structured teams. However, in all other aspects, the flexibility afforded by less formally structured teams typically leads to better team performance.

In Chapter 11 we discussed how individuals solve problems, and in Chapter 10 we discussed communication. To function as a team, a group of people must not only solve problems and communicate, but also coordinate with each other. The coordination problem highlights how important it is for each member of the team to share a common, accurate mental model of the task they must perform (e.g., Cooke & Gorman, 2006; Salas et al., 2012). The mental model allows the individual team members to maintain good situational awareness of the constantly changing conditions of the task (see Chapter 10). When all team members have an accurate representation of what has happened so far and what is supposed to happen next, they can effectively plan, communicate, and coordinate among themselves.

Team cognition is not the collection of the mental processes that each individual performs, but is something that emerges from the team members' interactions with each other (e.g., Kiekel & Cooke, 2011). Cooke, Gorman, Myers, and Duran (2013) argued that "teams are cognitive (dynamical) systems in which cognition emerges through interactions" (p. 255), a view that emphasizes how team cognition is an activity that should be studied at the team level. Working within a group can change how a team member thinks and feels about different issues, and this in turn can change how she processes information in the course of her teamwork. Social psychologists have studied these kinds of changes, and you can read about them in any introductory psychology textbook: the "risky shift" and "groupthink" are two well-studied phenomena arising from the interactions between members of a group. While human factors experts are well-trained and quite familiar with design issues that arise as a result of how individuals think and learn, they must also consider the cognition and performance of the team.

Computer technology and the Internet have changed how team members interact with each other. No longer do all team members have to work out of the same office or even the same city. "Virtual teams" are teams whose members may be located anywhere around the world (Gilson, Maynard, Young, Vartiainen, & Hakonen, 2015). Virtual teams collaborate with the help of applications generically referred to as *groupware* (e.g., Andriessen, 2012) or *Teamware* (e.g., van Tilburg & Briggs, 2005). Groupware consists of a collection of tools that help various aspects of teamwork, particularly communication, information exchange, and decision making. While many of these tools are the ones that computer users rely on every day, like electronic mail or instant messaging, groupware also gives each team member an interface to shared data and a common computer environment (e.g., Bontcheva et al., 2013).

There are three categories of groupware applications: those that facilitate communication, conferencing, and coordination. Communication tools like e-mail, instant messaging, Dropbox, and Google Docs permit team members to send files and data and other information to each other. Conferencing tools allow people to work together despite wide geographic distances. For example, there are several applications, such as Citrix® GoToMeeting™, Cisco WebEx, and Microsoft SharePoint™, which make use of the Web to hold online meetings. Many people can log in from their personal computers and meet together, watch and prepare presentations, and work on the same document.

Each of these collaboration applications permits the team to customize the tools for specific team needs. Team members can be assigned different roles (reader, contributor, designer, administrator), and the application will provide appropriate tools for different user roles.

Collaboration software can accomplish more than the editing of documents or showing presentations. In September of 2005, two bioengineers from the University of California Irvine reported that they had successfully performed microscopic laser surgery on cells in their University of California Irvine lab using a robotic microscope. This in itself was unremarkable, as such surgeries have been performed routinely for many decades. What is remarkable about their achievement is that they were in Brisbane, Australia at the time (Botvinick & Berns, 2005). Using a popular collaborative tool (LogMeIn, Inc.), they logged in to their lab computer (once from their hotel room), prepared slides and performed the surgeries online with no time delays. The tasks they performed required them to "trap" moving cells, so had there been any problems with connectivity or transmission, their experiments would have failed.

Despite successes like these, some critics complain that groupware has not lived up to its promise (Driskell & Salas, 2006). Almost anyone who has engaged in online conferences will have experienced technical failures, difficulty in comprehending speech, and so on. Grudin (2002) illustrates the drawbacks of online conferences by noting that although a 1990 meeting on the future of groupware was held in an electronic meeting room, no digital technology was used for a 2001 meeting on the same topic. The implication is that even the advocates of groupware find it limiting. These critics suggest that the development of improved groupware applications requires more consideration of group dynamics and the issues involved in team performance and cognition.

A study by Antunes, Ferreira, Zurita, and Baloion (2011) provides an illustration of how this can be done. They had computer science students carry out group design projects requiring the use of Google Maps in collaborative mode to identify places where a technological application might be beneficial. Perhaps not too surprisingly, the number of difficulties encountered by group members increased as group size increased, and the authors concluded that Google Maps improved collaborative activities primarily in small groups. However, even participants in larger groups perceived the collaboration enabled by Google Maps as useful for generating diverse ideas, providing feedback about ideas, and so on.

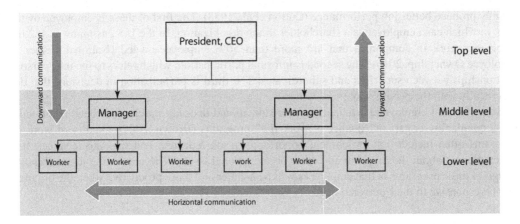

FIGURE 18.12 An organizational hierarchy, with arrows indicating upward, downward, and horizontal communication.

We have already mentioned that the effectiveness of centralized and decentralized communication networks depends on the jobs the employees are performing. It is also true that both job performance and satisfaction are correlated with the communication network (Shaw, 1981). Better performance is associated with centralized networks when the tasks making up the job are simple. In contrast, when the tasks making up the job are complicated, a decentralized network leads to better performance. Job satisfaction is better under decentralized communication networks, but people who report the highest levels of job satisfaction under either kind of network are those people who are the source of information.

Many organizations make use of "virtual workplaces" where employees work from various locations, often using flextime, away from their employers' offices. You might expect that communications would be more difficult under these conditions, and, consequently, employees might be less satisfied with their organization's communication network. However, "virtual" office workers report higher levels of satisfaction with the communication climate than do traditional office workers (Akkirman & Harris, 2005). So, even though the potential for communication "breakdown" exists in the virtual office, proper design of communication structure and practices apparently is sufficient to prevent communication problems.

EMPLOYEE PARTICIPATION

An organization's communication network will depend on the managerial style. There are four such styles: exploitative authoritative, benevolent authoritative, consultative, and participative (Likert, 1961). For the authoritative styles, organizational decisions are made at the higher levels of the hierarchy, and information flow is downward. Whereas an exploitative authority generally disregards suggestions from the lower levels of the hierarchy and uses fear to motivate employees, a benevolent authority takes into account suggestions of subordinates and uses both rewards and punishments for motivation. The consultative style allows some decision making by employees lower in the hierarchy and uses rewards as the primary source of motivation. With the participative style, employees throughout all levels of the organization are involved in the organization's decisions. Motivation under participative management uses economic rewards for employees who contribute to positive organizational changes.

Employee participation, sometimes called *participatory ergonomics*, is one of the hallmarks of macroergonomics (Brown, 2002). Employee satisfaction will usually be highest and performance will usually be best with the participative style of management, because people are more accepting of decisions which they have been involved in making. However, only certain forms of participation reliably produce better job performance (Cotton et al., 1988). The first of these is employee ownership, in which each employee is a shareholder in the organization. In the U.S., as many as 182,000 employees work in companies that are more than 50% employee-owned (National Center for Employee Ownership, 2016). The second is informal participation, which refers to the interpersonal relationships between superiors and subordinates. The third is participation in decisions that deal specifically with the person's job.

Some form of employee participation in organizational decision making has been instituted in many companies as part of "quality of work life" programs (Brown, 2002). Quality of work life programs often include job enrichment opportunities, job redesign, and ongoing feedback from the employees about the program and other organizational issues. The motivating idea behind the design of these programs is that employees will be happier and more productive when the organization is responsive to their personal needs.

ORGANIZATIONAL DEVELOPMENT

In the course of normal operations, managers often discover flaws in their organization's policies that adversely impact productivity and profits. Keeping the profit margin in mind, every organization has a structure and a set of goals. Associated with its structure and goals is the organization's effectiveness, or how well it achieves its goals. There are many measures of organizational effectiveness, such as the aforementioned profitability, and others such as stability. Organizational development is the improvement of organizational effectiveness through the deliberate change of structure and goals.

Organizational structure has three components (Hendrick & Kleiner, 2001; Robbins, 1983): complexity, formalization, and centralization. Complexity is the level of differentiation of the organization's activities; for example, the number of divisions, and the way that information is passed from one division to another. Vertical differentiation is the number of levels in the hierarchy between the chief executive and the employees directly responsible for the organization's production. Horizontal differentiation is the degree of specialization within a level of a hierarchy and the number of departments and divisions. For example, some universities have a single Arts and Sciences college, whereas others have a Liberal Arts college and a Science college. At the college level, the former would have less differentiation than the latter. Increases in differentiation produce increases in organizational complexity. The organization's communication network integrates the different divisions and levels in the organization's hierarchy.

Formalization refers to the rules and procedures that guide the behavior of the people in the organization. The more formalized an organization, the more standardized are its procedures. Generally, the higher the level of training the employees have, the less formalized the organizational structures are. An organization (e.g., a hospital) hires highly trained employees (physicians and nurses) for their problem-solving skills, and the organizational structure must be flexible and informal enough to let them exercise the skills for which they were hired.

Centralization is the degree to which authority is distributed through the organizational hierarchy. Does authority originate at a single level, or is it distributed across lower levels? The optimal degree of centralization will vary as a function of such things as the predictability of the organizational environment and the amount of coordinated, strategic planning that is needed to meet the organization's goals. A hospital, for example, is a fairly unpredictable place, and so there are layers of authority throughout the organization: the nursing staff, the hierarchy of physicians, housekeeping, dietary staff, and so forth. A factory, on the other hand, is very predictable, and so authority is concentrated in the management staff.

Whereas the structure of an organization defines its rules of operation, goals define what the organization is trying to achieve. Goals differ according to their time frame, focus, and criteria (Szilagyi & Wallace, 1983). Goals may be short, intermediate, or long term, and the action taken to achieve a goal may be one of maintenance, improvement, or development. Goals may include such things as increased productivity, improved resources and innovation, and greater profitability.

Organizational development is a change in structure and goals to improve organizational effectiveness. The process of change involves:

1. Identifying the system's purpose or goals;
2. Making explicit the relevant measures of organizational effectiveness; weighting them, and subsequently utilizing these organizational effectiveness measures as criteria for evaluating feasible alternative structures;
3. Systematically developing the design of the three major components of organizational structure;
4. Systematically considering the system's technology, psychosocial, and relevant external environment variables as moderators of organizational structure; and
5. Deciding on the general type of organizational structure for the system. (Hendrick, 1987, p. 472)

Although organizational development is usually initiated by top-level management, the agent of change is very often an outside management consultant. The immediate stimulus for change is often a problem, such as high turnover rates among employees or poor labor/management relations. However, change may also arise from the success of the organization; say, through the need to reorganize as the organization grows and expands.

Organizational development proceeds in much the same manner as the development and evaluation of any other system. We follow a series of steps from the initial perception of a problem or opportunity to the implementation and assessment of change. Figure 18.13 shows an ideal model of organizational development proposed by Lewin (1951). After the organization perceives and diagnoses a problem, it must develop a plan for implementing specific changes. After implementing the plan, the organization must collect data to evaluate the impact of the changes and use this information for further planning.

As Jewell (1998) notes, this model requires a great many resources that many organizations may lack, such as time, money, and outside expertise. It also fosters a dependency of the organization on the change agent, whoever that may be. Thus, in many cases, this model may not be adhered to rigidly or may not even be appropriate.

SUMMARY

The performance and productivity of any organization are a function of how well the organization manages human resources. In this chapter, we have discussed social and managerial issues that bear either directly or indirectly on employee well-being. This in turn influences productivity, which determines whether or not an organization will effectively achieve its goals.

We began by discussing job analysis, which provides a description of the tasks that a worker must perform. It can serve as the basis for job design, employee selection, and training, among other things. Implementing work schedules is part of job design. Work schedules interact with performance and job satisfaction. There are a variety of schedules that can be considered for use in specific situations and, when possible, adjusted to fit worker preferences.

The design of workplaces depends on concepts we referred to collectively as macroergonomics. These are the social influences that play an important role in job satisfaction and performance. We can design good workplaces by accommodating the types of social interactions that will occur in them. We also must consider the organizational structure. The way that information is transmitted to individuals in the organization, and the extent to which employees are involved in organizational decisions, have a direct impact on organizational effectiveness. If we view the entire organization from the systems perspective, we ensure that these macroergonomic factors, as well as such microergonomic factors as workspace design, are given appropriate consideration. As human factors experts possessing sophisticated knowledge about human–machine systems, we are in a unique position to aid in organizational development and job design.

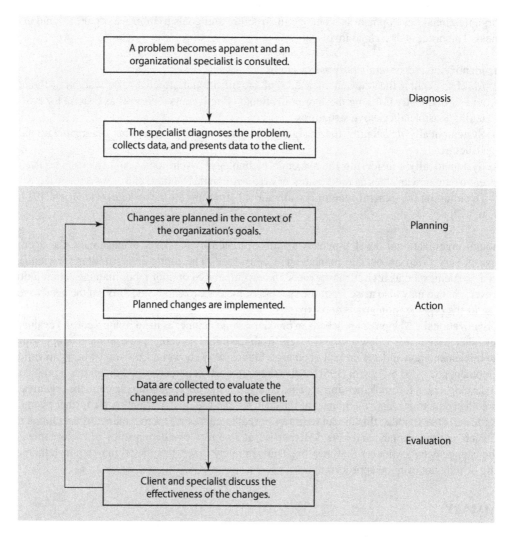

FIGURE 18.13 Stages in organizational development.

RECOMMENDED READINGS

Burke, W. W. (2013). *Organization Change: Theory and Practice* (3rd ed.). Thousand Oaks, CA: Sage.

Gifford, R. (2014). *Environmental Psychology: Principles and Practice* (5th ed.). Colville, WA: Optimal Books.

Goldstein, I. L., & Ford, J. K. (2002). *Training in Organizations* (4th ed.). Belmont, CA: Wadsworth.

Grandjean, E. (1987). *Ergonomics in Computerized Offices*. London: Taylor & Francis.

Foster, R. G., & Kreitzman, L. (2004). *Rhythms of Life: The Biological Clocks that Control the Daily Lives of Every Living Thing*. New Haven, CT: Yale University Press.

Hendrick, H. W., & Kleiner, B. M. (2001). *Macroergonomics: An Introduction to Work System Design*. Santa Monica, CA: Human Factors and Ergonomics Society.

Smither, R. D. (1998). *The Psychology of Work and Human Performance* (3rd ed.). Boston, MA: Addison-Wesley.

19 The Practice of Human Factors

We do not have to experience confusion or suffer from undiscovered errors. Proper design can make a difference in our quality of life ... Now you are on your own. If you are a designer, help fight the battle for usability. If you are a user, then join your voice with those who cry for usable products.

D. Norman
2013

INTRODUCTION

Our fundamental premise is that system performance, safety, and satisfaction can be improved by designing for human use. Objects as simple as a hammer or as complex as heavy construction equipment, or the complicated interactions arising between people and machines on a factory floor, or between people and their electronic devices, can benefit from a human factors analysis. Armchair evaluations, or "common sense" approaches, to most of the design issues discussed in this book will not ensure ergonomically appropriate designs. If common sense were all that was necessary to design safe and usable products, then everyone would be able to use their Blu-ray and DVD players, pilot error would not be the cause of many air-traffic accidents, secretaries would not complain about their computer workstations, and there would be no human factors science and profession.

Physical and psychological aspects of human performance in laboratory and work environments have been studied for more than 150 years. Consequently, we know a lot about factors that influence human performance and methods for evaluating performance under many different conditions. In this book, we have examined perceptual, cognitive, and motoric aspects of performance as well as some environmental and social factors, retaining throughout the conception of the human as an information-processing system. The value of this viewpoint is that both the human and machine components within the larger system can be evaluated using similar criteria.

The system concept is a framework for studying the influence of design variables (Pew & Mavor, 2007). Within this framework we evaluate the performance of the components, as well as overall system performance, relative to the system goals. Without such a framework, human factors would consist of an uncountable number of unrelated facts, and the way we apply these facts to specific design problems would be unclear. We would know that users prefer entering data with one software interface rather than another, or that operators of a control panel respond faster and more accurately when a critical switch is located on the left rather than the right. However, we would be unable to use this information to generalize to novel tasks or environments. Each time a new design problem surfaced, we would have to begin from scratch.

The body of design-related knowledge provided by human factors research, called *human–systems interface technology*, can be divided into five categories (Hendrick, 2000):

- Human–machine interface technology: design of interfaces for a range of human–machine systems to enhance usability and safety;
- Human–environment technology: design of physical environments to enhance comfort and performance;
- Human–software technology: design of computer software for compatibility with human cognitive capabilities;
- Human–job interface technology: designing work and jobs to improve performance and productivity;
- Human–organization interface technology: a sociotechnological systems approach in which the larger organizational system in which a person operates is taken into consideration.

We take the scientific knowledge behind each of these technologies and apply it to specific design problems. You now have read a lot about these kinds of problems and the techniques used to solve them. You also probably realize now that designing for human use requires contributions from many different people with different points of view. In fact, human factors/ergonomics is "a multidisciplinary endeavor that involves the design and engineering of systems for human use" (Dempsey, Wogalter, & Hancock, 2000, p. 6). In this final chapter, we will examine the issues that arise in the practice of human factors, and the interactions that human factors specialists have with other members of a design team.

The human factors specialist plays, or should play, an active role in many stages of the development process for systems and products (McBride & Newbold, 2016; Meister & Enderwick, 2002). Often the first step in this process is convincing management that the benefits of human factors analyses outweigh their costs. When everyone agrees that such analyses are necessary, the human factors expert needs to be careful not to make vague prescriptions, such as "Don't put the control too far away." When possible, she must provide the other members of the design team with quantitative predictions of performance for different design alternatives, and this is not a trivial task, as we have seen. The most detailed model of human performance may not be formulated for the application that is the target of the design team. Consequently, the human factors expert can use an engineering model, developed to make "ballpark" predictions for specific applications, or develop a more refined prediction from an integrative cognitive framework.

After the design phase is over and products are ready to go to market, the organization that made them will be concerned with safety and liability. If the product causes an accident or injury, or if a consumer is using the product when an accident or injury occurs, the organization may be held liable. Litigation may arise over issues of usability engineering, such as whether the product presented an unreasonable hazard to the user while performing the task for which it was intended.

We will examine each of these issues in turn in this chapter.

SYSTEM DEVELOPMENT

Although interest in understanding the role of humans in systems and accommodating that role in design has a history of more than 60 years, there has been a continuing concern that, in each phase of development, the human element is not sufficiently considered along with hardware and software elements. When information about the performance characteristics and preferences of operators and users is not introduced early enough in the process, there are higher risks for both simple and catastrophic failures in the systems that are produced.

R. W. Pew & A. S. Mavor
2007

MAKING THE CASE FOR HUMAN FACTORS

The consideration of the human element early in each phase of the design process advocated by Pew and Mavor (2007), consideration that is necessary to ensure that a system will operate as intended, is often not obvious to design team members who are not human factors specialists. The human factors specialists will have to convince managers, engineers, and other organizational authorities that the money invested in a human factors program is well spent. This will not always be easy: the costs of human factors analyses in both time and resources are readily apparent to management, but the benefits are often not as immediate and, in some cases, are difficult to express in tangible monetary values (Rensink & van Uden, 2006; Rouse & Boff, 2012). However, it should be obvious by now that human factors analyses improve safety and performance, which in turn translates into financial gains (Karat, 2005). Benefits arise from both improvements in equipment, facilities, and procedures within the organization, and improved usability of products produced by the organization.

An ergonomics program within an organization can increase productivity and decrease overhead, increase reliability, reduce maintenance costs, and increase safety, as well as improve employees' work satisfaction and health (Dillon, 2006; Rensink & van Uden, 2006). An ergonomic approach to product design reduces cost by identifying design problems in the early development stages, before a product is developed and tested. The final product will have reduced training expenses, greater user productivity and acceptance, reduced user errors, and decreased maintenance and support costs (e.g., Marcus, 2005). The benefits of ensuring usability can be particularly substantial for design of website, because poorly designed sites will force users to use a competitor's instead (Mayhew & Bias, 2003; Richeson, Bertus, Bias, & Tate, 2011).

Making the case for early consideration of human factors is easier when human factors specialists understand the perspective of managers in an organization and how human factors relates to the organization's strategic goals (Village, Salustri, & Neumann, 2013). A cost-benefit analysis is an effective way to communicate with management and convince them of the need to support ergonomics programs and usability engineering. There are several ways to conduct such an analysis (Rouse & Boff, 2012). The results of this analysis can then be presented in terms of the amount of money that the company will save through supporting such programs, an approach to argument based on return on investment (ROI; Richeson et al., 2011).

This approach was used by human factors engineers at the Shell Netherlands Refinery and Chemical complex. They developed a systematic cost-benefit method for helping to determine the costs and benefits for an ergonomically sound plant design (Rensink & van Uden, 2006). The method allowed designers "to visualize the potential benefits of ergonomic design and to serve as an aid to process technicians, human factors engineers and project managers who have to make decisions about the design in new construction of improvement projects" (Rensink & van Uden, 2006, p. 2582). The procedure resulted in a table with labeled columns for each of eight areas that may benefit from improved design (e.g., worker health). The table rows are specific benefits that may yield gains for more than one of the areas (e.g., reducing on-the-job accident rates; see Figure 19.1). With such a table it was easy to see the impact that a specific benefit might have across a range of areas, which in turn made it easier to assign values to the benefits. This meant that designers and managers could more easily identify cost savings and intangible benefits for safety and health.

By performing cost-benefit analyses, human factors specialists not only justify their own funding, but also:

1. Educate themselves and others about the financial contribution of human factors,
2. Support product management decisions about the allocation of resources to groups competing for resources on a product,
3. Communicate with product managers and senior management about shared fiscal and organizational goals,
4. Support other groups within the organization (e.g., marketing, education, maintenance), and
5. Get feedback about how to improve the use of human factors resource from individual to organization-wide levels. (Karat, 1992, 2005)

There are three steps to a basic cost-benefit analysis (Karat, 2005). First, identify those variables associated with costs and benefits for a project, along with their associated values (Richeson et al., 2011). Costs include salaries for human factors personnel, building and equipping facilities for conducting usability tests, payments for participants in the tests, and so on. Benefits, as noted, include things such as reduced training time and increased sales. Second, analyze the relationship between the costs and benefits, estimating the ROI for the cost. At this stage, you may develop a number of alternative usability proposals and compare them with each other. Finally, someone must decide how much money and resources will be invested in the human factors analysis, and the ROI provides a metric that anyone can understand.

Estimating the costs and benefits associated with implementing an occupational ergonomics program or a human factors program in system and product development is difficult (Beevis, 2003).

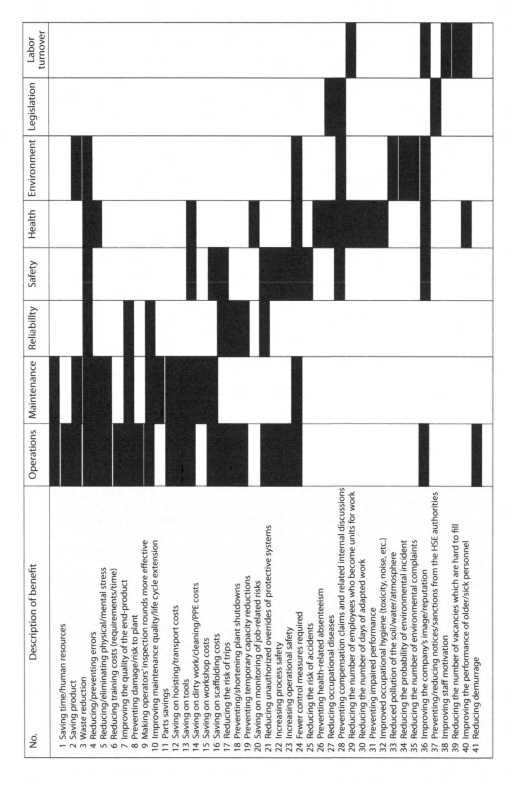

FIGURE 19.1 Cross-reference quick-scan benefit table.

However, it can be done effectively and, as illustrated below, will usually indicate the value of the program.

Occupational Ergonomics Programs

We will assess many different cost outcomes when we implement an ergonomics program for redesigning work (Macy & Mirvis, 1976). These include absenteeism, labor turnover, tardiness, human error, accidents, grievances and disputes, learning rate, productivity rate, theft and sabotage, inefficiency or yields, cooperative activities, and maintenance. Good work conditions also provide commercial and personal benefits by promoting increased comfort, satisfaction, and positive attitudes toward work (Corlett, 1988).

A case study of the implementation of an ergonomics program at a wood processing plant illustrates the benefits derived from such a program (Lahiri, Gold, & Levenstein, 2005). The company conducted ergonomic evaluations for the jobs performed by forklift operators, machine operators, crane operators, technicians, and general production workers. Based on these evaluations, they introduced workstation modifications that included adjustable chairs, lift tables, conveyors, grabbers, floor matting, and catwalks to be used instead of ladders. In addition, the company hired a physical therapist to teach the workers exercises that would help prevent musculoskeletal disorders. The company reported such benefits as a reduction in the number of cases of lower back pain (and the resulting loss of productive work time) and a 10% increase in productivity for all workers. They estimated the total financial benefit to be approximately 15 times greater than the cost of the program.

Another company reported similar benefits. A leading brewery instituted a manual handling ergonomics program for beer delivery personnel (Butler, 2003). Most of their beer deliverers start working for the company while in their 20s and work continuously until retirement. The daily load of beer deliveries that each person handles is large, and it has increased over the years as the company has added products to its offerings. The heavy physical demands of the delivery job force the delivery personnel to retire at the very early age of 45.

The development and implementation of a manual handling ergonomics program began in 1991. The company conducted task analyses for all manual materials activities performed by the delivery personnel, and they examined delivery sites for possible physical changes that would reduce handling difficulty. Some of the changes they implemented included lowering the loading height of the beer delivery vehicles and supplying cellar lifts for sites where the beer had to be lowered into a cellar. They also developed a training program. Everyone involved in beer delivery received 3 days of training on proper lifting and handling methods. The company reported a substantial decrease in work-related insurance claims and manual handling accident rates. They estimated the costs of the ergonomics program at $37,500, whereas the benefits were approximately $2.4 million.

Many companies may not have the resources for an occupational ergonomics program. However, some providers of office equipment supply their customers with ergonomic services at the time that the equipment is purchased (IBM, 2006; Sluchak, 1990). These include consultation about workstation design, assistance in conducting in-store evaluations of equipment, assistance in implementing employee training programs, recommendations for design of interfaces and website, and briefings on topics such as cumulative trauma disorders.

System and Product Development

The costs associated with incorporation of human factors into the system and product development process include wages for the human factors personnel. In addition, there are several distinct costs associated specifically with the human factors process (Mantei & Teorey, 1988, p. 430). These include expenses involved with evaluating the concerns of the intended user population in preliminary studies, constructing product or system mock-ups, designing and modifying prototypes, creating a laboratory and conducting advanced user studies, and conducting user surveys.

With regard to products such as computer software, the cost-benefit ratio will depend on the number of users affected by the ergonomic changes. Karat (1990) performed cost-benefit analyses for two software development projects that had incorporated human factors concerns. One of these projects was of small scale and the other was of large scale. She estimated the savings-to-cost ratio to be 2:1 for the small project and 100:1 for the large project. The savings that arise from human factors testing increase dramatically as the size of the user population increases (Mantei & Teorey, 1988). For smaller projects, a complete human factors testing program will not be cost-effective. Here again we see the importance of a cost-benefit analysis: not only can it justify human factors research, but it will also make the investment in such research commensurate with the expected savings-to-cost ratio.

Even in the military, we have to consider the benefits of investing in new systems to enhance human and system performance relative to the cost of such investments (Rouse & Boff, 2012). Benefits may include more precise and efficient weapons systems, increased operability of the system, improved design using new techniques, and new opportunities for military strategists. Costs are those associated with the initial research and development, recurring operating expenses, and development time. These costs and benefits accrue to other organizations as well, including the developers (the contracting companies who stand to benefit from receiving research and development funds) and the public (who stand to benefit from increased performance of the military's missions).

THE SYSTEM DEVELOPMENT PROCESS

Phases

The development of a system is driven by the system's requirements (Meister, 1989; Meister & Enderwick, 2002), and the primary goal of the development team is to produce a system that meets or exceeds those requirements. System development begins with a broadly defined goal and proceeds to more and smaller tasks and subtasks. Most system requirements do not include human performance objectives; initially, the requirements specify only how the physical system is to perform. Consequently, the human factors specialist must determine what the user will need from those physical requirements.

We have mentioned several times the importance of including human factors specialists from the outset of a project. The design decisions made at the early stages of system development have consequences for the later stages. From the initial decisions onward, there are human factors criteria, as well as physical performance criteria, that must be met if the system is to perform optimally. The U.S. military is well aware of this. In 1986, the U.S. Army initiated the MANPRINT program (Booher, 1990), which forces designers to deal with human factors concerns from the outset of the system development process.

The MANPRINT program is now called *Human Systems Integration*. It and other programs like it were established because failures to consider human factors concerns before initial design decisions were made had resulted in the production of equipment that could not be used effectively or meet its performance goals. For example, the U.S. Army's Stinger anti-aircraft missile system, designed without human factors considerations, was supposed to be capable of successfully striking an incoming enemy aircraft 60% of the time. However, because the designers did not consider the skill and training required of the soldier operating it, its actual performance was closer to 30% (Booher, 2003b). Julia Ruck, of the U.S. Army, said in 2014, "As a former soldier who spent years at the tactical edge, I can honestly say that the MANPRINT program, with its focus on integrating that human element, makes the difference between a material solution being used or sitting on a shelf" (Conant, 2014).

System development proceeds in a series of phases (Czaja & Nair, 2012; McBride & Newbold, 2016; Meister, 2006b). The first phase is *system planning*. During this phase, we will identify the need for the system, that is, what the system is to accomplish, and make assessments of its feasibility.

The second phase is *preliminary design*, or initial design, during which we will consider alternative designs for the system. In this phase we will begin testing, construct prototypes, and create a plan for future testing and evaluation of the system. The third phase is *detail design*, during which we will complete the development and testing of the system and make plans for production. The final phase is *design verification,* when we produce the system and then evaluate it in operation. Data about how effective the system is, its strengths and weaknesses, are used to improve the design in subsequent generations.

Several questions about human performance will arise at each phase of system development. These questions are shown in Table 19.1. At the system planning phase, the human factors specialist evaluates the changes in the task requirements, the personnel that will be required, and the nature and amount of training needed for the new system relative to its predecessor. She ensures that human factors issues are addressed in the system design goals and requirements.

System design, in both the preliminary and the detail phases, is concerned with generating and evaluating alternative design concepts. The human factors specialist focuses on issues such as allocation of function to machines or humans, task analysis, job design, interface design, and so on (Czaja & Nair, 2012). During preliminary design, the specialist will judge the alternative design concepts in terms of their usability. He will recommend designs that minimize the probability of human error. When development moves from the preliminary to the detail phase, many of the questions addressed in the preliminary phase will be revisited. The final system design will be engineered to accommodate human performance limitations.

Human factors activities in the design verification phase will help determine whether there are any deficiencies left in the final design. We will conduct tests on the final system in an environment that closely approximates the operational conditions to which the system will be subjected. We may conclude from these tests that there are design features that need to be changed before the product or system is distributed.

TABLE 19.1
Behavioral Questions Arising in System Development

System Planning
1. Assuming a predecessor system, what changes in the new system from the configuration of the predecessor system mandate changes in numbers and types of personnel, their selection, their training, and methods of system operation?

Preliminary Design
2. Which of the various design options available at this time is more effective from a behavioral standpoint?
3. Will system personnel in these options be able to perform all required functions effectively without excessive workload?
4. What factors are potential sources of difficulty or error, and can these be eliminated or modified in the design?

Detail Design
5. Which is the better or best of the design options proposed?
6. What level of personnel performance can one achieve with each design configuration, and does this level satisfy system requirements?
7. Will personnel encounter excessive workload, and what can be done about it?
8. What training should be provided to personnel to achieve a specified level of performance?
9. Are (a) hardware/software, (b) procedures, (c) technical data, and (d) total job design adequately engineered from the human point of view?

Design Verification
10. Are system personnel able to do their jobs effectively?
11. Does the system satisfy its personnel requirements?
12. Have all system dimensions affected by behavioral variables been properly engineered from the human point of view?
13. What design inadequacies must be corrected?

In each phase the human factors professional will provide four areas of support to the design team. He will provide input to the design of the system hardware, software, and operating procedures with the goal of optimizing human performance. He will also make recommendations regarding how system personnel should be selected and recruited. The third area for which the human factors specialist will provide support concerns issues of training: What type should be given, and how much is needed? Finally, he will conduct studies to evaluate the effectiveness of the entire system, and more specifically, of the human subsystem.

The systematic application of human performance data, principles, and methods at all phases of system development ensures that the design of the system will be optimized for human use. This optimization results in increased safety, utility, and productivity, and ultimately benefits everyone: the system managers, operators, and, ultimately, consumers.

Facilitating Human Factors Inputs

A central concern in human factors and ergonomics is how to get human factors experts involved in the design process, particularly in the early phases of development during which the initial design decisions are made. The design team often works under the pressure of a deadline, and they will focus primarily on developing a system or product that will achieve its primary development and operational goals. Consequently, the team may view incorporation of human factors methods and user/usability testing as a costly option that is not as important as other factors (Shepherd & Marshall, 2005; Steiner & Vaught, 2006).

Where should a human factors program be placed in an organization's structure, when the organization implements one? Most people agree that the human factors specialists should be in a single, centralized group or department, under a manager who is sensitive to human factors organizational issues (Hawkins, 1990; Hendrick, 1990). A centralized group has several advantages that allow the human factors specialists to maximize their contributions to projects. The manager can champion human factors concerns at higher organizational levels, which is essential for creating an environment in which human factors will flourish. By establishing a rapport with persons in authority and increasing their awareness about the role of human factors in system design, the manager and group can ensure that their efforts will be supported within the organization. Further, financial support for laboratories and research facilities will be more reliable if there is a human factors group or department, rather than single individuals scattered throughout different departments. A stable human factors group also helps to establish credibility with system designers and engineers. Project managers are more likely to seek advice from the human factors specialists, because of their credibility and visibility. Finally, the group will foster a sense of professional identity that will boost morale and help in the recruitment of other human factors specialists.

It is an unfortunate fact that many engineering designers do not fully appreciate human factors, or believe that human factors can be addressed by anyone with knowledge of their projects. Therefore, it is important for human factors experts to raise awareness of the fact that more than just common sense is required to properly address human factors issues in the design process (Helander, 1999).

Apart from being welcomed on the design team, another problem human factors experts face is that everyone involved in the design of a system will view the problem from the standpoint of their discipline. Each will discuss problems using the vocabulary with which they are familiar and attempt solutions to problems using discipline-specific methods (Rouse, Cody, & Boff, 1991). A designer may not know what questions to ask the human factors specialist, or how to interpret the answer that she provides. Communication difficulties may result in loss of human factors information, and so the recommendations provided by the human factors specialist may have little impact on the system development process.

To prevent information loss, the human factors specialist has the responsibility of knowing at least something of the core design area (e.g., automobile instrument displays) to which she is contributing. Similarly, designers, engineers, and managers need to learn about human factors and ergonomics. Blackwell and Green (2003) suggest that human factors experts, designers, and users all learn a common set of cognitive dimensions along which design alternatives may differ. The structure provided by these dimensions will provide a common ground for communication about usability issues.

We have noted several times that it is difficult to get appropriate human factors input in the early planning and design phases of system development. In fact, designers frequently wait to worry about human factors issues until late in the detail design phase, well after many crucial decisions have already been made (Rouse & Cody, 1988). Consequently, the contribution of the human factors specialist is diminished by the necessity of working around the established features of the designed system. As Shepherd and Marshall (2005) emphasize, "Human factors professionals must continue to address the question of how best to support organizations so that significant human factors issues can be taken into account during system development in time for problems to be identified and solved with minimum expense and inconvenience" (p. 720).

One way we can solve this problem is through an approach called *scenario-based design* (Haimes, Jung, & Medley, 2013), in which the human factors professional develops scenarios depicting possible uses of a product, tool, or system (see Box 19.1). Another way is the participatory design approach, using methods such as focus groups to obtain input from intended users about their wants and needs for the product or system under development (Clemensen, Larsen, Kyng, Morten, & Kirkevold, 2007). Finally, we may implement system models that allow the human factors specialist and the designer to collaborate in evaluating alternative designs before prototypes have been developed. Integrative and engineering models of human performance and human motion, described in the next section, bring existing knowledge to bear on initial design decisions.

COGNITIVE AND PHYSICAL MODELS OF HUMAN PERFORMANCE

Our discussions of human information processing and basic anthropometric characteristics in the earlier chapters emphasized how human performance is limited by characteristics of tasks and work environments. We have explored basic principles, such as the fact that a person's performance deteriorates when his working memory is overloaded and that his movement time is a linear function of movement difficulty (Fitts's law). We have also discussed many theories that can explain these phenomena. This foundation, formed from data and theory, must be understood by anyone who wishes to incorporate human factors and ergonomics into design decisions.

When faced with a specific design problem, you might begin by searching for information about how similar problems have been solved before. You can consult a variety of sources (Rouse & Cody, 1989): human factors textbooks like the present one; textbooks that cover specific content areas in more detail (e.g., attention; Wickens & McCarley, 2008); handbooks that provide detailed treatments of topics and prescribe specifications (e.g., Salvendy, 2012); journal articles (e.g., from *Human Factors*); and standards and guidelines (e.g., Karwowski, 2006b). Unfortunately, there is no easy way to determine exactly what factors will be critical for your specific problem and how they may interact with each other.

This is where models of human performance come in. Quantitative and computational models have played a significant role in human factors and ergonomics throughout the existence of the field (Byrne & Gray, 2003). We have encountered many such models throughout this book. However, a lot of these models were formulated to explore very narrow problems (e.g., the "tapping" or aimed movements that are the focus of Fitts's law), and so they may not be very useful for human factors engineering problems. Some researchers have made a greater effort to develop "general-purpose" models, which focus on information processing in human performance (e.g., Byrne, in press) and physical models of human motion (e.g., Haupt & Parkinson, 2015).

As an example, consider the problem of attention. In this book, we have talked about how attention has been studied in the laboratory, and there are a lot of models of how attention works. Logan (2004) reviewed formal theories of attention and concluded that two classes, one based on

BOX 19.1　SCENARIO-BASED DESIGN

Scenario-based design is an alternative approach for incorporating human factors into the design process (Carroll, 2006). Scenario-based design has been used extensively in the area of human–computer interaction, so much so that Carroll (2002) says that it "is now paradigmatic" (p. 621); that is, accepted practice. However, scenario-based design has not been adopted as widely within system design more generally (van den Anker & Schulze, 2006).

The human factors expert using a scenario-based approach generates narratives depicting various ways that a person might use a software tool or product. These narratives are then used to guide the design process, from addressing human factors requirements through the testing and evaluation of the tool. By exploring possible scenarios, the designer will discover potential difficulties that users may encounter and identify functions that would be beneficial for specific purposes.

Scenario-based design is important because it addresses several challenges in the design of technology (Carroll, 2006). First, scenarios require the designer to reflect on the purposes for which a person would be using the product or system, and the reasons why they might be using it. Scenarios focus the designer's attention on the context in which the product will be used. Second, scenarios make the task environment concrete: They describe specific situations that can be easily visualized. This in turn means that the designer will be able to view the problem from a number of different perspectives, and visualize and consider alternative solutions.

Third, because scenarios are oriented toward the work that people will perform, they tend to promote work-oriented communication among the designers and the people who will use the product. Fourth, specialized scenarios can be abstracted from more general scenario categories. The way that the designer implements these more specific scenarios can rely on any prior knowledge that was used to implement the more general scenario. From this perspective, particular design problems can be solved by first classifying the problem according to what kind of scenario it is.

Scenarios can differ in their form and content (van den Anker & Schulze, 2006). They are most often narrative descriptions or stories. As these narratives become more refined, they can be depicted visually in storyboard drawings and graphics, and even in the form of simulated or virtual environments. They can focus on the activities of an individual user or on collaborative activities between multiple users. The computer-supported cooperative work described in Chapter 18 is an example of collaborative activities that might lend itself well to a scenario-based design. Furthermore, their level of abstraction can vary greatly; general scenarios with little detail can guide early design decisions, and as the design begins to take shape, the scenarios may include a great level of detail.

Designers can develop and apply scenarios in a variety of ways. At later design stages, as in a participatory design approach, people who represent the end users may contribute to the process. Often designers use "tabs" (like post-it notes) on a storyboard that represent the interface control and display functions that different users will perform (Bonner, 2006). Users then pull off the appropriate tab from the board and place it on a mock product to indicate the action that they would perform at a particular point in a task.

A case study employing scenario-based design focused on how agencies within the U.S. Department of Homeland Security (DHS) monitor and respond to emergencies (Lacava & Mentis, 2005). Engineers at Lockheed Martin wanted to design a command and control system for DHS. This particular problem was well suited to scenario-based design, because the software engineers didn't even know who would be using their system. Other problems they faced included not really knowing what kinds of problems a DHS agency might be faced with, and the agencies themselves couldn't say exactly how they would carry out assignments.

The designers, focusing on the Coast Guard, began by devising scenarios. Each scenario had a setting, actors working to achieve specific goals within that setting, and a plot detailing the sequence of actions taken by the actors in response to events within the setting. Issues that they encountered in the design process included how information was shared between different DHS agencies, how the information flowed from top-level intelligence to the lower-level individuals within the Coast Guard (and back up again), who within the Coast Guard would be responding to the information, and how the information needed to be displayed to the people at all levels of information flow. Constant refining of these scenarios resulted in a prototype system that they were able to present to the Coast Guard. Once the Coast Guard realized that the design team had accurately identified the problems involved with managing first response teams and had some concrete solutions, the designers convinced the agency to work with them more closely. With a group of actual users, the design team then proceeded with a more traditional user-centered design.

signal-detection theory (see Chapter 4) and another on what is called *similarity-choice theory*, provide the best overall accounts of a range of attention phenomena. Logan states, "Their mathematical structure allows strong inferences and precise predictions, [and] they make sense of diverse phenomena" (p. 230). As important as these models and theories are for people who study human attention, they may not tell you, the human factors expert, what you need to know for your design problem.

Although many information-processing models are not directly applicable to design issues, modeling is valuable to design engineers for several reasons (Gray & Altmann, 2006; Rouse & Cody, 1989). A model forces rigor and consistency in the analyses. It also serves as a framework for organizing information and indicating what additional information is needed. A model is also capable of providing an explanation for why a particular result occurs. Perhaps most importantly, designers can incorporate the quantitative predictions provided by a model into design decisions, but this is more difficult to do with only vague recommendations derived from guidelines and other sources.

The benefits provided by formal models are so considerable that many people have worked to develop general frameworks and models that allow a designer or modeler to predict human performance in specific task contexts (Elkind, Card, Hochberg, & Huey, 1990; Gluck & Pew, 2005; McMillan et al., 1989; Pew & Mavor, 1998). We summarize several approaches in the following sections.

ENGINEERING MODELS OF HUMAN PERFORMANCE

The primary purpose of engineering models of human performance is to provide "ballpark" values of some aspect of performance, for example time to perform a task, in a simple and direct manner. Engineering models of human performance should satisfy three criteria (Card, Moran, & Newell, 1983). First, the models should be based on the view of the person as an information processor. Second, the models should emphasize approximate calculations based on a task analysis. The task analysis determines those information-processing operations that might be used for achieving the task goals. Third, the models should allow performance predictions for systems while they are still in the design phase of development, before they have been built.

In sum, an engineering model of human performance should make it easy for a designer to provide approximate quantitative predictions of performance for design alternatives. We will describe two types of engineering models of human performance that satisfy these criteria: cognitive models developed primarily from research in cognitive psychology, and digital human models developed primarily from research in anthropometrics and biomechanics.

Cognitive Models

The most widely used cognitive engineering models are based on a framework developed initially by Card et al. (1983) for application to the domain of human–computer interaction. This framework, described briefly in Box 3.1, has two components. The first is a general architecture of the human information processing system called the *Model Human Processor*. It consists of a perceptual processor, a cognitive processor, and a motor processor, as well as a working memory (with separate visual and auditory image stores) and a long-term memory (see Figure 19.2). Each processor has one quantitative parameter, the cycle time (time to process the smallest unit of information), and each memory has three parameters: the storage capacity (in chunks), the decay time (in seconds), and the code type (acoustic or visual). These parameters are presumed to be context-free; that is, their values will be the same regardless of the task being performed. Estimates of their values are determined from basic human performance research and "plugged in."

Table 19.2 summarizes the principles of operation of the Model Human Processor. Many of these principles are based on fundamental laws of human performance that we described in earlier chapters. The most fundamental of these is the rationality principle.

The rationality principle is the assumption that the user acts rationally to attain goals. If an individual acts irrationally, analyzing the goal structure of the task would not serve any useful purpose. The rationality principle justifies the second major criterion of the engineering model: a task analysis framed in terms of goals and requirements. In the Model Human Processor, the task analysis determines the Goals, Operators, Methods, and Selection rules (GOMS) that characterize a task, as we described in Chapter 13.

After the GOMS analysis determines the goal structure, we can specify the information-processing sequence by defining the methods for achieving the goals, the elementary operations from which the methods are composed, and the selection rules for choosing between alternative methods. The end result is an information-processing model that describes the sequence of operations executed to achieve goals pursuant to performance of the task. Table 19.3 shows an example model for deciphering vowel-deletion abbreviations that describes the goal structure at the keystroke level. By specifying cycle times for the execution of the elementary operations, the model will generate a prediction for the time it will take to perform the task.

To illustrate how such an approach can be used, consider an experiment conducted by Ramkumar et al. (2016) that examined interactive medical image segmentation. This is a process that uses

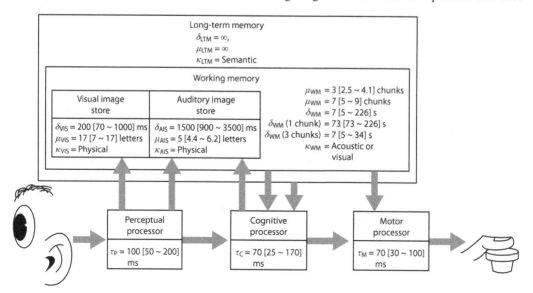

FIGURE 19.2   The Model Human Processor.

TABLE 19.2

The Model Human Processor—Principles of Operation

P0. **Recognize-Act Cycle of the Cognitive Processor**. On each side of the Cognitive Processor, the contents of Working Memory initiate actions associatively linked to them in Long-Term Memory; these actions in turn modify the contents of Working Memory.

P1. **Variable Perceptual Processor Rate Principle**. The Perceptual Processor cycle time τp varies inversely with stimulus intensity.

P2. **Encoding Specificity Principle**. Specific encoding operations performed on what is perceived determine what is stored, and what is stored determines what retrieval cues are effective in providing access to what is stored.

P3. **Discrimination Principle**. The difficulty of memory retrieval is determined by the candidates that exist in the memory, relative to the retrieval cues.

P4. **Variable Cognitive Processor Rate Principle**. The Cognitive Processor cycle time τ_c is shorter when greater effort is induced by increased task demands or information loads; it also diminishes with practice.

P5. **Fitts's Law**. The time T_{pos} to move the hand to a target of size S which lies a distance D away is given by:

$$T_{\text{pos}} = I_M \log_2\left(\frac{2D}{S+0.5}\right),$$

where $70 < I_M < 120$ (approximately), and we may fix $I_M = 100$ in most circumstances.

P6. **Power Law of Practice**. The time T_n to perform a task on the nth trial follows a power law:

$$T_n = T_1 n^{-\alpha},$$

where $0.2 < -\alpha < 0.6$ (approximately), and we may fix $-\alpha = 0.4$ in most circumstances.

P7. **Uncertainty Principle**. Decision time T increases with uncertainty about the judgment or decision to be made:

$$T = I_C H,$$

where H is decision uncertainty (in bits), and $0 < I_C < 157$ (approximately), and we may fix $I_C = 150$ in most circumstances.

For n alternatives with different probabilities, p_i, of occurrence,

$$H = \sum_{i=1}^{n} p_i \log_2\left(\frac{1}{p_i + 1}\right).$$

P8. **Rationality Principle**. A person acts so as to attain his goals through rational action, given the structure of the task and his inputs of information and bounded by limitations on his knowledge and processing ability:

$$\text{Goals} + \text{Task} + \text{Operators} + \text{Inputs} + \text{Knowledge} + \text{Process} - \text{limits} \rightarrow \text{Behavior}.$$

P9. **Problem Space Principle**. The rational activity in which people engage to solve a problem can be described in terms of (1) a set of states of knowledge, (2) operators for changing one state into another, (3) constraints on applying operators, and (4) control knowledge for deciding which operator to apply next.

---

image manipulation software to partition a digital image (such as an X-ray) into nonoverlapping regions to aid in medical diagnoses and planning treatments. They had three physicians segment images of body organs in preparation for radiation therapy. They used two task prototypes, one that required the physician to draw contours of an anatomical structure to segment it, and one that required the physician to draw strokes that indicated the desired foreground and background of the organ, from which an algorithm created the segment. Each physician segmented images of four organs (the spinal cord, the lungs, the heart, and the trachea) using both prototypes.

Using a GOMS analysis of videos of the physicians' performance, Ramkumar et al. (2016) identified 16 operators and 10 methods that were used to achieve the goal of segmentation of an organ. Operators included moving the cursor from the drawing region to a panel to select a tool, drawing a contour, and so on. The methods included combinations of the operators that were often performed together; for example, segmenting a single region by executing a click paint (the target region)

TABLE 19.3

GOMS Algorithm for Figuring out Vowel-Deletion Abbreviations

Algorithm	Operator Type
BEGIN	
Stimulus ← Get-Stimulus("Command")	Perceptual
Spelling ← Get-Spelling(Stimulus)	Cognitive
Initiate-Response(Spelling[First-Letter])	Cognitive
Execute-Response(Spelling[First-Letter])	Motor
Next-Letter ← Get-Next-Letter(Spelling)	Cognitive
REPEAT BEGIN	
IF-SUCCEEDED Is-Consonant?(Next-Letter)	Cognitive
THEN BEGIN	
Initiate-Response(Next-Letter)	Cognitive
Execute-Response(Next-Letter)	Motor
Next-Letter ← Get-Next-Letter(Spelling)	Cognitive
END	
ELSE IF-SUCCEEDED Is-Vowel?(Next-Letter)	Cognitive
THEN Next-Letter ← Get-Next-Letter(Spelling)	Cognitive
END	
UNTIL Null?(Next-Letter)	
IF-SUCCEEDED Null?(Next-Letter)	Cognitive
THEN BEGIN	
Initiate-Response("Return")	Cognitive
Execute-Response("Return")	Motor
END	
END	

---

operator, followed in succession by mouse move and draw operators. The researchers showed that the strokes approach was faster than the contour approach for large organs like the lungs, though not for smaller ones. However, though faster in some cases, the strokes approach also led to more errors than the contour approach, a finding that the researchers attributed to the fact that the strokes approach required more switching between tools than the contour approach.

The original GOMS and Model Human Processor framework has a number of shortcomings, which limit the accuracy of its predictive ability. It does not provide an account of performance changes that occur as skill is acquired, does not predict errors, assumes strictly serial processing, and does not address the effects of mental workload (Olson & Olson, 1990). However, extensions of the framework addressed issues of learning and errors (e.g., Lerch, Mantei, & Olson, 1989; Polson, 1988), and a family of GOMS models has been developed that has been successful at predicting efficiency of performance for a variety of tasks involving human–computer interaction (John, 2003; Olson & Olson, 1990). These include the use of text editors, graphics programs and spreadsheets, entering different kinds of keyboard commands, and manipulating files. GOMS models have also been used to generate a range of stimulus-response compatibility effects (see Chapter 13; Laird, Rosenbloom, & Newell, 1986) and to analyze the tasks performed by pilots with the flight management computer of a commercial aircraft (Irving, Polson, & Irving, 1994).

A variation of a GOMS analysis was used to design workstations for telephone toll and assistance operators (Gray, John, & Atwood, 1993). The human factors experts modeled operators' performance at a new workstation, which a telephone company was considering for purchase, and compared their predictions with the operators' performance at their old workstations. According to the workstation designers, the new workstation would reduce the average time each operator spent per call and so save the company money. However, the GOMS analysis predicted that the time per call would actually be longer with the new station than with the old. The researchers confirmed this prediction in a subsequent field study.

Digital Human Models

Digital human models are software design tools intended primarily for physical ergonomics, which deals with positions adopted by the human body and the loads imposed on it (Chaffin, 2005; Duffy, 2009). These tools allow a designer to create a virtual human being with specific physical attributes. The designer can then place the digital person in various environments and program it to perform specific tasks, like getting into an automobile or using a tool in a particular work environment. This enables the designer to evaluate the physical advantages and disadvantages of alternative designs relatively quickly and easily. More detailed aspects of performance, such as time and motion, field of view, work posture, and reach, are also available for additional analysis.

Any software system for creating digital humans must incorporate five elements (Seidl & Bubb, 2006). (1) The design of the digital human must take into account the number and mobility of the joints and accurately depict clothing; (2) the software must integrate anthropometric databases to assist in generation of digital humans with specific anthropometric characteristics; (3) the software must simulate posture and movement; (4) the software must include a way to analyze attributes relevant to the product or system being developed; (5) it should be possible to integrate the digital human model into a virtual world representing the design environment. Different digital modeling systems will vary with respect to the anthropometric database used, the algorithms used to simulate motion, and the analysis tools that are available.

Digital models cannot do everything. They have only a limited ability to incorporate differences in human sizes and shapes, to reproduce human body posture, and to predict human motion patterns (Woldstad, 2006). Also, because it is not always clear how these packages choose the algorithms they use to construct the models, it can be difficult for the designer to judge the accuracy of the "people" they produce.

JACK and RAMSIS are two digital human modeling tools used by designers in the analysis of automobile interiors (Hanson, Blomé, Dukic, & Högberg, 2006; Seidl & Bubb, 2006). To increase the applicability of these tools and to make them widely available within an organization (Saab Automobile, Sweden), Hanson et al. developed an internal Web-based usage guide and documentation system accessible by all members of a design team. The guide outlined a series of steps beginning with identifying the goal of modeling, how to prepare and use the modeling tool (modeling the people, physical environment, and task), and how to formulate recommendations (including results and discussion). They stored the data from all analyses (even those still underway) in a centralized data base. The result was an efficient system for organizing, conducting, and documenting simulation projects within the organization.

INTEGRATIVE COGNITIVE ARCHITECTURES

The engineering models that we have discussed so far are not very precise, although they are adequate for many purposes. In some cases, however, we may need more accurate predictions. These can be provided by integrative cognitive architectures. An integrative cognitive architecture is a relatively complete information-processing system, or unified theory, intended to provide a basis for developing computational models of performance in a range of specific tasks. This approach was introduced in Box 4.1, in which we mentioned three of the most prominent cognitive architectures: ACT-R (Adaptive Control of Thought–Rational; Boorst & Anderson, 2015), Soar (States, Operators, and Results; Howes & Young, 1997), and EPIC (Executive Process Interactive Control; Kieras, 2017).

All of these architectures are production systems (see Chapter 11), which rely on production rules (IF ... THEN statements that specify actions that occur when certain conditions are met) and a memory representation of the task to model cognitive processing. When the conditions for a given production are present in working memory, the production will "fire," and this produces a mental or physical action. The architectures differ in details, such as the extent to which processing is serial or parallel, and whether the architecture is more applicable to higher-level cognitive tasks, such as language learning and problem solving, or to perceptual-motor tasks, such as responding to two simultaneous stimuli.

The first applications of ACT and Soar architectures were cognitive tasks that required problem solving, learning, and memory. In contrast, because EPIC includes perceptual and motor processors, EPIC could simulate many aspects of multiple-task performance (for instance, driving; see Chapter 9). The most recent version of ACT-R, version 7.3 (Bothell, 2017), also includes perceptual (vision and audition modules) and motor (motor and speech modules) processors, and can model multiple-task performance. Although all these architectures were first developed to provide comprehensive accounts of basic cognitive phenomena, they have been used for applied problems in areas such as driving and interactions with digital devices.

Developing models within one of these architectures is a complex, time-consuming process that will require training. Even a skilled modeler will be challenged to develop models for a specific design problem, such as determining which of several alternative interfaces is best for a system. Consequently, some modelers have tried to simplify the modeling process (e.g., Salvucci & Lee, 2003).

Some engineering models can simulate human performance in complex human–machine systems. As Henninger and Whitaker (2015, p. 86) note, "Human behavioral modeling is motivated by the need to understand how people will react to a variety of possible environmental stimuli. It is used in war gaming … to understand enemy reactions, for marketing and product development decisions, in policy development for understanding policy alternatives and by organizational analysts to support organizational decisions." One model of this type is called the *Human Operator Simulator* (Harris, Iavecchia, & Dick, 1989; Pew, 2008), which helps design interfaces for weapon systems. It is a software system consisting of a resident Human Operator Model and a language that the designer uses to specify equipment characteristics and operator procedures. Much like other cognitive architectures, the Human Operator Model contains information-processing submodels for performing different subtasks ("micro-actions"). The major process submodels in the Human Operator Simulator are shown in Figure 19.3.

For simulating performance in a variety of weapons and flight systems, the designer must specify three major components of the task (see Figure 19.4; Harris et al., 1989, p. 286):

1. Environment (e.g., number, location, speed, and bearing of targets);
2. Hardware system (e.g., sensors, signal processors, displays, and controls);
3. Operator procedures and tactics for interacting with the system and for accomplishing mission goals.

The designer must also specify the interfaces between the three components: how information is passed from one component to the others. These interfaces determine such things as how well the hardware will detect changes in the environment, how heat and other environmental stressors will affect performance, and how difficult it will be for the operator to perform the required tasks. The simulation will produce timelines and accuracy predictions for task and system performance. The Human Operator Simulator is well suited to analyzing effects of control/display design, workstation layout, and task design.

Another human performance modeling technology used by the military is called *task network modeling*, or *discrete event simulation*. This modeling strategy is incorporated in commercially available applications like Micro Saint Sharp (Schunk, Bloechle, & Laughery, 2002). To use a task network model, the designer must first conduct a task analysis to decompose a person's functions into tasks, and then construct a network depicting the task sequence. After the initial task analysis,

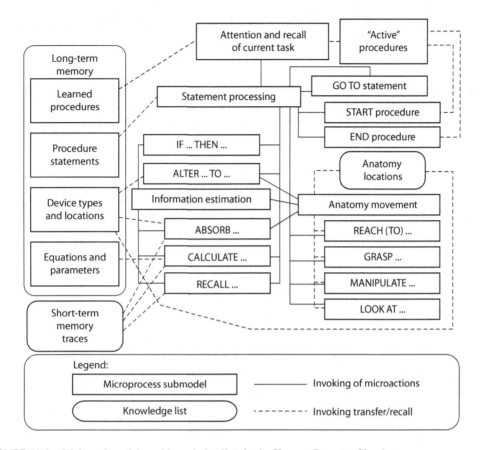

FIGURE 19.3   Major submodels and knowledge lists in the Human Operator Simulator.

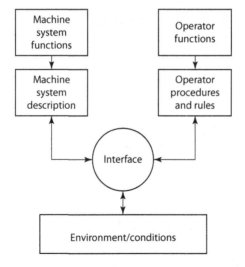

FIGURE 19.4   The three major Human Operator Simulator simulation components that are connected through the interface.

task network modeling is relatively easy to do and understand. It can include hardware and software models (which "plug in" at the appropriate points in the task network), which means that the complete human–machine system can be represented in the model (Dahn & Laughery, 1997). Another commercial application, the Integrated Performance Modeling Environment, combines the network modeling capabilities of Micro Saint Sharp with the modeling of the human information processing provided by the Human Operator Simulator.

Other widely used integrative architectures developed primarily for design purposes include (Pew & Mavor, 1998): COGNET (COGnition as a NETwork of tasks), which is used for building user models for intelligent interfaces; MIDAS (Man–machine Integrated Design and Analysis System), developed to model human performance in the context of aviation; and OMAR (Operator Model Architecture), intended to evaluate the procedures of operators in complex systems. An array of integrative architectures is available, and they are continually being developed and revised. A designer must examine the specific details of each modeling architecture relative to her needs and concerns, and make an informed choice as to which is best to use for her specific purpose.

CONTROL THEORY MODELS

Control theory models have a long history of use in human factors (Jagacinski & Flach, 2003). They are specialized for certain tasks, such as piloting an aircraft, that require monitoring and controlling operations of complex systems. Control theory models view the operator as a control element in a closed-loop system (see Figure 19.5). They assume that operators approximate the characteristics of good electromechanical control systems, subject to the limitations inherent in human information processing. Early models of this kind were limited in what they could do; they were only useful for dynamic systems involving one or more manual control tasks. Now, we have comprehensive models that span the range of supervisory activities engaged in by the operators of a complex system.

Some fundamental requirements have driven the development of comprehensive, multitask control theory models (Baron & Corker, 1989). First, a system model must represent the operators together with all of the nonhuman aspects of the system. Second, the cognitive and decision processes that characterize human performance in the complex system environment must be articulated clearly. Finally, communication between crew members and between operator and machine must be modeled, as should each crew member's mental model about the state of the system, goals, and so on.

As one example, the Procedure-Oriented Crew Model (PROCRU) was developed to evaluate the effects of changes in system design and procedures on the safety of landing approaches of aircraft (Baron & Corker, 1989; Baron et al., 1990). This application illustrates how the control-theoretic

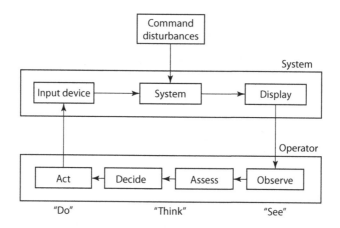

FIGURE 19.5 Closed-loop, control theory view of a human–machine system.

approach can help the designers develop comprehensive models of very complex systems. PROCRU is a closed-loop system that has separate models for the air-traffic controller, landing aids provided by the air-traffic control system, the aircraft, and crew members (a pilot and co-pilot; Vidulich, Tsang, & Flach, 2016).

The pilot models are, like the other components of the system, based on a control-theoretic information-processing structure. The pilots are assumed to have a set of tasks, or procedures, to be performed. The selection of the particular procedure to perform initially and when the previous one has been completed is based on the "expected gain" associated with performing each remaining task. Expected gain is a function of task priorities established by the flight mission and an estimate of the perceived urgency of performing particular tasks. When a procedure is chosen for execution, no other procedure will be considered during the time required to accomplish the chosen task.

PROCRU and other comprehensive control theory models do more than just predict how fast or accurate a person will be. They produce dynamic output: a continuous simulation of how the system will function over time. The simulation will vary as the representation of the situation evolves (Baron et al., 1990). Although we know that some aspects of control theory models work well in many contexts, we cannot say that they are true explanations of the way that complex systems operate over time. We have no empirical validation for comprehensive models such as PROCRU, and so we cannot say that they accurately represent what happens in the course of system operation.

FORENSIC HUMAN FACTORS

The decisions we make as designers, whether they are based on data or something else, determine the usability and safety of the final product our company sells. When something goes wrong, if people using the product get hurt, and if the human factors expert has been involved in the product design, then he or she must share the responsibility for design imperfections. Even human factors experts who were not involved in the development of a product may be asked to evaluate the product and its development process to determine what went wrong. The involvement of human factors considerations in the legal system is called *forensic human factors and ergonomics* (Dror, 2013; Noy & Karwowski, 2005).

LIABILITY

An organization is responsible for the safety of many people. Primarily, these are the people who use the products or services produced by the organization and the workers who are employed by the organization. If an organization fails to meet this responsibility, it can be held liable in a court of law. Thus, an organization must maintain safe practices and ensure that these practices can be justified if someone calls them into question.

In the U.S., safety in the workplace is governed by the directives of the Occupational Safety and Health Administration (OSHA) and the National Institute for Occupational Safety and Health (NIOSH). Some of the guidelines we discussed in Chapter 16 were determined by OSHA and NIOSH. OSHA was created by the passage of the Occupational Safety and Health Act of 1970 to ensure a safe work environment. It is responsible for safety and health regulations, and required employers to reduce workplace hazards and implement safety and health programs that informed and trained their employees (www.osha.gov/). An organization that knowingly or unknowingly violates these standards may be subject to citations and fines, which are leveled by OSHA.

NIOSH was established in conjunction with OSHA to provide the research and information on which the OSHA regulations are based, as well as education and training in occupational safety and health. Human factors specialists contribute to the development and evaluation of the standards. Human factors specialists also devise safety and training guidelines that keep the employer in compliance with OSHA regulations and ensure that employees will follow the safety procedures.

When an employee (or visitor) is injured or dies in an organization's workplace, the organization may be responsible. This responsibility also extends to the people outside of the organization who bought or sold the organization's products or services. If the organization is responsible, the law says that the organization has been negligent. Negligence is either criminal or civil. Criminal negligence occurs when the organization willfully violates the laws established to ensure safe products and safe work environments.

If the organization has not been criminally negligent, it may nonetheless have breached its civil responsibility ("duty of care") to its employees or customers. The law distinguishes between product liability (Wardell, 2005) and service liability cases; both arise from the failure of performance, in the first case of a product and in the second case of a person. When someone is injured as a result of such failures, he or his family may undertake litigation to prove negligence and, if appropriate, get compensation for his losses. The law will decide negligence by evaluating whether "reasonable care" was taken in the design and maintenance of products and equipment (Cohen & LaRue, 2006).

A now-famous skit aired on *Saturday Night Live* in 1976. It starred Dan Akroyd as sleazy toy manufacturer "Irwin Mainway," who attempted to justify the extreme danger of children's toys like "Bag O' Glass" and "Johnny Switchblade" to an incredulous consumer advocate (played by Candice Bergen). The skit is funny even decades later, because the product (a doll with spring-loaded knives under his arms) was obviously inappropriate for its intended users, regardless of the wild justifications offered by Mr. Mainway. Such mismatches between product design and user capabilities create hazards, risks, and dangers. A hazard is a situation for which there is the potential for injury or death; risk is the probability of injury or death occurring; danger is the combination of hazard and risk. A danger exists when there is a hazard for which there is a significant risk probability. Hence, we see immediately that "Bag O' Glass" and "Johnny Switchblade" are unreasonably dangerous toys.

Unlike "Johnny Switchblade," it is usually very difficult to determine whether a product is unreasonably dangerous. High frequency of injury does not necessarily mean that the risk associated with a product is unreasonable, nor does absence of injury indicate that the risk is reasonable (Statler, 2005). For example, chainsaws are dangerous, but at least some of the risk associated with their use is inherent in the product itself. Several criteria that form the basis of the "unreasonable danger" test are as follows (Weinstein, Twerski, Piehler, & Donaher, 1978):

1. The usefulness and functionality of the product.
2. The availability of similar but safer products that serve the same purposes.
3. The likelihood and seriousness of injury.
4. How obvious the danger is.
5. Common knowledge and normal public expectation of the danger.
6. Whether injury can be avoided by being careful in use of the product (including the effect of instructions or warnings).
7. Whether the product could be redesigned without impairing the usefulness of the product or making it too expensive.

Standards, such as those published by the American National Standards Institute (ANSI), make contractual agreements about and identification of mass-produced products possible. Standards are intended to guarantee uniformity in mass-produced goods, and safety is only one of many concerns that published standards are intended to address. However, adherence to published safety standards is not sufficient to ensure a safe product and does not absolve a manufacturer of liability. ANSI and the courts regard standards as only the minimum requirements for a reasonable product. The criteria for the standards may be outdated, standards published by different institutions may be inconsistent, the risk allowed by the standards may still be significant, and many aspects of product design will not be covered by standards. Generally, standards may not be good enough, and the time and money spent trying to meet minimum requirements set forth in the standards could be better spent in research and design (Peters, 1977).

An example case study illustrating the inadequacy of industry standards is seen in a certain kind of riding mower accidents (Statler, 2005). A significant number of accidents with riding mowers involve backing over someone, usually a child, while the mower blade is moving. These accidents occur because a small child is difficult to see behind the mower, and the driver also needs to look forward at the mower controls while backing up. In the 1970s, the U.S. Consumer Product Safety Commission (CPSC) urged manufacturers to stop the blade when the mowers are put in reverse, and a few companies added no-mow-in-reverse devices to their riding lawnmowers. Even though this safer design was economically feasible, most companies did not make the change, and industry standards have not been modified to require the design change. Stuart Statler (2005), former Commissioner of the CPSC, notes, "Not unexpectedly, the industry standard for riding mowers in effect over the past two decades represents, more or less, a low point for safety for a product that, by its very nature, engendered the highest degree of risk to life, namely, severe injury or death from backover" (p. 25).

The "reasonable danger" test was refashioned by Weinstein et al. (1978) into a set of criteria that a designer can apply to ensure that the danger is reasonable:

1. Delineate the scope of product uses.
2. Identify the environments within which the product will be used.
3. Describe the user population.
4. Postulate all possible hazards, including estimates of probability of occurrence and seriousness of resulting harm.
5. Delineate all alternative design features or production techniques, including warnings and instructions, that can be expected to effectively mitigate or eliminate the hazards.
6. Evaluate such alternatives relative to the expected performance standards of the product, including the following:
   a. Other hazards that may be introduced by the alternatives.
   b. Their effect on the subsequent usefulness of the product.
   c. Their effect on the ultimate cost of the product.
   d. A comparison to similar products.
7. Decide which features to include in the final design. (p. 140)

When an injury occurs for which a product is implicated as a possible cause, a legal complaint may be filed in civil court. The plaintiff (victim) must not only prove that the product was the likely cause of her injury, but she must also establish that a legal responsibility to the consumer was not met by the manufacturer of the product. A manufacturer may fail to meet her or his legal responsibilities in one of three ways: negligence, strict liability, or breach of warranty (Moll, Robinson, & Hobscheid, 2005). Negligence, as we have been discussing, is focused on the behavior of the defendant (the manufacturer), in that the defendant failed to take reasonable actions that would have prevented the accident. If the defendant is accused of engaging in reckless and wanton misconduct, the charge is one of gross negligence (which may also be criminal negligence).

Strict liability focuses on the product and not the defendant. Although the manufacturer need not have been in any way negligent, the manufacturer can be held liable for any product defect if that defect was the cause of the injury. Under strict liability, it is not only the manufacturer that can be held liable. The manufacturer must have sold the product either to the plaintiff or to one of many members of a distributive chain, all of whom may be named as defendants in the trial (Weinstein et al., 1978): (1) the producer of the raw material; (2) the maker of a component part; (3) the assembler or subassembler; (4) the packager of the final product; (5) the wholesaler, distributor, or middleman; (6) the person who holds the product out to be his or her own; and (7) the retailer. Any or all of these members can be held liable if it can be proven that a product was defective when it left their possession.

Breach of warranty occurs when a product fails to function as the defendant stated it would. Express warranty is an explicit statement in an oral or written contract. Implied warranties are not explicitly stated but are ones that a person could reasonably infer, for example, in the advertised uses of a product or in the product's name. For example, the drug Rogaine, marketed as a baldness cure in the U.S. and other countries, is sold under the name Regaine in the U.K. Under U.S. product liability law, the name Regaine provides an express warranty that the product will cure baldness (if you use it you will "regain" your hair).

EXPERT TESTIMONY

Human factors specialists are called upon in the development of a product or system to improve the product and so reduce a manufacturer's risk of liability. They may also be hired to provide expert testimony during litigation about human limitations and the product in question (Cohen & LaRue, 2006). In the role of an expert witness, the forensic human factors consultant will first be contacted by an attorney, either the plaintiff's or the defendant's. The consultant must make sure that the issues involved in the case fall within her areas of expertise, that she has no apparent conflict of interest, and that she will be able to work with the attorney (Hess, 2005). After she and the attorney reach an agreement, she will examine all of the information in the case to determine the relevant facts (Askren & Howard, 2005). She will also inspect the product or the location where the accident or injury occurred. She may also need to conduct some research. This could involve reading standards, guidelines, and relevant scientific literature, and possibly even conducting an experiment.

After all this, the consultant will write a report for the attorney, in which she integrates the information from all these different sources and summarizes her opinions relevant to the case. If the consultant is called to provide testimony as an expert witness, this will take place in two stages. First, she provides her opinion and answers questions from the opposing attorney in a deposition. If the consultant's evidence is strong, the case will often end here, because the opposing attorney will be unwilling to let a jury listen to evidence detrimental to his client's case. If it does not end with the deposition, the consultant will have to testify in court in front of a jury, answering questions posed both by her attorney and by the opposing attorney.

An example case of some notoriety involved incidents of "unintended acceleration" of the Audi 5000 automobile. As we discussed in Chapter 15, unintended acceleration incidents in vehicles with automatic transmissions have been reported since the 1940s (Schmidt, 1989). Such incidents are relatively rare and are not limited to any particular make or model. However, in the late 1980s, a number of people charged that the Audi 5000 was involved in an unusually high number of unintended acceleration accidents.

These charges peaked as a result of an incident in February 1986 in which a woman driving an Audi 5000 struck and killed her 6-year-old son when the car accelerated out of control. Her lawsuit was filed in April of that year, in which she claimed that a design defect in the Audi transmission was the cause of the unintended acceleration. The case received considerable media coverage, culminating in November with an expose on the CBS investigative reporting program, "60 Min." Following this program, a flood of claims were made alleging incidents of unintended acceleration involving the Audi 5000.

The litigation against Audi's parent company, Volkswagen of America, proceeded in two phases (Huber, 1991). In the first phase, the plaintiffs insisted that there was a flaw in the Audi transmission, as in the initial case described above. However, the evidence overwhelmingly indicated that the unintended acceleration was due to foot placement errors and not mechanical failure, leading the jury in the initial case to return a verdict in favor of the defendants in June, 1988 (Baker, 1989). At that point, at least one plaintiff returned to court, charging that the sudden acceleration was in fact due to foot placement errors that would not have occurred if the pedals had been designed differently.

Many product liability cases hinge on human capabilities and design, and this is apparent in the case of the Audi 5000. In the first phase of Audi's litigation, the human factors expert could have testified how likely it was that an instance of unintended acceleration was due to an undetected foot placement error. This same testimony, together with information about the sizes and locations of the pedals, could have been used by opposing counsel during the second phase of Audi's litigation. However, to make such a case against Audi, it would have to be shown that unintended acceleration incidents were in fact greater for the Audi than for other automobiles and that the pedals were placed in such a way that the likelihood of foot placement errors was greater in the Audi than for other automobiles. Because neither of these claims is true, the defense could use the testimony of a human factors specialist to prevent Audi from being unjustly found negligent.

Despite the fact that Audi was not found negligent in any of the cases, and evidence has indicated that virtually all instances of unintended acceleration are due to the driver mistakenly stepping on the gas pedal, unintended acceleration cases continue to be filed in courts. On August 6, 2006, a jury awarded $18 million to the driver of an SUV who charged that a defective speed control system was responsible for her crash on an interstate highway (Alongi & Davis, 2006). The high stakes involved for automobile manufacturers and plaintiffs in such cases suggest that human factors experts will continue to play an important role in product liability cases.

The issues described in the Audi 500 case are illustrative of the types of questions that arise in legal proceedings for which the testimony of a human factors specialist may be of value. During litigation, a human factors specialist can provide information pertinent to the following questions:

1. Was the product design, service, or process appropriate for the knowledge, skills, and abilities (KSAs) to be expected of normally functioning users (or clients) in the expected operational environment?
2. If not, could the service or the product design have been modified so that it would have been appropriate to the KSA of the anticipated user population?
3. If there was less than an optimal match between product design and the KSA of the expected user population, was an attempt made to modify the user population KSA by adequate selection procedures and/or by providing appropriate information by means of adequate training, instructions, and warnings?
4. If not, was it technically feasible to have provided such selection procedures and/or information transfer?
5. If [testimony indicates] that the information provided was not appropriate to the idiosyncrasies of the injured party, was it technically possible for the design of the product, selection, and/or information exchange to have been altered to accommodate those idiosyncrasies? (Kurke, 1986, p. 13)

Because litigation is adversarial, attorneys for both the defendant and the plaintiff are legally obligated to use any (ethical) means possible to win the case for their clients. Consequently, rendering expert opinion is rarely a pleasant experience. During cross examination, the human factors expert may be subjected to what is essentially a personal attack. The expert will be called upon to defend his credentials and the basis of his opinion. He will be asked misleading questions and may have his or her testimony restricted to exclude possibly relevant information on the basis of opposing counsel's objections.

The expert witness is in a position of authority regarding the issues on which she testifies. The expert witness is also paid, often lavishly, for her time by one of the interested parties. The combination of unquestioned authority and monetary compensation puts the expert witness, as well as the field of human factors, in a position where professional and scientific integrity come into question. For this reason, the Human Factors and Ergonomics Society has a section on principles of Forensic Practice in its code of ethics (Human Factors and Ergonomics Society, 2005). These principles outline behaviors that ensure that the expert witness is unbiased and not motivated by personal gain, that the witness adheres to high scientific and personal standards, and that the witness does not abuse her position of authority and so damage the reputation of the human factors profession.

HUMAN FACTORS AND SOCIETY

As the field of human factors emerged from World War II, its emphasis was on the "lights and buttons" systems so often encountered in the military. Since that time, the field has rapidly expanded. It now includes a wide range of domains covering both the military and the private sectors. Many forces have led to the rise of human factors, the most compelling being the rapid growth of high technology systems in which human performance is often the variable that limits the performance of the system. With each new technological development, a host of specific human factors issues arise that are unique to that technology, though the basic and applied principles of human performance acquired through years of research remain applicable. Other pressures that have led to increasing emphasis on human factors include greater concern with workers' health and safety, demands from consumers for products that are easier to use, and the financial benefits that arise from improvements in the match between the human and the machine.

As the field of human factors and ergonomics has grown, the range of disciplines that interact to form its knowledge base has also grown. Participants in the field include graduates not only of human factors programs, but also from such fields as psychology, industrial engineering, civil engineering, biomechanics, physiology, medicine, cognitive sciences, machine intelligence, computer science, anthropology and sociology, and education. The highly interdisciplinary nature of the profession encourages communication across discipline boundaries. Such interdisciplinary communication provides a basis for fundamental advances in scientific understanding that contribute to society through more usable, safer products and services.

An immediate application of human factors research is the design of equipment and environments for the very young, the aged, and the handicapped. In recent years, our society has become more aware of the challenges that face such special populations. One challenge to human factors is to improve the quality of life for these populations through designs that allow them to attain personal goals and fulfillment with the same ease as those not so challenged. Human factors experts have a responsibility to see that products intended for use by special populations are more than just modifications of products designed for the population at large.

With the development of the Internet and the World Wide Web, and the many computer-mediated activities in which we now engage, we often talk about a concept of "universal design" (Stephanidis & Akoumianakis, 2011). Universal design ensures that anyone will have access to information and services at all places and at all times. One goal of the proponents of universal design is the development of a code of practice. This code of practice is intended to see that considerations of usability during product and system development are not restricted to just an average, able user but to the larger population of users of diverse abilities. In advocating that systems should work well for all users, Vanderheiden (2005) emphasizes, "Web content that is more usable by individuals who have disabilities is also more usable by individuals with mobile technologies and, often, more understandable and usable by all users" (p. 281). We can make this claim for virtually all aspects of product and system design.

Because the forces that led to the founding and expansion of the human factors profession continue to exert their influence, human factors will continue to grow. Moreover, since technology is moving forward in leaps and bounds, providing us with new, complex machines whose effective use requires that we be able to interact with them intuitively and naturally, there will continue to be new frontiers for application of knowledge concerning human factors and ergonomics. Our efforts to provide a better integration of the basic facts of human performance and the applied concerns of system and product designers emphasize how usability engineering is a fundamental component of any design process.

RECOMMENDED READINGS

Bias, R. G., & Mayhew, D. J. (Eds.) (2005). *Cost-Justifying Usability: An Update for the Internet Age*. San Francisco, CA: Morgan Kaufman.

Card, S. K., Moran, T. P., & Newell, A. (1983). *The Psychology of Human-Computer Interaction*. Hillsdale, NJ: Erlbaum.

Gluck, K. A., & Pew, R. W. (Eds.) (2005). *Modeling Human Behavior with Integrated Cognitive Architectures: Comparison, Evaluation, and Validation*. Mahwah, NJ: Erlbaum.

Meister, D., & Enderwick, T. P. (2002). *Human Factors in System Design, Development, and Testing*. Mahwah, NJ: Erlbaum.

Noy, Y. I., & Karwowski, W. (Eds.) (2005). *Handbook of Human Factors in Litigation*. Boca Raton, FL: CRC.

Appendix I: Areas under the Standard Normal Curve

FIGURE AI.1

z	0	0.01	0.02	0.03	0.04	0.05	0.06	0.07	0.08	0.09
0.0	0.5000	0.4960	0.4920	0.4880	0.4840	0.4801	0.4761	0.4721	0.4681	0.4641
0.1	0.4602	0.4562	0.4522	0.4483	0.4443	0.4404	0.4364	0.4325	0.4286	0.4247
0.2	0.4207	0.4168	0.4129	0.4090	0.4052	0.4013	0.3974	0.3936	0.3897	0.3859
0.3	0.3821	0.3783	0.3745	0.3707	0.3669	0.3632	0.3594	0.3557	0.3520	0.3483
0.4	0.3446	0.3409	0.3372	0.3336	0.3300	0.3264	0.3228	0.3192	0.3156	0.3121
0.5	0.3085	0.3050	0.3015	0.2981	0.2946	0.2912	0.2877	0.2843	0.2810	0.2776
0.6	0.2743	0.2709	0.2676	0.2643	0.2611	0.2578	0.2546	0.2514	0.2483	0.2451
0.7	0.2420	0.2389	0.2358	0.2327	0.2296	0.2266	0.2236	0.2206	0.2177	0.2148
0.8	0.2119	0.2090	0.2061	0.2033	0.2005	0.1977	0.1949	0.1922	0.1894	0.1867
0.9	0.1841	0.1814	0.1788	0.1762	0.1736	0.1711	0.1685	0.1660	0.1635	0.1611
1.0	0.1587	0.1562	0.1539	0.1515	0.1492	0.1469	0.1446	0.1423	0.1401	0.1379
1.1	0.1357	0.1335	0.1314	0.1292	0.1271	0.1251	0.1230	0.1210	0.1190	0.1170
1.2	0.1151	0.1131	0.1112	0.1093	0.1075	0.1056	0.1038	0.1020	0.1003	0.0985
1.3	0.0968	0.0951	0.0934	0.0918	0.0901	0.0885	0.0869	0.0853	0.0838	0.0823
1.4	0.0808	0.0793	0.0778	0.0764	0.0749	0.0735	0.0721	0.0708	0.0694	0.0681
1.5	0.0668	0.0655	0.0643	0.0630	0.0618	0.0606	0.0594	0.0582	0.0571	0.0559
1.6	0.0548	0.0537	0.0526	0.0516	0.0505	0.0495	0.0485	0.0475	0.0465	0.0455
1.7	0.0446	0.0436	0.0427	0.0418	0.0409	0.0401	0.0392	0.0384	0.0375	0.0367
1.8	0.0359	0.0351	0.0344	0.0336	0.0329	0.0322	0.0314	0.0307	0.0301	0.0294
1.9	0.0287	0.0281	0.0274	0.0268	0.0262	0.0256	0.0250	0.0244	0.0239	0.0233
2.0	0.0228	0.0222	0.0217	0.0212	0.0207	0.0202	0.0197	0.0192	0.0188	0.0183
2.1	0.0179	0.0174	0.0170	0.0166	0.0162	0.0158	0.0154	0.0150	0.0146	0.0143
2.2	0.0139	0.0136	0.0132	0.0129	0.0125	0.0122	0.0119	0.0116	0.0113	0.0110
2.3	0.0107	0.0104	0.0102	0.0099	0.0096	0.0094	0.0091	0.0089	0.0087	0.0084
2.4	0.0082	0.0080	0.0078	0.0075	0.0073	0.0071	0.0069	0.0068	0.0066	0.0064
2.5	0.0062	0.0060	0.0059	0.0057	0.0055	0.0054	0.0052	0.0051	0.0049	0.0048
2.6	0.0047	0.0045	0.0044	0.0043	0.0041	0.0040	0.0039	0.0038	0.0037	0.0036
2.7	0.0035	0.0034	0.0033	0.0032	0.0031	0.0030	0.0029	0.0028	0.0027	0.0026
2.8	0.0026	0.0025	0.0024	0.0023	0.0023	0.0022	0.0021	0.0021	0.0020	0.0019
2.9	0.0019	0.0018	0.0018	0.0017	0.0016	0.0016	0.0015	0.0015	0.0014	0.0014
3.0	0.0013	0.0013	0.0013	0.0012	0.0012	0.0011	0.0011	0.0011	0.0010	0.0010
3.1	0.0010	0.0009	0.0009	0.0009	0.0008	0.0008	0.0008	0.0008	0.0007	0.0007
3.2	0.0007	0.0007	0.0006	0.0006	0.0006	0.0006	0.0006	0.0005	0.0005	0.0005
3.3	0.0005	0.0005	0.0005	0.0004	0.0004	0.0004	0.0004	0.0004	0.0004	0.0003
3.4	0.0003	0.0003	0.0003	0.0003	0.0003	0.0003	0.0003	0.0003	0.0003	0.0002
3.5	0.0002	0.0002	0.0002	0.0002	0.0002	0.0002	0.0002	0.0002	0.0002	0.0002
3.6	0.0002	0.0002	0.0001	0.0001	0.0001	0.0001	0.0001	0.0001	0.0001	0.0001
3.7	0.0001	0.0001	0.0001	0.0001	0.0001	0.0001	0.0001	0.0001	0.0001	0.0001
3.8	0.0001	0.0001	0.0001	0.0001	0.0001	0.0001	0.0001	0.0001	0.0001	0.0001
3.9	0.0000	0.0000	0.0000	0.0000	0.0000	0.0000	0.0000	0.0000	0.0000	0.0000

Appendix II: Information Theory

A methodological tool that played a key role in the rise of the human information-processing approach is *information theory*, developed by Claude Shannon (1948), which Gleick (2011) characterizes as one of the most important technological developments of the mid-20th century:

> Shannon's theory made a bridge between information and uncertainty; between information and entropy and between information and chaos. It led to compact discs and fax machines, computers, and cyberspace ... and all the world's Silicon Alleys. Information processing was born, along with information storage and information retrieval.

Information theory was developed by communication engineers to quantify the flow of information through communication channels such as telephone lines and computer systems. In the 1950s, contemporaneous with the development of signal-detection theory, psychologists began to apply the concepts of information theory to human performance (Fitts & Posner, 1967; Garner, 1962).

Information theory does not play the prominent role in human factors today that it once did, but it is still useful in many circumstances. As one example, Kang and Seong (2001) used an information-theoretic approach to quantify the perceived complexity of control room interfaces for nuclear power plants and to estimate the extent to which the interface would overload the operator's capacity for processing information. As another, Strange et al. (2005) used information theory to perform a quantitative analysis showing that activity in a part of the brain involved in visual perception, the hippocampus, is a function of event uncertainty. More generally, Castro (2009) stated with respect to human factors in driving,

> We believe a description of the road environment in information theory terms—a theory in decline over the past decades—is enriching and useful in order to understand more about driver limitations. It enables us to quantify and assess the usefulness of traffic elements with regard to the amount of information they can transmit, so that we can estimate the guarantees of such information being received. (p. 7)

Information theory is not a scientific theory. It is a system of measurement for quantifying information, as implied in the last sentence of Castro's (2009) quote. The amount of information conveyed by the occurrence of an event (a stimulus, response, or the like) is a function of the number of possible events and their probabilities of occurring. If an event is sure to occur, then its occurrence conveys no information. For example, if I know that my car's engine is not working, then I gain no information by turning the key in the ignition and observing that the car will not start. On the other hand, if I am uncertain about whether the engine is working, say, on a cold winter morning, then I gain information from the same event. The uncertainty of the event is the amount of information that we gain by observing it.

The general idea behind information theory is that the most efficient way to uniquely identify one of a set of events is to ask a series of binary questions. For example, if I told you that I was thinking of a number between 1 and 16, and you were to identify that number by asking me questions, you could proceed in several ways. One way would be to guess randomly each of the numbers until I said *yes*. Though you occasionally might guess the number the first time, on the average it would take eight questions to determine the correct number.

It would make more sense to systematically restrict the number of possibilities by asking *yes-no* questions. There are many ways that you could do this, but the most efficient would be to ask the questions in such a way that each reduced the number of possible alternatives by half. For identifying one of 16 numbers, your first question might be, "Is it between 1 and 8?" If my answer is *yes*, your next question should be, "Is it between 1 and 4?" Proceeding in this manner, you always would identify the correct number with four questions. In fact, of all possible guessing strategies you could use, four is the minimum number of questions that would have to be asked on average to correctly identify the number.

This idea of binary questions underlies the information theory definition of information. The number of binary questions required to decode a message provides the measure of information.

When all alternatives are equally likely, the amount of information (H) is given by

$$H = \log_2 N,$$

where N is the number of possible events. The basic unit of information is the bit, or binary digit. Thus, an event conveys 1 bit of information when there are 2 equally likely possibilities, 2 bits when there are 4 possibilities, 3 bits when there are 8 possibilities, and, as we have demonstrated, 4 bits when there are 16 possibilities. In other words, each item in a set of 16 can be represented by a unique 4-digit binary code.

The amount of uncertainty, and thus the average information conveyed by the occurrence of one of N possible events, is a function of the probability for each event. The maximum amount of information is conveyed when the N events are equally likely. The average amount of information is less when the events are not equally likely. To understand this, think back to the problem of knowing whether your car's engine is working on a cold morning. If you know that your car has problems in cold weather, so that the probability of it not starting is greater than the probability of it starting, then the car's failure to start when you turn the key does not transmit as much information as it might have.

The uncertainty of a single event i that occurs with probability p_i is $-\log_2 p_i$; thus, the average uncertainty over all possible events is

$$H = -\sum_{i=1}^{N} p_i \log_2 p_i$$

The equation for H when all events are equally likely, that is, $p_i = 1/N$, can be easily derived from the more general equation by noting that $-\log_2 p_i = \log_2(1/p_i)$.

The importance of information theory is in analyzing the amount of information transmitted through a system. Because a person can be regarded as a communication system, computing the information input $H(S)$ (stimulus information) and the information output $H(R)$ (response information) will tell us about the properties of the human system. Suppose that a person's task is to identify a letter spoken over headphones. If four equally likely stimuli can occur (say, the letters A, B, C, and D), then there are 2 bits of stimulus information. If there are four response categories, again A, B, C, and D, each used equally often, then there are 2 bits of response information.

In most communication systems, we are interested in the output that results from a particular input. Given that, say, the stimulus A is input into the system, we can record the number of times that the responses A, B, C, and D are output. The frequency of each stimulus-response pair can be counted, thus forming a bivariate frequency distribution (see Table 13A.1). From such a table, the joint information can be computed by the equation

$$H(S,R) = -\sum_{j=1}^{N} \sum_{i=1}^{N} p_{ij} \log_2 p_{ij}$$

where p_{ij} equals the relative frequency of response j to stimulus i.

Using the joint information in a system, we can determine the amount of information transmitted through the system, or the ability of the system to carry information. If the responses correlate perfectly with the stimuli, for example if the stimulus A is classified as A every time, then all the information in the stimuli is maintained in the responses, and the information transmitted is 2 bits (see top panel of Table AII.1). Although the example in the table shows all of the responses as correct, note that the responses only have to be consistent. In other words, if stimulus A were always identified as B, and vice versa, the information transmitted would still be the same. If the responses are distributed equally across the four stimuli, as in the center panel of Table AII.1, then no information is transmitted. When there is a less than perfect, nonzero correlation between stimuli and responses, as in the bottom panel of Table AII.1, then the information transmitted is between 0 and 2 bits. To determine the amount of information transmitted, we must calculate the stimulus information, response information, and joint information. Transmitted information is then given by

$$T(S,R) = H(S) + H(R) - H(S,R)$$

For the data in the bottom panel of Table AII.1, the amount of transmitted information is computed as follows (see Table AII.2). By summing across the frequencies of the responses to each stimulus, we can determine that each stimulus was presented 24 times. Because the four stimuli were equally likely, we compute the stimulus information to be $\log_2 4$ or 2 bits (see Appendix III for values of $\log_2 N$ and $p\log_2 p$). By summing across the stimuli, we can determine that the responses were not made equally often. Thus, we must use the second equation, where p_i is the relative frequency of response i, to calculate the response information to be 1.92 bits. Similarly, the joint information is found by using the third equation to be 3.64 bits. The information transmitted then can be found by adding the stimulus information and response information ($2.00 + 1.92 = 3.92$) and subtracting the joint information from the result ($3.92 - 3.64 = 0.28$). Thus, in this example, the transmitted information is 0.28 bits.

Among other things, information theory has been applied to the measurement of the human's ability to make absolute judgments, like the letter identification task we used for our example. This

TABLE AII.1
Stimulus-Response Matrices for Three Amounts
of Information Transmission

Stimulus	Response			
	A	B	C	D
Perfect Information Transmission				
A	24	—	—	—
B	—	24	—	—
C	—	—	24	—
D	—	—	—	24
No Information Transmission				
A	6	6	6	6
B	6	6	6	6
C	6	6	6	6
D	6	6	6	6
Partial Information Transmission				
A	9	8	3	4
B	3	15	2	4
C	4	4	8	8
D	0	5	3	16

TABLE AII.2

Calculating Transmitted Information

Stimulus	Response				Stimulus Frequency
	A	B	C	D	
A	9	8	3	4	24
B	3	15	2	4	24
C	4	4	8	8	24
D	0	5	3	16	24
Response Frequency	16	32	16	32	

Stimulus Information:

$$H(S) = \log_2 4 = 2.00 \text{ bits}$$

Response Information:

$$H(R) = -\sum_{i=1}^{4} p_i \log_2 p_i$$

$$= \left[\frac{16}{96} \log_2 \frac{96}{16} + \frac{32}{96} \log_2 \frac{96}{32} \right] \times 2$$

$$= 1.92 \text{ bits}$$

Joint Information:

$$H(S,R) = -\sum_{j=1}^{4} \sum_{i=1}^{4} p_{ij} \log_2 p_{ij}$$

$$= \frac{9}{96} \log_2 \frac{96}{9} + \frac{8}{96} \log_2 \frac{96}{8} + \dots + \frac{16}{96} \log_2 \frac{96}{16}$$

$$= 3.64 \text{ bits}$$

Transmitted Information:

$$T(S,R) = H(S) + H(R) - H(S,R) = 2.00 + 1.92 - 3.64 = 0.28 \text{ bits}$$

ability is important when an operator needs to identify displayed signals accurately. Usually, as the amount of stimulus information increases, the amount of information a person transmits increases and then levels off. This asymptotic value of information transmitted can be seen as the channel capacity of the human information-processing system. For example, the channel capacity for discriminating distinct pitches of tones is approximately 2.3 bits, or five pitches (Pollack, 1952). This means that in situations that require a listener to distinguish between six or more pitches, the listener will make mistakes. Across a variety of sensory dimensions, the channel capacity is approximately 2.5 bits of information. This point was stressed in a classic article by George Miller (1956) on limitations in perception and memory, called "The Magical Number Seven, Plus or Minus Two."

Perhaps of most concern to human factors specialists is the fact that this limit in the number of stimuli that a person can identify accurately applies only to stimuli that vary on one dimension. When two dimensions, for example pitch and location, are varied simultaneously, the capacity for transmitting information increases. So, if we present a listener with a series of tones to identify, her channel capacity will probably not exceed 2.3 bits. But if we present the same tones, half originating from the left headphone and half originating from the right, her channel capacity will increase.

This tells us that we should use multidimensional stimuli in situations where more than just a few potential signals can occur.

As discussed in Chapters 13 and 14, information theory has been used to describe the relationship between uncertainty and response time, as well as movement time (Schmidt & Lee, 2011). However, in recent years, research in human information processing has become less concerned with information theory and more concerned with information flow. Modern research emphasizes developing models of the processes and representations that intervene between stimuli and responses, rather than just looking at the correspondences between them. Nevertheless, information theory's emphasis on uncertainty continues to play an important role in contemporary human performance.

Appendix III: Values of $\log_2 n$ and $-p\log_2 p$

n or p*	$\log_2 n$	$-p\log_2 p$	n or p	$\log_2 n$	$-p\log_2 p$	n or p	$\log_2 n$	$-p\log_2 p$
1	0.000	0.066	41	5.358	0.527	81	6.340	0.246
2	1.000	0.113	42	5.392	0.526	82	6.358	0.235
3	1.585	0.152	43	5.426	0.524	83	6.375	0.223
4	2.000	0.186	44	5.459	0.521	84	6.392	0.211
5	2.322	0.216	45	5.492	0.518	85	6.409	0.199
6	2.585	0.244	46	5.524	0.515	86	6.426	0.187
7	2.807	0.269	47	5.555	0.512	87	6.443	0.175
8	3.000	0.292	48	5.585	0.508	88	6.459	0.162
9	3.170	0.313	49	5.615	0.504	89	6.476	0.150
10	3.322	0.332	50	5.644	0.500	90	6.492	0.137
11	3.459	0.350	51	5.672	0.495	91	6.508	0.124
12	3.585	0.367	52	5.700	0.491	92	6.524	0.111
13	3.700	0.383	53	5.728	0.485	93	6.539	0.097
14	3.807	0.397	54	5.755	0.480	94	6.555	0.084
15	3.907	0.411	55	5.781	0.474	95	6.570	0.070
16	4.000	0.423	56	5.807	0.468	96	6.585	0.057
17	4.087	0.435	57	5.833	0.462	97	6.600	0.043
18	4.170	0.445	58	5.858	0.456	98	6.615	0.029
19	4.248	0.455	59	5.883	0.449	99	6.629	0.014
20	4.322	0.464	60	5.907	0.442	100	6.644	0.000
21	4.392	0.473	61	5.931	0.435			
22	4.459	0.481	62	5.954	0.428			
23	4.524	0.488	63	5.977	0.420			
24	4.585	0.494	64	6.000	0.412			
25	4.644	0.500	65	6.022	0.404			
26	4.700	0.505	66	6.044	0.396			
27	4.755	0.510	67	6.066	0.387			
28	4.807	0.514	68	6.087	0.378			
29	4.858	0.518	69	6.109	0.369			
30	4.907	0.521	70	6.129	0.360			
31	4.954	0.524	71	6.150	0.351			
32	5.000	0.526	72	6.170	0.341			
33	5.044	0.528	73	6.190	0.331			
34	5.087	0.529	74	6.209	0.321			
35	5.129	0.530	75	6.229	0.311			
36	5.170	0.531	76	6.248	0.301			
37	5.209	0.531	77	6.267	0.290			
38	5.248	0.530	78	6.285	0.280			
39	5.285	0.530	79	6.304	0.269			
40	5.322	0.529	80	6.322	0.258			

*p is tabled in hundredths

Glossary

abduction: a form of reasoning in which hypotheses are generated to explain some observed phenomena, and the best explanation is the one that is accepted

absolute threshold: the minimum amount of physical energy in a stimulus necessary for a person to detect that stimulus

accommodation: the process by which the lens changes shape to keep images focused on the retina

acoustic reflex: the muscular reflex within the middle ear that restricts the movement of the bones of the middle ear, protecting the inner ear from loud sounds

active touch: the perception of an object through manipulation of the object

acuity: the ability to perceive fine detail

additive-factors logic: the notion that, if the effect of two variables on reaction time is additive (i.e., the effect of both variables together is equal to the sum of the effects of both variables alone), then the variables must influence different stages of information processing. Systematic application of additive-factors logic can give some idea of the stages of processing required for a task and how these stages are arranged

analogy: a problem-solving heuristic that relies on a comparison between an unfamiliar problem and a well-known problem

anchoring heuristic: an inductive heuristic from which the estimated frequency of an event is determined by the initial evidence presented about the event

anthropometrics: the measurement of human bodily characteristics

apparent motion: perceived motion produced by discrete changes in location of stimulation

archival data: preexisting data that have been collected for some other purpose, such as medical records

arithmetic mean: the sum of all values of a dependent variable divided by the number of such values

articulation index: a measure of speech intelligibility, used especially for situations with background noise

assembly error: see *manufacturing error*

associative phase: the intermediate phase of skill acquisition in which associations between task elements are being formed

astigmatism: irregularities in the shape of the cornea that blur contours of the image that are in certain orientations

auditory canal: the canal of the outer ear that is located between the pinna and the tympanic membrane

autonomous phase: the final phase of skill acquisition in which task execution becomes automatic

availability heuristic: an inductive heuristic used to estimate probabilities of events according to the ease with which the events can be remembered

backlash: insensitivity to control movement that is present at any control position

basilar membrane: an organ in the inner ear that contains the auditory sensory receptors

behavioral variables: aspects of human action, such as time to respond, that can be measured

binocular depth cues: cues to the distance of an object in an image based on slight differences in the two images that each eye receives

binocular disparity: the retinal distance between corresponding points in the images received by each eye

biomechanics: the mechanical properties of the moving body, including the forces applied by muscles

blind spot: the location on the retina where the optic nerve leaves the eye and, hence, there are no sensory receptors

breach of express warranty: the failure of a product to function as its manufacturer stated or implied that it would

brightness: the sensation corresponding mainly to the intensity of light waves

carpal tunnel syndrome: a cumulative trauma disorder characterized by pain and tingling of the fingers and hand, caused by compression of the median nerve in the carpal tunnel of the wrist

carryover effects: a problem that arises in a within-subject experimental design where performance in one treatment condition is affected by previously received treatments

categorical perception: the tendency to perceive stimuli in discrete categories, rather than as varying along continua

central tendency: a value around which a distribution of numbers (scores or measurements, for example) tends to cluster

certainty effect: gambles with highly probable outcomes tend to be selected over gambles with improbable outcomes of higher value

change blindness: a salient change in a display often goes unnoticed

check reading: a systematic inspection of each of several dials to verify that all register normal operating values

choice reaction time: the amount of time required to select an appropriate response to the onset of a stimulus from two or more alternative responses

circadian rhythms: biological oscillations of the body with periods of approximately 24 hours

closed-loop systems: systems that make use of feedback

cochlea: a bony, fluid-filled coiled cavity in the inner ear that contains the basilar membrane

cognitive architecture: a relatively completely specified information-processing system intended to provide a basis for developing computational models of performance in a range of specific tasks

cognitive phase: the initial phase of skill acquisition, in which performance of a task relies on rules and instructions

color circle: the color appearance system created by connecting the short- and long-wavelength ends of the visual spectrum with nonspectral purple

comfort zone: the temperature and humidity combinations that are comfortable for most people

communication error: inaccurate transmission of information between members of a team

computational method: a method of human reliability analysis that calculates the probability of system success from tabled data giving the probabilities of relevant human and machine errors

cones: the sensory receptors responsible for color vision and perception of detail

connectionist models: models of cognitive function that store information as connections between "nodes" which represent cognitive or neural structures that may be found in the brain

conspicuity: the ability of a display to attract attention, or how conspicuous it is

contextual interference: difficulty in remembering an item due to the context in which it is presented

continuous controls: controls that can be set to any value along a continuum

contrast sensitivity function: a graph expressing sensitivity to contrast as a function of the spatial frequency of a sine-wave grating

control–display ratio: the ratio of the magnitude of control adjustment to the magnitude of the change in a display indicator

control knowledge: knowledge of how to structure and coordinate a problem to achieve a solution

control order: the relationship between the position of a control and the position, velocity, or acceleration of a display or system

control procedures: systematic methods used to reduce the influence of extraneous variables in a study. Control procedures help ensure that the effects observed on the dependent variables are due to the independent variables and nothing else

control structure: the collection of programs that drive a knowledge-based software system

cooperative principle: the assumption that a speaker is being cooperative and sincere to further the purpose of a conversation

correct rejection: correctly responding that a signal is not present

cost-benefit analysis: calculating the costs associated with some implementation, such as a usability study, relative to the benefits that would be obtained

cost of concurrence: the difference between the performance level on a task when it is performed alone versus when it is performed with another task to which no attentional resources are devoted

counterbalancing procedures: procedures used in within-subjects designs to minimize the effects of practice and fatigue, involving the presentation of treatment conditions in different orders

critical bandwidth: the range of frequencies contained in a complex tone outside which inclusion of additional frequencies increases the loudness

critical flicker frequency: the highest rate of flicker at which a stimulus can still be perceived as flickering. Flicker frequencies higher than this critical frequency result in the perception of a continuous stimulus

crowding: a psychological experience associated with a high population density

cumulative trauma disorders: a family of syndromes arising from repeated physical stress on a joint

dark adaptation: the process of improvement in sensitivity to light energy under conditions of low illumination

data-limited processing: limitations of human information processing attributable to impoverished input

deadspace: the amount of control movement around a neutral position that can occur with no effect on the system

decision analysis: the reduction of a complex decision problem into a series of smaller, simpler component problems

decision-support system: a computer program that guides the decision-making process

declarative knowledge: knowledge that is available for verbalization

deduction: reasoning about the solution to a problem based on formal logic applied to conditions of the problem

dependent variable: a variable representing the phenomenon of interest that is measured as a function of the independent variables

depth of field: the extent of the area before and beyond a fixated object in which other objects are also in focus

descriptive models: models of decision making that capture the ways that people think and decide

descriptive statistics: methods of condensing data to allow the description or summary of research results

design error: an error in machine design that makes operation difficult or error-prone

detail design: the third phase of system development, in which the initial preliminary design is developed further and plans are made for production

detectability: the degree to which the presence or absence of a stimulus can be determined

dichromatic vision: color blindness in which one of the three types of cone photopigments is missing

difference threshold: the minimum amount of difference between the physical energies in two stimuli necessary to detect a difference between the stimuli

differential research: experiments that use subject variables as independent variables to evaluate the effects of individual differences on other variables of interest

digital human models: software design tools that allow a designer to create a virtual human with specific physical attributes who can be inserted into environments with various dimensions and properties

disability glare: glare that reduces the detectability, legibility, and readability of display characters, which in turn impairs performance

discomfort glare: glare that causes visual discomfort when a work surface is viewed for a period of time

discrete controls: controls that can be set to one of a fixed number of states

discriminability: the degree to which a difference between two stimuli can be detected

distinctiveness: the degree to which one remembered item stands apart from other remembered items

distributed practice: performance of a task for periods interspersed with periods of rest

divided attention: the act of focusing attention on several sources of input at once

dorsal stream: visual pathway in the brain that processes information about where objects are located and how to respond to them

dynamic acuity: the ability to resolve detail for moving stimuli

dynamic displays: displays that change over time, such as altimeters

echoic memory: the sensory store for the auditory system

ecological interface design: an approach to interface design that is based on a description of the work domain at different levels of abstraction and on the skills-rule-knowledge framework (see glossary):

ecological validity: the extent to which the effects observed in a research setting can be applied to a real-world setting

elaborative rehearsal: constructing relationships among items in short-term memory to enhance long-term retention

elastic resistance: the resistance felt in a spring-loaded control, which causes the control to return to a neutral position when released

elimination by aspects: a descriptive decision-making heuristic by which the decision-making process occurs through a systematic elimination of features for comparison

empiricism: evaluating scientific hypotheses through the collection of data based on controlled observations

encoding specificity principle: the ability to remember an item will depend on the match between the context in which it is retrieved and the context in which it is encoded

engineering anthropometry: the use of anthropometric data in the design of equipment

engineering models of human performance: models intended to produce quick, approximate predictions of human functioning that can be used to make design decisions

engineering psychology: see *human factors*

episodic memory: memory for specific events

equal loudness contours: the intensity levels across tones of varying frequencies that result in equal perceived loudness

equal pitch contours: the frequencies across tones of varying intensity levels that result in equal perceived pitch

ergonomics: see *human factors*

error of commission: the performance of an incorrect action

error of omission: the failure to perform a necessary action

ethnographic methods: research methods that provide qualitative descriptions of human behavioral and social phenomena based on field observations

executive control: processes involved in the coordination of more basic cognitive functions such as direction of attention, rehearsal of information, and so on

expected-utility theory: a normative theory of decision making in which choices are based on the average utility of different objects or outcomes

expert system: a knowledge-based software system intended to perform as an expert consultant

external validity: the extent to which the results obtained in a study generalize to other situations

fact base: the data base and models used by a knowledge-based software system

false alarm: incorrectly responding that a signal was present when it was not

far point: the point beyond which increasing the distance of a fixated object requires no further change in accommodation to keep its image in focus

fatigue effects: decrements in performance attributable only to the amount of time spent at a task

feature-comparison models: models of memory that assume concepts are stored as lists of features

Fechner's law: the magnitude of a sensation is proportional to the logarithm of the physical intensity of a stimulus

figure–ground organization: the segregation of parts of an image into objects against a background

filter-attenuation model: a model of attention similar to filter theory that presumes that several sources of input are differentially weighted, which allows some information from unattended sources to enter the central processing channel

filter theory: a model of attention that presumes the existence of a central processing channel that can act on input from only a single source at one time

fine adjust time: after the travel time, the time required to adjust the position of a control precisely

Fitts's law: movement time is a linear function of the index of difficulty

focus group: a small group of people selected from a larger population to discuss opinions and views on a topic or product

Fourier analysis: a method of decomposing a complex waveform into its component sinusoids

fovea: that region on the retina containing only cone receptors. Acuity is highest in this region

framing: how a decision-making problem is presented

frequency distribution: a plot of the number of times each value of a dependent variable was observed

frequency of use: a design principle that the most frequently used and important displays or controls should be located in the central visual field

frequency theory: a theory of pitch perception suggesting that the frequency of vibration of the basilar membrane is represented by a pattern of neural firing that occurs at the same frequency

frictional resistance: resistance encountered at any point during the movement of a control due to the mechanical properties of the control

functional equivalence: the extent to which the tasks performed in a simulated environment mimic those of the real world

gain: a measure of the responsiveness of a control, inversely related to the control–display ratio

gaze-contingent multiresolution display: a display for which a region of the image around fixation is of higher resolution than the remainder of the display. The high-resolution region shifts along with shifts in fixation

general adaptation syndrome: a physiological response to stress characterized by swollen adrenal glands, atrophied thymus glands, and stomach ulcers when exposure to stress has been prolonged and severe

gestalt grouping: the tendency for individual elements to be grouped into a larger whole on the basis of principles of proximity, similarity, and so on

given-new strategy: the fact that sentences in a meaningful conversation contain both old and new information

glare: a high-intensity light that interferes with the perception of objects of lower intensity

go-no go reaction time: the amount of time required to execute a single response to the onset of a particular subset of the possible stimuli

gross negligence: reckless and wanton disregard by the manufacturer of a product of the manufacturer's legal responsibilities

groupware: computer software developed to support interactions among group and team members on projects

harmonics: integer multiples of the fundamental frequency of a complex tone

Hawthorne effect: changes in performance or productivity that can be traced to any alteration of the workplace environment and not to any specific variable that was manipulated to effect the alteration

head-up display: a display on the windshield of an aircraft, automobile, or other vehicle that allows the operator to read the display without having to direct his gaze away from the outside world

helmet-mounted display: a display mounted on a helmet worn by a person that enables the display to be visible regardless of the direction in which the person is looking

Hick–Hyman law: choice reaction time is a linear function of the amount of information transmitted

hit: correctly responding that a signal is present

human–computer interaction: that area of human factors concerned with the design of computer workstations and software interfaces to optimize performance of computer-based tasks

human–computer interaction (1): the subfield of human factors and ergonomics concerned with designing usable interfaces for people to interact with computerized systems

human error: a decision or action made by a person that has undesirable consequences for the operation of a system or use of a product

human factors: the study of human cognitive, behavioral, and biological characteristics that influence the efficiency with which a human can interact with the inanimate components of a human–machine system

human information processing: the view that human perception, cognition, and action are based on a systematic processing of information from the environment

human–machine system: an entity consisting of a human operator and a machine that work together to achieve some goal

human reliability: the probability that an operator makes no errors while interacting as part of a human–machine system

human–systems integration: a term that refers to the consideration and integration of human issues across an entire system. It is a somewhat broader term than human factors, because human issues might encompass, in addition to engineering psychology or ergonomics/human factors, sociological, economic, political, and psychological concerns

hyperopia: farsightedness, or the inability to see close objects

hypervigilance: a state of panic in which thinking becomes overly simplistic, resulting in hasty, poor decision making

hypothesis: a tentative and testable statement about the cause of some phenomenon

iconic memory: the sensory store for the visual system

identification acuity: acuity as measured by a Snellen eye chart; the distance at which an observer could identify letters that an observer with normal vision could identify at a standard distance

illuminance: the amount of light falling on a surface

independence point: the point in the performance operating characteristic space indicated by the performance level of each task when performed alone

independent variable: a variable that is overtly changed in an experiment to determine whether it affects a dependent variable

index of accessibility: a measure of the ease with which frequently used controls on a panel can be reached

index of difficulty: a measure of the difficulty of an aimed movement, given by the logarithm to the base 2 of the ratio of twice the distance to the target divided by the target width

induced motion: perceived movement of a stationary element induced by motion of its frame of reference

induction: reasoning in which a general solution to a problem is generated from the particular conditions of the problem

inertial resistance: a control resistance that decreases as control acceleration increases

information theory: quantifying the information in a set of events by the average minimum number of binary questions required to determine the identity of an item in the set

input error: an error that occurs during the perception of a stimulus

installation error: an error in the installation of a machine that leads to system failure

interaural intensity differences: differences in the intensity of a sound at each ear, due to a sound shadow created by the head, that provide cues to position

interaural time differences: differences in the time at which a tone reaches each ear that provide cues to positions

internal validity: the degree to which effects observed in a study can be attributed to the variables of interest

inverse square law: the intensity of an auditory signal is inversely related to the squared distance of the sound source

isolation effect: more attention is focused on features that are unique to different choices rather than on features that the choices have in common

isometric control: a fixed control that responds according to the amount of force exerted on it

isotonic control: a movable control that responds according to its amount of displacement

iterative corrections model: a theory of movement control that assumes that an aimed movement is composed of a series of discrete submovements, each traversing a fixed proportion of the distance to the target

job analysis: an analysis of a position (job): to determine the tasks and responsibilities of a worker in that position, the conditions under which that worker must perform, and the skills and training that the position requires

job design: the act of structuring tasks and assigning them to positions

kinesthesis: sensory information about the location of the limbs during movement

knowledge-based behavior: a mode of behavior in which the person must solve problems for which they are not trained and have not learned rules for action

knowledge elicitation: methods for drawing out the knowledge that an expert or user possesses about a domain or task

knowledge of performance: detailed feedback concerning the performance of a movement

knowledge of results: feedback concerning the success or failure of a movement

latent semantic analysis: an analysis that, when applied to a sample of text, produces a semantic space that depicts the relationships between concepts

lateral inhibition: the inhibition of a cell's firing rate due to the activity of neighboring cells

late-selection model: a model of attention that presumes that information from all input channels is identified, but that only the information from the attended input source is acted on

legibility: the ease with which symbols and letters can be discerned

level of processing: the degree of elaborative or semantic processing performed on information in short-term memory

lightness: the perceived reflectance of an object, or how dark or light the object appears on a scale from black to white

lightness constancy: maintenance of perceived relative lightness under different levels of illumination

lightness contrast: changes in the lightness of an object with changes in the intensity of the surrounding area

likelihood alarm: a warning, caution, or advisory signal that also presents information about the likelihood of an event

link analysis: an analysis of display panel design based on connections between displays, defined in terms of frequency and sequence of use. Link analysis can also be used to analyze control panels and to aid in the design of workstations

loading task paradigm: a method of measuring mental workload in a dual-task situation in which the emphasis is placed on the secondary task, and mental workload is estimated from performance on the primary task

long-term store: an unlimited-capacity memory system that retains information for an indefinite period of time

luminance: the amount of light generated by a surface

macroergonomics: an approach to human factors that stresses the organizational and social environment in which the human–machine system functions

maintenance error: an error during routine maintenance of a machine that leads to system failure

maintenance rehearsal: covert repetition of material held in short-term memory

manufacturing error: an error in the fabrication of a machine that leads to system failure

masked threshold: the amount of physical energy in a stimulus necessary to detect that stimulus when it is presented in a noisy background

masking: the interference between the presentation of one stimulus and the perception of another presented in close spatial and/or temporal proximity

massed practice: continuous performance of a task for an extended period of time

median: that value of a dependent variable below which and above which 50% of all values fall; the value with a percentile rank of 50%

mediation error: an error that occurs during cognition that is not attributable to misperception of a stimulus or incorrect execution of an intended action

mental effort: the amount of cognitive work required to perform a task

mental model: a dynamic representation or simulation of a problem held in working memory

mental workload: an estimate of the attentional demands of a task

method of constant stimuli: a method to determine a threshold that presents a large number of stimulus intensities in random order

method of limits: a method to determine a threshold that presents stimulus intensities in increasing or decreasing increments

miss: incorrectly responding that a signal was not present

mistakes: errors that arise in the planning of an action

mnemonics: mental strategies used to organize and aid memory for information

mode: the most frequently occurring value of a dependent variable

monochromatic vision: a kind of color blindness in which an individual has either no cones or only one type of cone

monocular depth cues: cues to the depth of an object in an image that are available to a monocular viewer. Stationary monocular cues are those used to portray depth in still paintings. Additional monocular cues are provided when an observer moves

Monte Carlo method: a method of human reliability analysis in which system performance is predicted by simulating a model system

motion contrast: apparent motion of a stationary texture induced by motion of a surrounding texture

motor program: an abstract plan thought to control specific classes of movements

motor unit: a small group of muscle fibers innervated by a single motor neuron

multiple-resource model: a model of attention that presumes the existence of several pools of mental resources, each appropriate to different kinds of stimuli, processing, and response modalities

myopia: nearsightedness, or the inability to see distant objects

naturalistic research: the observation of behavior in real-world settings without manipulation of any independent variables

near point: the point at which moving an object closer produces no further accommodation

negligence: the failure of a manufacturer to engage in reasonable actions to meet his or her legal responsibilities

network models: models of memory in which concepts are represented as connections between functionally related neural units

neuron: a cell that transmits an electrochemical signal within the nervous system

normative models: models of decision making that predict the choices that would be made by an optimal decision maker

null hypothesis: the proposal that a treatment had no effect on the dependent variable

observational learning: learning to perform a task by watching another performer

occupational ergonomics program: a plan for redesigning the work environment and practices to conform to ergonomic principles

occupational stress: stress that arises from the work environment

oculomotor depth cues: cues to the depth of an object in an image based on proprioceptive feedback from the muscles in the eye

olfactory cilia: the likely sensory receptors for olfaction

olfactory epithelium: that area of the nasal cavity that contains the olfactory sensory receptors

open-loop systems: systems that do not make use of feedback

operating error: an inappropriate use or operation of a machine

operational definition: the definition of a concept in terms of the methods by which it is measured

opponent process theory: a theory of color vision that proposes that neural mechanisms code blue and yellow together and red and green together so that one color of a pair can be signaled, but not both

optimized initial impulse model: a model of movement control that combines elements of the iterative corrections and impulse variability models

organizational development: changes in the structure and goals of an organization, designed to improve organizational effectiveness

ossicles: the three small bones in the inner ear that transmit pressure changes from the tympanic membrane to the oval window

output error: the selection and execution of an inappropriate action

oval window: a membrane that receives vibrations from the ossicles and produces waves in the fluid around the basilar membrane

parallel components: system components that receive input and commence operation simultaneously

part-whole transfer: the extent to which practice with the components of a task improves performance of the entire task

passive touch: the perception of a texture pressed against the skin

pay for performance: a pay schedule in which salary depends on a worker's level of productivity

percentile rank: a measurement given to a particular value of a dependent variable that specifies the percentage of scores that fall below it

perceptual organization: the way that relationships are formed among the different elements of an image to produce a percept

performance appraisal: the formal evaluation of an employee's performance

performance efficiency: a measure of how efficiently two tasks can be performed together, defined as the smallest distance between the performance operating characteristic curve and the independence point

performance operating characteristic: a plot of performance for a divided-attention situation, by which the performance of one task is plotted as a function of the performance on another task under several levels of relative task emphasis

personal space: the area immediately surrounding one's body

personnel selection: choosing employees for a job on the basis of the match between their characteristics or qualifications and the job requirements

phoneme: the smallest unit of speech that, when changed, changes the meaning of an utterance

photometry: measurement of the functional amount of light energy for human vision

photopic vision: vision under conditions of bright light, controlled primarily by cones

pinna: the outer, visible part of the ear

place theory: a theory of pitch perception that proposes that the perception of pitch is determined by the location of the active receptors on the basilar membrane and the neurons that they innervate

population stereotype: an intuitive association between a control motion and its associated effect

positive misaccommodation: a problem that arises in the use of head-up displays in which an observer's eyes accommodate for a distance closer than the far point. This results in poor size and depth perception

power law of practice: the empirical finding that performance (as measured by response time or accuracy) improves as a power function of the amount of time spent practicing a task

practice effects: improvements in performance attributable only to the amount of time spent performing a task

preference reversals: a change in the most preferred object under changes in the context in which the choice is presented

preferred noise criterion: a level of background noise intensity and frequency that is optimal for a given task environment

preliminary design: the second phase of system development, in which alternative designs are considered, resulting in an initial, tentative design

presbyopia: a loss of accommodative ability that comes with age

proactive interference: forgetting of information that occurs because of the memory of previously presented information

probability: a number from 0 to 1 that indicates the likelihood of a random event. Usually, the number of times that an event of interest is observed divided by the total number of observations made

probability density function: the continuous analogue to the probability distribution; used to assign probabilities to continuous events (e.g., time)

probability distribution: a relative frequency distribution over an entire set of discrete events, describing the proportion of times that each event occurs relative to all other events

problem space hypothesis: a conception of problem solving as a mental space in which the problem solver must move along a solution path from a start state to a goal state

procedural knowledge: knowledge of how tasks are performed that is not available for verbalization

production and development phase: the final phase of system development, in which the system is actually built, tested, and evaluated

production system: a data base, control system, and set of if-then rules that can be used to solve simple or complex problems

proprioception: sensory information about the position of the limbs

proxemics: the way that people manage the space around them and their distances from other people

psychological refractory period effect: increases in response time for the second response when two tasks must be performed in rapid succession

psychophysical scaling: a mathematical expression relating the physical intensity of a stimulus to its perceived magnitude

psychophysics: the study of the relation between physical stimulus properties and psychological experience

pupillometry: the measurement of the diameter of the pupil

Purkinje shift: the relatively greater perceived brightness of objects of short wavelength under scotopic viewing conditions

quantitative error: an action that fails by being either insufficient or excessive

radiometry: measurement of light energy

random walk: a continuous model of information processing that assumes that evidence is accumulated over time toward alternative responses

reach envelope: an area in which controls and other objects should be located to ensure that some large percentage of the population will be able to reach them

reactivity: changes in a mental process due to concurrent verbalization of that process

readability: the degree to which a display of letters or characters allows fast and accurate recognition of information

receiver operating characteristic: a plot of the proportion of hits as a function of the proportion of false alarms under several levels of response bias

receptive field: the area of sensory receptors that, when stimulated, affects the firing rate of a particular neuron

recommended weight limit: the weight of a load that a healthy individual can lift for as much as 8 hours per day without increased risk of lower back pain

reflection effect: when expected utilities are positive, the high probability outcomes are preferred even when their expected utility is low. When expected utilities are negative, low-probability outcomes are preferred

relative frequency distribution: a plot of the proportion of times that a value of a dependent variable was observed

reliability: the probability that a system, subsystem, or component does not fail

representativeness heuristic: an inductive heuristic used to assign probabilities to events according to their perceived similarity between some representative outcome

resolution acuity: the ability to distinguish between a field of varying contrast and a field of uniform intensity

resource-limited processing: limitations in human information processing attributable to a lack of cognitive resources; for example, attention or working memory

response bias: a tendency to prefer one response over others, regardless of the stimulus conditions

retina: a two-dimensional grid of sensory receptors and associated neurons lining the back wall of the eye

risk analysis: a comprehensive analysis of the costs of system failure, taking into account system and human reliability and the risks that accompany specific failures

rods: the sensory receptors responsible for vision under conditions of low illumination

rule-based behavior: a mode of behavior in which a person's skills are not applicable and she must retrieve previous learned rules from memory

scenario-based design: narratives are developed that depict ways that a person might use a software tool or product, and these narratives are used to guide the design process

schema: an abstract mental representation, similar to a mental model, for organizing sequences of events

schema theory: a theory of motor skill that assumes the existence of a generalized motor program, the parameters of which are determined by schemas acquired through practice

scientific method: the process by which alternative hypotheses concerning the cause of some phenomenon are evaluated. This evaluation is based on the outcomes of controlled observations

scotopic vision: vision under conditions of low illumination, primarily controlled by the rods

selection error: an action performed with the wrong control

selective attention: the act of focusing on one source of information and ignoring all others

semantic context: the effect of the meaning of a context on the perception of a stimulus

semantic memory: long-term memory for general knowledge

sensory receptors: specialized cells in a sensory system that convert physical energy into nervous impulses

sensory store: a buffer that retains sensory information briefly

sequence error: the performance of an action at the wrong position within a sequence of actions

sequence of use: a design principle that states that, if displays must be scanned in a fixed sequence, the displays should be arranged in that sequence

serial components: an arrangement of system components in which each component receives as input the output of a previous component and delivers its output as input to the following component

shape constancy: the tendency to perceive an object as having the same shape regardless of its slant or tilt

short-term store: a limited-capacity memory system in which information is retained through rehearsal

sick building syndrome: a condition in which many occupants of a building experience chronic respiratory symptoms, headaches, and eye irritation

signal-detection theory: a theory that assumes that binary decisions concerning the presence or absence of a signal are based on discriminability of the signal and a response criterion

simple reaction time: the amount of time required to react with a single response to the onset of any stimulus event

situation awareness: consciousness of the objects in the environment, what they mean, and their future status

size constancy: the tendency to perceive an object as having the same size regardless of its visual angle

skill-based behavior: a mode of behavior in which the person is engaged in highly overlearned activities for which she has been trained

skill-rule-knowledge framework: a framework of cognitive behavior in which behaviors are classified according to the level of skill involved

slips: errors arising in the execution of an action

sociotechnical system: an organizational system comprised of a technical subsystem and a personnel subsystem

somesthetic senses: those senses associated with skin, joints, muscles, and tendons, including touch, pressure, temperature, pain, vibration, and proprioception

span of apprehension: the number of briefly displayed visual stimuli that can be reported without error

speech spectrogram: a plot of the frequencies that appear in a speech signal over time

speed–accuracy tradeoff: for performance of tasks, a person can respond faster and less accurately or slower and more accurately

spinal reflex: simple actions controlled by the spinal cord

standard deviation: the square root of the variance of a dependent variable

static displays: displays that do not change over time, such as road signs

Stevens's law: the magnitude of sensation provided by a stimulus is directly proportional to some power of the physical intensity of the stimulus when sensation is scaled using magnitude estimation procedures

stimulus-response compatibility: the ease with which a response to a stimulus can be selected based on the assignment of stimuli to responses

stimulus variables: environmental factors that affect behavior

strict liability: a manufacturer's responsibility for any product defect

stroboscopic motion: the perception of movement arising from the sequential illumination of two or more spatially separated lights in close succession

strong methods: methods of problem solving based on an expert's knowledge of a domain

structurally limited processing: limitations in human information processing that arise when one structure is called on to perform more than one task

subject variables: individual differences such as physical characteristics, mental abilities, and training

subjective assessment techniques: measurements obtained through an operator's evaluation of some aspect of a task or procedure. These techniques are commonly used to measure mental workload

subsidiary task paradigm: a method of measuring mental workload using a dual-task situation in which emphasis is placed on the primary task and mental workload is estimated from performance on the secondary task

subtractive logic: the notion that the time to perform a mental event can be found by measuring the reaction time in a task that requires that event and in a task that requires everything except that event and then subtracting one from the other

syllogism: a list of premises and a conclusion drawn from them

syntactic context: the effect of grammatical context on the perception of a stimulus

system: a collection of components that act together to achieve a goal that could not be achieved by any single component alone

system planning: the first phase of system development, in which the need for a system is identified

systems engineering: an interdisciplinary approach to the design of complex systems that bases design decisions on achieving system goals

task analysis: the analysis of a task in terms of its perceptual, cognitive, and motor components

task environment: the objects and allowable actions that may be used to achieve a solution to a problem

taste buds: groups of sensory receptors on the tongue

team performance: study of the actions of functioning of two or more people as a team rather than the functioning of a single person

teleoperators: general-purpose, dextrous human–machine systems that augment the physical skills of the operator by allowing him to pick up and manipulate objects from a remote location

territoriality: behavior patterns oriented toward occupying and controlling physical spaces

theory: an organized framework of causal statements that allows the understanding, prediction, and control of some phenomena

threshold shift: a decrease in auditory sensitivity due to exposure to high noise levels

timbre: the texture of a complex tone, which is determined by such factors as the relative intensities of its harmonics

time-and-motion study: an analysis of the movements required to perform a job and the time required for each movement

timing error: the performance of an action at the wrong time

tracking task: a task that requires matching a dynamic stimulus signal with an identical output signal

transfer-appropriate processing: the ability to remember an item encoded in a particular way depends on the way in which the item is tested

transmitted information: the amount of information (in bits) passing through a communication channel, as derived from the amount of information in the input and the amount of information in the output

travel time: the time required to move a control into the vicinity of a desired position

trichromatic color theory: a theory of color vision that proposes that color is perceived as a function of the relative activity in the blue, green, and red color systems

two-point thresholds: the minimum distance between two points of stimulation on the skin that allows the perception of two distinct stimuli

tympanic membrane: a delicate membrane that vibrates with changes in air pressure created by an auditory stimulus. It is also called the eardrum

unitary-resource model: a model of attention that views attention as a single pool of resources reserved for mental activities

user interface: the component of a software system responsible for presenting output to and receiving input from the user

utility: the subjective worth of an object or event

validity: the degree to which a test or some other measurement device measures what it is supposed to measure

variability: a measure that indicates the degree of "spread" in a distribution of numbers from a central point. Usually, the variance

variability of practice: the extent to which the specific movements executed during practice of a motor skill differ from each other

variables: critical events or objects that change or can be changed

variance: the sum of all squared differences between the values of a dependent variable and their mean, divided by the total number of such values minus 1

ventral stream: visual pathway in the brain that processes information about what an object is

verbal protocol analysis: a method for organizing verbal reports obtained as a person describes what she is thinking of while performing a task

vergence: rotations of the eyes inward or outward with changes in the point of fixation

vernier acuity: the ability to discriminate between a broken and unbroken line

vestibular sense: the sense associated with the perception of bodily motion and balance

vigilance decrement: a decline in the hit rate over time in the performance of a vigilance task

vigilance task: a task characterized by the requirement of detecting small, infrequent changes in the environment over long periods of time

virtual reality environments: computerized "worlds" intended to provide the experience of moving about and interacting with objects in a three-dimensional space

viscous resistance: control resistance that increases with control velocity

visibility: how well a display can be seen, or how visible it is

visual angle: a measure of the size of the retinal image of an object

visual cortex: the primary receiving area of the cortex in which visual signals are processed and recombined

visual dominance: the priority that visual information receives when information arrives from the visual and other systems simultaneously

Warrick's principle: the pointer of a display should move in the same direction as the side of the control nearest the display

weak methods: methods of problem solving of broad applicability used to solve unfamiliar problems when the correct way to proceed is unknown

Weber's law: the smallest detectable change in the magnitude of a stimulus is a constant proportion of the magnitude of the original stimulus

work tolerance: the ability of an operator to perform well while maintaining physical and emotional health

working memory: another name for short-term memory that emphasizes the operations that occur on information in short-term memory

Yerkes–Dodson law: performance is an inverted U-shaped function of arousal, with best performance at intermediate levels of arousal

References

AAIB Bulletin: 12/2015 N103CD EW/C2014/10/01. Downloaded 1/3/2016 from https://assets.digital.cabinet-office.gov.uk/media/5669383fed915d035f000000/Gulfstream_III_G-1159A_N103CD_12-15.pdf

Aaronson, D. & Scarborough, H. 1976. Performance theories for sentence coding: Some qualitative evidence. *Journal of Experimental Psychology: Human Perception and Performance*, 2, 56–70.

Abendroth, B. & Landau, K. 2006. Ergonomics of cockpits in cars. In W. Karwowski (Ed.), *International Encyclopedia of Ergonomics and Human Factors* (2nd ed., Vol.2, pp. 1626–1635). Boca Raton, FL: CRC Press.

Abernethy, B., Kippers, V., Hanrahan, S., Pandy, M., McManus, A. & Mackinnon, L. 2013. *Biophysical Foundations of Human Movement* (3rd ed.). Champaign, IL: Human Kinetics.

Adams, J.A. 1967. Engineering psychology. In H. Helson & W. Bevan (Eds.), *Contemporary Approaches to Psychology* (pp. 345–383). Princeton, NJ: Van Nostrand Reinhold.

Adams, J.A. 1972. Research and the future of engineering psychology. *American Psychologist*, 27, 615–622.

Adams, J.A. 1984. Learning of movement sequences. *Psychological Bulletin*, 96, 3–28.

Adams, S.K. 2006. Input devices and controls: Manual, foot, and computer. In W. Karwowski (Ed.), *International Encyclopedia of Ergonomics and Human Factors* (2nd ed., Vol. 1, pp. 1419–1439). Boca Raton, FL: CRC Press.

Ahlstrom, U. & Suss, J. 2015. Change blindness in pilot perception of METAR symbology. *International Journal of Industrial Ergonomics*, 46, 44–58.

Aiello, J.R., DeRisi, D.T., Epstein, Y.M. & Karlin, R.A. 1977. Crowding and the role of interpersonal distance preference. *Sociometry*, 40, 271–282.

Åkerstedt, T. & Gillberg, M. 1982. Displacement of the sleep period and sleep deprivation: Implications for shift work. *Human Neurobiology*, 1, 163–171.

Akhmadeeva, L., Tukhvatullin, I. & Veytsman, B. 2012. Do serifs help in comprehension of printed text? An experiment with Cyrillic readers. *Vision Research*, 65, 21–24.

Akkirman, A.D. & Harris, D.L. 2005. Organizational communication satisfaction in the virtual workplace. *Journal of Management Development*, 24, 397–409.

Al Amin, M.S., Nuradilah, Z., Isa, H., Nor Akramin, M. & Febrian, I. 2013. A review on ergonomics risk factors and health effects associated with manual materials handling. In *Advanced Engineering Forum* (Vol. 10, pp. 251–256). Trans Tech Publications.

Aldrich, F.K. & Parkin, A.J. 1987. Tangible line graphs: An experimental investigation of three formats using capsule paper. *Human Factors*, 20, 301–309.

Aldrich, T.B. & Szabo, S.M. 1986. A methodology for predicting crew workload in new weapon systems. In *Proceedings of the Human Factors Society 30th Annual Meeting* (pp. 633–637). Santa Monica, CA: Human Factors Society.

Alexander, T., Schlick, C., Sievert, A. & Leyk, D. 2008. Assessing human mobile computing performance by Fitts' law. In J.Lumsden, J. Lumsden (Eds.), *Handbook of Research on User Interface Design and Evaluation for Mobile Technology* (Vols. 1 and 2; pp. 830–846). Hershey, PA: Information Science Reference/IGI Global.

Algvere, P.V., Marshall, J. & Seregard, S. 2006. Age-related maculopathy and the impact of blue light hazard. *Acta Ophthalmologica Scandinavica*, 84, 4–15.

Allami, N., Paulignan, Y., Brovelli, A. & Boussaoud, D. 2008. Visuo-motor learning with combination of different rates of motor imagery and physical practice. *Experimental Brain Research*, 184, 105–113.

Allen, T.J. & Gerstberger, P.G. 1973. A field experiment to improve communications in a product engineering department: The nonterritorial office. *Human Factors*, 15, 487–498.

Allport, D.A., Antonis, B. & Reynolds, P. 1972. On the division of attention: A disproof of the single channel hypothesis. *Quarterly Journal of Experimental Psychology*, 24, 225–235.

Alongi, P. & Davis, J. 2006. Cruise control led to crash, jury says. *The Greenville News*. Downloaded on August 7, 2006, from http://www.greenvillenews.com/apps/pbcs.dll/article? AID¼=/20060807/NEWS01/60807 0315&SearchID=73253038864740

Altinsoy, M.E. & Hempel, T. 2011. Multimodal user interfaces: Designing media for the auditory and tactile channels. In K.-P.L. Vu & R.W. Proctor (Eds.), *Handbook of Human Factors in Web Design* (2nd ed.; pp. 127–151). Boca Raton, FL: CRC Press.

Altman, I. & Chemers, M. 1980. *Culture and Environment.* Monterey, CA: Brooks/Cole.

Altmann, C.F., Uesaki, M., Ono, K., Matsuhashi, M., Mima, T. & Fukuyama, H. 2014. Categorical speech perception during active discrimination of consonants and vowels. *Neuropsychologia,* 64, 13–23.

Alvarez, G.A., Horowitz, T.S., Arsenio, H.C., DiMase, J.S. & Wolfe, J.M. 2005. Do multielement visual tracking and visual search draw continuously on the same visual attention resources? *Journal of Experimental Psychology: Human Perception and Performance,* 31, 643–667.

Amar, J. 1920. *The Human Motor.* New York: Dutton.

Amell, T.K. & Kumar, S. 1999. Cumulative trauma disorders and keyboarding work. *International Journal of Industrial Ergonomics,* 25, 69–78.

American Medical Association 2016. *Human and Environmental Effects of Light Emitting Diode (LED) Community Lighting.* Report of the Council on Science and Public Health (CSAPH Report 2-A-16). Chicago, IL: American Medical Association.

American Society of Heating, Refrigerating, and Air Conditioning Engineers (ASHRAE). 1985. *ASHRAE Handbook 1985: Fundamentals.* Atlanta, GA: ASHRAE.

Anderson, D.I. 1999. The discrimination, acquisition, and retention of aiming movements made with and without elastic resistance. *Human Factors,* 41, 129–138.

Anderson, D.I., Magill, R.A., Sekiya, H. & Ryan, G. 2005. Support for an explanation of the guidance effect in motor skill learning. *Journal of Motor Behavior,* 37, 231–238.

Anderson, J.R. 1983. *The Architecture of Cognition.* Cambridge, MA: Harvard University Press.

Anderson, J.R., Bothell, D., Byrne, M.D., Douglass, S., Lebiere, C. & Qin, Y. 2004. An integrated theory of the mind. *Psychological Review,* 111, 1036–1060.

Anderson, J.R., Douglass, S. & Qin, Y. 2005b. How should a theory of learning and cognition inform instruction? In A.F. Healy (Ed.), *Experimental Cognitive Psychology and Its Applications* (pp. 47–58). Washington, DC: American Psychological Association.

Anderson, J.R., Matessa, M. & Lebiere, C. 1997. ACT-R: A theory of higher level cognition and its relation to visual attention. *Human–Computer Interaction,* 12, 439–462.

Anderson, J.R., Taatgen, N.A. & Byrne, M.D. 2005c. Learning to achieve perfect timesharing: Architectural implications of Hazeltine, Teague, and Ivry (2002). *Journal of Experimental Psychology: Human Perception and Performance,* 31, 749–761.

Andersson, G.B.J. 1986. Loads on the spine during sitting. In N. Corlett, J. Wilson & I. Manenica (Eds.), *The Ergonomics of Working Postures* (pp. 309–318). London: Taylor & Francis.

Andre, J. 2003. Controversies concerning the resting state of accommodation: Focusing on Leibowitz. In J. Andre, D.A. Owens & L.O. Harvey, Jr. (Eds.), *Visual Perception: The Influence of H.W. Leibowitz* (pp. 69–79). Washington, DC: American Psychological Association.

Andre, J.T. & Owens, D.A. 1999. Predicting optimal accommodative performance from measures of the dark focus of accommodation. *Human Factors,* 4, 139–145.

Andreoni, G., Piccini, L. & Maggi, L. 2006. Active customized control of thermal comfort. In W. Karwowski (Ed.), *International Encyclopedia of Ergonomics and Human Factors* (2nd ed., Vol. 2, pp. 1761–1766). Boca Raton, FL: CRC Press.

Andres, R.O. & Hartung, K.J. 1989. Prediction of head movement time using Fitts' law. *Human Factors,* 31, 703–713.

Andriessen, J.E. 2012. *Working with Groupware: Understanding and Evaluating Collaboration Technology.* New York: Springer Science & Business Media.

Ang, B. & Harms-Ringdahl, K. 2006. Neck pain and related disability in helicopter pilots: A survey of prevalence and risk factors. *Aviation, Space, and Environmental Medicine,* 77, 713–719.

Annett, J. 2004. Hierarchical task analysis. In D. Diaper & N. Stanton (Eds.), *The Handbook of Task Analysis for Human–Computer Interaction* (pp. 67–82). Mahwah, NJ: Lawrence Erlbaum.

Anonymous. 2000. Basic ergonomics for hand tool users. *Occupational Health and Safety,* 69(7), 68–74.

Anson, G., Elliott, D. & Davids, K. 2005. Information processing and constraints-based views of skill acquisition: Divergent or complementary? *Motor Control,* 9, 217–241.

Ansorge, U. & Wühr, P. 2004. A Response-Discrimination Account of the Simon Effect. *Journal of Experimental Psychology: Human Perception and Performance,* 30, 365–377.

Anthony, W.P. & Harrison, C.W. 1972. Tympanic membrane perforations: Effect on audiograms. *Archives of Otolaryngology,* 95, 506–510.

Antin, J.F., Lauretta, D.J. & Wolf, L.D. 1991. The intensity of auditory warning tones in the automobile environment: Detection and preference evaluations. *Applied Ergonomics*, 22, 13–19.

Antunes, P., Ferreira, A., Zurita, G. & Baloian, N. 2011. Analyzing the support for large group collaborations using Google maps. In *Computer Supported Cooperative Work in Design (CSCWD, 2011 15th International Conference on)* (pp. 748–755). Piscataway, NJ: IEEE.

Anzai, Y. 1984. Cognitive control of real-time event-driven systems. *Cognitive Science*, 8, 221–254.

Arditi, A. & Cho, J. 2005. Serifs and font legibility. *Vision Research*, 45, 2926–2933.

Arkes, H.R., Shaffer, V.A. & Medow, M.A. 2007. Patients derogate physicians who use a computer-assisted diagnostic aid. *Medical Decision Making*, 27, 189–202.

Arevalillo-Herráez, M., Arnau, D. & Marco-Giménez, L. 2013. Domain-specific knowledge representation and inference engine for an intelligent tutoring system. *Knowledge-Based Systems*, 4, 997–105.

Armstrong, J., McPherson, K. & Nayar, S. 2012. External memory aid training after traumatic brain injury: 'making it real'. *The British Journal of Occupational Therapy*, 75, 541–548.

Armstrong, T. J. 2010. Hand tool selection, design and prevention of Work Related Musculoskeletal Disorders, WMSDs. In *Proceedings of the Human Factors and Ergonomics Society 54th Annual Meeting* (pp. 1145–1146). Santa Monica, CA: HFES.

Armstrong, T.J., Foulke, J.A., Joseph, B.S. & Goldstein, S.A. 1982. Investigation of cumulative trauma disorders in a poultry processing plant. *American Industrial Hygiene Association Journal*, 43, 103–116.

Armstrong, T.J., Ulin, S. & Ways, C. 1990. Hand tools and control of cumulative trauma disorders of the upper limb. In C.M. Haslegrave, J.R. Wilson, E.N. Corlett, & I. Manenica (Eds.), *Work Design in Practice* (pp. 43–50). London: Taylor & Francis.

Arnaut, L.Y. & Greenstein, J.S. 1990. Is display/control gain a useful metric for optimizing an interference? *Human Factors*, 32, 651–663.

Ascar, B.S. & Weekes, A.M. 2005. Design guidelines for pregnant occupant safety. *Journal of Automobile Engineering*, 219, 857–867.

Askey, S. & Sheridan, T. 1996. Safety of high-speed ground transportation systems—Human factors phase II: Design and evaluation of decision aids for control of high-speed trains: Experiments and model. Report No. DOT=FRA=ORD-96=09. Washington, DC: U.S. Department of Transportation.

Askren, W.B. & Howard, J.M. 2005. A road map for the practice of forensic human factors and ergonomics. In Y.I. Noy & W. Karwowski (Eds.), *Handbook of Human Factors in Litigation* (pp. 5-1–5-16). Boca Raton, FL: CRC Press

Associated Press wire story 1990a. Drinking on the job. April 29, 1990a.

Associated Press wire story 1990b. Failure to enunciate causes bus chaos. July 31, 1990b.

Atkinson, R.C. & Shiffrin, R.M. 1968. Human memory: A proposed system and its control processes. In K.W. Spence (Ed.), *The Psychology of Learning and Motivation* (Vol. 2, pp. 89–195). New York: Academic Press.

Atkinson, R.C., Holmgren, J.E. & Juola, J.F. 1969. Processing time as influenced by the number of elements in a visual display. *Perception & Psychophysics*, 6, 321–326.

Atwood, M.E. & Polson, P.G. 1976. A process model for water jug problems. *Cognitive Psychology*, 8, 191–216.

Averbach, E. & Coriell, A.S. 1961. Short-term memory in vision. *Bell Systems Technical Journal*, 40, 309–328.

Awater, H., Kerlin, J.R., Evans, K.K. & Tong, F. 2005. Cortical representation of space around the blind spot. *Journal of Neurophysiology*, 94, 3314–3324.

Ayas, N.T., Barger, L.K., Cade, B.E., et al. 2006. Extended work duration and the risk of self-reported percutaneous injuries in interns. *Journal of the American Medical Association*, 296, 1055–1062.

Babbage, C. 1832. *On the Economy of Machinery and Manufactures*. Philadelphia, PA: Carey & Lea.

Baber, C. 2006. Cognitive aspects of tool use. *Applied Ergonomics*, 37, 3–15.

Baber, C. & Wankling, J. 1992. An experimental comparison of text and symbols for in-car reconfigurable displays. *Applied Ergonomics*, 23, 255–262.

Baddeley, A. 1986. *Working Memory*. New York: Oxford University Press.

Baddeley, A. 1999. *Essentials of Human Memory*. Hove, UK: Psychology Press.

Baddeley, A. 2000. The episodic buffer: A new component of working memory? *Trends in Cognitive Sciences*, 4, 417–423.

Baddeley, A. 2003. Working memory: Looking back and looking forward. *Nature Reviews: Neuroscience*, 4, 829–839.

Baddeley, A. 2016. Working memory. In R. J. Sternberg, S. T. Fiske, & D. J. Foss (Eds.), *Scientists Making a Difference: One Hundred Eminent Behavioral and Brain Scientists Talk about their Most Important Contributions* (pp. 119–122). New York: Cambridge University Press.

Baddeley, A.D. & Ecob, J.R. 1973. Reaction time and short-term memory: A trace strength alternative to the high-speed scanning hypothesis. *Quarterly Journal of Experimental Psychology*, 25, 229–240.

Baddeley, A. & Hitch, G.J. 1974. Working memory. In G.H. Bower (Ed.), *The Psychology of Learning and Motivation* (Vol. 8, pp. 47–89). New York: Academic Press.

Badets, A. & Blandin, Y. 2004. The role of knowledge of results frequency in learning through observation. *Journal of Motor Behavior*, 36, 62–70.

Baek, Y., Cha, O. & Chong, S. C. 2012. Characteristics of the filled-in surface at the blind spot. *Vision Research*, 58, 33–44.

Bahr, N.J. 2017. *System Safety Engineering and Risk Assessment: A Practical Approach.* Second Edition. Boca Raton, FL: CRC Press.

Bahrick, H.P. 1957. An analysis of stimulus variables influencing the proprioceptive control of movements. *Psychological Review*, 64, 324–328.

Bailey, R.W. 1996. *Human Performance Engineering: Designing High Quality Professional User Interfaces for Computer Products, Applications, and Systems* (3rd ed.). Englewood Cliffs, NJ: Prentice-Hall.

Baird, K.M., Hoffmann, E.R. & Drury, C.G. 2002. The effects of probe length on Fitts' law. *Applied Ergonomics*, 33, 9–14.

Baker, C.H. 1963. Signal duration as a factor in vigilance tasks. *Science*, 141, 1196–1197.

Baker, D.S. 1989. Major defense verdicts of 1988. *ABA Journal*, November, 82–86.

Baldwin, C.L. 2003. Neuroergonomics of mental workload: New insights from the convergence of brain and behaviour in ergonomics research. *Theoretical Issues in Ergonomics Science*, 4, 132–141.

Baldwin, C.L. 2012. *Auditory Cognition and Human Performance: Research and Applications.* Boca Raton, FL: CRC Press.

Baltes, B.B., Briggs, T.E., Huff, J.W., Wright, J.A. & Neuman, G.A. 1999. Flexible and compressed workweek schedules: A meta-analysis of their effects on work-related criteria. *Journal of Applied Psychology*, 84, 496–513.

Balthazar, B. 1998. Imagine it's Friday and you've got to hurry off to work, where you taste whiskey for Jack Daniel Distillery. Life is good. Retrieved on July 28, 2002, from: http://www.beveragebusiness. Com/art-arch/baltha10.html

Bandura, A. 1986. *Social Foundations of Thought and Action: A Social Cognitive Theory.* Englewood Cliffs, NJ: Prentice-Hall.

Banks, W.P. & Prinzmetal, W. 1976. Configurational effects in visual information processing. *Perception & Psychophysics*, 19, 361–367.

Banks, W.W. & Boone, M.P. 1981. A method for quantifying control accessibility. *Human Factors*, 23, 299–303.

Barber, A.V. 1990. Visual mechanisms and predictors of far field visual task performance. *Human Factors*, 32, 217–234.

Barber, P. & O'Leary, M. 1997. The relevance of salience: Towards an activation account of irrelevant stimulus–response compatibility effects. In B. Hommel & W. Prinz (Eds.), *Theoretical Issues in Stimulus–Response Compatibility* (pp. 135–172). Amsterdam: North-Holland.

Barcroft, J. 2013. Mnemonics. In C. A. Chapelle (Ed.), *The Encyclopedia of Applied Linguistics* (pp. 1–9). Malden, MA: Blackwell.

Barger, L.K., Cole, B.E., Ayas, N.T., et al. 2005. Extended work shifts and the risk of motor vehicle crashes among interns. *The New England Journal of Medicine*, 352, 125–134.

Barlow, R.E. & Proschan, F. 1965. *Mathematical Theory of Reliability.* New York: Wiley.

Baron, J. 2008. *Thinking and Deciding* (4th ed.). New York: Cambridge University Press.

Baron, R. & Rodin, J. 1978. Personal control as a mediator of crowding. In A. Baum, J.E. Singer & S. Valins (Eds.), *Advances in Environmental Psychology* (Vol. 1: *The Urban Environment*, pp. 145–190). Hillsdale, NJ: Lawrence Erlbaum.

Baron, S. & Corker, K. 1989. Engineering-based approaches to human performance modeling. In G.R. McMillan, D. Beevis, E. Salas, M.H. Strub, R. Sutton, & L. Van Breda (Eds.), *Applications of Human Performance Models to System Design* (pp. 203–217). New York: Plenum Press.

Baron, S., Kruser, D.S. & Huey, B.M. (Eds.) 1990. *Quantitative Modeling of Human Performance in Complex, Dynamic Systems.* Washington, DC: National Academy Press.

Barrett, G.V. 2000. Personnel selection: Selection and the law. In A.E. Kazdin (Ed.), *Encyclopedia of Psychology* (Vol. 6, pp. 156–160). Washington, DC: American Psychological Association.

Barrett, H.H. & Swindell, W. 1981. *Radiological Imaging: The Theory of Image Formation, Detection and Processing.* New York: Academic Press.

Barrett, R.S. & Dennis, G.J. 2005. Ergonomic issues in team lifting. *Human Factors and Ergonomics in Manufacturing*, 15, 293–307.

Bashford, J.A., Jr., Riener, K.R. & Warren, R.M. 1992. Increasing the intelligibility of speech through multiple phonemic restorations. *Perception & Psychophysics*, 51, 211–217.

Bassok, M. 2003. Analogical transfer in problem solving. In J.E. Davidson & R.J. Sternberg (Eds.), *The Psychology of Problem Solving* (pp. 343–369). New York: Cambridge University Press.

Bastian, A.J. 2006. Learning to predict the future: The cerebellum adapts feedforward movement control. *Current Opinion in Neurobiology*, 16, 645–649.

Battig, W.F. 1979. The flexibility of human memory. In L.S. Cermak & F.I.M. Craik (Eds.), *Levels of Processing in Human Memory* (pp. 23–44). Hillsdale, NJ: Lawrence Erlbaum.

Baudhuin, E.A. 1987. The design of industrial and flight simulators. In S.M. Cormier & J.D. Hagman (Eds.), *Transfer of Learning* (pp. 217–237). New York: Academic Press.

Baum, A. & Paulus, P.B. 1987. Crowding. In D. Stokols &I. Altman (Eds.), *Handbook of Environmental Psychology* (Vol. 1, pp. 533–570). New York: Wiley.

Bausenhart, K.M., Rolke, B., Hackley, S.A. & Ulrich, R. 2006. The locus of temporal preparation effects: Evidence from the psychological refractory period paradigm. *Psychonomic Bulletin & Review*, 13, 536–542.

Bazovsky, I. 1961. *Reliability: Theory and Practice*. Englewood Cliffs, NJ: Prentice-Hall.

Beaman, C.P. & Morton, J. 2000. The effects of rime on auditory recency and the suffix effect. *European Journal of Cognitive Psychology*, 12, 223–242.

Beatty, J. 1982. Task-evoked pupillary responses, processing load, and the structure of processing resources. *Psychological Bulletin*, 91, 276–292.

Beck, J. 1966. Effects of orientation and of shape similarity on perceptual grouping. *Perception & Psychophysics*, 1, 300–302.

Becker, A.B., Warm, J.S., Dember, W.N. & Hancock, P.A. 1991. Effects of feedback on perceived workload in vigilance performance. In *Proceedings of the Human Factors Society 35th Annual Meeting* (Vol. 2, pp. 1491–1494). Santa Monica, CA: Human Factors Society.

Becker, A.B., Warm, J.S., Dember, W.N. & Hancock, P.A. 1995. Effects of jet engine noise and performance feedback on perceived workload in a monitoring task. *International Journal of Aviation Psychology*, 5, 49–62.

Bedford, T. & Cooke, R. 2001. *Probabilistic Risk Analysis: Foundations and Methods*. Cambridge, UK: Cambridge University Press.

Bednar, J.A. & Miikkulainen, R. 2000. Tilt aftereffects in a self-organizing model of the primary visual cortex. *Neural Computation*, 12, 1721–1740.

Beevis, D. 2003. Ergonomics—Costs and benefits revisited. *Applied Ergonomics*, 34, 491–496.

Behar, A., Chasin, M. & Cheesman, M. 2000. *Noise Control: A Primer*. San Diego,CA: Singular.

Beith, B.H. 2006. The ABCs of career development: Attitude, basics, and communication. In *Proceedings of the Human Factors and Ergonomics Society 50th Annual Meeting*(pp. 2302–2303). Santa Barbara, CA: Human Factors and Ergonomics Society.

Békésy, G. von. 1960. *Experiments in Hearing*. New York: McGraw-Hill.

Bell, P.A. 1978. Effects of heat and noise stress of primary and subsidiary task performance. *Human Factors*, 20, 749–752.

Belz, S.M., Robinson, G.S. & Casali, J.G. 1999. A new class of auditory warning signals for complex systems: Auditory icons. *Human Factors*, 41, 608–618.

Benjamin, A.S. & Bjork, R.A. 2000. On the relationship between recognition speed and accuracy for words rehearsed via rote versus elaborative rehearsal. *Journal of Experimental Psychology: Learning, Memory, & Cognition*, 26, 638–648.

Bennett, K.B. & Flach, J.M. 2011. *Display and Interface Design Subtle Science, Exact Art*. Boca Raton, FL: CRC Press.

Bennett, K.B., Flach, J.M., Edman, C., Holt, J. & Lee, P. 2015. Ecological interface design: A selective overview. In R. R. Hoffman, P. A. Hancock, M. W. Scerbo, R. Parasuraman, & J. L. Szalma (Eds.), *The Cambridge Handbook of Applied Perception rResearch* (pp. 669–691). New York: Cambridge University Press.

Beranek, L.L. 1989. Balanced noise-criterion (NCB) curves. *Journal of the Acoustical Society of America*, 86, 650–664.

Berglund, B. 1991. Quality assurance in environmental psychophysics. In S.J. Bolanowski, Jr. & G.A. Gescheider (Eds.), *Ratio Scaling of Psychological Magnitude* (pp. 140–162). Hillsdale, NJ: Lawrence Erlbaum.

Berglund, B., Berglund, U. & Lindvall, T. 1974. A psychological detection method in environmental research. *Environmental Research*, 7, 342–352.

Bergman, M., Blumenfeld, V.G., Cascardo, D., Dash, B., Levitt, H. & Margulies, M.K. 1976. Age-related decrement in hearing for speech: Sampling and longitudinal studies. *Journal of Gerontology*, 31, 533–538.

Beringer, D.B., Williges, R.C. & Roscoe, S.N. 1975. The transition of experienced pilots to a frequency-separated aircraft attitude display. *Human Factors*, 17, 401–414.

Berkowitz, P. & Casalli, S.P. 1990. Influence of age on the ability to hear telephone ringers of different spectral content. In *Proceeding of the Human Factors Society 34th Annual Meeting* (Vol. 1, pp. 132–136). Santa Monica, CA: Human Factors Society.

Berliner, D.C., Angell, D. & Shearer, J.W. 1964. Behaviors, measures and instruments for performance evaluation in simulated environments. Paper presented at a Symposium and Workshop on Quantification of Human Performance, Albuquerque, NM.

Bernstein, I.H. 2011. A brief history of computers and the internet. In K.-P.L. Vu & R.W. Proctor (Eds.), *Handbook of Human Factors in Web Design* (2nd ed., pp. 13–34). Mahwah, NJ: Lawrence Erlbaum.

Bernstein, N. 1967. *The Coordination and Regulation of Movements*. New York: Pergamon Press.

Berryhill, M.E., Kveraga, K., Webb, L. & Hughes, H.C. 2005. Effect of uncertainty on the time course for selection of verbal name codes. *Perception & Psychophysics*, 67, 1437–1445.

Best, P.S., Littleton, M.H., Gramopadhye, A.K. & Tyrrell, R.A. 1996. Relations between individual differences in oculomotor resting states and visual inspection performance. *Ergonomics*, 39, 35–40.

Beveridge, M. & Parkins, E. 1987. Visual representation in analogical problem solving. *Memory & Cognition*, 15, 230–234.

Beyer, H. & Holtzblatt, K. 1998. *Contextual Design: A Customer Centered Approach to Systems Designs*. San Francisco, CA: Morgan Kaufmann.

Bezdjian, S., Baker, L. A., Lozano, D. I. & Raine, A. 2009. Assessing inattention and impulsivity in children during the Go/NoGo task. *British Journal of Developmental Psychology*, 27, 365–383.

Bhise, V.D. 2006. Incorporating hard disks in vehicles—Uses and challenges. *SAE Technical Paper Series* (2006-01-0814). Warrendale, PA: SAE International.

Biederman, I., Glass, A.L. & Stacy, E.W., Jr. 1973. Searching for objects in real-world scenes. *Journal of Experimental Psychology*, 97, 22–27.

Biederman, I., Mezzanotte, R.J., Rabinowitz, J.C., Francolini, C.M. & Plude, D. 1981. Detecting the unexpected in photo interpretation. *Human Factors*, 23, 153–164.

Biehal, G. & Chakravarti, D. 1989. The effects of concurrent verbalization on choice processing. *Journal of Marketing Research*, 26, 84–96.

Billinghurst, S, & Vu, K.-P.L. 2015. Touch screen gestures for Web browsing tasks. *Computers in Human Behavior*, 53, 71–81.

Birolini, A. 1999. *Reliability Engineering: Theory and Practice* (7th ed.). New York: Springer.

Bisantz, A.M. & Drury, C.G. 2005. Applications of archival and observational data. In J.R. Wilson & N. Corlett (Eds.), *Evaluation of Human Work* (3rd ed., pp. 61–82). Boca Raton, FL: CRC Press.

Bizzi, E. & Abend, W. 1983. Posture control and trajectory formation in single- and multi-joint arm movements. In J.E. Desmedt (Ed.), *Motor Control Mechanisms in Health and Disease* (pp. 31–45). New York: Raven Press.

Bjorn, V. 2004. New twists to an old design. *Air Force Research Technology Horizons*, Human Effectiveness Directorate, Technical Article HE-03–03. Downloaded on August 19, 2006, from http://www.afrlhorizons.com=Briefs=Apr04=HE0303.html

Blackwell, A. & Green, T. 2003. Notational systems—The cognitive dimensions of notations framework. In J.M. Carroll (Ed.), *HCI Models, Theories, and Frameworks: Toward a Multidisciplinary Science* (pp. 103–133). San Francisco, CA: Morgan Kaufman.

Blair, S.M. 1991. The Antarctic experience. In A.A. Harrison, Y.A. Clearwater & C.P. McKay (Eds.), *From Antarctica to Outer Space: Life in Isolation and Confinement* (pp. 57–64). New York: Springer.

Blankenship, E. 2003. An introduction to designing user interface controls at SAP. SAP Design Guild. Downloaded on August 10, 2006, from http://www.sapdesignguild.org/community/design/controls.asp

Blauert, J. 1997. *Spatial Hearing. The Psychophysics of Human Sound Localization*. Cambridge, MA: MIT Press.

Blazier, W.E. 1981. Revised noise criteria for application in the acoustical design and rating of HVAC systems. *Noise Control Engineering*, 16, 64–73.

Blazier, W.E. 1997. RC Mark II: A refined procedure for rating the noise of heating, ventilating and air-conditioning (HVAC) systems in buildings. *Noise Control Engineering Journal*, 45, 243–250.

Bliss, J.C., Crane, H.D., Mansfield, P.K. & Townsend, J.T. 1966. Information available in brief tactile presentations. *Perception & Psychophysics*, 1, 273–283.

Boer, L.C., Harsveld, M. & Hermans, P.H. 1997. The selective-listening task as a test for pilots and air traffic controllers. *Military Psychology*, 9, 136–149.

Boersema, T., Zwaga, H. J. G. & Adams, A. S. 1989. Conspicuity in realistic scenes: An eye-movement measure. *Applied Ergonomics*, 20, 267–273.

Boff, K.R. & Lincoln, J.E. (Eds.) 1988. Engineering Data Compendium: Human Perception and Performance. Wright-Patterson Air Force Base, OH: Harry G. Armstrong Aerospace Medical Research Laboratory.

Boles, D.B., Bursk, J.H., Phillips, J.B. & Perdelwitz, J.R. 2007. Predicting dual-task performance with the Multiple Resources Questionnaire (MRQ). *Human Factors*, 49, 32–45.

Boles, D.B. & Dillard, M.B. 2015. The measurement of perceptual resources and workload. In R.R. Hoffman, P.A. Hancock, M.W. Scerbo, R. Parasuraman, & J.L. Szalma (Eds.), *The Cambridge Handbook of Applied Perception Research* (Vol. 1, pp. 39–59). New York: Cambridge University Press.

Boulton, L. & Cole, J. 2016. Adaptive Flexibility: Examining the Role of Expertise in the Decision Making of Authorized Firearms Officers During Armed Confrontation, *Journal of Cognitive Engineering and Decision Making*, (Vol. 10(3), pp. 291–308). DOI 10.1177/1555343416646684.

Bongers, P.M., Hulshof, C.T.J., Dijkstra, L., Boshuizen, H.C., Groenhout, H.J.M. & Valken, E. 1990. Back pain and exposure to whole body vibration in helicopter pilots. *Ergonomics*, 33, 1007–1026.

Bonner, J.V.H. 2006. Human factors design tools for consumer-product interfaces. In W. Karwowski (Ed.), *International Encyclopedia of Ergonomics and Human Factors* (2nd ed., Vol. 3, pp. 3203–3206). Boca Raton, FL: CRC Press.

Bonnet, M., Decety, J., Jeannerod, M. & Requin, J. 1997. Mental simulation of an action modulates the excitability of spinal reflex pathways in man. *Cognitive Brain Research*, 5, 221–228.

Booher H.R. (Ed.) 1990. *Manprint: An Approach to Systems Integration*. New York: Van Nostrand Reinhold.

Booher, H.R. (Ed.) 2003a. *Handbook of Human Systems Integration*. Hoboken, NJ: Wiley.

Booher, H.R. 2003b. Introduction: Human systems integration. In H.R. Booher (Ed.), *Handbook of Human Systems Integration* (pp. 1–30). Hoboken, NJ: Wiley.

Boorst, J. P. & Anderson, J. R. 2015. Using the ACT-R cognitive architecture in combination with fMRI data. In B. U. Forstmann E.-J., (Eds.), *An Introduction to Model-based Cognitive Neuroscience* (pp. 339–352). New York: Springer Science+Business Media.

Borelli, G.A. 1679/1989. *On the Movement of Animals* (P. Maquet, Trans.). New York: Springer.

Boring, E.G. 1942. *Sensation and Perception in the History of Experimental Psychology*. New York: Appleton-Century-Crofts.

Boring, R.L. 2012, June. Fifty years of THERP and human reliability analysis. In *Proceedings of the Probabilistic Safety Assessment and Management and European Safety and Reliability Conference (PSAM 11 & ESREL 2012)*. Helsinki, Finland.

Borsky, P.N. 1965. Community reactions to sonic booms in the Oklahoma City area. USAF Aerospace Medical Research Laboratory Rep. AMRL-TR-65–37. OH: Wright-Patterson Air Force Base.

Boswell, W.R. & Boudreau, J.W. 2002. Separating the developmental and evaluative performance appraisal uses. *Journal of Business and Psychology*, 16, 391–412.

Bothell, D. 2017. ACT-R 7 reference manual. Downloaded 2/19/2017 from http://act-r.psy.cmu.edu/software/.

Botvinick, E.L. & Berns, M.W. 2005. Internet-based robotic laser scissors and tweezers microscopy. *Microscopy Research and Technique*, 68, 65–75.

Bouchard, S., Dumoulin, S., Talbot, J., Ledoux, A., Phillips, J., Monthuy-Blanc, J. & Renaud, P. 2012. Manipulating subjective realism and its impact on presence: Preliminary results on feasibility and neuroanatomical correlates. *Interacting with Computers*, 24, 227–236.

Boucsein, W. & Backs, R.W. 2000. Engineering psychophysiology as a discipline: Historical and theoretical aspects. In R.W. Backs & W. Boucsein (Eds.), *Engineering Psychophysiology: Issues and Applications* (pp. 3–30). Mahwah, NJ: Lawrence Erlbaum.

Bourne, L.J., Kole, J.A. & Healy, A.F. 2014. Expertise: Defined, described, explained. *Frontiers in Psychology*, 5, 1–3.

Boutis, K., Pecaric, M., Seeto, B. & Pusic, M. 2010. Using signal detection theory to model changes in serial learning of radiological image interpretation. *Advances in Health Sciences Education*, 15, 647–658.

Bower, G.H., Clark, M.C., Lesgold, A.M. & Winzenz, D. 1969. Hierarchical retrieval schemes in recall of categorized word lists. *Journal of Verbal Learning and Verbal Behavior*, 8, 323–343.

Bowers, C., Jentsch, F. & Morgan, B.B., Jr. 2006. Team performance. In W. Karwowski (Ed.), *International Encyclopedia of Ergonomics and Human Factors* (2nd ed., Vol. 2, pp. 2390–2393). Boca Raton, FL: CRC Press.

Bowling, B. 2016. *Kanski's Clinical Ophthalmology: A Systematic Approach* (8th ed.). New York: Elsevier.

Boyce, P.R. 2014. *Human Factors in Lighting* (3rd ed.). Boca Raton, FL: CRC Press.

Bradley, D.R. & Abelson, S.B. 1995. Desktop flight simulators: Simulation fidelity and pilot performance. *Behavior Research Methods, Instruments and Computers*, 27, 152–159.

Bradley, J.V. 1967. Tactual coding of cylindrical knobs. *Human Factors*, 9, 483–496.

Bradley, J.V. 1969a. Desirable dimensions for concentric controls. *Human Factors*, 11, 213–226.

Bradley, J.V. 1969b. Optimum knob crowding. *Human Factors*, 11, 227–238.

Bradley, J.V. & Wallis, R.A. 1958. Spacing of on–off controls. 1: Pushbuttons (WADC-TR 58-2). Wright-Patterson Air Force Base, OH: Wright Air Development Center.

Bradley, J.V. & Wallis, R.A. 1960. Spacing of toggle switch on–off controls. *Engineering and Industrial Psychology*, 2, 8–19.

Bradley, M. M., Miccoli, L., Escrig, M. A. & Lang, P. J. 2008. The pupil as a measure of emotional arousal and autonomic activation. *Psychophysiology*, 45, 602–607.

Bragança, S., Arezes, P.M. & Carvalho, M. 2015. An overview of the current three-dimensional body scanners for anthropometric data collection. In P. M. Arezes et al. (Eds.), *Occupational Safety and Hygiene III* (pp. 149–153). Boca Raton, FL: CRC Press.

Brannick, M.T. & Levine, E. 2007. *Job Analysis: Methods, Research, and Applications for Human Resource Management* (2nd ed.). Thousand Oaks, CA: Sage.

Bransford, J.D. & Johnson, M.K. 1972. Contextual prerequisites for understanding: Some investigations of comprehension and recall. *Journal of Verbal Learning and Verbal Behavior*, 11, 717–726.

Brauner, T., Sterzing, T., Wulf, M. & Horstmann, T. 2014. Leg stiffness: Comparison between unilateral and bilateral hopping tasks. *Human Movement Science*, 33, 263–272.

Brebner, J., Shephard, M. & Cairney, P.T. 1972. Spatial relationships and S–R compatibility, *Acta Psychologica*, 36, 1–15.

Bregman, A.S., Colantonio, C. & Ahad, P.A. 1999. Is a common grouping mechanism involved in the phenomena of illusory continuity and stream segregation? *Perception & Psychophysics*, 61, 195–205.

Bregman, A.S. & Rudnicky, A.I. 1975. Auditory segregation: Stream or streams? *Journal of Experimental Psychology: Human Perception and Performance*, 1, 263–267.

Breitmeyer, B.G. & Ganz, L. 1976. Implications of sustained and transient channels for theories of visual pattern masking, saccadic suppression, and information processing. *Psychological Review*, 83, 1–36.

Breitmeyer, B.G., Kropfl, W. & Julesz, B. 1982. The existence and role of retinotopic and spatiotopic forms of visual persistence. *Acta Psychologica*, 52, 175–196.

Breitmeyer, B.G. & Ögmen, H. 2000. Recent models and findings in visual backward masking: A comparison, review, and update. *Perception & Psychophysics*, 62, 1572–1595.

Breitmeyer, B.G. & Ögmen, H. 2006. *Visual Masking: Time Slices through Conscious and Unconscious Vision*. Oxford: Oxford University Press.

Brennan, A., Chugh, J.S. & Kline, T. 2002. Traditional versus open office design: A longitudinal study. *Environment and Behavior*, 34, 279–299.

Brenner, M.D., Fairris, D. & Ruser, J. 2004. "Flexible" work practices and occupational safety and health: Exploring the relationship between cumulative trauma disorders and workplace transformation. *Industrial Relations*, 43, 242–266.

Breuer, K., Nieken, P. & Sliwka, D. 2013. Social ties and subjective performance evaluations: An empirical investigation. *Review of Managerial Science*, 7, 141–157.

Brewer, W.F. & Lichtenstein, E.H. 1981. Event schemas, story schemas, and story grammars. In J.L. Long & A. Baddeley (Eds.), *Attention and Performance IX* (pp. 363–379). Hillsdale, NJ: Lawrence Erlbaum.

Bontcheva, K., Cunningham, H., Roberts, I., Roberts, A., Tablan, V., Aswani, N. & Gorrell, G. 2013. GATE Teamware: A web-based, collaborative text annotation framework. *Language Resources and Evaluation*, 47, 1007–1029.

Brewster, S. & Murray-Smith, R. (Eds.) 2001. *Haptic Human–Computer Interaction*. New York: Springer.

Brick, J. 2001. Biobehavioral factors: In unintended acceleration. *Forensic Examiner*, 10(5–6), 26–31.

Bridgeman, B. 1995. Extraretinal signals in visual orientation. In W. Prinz & B. Bridgeman (Eds.), *Handbook of Perception and Action* (Vol. 1: *Perception*, pp. 191–223). San Diego, CA: Academic Press.

Bridgeman, B. & Delgado, D. 1984. Sensory effects of eyepress are due to efference. *Perception & Psychophysics*, 36, 482–484.

Broadbent, D. E. 1958. *Perception and Communication*. New York: Pergamon Press.

Broadbent, D.E. & Gregory, M. 1965. Effects of noise and of signal rate upon vigilance analyzed by means of decision theory. *Human Factors*, 7, 155–162.

Broner, N. 2005. Rating and assessment of noise. *EcoLibrium, March*, 21–25.

Bronkhorst, A. W. 2015. The cocktail-party problem revisited: Early processing and selection of multi-talker speech. *Attention, Perception, & Psychophysics*, 77, 1465–1487.

Brooke, R.E., Isherwood, S., Herbert, N.C., Raynor, D.K. & Knapp, P. 2012. Hearing aid instruction booklets: Employing usability testing to determine effectiveness. *American Journal of Audiology*, 21, 206–214.

Brookes, M.J. & Kaplan, A. 1972. The office environment: Space planning and affective behavior. *Human Factors*, 14, 373–391.

Brookhuis, K.A. & De Waard, D. 2002. On the assessment of (mental) workload and other subjective qualifications. *Ergonomics*, 45, 1026–1030.

Brooks, L.R. 1968. Spatial and verbal components of the act of recall. *Canadian Journal of Psychology*, 22, 349–368.

Brown, J. 1958. Some tests of the decay theory of immediate memory. *Quarterly Journal of Experimental Psychology*, 10, 12–21.

Brown, J.L. 1965. Flicker and intermittent stimulation. In C.H. Graham (Ed.), *Vision and Visual Perception* (pp. 251–320). New York: Wiley.

Brown, O., Jr. 2002. Macroergonomic methods: Participation. In H.W. Hendrick & B.M. Kleiner (Eds.), *Macroergonomics: Theory, Methods, and Applications* (pp. 111–131). Mahwah, NJ: Lawrence Erlbaum.

Browne, R.C. 1949. The day and night performance of teleprinter switchboard operators. *Journal of Occupational Psychology*, 23, 121–126.

Bruins, J. & Barber, A. 2000. Crowding, performance, and affect: A field experiment investigating mediational processes. *Journal of Applied Social Psychology*, 30, 1268–1280.

Brush, S.G. 1989. Prediction and theory evaluation: The case of light bending. *Science*, 246, 1124–29.

Bryan, W.L. & Harter, N. 1899. Studies of the telegraphic language. The acquisition of a hierarchy of habits. *Psychological Review*, 6, 345–375.

Bubb, H. 2004. Challenges in the application of anthropometric measurements. *Theoretical Issues in Ergonomics Science*, 5, 154–168.

Bubb, H. 2005. Human reliability: A key to improved quality in manufacturing. *Human Factors and Ergonomics in Manufacturing*, 15, 353–368.

Buchanan, B.G., Davis, R. & Feigenbaum, E.A. 2007. Expert systems: A perspective from computer science. In K.A. Ericsson, N. Charness, P.J. Feltovich & R.R. Hoffman (Eds.), *Cambridge Handbook of Expertise and Expert Performance* (pp. 87–103). Cambridge, UK: Cambridge University Press.

Buck, J.R., Zellers, S.M. & Opar, M.E. 2000. Control error statistics. *Ergonomics*, 43, 1–16.

Bucker, B. & Theeuwes, J. 2014. The effect of reward on orienting and reorienting in exogenous cuing. *Cognitive, Affective & Behavioral Neuroscience*, 14, 635–646.

Budinger, E. 2005. Introduction: Auditory cortical fields and their functions. In R. König, P. Heil, E. Budinger & H. Schleich (Eds.), *The Auditory Cortex: A Synthesis of Human and Animal Research* (pp. 3–6). Mahwah, NJ: Lawrence Erlbaum.

Buhr, E.D., Yoo, S.-H. & Takahashi, J.S. 2010. Temperature as a universal resetting cue for mammalian circadian oscillators. *Science*, 330, 379–385.

Bureau of Labor Statistics. 2015, Nov. Nonfatal Occupational Injuries and Illnesses Requiring Days away from Work, 2014 (USDL 15-2205). Washington, DC: U.S. Department of Labor.

Buetti, S., Lleras, A. & Moore, C.M. 2014. The flanker effect does not reflect the processing of "task-irrelevant" stimuli: Evidence from inattentional blindness. *Psychonomic Bulletin & Review*, 21, 1231–1237.

Bugarski, V., Bačkalić, T. & Kuzmanov, U. 2013. Fuzzy decision support system for ship lock control. *Expert Systems with Applications*, 40, 3953–3960.

Bullinger, H.-J., Kern, P. & Braun, M. 1997. Controls. In G. Salvendy (Ed.), *Handbook of Human Factors* (2nd ed., pp. 697–728). New York: Wiley.

Burnett, G.E. & Donkor, R.A. 2012. Evaluating the impact of head-up display complexity on peripheral detection performance: A driving simulator study. *Advances in Transportation Studies*, 28, 5–16.

Buschman, T.J. & Miller, E.K. 2010. Shifting the spotlight of attention: Evidence for discrete computations in cognition. *Frontiers in Human Neuroscience*, 4, doi:10.3389/fnhum.2010.00194

Butler, M.P. 2003. Corporate ergonomics programme at Scottish & Newcastle. *Applied Ergonomics*, 34, 35–38.

Byrne, M.D. 2015. Human performance modeling. In D.A. Boehm-Davis, F.T. Durso, & J.D. Lee (Eds.), *APA Handbook of Human Systems Integration* (pp. 345–358). Washington, DC: American Psychological Association.

Byrne, M.D. & Gray, W.D. 2003. Steps toward building mathematical and computer models from cognitive task analyses. *Human Factors*, 45, 1–4.

Cacciabue, P.C. 1997. Methodology of human factors analysis for systems engineering: Theory and applications. *IEEE Transactions on Systems, Man, and Cybernetics Part A: Systems and Humans*, 27, 325–339.

Cacha, C.A. 1999. *Ergonomics and Safety in Hand Tool Design*. Boca Raton, FL: Lewis Publishers.

Caelli, T. & Porter, D. 1980. On difficulties in localizing ambulance sirens. *Human Factors*, 22, 719–724.

Cal/OSHA Consultation Service and NIOSH 2004. Easy ergonomics: A guide to selecting non-powered hand tools. DHHS (NIOSH) Publication No. 2004–164). Cincinnati, OH: National Institute for Occupational Safety and Health.

Calhoun, G.L. 2006. Gaze-based control. In W. Karwowski (Ed.), *International Encyclopedia of Ergonomics and Human Factors* (2nd ed., Vol. 1, pp. 352–355). Boca Raton, FL: CRC Press.

Campbell, F.W. & Westheimer, G. 1960. Dynamics of accommodation responses in the human eye. *Journal of Physiology*, 151, 285–295.

Canas, J.J., Salmeron, L., Antoli, A., Fajardo, I., Chisalita, C. & Escudero, J.T. 2003. Differential roles for visuospatial and verbal working memory in the construction of mental models of physical systems. *International Journal of Cognitive Technology*, 8, 45–53.

Capps, K. 2016. America's sudden U-turn on highway fonts: Clearview is out, Highway Gothic is (back) in. Critics want to know why. *The Atlantic CityLab*, January 27. http://www.citylab.com/commute/2016/01/official-united-states-highway-sign-font-clearview/427068/

Cao, A., Chintamani, K.K., Pandya, A.K. & Ellis, R.D. 2009. NASA TLX: Software for assessing subjective mental workload. *Behavior Research Methods*, 41, 113–117.

Carayon, P., Kianfar, S., Li, Y. & Wooldridge, A. 2015. Organizational design: Macroergonomics as a foundation for human systems integration. In D.A. Boehm-Davis & F. T. Durso (Eds.), *APA Handbook of Human Systems Integration* (pp. 573–589). Washington, DC: American Psychological Association.

Carayon, P., Wetterneck, T.B., Rivera-Rodriguez, A.J., Hundt, A.S., Hoonakker, P., Holden, R. & Gurses, A.P. 2014. Human factors systems approach to healthcare quality and patient safety. *Applied Ergonomics*, 45, 14–25.

Carayon, P. & Xie, A. 2012. Decision making in healthcare system design: When human factors engineering meets health care. In R.W. Proctor, S.Y. Nof, Y. Yih (Eds.), *Cultural Factors in Systems Design: Decision Making and Action* (pp. 219–238). Boca Raton, FL: CRC Press.

FY 011 Department of Defense Human Systems Integration Management Plan. 2011. Washington, DC: DDRE, Director, Mission Assurance, and Director of Human Performance, Training & Biosystems.

Card, S.K., Moran, T.P. & Newell, A. 1983. *The Psychology of Human–Computer Interaction*. Hillsdale, NJ: Lawrence Erlbaum.

Caroux, L. & Isbister, K. 2016. Influence of head-up displays' characteristics on user experience in video games. *International Journal of Human-Computer Studies*, 87, 65–79.

Carroll, J.M. 2002. Making use is more than a matter of task analysis. *Interacting with Computers*, 14, 619–627.

Carroll, J.M. (Ed.) 2003. *HCI Models, Theories, and Frameworks: Toward a Multidisciplinary Science*. San Francisco, CA: Morgan Kaufman.

Carroll, J.M. 2006. Scenario-based design. In W. Karwowski (Ed.), *International Encyclopedia of Ergonomics and Human Factors* (2nd ed., Vol. 1, pp. 198–202). Boca Raton, FL: CRC Press.

Carroll, W.R. & Bandura, A. 1982. The role of visual monitoring in observational learning of action patterns: Making the unobservable observable. *Journal of Motor Behavior*, 14, 153–167.

Carroll, W.R. & Bandura, A. 1985. Role of timing of visual monitoring and motor rehearsal in observational learning of action patterns. *Journal of Motor Behavior*, 17, 269–281.

Carroll, W.R. & Bandura, A. 1987. Translating cognition into action: The role of visual avoidance in observational learning. *Journal of Motor Behavior*, 19, 385–398.

Carroll, W.R. & Bandura, A. 1990. Representational guidance of action production in observational learning: A causal analysis. *Journal of Motor Behavior*, 22, 85–97.

Carron, A.V. 1969. Performance and learning in a discrete motor task under massed versus distributed practice. *Research Quarterly*, 40, 481–4989.

Carter, M.J. & Ste-Marie, D.M. 2017. An interpolated activity during the knowledge-of-results delay interval eliminates the learning advantages of self-controlled feedback schedules. *Psychological Research*, 81, 399–406.

Caruso, C.C. 2012. Running on empty: Fatigue and healthcare professionals. NIOSH: Workplace Safety and Health. Medscape and NIOSH.

Casali, J.G. 2006. Sound and noise. In G. Salvendy (Ed.), *Handbook of Human Factors and Ergonomics* (3rd ed., pp. 612–642). Hoboken, NJ: Wiley.

Casali, J.G., Robinson, G.S., Dabney, E.C. & Gauger, D. 2004. Effect of electronic ANR and conventional hearing protectors on vehicle backup alarm detection in noise. *Human Factors*, 46, 1–10.

Casey, S.M. 1989. Anthropometry of farm equipment operators. *Human Factors Society Bulletin*, 32(7), 1–3.

Casey, S.M. & Kiso, J.L. 1990. The acceptability of control locations and related features in agricultural tractor cabs. *Proceedings of the Human Factors Society 34th Annual Meeting* (pp. 743–747). Santa Monica, CA: Human Factors Society.

Castiello, U., Bennett, K.M. & Stelmach, G.E. 1993. The bilateral reach to grasp movement. *Behavioural Brain Research*, 56, 43–57.

Castro, C. 2009. Visual demands and driving. In C. Castro (Ed.), *Human Factors of Visual and Cognitive Performance in Driving*. Boca Raton, FL: CRC Press.

Cattell, J.M. 1886. The time it takes to see and name objects. *Mind*, 11, 63–65.

Cavallo, V. & Pinto, M. 2012. Are car daytime running lights detrimental to motorcycle conspicuity? *Accident Analysis and Prevention*, 49, 78–85.

Cavanagh, J.P. 1972. Relation between immediate memory span and the memory search rate. *Psychological Review*, 79, 525–530.

Cervero, F. 2012. *Understanding Pain*. Cambridge, MA: MIT Press.

Chaffin, D.B. 1973. Localized muscle fatigue-definition and measurement. *Journal of Occupational Measurement*, 15, 346–334.

Chaffin, D.B. 1997. Biomechanical aspects of workplace design. In G. Salvendy (Ed.), *Handbook of Human Factors* (2nd ed., pp. 772–789). New York: Wiley.

Chaffin, D.B. 2005. Improving digital human modelling for proactive ergonomics in design. *Ergonomics*, 48, 478–491.

Chaffin, D.B., Andersson, G.B.J. & Martin, B.J. 2006. *Occupational Biomechanics* (4th ed.). New York: Wiley.

Chaffin, D.B., Andres, R.O. & Garg, A. 1983. Volitional postures during maximal push/pull exertions in the sagittal plane. *Human Factors*, 25, 541–550.

Chan, J., Paletz, S.F. & Schunn, C.D. 2012. Analogy as a strategy for supporting complex problem solving under uncertainty. *Memory & Cognition*, 40, 1352–1365.

Chang, C.M. 2010. *Service Systems Management and Engineering: Creating Strategic Differentiation and Operational Excellence*. Hoboken, NJ: John Wiley.

Chapanis, A., Garner, W.R. & Morgan, C.T. 1949. *Applied Experimental Psychology: Human Factors in Engineering Design*. New York: John Wiley.

Chapanis, A. & Kinkade, R.G. 1972. Design of controls. In H.P. Van Cott & R.G. Kinkade (Eds.), *Human Engineering Guide to Equipment Design* (pp. 345–379). Washington, DC: U.S. Superintendent of Documents.

Chapanis, A. & Lindenbaum, L.E. 1959. A reaction time study of four control-display linkages. *Human Factors*, 1, 1–7.

Chapanis, A. & Moulden, J.V. 1990. Short-term memory for numbers. *Human Factors*, 32, 123–138.

Chapanis, A., Parrish, R.N., Ochsman, R.B. & Weeks, G.D. 1977. Studies in interactive communication: II. The effects of four communication modes on the linguistic performance of teams during cooperative problem solving. *Human Factors*, 19, 101–126.

Chaparro, B.S., Shaikh, D. & Chaparro, A. 2006. The legibility of cleartype fonts. *Proceedings of the Human Factors and Ergonomics Society 30th Annual Meeting* (pp. 1829–1833). Santa Monica, CA: Human Factors and Ergonomics Society.

Charles, J.M. 2002. *Contemporary Kinesiology*. Champaign, IL: Stipes Publisher.

Charlton, S.G. 2002. Questionnaire techniques for test and evaluation. In T.G. O'Brien & S.G. Charlton (Eds.), *Handbook of Human Factors Testing and Evaluation* (2nd ed., pp. 225–246). Mahwah, NJ: Lawrence Erlbaum.

Chase, W.G. 1983. Spatial representations of taxi drivers. In D.R. Rogers & J.H. Sloboda (Eds.), *Acquisition of Symbolic Skills* (pp. 391–405). New York: Plenum Press.

Chase, W.G. & Ericsson, K.A. 1981. Skilled memory. In J.R. Anderson (Ed.), *Cognitive Skills and Their Acquisition* (pp. 141–189). Hillsdale, NJ: Lawrence Erlbaum.

Chase, W.G. & Simon, H.A. 1973. Perception in chess. *Cognitive Psychology*, 4, 55–81.

Chatterjee, D.S. 1987. Repetition strain injury—A recent review. *Journal of the Society of Occupational Medicine*, 37, 100–105.

Chebat, D.R., Maidenbaum, S. & Amedi, A. 2015. Navigation using sensory substitution in real and virtual mazes. *PloS One*, 10(6), e0126307.

Chen, S.Y. & Macredie, R. 2010. Web-based interaction: A review of three important human factors. *International Journal of Information Management*, 30, 379–387.

Cheng, H., Yang, L. & Liu, Z. 2016. Survey on 3D Hand Gesture Recognition. *IEEE Transactions on Circuits and Systems for Video Technology*, 26(9), 1659–1673.

Cheng, P.W. & Holyoak, K.J. 1985. Pragmatic reasoning schemas. *Cognitive Psychology*, 17, 391–416.

Cheng, P.W., Holyoak, K.J., Nisbett, R.E. & Oliver, L.M. 1986. Pragmatic versus Syntactic Approaches to Training Deductive Reasoning. *Cognitive Psychology*, 18, 293–328.

Chengalur, S.N., Rodgers, S.H. & Bernard, T.E. 2004. *Kodak's Ergonomic Design for People at Work*. Hoboken, NJ: Wiley.

Chernikoff, R., Birmingham, H.P. & Taylor, F.V. 1956. A comparison of pursuit and compensatory tracking in a simulated aircraft control loop. *Journal of Applied Psychology*, 40, 47–52.

Cherry, E.C. 1953. Some experiments on the recognition of speech, with one and with two ears. *Journal of the Acoustical Society of America*, 25, 975–979.

Cherry, E.C. & Taylor, W.K. 1954. Some further experiments on the recognition of speech with one and two ears. *Journal of the Acoustical Society of America*, 26, 554–559.

Chervinskaya, K.R. & Wasserman, E.L. 2000. Some methodological aspects of tacit knowledge elicitation. *Journal of Experimental and Theoretical Artificial Intelligence*, 12, 43–55.

Chery, S. & Vicente, K.J. 2006. Ecological interface design: Applications. In W. Karwowski (Ed.), *International Encyclopedia of Ergonomics and Human Factors* (2nd ed., Vol. 3, pp. 3101–3106). Boca Raton, FL: CRC Press.

Chi, M.T.H., Feltovich, P.J. & Glaser, R. 1981. Categorization and representation of physics problems by experts and novices. *Cognitive Science*, 5, 121–125.

Chichilnisky, E.J. & Wandell, B.A. 1999. Trichromatic opponent color classification. *Vision Research*, 39, 3444–3458.

Chignell, M.H. & Peterson, J.G. 1988. Strategic issues in knowledge engineering. *Human Factors*, 30, 381–394.

Cho, Y.S. & Proctor, R.W. 2003. Stimulus and response representations underlying orthogonal stimulus–response compatibility effects. *Psychonomic Bulletin & Review*, 10, 45–73.

Christ, R.E. 1975. Review and analysis of color coding research for visual displays. *Human Factors*, 17, 542–570.

Christ, R.E. & Corso, G.M. 1983. The effects of extended practice on the evaluation of visual display codes. *Human Factors*, 25, 71–84.

Chronicle, E.P., MacGregor, J.N. & Ormerod, T.C. 2004. What makes an insight problem? The roles of heuristics, goal conception, and solution recoding in knowledge-lean problems. *Journal of Experimental Psychology: Learning, Memory, and Cognition*, 30, 14–27.

Chui, T.Y.P., Yap, M.K.H., Chan, H.H.L. & Thibos, L.N. 2005. Retinal stretching limits peripheral visual acuity in myopia. *Vision Research*, 45, 593–605.

Ciampolini, A. & Torroni, P. 2004. Using abductive logic agents for modeling the judicial evaluation of criminal evidence. *Applied Artificial Intelligence*, 18, 251–275.

Cicerone, C.M. & Nerger, J.L. 1989. The density of cones in the fovea centralis of the human dichromat. *Vision Research*, 24, 1587–1595.

Ciriello, V.M. 2004. Comparison of two techniques to establish maximum acceptable forces of dynamic pushing for female industrial workers. *International Journal of Industrial Ergonomics*, 34, 93–99.

Clark, H.H. & Clark, E.V. 1968. Semantic distinctions and memory for complex sentences. *Quarterly Journal of Experimental Psychology*, 20, 129–138.

Clark, T. S. & Corlett, E. N. 1984. *The ergonomics of workspaces and machines: A design manual*. London: Taylor & Francis.

Clegg, C.W. 2000. Sociotechnical principles for system design. *Applied Ergonomics*, 31, 463–477.

Clemensen, J., Larsen, S.B., Kyng, M. & Kirkevold, M. 2007. Participatory design in health sciences: Using cooperative experimental methods in developing health services and computer technology. *Qualitative Health Research*, 17, 122–130.

Clifton, C., Jr. & Duffy, S.A. 2001. Sentence and text comprehension: Roles of linguistic structure. *Annual Review of Psychology*, 52, 167–196.

Cloutier, R., Baldwin, C. & Bone, M.A. 2015. *Systems Engineering Simplified*. Boca Raton, FL: CRC Press.

Cockton, G., Woolrych, A., Hornbæk, K. & Frøkjær, E. 2012. Inspection-based evaluations. In J. Jacko (Ed.), *The Human-Computer Interaction Handbook: Fundamentals, Evolving Technologies, and Emerging Applications* (3rd ed., pp. 1279–1298). Boca Raton, FL: CRC Press.

Cohen, D., Otakeno, S., Previc, F.H. & Ercoline, W.R. 2001. Effects of "inside-out" and "outside-in" attitude displays on off-axis tracking in pilots and nonpilots. *Aviation, Space, and Environmental Medicine*, 72, 170–176.

Cohen, H.H. & LaRue, C.A. 2006. Forensic human factors/ergonomics. In W. Karwowski (Ed.), *International Encyclopedia of Ergonomics and Human Factors* (2nd ed., Vol. 3, pp. 2909–2916). Boca Raton, FL: CRC Press.

Cohen, H.S., Kimball, K.T., Mulavara, A.P., Bloomberg, J.J. & Paloski, W.H. 2012. Posturography and locomotor tests of dynamic balance after long-duration spaceflight. *Journal of Vestibular Research: Equilibrium & Orientation*, 22, 191–196.

Cohen, M.M. 2000. Perception of facial features and face-to-face communications in space. *Aviation, Space, and Environmental Medicine*, 71(9, Sect. 2, Suppl.), A51–A57.

Coles, M.G.H., Gratton, G., Bashore, T.R., Eriksen, C.W. & Donchin, E. 1985. A psychophysiological investigation of the continuous flow model of human information processing. *Journal of Experimental Psychology: Human Perception and Performance*, 11, 529–553.

Colle, H.A. & Reid, G.B. 1998. Context effects in subjective mental workload ratings. *Human Factors*, 40, 591–600.

Colley, A.M. & Beech, J.R. 1989. Acquiring and performing cognitive skills. In A.M. Colley & J.R. Beech (Eds.), *Acquisition and Performance of Cognitive Skills* (pp. 1–16). New York: Wiley.

Collier, S., Ludvigsen, J.T. & Svengren, H. 2004. Human rreliability data from simulator experiments: Principles and context-sensitive analysis. Paper presented at PSAM 7-ESREL'04 International Conference, Berlin.

Collins, A.M. & Loftus, E.F. 1975. A spreading-activation theory of semantic memory. *Psychological Review*, 82, 407–428.

Colombini, D., Occhipinti, E., Molteni, G. & Grieco, A. 2006. Evaluation of work chairs. In W. Karwowski (Ed.), *International Encyclopedia of Ergonomics and Human Factors* (2nd ed., Vol. 2, pp. 1636–1642). Boca Raton, FL: CRC Press.

Colquhoun, W.P. 1971. *Biological Rhythms and Human Performance*. New York: Academic Press.

Colquhoun, W.P., Blake, M.J.F. & Edwards, R.S. 1968. Experimental studies of shift-work II: Stabilized eight-hour shift systems. *Ergonomics*, 11, 527–546.

Coltheart, M. 1980. Iconic memory and visible persistence. *Perception & Psychophysics*, 27, 183–228

Coltheart, M. 2009. Memory, iconic. In T. Bayne, A. Cleeremans, & P. Wilkens (Eds.), *The Oxford Companion to Consciousness* (pp. 427–430). Oxford, UK: Oxford University Press.

Conant, J.M. 2014. MANPRINT program focuses on integrating human element. Downloaded from https://www.arl.army.mil/www/?article=2470.

Conrad, R. 1964. Acoustic confusion in immediate memory. *British Journal of Psychology*, 55, 75–84.

Cook, N.L. 1989. The applicability of verbal mnemonics for different populations: A review. *Applied Cognitive Psychology*, 3, 3–22.

Cooke, N.J. & Gorman, J.C. 2006. Assessment of team cognition. In W. Karwowski (Ed.), *International Encyclopedia of Ergonomics and Human Factors* (2nd ed., Vol. 1, pp. 271–276). Boca Raton, FL: CRC Press.

Cooke, N.J., Gorman, J.C., Myers, C.W. & Duran, J.L. 2013. Interactive team cognition. *Cognitive Science*, 37, 255–285.

Cooper, G.E. & Harper, R.P., Jr. 1969. *The Use of Pilot Rating in the Evaluation of Aircraft Handling Qualities*. (NASA TN-D-5153). Washington, DC: National Aeronautics and Space Administration.

Cooper, W.E. 1983. Introduction. In W.E. Cooper (Ed.), *Cognitive Aspects of Skilled Typewriting* (pp. 1–38). New York: Springer.

Corballis, M.C., Kirby, J. & Miller, A. 1972. Access to elements of a memorized list. *Journal of Experimental Psychology*, 94, 185–190.

Coren, S. & Girgus, J.S. 1978. *Seeing Is Deceiving: The Psychology of Visual Illusions*. Mahwah, NJ: Lawrence Erlbaum.

Coren, S., Ward, L.M. & Enns, J.T. 2004. *Sensation and Perception* (6th ed.). Hoboken, NJ: Wiley.

Corey, G.A. & Merenstein J.H. 1987. Applying the acute ischemic heart disease predictive instrument. *Journal of Family Practice*, 25, 127–133.

Corlett, E.N. 1988. Cost–benefit analysis of ergonomic and work design changes. In D.J. Oborne (Ed.), *International Reviews of Ergonomics* (Vol. 2, pp. 85–104). London: Taylor & Francis.

Corlett, E.N. 2005. The evaluation of industrial seating. In J.R. Wilson & N. Corlett (Eds.), *Evaluation of Human Work* (3rd ed., pp. 729–741). Boca Raton, FL: CRC Press.

Corlett, J. 1992. The role of vision in the planning, guidance and adaptation of locomotion through the environment. In D. Elliott & L. Proteau (Eds.), *Vision and Motor Control* (pp. 375–397). Amsterdam: North-Holland.

Cormier, S.M. & Hagman, J.D. (Eds.) 1987. *Transfer of Learning: Contemporary Research and Applications*. San Diego, CA: Academic Press.

Cornsweet, T.N. 1970. *Visual Perception*. New York: Academic Press.

Cotten, D.J., Thomas, J.R., Spieth, W.R. & Biasiotto, J. 1972. Temporary fatigue effects in a gross motor skill. *Journal of Motor Behavior*, 4, 217–222.

Cotton, J.L., Vollrath, D.A., Froggatt, K.L., Lengnick-Hall, M.L. & Jennings, K.R. 1988. Employee participation: Diverse forms and different outcomes. *Academy of Management Review*, 13, 8–22.

Courtney, A.J. & Chow, H.M. 2001. A study of the discriminability of shape symbols by the foot. *Ergonomics*, 44, 328–338.

Cowan, N. 1997. *Attention and Memory: An Integrated Framework*. New York: Oxford University Press.

Cowan, N. 2008. Sensory memory. In H.L. Roediger, III (Ed.), *Cognitive Psychology of Memory* (pp. 23–32). Oxford, UK: Elsevier.

Cowan, N., Chen, Z. & Rouder, J.N. 2004. Constant capacity in an immediate serial-recall task: A logical sequel to Miller 1956. *Psychological Science*, 15, 634–640.

Cox, B.D. 1997. The rediscovery of the active learner in adaptive contexts: A developmental-historical analysis of transfer of training. *Educational Psychologist*, 32, 41–55.

Craig, J. 1980. *Designing with Type* (rev. ed.). New York: Watson-Guptill.

Craik, F.I.M. 2002. Levels of processing: Past, present, and future? *Memory*, 10, 305–318.

Craik, F.I.M. & Bialystock, E. 2006. Cognition through the lifespan: Mechanisms of change. *Trends in Cognitive Sciences*, 10, 131–138.

Craik, F.I.M. & Lockhart, R.S. 1972. Levels of processing: A framework for memory research. *Journal of Verbal Learning and Verbal Behavior*, 11, 671–684.

Craik, F.I.M. & Tulving, E. 1975. Depth of processing and the retention of words in episodic memory. *Journal of Experimental Psychology: General*, 104, 268–294.

Cranor, L.F., Guduru, P. & Arjula, M. 2006. User interfaces for privacy agents. *ACM Transactions on Computer–Human Interaction*, 13, 135–178.

Crawford, J. & Neal, A. 2006. A review of the perceptual and cognitive issues associated with the use of head-up displays in commercial aviation. *International Journal of Aviation Psychology*, 16, 1–19.

Crossman, E.R.F.W. 1959. A theory of the acquisition of speed-skill. *Ergonomics*, 2, 153–156.

Crossman, E.R.F.W. & Goodeve, P.J. 1983. Feedback control of hand-movement and Fitts' Law. *Quarterly Journal of Experimental Psychology*, 35A, 251–278. (Originally presented at a meeting of the Experimental Psychology Society, Oxford, England, 1963.)

Crowder, R.G. & Surprenant, A.M. 2000. Sensory stores. In A.E. Kazdin (Ed.), *Encyclopedia of Psychology* (Vol. 7, pp. 227–229). Washington, DC: American Psychological Association.

Cruikshank, P.J. 1985. Patient ratings of doctors using computers. *Social Science Medicine*, 21, 615–622.

Cuijpers, R.H., Smeets, J.B.J. & Brenner, E. 2004. On the relation between object shape and grasping kinematics. *Journal of Neurophysiology*, 91, 2598–2606.

Cullinane, T.P. 1977. Minimizing cost and effort in performing a link analysis. *Human Factors*, 19, 151–156.

Culver, C.C. & Viano, D.C. 1990. Anthropometry of seated women during pregnancy: Defining a fetal region for crash protection research. *Human Factors*, 32, 625–636.

Cunningham, G. 1990. Air quality. In N.C. Ruck (Ed.), *Building Design and Human Performance* (pp. 29–39). New York: Van Nostrand Reinhold.

Cushman, W.H. & Crist, B. 1987. Illumination. In G. Salvendy (Ed.), *Handbook of Human Factors* (pp. 670–695). New York: Wiley.

Cushman, W.H., Dunn, J.E. & Peters, K.A. 1984. Workplace luminance ratios: Do they affect subjective fatigue and performance? *Proceedings of the Human Factors Society 28th Annual Meeting* (p. 991). Santa Monica, CA: Human Factors Society.

Cyr, D., Head, M. & Larios, H. 2010. Colour appeal in website design within and across cultures: A multimethod evaluation. *International Journal of Human-Computer Studies*, 68, 1–21.

Czaja, S.J. & Nair, S.N. 2012. Human factors engineering and systems design. In G. Salvendy (Ed.), *Handbook of Human Factors and Ergonomics* (4th ed., pp. 38–56). Hoboken, NJ: Wiley.

Czichos, H. 2013. Physics of failure. In H. Czichos (Ed.), *Handbook of Technical Diagnostics* (pp. 23–40). Berlin Heidelberg: Springer.

Daanen, H.A.M., van de Vliert, E. & Huang, X. 2003. Driving performance in cold, warm, and thermoneutral environments. *Applied Ergonomics*, 34, 597–602.

Dahlgren, K. 1988. Shiftwork scheduling and their impact upon operators in nuclear power plants. *IEEE Fourth Conference on Human Factors in Power Plants* (pp. 517–521). Piscataway, NJ: IEEE.

Dahn, D. & Laughery, K.R. 1997. The integrated performance modeling environment—Simulating human-system performance. In S. Andradóttir, K.J. Healy, D.H. Withers & B.L. Nelson (Eds.), *Proceedings of the 1997 Winter Simulation Conference* (pp. 1141–1145).

Dail, T.E. & Christina, R.W. 2004. Distribution of practice and metacognition in learning and long-term retention of a discrete motor task. *Research Quarterly for Exercise and Sport*, 75, 148–155.

Danckert, S.L. & Craik, F.M. 2013. Does aging affect recall more than recognition memory? *Psychology and Aging*, 28, 902–909.

Dar, S. & Hall, J.W., III. 2012. *Otoacoustic Emissions: Principles, Procedures, and Protocols*. San Diego, CA: Plural Publishing.

Darwin, C.J., Turvey, M.T. & Crowder, R.G. 1972. An auditory analogue of the Sperling partial report proce-dure: Evidence for brief auditory storage. *Cognitive Psychology*, 3, 255–267.

Davies, B.T. & Watt, J.M., Jr. 1969. Preliminary investigation of movement time between brake and accelerator pedals in automobiles. *Human Factors*, 11, 407–410.

Davies, D.R. & Tune, G.S. 1969. *Human Vigilance Performance*. New York: Elsevier.

Davis, D.N. 2001. Control states and complete agent architectures. *Computational Intelligence*, 17, 621–650.

Davis, G. & Altman, I. 1976. Territories at the work-place: Theory into design guidelines. *Man–Environment Systems*, 6, 46–53.

Davis, K. & Marras, W. 2005. Load spatial pathway and spine loading: How does lift origin and destination influence low back response? *Ergonomics*, 48, 1031–1046.

Davis, K., Stremikis, K., Schoen, C. & Squires, D. 2014, June. Mirror, mirror on the wall, 2014 update: How the U.S. health care system compares internationally. The Commonwealth Fund.

Davis, L.E. & Wacker, G.J. 1987. Job design. In G. Salvendy (Ed.), *Handbook of Human Factors* (pp. 431–452). New York: Wiley.

Davis, S. (Ed.) 2000. *Color Perception: Philosophical, Psychological, Artistic, and Computational Perspectives*. New York: Oxford University Press.

Davranche, K. & Pichon, A. 2005. Critical flicker frequency threshold increment after an exhausting exercise. *Journal of Sport and Exercise Psychology*, 27, 515–520.

Dawson, E., Gilovich, T. & Regan, D.T. 2002. Motivated reasoning and performance on the Wason selection task. *Personality and Social Psychology Bulletin*, 28, 1379–1387.

Deatherage, B.H. 1972. Auditory and other sensory forms of information presentation. In H.P. Van Cott & R.G. Kinkade (Eds.), *Human Engineering Guide to Equipment Design* (revised edition, pp. 123–160). Washington, DC: U.S. Government Printing Office.

de Freitas Araujo, S. 2016. *Wundt and the Philosophical Foundations of Psychology: A Reappraisal*. Cham, Switzerland: Springer.

De Grip, A. & Sauermann, J. 2013. The effect of training on productivity: The transfer of on-the-job training from the perspective of economics. *Educational Research Review*, 8, 28–36.

de Groot, A. 1966. Perception and memory versus thought: Some old ideas and recent findings. In B. Kleinmuntz (Ed.), *Problem Solving* (pp. 19–50). New York: Wiley.

De Houwer, J. 2003. On the role of stimulus–response and stimulus–stimulus compatibility in the Stroop effect. *Memory & Cognition*, 31, 353–359.

De Keyser, V. & Javaux, D. 2000. Mental workload and cognitive complexity. In N.B. Sarter & R. Amalberti (Eds.), *Cognitive Engineering in the Aviation Domain* (pp. 43–63). Mahwah, NJ: Lawrence Erlbaum.

de Lange, H. 1958. Research into the dynamic nature of human fovea-cortex systems with intermittent and modulated light. I. Attenuation characteristics with white and colored light. *Journal of the Optical Society of America*, 48, 777–784.

de Lussanet, M.H.E, Smeets, J.B.J. & Brenner, E. 2002. Relative damping improves linear mass-spring models of goal-directed movements. *Human Movement Science*, 21, 85–100.

De Valois, R.L. & De Valois, K.K. 1980. Spatial vision. *Annual Review of Psychology,* 31, 309–341.

Deininger, R.L. 1960. Human factors engineering studies of the design and use of push button telephone tests. *Bell System Technical Journal*, 39, 995–1012.

Dekker, S.W.A. 2005. *Ten Questions about Human Error: A New View of Human Factors and System Safety*. Mahwah, NJ: Lawrence Erlbaum.

Delleman, N.J. & Dul, J. 1990. Ergonomic guidelines for adjustment and redesign of sewing machine work-places. In C.M. Haslegrave, J.R. Wilson, E.N. Corlett & I. Manenica (Eds.), *Work Design in Practice* (pp. 155–160). London: Taylor & Francis.

Dellis, E. 1988. Automotive head-up displays: Just around the corner. *Automotive Engineering*, 96, 107–110.

DeLucia, P.R., Mather, R.D., Griswold, J.A. & Mitra, S. 2006. Toward the improvement of image- guided inter-ventions for minimally invasive surgery: Three factors that affect performance. *Human Factors*, 48, 23–38.

DeMonasterio, F.M. 1978. Center and surround mechanisms of opponent-color X and Y ganglion cells of retina of macaques. *Journal of Neurophysiology*, 41, 1418–1434.

Dempsey, P.G., Wogalter, M.S. & Hancock, P.A. 2000. What's in a name? Using terms from definitions to examine the fundamental foundation of human factors and ergonomics science. *Theoretical Issues in Ergonomics Science*, 1, 3–10.

Dempsey, P.G., Wogalter, M.S. & Hancock, P.A. 2006. Defining ergonomics/human factors. In W. Karwowski (Ed.), *International Encyclopedia of Ergonomics and Human Factors* (2nd ed., Vol. 1, pp. 32–35). Boca Raton, FL: CRC Press.

DeNisi, A. & Smith, C. E. 2014. Performance appraisal, performance management, and firm-level performance: A review, a proposed model, and new directions for future research. *The Academy of Management Annals*, 8, 127–179.

Dettinger, K.M. & Smith, M.J. 2006. Human factors in organizational design and management. In G. Salvendy (Ed.), *Handbook of Human Factors and Ergonomics* (3rd ed., pp. 513–527). New York: Wiley.

Deutsch, J.A. & Deutsch, D. 1963. Attention: Some theoretical considerations. *Psychological Review*, 70, 80–90.

Dewar, R.E. 2006. Road warnings with traffic control devices. In M.S. Wogalter (Ed.), *Handbook of Warnings* (pp. 177–185). Mahwah, NJ: Lawrence Erlbaum.

Dewey, J. 1916. *Democracy and Education: An Introduction to the Philosophy of Education*. New York: Macmillan.

Dhillon, B.S. 1999. *Design Reliability: Fundamentals and Applications*. Boca Raton, FL: CRC Press.

Dhillon, B.S. 2009. *Human Reliability, Error, and Human Factors in Engineering Maintenance with Reference to Aviation and Power Generation*. Boca Raton, FL: CRC Press.

Di Lorenzo, P.M. & Youngentob, S.L. 2013. Olfaction and taste. In R.J. Nelson & S.J.Y. Mizumori (Eds.), *Biological Psychology* (Vol. 3: *Handbook of Psychology*, pp. 272–305). Hoboken, NJ: Wiley.

Diaper, D. & Stanton, N. (Eds.) 2004. *The Handbook of Task Analysis for Human–Computer Interaction*. Mahwah, NJ: Lawrence Erlbaum.

Diederich, A. & Busemeyer, J.R. 2003. Simple matrix methods for analyzing diffusion models of choice probability, choice response time, and simple response time. *Journal of Mathematical Psychology*, 47, 304–322.

Dijkstra, J.J. 1999. User agreement with incorrect expert system advice. *Behaviour and Information Technology*, 18, 299–411.

Dillon, C. 2006. Management perspectives for workplace ergonomics. In W. Karwowski (Ed.), *International Encyclopedia of Ergonomics and Human Factors* (2nd ed., Vol. 3, pp. 2930–2935). Boca Raton, FL: CRC Press.

Dillon, C. & Sanders, M. 2006. Diagnosis of work-related musculoskeletal disorders. In W. Karwowski (Ed.), *International Encyclopedia of Ergonomics and Human Factors* (2nd ed., Vol. 3, pp. 2584–2588). Boca Raton, FL: CRC Press.

Dimitriou, M., Wolpert, D.M. & Franklin, D.W. 2013. The temporal evolution of feedback gains rapidly update to task demands. *The Journal of Neuroscience*, 33, 10898–10909.

Dingus, T.A. 1995. A meta-analysis of driver eye-scanning behavior while navigating. *Proceedings of the Human Factors and Ergonomics Society 39th Annual Meeting* (pp. 1127–1131). Santa Monica, CA: HFES.

DiVita, J., Obermayer, R., Nugent, W. & Linville, J.M. 2004. Verification of the change blindness phenomenon while managing critical events on a combat information display. *Human Factors*, 46, 205–218.

Dix, A.J., Finlay, J.E., Abowd, G.D. & Beale, R. 2003. *Human–Computer Interaction* (3rd ed.). Upper Saddle River, NJ: Prentice-Hall.

Dixon, P. 1982. Plans and written directions for complex tasks. *Journal of Verbal Learning and Verbal Behavior*, 21, 70–84.

Donchin, E. 1981. Event-related brain potentials: A tool in the study of human information processing. In H. Begleiter (Ed.), *Evoked Potentials in Psychiatry*. New York: Plenum Press.

Donders, F.C. 1868/1969. On the speed of mental processes (W.G. Koster, Trans.). *Acta Psychologica*, 30, 412–431.

Donner, Y. & Hardy, J.L. 2015. Piecewise power laws in individual learning curves. *Psychonomic Bulletin & Review*, 22, 1308–1319.

Donovan, J.J. & Radosevich, D.J. 1999. A meta-analytic review of the distribution of practice effect: Now you see it, now you don't. *Journal of Applied Psychology*, 84, 795–805.

Dougherty, D.J., Emery, J.H. & Curtin, J.G. 1964. Comparison of perceptual workload in flying standard instrumentation and the contact analogy vertical display (JANAIR-D278-421-019). Fort Worth, TX: Bell Helicopter Co. (DTIC NO. 610617).

Dougherty, E. 1997. Human reliability analysis—Where shouldst thou turn? *Reliability Engineering and System Safety*, 29, 283–299.

Dougherty, E.M. & Fragola, J.R. 1988. *Human Reliability Analysis*. New York: Wiley.

Doya, K. 2000. Complementary roles of basal ganglia and cerebellum in learning and motor control. *Current Opinion in Neurobiology*, 10, 732–739.

Drakopoulos, D. & Mann, D.D. 2008. A mathematical equation for quantifying control functionality in agricultural tractors. *Journal of Agricultural Safety and Health*, 14, 377–389.

Draper, J.V., Handel, S. & Hood, C.C. 1990. Fitts' task by teleoperator: Movement time, velocity, and accelera- tion. *Proceedings of the Human Factors Society 34th Annual Meeting* (pp. 127–131). Santa Monica, CA: Human Factors Society.

Drews, F.A. 2012. Human error in health care. In P. Carayon (Ed.), *Handbook of Human Factors and Ergonomics in Health Care and Patient Safety* (2nd ed., pp. 323–340). Boca Raton, FL: CRC Press.

Drillis, R.J. 1963. Folk norms and biomechanics. *Human Factors*, 5, 427–441.

Driskell, J.E. & Salas, E. 2006. Groupware, group dynamics, and team performance. In C. Bowers, E. Salas & F. Jentsch (Eds.), *Creating High-Tech Teams: Practical Guidance on Work Performance and Technology* (pp. 11–34). Washington, DC: American Psychological Association.

Driskell, J.E., Copper, C. & Moran, A. 1994. Does mental practice enhance performance? *Journal of Applied Psychology*, 79, 481–492.

Drolet, B.C., Schwede, M., Bishop, K.D. & Fischer, S.A. 2013. Compliance and falsification of duty hours: Reports from residents and program directors. *Journal of Graduate Medical Education*, September, 368–373.

Dror, I.E. 2013. Practical solutions to cognitive and human factor challenges in forensic science. *Forensic Science Policy & Management: An International Journal*, 4(3–4), 105–113.

Dror, I.E. 2016. A hierarchy of expert performance. *Journal of Applied Research in Memory and Cognition*, 5, 121–127.

Druckman, D. & Bjork, R.A. (Eds.) 1991. *In the Mind's Eye: Enhancing Human Performance*. Washington, DC: National Academy Press.

Druckman, D. & Swets, J.A. (Eds.) 1988. *Enhancing Human Performance: Issues, Theories, and Techniques*. Washington, DC: National Academy Press.

Drury, C. 1975. Application of Fitts' Law to foot-pedal design. *Human Factors*, 17, 368–373.

Drury, C.G. & Coury, B.G. 1982. Container and handle design for manual handling. In R. Easterby, K.H.E. Kroemer & D.B. Chaffin (Eds.), *Anthropometry and Biomechanics: Theory and Application* (pp. 259– 268). New York: Plenum Press.

Duda, K.R., Vasquez, R.A., Middleton, A.J., Hansberry, M.L., Newman, D.J., Jacobs, S.E. & West, J.J. 2015. The Variable Vector Countermeasure Suit (V2Suit) for space habitation and exploration. *Frontiers in Systems Neuroscience*, 9.

Duffy, V.G. (Ed.) 2009. *Handbook of Digital Human Modeling*. Boca Raton, FL: CRC Press.

Dukic, T., Hanson, L. & Falkmer, T. 2006. Effect of drivers' age and push button locations on visual time off road, steering wheel deviation and safety perception. *Ergonomics*, 49, 78–92.

Dumas, J. S. & Fox, J. E. 2012. Usability testing. In J. Jacko (Ed.), *The Human-Computer Interaction Handbook: Fundamentals, Evolving Technologies, and Emerging Applications* (3rd ed., pp. 1221–1241). Boca Raton, FL: CRC Press.

Dumesnil, C.D. 1987. Office case study: Social behavior in relation to the design of the environment. *The Journal of Architectural and Planning Research*, 4, 7–13.

Dunbar, K. & Blanchette, I. 2001. The in vivo/in vitro approach to cognition: The case of analogy. *Trends in Cognitive Sciences*, 5, 334–339.

Duncan, J. 1977. Response selection rules in spatial choice reaction tasks. In S. Dornic (Ed.), *Attention and Performance IV* (pp. 49–61). Hillsdale, NJ: Lawrence Erlbaum.

Duncan, J. 1980. The locus of interference in the perception of simultaneous stimuli. *Psychological Review*, 87, 272–300.

Duncan-Johnson, C.C. & Donchin, E. 1977. On quantifying surprise: The variation in event-related potentials with subjective probability. *Psychophysiology*, 14, 456–467.

Duncker, K. 1945. On problem solving. *Psychological Monographs*, 58 (5, Whole No. 270).

Durlach, P.J. 2004. Change blindness and its implications for complex monitoring and control systems design and operator training. *Human–Computer Interaction*, 19, 423–451.

Durso, F.T., Boehm-Davis, D.A. & Lee, J.D. 2015. A view of human systems integration from the academy. In D. A. Boehm-Davis, F. T. Durso, & J. D. Lee (Eds.), *APA Handbook of Human Systems Integration* (pp. 5–19). Washington, DC: American Psychological Association.

Durso, F.T. & Dattel, A.R. 2004. SPAM: The real-time assessment of SA. In S. Banbury & S. Tremblay (Eds.), *A Cognitive Approach to Situation Awareness: Theory, Measurement and Application* (pp. 137–154). Burlington, VT: Ashgate.

Dutta, A. & Proctor, R.W. 1992. Persistence of stimulus–response compatibility effects with extended practice. *Journal of Experimental Psychology: Learning, Memory, and Cognition*, 18, 801–809.

Dzhafarov, E.N. & Colonius, H. 2006. Reconstructing distances among objects from their discriminability. In H. Colonius & E.N. Dzhafarov (Ed.), *Measurement and Representation of Sensations* (pp. 47–88). Mahwah, NJ: Lawrence Erlbaum.

Easterbrook, J.A. 1959. The effect of emotion on cue utilization and the organization of behavior. *Psychological Review*, 66, 183–201.

Easterby, R.S. 1967. Perceptual organization in static displays for man/machine systems. *Ergonomics*, 10, 193–205.

Easterby, R.S. 1970. The perception of symbols for machine displays. *Ergonomics*, 13, 149–158.

Eastman Kodak Company. 1983. *Ergonomic Design for People at Work* (Vol. 1). New York: Van Nostrand Reinhold.

Ebbinghaus, H. 1885/1964. *Memory* (H.A. Ruger & C.E. Bussenius, Trans.). New York: Dover Press.

Eberts, R., Lang, G.T. & Gabel, M. 1987. Expert/novice differences in designing with a CAD system. *Proceedings of 1987 IEEE Conference on Systems, Man, and Cybernetics* (pp. 985–989). New York: IEEE.

Eberts, R.E. & MacMillan, A.G. 1985. Misperception of small cars. In R.E. Eberts & C.G. Eberts (Eds.), *Trends in Ergonomics/Human Factors II* (pp. 33–39). Amsterdam: North-Holland.

Eberts, R.E. & Posey, J.W. 1990. The mental model in stimulus–response compatibility. In R.W. Proctor & T.G. Reeve (Eds.), *Stimulus–Response Compatibility: An Integrated Perspective* (pp. 389–425). Amsterdam: North-Holland.

Eberts, R.E. & Schneider, W. 1985. Internalizing the system dynamics for a second-order system. *Human Factors*, 27, 371–393.

Edwards, J.R., Caplan, R.D. & van Harrison, R. 1998. Person–environment fit theory. In C.L. Cooper (Ed.), *Theories of Organizational Stress* (pp. 28–67). Oxford: Oxford University Press.

Edwards, W. 1998. Hailfinder: Tools for and experiences with Bayesian normative modeling. *American Psychologist*, 53, 416–428.

Edworthy, J. & Hellier, E.J. 2006. Auditory warnings. In W. Karwowski (Ed.), *International Encyclopedia of Ergonomics and Human Factors* (2nd ed., Vol. 1, pp. 1026–1028). Boca Raton, FL: CRC Press.

Egan, J.P., Carterette, E.C. & Thwing, E.J. 1954. Some factors affecting multichannel listening. *Journal of the Acoustical Society of America*, 26, 774–782.

Eggemeier, F.T. 1988. Properties of workload assessment techniques. In P.A. Hancock & N. Meshkati (Eds.), *Human Mental Workload* (pp. 41–62). Amsterdam: North-Holland.

Eggleston, R.G. & Quinn, T.J. 1984. A preliminary evaluation of a projective workload assessment procedure. In *Proceedings of the Human Factors Society 28th Annual Meeting* (pp. 695–699). Santa Monica, CA: Human Factors Society.

Einstein, G.O. & Hunt, R.R. 1980. Levels of processing and organization: Additive effects of individual-item and relational processing. *Journal of Experimental Psychology: Human Learning and Memory*, 6, 588–598.

Eklund, J.A.E. & Corlett, E.N. 1986. Experimental and biomechanical analysis of seating. In N. Corlett, J.W. Wilson & I. Manenica (Eds.), *The Ergonomics of Working Postures* (pp. 319–330). London: Taylor & Francis.

ElBardissi, A.W., Wiegmann, D.A., Dearani, J.A., Daly, R.C. & Sundt, T.M. 2007. Application of the human factors analysis and classification system methodology to the cardiovascular surgery operating room. *The Annals of Thoracic Surgery*, 83, 1412–1419.

Elkind, J.I., Card, S.K., Hochberg, J. & Huey, B.M. (Eds.) 1990. *Human Performance Models for Computer-Aided Engineering*. New York: Academic Press.

Elliott, D., Helsen, W.F. & Chua, R. 2001. A century later: Woodworth's 1899 two-component model of goal-directed aiming. *Psychological Bulletin*, 127, 342–357.

Ellis, R.D., Cao, A., Pandya, A., Composto, A., Chacko, M., Klein, M. & Auner, G. 2004. Optimizing the surgeon–robot interface: The effect of control-display gain and zoom level on movement time. *Proceedings of the Human Factors and Ergonomics Society 48th Annual Meeting* (pp. 1713–1717). Santa Monica, CA: Human Factors and Ergonomics Society.

Ellis, S.E. 2000. Collision in space. *Ergonomics in Design*, 8(1), 4–9.

Ells, J.G. & Dewar, R.E. 1979. Rapid comprehension of verbal and symbolic traffic sign messages. *Human Factors*, 21, 161–168.

Elsbach, K.D. 2003. Relating physical environment to self-categorizations: Identity threat and affirmation in a non-territorial office space. *Administrative Science Quarterly*, 48, 622–654.

Embrey, D.E. 1986. SHERPA: A systematic human error reduction and prediction approach. Paper presented at the International Meeting on Advances in Nuclear Power Systems, Knoxville, TN.

Emery, F.E. & Trist, E.L. 1960. Sociotechnical systems. In C.W. Churchman & M. Verhulst (Eds.), *Management Science, Models and Techniques II* (pp. 83–97). London: Pergamon Press.

Endsley, M.R. 1988. Design and evaluation for situation awareness enhancement. *Proceedings of the Human Factors Society 30th Annual Meeting* (pp. 97–101). Santa Monica, CA: Human Factors Society.

Endsley, M.R. & Jones, D.G. 2012. *Designing for Situation Awareness: An Approach to User-Centered Design.* Boca Raton, FL: CRC Press.

English, H.B. 1942. How psychology can facilitate military training—A concrete example. *Journal of Applied Psychology*, 26, 3–7.

Engst, C., Chhokar, R., Miller, A., Tate, R.B. & Yassi, A. 2005. Effectiveness of overhead lifting devices in reducing the risk of injury to care staff in extended care facilities. *Ergonomics*, 48, 187–199.

Entwistle, M.S. 2003. The performance of automated speech recognition systems under adverse conditions of human exertion. *International Journal of Human–Computer Interaction*, 16, 127–140.

Epstein, W., Park, J. & Casey, A. 1961. The current status of the size-distance hypothesis. *Psychological Bulletin*, 58, 491–514.

Ercoline, W. 2000. The good, the bad, and the ugly of head-up displays. *IEEE Engineering in Medicine and Biology Magazine*, 19(2), 66–70.

Ergai, A., Cohen, T., Sharp, J., Wiegmann, D., Gramopadhye, A. & Shappell, S. 2016. Assessment of the Human Factors Analysis and Classification System (HFACS): Intra-rater and inter-rater reliability. *Safety Science*, 82, 393–398.

Ericsson, K.A. 2005. Recent advances in expertise research: A commentary on the contributions to the special issue. *Applied Cognitive Psychology*, 19, 233–241.

Ericsson, K.A. 2006a. An introduction to *Cambridge Handbook of Expertise and Expert* Performance: Its development, organization, and content. In K.A. Ericsson, N. Charness, P.J. Feltovich & R.R. Hoffman (Eds.), *Cambridge Handbook of Expertise and Expert Performance* (pp. 3–19). Cambridge: Cambridge University Press.

Ericsson, K.A. 2006b. The influence of experience and deliberate practice on the development of superior expert performance. In K.A. Ericsson, N. Charness, P.J. Feltovich & R.R. Hoffman (Eds.), *The Cambridge Handbook of Expertise and Expert Performance* (pp. 683–703). New York: Cambridge University Press.

Ericsson, K.A. & Polson, P.G. 1988. A cognitive analysis of exceptional memory for restaurant orders. In M.T.H. Chi, R. Glaser & M.J. Farr (Eds.), *The Nature of Expertise* (pp. 23–70). Hillsdale, NJ: Lawrence Erlbaum.

Ericsson, K.A. & Simon, H.A. 1993. *Protocol Analysis: Verbal Reports as Data* (rev. ed.). Cambridge, MA: MIT Press.

Ericsson, K.A., Krampe, R.T. & Tesch-Romer, C. 1993. The role of deliberate practice in the acquisition of expert performance. *Psychological Review*, 100, 363–406.

Eriksen, B.A. & Eriksen, C.W. 1974. Effects of noise letters upon the identification of a target letter in a non-search task. *Perception & Psychophysics*, 16, 143–149.

Eriksen, C.W. & Collins, J.F. 1967. Some temporal characteristics of visual pattern perception. *Journal of Experimental Psychology*, 74, 476–484.

Eriksen, C.W. & Schultz, D.W. 1979. Information processing in visual search: A continuous flow conception and experimental results. *Perception & Psychophysics*, 25, 249–263.

Eriksen, C.W. & St. James, J.D. 1986. Visual attention within and around the field of focal attention: A zoom lens model. *Perception & Psychophysics*, 40, 225–240.

Estes, W.K. 1972. An associative basis for coding and organization in memory. In A.W. Melton & E. Martin (Eds.), *Coding Processes in Human Memory* (pp. 161–190). Washington, DC: Winston.

Evangelopoulos, N.E. 2013. Latent semantic analysis. *Wires Cognitive Science*, 4, 683–692.

Evans, D.W. & Ginsburg, A.P. 1982. Predicting age-related differences in discriminating road signs using contrast sensitivity. *Journal of the Optical Society of America*, 72, 1785–1786 (Abstract).

Evans, G.W. & Johnson, D. 2000. Stress and open-office noise. *Journal of Applied Psychology*, 85, 779–783.

Evans, G.W., Lepore, S.J. & Allen, K.M. 2000. Cross-cultural differences in tolerance for crowding: Fact or fiction? *Journal of Personality and Social Psychology*, 79, 204–210.

Evans, J. St. B.T. 1989. *Bias in Human Reasoning: Causes and Consequences.* Hillsdale, NJ: Lawrence Erlbaum.

Evans, J. St. B.T. 1998. Matching bias in conditional reasoning: Do we understand it after 25 years? *Thinking and Reasoning*, 4, 45–82.

Evans, J. St. B.T. 2002. Logic and human reasoning: An assessment of the deduction paradigm. *Psychological Review*, 128, 978–996.

Eysenck, M.W. 1979. Depth, elaboration, and distinctiveness. In L. S. Cermak & F. I. M. Craik (Eds.), *Levels of Processing in Human Memory* (pp. 89–118). Hillsdale, NJ: Lawrence Erlbaum.

Fan, J., Han, F. & Liu, H. 2014. Challenges of big data analysis. *National Science Review*, 1 (2), 293–314.

Fanger, P.O. 1977. Local discomfort to the human body caused by nonuniform thermal environments. *Annals of Occupational Hygiene*, 20, 285–291.

Farell, T.R., Weir, R.F., Heckathorne, C.W. & Childress, D.S. 2005. The effects of static friction and backlash on extended physiological proprioception control of a power prosthesis. *Journal of Rehabilitation Research and Development*, 42, 327–342.

Fechner, G.T. 1860/1966. *Elements of Psychophysics* (Vol. 1; E.G. Boring & D.H. Howes, Eds.; H.E. Adler, Trans.). New York: Holt, Rinehart & Winston.

Feltz, D. & Landers, D.M. 1983. The effects of mental practice on motor skill learning and performance: A meta-analysis. *Journal of Sport Psychology*, 5, 25–57.

FEMA 2009. Emergency vehicle visibility and conspicuity study. Report FA-323. Emmitsburg, MD: Department of Homeland Security.

Ferguson, S.A., Marras, W.S. & Burr, D.L. 2004. The influence of individual low back health status on workplace trunk kinematics and risk of low back disorder. *Ergonomics*, 47, 1226–1237.

Ferreira, J. & Hignett, S. 2005. Reviewing ambulance design for clinical efficiency and paramedic safety. *Applied Ergonomics*, 36, 97–105.

Feuerstein, M., Huang, G.D. & Pransky, G. 1999. Low back pain: An epidemic in industrialized countries. In R.J. Gatchel & D.C. Turk (Eds.), *Psychosocial Factors in Pain: Critical Perspectives* (pp. 175–192). New York: Guilford Press.

Fincannon, T., Keebler, J.R. & Jentsch, F. 2014. Examining external validity issues in research with human operation of unmanned vehicles. *Theoretical Issues in Ergonomics Science*, 15, 395–414.

Fine, S.A. 1974. Functional job analysis: An approach to a technology for manpower planning. *Personnel Journal*, 53, 813–818.

Finomore, V.J., Shaw, T.H., Warm, J.S., Matthews, G. & Boles, D.B. 2013. Viewing the workload of vigilance through the lenses of the NASA-TLX and the MRQ. *Human Factors*, 55, 1044–1063.

Fiore, S.M., Cuevas, H.M. & Salas, E. 2003. Putting working memory to work: Integrating cognitive science theories with cognitive science research. In *Proceedings of the Human Factors and Ergonomics Society 47th Annual Meeting* (pp. 508–512). Santa Monica, CA: HFES.

Fiorentini, A. 2003. Brightness and lightness. In L.M. Chalupa & J.S. Werner (Eds.), *The Visual Neurosciences* (pp. 881–891). Cambridge, MA: MIT Press.

Fiscus, J. G., Ajot, J., Garofolo, J. S., & Doddingtion, G. 2007. Results of the 2006 spoken term detection evaluation. In *Proceedings of the SIGIR 2007 Workshop* (Vol. 7, pp. 51–57). Amsterdam, The Netherlands.

Fisher, D.L., Coury, B.G., Tengs, T.O. & Duffy, S.A. 1989. Minimizing the time to search visual displays: The role of highlighting. *Human Factors*, 31, 167–182.

Fisher, D.L. & Tan, K.C. 1989. Visual displays: The highlighting paradox. *Human Factors*, 31, 17–30.

Fitchett, S. & Cockburn, A. 2015. An empirical characterisation of file retrieval. *International Journal of Human-Computer Studies*, 74, 1–13.

Fitts, D. 2000. An overview of NASA ISS human engineering and habitability: Past, present, and future. *Aviation, Space, and Environmental Medicine*, 71(9, Sect. 2, Suppl.), A112–A116.

Fitts, P.M. 1954. The information capacity of the human motor system in controlling the amplitude of movement. *Journal of Experimental Psychology*, 47, 381–391.

Fitts, P.M. 1964. Perceptual-motor skill learning. In A.W. Melton (Ed.), *Categories of Human Learning* (pp. 243–285). New York: Academic Press.

Fitts, P.M. & Crannell, C.W. 1953. Studies in location discrimination. Wright Air Development Center Technical Report.

Fitts, P.M. & Deininger, R.L. 1954. S–R compatibility: Correspondence among paired elements within stimulus and response codes. *Journal of Experimental Psychology*, 48, 483–491.

Fitts, P.M. & Jones, R.E. 1947. Analysis of factors contributing to 460 "pilot-error" experiences in operating aircraft controls. Report TSEAA-694-12, Air Material Command, Wright Patterson Air Force Base. Reprinted in H.W. Sinaiko (Ed.) 1961. *Selected Papers on Human Factors in the Design and Use of Control Systems*. New York: Dover.

Fitts, P.M., Jones, R.E. & Milton, J.L. 1950. Eye movements of aircraft pilots during instrument-landing approaches. *Aeronautical Engineering Review*, 9, 1–16.

Fitts, P.M. & Posner, M.I. 1967. *Human Performance*. Belmont, CA: Brooks/Cole.

Fitts, P.M. & Seeger, C.M. 1953. S–R compatibility: Spatial characteristics of stimulus and response codes. *Journal of Experimental Psychology*, 46, 199–210.

Fitts, R.H., Riley, D.R. & Widrick, J.J. 2000. Invited review: Microgravity and skeletal muscle. *Journal of Applied Physiology*, 89, 823–839.

Fitzgerald, K. 1989. Probing Boeing's crossed connections. *IEEE Spectrum, May*, 30–35.

Fleming, R.A. 1970. The processing of conflicting information in a simulated tactical decision-making task. *Human Factors*, 12, 275–285.

Fletcher, J.L. & Riopelle, A.J. 1960. Protective effect of the acoustic reflex for impulsive noises. *Journal of the Acoustical Society of America*, 32, 401–404.

Flynn, J.E. 1977. A study of subjective responses to low energy and nonuniform lighting systems. *Lighting Design and Application*, 7(2), 6–15.

Fodor, J.A. & Garrett, M. 1967. Some syntactic determinants of sentential complexity. *Perception & Psychophysics*, 2, 289–296.

Folds, D.J. 2015. Systems engineering perspective on human systems integration. In D.A. Boehm-Davis, F.T. Durso, & J.D. Lee (Eds.), *APA Handbook of Human Systems Integration* (pp. 21–35). Washington, DC: American Psychological Association.

Folkard, S. 1975. Diurnal variation in logical reasoning. *British Journal of Psychology*, 66, 1–8.

Folkard, S. & Lombardi, D.A. 2006. Modeling the impact of the components of long work hours on injuries and "accidents". *American Journal of Independent Medicine*, 49, 953–963.

Folkard, S. & Monk, T.H. 1980. Circadian rhythms in human memory. *British Journal of Psychology*, 71, 295–307.

Fong, G.T., Krantz, D.H. & Nisbett, R.E. 1986. The effects of statistical training on thinking about everyday problems. *Cognitive Psychology*, 18, 253–292.

Fonseca, D.J., Bisen, K.B, Midkiff, K.C. & Moynihan, G.P. 2006. An expert system for lighting energy management in public school facilities. *Expert Systems: International Journal of Knowledge Engineering and Neural Networks*, 23, 194–211.

Forgery Warning I (Oct. 1999). Downloaded on July 27, 2002, from http://www.s-line.de/homepages/ronald.frank/PGP/forgery_1.html

Fosse, P. & Valberg, A. 2004. Lighting needs and lighting comfort during reading with age-related macular degeneration. *Journal of Visual Impairment and Blindness, July*, 389–409.

Fowler, C.A. & Galantucci, B. 2005. The relation of speech perception and speech production. In D.B. Pisoni & R.E. Remez (Eds.), *The Handbook of Speech Perception* (633–652). Malden, MA: Blackwell.

Foyle, D.C., Dowell, S.R. & Hooey, B.L. 2001. Cognitive tunneling in head-up display (HUD) superimposed symbology: Effects of information location. In R.S. Jensen, L. Chang & K. Singleton (Eds.), *Proceedings of the Eleventh International Symposium on Aviation Psychology* (pp. 143:1–143:6). Columbus, OH: Ohio State University.

Francis, G. 2000. Quantitative theories of metacontrast masking. *Psychological Review*, 107, 768–785.

Frane, A. 2015. A call for considering color vision deficiency when creating graphics for psychology reports. *The Journal of General Psychology*, 142, 194–211.

Frazier, L. & Clifton, C., Jr. 1996. *Construal*. Cambridge, MA: MIT Press.

Frazier, L., Carlson, K. & Clifton, C. Jr. 2006. Prosodic phrasing is central to language comprehension. *Trends in Cognitive Sciences*, 10, 244–249.

Freund, L.E. & Sadosky, T.L. 1967. Linear programming applied to optimization of instrument panel and workplace layout. *Human Factors*, 9, 295–300.

Friesen, L. & Earl, P.E. 2015. Multipart tariffs and bounded rationality: An experimental analysis of mobile phone plan choices. *Journal of Economic Behavior & Organization*, 116, 239–253.

Frishman, L.J. 2001. Basic visual processes. In E.B. Goldstein (Ed.), *Blackwell Handbook of Perception* (pp. 53–91). Malden, MA: Blackwell.

Frost, N. 1972. Encoding and retrieval in visual memory tasks. *Journal of Experimental Psychology*, 95, 317–326.

Fryar, C.D., Gu, Q. & Ogden, C.L. 2012. Anthropometric reference data for children and adults: United States, 2007–2010. *Vital and Health Statistics. Series 11, Data from the National Health Survey*, (252), 1–48.

Fucigna, J.T. 1967. The ergonomics of offices. *Ergonomics*, 10, 589–604.

Fulton-Suri, J. 1999. The next 50 years: Future challenges and opportunities. *The Ergonomics Society Annual Conference 1999*. University of Leicester, April 7–9, 1999.

Gajilan, A.C. 2006. Today's medical training—Better or worse for patients. *Health, CNN.com*, September 6, 2006. Retrieved on September 24, 2006, from http://www.cnn.com/006/HEALTH/09/06/doctors.hours.cnn/

Galinsky, T.L., Warm, J.S., Dember, W.N., Weiler, E.M. & Scerbo, M.W. 1990. Sensory alternation and vigilance performance: The role of pathway inhibition. *Human Factors*, 32, 717–728.

Gallant, S.I. 1988. Connectionist expert systems. *Communications of the ACM*, 31, 152–169.

Gallaway, G. 2007. Aviation psychology research—Extending scientific method to incorporate value and application. *Proceedings of the 14th International Symposium on Aviation Psychology* (pp. 224–229). Dayton, OH (CD-ROM).

Gallimore, J.J. & Stouffer, J.M. 2001. Research and development toward automatic luminance control of electronic displays. *International Journal of Aviation Psychology*, 11, 149–168.

Gao, S. 2013. Mobile decision support systems research: A literature analysis. *Journal of Decision Systems*, 22, 10–27.

Garcia, R. & Barnes, L. 2010. Multi-UAV simulator utilizing X-Plane. *Journal of Intelligent & Robotic Systems*, 57, 393–406.

García-Izquierdo, A.L., Vilela, L.D. & Moscoso, S. 2015. Work analysis for personnel selection. In I. Nikolaou & J.K. Oostrom (Eds.), *Employee Recruitment, Selection, and Assessment: Contemporary Issues for Theory and Practice* (pp. 9–26). New York: Psychology Press.

Gardin, H., Kaplan, K.J., Firestone, I. & Cowan, G. 1973. Proxemic effects on cooperation, attitude, and approach-avoidance in a prisoner's dilemma game. *Journal of Personality and Social Psychology*, 27, 13–18.

Gardiner, P.C. & Edwards, W. 1975. Public values: Multiattribute utility measurement for social decision-making. In M.F. Kaplan & S. Schwartz (Eds.), *Human Judgment and Decision Processes* (pp. 1–37). New York: Academic Press.

Garland, D.J., Stein, E.S. & Muller, J.K. 1999. Air traffic controller memory: Capabilities, limitations, and volatility. In D.J. Garland, J.A. Wise & V.D. Hopkin (Eds.), *Handbook of Aviation Human Factors* (pp. 455–496). Mahwah, NJ: Lawrence Erlbaum.

Garner, W.R. 1962. *Uncertainty and Structure as Psychological Concepts*. New York: Wiley.

Garner, W.R. 1974. *The Processing of Information and Structure*. Hillsdale, NJ: Lawrence Erlbaum.

Garofolo, J.P. & Polatin, P. 1999. Low back pain: An epidemic in industrialized countries. In R.J. Gatchel & D.C. Turk (Eds.), *Psychosocial Factors in Pain: Critical Perspectives* (pp. 164–174). New York: Guilford Press.

Garvey, P.G., Pietrucha, M.T. & Meeker, D.T. 1998. Clearer road signs ahead. *Ergonomics in Design*, 6(3), 7–11.

Gaspar, J.G., Ward, N., Neider, M.B., Crowell, J., Carbonari, R., Kaczmarski, H. & Loschky, L.C. 2016. Measuring the useful field of view during simulated driving with gaze-contingent displays. *Human Factors*, 58, 630–641.

Gawron, V.J. 2008. *Human Performance, Workload, and Situational Awareness Measures Handbook* (2nd ed.). Boca Raton, FL: CRC Press.

Geiselman, R.E., McCloskey, B.P., Mossler, R.A. & Zielan, D.S. 1984. An empirical evaluation of mnemonic instruction for remembering names. *Human Learning*, 3, 1–7.

Gelb, A. 1929. Die "Farbenkonstanz" der Sehding. *Handbook of Normal Pathological Physiology*, 12, 594–678.

Geldard, F.A. 1972. *The Human Senses* (2nd ed.). New York: Wiley.

Gell, N., Werner, R.A., Franzblau, A., Ulin, S.S. & Armstrong, T.J. 2005. A longitudinal study of industrial and clerical workers: Incidence of carpal tunnel syndrome and assessment of risk factors. *Journal of Occupational Rehabilitation*, 15, 47–55.

Gemperle, F., Ota, N. & Siewiorek, D. 2001. Design of a wearable tactile display. In *Proceedings of the. Fifth International Symposium on Wearable Computers*, 2001. (pp. 5–12). Piscataway, NJ: IEEE.

Gentaz, E. & Tschopp, C. 2002. The oblique effect in the visual perception of orientations. In S.P. Shohov (Ed.), *Advances in Psychology Research* (Vol. 10, pp. 3–28). Huntington, NY: Nova Science Publishers.

Gentner, D. & Stevens, A.L. (Eds.) 1983. *Mental Models*. Hillsdale, NJ: Lawrence Erlbaum.

Gentner, D.R. 1983. Keystroke timing in transcription typing. In W.E. Cooper (Ed.), *Cognitive Aspects of Skilled Typewriting* (pp. 95–120). New York: Springer.

Gentner, D.R., Larochelle, S. & Grudin, J. 1988. Lexical, sublexical, and peripheral effects in skilled typewriting. *Cognitive Psychology*, 20, 524–548.

Gescheider, G.A. 1997. *Psychophysics: The Fundamentals* (3rd ed.). Mahwah, NJ: Lawrence Erlbaum.

Giarratano, J.C. & Riley, G.D. 2004. *Expert Systems: Principles and Programming* (4th ed.). Boston, MA: Course Technology.

Gibson, J.J. 1950. *The Perception of the Visual World*. Boston, MA: Houghton Mifflin.

Gick, M.L. & Holyoak, K.J. 1980. Analogical problem solving. *Cognitive Psychology*, 12, 306–355.

Gick, M.L. & Holyoak, K.J. 1983. Schema induction and analogical transfer. *Cognitive Psychology*, 15, 1–38.

Gielen, S., van Bolhuis, B. & Vrijenhoek, E. 1998. On the number of degrees of freedom in biological limbs. In M.L. Latash (Ed.), *Progress in Motor Control* (Vol. 1, pp. 173–190). Champaign, IL: Human Kinetics.

Gies, J. 1991. Automating the worker. *American Heritage of Invention and Technology*, 6(3), 56–63.

Gifford, R. 2014. *Environmental Psychology: Principles and Practice* (5th ed.). Colville, WA: Optimal Books.

Gilbreth, F.B. 1909. *Bricklaying System*. New York: Clark.

Gilbreth, F.B. & Gilbreth, L.M. 1924. Classifying the elements of work. *Management and Administration*, 8, 151–154.

Gilchrist, A.L. 1977. Perceived lightness depends on perceived spatial arrangement. *Science*, 195, 185–187.

Gilchrist, A.L. 2006. *Seeing Black and White*. Oxford: Oxford University Press.

Gill, N.F. & Dallenbach, K.M. 1926. A preliminary study of the range of attention. *American Journal of Psychology*, 37, 247–256.

Gillam, B. 2001. Varieties of grouping and its role in determining surface layout. In T.F. Shipley & P.J. Kellman (Eds.), *From Fragments to Objects: Segmentation and Grouping in Vision* (pp. 247–264). Amsterdam: North-Holland.

Gillan, D.J., Burns, M.J., Nicodemus, C.L. & Smith, R.L. 1986. The space station: Human factors and productivity. *Human Factors Society Bulletin*, 29(11), 1–3.

Gillespie, R. 1991. *Manufacturing Knowledge: A History of the Hawthorne Experiments*. New York: Cambridge University Press.

Gilson, L. L., Maynard, M. T., Young, N. C. J., Vartiainen, M. & Hakonen, M. 2015. Virtual teams research 10 years, 10 themes, and 10 opportunities. *Journal of Management*, 41, 1313–1337.

Ginsburg, A.P., Evans, D.W., Sekuler, R. & Harp, S.A. 1982. Contrast sensitivity predicts pilots' performance in aircraft simulators. *American Journal of Optometry and Physiological Optics*, 59, 105–108.

Gkikas, N. (Ed.) 2013. *Automotive Ergonomics: Driver-Vehicle Interaction*. Boca Raton, FL: CRC Press.

Glaser, R. & Chi, M.T.H. 1988. Overview. In M.T.H. Chi, R. Glaser & M.J. Farr (Eds.), *The Nature of Expertise* (pp. xv–xxviii). Hillsdale, NJ: Lawrence Erlbaum.

Glass, S. & Suggs, C. 1977. Optimization of vehicle-brake foot pedal travel time. *Applied Ergonomics*, 8, 215–218.

Glavin, R.J. & Maran, N.J. 2003. Integrating human factors into the medical curriculum. *Medical Education*, 37(Suppl. 1), 59–74.

Gleick, J. 2011. *The Information: A History, A Theory, A Flood*. New York: Pantheon

Gluck, K.A. & Pew, R.W. (Eds.) 2005. *Modeling Human Behavior with Integrated Cognitive Architectures: Comparison, Evaluation, and Validation*. Mahwah, NJ: Lawrence Erlbaum.

Gluck, M.A. & Bower, G.H. 1988. Evaluating an adaptive network model of human learning. *Journal of Memory and Language*, 27, 166–195.

Gobet, F. & Charness, N. 2007. Expertise in chess. In K.A. Ericsson, N. Charness, P.J. Feltovich & R.R. Hoffman (Eds.), *Cambridge Handbook of Expertise and Expert Performance* (pp. 523–538). Cambridge: Cambridge University Press.

Godwin, M.A. & Schmidt, R.A. 1971. Muscular fatigue and discrete motor learning. *Research Quarterly*, 42, 374–383.

Goh, H., Gordon, J., Sullivan, K.J. & Winstein, C.J. 2014. Evaluation of attentional demands during motor learning: Validity of a dual-task probe paradigm. *Journal of Motor Behavior*, 46, 95–105.

Goldberg, J.M., Wilson, V.J., Cullen, K.E., Angelaki, D.E., Broussard, D.M., Buttner-Ennever, J., Fukushima, K. & Minor, L.B. 2012. *The Vestibular System: A Sixth Sense*. New York: Oxford University Press.

Goldstein, E. B. 2016. *Sensation and Perception* (10th ed.). Belmont, CA: Wadsworth.

Goldstein, I.L. & Ford, J.K. 2002. *Training in Organizations* (4th ed.). Belmont, CA: Wadsworth.

Goodwin, G.M., McCloskey, D.J. & Matthews, P.B.C. 1972. The contribution of muscle afferents to kinaesthesia shown by vibration-induced illusions of movement and by the effect of paralysing joint afferents. *Brain*, 95, 705–748.

Gopher, D. & Kahneman, D. 1971. Individual differences in attention and the prediction of flight criteria. *Perceptual and Motor Skills*, 33, 1335–1342.

Gopher, D., Karis, D & Koenig, W. 1985. The representation of movement schemas in long-term memory: Lessons from the acquisition of a transcription skill. *Acta Psychologica*, 60, 105–134.

Gordon, C.C., Churchill, T., Clauser, C.C., Bradtmiller, B., McConville, J.T., Tebbets, I. & Walker, R.A. 1989. 1988 anthropometric survey of U.S. Army personnel: Summary statistics interim report. Natick TR-89=027. Matcik, MA: U.S. Army Natick Research, Development and Engineering Center.

Gore, J., Flin, R., Stanton, N. & Wong, B.W. 2015. Applications for naturalistic decision-making. *Journal of Occupational and Organizational Psychology*, 88, 223–230.

Gorea, A. 2015. A refresher of the original Bloch's law paper (Bloch, July 1885). *I-Perception*, 6(4), 1–6.

Goswami, A. 1997. Anthropometry of people with disability. S. Kumar (Ed.), *Perspectives in Rehabilitation Ergonomics* (pp. 339–359). London: Taylor & Francis.

Goteman, O., Smith, K. & Dekker, S. 2007. HUD with a velocity (flight-path) vector reduces lateral error during landing in restricted visibility. *International Journal of Aviation Psychology*, 17, 91–108.

Grant, E.R. & Spivey, M.J. 2003. Eye movements and problem solving: Guiding attention guides thought. *Psychological Science*, 14, 462–466.

Gray, W.D. 2000. The nature and processing of errors in interactive behavior. *Cognitive Science*, 24, 205–248.

Gray, W.D. & Altmann, E.M. 2006. Cognitive modeling in human–computer interaction. In W. Karwowski (Ed.), *International Encyclopedia of Ergonomics and Human Factors* (2nd ed., Vol. 1, pp. 609–614). Boca Raton, FL: CRC Press.

Gray, W.D., John, B.E. & Atwood, M.E. 1993. Project Ernestine: Validating a GOMS analysis for predicting and explaining real-world performance. *Human–Computer Interaction*, 8, 237–309.

Green, D.M. & Swets, J.A. 1966. *Signal Detection Theory and Psychophysics*. New York: Wiley.

Green, R.J., Self, H.C. & Ellifritt, T.S. (Eds.) 1995. *50 Years of Human Engineering: History and Cumulative Bibliography of the Fitts Human Engineering Division*. Wright-Patterson Air Force Base, OH: Armstrong Laboratory.

Greenberg, L. & Chaffin, D. 1978. *Workers and Their Tools: A Guide to the Ergonomic Design of Handtools and Small Presses* (rev. ed.). Midland, MI: Pendell Publishing.

Greenwald, A.G. 1970. A choice reaction time test of ideomotor theory. *Journal of Experimental Psychology*, 86, 20–25.

Greenwald, A.G. & Shulman, H.G. 1973. On doing two things at once: II. Elimination of the psychological refractory period effect. *Journal of Experimental Psychology*, 101, 70–76.

Gregory, R. L. 2015. *Eye and Brain: The Psychology of Seeing* (5th ed.). Princeton, NJ: Princeton University Press.

Grether, W.F. & Baker, C.A. 1972. Visual presentation of information. In H.A. Van Cott & R.G. Kinkade (Eds.), *Human Engineering Guide to Equipment Design* (rev. ed., pp. 41–121). Washington, DC: U.S. Government Printing Office.

Grice, H.P. 1975. Logic and conversation. In P. Cole & J.L. Morgan (Eds.), *Syntax and Semantics* (Vol. 3: *Speech Acts*, pp. 41–58). New York: Seminar Press.

Grier, R.A., Warm, J.S., Dember, W.N., Matthews, G., Galinsky, T.L., Szalma, J.L. & Parasuraman, R. 2003. The vigilance decrement reflects limitations in effortful attention, not mindlessness. *Human Factors*, 45, 349–359.

Griffin, M.J. 2006. Vibration and motion. In G. Salvendy (Ed.), *Handbook of Human Factors and Ergonomics* (3rd ed., pp. 590–611). Hoboken, NJ: Wiley.

Griffin, M.J. & Bovenzi, M. 2002. The diagnosis of disorders caused by hand-transmitted vibration: Southampton workshop 2000. *International Archives of Occupational and Environmental Health*, 75(1–2), 1–5.

Griggs, R.A. & Cox, J.R. 1982. The elusive thematic-materials effect in Wason's selection task. *British Journal of Psychology*, 73, 407–420.

Griggs, R.A. & Newstead, S.E. 1982. The role of problem structure in deductive reasoning. *Journal of Experimental Psychology: Learning, Memory, and Cognition*, 8, 297–307.

Grigoriev, A.I. & Potapov, A.N. 2013. From the flight of Yu. A. Gagarin to the contemporary piloted space flights and exploration missions. *Human Physiology*, 39, 675–686.

Grillner, S. 1975. Locomotion in vertebrates: Central mechanisms and reflex interaction. *Physiological Reviews*, 55, 247–304.

Grimes, J. 1996. On the failure to detect changes in scenes across saccades. In K. Akins (Ed.), *Vancouver Studies in Cognitive Science* (Vol. 5: *Perception*, pp. 89–110). New York: Oxford University Press.

Grobelny, J. & Karwowski, W. 2006. Facility and workspace layout problems in ergonomic design. In W. Karwowski (Ed.), *International Encyclopedia of Ergonomics and Human Factors* (2nd ed., Vol. 2, pp. 1643–1654). Boca Raton, FL: CRC Press.

Groome, D., Eysenk, M.W. 2016. *Applied Cognitive Psychology* (2nd ed.). New York: Routledge.

Grosslight, J.H., Fletcher, H.J., Masterton, R.B. & Hagen, R. 1978. Monocular vision and landing performance in general aviation pilots: Cyclops revisited. *Human Factors*, 20, 27–33.

Grossman, L. & Eagle, M. 1970. Synonymity, antonymity, and association in false recognition responses. *Journal of Experimental Psychology*, 83, 244–248.

Grudin, J. 2002. Group dynamics and ubiquitous computing. *Communications of the ACM*, 45(12), 74–78.

Guiard, Y. 1983. The lateral coding of rotations: A study of the Simon effect with wheel-rotation responses. *Journal of Motor Behavior*, 15, 331–342.

Guiard, Y. & Beaudouin-Lafon, M. 2004. Fitts' law 50 years later: Applications and contributions from human–computer interaction. *International Journal of Human–Computer Studies*, 61, 747–750.

Günther, F., Dudschig, C. & Kaup, B. 2015. LSAfun—An R package for computations based on Latent Semantic Analysis. *Behavior Research Methods*, 47, 930–944.

Guo, C., Wang, W., Guo, B. & Si, X. 2013. A maintenance optimization model for mission-oriented systems based on Wiener degradation. *Reliability Engineering and System Safety*, 111, 183–194.

Gupta, N., Bisantz, A.M. & Singh, T. 2002. The effects of adverse condition warning system characteristics on driver performance: An investigation of alarm signal type and threshold level. *Behaviour and Information Technology*, 21, 235–248.

Guth, S.K. 1963. A method for the evaluation of discomfort glare. *Illumination Engineering*, 58, 351.

Gwóz´dz, B.M. 2006. Occupational stress mechanisms. In W. Karwowski (Ed.), *International Encyclopedia of Ergonomics and Human Factors* (2nd ed., Vol. 1, pp. 852–853). Boca Raton, FL: CRC Press.

Haas, E. & Edworthy, J. 2006. An introduction to auditory warnings and alarms. In M.S. Wogalter (Ed.), *Handbook of Warnings* (pp. 189–198). Mahwah, NJ: Lawrence Erlbaum.

Haas, E.C. & Edworthy, J. 1996. Designing urgency into auditory warnings using pitch, speed and loudness. *Computing and Control Engineering*, 193–198.

Haber, R.N. & Standing, L.G. 1970. Direct estimates of the apparent duration of a flash. *Canadian Journal of Psychology*, 24, 216–229.

Hackman, J. & Oldham, G. 1976. Motivation through the design of work: Test of a theory. *Organizational Behavior and Human Performance*, 15, 250–279.

Hagman, J.D. & Smith, M.D. 1996. Device-based prediction of tank gunnery performance. *Military Psychology*, 8, 59–68.

Haig, B. D. 2009. Inference to the best explanation: A neglected approach to theory appraisal in psychology. *American Journal of Psychology*, 122, 219–234.

Haig, B. D. 2014. *Investigating the Psychological World: Scientific Method in the Behavioral Sciences.* Cambridge, MA: MIT Press.

Haimes, P., Jung, J.H. & Medley, S. 2013. Bridging the gap: Scenario-based design as a solution for delayed access to users. In Proceedings of Australian Council of University Art and Design Schools (ACUADS). Perth, Western Australia: ACUADS.

Hall, E.T. 1959. *The Silent Language.* New York: Doubleday.

Hall, E.T. 1966. *The Hidden Dimension.* New York: Doubleday.

Hamburger, H. & Booker, L.D. 1989. Managing uncertainty in expert systems: Rationale, theory, and techniques. In J. Liebowitz & D.A. De Salvo (Eds.), *Structuring Expert Systems: Domain Design, and Development* (pp. 241–271). Englewood Cliffs, NJ: Yourdon Press.

Hammerton, M. 1989. Tracking. In D.H. Holding (Ed.), *Human Skills* (2nd ed., pp. 171–195). New York: Wiley.

Han, J., Waddington, G., Anson, J. & Adams, R. 2013. Does elastic resistance affect finger pinch discrimination? *Human Factors*, 55, 976–984.

Hancock, P.A. & Ganey, H.N. 2003. From the inverted-U to the extended-U: The evolution of a law of psychology. *Journal of Human Performance in Extreme Environments*, 7, 5–14.

Handy, T.C., Soltani, M. & Mangun, G.R. 2001. Perceptual load and visuocortical processing: Event-related potentials reveal sensory-level selection. *Psychological Science*, 12, 213–218.

Hanisch, K.A., Kramer, A.F. & Hulin, C.L. 1991. Cognitive representations, control, and understanding of complex systems: A field study focusing on components of users' mental models and expert/novice differences. *Ergonomics*, 34, 1129–1145.

Hankey, J.M. & Dingus, T.A. 1990. A validation of SWAT as a measure of workload induced by changes in operator capacity. *Proceedings of the Human Factors Society 34th Annual Meeting* (pp. 112–115). Santa Monica, CA:Human Factors Society.

Hannaman, G.W., Spurgin, A.J. & Lukic, Y. 1985. A model for assessing human cognitive reliability in PRA studies. *1985 IEEE Third Conference on Human Factors and Nuclear Safety* (pp. 343–353). New York: Institute of Electrical and Electronics Engineers.

Hanne, K.-H. & Hoepelman, J. 1990. Natural language and direct manipulation interfaces for expert systems (multimodal communication). In D. Berry & A. Hart (Eds.), *Expert Systems: Human Issues* (pp. 156–168). Cambridge, MA: MIT Press.

Hanoch, Y. & Vitouch, O. 2004. When less is more: Information, emotional arousal and the ecological reframing of the Yerkes–Dodson law. *Theory and Psychology*, 14, 427–452.

Hanson, L., Blomé, M., Dukic, T. & Högberg, D. 2006. Guide and documentation system to support digital human modeling applications. *International Journal of Industrial Ergonomics*, 36, 17–24.

Harm, D.L. 2002. Motion sickness neurophysiology, physiological correlates, and treatment. In K.M. Stanney (Ed.), *Handbook of Virtual Environments: Design, Implementation, and Applications* (pp. 637–661). Mahwah, NJ: Lawrence Erlbaum.

Harmon, L.D. & Julesz, B. 1973. Masking in visual recognition: Effects of two-dimensional filtered noise. *Science*, 180, 1194–1197.

Harrigan, J.E. 1987. Architecture and interior design. In G. Salvendy (Ed.), *Handbook of Human Factors* (pp. 742–764). New York: Wiley.

Harrigan, J.A. 2005. Proxemics, kinesics, and gaze. In J.A. Harrigan, R. Rosenthal & K.R. Scherer (Eds.), *The New Handbook of Methods in Nonverbal Behavior Research* (pp. 137–198). New York: Oxford University Press.

Harris, D.F. 2014. *The Complete Guide to Writing Questionnaires: How to get Better Information for Better Decisions.* US: I&M Press.

Harris, R., Iavecchia, H.P. & Dick, A.O. 1989. The human operator simulator (HOS-IV). In G.R. McMillan, D. Beevis, E. Salas, M.H. Strub, R. Sutton & L. Van Breda (Eds.), *Applications of Human Performance Models to System Design* (pp. 275–280). New York: Plenum Press.

Harrison, A.A. 2001. *Spacefaring: The Human Dimension.* Berkeley, CA: University of California Press.

Harrison, A.A., Clearwater, Y.A. & McKay, C.P. (Eds.) 1991. *From Antarctica to Outer Space: Life in Isolation and Confinement.* New York: Springer.

Hart, S.G. & Staveland, L.E. 1988. Development of NASA-TLX (Task Load Index): Results of empirical and theoretical research. In P.A. Hancock & N. Meshkati (Eds.), *Human Mental Workload* (pp. 139–183). Amsterdam: North-Holland.

Hasher, L., Goldstein, D. & May, C.P. 2005. It's about time: Circadian rhythms, memory, and aging. In C. Izawa & N. Ohta (Eds.), *Human Learning and Memory: Advances in Theory and Application* (pp. 199–217). Mahwah, NJ: Lawrence Erlbaum.

Haslegrave, C.M. 2005. Auditory environment and noise assessment. In J.R. Wilson & N. Corlett (Eds.), *Evaluation of Human Work* (3rd ed., pp. 693–713). Boca Raton, FL: CRC Press.

Haupt, D.R. & Parkinson, M.B. 2015. Digital modeling of physical constraints. In D.A. Boehm-Davis, F.T. Durso, & J.D. Lee (Eds.), *APA Handbook of Human Systems Integration* (pp. 133–148). Washington, DC: American Psychological Association.

Haviland, S. & Clark, H.H. 1974. What's new? Acquiring new information as a process in comprehension. *Journal of Verbal Learning and Verbal Behavior*, 13, 512–521.

Hawkins, W.H. 1990. Where does human factors fit in R&D organizations? *IEEE Aerospace and Electronic Systems Magazine*, 5(9), 31–33.

Health & Safety Executive. 2002. Noise and vibration. Offshore Technology Report 2001=068. Norwich: HSE Books.

Healy, A.F. & Bourne, L.E., Jr. 2012. *Training Cognition: Optimizing Efficiency, Durability, and Generalizability.* New York: Psychology Press.

Healy, A.F. & Wohldmann, E.L. 2012. Specificity and transfer of learning. In B.H. Ross (Ed.), *The psychology of learning and motivation* (Vol 57, pp. 227–253). San Diego, CA: Academic Press.

Heath, M., Rival, C., Westwood, D.A. & Neely, K. 2005. Time course analysis of closed- and open-loop grasping of the Muller–Lyer illusion. *Journal of Motor Behavior*, 37, 179–185.

Heathcote, A. & Hayes, B. 2012. Diffusion versus linear ballistic accumulation: Different models for response time with different conclusions about psychological mechanisms? *Canadian Journal of Experimental Psychology*, 66, 125–136.

Hebb, D.O. 1961. Distinctive features of learning in the higher animal. In J.F. Delafresnaye (Ed.), *Brain Mechanisms and Learning* (pp. 37–46). London: Oxford University Press.

Hecht, E. 2016. *Optics* (5th ed.). Boston, MA: Addison-Wesley.

Hecht, H. & Salvesbergh, G.J.P. (Eds.) 2004. *Time-to-Contact.* Amsterdam: Elsevier.

Hedge, A. 2006. Environmental ergonomics. In W. Karwowski (Ed.), *International Encyclopedia of Ergonomics and Human Factors* (2nd ed., Vol. 2, pp. 1770–1775). Boca Raton, FL: CRC Press.

Hedge, A., Sims, W.R. & Becker, F.D. 1995. Effects of lensed indirect and parabolic lighting on the satisfaction, visual health, and productivity of office workers. *Ergonomics*, 38, 260–280.

Hedge, J.W. & Borman, W.C. 2006. Personnel selection. In G. Salvendy (Ed.), *Handbook of Human Factors and Ergonomics* (3rd ed., pp. 458–471). Hoboken, NJ: Wiley.

Hegarty, M. 2011. The cognitive science of visual-spatial displays: Implications for design. *Topics in Cognitive Science*, 3, 446–474.

Heise, G.A. & Miller, G.A. 1951. An experimental study of auditory patterns. *American Journal of Psychology*, 64, 68–77.

Heister, G., Schroeder-Heister, P. & Ehrenstein, W.H. 1990. Spatial coding and spatio-anatomical mapping: Evidence for a hierarchical model of spatial stimulus–response compatibility. In R.W. Proctor & T.G. Reeve (Eds.), *Stimulus–Response Compatibility: An Integrated Perspective* (pp. 117–143). Amsterdam: North-Holland.

Heitman, R.J., Stockton, C.A. & Lambert, C. 1987. The effects of fatigue on motor performance and learning in mentally retarded individuals. *American Corrective Therapy Journal*, 41, 40–43.

Helander, M.G. 1987. Design of visual displays. In G. Salvendy (Ed.), *Handbook of Human Factors* (pp. 507–549). New York: Wiley.

Helander, M.G. 1999. Seven common reasons to not implement ergonomics. *International Journal of Industrial Ergonomics*, 25, 97–101.

Helander, M.G. 2003. Forget about ergonomics in chair design? Focus on aesthetics and comfort! *Ergonomics*, 46, 1306–1319.

Hellier, E. & Edworthy, J. 1999. On using psychophysical techniques to achieve urgency mapping in auditory warnings. *Applied Ergonomics*, 30, 167–171.

Hellier, E., Edworthy, J., Weedon, B., Walters, K. & Adams, A. 2002. The perceived urgency of speech warnings: Semantics versus acoustics. *Human Factors*, 44, 1–17.

Helmholtz, H. von. 1852. On the theory of compound colors. *Philosophical Magazine*, 4, 519–534.

Helmholtz, H. von. 1867. *Handbook of Physiological Optics* (Vol. 3). Leipzig: Voss.

Helvacioglu, S. & Insel, M. 2005. A reasoning method for a ship design expert system. *Expert Systems*, 22, 72–77.

Henderson, L. & Dittrich, W.H. 1998. Preparing to react in the absence of uncertainty: I. New perspectives on simple reaction time. *British Journal of Psychology*, 89, 531–554.

Hendrick, H.W. 1987. Organizational design. In G. Salvendy (Ed.), *Handbook of Human Factors* (pp. 470–494). New York: Wiley.

Hendrick, H.W. 1990. Factors affecting the adequacy of ergonomic efforts on large-scale-system development programs. *Ergonomics*, 33, 639–642.

Hendrick, H.W. 1991. Ergonomics in organizational design and management. *Ergonomics*, 34, 743–756.

Hendrick, H.W. 2000. The technology of ergonomics. *Theoretical Issues in Ergonomics Science*, 1, 22–33.

Hendrick, H.W. & Kleiner, B.M. 2001. *Macroergonomics: An Introduction to Work System Design*. Santa Monica, CA: Human Factors and Ergonomics Society.

Hendrick, H.W. & Kleiner, B.M. (Eds.) 2002. *Macroergonomics: Theory, Methods, and Applications*. Mahwah, NJ: Lawrence Erlbaum.

Henninger, A.E. & Whitaker, E.T. 2015. Modeling human behavior. In M.L. Loper (Ed.), *Modeling and Simulation in the Systems Engineering Life Cycle* (pp. 75–87). London: Springer.

Henry, F.M. & Rogers, D.E. 1960. Increased response latency for complicated movements and a "memory drum" theory of neuromotor reaction. *Research Quarterly*, 31, 448–458.

Herbart, J.F. 1816/1891. *A Textbook in Psychology: An Attempt to Found the Science of Psychology on Experience, Metaphysics, and Mathematics* (2nd ed., W.T. Harris, Ed.; M.K. Smith, Trans.). New York: Appleton.

Herrmann, D.J. & Petros, S.J. 1990. Commercial memory aids. *Applied Cognitive Psychology*, 4, 439–450.

Hess, A.K. 2005. Practical ethics for the expert witness in ergonomics and human factors forensic cases. In Y.I. Noy & W. Karwowski (Eds.), *Handbook of Human Factors in Litigation* (pp. 4-1–4-11). Boca Raton, FL: CRC Press.

Hesse-Biber, S. & Johnson, R. B. (Eds.) 2015. *The Oxford Handbook of Multimethod and Mixed Methods Research Inquiry*. New York: Oxford University Press.

Heuer, H., Hollendiek, G., Kroger, H. & Romer, T. 1989. The resting position of the eyes and the influence of observation distance and visual fatigue on VDT work. *Zeitschrift für Experimentelle und Angewandte Psychologie*, 36, 538–566.

Heuer, H., Manzey, D., Lorenz, B. & Sangals, J. 2003. Impairments of manual tracking performance during spaceflight are associated with specific effects of microgravity on visuomotor transformations. *Ergonomics*, 46, 920–934.

Hick, W.E. 1952. On the rate of gain of information. *Quarterly Journal of Experimental Psychology*, 4, 11–26.

Highstein, S.M., Fay, R.R. & Popper, A.N. (Eds.) 2004. *The Vestibular System*. New York: Springer.

Hitt, J.D. 1961. An evaluation of five different abstract coding methods—Experiment IV. *Human Factors*, 3, 120–130.

Hobson, D.A. & Molenbroek, J.F.M. 1990. Anthropometry and design for the disabled: Experience with seating design for the cerebral palsy population. *Applied Ergonomics*, 21, 43–54.

Hochberg, J.E. 1988. Visual perception. In R.C. Atkinson, R.J. Herrnstein, G. Lindzey & R.D. Luce (Eds.), *Stevens' Handbook of Experimental Psychology* (2nd ed., Vol. 1: *Perception and Motivation*, pp. 195–276). New York: Wiley.

Hockey, G.R.J. 1986. Changes in operational efficiency as a function of environmental stress, fatigue, and circadian rhythms. In K.R. Boff, L. Kaufman & J.P. Thomas (Eds.), *Handbook of Perception and Human Performance* (Vol. II: *Cognitive Processes and Performance*, pp. 44-1–44-49). New York: Wiley.

Hoffmann, E.R. 1990. Strength of component principles determining direction-of-turn stereotypes for horizontally moving displays. *Proceedings of the Human Factors Society 34th Annual Meeting* (Vol. 1, pp. 457–461). Santa Monica, CA: Human Factors Society.

Hoffmann, E.R. 1997. Strength of component principles determining direction of turn stereotypes–linear displays and rotary controls. *Ergonomics*, 40, 199–222.

Hoffmann, E.R. 2016. Fitts' law with an average of two or less submoves? *Journal of Motor Behavior*, 48, 318–331.

Hoffmann, E.R. & Chan, A.S. 2013. The Worringham and Beringer 'visual field' principle for rotary controls. *Ergonomics*, 56, 1620–1624.

Hoffman, R.R. 2008. Human factors contributions to knowledge elicitation. *Human Factors*, 50, 481–488.

Hofstetter, H.W., Griffin, J.R., Berman, M.S. & Everson, R.W. 2000. *Dictionary of Visual Science and Related Clinical Terms* (5th ed.). Boston, MA: Butterworth-Heinemann.

Holcomb, H.R., III. 1998. Testing evolutionary hypotheses. In C. Crawford & D.L. Krebs (Eds.), *Handbook of Evolutionary Psychology: Ideas, Issues, and Applications* (pp. 303–334). Mahwah, NJ: Lawrence Erlbaum.

Holick, A. J., Chrysler, S. T., Park, E. & Carlson, P. J. 2006. Evaluation of the Clearview font for negative contrast traffic signs. Texas Transportation Research Institute, Texas A & M University, Report number FHWA/TX-06/0-4984-1. http://tti.tamu.edu/documents/0-4984-1.pdf

Holladay, C.L. & Quinones, M.A. 2003. Practice variability and transfer of training: The role of self-efficacy generality. *Journal of Applied Psychology*, 88, 1094–1103.

Holland, J.H., Holyoak, K.J., Nisbett, R.E. & Thagard, P.R. 1986. *Induction: Processes of Inference, Learning, and Discovery*. Cambridge, MA: MIT Press.

Hollingworth, A. 2004. Constructing visual representations of natural scenes: The roles of short-and long-term visual memory. *Journal of Experimental Psychology: Human Perception and Performance*, 30, 519–537.

Hollnagel, E. 1998. *Cognitive Reliability and Error Analysis Method*. London: Elsevier.

Hollnagel, E. 2000. Looking for errors of omission and commission of *The Hunting of the Snark revisited*. *Reliability Engineering and System Safety*, 68, 135–145.

Holman, D., Clegg, C. & Waterson, P. 2002. Navigating the territory of job design. *Applied Ergonomics*, 33, 197–205.

Holman, G.T., Carnahan, B.J. & Bulfin, R.L. 2003. Using linear programming to optimize control panel design from an ergonomics perspective. *Proceedings of the Human Factors and Ergonomics Society 47th Annual Meeting* (pp. 1317–1321). Santa Monica, CA: HFES.

Holton, G. & Brush, S.G. 2000. *Physics, the Human Adventure*. New Brunswick, NJ: Rutgers University Press.

Holway, A.H. & Boring, E.G. 1941. Determinants of apparent visual size with distance variant. *American Journal of Psychology*, 54, 21–37.

Holyoak, K.J. & Koh, K. 1987. Surface and structural similarity in analogical transfer. *Memory & Cognition*, 15, 332–340.

Holyoak, K.J. & Morrison, R.G. (Eds.) 2012. *The Oxford Handbook of Thinking and Reasoning*. New York: Oxford University Press.

Holyoak, K.J. & Nisbett, R.E. 1988. Induction. In R.J. Sternberg & E.E. Smith (Eds.), *The Psychology of Human Thought* (pp. 50–91). New York: Cambridge University Press.

Hommel, B. 1998. Automatic stimulus–response translation in dual-task performance. *Journal of Experimental Psychology: Human Perception and Performance*, 24, 1368–1384.

Hommel, B. 2004. Event files: Feature binding in and across perception and action. *Trends in Cognitive Science*, 11, 494–500.

Hommel, B. 2011. The Simon effect as tool and heuristic. *Acta Psychologica*, 136, 189–202.

Hommel, B., Müsseler, J., Aschersleben, G. & Prinz, W. 2001. The theory of event-coding (TEC): A framework for perception and action planning. *Behavioral and Brain Sciences*, 24, 849–878.

Hood, P.C. & Finkelstein, M.A. 1986. Sensitivity to light. In K.R. Boff, L. Kaufman & J.P. Thomas (Eds.), *Handbook of Perception and Human Performance* (Vol. I: *Sensory Processes and Perception*, pp. 5–66). New York: Wiley.

Hopkinson, R.G. & Longmore, J. 1959. Attention and distraction in the lighting of work places. *Ergonomics*, 2, 321–333.

Horne, J.A., Brass, C.G. & Pettitt, A.N. 1980. Circadian performance differences between morning and evening "types." *Ergonomics*, 23, 29–36.

Hotta, A., Takahashi, T., Takahashi, K. & Kogi, K. 1981. Relations between direction-of-motion stereotypes for control in living space. *Journal of Human Ergology*, 10, 73–82.

Houck, D. 1991, March. Fighter pilot display requirements for post-stall maneuvers. *The Visual Performance Group Technical Newsletter*, 13(1), 1–4.

Houvet, P. & Obert, L. 2013. Upper limb cumulative trauma disorders for the orthopaedic surgeon. *Orthopaedics & Traumatology: Surgery & Research*, 99, S104–S114.

Howard, I.P. 2002. Depth perception. In S. Yantis (Ed.), *Stevens' Handbook of Experimental Psychology* (3rd ed., Vol. 1: *Sensation and Perception*, pp. 77–120). New York: Wiley.

Howard, I.P. (Ed.) 2012. *Perceiving in depth (Vol 1: Basic mechanisms)*. New York: Oxford University Press.

Howes, A. & Young, R.M. 1997. The role of cognitive architecture in modeling the user: Soar's learning mechanism. *Human–Computer Interaction*, 12, 311–343.

Howland, D. & Noble, M.E. 1953. The effect of physical constants of a control on tracking performance. *Journal of Experimental Psychology*, 46, 353–360.

HSE. 2004. Getting to grips with manual materials handling: A short guide. Leaflet INDG143REV2. Sudbury: HSE Books.

Hsiao, H., Whitestone, J., Bradtmiller, B., Whisler, R., Zwiener, J., Lafferty, C., Kau, T.-Y. & Gross, M. 2005. Anthropometric criteria for the design of tractor cabs and protection frames. *Ergonomics*, 48, 323–352.

Hsiao, S.-W. & Chou, J.-R. 2006. A Gestalt-like perceptual measure for home page design using a fuzzy entropy approach. *International Journal of Human–Computer Studies*, 64, 137–156.

Hubbard, C., Naqvi, S.A. & Capra, M. 2001. Heavy mining vehicle controls and skidding accidents. *International Journal of Occupational Safety and Ergonomics*, 7, 211–221.

Hubel, D.H. & Wiesel, T.N. 1979. Brain mechanisms of vision. *Scientific American*, 241, 150–163.

Huber, P.W. 1991. *Galileo's Revenge: Junk Science in the Courtroom*. New York: Basic Books.

Huettel, S.A., Song, A.W. & McCarthy, G. 2014. *Functional Magnetic Resonance Imaging* (2nd ed.). Sunderland, MA: Sinauer Associates.

Hughes, D.G. & Folkard, S. 1976. Adaptation to an 8-hr shift in living routine by members of a socially isolated community. *Nature*, 264, 432–434.

Hughes, J. & Parkes, S. 2003. Trends in use of verbal protocol analysis in software engineering research. *Behaviour and Information Technology*, 22, 127–140.

Human Factors and Ergonomics Society. 2005. Human Factors and Ergonomics Society Code of Ethics. https://www.hfes.org/web/AboutHFES/ethics.html

Humphreys, P.C. & McFadden, W. 1980. Experiences with MAUD: Aiding decision structuring versus bootstrapping the decision maker. *Acta Psychologica*, 45, 51–69.

Humphries, M.D., Stewart, R.D. & Gurney, K.M. 2006. A physiologically plausible model of action selection and oscillatory activity in the Basal Ganglia. *Journal of Neuroscience*, 26(50), 12921–12942.

Hung, F. & Govindasamy, S. 2015. "Wow, pulled back wrong throttle:" Captain of crashed TransAsia plane. Reuters (Edition U.S.), July 2, 2015. Downloaded from: http://www.reuters.com/article/2015/07/02/us-taiwan-airplane-idUSKCN0PC05L20150702

Hung, Y.H., Lin, C.F. & Chang, R.I. 2015. Developing a dynamic inference expert system to support individual learning at work. *British Journal of Educational Technology*, 46, 1378–1391.

Hunt, D.P. 1953. The coding of aircraft controls (Report No. 53-221). Wright Air Development Center: U.S. Air Force.

Hunt, E. 1989. Connectionist and rule-based representations of expert knowledge. *Behavior Research Methods, Instruments, and Computers*, 21, 88–95.

Hunting, W. & Grandjean, E. 1976. Hunting and Grandjean high back. *Design*, 333, 34–35.

Huxley, J. 1934. Science and industry. *The Human Factor*, 8, 83–86.

Huysmans, M.A., de Looze, M.P., Hoozemans, M.J.M., van der Beek, A.J. & van Dieen, J.H. 2006. The effect of joystick handle size and gain at two levels of required precision on performance and physical load on crane operators. *Ergonomics*, 49, 1021–1035.

Hyde, T.S. & Jenkins, J.J. 1973. Recall for words as a function of semantic, graphic, and syntactic orienting tasks. *Journal of Verbal Learning and Verbal Behavior*, 12, 471–480.

Hyman, R. 1953. Stimulus information as a determinant of reaction time. *Journal of Experimental Psychology*, 45, 188–196.

IBM, 2006. IBM Global Services consulting: With the right help you can beat the traffic. Downloaded on July 31, 2006, from http://www-306.ibm.com/ibm/easy/eou_ext.nsf/publish/1601

Inglis, E.A., Szymkowiak, A., Gregor, P., Newell, A.F., Hine, N., Wilson, B.A., Evans, J. & Shah, P. 2004. Usable technology? Challenges in designing a memory aid with current electronic devices. *Neuropsychological Rehabilitation*, 14, 77–87.

Intons-Peterson, M.J. & Fournier, J. 1986. External and internal memory aids: When and how often do we use them? *Journal of Experimental Psychology: General*, 115, 267–280.

Irving, S., Polson, P. & Irving, J.E. 1994. A GOMS analysis of the advanced automated cockpit. *Proceedings of the SIGCHI Conference on Human Factors in Computing Systems: Celebrating Interdependence* (pp. 344–350). New York: ACM Press.

Isreal, J.B., Wickens, C.D., Chesney, G.L. & Donchin, E. 1980. The event-related brain potential as an index of display-monitoring workload. *Human Factors*, 22, 211–224.

Ivergard, T. 2006. Manual control devices. In W. Karwowski (Ed.), *International Encyclopedia of Ergonomics and Human Factors* (2nd ed., Vol. 1, pp. 1457–1462). Boca Raton, FL: CRC Press.

Ivry, R.B., Spencer, R.M., Zelaznik, H.N. & Diedrichsen, J. 2002. The cerebellum and event timing. In S. Highstein & W. Thach (Eds.), *The Cerebellum: Recent Developments in Cerebellar Research* (pp. 302–317). New York: New York Academy of Sciences.

Jacko, J. (Ed.). 2012. *Human Computer Interaction Handbook: Fundamentals, Evolving Technologies, and Emerging Applications* (3rd ed.). Boca Raton, FL: CRC Press.

Jacko, J.A., Yi, J.S., Sainfort, F. & McClellan, M. 2012. Human factors and ergonomic methods. In G. Salvendy (Ed.), *Handbook of Human Factors and Ergonomics* (4th ed., pp. 298–329). Hoboken, NJ: John Wiley.

Jagacinski, R.J. & Flach, J.M. 2003. *Control Theory for Humans: Quantitative Approaches to Modeling Performance.* Mahwah, NJ: Lawrence Erlbaum.

Jagacinski, R.J. & Monk, D.L. 1985. Fitts' law in two dimensions with hand and head movements. *Journal of Motor Behavior,* 17, 77–95.

Jagacinski, R., Miller, D. & Gilson, R. 1979. A comparison of kinesthetic-tactual displays via a critical tracking task. *Human Factors,* 21, 79–86.

Jahweed, M.M. 1994. Muscle structure and function. In A.E. Nicogossian, C.L. Huntoon & S.L. Pool (Eds.), *Space Physiology and Medicine* (3rd ed.). Malvern, PA: Lea & Febiger.

James, J.T. 2013. A new, evidence-based estimate of patient harms associated with hospital care. *Journal of Patient Safety,* 9 (3), 122–128.

James, W. 1890/1950. *The Principles of Psychology* (Vol. 1). New York: Dover Press.

Janelle, C.M., Singer, R.N. & Williams, A.M. 1999. External distraction and attentional narrowing: Visual search evidence. *Journal of Sport and Exercise Psychology,* 21, 70–91.

Jang, R., Molesworth, B.R., Burgess, M. & Estival, D. 2014. Improving communication in general aviation through the use of noise cancelling headphones. *Safety Science,* 62, 499–504.

Janis, I.L. & Mann, L. 1977. *Decision Making: A Psychological Analysis of Conflict, Choice, and Commitment.* New York: Free Press.

Jaschinski, W., Jainta, S., Hoormann, J. & Walper, N. 2007. Objective vs subjective measurements of dark vergence. *Ophthalmic and Physiological Optics,* 27, 85–92.

Jaschinski-Kruza, W. 1991. Eyestrain in VDU users: Viewing distance and the resting position of ocular muscles. *Human Factors,* 33, 69–83.

Jastrzębowski, W.B. 1857. An outline of ergonomics of the science of work based upon the truths drawn from the science of nature, Part I. *Nature and Industry,* 29, 227–231.

Jaynes, L. S. & Boles, D. B. 1990. Effects of symbols on ewarning compliance. *Proceedings of the Human Factors Society 34th Annual Meeting* (pp. 984–987). Santa Monica, CA: HFES.

Jeannerod, M. 1981. Intersegmental coordination during reaching at natural objects. In J.L. Long & A. Baddeley (Eds.), *Attention and Performance IX* (pp. 153–169). Hillsdale, NJ: Lawrence Erlbaum.

Jeannerod, M. 1984. The timing of natural prehension movement. *Journal of Motor Behavior,* 26, 235–254.

Jebaraj, D., Tyrrell, R.A. & Gramopadhye, A.K. 1999. Industrial inspection performance depends on both viewing distance and oculomotor characteristics. *Applied Ergonomics,* 30, 223–228.

Jenkins, W.O. 1946. Investigation of shapes for use in coding aircraft control knobs. USAF Air Materiel Command Memorandum Report No. TSEAA-694-4.

Jewell, L.N. 1998. *Contemporary Industrial/Organizational Psychology* (3rd ed.). Belmont, CA: Wadsworth.

Joe, J.C. & Boring, R.L. 2014. Individual differences in human reliability analysis. *Probabilistic Safety Assessment and Management PSAM 12,* June 2014, Honolulu, Hawaii.

John, B.E. 2003. Information processing and skilled behavior. In J.M. Carroll (Ed.), *HCI Models, Theories, and Frameworks: Toward a Multidisciplinary Science* (pp. 55–101). San Francisco, CA: Morgan Kaufmann.

John, B.E. & Newell, A. 1987. Predicting the time to recall computer command abbreviations. In *CHI & GI 1987 Conference Proceedings: Human Factors in Computing Systems and Graphics Interface* (pp. 33–40). New York: ACM.

John, B.E. & Newell, A. 1990. Toward an engineering model of stimulus–response compatibility. In R.W. Proctor & T.G. Reeve (Eds.), *Stimulus–Response Compatibility: An Integrated Perspective* (pp. 427–479). Amsterdam: North-Holland.

John, B.E., Rosenbloom, P.S. & Newell, A. 1985. A theory of stimulus–response compatibility applied to human–computer interaction. In *CHI '85 Conference Proceedings: Human Factors in Computing Systems* (pp. 213–219). New York: ACM.

Johnsen, E.G. & Corliss, W.R. 1971. *Human Factors Applications in Teleoperator Design and Operation.* New York: Wiley.

Johnson, A. & Proctor, R.W. 2004. *Attention: Theory and Practice.* Thousand Oaks, CA: Sage.

Johnson, A. & Proctor, R.W. (Eds.) 2013. *Neuroergonomics: A Cognitive Neuroscience Approach to Human Factors and Ergonomics.* Basingstoke, UK: Palgrave Macmillan.

Johnson, A. & Proctor, R.W. 2017. *Skill Acquisition and Training: Achieving Expertise in Simple and Complex Tasks*. New York: Routledge.

Johnson, E.J., Bellman, S. & Lohse, G.L. 2003. Cognitive lock-in and the power law of practice. *Journal of Marketing*, 67, 62–75.

Johnson, M.K., Bransford, J.D. & Solomon, S.K. 1973. Memory for tacit implications of sentences. *Journal of Experimental Psychology*, 98, 203–205.

Johnson, P.J., Forester, J.A., Calderwood, R. & Weisgerber, S.A. 1983. Resource allocation and the attentional demands of letter encoding. *Journal of Experimental Psychology: General*, 112, 616–638.

Johnson-Laird, P.N. 1983. *Mental Models*. Cambridge, MA: Harvard University Press.

Johnson-Laird, P.N. 1989. Mental models. In M.I. Posner (Ed.), *Foundations of Cognitive Science* (pp. 469–499). Cambridge, MA: MIT Press.

Johnson-Laird, P.N., Khemlani, S.S. & Goodwin, G.P. 2015. Logic, probability, and human reasoning. *Trends in Cognitive Sciences*, 19, 201–214.

Johnson-Laird, P.N., Legrenzi, P. & Legrenzi, M.S. 1972. Reasoning and a sense of reality. *British Journal of Psychology*, 63, 395–400.

Johnston, W.A. & Heinz, S.P. 1978. Flexibility and capacity demands of attention. *Journal of Experimental Psychology: General*, 107, 420–435.

Jokinen, J. P. 2015. Emotional user experience: Traits, events, and states. *International Journal of Human-Computer Studies*, 76, 67–77.

Jones, D.M. & Broadbent, D.E. 1987. Noise. In G. Salvendy (Ed.), *Handbook of Human Factors* (pp. 623–649). New York: Wiley.

Jones, D.M., Morris, N. & Quayle, A.J. 1987. The psychology of briefing. *Applied Ergonomics*, 18, 335–339.

Jones, R.M., Laird, J.E., Nielsen, P.E., Coulter, K.J., Kenny, P. & Koss, F.V. 1999. Automated intelligent pilots for combat flight simulation. *AI Magazine*, 20(1), 27–41.

Jonides, J., Lacey, S.C. & Nee, D.E. 2005. Processes of working memory in mind and brain. *Current Directions in Psychological Science*, 14, 2–5.

Jorgensen, Z., Chen, J., Gates, C. S., Li, N., Proctor, R.W. & Yu, T. 2015. Dimensions of risk in mobile applications: A user study. In *Proceedings of the Fifth ACM Conference on Data and Application Security and Privacy* (pp. 49–60). New York: Association for Computing Machinery.

Julesz, B. 1971. *Foundations of Cyclopean Perception*. Chicago, IL: University of Chicago Press.

Juslén, H. & Tenner, A. 2005. Mechanisms involved in enhancing human performance by changing the lighting in the industrial workplace. *International Journal of Industrial Ergonomics*, 35, 843–855.

Juslén, H., Wouters, M. & Tenner, A. 2007. The influence of controllable task-lighting on productivity: A field study in a factory. *Applied Ergonomics*, 38, 39–44.

Kaber, D.B., Riley, J.M. & Tan, K.-W. 2002. Improved usability of aviation automation through direct manipulation and graphical user interface design. *International Journal of Aviation Psychology*, 12, 153–178.

Kahana, M.J. & Wingfield, A. 2000. A functional relation between learning and organization in free recall. *Psychonomic Bulletin & Review*, 7, 516–521.

Kahane, C.J. 1998. The long-term effectiveness of center high mounted stop lamps in passenger cars and light trucks. NHTSA Technical Report Number DOT HS 808 696.

Kahneman, D. 1973. *Attention and Effort*. Englewood Cliffs, NJ: Prentice-Hall.

Kahneman, D. & Tversky, A. 1972. Subjective probability: A judgment of representativeness. *Cognitive Psychology*, 3, 430–454.

Kahneman, D. & Tversky, A. 1973. On the psychology of prediction. *Psychological Review*, 80, 237–251.

Kahneman, D., Norman, J. & Kubovy, M. 1967. Critical duration for the resolution of form: Centrally or peripherally determined? *Journal of Experimental Psychology*, 73, 323–327.

Kahneman, D., Ben-Ishai, R. & Lotan, M. 1973. *Journal of Applied Psychology*, 58, 113–1155.

Kahneman, D., Slovic, P. & Tversky, A. 1982. *Judgment under Uncertainty: Heuristics and Biases*. New York: Cambridge University Press.

Kahneman, D., Treisman, A. & Gibbs, B.J. 1992. The reviewing of object files: Object-specific integration of information. *Cognitive Psychology*, 24, 175–219.

Kaminaka, M.S. & Egli, E.A. 1984. Determination of stereotypes for lever control of object motions using computer generated graphics. *Ergonomics*, 27, 918–995.

Kammler, D. W. 2007. *A First Course in Fourier Analysis* (2nd ed.). Cambridge, UK: Cambridge University Press.

Kang, H.G. & Seong, P.H. 2001. Information theoretic approach to man–machine interface complexity evaluation. *IEEE Transactions on Systems, Man, and Cybernetics—Par A: Systems and Humans*, 31, 163–171.

Kanis, H. 2002. Can design supportive research be scientific? *Ergonomics*, 45, 1037–1041.

Kanis, H. 2014. Reliability and validity of findings in ergonomics research. *Theoretical Issues in Ergonomics Science*, 15, 1–46.

Kantowitz, B.H. 1987. Mental workload. In P.A. Hancock (Ed.), *Human Factors Psychology* (pp. 81–121). Amsterdam: North-Holland.

Kantowitz, B.H. 1989. The role of human information processing models in system development. *Proceedings of the Human Factors Society 33rd Annual Meeting* (pp. 1059–1063). Santa Monica, CA: Human Factors Society.

Kantowitz, B.H. & Elvers, G.C. 1988. Fitts' law with an isometric controller: Effects of order of control and control-display gain. *Journal of Motor Behavior*, 20, 53–66.

Kantowitz, B.H., Triggs, T.J. & Barnes, V.E. 1990. Stimulus–response compatibility and human factors. In R.W. Proctor & T.G. Reeve (Eds.), *Stimulus–Response Compatibility: An Integrated Perspective* (pp. 365–388). Amsterdam: North-Holland.

Karat, C.-M. 1990. Cost–benefit analyses of usability engineering techniques. *Proceedings of the Human Factors Society 34th Annual Meeting* (pp. 839–843). Santa Monica, CA: Human Factors Society.

Karat, C.-M. 1992. Cost-justifying human factors support on software development projects. *Human Factors Society Bulletin*, 35(11), 1–4.

Karat, C.-M. 2005. A business case approach to usability cost justification for the Web. In R.G. Bias & D.J. Mayhew (Eds.), *Cost-Justifying Usability: An Update for the Internet Age* (pp. 103–141). San Francisco, CA: Morgan Kaufman.

Karn, J.S. & Cowling, A.J. 2006. Using ethnographic methods to carry out human factors research in software engineering. *Behavior Research Methods*, 38, 495–503.

Karwowski, W. 2006a. The discipline of ergonomics and human factors. In G. Salvendy (Ed.), *Handbook of Human Factors and Ergonomics* (3rd ed., pp. 3–31). Hoboken, NJ: Wiley.

Karwowski, W. (Ed.) 2006b. Handbook of Standards and Guidelines in Ergonomics and Human Factors. Mahwah, NJ: Lawrence Erlbaum.

Katsikopoulos, K. V. & Gigerenzer, G. 2013. Modeling decision heuristics. In J. D. Lee & A. (Eds.), *The Oxford Handbook of Cognitive Engineering* (pp. 490–500). New York: Oxford University Press.

Kausel, E.E., Culbertson, S.S. & Madrid, H.P. 2016. Overconfidence in personnel selection: When and why unstructured interview information can hurt hiring decisions. *Organizational Behavior and Human Decision Processes*, 137, 27–44.

Kawabata, N. 1984. Perception at the blind spot and similarity grouping. *Perception & Psychophysics*, 36, 151–158.

Kawamoto, A.H., Farrar, W.T. & Kello, C.T. 1994. When two meanings are better than one: Modeling the ambiguity advantage using a recurrent distributed network. *Journal of Experimental Psychology: Human Perception and Performance*, 20, 1233–1247.

Keele, S.W. 1968. Movement control in skilled motor performance. *Psychological Bulletin*, 70, 387–403.

Keele, S.W. & Posner, M.I. 1968. Processing of visual feedback in rapid movements. *Journal of Experimental Psychology*, 77, 155–158.

Keele, S.W., Cohen, A. & Ivry, R. 1990. Motor programs: Concepts and issues. In M. Jeannerod (Ed.), *Attention and Performance XIII* (pp. 77–110). Hillsdale, NJ: Lawrence Erlbaum.

Keen, P.G.W. & Scott-Morton, M.S. 1978. *Decision Support Systems: An Organizational Perspective*. Reading, MA: Addison Wesley.

Keetch, K.M., Schmidt, R.A., Lee, T.D. & Young, D.E. 2005. Especial skills: Their emergence with massive amounts of practice. *Journal of Experimental Psychology: Human Perception and Performance*, 31, 970–978.

Keil, M.S. 2006. Smooth gradient representations as a unifying account of Chevreul's illusion, Mach bands, and a variant of the Ehrenstein disk. *Neural Computation*, 18, 871–903.

Keir, P.J., Bach, J.M. & Rempel, D. 1999. Effects of computer mouse design and task on carpal tunnel pressure. *Ergonomics*, 42, 1350–1360.

Kelley, C.R. 1962. Predictor instruments look into the future. *Control Engineering*, 9, May, 86–90.

Kelly, P.L. & Kroemer, K.H.E. 1990. Anthropometry of the elderly: Status and recommendations. *Human Factors*, 32, 571–595.

Kelso, J.A.S., Putnam, C.A. & Goodman, D. 1983. On the space-time structure of human interlimb coordination. *Quarterly Journal of Experimental Psychology*, 35A, 347–375.

Kelso, J.A.S., Southard, D.L. & Goodman, D. 1979. On the nature of human interlimb coordination. *Science*, 203, 1029–1031.

Kemper, H.C.G., van Aalst, R., Leegwater, A., Maas, S. & Knibbe, J.J. 1990. The physical and physiological workload of refuse collectors. *Ergonomics*, 33, 1471–1486.

Kenshalo, D.R. 1972. The cutaneous senses. In J.W. Kling & L.A. Riggs (Eds.), *Woodworth and Schlossberg's Experimental Psychology* (3rd ed.). New York: Holt, Rinehart & Winston.

Keppel, G. & Underwood, B.J. 1962. Proactive inhibition in short-term retention of single items. *Journal of Verbal Learning and Verbal Behavior*, 1, 153–161.

Kerlinger, F.N. & Lee, H.B. 2000. *Foundations of Behavioral Research* (4th ed.). Fort Worth, TX: Harcourt Brace.

Kerr, B.A. & Langolf, G.D. 1977. Speed of aiming movements. *Quarterly Journal of Experimental Psychology*, 29, 475–481.

Kerr, R. 1973. Movement time in an underwater environment. *Journal of Motor Behavior*, 5, 175–178.

Kershaw, T.C. & Ohlsson, S. 2004. Multiple causes of difficulty in insight: The case of the nine-dot problem. *Journal of Experimental Psychology: Learning, Memory, and Cognition*, 30, 3–13.

Kheddar, A., Chellali, R. & Coiffet, P. 2002. Virtual environment—Assisted teleoperation. In K.M. Stanney (Ed.), *Handbook of Virtual Environments: Design, Implementation, and Applications* (pp. 959–997). Mahwah, NJ: Lawrence Erlbaum.

Khuu, S. K. & Kalloniatis, M. 2015. Spatial summation across the central visual field: Implications for visual field testing. *Journal of Vision*, 15(1):6, 1–15.

Kiekel, P.A. & Cooke, N.J. 2011. Human factors aspects of team cognition. In K.-P.L. Vu & R.W. Proctor (Eds.), *Handbook of Human Factors in Web Design* (2nd ed., pp. 107–123). Boca Raton, FL: CRC Press.

Kieras, D. 2004. GOMS models for task analysis. In D. Diaper & N.A. Stanton (Ed.), *The Handbook of Task Analysis for Human–Computer Interaction* (pp. 83–116). Mahwah, NJ: Lawrence Erlbaum.

Kieras, D.E. 2017. A summary of the EPIC cognitive architecture. In S. E. F. Chipman (Ed.), *The Oxford Handbook of Cognitive Science*. New York: Oxford University Press.

Kieras, D.E. & Meyer, D.E. 1997. An overview of the EPIC architecture for cognition and performance with application to human–computer interaction. *Human–Computer Interaction*, 12, 391–438.

Kieras, D.E. & Meyer, D.E. 2000. The role of cognitive task analysis in the application of predictive models of human performance. In J.M. Schraagen, S.F. Chipman & V.L. Shalin (eds.), *Cognitive Task Analysis* (pp. 237–260). Mahwah, NJ: Lawrence Erlbaum.

Kieras, D.E., Wakefield, G.H., Thompson, E.R., Iyer, N. & Simpson, B.D. 2016. Modeling two-channel speech processing with the epic cognitive architecture. *Topics in Cognitive Science*, 8, 291–304

Kim, G.J. 2015. *Human-Computer Interaction: Fundamentals and Practice*. Boca Raton, FL: CRC Press.

Kim, I.S. 2001. Human reliability analysis in the man–machine interface design review. *Annals of Nuclear Energy*, 28, 1069–1081.

Kim, J. & de Dear, R. 2013. Workspace satisfaction: The privacy-communication trade-off in open-plan offices. *Journal of Environmental Psychology*, 36, 18–26.

Kimchi, R., Behrmann, M. & Olson, C.R. 2003. *Perceptual Organization in Vision: Behavioral and Neural Perspectives*. Mahwah, NJ: Lawrence Erlbaum.

King, R.B. & Oldfield, S.R. 1997. The impact of signal bandwidth on auditory localization: Implications for the design of three-dimensional audio displays. *Human Factors*, 39, 287–295.

Kingdom, F.A.A. & Prins, N. 2010. *Psychophysics: A Practical Introduction*. London: Academic Press.

Kingery, D. & Furuta, R. 1997. Skimming electronic newspaper headlines: A study of typeface, point size, screen resolution, and monitor size. *Information Processing and Management*, 33, 685–696.

Kintsch, W. & Keenan, J. 1973. Reading rate and retention as a function of the number of propositions in the base structure of sentences. *Cognitive Psychology*, 5, 257–274.

Kirwan, B. 1988. A comparative evaluation of five human reliability assessment techniques. In B.A. Sayers (Ed.), *Human Factors and Decision Making* (pp. 87–109). New York: Elsevier Applied Sciences.

Kirwan, B. 1994. *A Guide to Practical Human Reliability Assessment*. London: Taylor & Francis.

Kittusamy, N.K. & Buchholz, B. 2004. Whole-body vibration and postural stress among operators of construction equipment: A literature review. *Journal of Safety Research*, 35, 255–261.

Kitsinelis, S. 2015. *Light Sources: Basics of Lighting Technologies and Applications* (2nd ed.). Boca Raton, FL: CRC Press.

Kivetz, R. & Simonson, I. 2000. The effects of incomplete information on consumer choice. *Journal of Marketing Research*, 37, 427–448.

Kivimäki, M. & Lindström, K. 2006. Psychosocial approach to occupational health. In G. Salvendy (Ed.), *Handbook of Human Factors and Ergonomics* (3rd ed., pp. 801–817). Hoboken, NJ: Wiley.

Klapp, S.T. 1977. Reaction time analysis of programmed control. *Exercise and Sport Science Reviews*, 5, 231–253.

Klatzky, R.L. & Lederman, S.J. 2013. In & A.F. Healy R.W. Proctor (Eds.), *Exeperimental Psychology* (pp. 152–178). Volume 4 in I.B. Weiner (Editor-in-Chief) *Handbook of Psychology* (2nd ed.). Hoboken, NJ: John Wiley.

Klauer, K.C. 2014. Random-walk and diffusion models. In J.W. Sherman, B. Gawronski, Y. Trope, (Eds.), *Dual-Process Theories of the Social Mind* (pp. 139–152). New York: Guilford Press.

Kleffner, D. A., & Ramachandran, V. S. 1992. On the perception of shape from shading. *Perception & Psychophysics*, 52, 18–36.

Klein, G.A. 1989. Recognition-primed decisions. In W.B. Rouse (Ed.), *Advances in Man–Machine Systems Research* (Vol. 5, pp. 47–92). Greenwich, CT: JAI Press.

Kleiner, B.M. 2004. Macroergonomics as a large work-system transformation technology. *Human Factors and Ergonomics in Manufacturing*, 14, 99–115.

Kleiner, B.M. 2006. Macroergonomics: Analysis and design of work systems. *Applied Ergonomics*, 37, 81–89.

Kleiner, B. M. 2008. Macroergonomics: Work system analysis and design. *Human Factors*, 50, 461–467.

Kleiner, B.M., Drury, C.G. & Christopher, C.L. 1987. Sensitivity of human tactile inspection. *Human Factors*, 29, 1–7.

Knapp, M.L. 1978. *Nonverbal Communication in Human Interaction*. New York: Holt, Rinehart & Winston.

Knauth, P. & Hornberger, S. 2003. Preventive and compensatory measures for shift workers. *Occupational Medicine*, 53, 109–116.

Knoblich, G. (Ed.) 2006. *Human Body Perception from Inside Out*. New York: Oxford University Press.

Knowles, E.S. 1983. Social physics and the effects of others: Tests of the effects of audience size and distance on social judgments and behavior. *Journal of Personality and Social Psychology*, 45, 1263–1279.

Knowles, W.B. 1963. Operator loading tasks. *Human Factors*, 5, 151–161.

Knowles, W.B. & Sheridan, T.B. 1966. The "feel" of rotary controls: Friction and inertia. *Human Factors*, 8, 209–216.

Ko, Y. & Miller, J. 2014. Locus of backward crosstalk effects on task 1 in a psychological refractory period task. *Experimental Psychology*, 61, 30–37.

Ko, Y.-C., Wu, C.-H. & Lee, M. 2006. Evaluation of the impact of SAMG on the level-2 PSA results of a pressurized water reactor. *Nuclear Technology*, 155, 22–33.

Koffka, K. 1935. *Principles of Gestalt Psychology*. New York: Harcourt, Brace & World.

Kohn, L., Corrigan, J. & Donaldson, M. (Eds.) 2000. *To Err Is Human: Building a Safer Health System*. Washington, DC: National Academy Press.

Kolodner, J.L. 1991. Improving human decision making through case-based decision aiding. *AI Magazine*, 12 (2), 52–68.

Konis, K. 2013. Evaluating daylighting effectiveness and occupant visual comfort in a side-lit open-plan office building in San Francisco, California. *Building and Environment*, 59, 662–677.

Konstantzos, I., Tzempelikos, A. & Chan, Y.C. 2015. Experimental and simulation analysis of daylight glare probability in offices with dynamic window shades. *Building and Environment*, 87, 244–254.

Konz, S. 1974. Design of hand tools. *Proceedings of the Human Factors Society 18th Annual Meeting* (pp. 292–300). Santa Monica, CA: Human Factors Society.

Koradecka, D. 2006. Wojciech Bogumił Jastrzębowski. In W. Karwowski (Ed.), *International Encyclopedia of Ergonomics and Human Factors* (2nd ed., Vol. 3, pp. 3447–3448). Boca Raton, FL: CRC Press.

Kornblum, S. 1973. Sequential effects in choice reaction time: A tutorial review. In S. Kornblum (Ed.), *Attention and Performance IV* (pp. 259–288). New York: Academic Press.

Kornblum, S. 1991. Stimulus–response coding in four classes of stimulus–response ensembles. In J. Requin & G.E. Stelmach (Eds.), *Tutorials in Motor Neuroscience* (pp. 3–15). Dordrecht, The Netherlands: Kluwer Academic.

Kornblum, S., Hasbroucq, T. & Osman, A. 1990. Dimensional overlap: Cognitive basis for stimulus–response compatibility—A model and taxonomy. *Psychological Review*, 97, 253–270.

Kornblum, S. & Lee, J.-W. 1995. Stimulus–response compatibility with relevant and irrelevant stimulus dimensions that do and do not overlap with the response. *Journal of Experimental Psychology: Human Perception and Performance*, 21, 855–875.

Koslowski, B. 2012. Inference to the best explanation (IBE) and the causal and scientific reasoning of nonscientists. In R. W. Proctor & E. J. Capaldi (Eds.), *Psychology of Science: Implicit and Explicit Processes* (pp. 112–136). New York: Oxford University Press.

Kosnik, W.D., Sekuler, R. & Kline, D.W. 1990. Self-reported visual problems of older drivers. *Human Factors*, 5, 597–608.

Kossiakoff, A., Sweet, W. N., Seymour, S. & Biemer, S. M. 2011. *Systems Engineering: Principles and Practice* (2nd ed.). Hoboken, NJ: John Wiley.

Kosslyn, S.M. 1975. Information representation in visual images. *Cognitive Psychology*, 7, 341–370.

Kosslyn, S.M. & Thompson, W.L. 2003. When is early visual cortex activated during visual mental imagery? *Psychological Bulletin*, 129, 723–746.

Kosslyn, S.M., Ball, T.M. & Reiser, B.J. 1978. Visual images preserve metric spatial information: Evidence from studies of image scanning. *Journal of Experimental Psychology: Human Perception and Performance*, 4, 47–60.

Kosslyn, S.M., Thompson, W.L. & Ganis, G. 2006. *The Case for Mental Imagery*. New York: Oxford University Press.

Kosten, C.W. & Van Os, G.J. 1962. Community reaction criteria for external noises. *National Physical Laboratory, Symposium No. 12* (pp. 373–387). London: Her Majesty's Stationary Office.

Kramer, A.F., Sirevaag, E.J. & Braune, R. 1987. A psychophysiological assessment of operator workload during simulated flight missions. *Human Factors*, 29, 145–160.

Kreifeldt, J., Parkin, L., Rothschild, P. & Wempe, T. 1976, May. Implications of a mixture of aircraft with and without traffic situation displays for air traffic management. *Twelfth Annual Conference on Manual Control* (pp. 179–200). Washington, DC: National Aeronautics and Space Administration.

Kring, J.P. 2001. Multicultural factors for international spaceflight. *Human Performance in Extreme Environments*, 5, 11–32.

Kristjánsson, Á., Moldoveanu, A., Jóhannesson, Ó.I., Balan, O., Spagnol, S., Valgeirsdóttir, V.V. & Unnthorsson, R. 2016. Designing sensory-substitution devices: Principles, pitfalls and potential. *Restorative Neurology and Neuroscience*, 34, 769–787.

Kristofferson, M.W. 1972. When item recognition and visual search functions are similar. *Perception & Psychophysics*, 12, 379–384.

Kroemer, K.H.E. 1971. Foot operation of controls. *Ergonomics*, 14, 333–361.

Kroemer, K.H.E. 1983a. Engineering anthropometry: Workspace and equipment to fit the user. In D.J. Oborne & M.M. Gruneberg (Eds.), *The Physical Environment at Work* (pp. 39–68). New York: Wiley.

Kroemer, K.H.E. 1983b. Isoinertial technique to assess individual lifting capability. *Human Factors*, 25, 493–506.

Kroemer, K.H.E. 1989. Cumulative trauma disorders: Their recognition and ergonomics measures to avoid them. *Applied Ergonomics*, 20, 274–280.

Kroemer, K.H.E. 1997. Anthropometry and biomechanics. In A.D. Fisk & W.A. Rogers (Eds.), *Handbook of Human Factors and the Older Adult* (pp. 87–124). San Diego, CA: Academic Press.

Kroemer, K.H.E. 2006a. *"Extra-Ordinary" Ergonomics: How to Accommodate Small and Big Persons, the Disabled and Elderly, Expectant Mothers, and Children*. Boca Raton, FL: CRC Press.

Kroemer, K.H.E. 2006b. Static and dynamic strength. In W. Karwowski (Ed.), *International Encyclopedia of Ergonomics and Human Factors* (2nd ed., Vol. 1, pp. 511–512). Boca Raton, FL: CRC Press.

Kroemer, K.H.E. & Grandjean, E. 1997. *Fitting the Task to the Human: A Textbook of Occupational Ergonomics* (5th ed.). London: Taylor & Francis.

Kroemer, K.H.E., Kroemer, H.J. & Kroemer-Elbert, K.E. 1997. *Engineering Physiology: Bases of Human Factors/Ergonomics* (3rd ed.). New York: Van Nostrand Reinhold.

Kroemer, K.H.E., Kroemer, H.J. & Kroemer-Elbert, K.E. 2010. *Engineering Physiology: Bases of Human Factors/Ergonomics* (4th ed.). Berlin, Heidelberg: Springer.

Krug, S. 2014. *Don't Make Me Think, Revisited: A Common Sense Approach to Web Usability* (3rd Ed.). Indianapolis, IN: New Riders.

Krueger, R.A. & Casey, M.A. 2015. *Focus Groups: A Practical Guide for Applied Research*. Thousand Oaks, CA: Sage.

Kryter, K.D. 1972. Speech communication. In H.P. Van Cott & R.G. Kinkade (Eds.), *Human Engineering Guide to Equipment Design* (pp. 161–226). Washington, DC: U.S. Government Printing Office.

Kryter, K.D. & Williams, C.E. 1965. Masking of speech by aircraft noise. *Journal of the Acoustical Society of America*, 37, 138–150.

Kumar, S. & Ng, B. 2001. Crowding and violence on psychiatric wards: Explanatory models. *Canadian Journal of Psychiatry*, 46, 433–437.

Kunde, W. 2001. Response-effect compatibility in manual choice reaction tasks. *Journal of Experimental Psychology: Human Perception and Performance*, 27, 387–394.

Kurke, M.I. 1986. Anatomy of product liability/personal injury litigation. In Kurke, M.I. & Meyer, R.G. (Eds.), *Psychology in Product Liability and Personal Injury Litigation* (pp. 3–15). Washington, DC: Hemisphere Publishing.

Kurlychek, R.T. 1983. Use of a digital alarm chronograph as a memory aid in early dementia. *Clinical Gerontologist*, 1(3), 93–94.

Kurmann, A., Tschan, F., Semmer, N.K., Seelandt, J., Candinas, D. & Beldi, G. 2012. Human factors in the operating room–The surgeon's view. *Trends in Anaesthesia and Critical Care*, 2, 224–227.

Kuronen, P., Sorri, M.J., Paakkonen, R. & Muhli, A. 2003. Temporary threshold shift in military pilots measured using conventional and extended high-frequency audiometry after one flight. *International Journal of Audiology*, 42, 29–33.

Kvälseth, T.O. 1980. Factors influencing the implementation of ergonomics: An empirical study based on a psychophysical scaling technique. *Ergonomics*, 23, 821–826.

Kveraga, K., Boucher, L. & Hughes, H.C. 2002. Saccades operate in violation of Hick's law. *Experimental Brain Research*, 146, 307–314.

La Sala, K.P. 1998. Human performance reliability: A historical perspective. *IEEE Transactions on Reliability*, 47(3-SP), 365–371.

LaBerge, D. 1983. Spatial extent of attention to letters and words. *Journal of Experimental Psychology: Human Perception and Performance*, 9, 371–379.

Lacava, D. & Mentis, H.M. 2005. Beginning design without a user: Application of scenario-based design. *Proceedings of the 11th International Conference on Human–Computer Interaction (HCII)*, Las Vegas, NV.

Lachman, R., Lachman, J.L. & Butterfield, E.C. 1979. *Cognitive Psychology and Information Processing: An Introduction*. Hillsdale, NJ: Lawrence Erlbaum.

Lackner, J.R. 1990. Sensory-motor adaptation to high force levels in parabolic flight maneuvers. In M. Jeannerod (Ed.), *Attention and Performance XIII: Motor Representation and Control* (pp. 527–548). Hillsdale, NJ: Lawrence Erlbaum.

Lackner, J.R. & DiZio, P. 2005. Vestibular, proprioceptive, and haptic contributions to spatial orientation. *Annual Review of Psychology*, 56, 115–147.

Lahiri, S., Gold, J. & Levenstein, C. 2005. Net-cost model for workplace interventions. *Journal of Safety Research*, 36, 241–255.

Laird, J. E. 2012. *The Soar Cognitive Architecture*. Cambridge, MA: MIT Press.

Laird, J., Rosenbloom, P. & Newell, A. 1986. *Universal Subgoaling and Chunking*. New York: Kluwer Academic.

Laita, L.M., López-Bravo, B., Roanes-Lozano, E., Laita, L. & de Ledesma, L. 2007. A logic and computer algebra based expert system for diagnosis and treatment of migraine. In A.R. Tyler (Ed.), *Expert Systems: Research Trends* (pp. 1–84). New York: Nova Science Publishers.

Lamberg, L. 2002. Long hours, little sleep: Bad medicine for physicians in training? *Journal of the American Medical Association*, 287, 303–306.

Landauer, T.K. 1998. Learning and representing verbal meaning: The latent semantic analysis theory. *Current Directions in Psychological Science*, 7, 161–164.

Landauer, T.K., Laham, D. & Foltz, P. 2003. Automatic essay assessment. *Assessment in Education: Principles, Policy and Practice*, 10, 295–308.

Landauer, T.K., McNamara, D.S., Dennis, S. & Kintsch, W. (Eds.) 2007. *The Handbook of Latent Semantic Analysis*. Mahwah, NJ: Lawrence Erlbaum.

Landrigan, C.P., Rothschild, J.M., Cronin, J.W., et al. 2004. Effect of reducing interns' work hours on serious medical errors in intensive care units. *The New England Journal of Medicine*, 351, 1838–1848.

Landrigan, C.P., Barger, L.K., Cade, B.E., et al. 2006. Interns' compliance with accreditation council for graduate medical education work-hour limits. *Journal of the American Medical Association*, 296, 1063–1070.

Landy, F.J. & Farr, J.L. 1980. Performance rating. *Psychological Bulletin*, 87, 72–107.

Lane, N.E. 1987. *Skill Acquisition Rates and Patterns: Issues and Training Implications*. New York: Springer.

Langendijk, E.H.A. & Bronkhorst, A.W. 2000. Fidelity of three-dimensional-sound reproduction using a virtual auditory display. *Journal of the Acoustical Society of America*, 107, 528–537.

Langham, M.P. & Moberly, N.J. 2003. Pedestrian conspicuity research: A review. *Ergonomics,* 46, 345–363.

Langolf, G. & Hancock, W.M. 1975. Human performance times in microscope work. *AIEE Transactions*, 7, 110–117.

Lashley, K.S. 1951. The problem of serial order in behavior. In L.A. Jefress (Ed.), *Cerebral Mechanisms in Behavior* (pp. 112–136). New York: Wiley.

Lathan, C., Tracey, M.R., Sebrechts, M.M., Clawson, D.M. & Higgins, G.A. 2002. Using virtual environments as training simulators: Measuring transfer. In K.M. Stanney (Ed.), *Handbook of Virtual Environments: Design, Implementation, and Applications* (pp. 403–414). Mahwah, NJ: Lawrence Erlbaum.

Lavery, J.J. 1962. Retention of simple motor skills as a function of type of knowledge of results. *Canadian Journal of Psychology*, 16, 300–311.

Lavie, N., Hirst, A., de Fockert, J.W. & Viding, E. 2004. Load theory of selective attention and cognitive control. *Journal of Experimental Psychology: General*, 133, 339–354.

Lawrence, C. & Andrews, K. 2004. The influence of perceived prison crowding on male inmates' perception of aggressive events. *Aggressive Behavior*, 30, 273–283.

Lazarus, R.S. & Folkman, S. 1984. *Stress, Appraisal, and Coping*. New York: Springer.

Leape, L.L. 1994. Error in medicine. *Journal of the American Medical Association*, 272, 1851–1857.

Leape, L.L. & Berwick, D.M. 2005. Five years after *To Err Is Human*: What have we learned? *The Journal of the American Medical Association*, 293, 2083–2090.

Lederman, S.J. & Campbell, J.I. 1982. Tangible graphs for the blind. *Human Factors*, 24, 85–100.

Ledford, M. (Ed.) 2014. *Carpal Tunnel Syndrome: Risk Factors, Symptoms and Treatment Options*. New York: Nova Biomedical.

Lee, A.T. 2005. *Flight Simulation: Virtual Environments in Aviation*. Burlington, VT: Ashgate.

Lee, D.N. 1976. A theory of visual control of braking based on information about time-to-collision. *Perception*, 5, 437–459.

Lee, J.Y., Bahn, S. & Nam, C.S. 2014. Use of reference frame and movement pattern in haptically enhanced 3d virtual environment. *International Journal of Human-Computer Interaction*, 30, 891–903.

Lee, T.D. & Genovese, E.D. 1988. Distribution of practice in motor skill acquisition: Learning and performance effects reconsidered. *Research Quarterly for Exercise and Sport*, 59, 277–287.

Lee, T.D. & Genovese, E.D. 1989. Distribution of motor skill acquisition: Different effects for discrete and continuous tasks. *Research Quarterly for Exercise and Sport*, 70, 59–65.

Lee, T.D., Magill, R.A. & Weeks, D.J. 1985. Influence of practice schedule on testing schema theory predictions in adults. *Journal of Motor Behavior*, 17, 283–299.

Lehman, J.F., Laird, J.E. & Rosenbloom, P. 1998. A gentle introduction to SOAR: An architecture for human cognition. In D. Scarborough & S. Sternberg (Eds.), *An Invitation to Cognitive Science* (2nd ed., Vol. 4: *Methods, Models, and Conceptual Issues*, pp. 211–253). Cambridge, MA: MIT Press.

Lehto, M.R., Nah, F.F.H. & Yi, J.S. 2012. Decision-making models, decision support, and problem solving. In G. Salvendy (Ed.), *Handbook of Human Factors and Eergonomics* (4th ed, pp. 192–242). Hoboken, NJ: John Wiley.

Leibowitz, H.W. 1996. The symbiosis between basic and applied research. *American Psychologist*, 51, 366–370.

Leibowitz, H.W. & Owens, D.A. 1986. We drive by night. *Psychology Today, January*, 55–58.

Leibowitz, H.W. & Post, R.B. 1982. The two modes of processing concept and some implications. In J. Beck (Ed.), *Organization and Representation in Perception* (pp. 343–363). Hillsdale, NJ: Lawrence Erlbaum.

Leibowitz, H.W., Post, R.B., Brandt, T. & Dichgans, J. 1982. Implications of recent developments in dynamic spatial orientation and visual resolution for vehicle guidance. In A.H. Wertheim, W.A. Wagenaar & H.W. Leibowitz (Eds.), *Tutorials on Motion Perception* (pp. 231–260). New York: Plenum Press.

Leider, P.C., Boschman, J.S., Frings-Dresen, M.H. & van der Molen, H.F. 2015. When is job rotation perceived useful and easy to use to prevent work-related musculoskeletal complaints? *Applied Ergonomics*, 51, 205–210.

Leighton, J.P. 2004a. Defining and describing reason. In J.P. Leighton & R.J. Sternberg (Eds.), *The Nature of Reasoning* (pp. 3–11). Cambridge: Cambridge University Press.

Leighton, J.P. 2004b. The assessment of logical reasoning. In J.P. Leighton & R.J. Sternberg (Eds.), *The Nature of Reasoning* (pp. 291–312). Cambridge: Cambridge University Press.

Lennard, S.H.C. & Lennard, H.L. 1977. Architecture: Effect of territory, boundary, and orientation on family functioning. *Family Process*, 16, 49–66.

Lennie, P. 2003. The physiology of color vision. In S.K. Shevell (Ed.), *The Science of Color* (2nd (Ed.), pp. 217–246). Amsterdam: Elsevier.

Lenz, M., Bartsch-Sporl, B., Burkhard, H.D. & Wess, S. (Eds.) 1998. *Case-Based Reasoning Technology: From Foundations to Applications*. Berlin: Springer.

Leonard, J.A. 1958. Partial advance information in a choice reaction task. *British Journal of Psychology*, 49, 89–96.

Leonard, J.A. 1959. Tactual choice reactions: I. *Quarterly Journal of Experimental Psychology*, 11, 76–83.

Lerch, F.J., Mantei, M.M. & Olson, J.R. 1989. Translating ideas into action: Cognitive analysis of errors in spreadsheet formulas. *Proceedings of the CHI '89 Conference on Human Factors in Computing Systems* (pp. 121–126). New York: Van Nostrand Reinhold.

Lessiter, J., Freeman, J., Davis, R. & Drumbreck, A. 2003. Helping viewers press the right buttons: Generating intuitive labels for digital terrestrial TV remote controls. *Psychology Journal*, 1, 355–377.

Levin, D.T. & Simons, D.J. 1997. Failure to detect changes to attended objects in motion pictures. *Psychonomic Bulletin & Review*, 4, 501–506.

Levin, I.P., McElroy, T., Gaeth, G.J., Hedgcock, W., Denburg, N.L. & Tranel, D. 2015. Studying decision processes through behavioral and neuroscience analyses of framing effects. In E.A. Wilhelms, & V.F. Reyna (Eds.), *Neuroeconomics, Judgment, and Decision Making* (pp. 131–156). New York: Psychology Press.

Levy, P.E. & Williams, J.R. 2004. The social context of performance appraisal: A review and framework for the future. *Journal of Management*, 30, 881–905.

Lewin, K. 1951. *Field Theory in Social Science*. New York: Harper.

Lewis, J.L. 1970. Semantic processing of unattended messages using dichotic listening. *Journal of Experimental Psychology*, 85, 225–228.

Lewis, R. 1990. Design economy in the creation of manned space systems. *Human Factors Society Bulletin*, 32(3), 5–6.

Li, C.-C., Hwang, S.-L. & Wang, M.-Y. 1990. Static anthropometry of civilian Chinese in Taiwan using computer-analyzed photography. *Human Factors*, 32, 359–370.

Li, J., Deng, L., Gong, Y. & Haeb-Umbach, R. 2014. An overview of noise-robust automatic speech recognition. *IEEE/ACM Transactions on Audio, Speech, and Language Processing*, 22, 745–777.

Li, W.-C. & Harris, D. 2005. HFACS analysis of ROC air force aviation accidents: Reliability analysis and cross-cultural comparison. *International Journal of Applied Aviation Studies*, 5, 65–81.

Lichtenstein, S. & Slovic, P. 1971. Reversal of preferences between bids and choices in gambling decision. *Journal of Experimental Psychology*, 89, 46–55.

Lichtenstein, S., Slovic, P., Fischoff, B., Layman, M. & Combs, B. 1978. Judged frequency of lethal events. *Journal of Experimental Psychology: Human Learning and Memory*, 4, 551–578.

Lie, I. 1980. Visual detection and resolution as a function of retinal locus. *Vision Research*, 20, 967–974.

Lieberman, M. D. & Cunningham, W. A. 2009. Type I and Type II error concerns in fMRI research: Re-balancing the scale. *Social Cognitive and Affective Neuroscience*, 4, 423–428.

Liebowitz, J. 1990. *The Dynamics of Decision Support Systems and Expert Systems*. Chicago, IL: Dryden Press.

Lien, M.-C. & Proctor, R.W. 2000. Multiple spatial correspondence effects on dual-task performance. *Journal of Experimental Psychology: Human Perception and Performance*, 26, 1260–1280.

Lien, M.-C., Proctor, R.W. & Allen, P.A. 2002. Ideomotor compatibility in the psychological refractory period effect: 29 years of oversimplification. *Journal of Experimental Psychology: Human Perception and Performance*, 28, 396–409.

Likert, R. 1961. *New Patterns of Management*. New York: McGraw-Hill.

Lim, V., Liu, S., Wang, W. & Joines, S. 2011. Assessment of the redesigned Asian wok handles. In *Proceedings of the Human Factors and Ergonomics Society 55th Annual Meeting* (pp. 1706–1710). Santa Monica, CA: HFES.

Lindahl, L.G. 1945. Movement analysis as an industrial training method. *Journal of Applied Psychology*, 29, 420–436.

Lindemann, P.G. & Wright, C.E. 1998. Skill acquisition and plans for actions: Learning to write with your other hand. In D. Scarborough & S. Sternberg (Eds.), *Methods, Models, and Conceptual Issues: An Invitation to Cognitive Science* (Vol. 4, pp. 523–584). Cambridge, MA: MIT Press.

Little, R.G., Manzanares, T. & Wallace, W.A. 2015. Factors influencing the selection of decision support systems for emergency management: An empirical analysis of current use and user preferences. *Journal of Contingencies and Crisis Management*, 23, 266–274.

Lipshitz, R., Klein, G., Orasanu, J. & Salas, E. 2001. Taking stock of naturalistic decision making. *Journal of Behavioral Decision Making*, 14, 331–352.

Liu, W. 2003. Field dependence–independence and sports with a preponderance of closed or open skill. *Journal of Sport Behavior*, 26, 285–297.

Lockhart, R.S. 2002. Levels of processing, transfer-appropriate processing, and the concept of robust encoding. *Memory*, 10, 397–403.

Lockhart, T.E., Atsumi, B., Ghosh, A., Mekaroonreung, H. & Spaulding, J. 2006. Effects of planar and non-planar driver-side mirrors on age-related discomfort-glare responses. *Safety Science*, 44, 187–195.

Loftus, E.F. & Loftus, G.R. 1980. On the permanence of stored information in the human brain. *American Psychologist*, 35, 409–420.

Loftus, G.R. & Irwin, D.E. 1998. On the relations among different measures of visible and informational persistence. *Cognitive Psychology*, 35, 135–199.

Loftus, G.R., Dark, V.J. & Williams, D. 1979. Short-term memory factors in ground controller/pilot communication. *Human Factors*, 21, 169–181.

Logan, G.D. 2004. Cumulative progress in formal theories of attention. *Annual Review of Psychology*, 55, 207–234.

Logan, G.D. & Schulkind, M.D. 2000. Parallel memory retrieval in dual-task situations: I. Semantic memory. *Journal of Experimental Psychology: Human Perception and Performance*, 26, 1260–1280.

Long, G.M. & Kearns, D.F. 1996. Visibility of text and icon highway signs under dynamic viewing conditions. *Human Factors*, 38, 690–701.

Long, G.M. & Zavod, M.J. 2002. Contrast sensitivity in a dynamic environment: Effects of target conditions and visual impairment. *Human Factors*, 44, 120–132.

LoPresti, E.F., Brienza, D.M. & Angelo, J. 2002. Head-operated computer controls: Effect of control method on performance for subjects with and without disability. *Interacting with Computers*, 14, 359–377.

Lorge, I. 1930. Influence of regularly interpolated time intervals upon subsequent learning (Teacher College Contributions to Education, No. 438). New York: Columbia University, Teachers College.

Loschky, L.C. & McConkie, G.W. 2002. Investigating spatial vision and dynamic attentional selection using a gaze-contingent multi-resolutional display. *Journal of Experimental Psychology: Applied*, 8, 99–117.

Loschky, L.C., McConkie, G.W., Yang, J. & Miller, M.E. 2005. The limits of visual resolution in natural scene viewing. *Visual Cognition*, 12, 1057–1092.

Louviere, A.J. & Jackson, J.T. 1982. Man–machine design for spaceflight. In T.S. Cheston & D.L. Winter (Eds.), *Human Factors of Outer Space Production* (pp. 97–112). Boulder, CO: Westview Press.

Lovasik, J.V., Matthews, S.M.L. & Kergoat, H. 1989. Neural, optical, and search performance in prolonged viewing of chromatic displays. *Human Factors*, 31, 273–289.

Loveless, N.E. 1962. Direction-of-motion stereotypes: A review. *Ergonomics*, 5, 357–383.

Lovie, A.D. 1983. Attention and behaviourism. *British Journal of Psychology*, 74, 301–310.

Lozinskii, E.I. 2000. Explaining by evidence. *Journal of Experimental and Theoretical Artificial Intelligence*, 12, 69–89.

Lu, C.-H. & Proctor, R.W. 1995. The influence of irrelevant location information on performance: A review of the Simon and spatial Stroop effects. *Psychonomic Bulletin & Review*, 2, 174–207.

Lu, J. & Hignett, S. 2009. Using task analysis in healthcare design to improve clinical efficiency. *Health Environments Research and Design Journal*, 2, 60–69.

Lu, M., Waters, T. R., Krieg, E. & Werren, D. 2014. Efficacy of the revised NIOSH lifting equation to predict risk of low-back pain associated with manual lifting: A one-year prospective study. *Human Factors*, 56, 73–85.

Lucas, J.L. & Heady, R.B. 2002. Flextime commuters and their driver stress, feelings of time urgency, and commute satisfaction. *Journal of Business and Psychology*, 16, 565–572.

Luce, P.A., Feustal, T.C. & Pisoni, D.B. 1983. Capacity demands in short-term memory for synthetic and natural speech. *Human Factors*, 25, 17–32.

Luck, S.J. 2014. *An Introduction to the Event-Related Potential Technique* (2nd ed.). Cambridge, MA: MIT Press.

Lukosch, S. & Schummer, T. 2006. Groupware development support with technology patterns. *International Journal of Human–Computer Studies*, 64, 599–610.

Luque-Casado, A., Perales, J. C., Cárdenas, D. & Sanabria, D. 2016. Heart rate variability and cognitive processing: The autonomic response to task demands. *Biological Psychology*, 113, 83–90.

Lusted, L.B. 1971. Signal detectability and medical decision-making. *Science*, 171, 1217–1219.

Lutz, M.C. & Chapanis, A. 1955. Expected locations of digits and letters on ten-button keysets. *Journal of Applied Psychology*, 39, 314–317.

Luursema, J., Verwey, W.B. & Burie, R. 2012. Visuospatial ability factors and performance variables in laparoscopic simulator training. *Learning and Individual Differences*, 22, 632–638.

Luximon, A. & Goonetilleke, R.S. 2001. Simplified subjective workload assessment technique. *Ergonomics*, 44, 229–243.

Lyman, S.L. & Scott, M.B. 1967. Territoriality: A neglected sociological dimension. *Social Problems*, 15, 235–249.

Lyons, M., Adams, S., Woloshynowych, M. & Vincent, C. 2004. Human reliability analysis in healthcare: A review of techniques. *International Journal of Risk and Safety in Medicine*, 16, 223–237.

Lysaght, R.S., Hill, S.G., Dick, A.O., Plamondon, B.D., Wherry, R.J., Zaklad, A.L. & Bittner, A.C. 1989. Operator workload: A comprehensive review of operator workload methodologies. Technical Report 851, U.S. Army Research Institute for the Social Sciences. Alexandria, VA.

MacDonald, L.W. 1999. Using color effectively in computer graphics. *IEEE Computer Graphics and Applications*, 19(4), 20–35.

Machado, G.M., Oliveira, M.M. & Fernandes, L.A. 2009. A physiologically-based model for simulation of color vision deficiency. *IEEE Transactions on Visualization and Computer Graphics*, 15, 1291–1298.

Mack, A. 1986. Perceptual aspects of motion in the frontal plane. In K.R. Boff, L. Kaufman & J.P. Thomas (Eds.), *Handbook of Perception and Human Performance* (Vol. I: *Sensory Processes and Perception*), pp. 17–1–17–38). New York: Wiley.

MacKay, D.G. 1981. The problem of rehearsal or mental practice. *Journal of Motor Behavior*, 13, 274–285.

Mackworth, N.H. 1950. Researches on the measurement of human performance (Special Report No. 268). London: Medical Research Council, Her Majesty's Stationary Office.

Mackworth, N. H. 1948. The breakdown of vigilance during prolonged visual search. *Quarterly Journal of Experimental Psychology*, 1, 6–21.

Mackworth, N.H. 1965. Visual noise causes tunnel vision. *Psychonomic Science*, 3, 67–68.

MacLeod, C.M. 1991. Half a century of research on the Stroop effect: An integrative review. *Psychological Review*, 109, 163–203.

Macmillan, N.A. & Creelman, C.D. 2005. *Detection Theory: A User's Guide* (2nd ed.). Mahwah, NJ: Lawrence Erlbaum.

Macy, B.A. & Mirvis, P.H. 1976. A methodology for assessment of quality of work life and organizational effectiveness in behavioural-economic terms. *Administrative Science Quarterly*, 21, 212–216.

Madni, A.M. 1988. The role of human factors in expert systems design and acceptance. *Human Factors*, 30, 395–414.

Maeda, T., Ando, H. & Sugimoto, M. 2005, March. Virtual acceleration with galvanic vestibular stimulation in a virtual reality environment. In IEEE *Proceedings. VR 2005. Virtual Reality, 2005.* (pp. 289–290). Piscataway, NJ: IEEE.

Magill, R.A. & Parks, P.F. 1983. The psychophysics of kinesthesis for positioning responses: The physical stimulus-psychological response relationship. *Research Quarterly for Exercise and Sport*, 54, 346–351.

Magill, R.A. & Wood, C.A. 1986. Knowledge of results precision as a learning variable in motor skill acquisition. *Research Quarterly for Exercise and Sport*, 57, 170–173.

Magnussen, S. & Kurtenbach, W. 1980. Linear summation of tilt illusion and tilt aftereffect. *Vision Research*, 20, 39–42.

Magnussen, S., Andersson, J., Cornoldi, C., De Beni, R., Endestad, T., Goodman, G.S., Helstrup, T., Koriat, A., Larsson, M., Melinder, A., Nilsson, L.-G., Ronnberg, J. & Zimmer, H. 2006. What people believe about memory. *Memory*, 14, 595–613.

Makhsous, M., Hendrix, R., Crowther, Z., Nam, E. & Lin, F. 2005. Reducing whole-body vibration and musculoskeletal injury with a new car seat design. *Ergonomics*, 48, 1183–1199.

Makous, J.C. & Middlebrooks, J.C. 1990. Two-dimensional sound localization by human listeners. *Journal of the Acoustical Society of America*, 87, 2188–2200.

Malacara, D. 2011, August. *Color vision and colorimetry: Theory and Applications* (2nd ed.). Bellingham, WA: SPIE.

Mallis, M.M. & DeRoshia, C.W. 2005. Circadian rhythms, sleep, and performance in space. *Aviation, Space, and Environmental Medicine*, 76, B94–B107.

Malone, T.B. 1986. The centered high-mounted brakelight: A human factors success story. *Human Factors Society Bulletin*, 29(10), 1–3.

Manchi, G., Gowda, S. & Hanspal, J.S. 2013. Study on cognitive approach to human error and its application to reduce the accidents at workplace. *International Journal of Engineering and Advanced Technology*, 2, 236–242.

Mantei, M.M. & Teorey, T.J. 1988. Cost/benefit analysis for incorporating human factors in the software cycle. *Communications of the ACM*, 31, 428–439.

Manzey, D. & Lorenz, B. 1998. Mental performance during short-term and long-term spaceflight. *Brain Research Reviews*, 28, 215–221.

Marakas, G.M. 2003. *Decision Support Systems in the 21st Century*. Upper Saddle River, NJ: Prentice-Hall.

Marcos, S., Moreno, E. & Navarro, R. 1999. The depth-of-field of the human eye from objective and subjective measurements. *Vision Research*, 39, 2039–2049.

Marcus, A. 2005. User interface design's return on investment: Examples and statistics. In R.G. Bias & D.J. Mayhew (Eds.), *Cost-Justifying Usability: An Update for the Internet Age* (pp. 17–39). San Francisco, CA: Morgan Kaufman.

Mardex, J. 2004. Auditory, visual, and physical distractions in the workplace. *Cornell University, Department of Design and Environmental Analysis*.

Marey, E.-J. 1902. *The History of Chronophotography*. Washington, DC: Smithsonian Institute.

Mariampolski, H. 2006. *Ethnography for Marketers: A Guide to Consumer Immersion*. Thousand Oaks, CA: Sage.

Marics, M.A. & Williges, B.H. 1988. The intelligibility of synthesized speech in data inquiry systems. *Human Factors*, 30, 719–732.

Marine, A., Ruotsalainen, J., Serra, C. & Verbeek, J. 2006. Preventing occupational stress in healthcare workers. *Cochrane Database of Systematic Reviews*, Issue 4. Art. No.: CD002892. DOI:10.1002/14651858. CD002892.pub2.

Marklin, R.W. & Simoneau, G.G. 2006. Biomechanics of the wrist in computer keyboarding. In W. Karwowski (Ed.), *International Encyclopedia of Ergonomics and Human Factors* (2nd ed., Vol. 2, pp. 1549–1554). Boca Raton, FL: CRC Press.

Marklin, R.W., Simoneau, G.G. & Monroe, J.F. 1999. Wrist and forearm posture from typing on split and vertically inclined computer keyboards. *Human Factors*, 41, 559–569.

Marks, L.E. & Gescheider, G.A. 2002. Psychophysical scaling. In H. Pashler & J. Wixted (Eds.), *Stevens' Handbook of Experimental Psychology (3rd ed.), Vol. 4: Methodology in Experimental Psychology* (pp. 91–138). Hoboken, NJ: John Wiley.

Marois, R. & Ivanoff, J. 2005. Capacity limits of information processing in the brain. *Trends in Cognitive Sciences*, 9, 296–305.

Marrao, C., Tikuisis, P., Keefe, A.A., Gil, V. & Giesbrecht, G.G. 2005. Physical and cognitive performance during long-term cold weather operations. *Aviation, Space, and Environmental Medicine*, 76, 744–752.

Marras, W.S. 2006. Basic biomechanics and workstation design. In G. Salvendy (Ed.), *Handbook of Human Factors and Ergonomics* (3rd ed., pp. 340–370). Hoboken, NJ: Wiley.

Marras, W.S. & Kroemer, K.H.E. 1980. A method to evaluate human factors/ergonomics design variables of distress signals. *Human Factors*, 22, 389–399.

Marras, W., Lavender, S., Leurgans, S., Rajulu, S., Allread, W., Fathallah, F. & Ferguson, S. 1993. The role of dynamic three-dimensional trunk motion in occupationally-related low back disorders, *Spine*, 18, 617–628.

Marsalek, P. & Kofranek, J. 2004. Sound localization at high frequencies and across the frequency range. *Neurocomputing, 58–60*, 999–1006.

Marsh, E.J. & Roediger, H.L., III 2013. Episodic and autobiographical memory. In A.F. Healy & R.W. Proctor (Eds.), *Experimental Psychology* (Vol. 4. of *Handbook of Psychology*, 2nd ed. pp. 472–494). Hoboken, NJ: John Wiley.

Marslen-Wilson, W.D. 1975. Sentence perception as an interactive parallel process. *Science*, 189, 226–228.

Marteniuk, R.G. 1986. Information processes in movement learning: Capacity and structural interference effects. *Journal of Motor Behavior*, 18, 55–75.

Martín-Arévalo, E., Chica, A.B. & Lupiáñez, J. 2016. No single electrophysiological marker for facilitation and inhibition of return: A review. *Behavioural Brain Research*, 300, 1–10.

Matsuoka, J., Berger, R. A., Berglund, L. J., & An, K. N. 2006. An analysis of symmetry of torque strength of the forearm under resisted forearm rotation in normal subjects. *The Journal of Hand Surgery*, 31, 801–805.

Matthis, J.S., Barton, S.L. & Fajen, B.R. 2015. The biomechanics of walking shape the use of visual information during locomotion over complex terrain. *Journal of Vision*, 15(3):10, 1–13.

Mathôt, S. & Van der Stigchel, S. 2015. New light on the mind's eye: The pupillary light response as active vision. *Current Directions in Psychological Science*, 24, 374–378.

Matin, L., Picoult, E., Stevens, J., Edwards, M. & MacArthur, R. 1982. Oculoparalytic illusion: Visual-field dependent spatial mislocations by humans partially paralyzed with curare. *Science*, 216, 198–201.

Matthews, G., Davies, D.R., Westerman, S.J. & Stammers, R.B. 2000. *Human Performance: Cognition, Stress and Individual Differences*. Hove: Psychology Press.

Matthews, G., Reinerman-Jones, L.E., Barber, D.J. & Abich IV, J. 2015. The psychometrics of mental workload: Multiple measures are sensitive but divergent. *Human Factors*, 57, 125–143.

Mattingly, I.G. & Studdert-Kennedy, M. (Eds.) 1991. *Modularity and the Motor Theory of Speech Perception*. Hillsdale, NJ: Lawrence Erlbaum.

Maula, H., Hongisto, V., Östman, L., Haapakangas, A., Koskela, H. & Hyönä, J. 2016. The effect of slightly warm temperature on work performance and comfort in open-plan offices-a laboratory study. *Indoor Air*, 26, 286–297.

May, J. & Barnard, P.J. 2004. Cognitive task analysis in interacting cognitive subsystems. In D. Diaper & N. Stanton (Eds.), *The Handbook of Task Analysis for Human–Computer Interaction* (pp. 291–325). Mahwah, NJ: Lawrence Erlbaum.

Mayhew, D.J. & Bias, R.G. 2003. Cost-justifying Web usability. In J. Ratner (Ed.), *Human Factors and Web Development* (2nd ed., pp. 63–87). Mahwah, NJ: Lawrence Erlbaum.

Mayr, S., Köpper, M. & Buchner, A. 2013. Comparing colour discrimination and proofreading performance under compact fluorescent and halogen lamp lighting. *Ergonomics*, 56, 1418–1429.

McAnany, J. J. & Alexander, K. R. 2008. Spatial frequencies used in Landolt C orientation judgments: Relation to inferred magnocellular and parvocellular pathways. *Vision Research*, 48, 2615–2624.

McBride, D.K. & Schmorrow, D. (Eds.) 2005. *Quantifying Human Information Processing*. Lanham, MA: Lexington Books.

McBride, S. & Newbold, S. 2016. Systems development life cycle for achieving meaningful use. In S. McBride & M. Tietze (Eds.), *Nursing Informatics for the Advanced Practice Nurse* (pp. 191–223). New York: Springer.

McCann, E. 2014. Deaths by medical mistakes hit records. Healthcare IT News. Downloaded from http://www.healthcareitnews.com/news/deaths-by-medical-mistakes-hit-records

McClelland, J.L. 1979. On the time relations of mental processes: An examination of systems of processes in cascade. *Psychological Review*, 86, 287–330.

McCloskey, M. 1983. Naive theories of motion. In D. Gentner & A.L. Stevens, *Mental Models* (pp. 299–324). Hillsdale, NJ: Lawrence Erlbaum.

McColl, S.L. & Veitch, J.A. 2001. Full-spectrum fluorescent lighting: A review of its effects on physiology and health. *Psychological Medicine*, 31, 949–964.

McConville, J.T., Robinette, K.M. & Churchill, T.D. 1981. An anthropometric database for commercial design applications (Phase I). Final Report NSF=BNS-81001 (PB 81–211070). Washington, DC: National Science Foundation.

McCormick, E.J., Jeanneret, P.R. & Mecham, R.C. 1972. A study of job characteristics and job dimensions as based on the Position Analysis Questionnaire (PAQ). *Journal of Applied Psychology*, 56, 347–368.

McCracken, J.H. & Aldrich, T.B. 1984. Analysis of selected LHX mission functions: Implications for operator workload and system automation goals (TNA AS1479-24–84). Fort Rucker, AL: Anacapa Sciences.

McDowd, J.M. 1986. The effects of age and extended practice on divided attention performance. *Journal of Gerontology*, 41, 764–769.

McFarland, R.A. 1946. *Human Factors in Air Transport Design*. New York: McGraw-Hill.

McGrath, J.J. 1963. Irrelevant stimulation and vigilance performance. In D.N. Buckner & J.J. McGrath (Eds.), *Vigilance: A Symposium*. New York: McGraw-Hill.

McIntire, J.P., Havig, P.R. & Geiselman, E.E. 2014. Stereoscopic 3D displays and human performance: A comprehensive review. *Displays*, 35(1), 18–26.

McKenzie, I. S. 2013. *Human-Computer Interaction: An Empirical Research Perspective*. Waltham, MA: Morgan Kaufman.

McKnight, A.J., Shinar, D. & Hilburn, B. 1991. The visual and driving performance of monocular and binocular heavy-duty truck drivers. *Accident Analysis and Prevention*, 23, 225–237.

McMillan, G.R., Beevis, D., Salas, E., Strub, M.H., Sutton, R. & Van Breda, L. 1989. *Applications of Human Performance Models to System Design*. New York: Plenum Press.

McNamara, T.P. 2013. Semantic memory and priming. In A.F. Healy & R.W. Proctor (Eds.), *Experimental psychology* (Vol. 4. of *Handbook of psychology*), 2nd ed., pp. 449–471. Hoboken, NJ: John Wiley.

McNaughton, G.B. 1985, October. The problem. Presented at a workshop on flight attitude awareness. Wright-Patterson Air Force Base, OH: Wright Aeronautical Laboratories and Life Support System Program Office.

Medina, J.M., Wong, W., Díaz, J.A. & Colonius, H. 2015. Advances in modern mental chronometry. *Frontiers in Human Neuroscience*, 9, 1–3.

Megaw, E.D. & Bellamy, L.J. 1983. Illumination at work. In D.J. Oborne & M.M. Gruneberg (Eds.), *The Physical Environment at Work* (pp. 109–141). New York: Wiley.

Mehler, B., Kidd, D., Reimer, B., Reagan, I., Dobres, J. & McCartt, A. 2016. Multi-modal assessment of on-road demand of voice and manual phone calling and voice navigation entry across two embedded vehicle systems. *Ergonomics*, 59, 344–367.

Mehrparvar, A.H., Mirmohammadi, S.J., Hashemi, S.H., Davari, M.H., Mostaghaci, M., Mollasadeghi, A. & Zare, Z. 2015. Concurrent effect of noise exposure and smoking on extended high-frequency pure-tone thresholds. *International Journal of Audiology*, 54, 301–307.

Meister, D. 1971. *Human Factors: Theory and Practice*. New York: Wiley.

Meister, D. 1985. *Behavioral Analysis and Measurement Methods*. New York: Wiley.

Meister, D. 1989. *Conceptual Aspects of Human Factors*. Baltimore, MD: Johns Hopkins University Press.

Meister, D. 1991. The definition and measurement of systems. *Human Factors Society Bulletin*, 34(2), 3–5.

Meister, D. 2006a. History of human factors in the United States. In W. Karwowski (Ed.), *International Encyclopedia of Ergonomics and Human Factors* (2nd ed., Vol. 1, pp. 98–101). Boca Raton, FL: CRC Press.

Meister, D. 2006b. Human factors system design. In W. Karwowski (Ed.), *International Encyclopedia of Ergonomics and Human Factors* (2nd ed., Vol. 2, pp. 1967–1971). Boca Raton, FL: CRC Press.

Meister, D. & Enderwick, T.P. 2002. *Human Factors in System Design, Development, and Testing*. Mahwah, NJ: Lawrence Erlbaum.

Meister, D. & Rabideau, G. 1965. *Human Factors Evaluation in System Development*. New York: Wiley.

Meltzer, J.E. & Moffitt, K. 1997. *Head-Mounted Displays: Designing for the User*. New York: McGraw-Hill.

Mendl, M. 1999. Performing under pressure: Stress and cognitive function. *Applied Animal Behaviour Science*, 65, 221–244.

Mertens, H.W. & Lewis, M.F. 1981. Effect of different runway size on pilot performance during simulated night landing approaches. U.S. Federal Aviation Administration Office of Aviation Medicine Technical Report (FAA-AH-81-6). Washington, DC: Federal Aviation Administration.

Mertens, H.W. & Lewis, M.F. 1982. Effects of approach lighting and visible runway length on perception of approach angle in simulated night landings. U.S. Federal Aviation Administration Office of Aviation Technical Report (FAA-AM-82-6). Washington, DC: Federal Aviation Administration.

Merton, P.A. 1972. How we control the contraction of our muscles. *Scientific American*, 226, 30–37.

Meshkati, N. 1988. Heart rate variability and mental workload assessment. In P.A. Hancock & N. Meshkati (Eds.), *Human Mental Workload* (pp. 101–115). Amsterdam: North-Holland.

Meso, P., Troutt, M.D. & Rudnicka, J. 2002. A review of naturalistic decision making research with some implications for knowledge management. *Journal of Knowledge Management*, 6, 63–73.

Metz, S., Isle, B., Denno, S. & Li, W. 1990. Small rotary controls: Limitations for people with arthritis. *Proceedings of the Human Factors Society 34th Annual Meeting* (pp. 137–140). Santa Monica, CA: Human Factors Society.

Meyer, D.E., Abrams, R.A., Kornblum, S., Wright, C.W. & Smith, J.E.K. 1988. Optimality in human motor performance: Ideal control of rapid aimed movements. *Psychological Review*, 95, 340–370.

Meyer, D.E. & Kieras, D.E. 1997a. A computational theory of executive cognitive processes and multiple-task performance: Part 2. Accounts of psychological refractory-period phenomena. *Psychological Review*, 104, 749–791.

Meyer, D.E. & Kieras, D.E. 1997b. A computational theory of executive cognitive processes and multiple-task performance: Part 1. Basic mechanisms. *Psychological Review*, 104, 3–65.

Meyer, D.E., Smith, J.E.K., Kornblum, S., Abrams, R.A. & Wright, C.E. 1990. Speed–accuracy tradeoffs in aimed movements: Toward a theory of rapid voluntary action. In M. Jeannerod (Ed.), *Attention and Performance XIII* (pp. 173–226). Hillsdale, NJ: Lawrence Erlbaum.

Meyer, J. 2006. Responses to dynamic warnings. In M.S. Wogalter (Ed.), *Handbook of Warnings* (pp. 221–229). Mahwah, NJ: Lawrence Erlbaum.

Michelson, G. & Mouly, V.S. 2004. Do loose lips sink ships? The meaning, antecedents and consequences of rumour and gossip in organisations. *Corporate Communications*, 9, 189–201.

Miles, J. D. & Proctor, R. W. 2012. Correlations between spatial compatibility effects: Are arrows more like locations or words? *Psychological Research*, 76, 777–791.

Miller, D.P. & Swain, A.D. 1987. Human error and human reliability. In G. Salvendy (Ed.), *Handbook of Human Factors* (pp. 219–250). New York: Wiley.

Miller, G.A. 1956. The magical number seven, plus or minus two: Some limits on our capacity for processing information. *Psychological Review*, 63, 81–97.

Miller, G.A. & Glucksberg, S. 1988. Psycholinguistic aspects of pragmatics and semantics. In R.C. Atkinson, R.J. Herrnstein, G. Lindzey & R.D. Luce (Eds.), *Stevens' Handbook of Experimental Psychology* (2nd ed., pp. 417–472). New York: Wiley.

Miller, G.A. & Isard, S. 1963. Some perceptual consequences of linguistic rules. *Journal of Verbal Learning and Verbal Behavior*, 2, 212–228.

Miller, J. 1988. Discrete and continuous models of human information processing: Theoretical distinctions and empirical results. *Acta Psychologica*, 67, 191–257.

Miller, J. 2006. Backward crosstalk effects in psychological refractory period paradigms: Effects of second-task response types on first-task response latencies. *Psychological Research*, 70, 484–493.

Miller, J., Atkins, S.G. & Van Nes, F. 2005. Compatibility effects based on stimulus and response numerosity. *Psychonomic Bulletin & Review*, 12, 265–270.

Miller, J., Beutinger, D. & Ulrich, R. 2009. Visuospatial attention and redundancy gain. *Psychological Research*, 73, 254–262.

Miller, J., & Schwarz, W. 2006. Dissociations between reaction times and temporal order judgments: A diffusion model approach. *Journal of Experimental Psychology: Human Perception and Performance*, 32, 394–412.

Miller, J. & Ulrich, R. 2003. Simple reaction time and statistical facilitation: A parallel grains model. *Cognitive Psychology*, 46, 101–151.

Miller, M.E. & Beaton, R.J. 1994. The alarming sounds of silence. *Ergonomics in Design, January*, 21–23.

Miller, R.J. & Penningroth, S. 1997. The effects of response format and other variables on comparisons of digital and dial displays. *Human Factors*, 39, 417–424.

Milligan, M.W. & Tennant, J.S. 1997. Enhancing the conspicuity of personal watercrafts. *Marine Technology, Society Journal*, 31(2), 50–55.

Mirka, G.A. & Marras, W.S. 1990. Lumbar motion response to a constant load velocity lift. *Human Factors*, 32, 493–501.

Miskewicz-Zastrow, A., Bishop, E., Zastrow, A., Cuevas, D. M. & Rainey, B. B. 2015. A standardized procedure and normative values for measuring binocular dynamic visual acuity. *Optometry & Visual Performance*, 3, 169–175.

Mistlberger, R.E. 2003. Circadian rhythms. In L. Nadel (editor-in-chief), *Encyclopedia of Cognitive Science* (Vol. 1, pp. 514–518). New York: Nature Publishing Group.

Mistrot, P., Donati, P., Galimore, J.P. & Florentin, D. 1990. Assessing the discomfort of the whole-body multi-axis vibration: Laboratory and field experiments. *Ergonomics*, 33, 1523–1536.

Mital, A. & Ramanan, S. 1985. Accuracy of check-reading dials. In R.E. Eberts & C.G. Eberts (Eds.), *Trends in Ergonomics/Human Factors II* (pp. 105–113). Amsterdam: North-Holland.

Mitsopoulos-Rubens, E. & Lenné, M.G. 2012. Issues in motorcycle sensory and cognitive conspicuity: The impact of motorcycle low-beam headlights and riding experience on drivers' decisions to turn across the path of a motorcycle. *Accident Analysis and Prevention*, 49, 86–95.

Mohrman, A.M., Jr., Resnick-West, S.M. & Lawler, E.E., II 1989. *Designing Performance Appraisal Systems*. San Francisco, CA: Jossey-Bass.

Møler, A.R. 2014. *Sensory Systems: Anatomy and Physiology* (3rd ed.). Dallas, TX: Aage R. Møler Publishing.

Moll, D., Robinson, P.A. & Hobscheid, H.M. 2005. Products liability law: What engineering experts need to know. In Y.I. Noy & W. Karwowski (Eds.), *Handbook of Human Factors in Litigation* (pp. 27-1–27-9). Boca Raton, FL: CRC Press.

Mollon, J.D. 2003. The origins of modern color science. In S.K. Shevell (Ed.), *The Science of Color* (2nd ed., pp. 1–39). Amsterdam: Elsevier.

Monk, A. 2003. Common ground in electronically mediated communication: Clark's theory of language use. In J.M. Carroll (Ed.), *HCI Models, Theories, and Frameworks: Toward a Multidisciplinary Science* (pp. 265–289). San Francisco, CA: Morgan Kaufman.

Monk, T.H. 1989. Human factors implications of shiftwork. *International Reviews of Ergonomics*, 2, 111–128.

Monk, T.H. 2006. Application of basic knowledge to the human body: Shiftwork. In W. Karwowski (Ed.), *International Encyclopedia of Ergonomics and Human Factors* (2nd ed., Vol. 2, pp. 2049–2055). Boca Raton, FL: CRC Press.

Monsell, S. & Driver, J. (Eds.) 2000. *Control of Cognitive Processes: Attention and Performance XVIII*. Cambridge, MA: MIT Press.

Mon-Williams, M. & Tresilian, J.R. 1999. Some recent studies on the extraretinal contribution to distance perception, *Perception*, 28, 167–181.

Mon-Williams, M. & Tresilian, J.R. 2000. Ordinal depth information from accommodation? *Ergonomics*, 43, 391–404.

Moore, M. & Kearsley, G. 2012. *Distance Education: A Systems View of Online Learning* (3rd ed.). Belmont, CA: Wadsworth.

Moore, T. G. 1974. Tactile and kinesthetic aspects of pushbuttons. *Applied Ergonomics*, 52, 66–71.

Moore, T.G. 1975. Industrial push-buttons. *Applied Ergonomics*, 6, 33–38.

Morahan, P., Meehan, J.W., Patterson, J. & Hughes, P.K. 1998. Ocular vergence measurement in projected and collimated simulator displays. *Human Factors*, 40, 376–385.

Moray, N. 1959. Attention in dichotic listening: Affective cues and the influence of instructions. *Quarterly Journal of Experimental Psychology*, 11, 56–60.

Moray, N. 1987. Intelligent aids, mental models, and the theory of machines. *International Journal of Man–Machine Studies*, 27, 619–629.

Moray, N. 1993. Designing for attention. In A. Baddeley & L. Weiskrantz (Eds.), *Attention: Selection, Awareness, and Control* (pp. 111–134). Oxford, UK: Oxford University Press.

Morgan, M.J., Watt, R.J. & McKee, S.P. 1983. Exposure duration affects the sensitivity of vernier acuity to target motion. *Vision Research*, 23, 541–546.

Morgeson, F.P., Medsker, G.J. & Campion, M.A. 2012. Job and team design. In G. Salvendy (Ed.), *Handbook of Human Factors and Ergonomics* (4th ed., pp. 441–474). Hoboken, NJ: Wiley.

Morin, R.E. & Grant, D.A. 1955. Learning and performance on a key-pressing task as a function of the degree of spatial stimulus–response correspondence. *Journal of Experimental Psychology*, 49, 39–47.

Morley, N.J., Evans, J. St. B.T. & Handly, S.J. 2004. Belief bias and figural bias in syllogistic reasoning. *Quarterly Journal of Experimental Psychology, 57A*, 666–692.

Morris, C.D., Bransford, J.D. & Franks, J.J. 1977. Levels of processing versus transfer appropriate processing. *Journal of Verbal Learning and Verbal Behavior*, 16, 519–533.

Morrison, D., Cordery, J., Girardi, A. & Payne, R. 2005. Job design, opportunities for skill utilization, and intrinsic job satisfaction. *European Journal of Work and Organizational Psychology*, 14, 59–79.

Morrison, J.D. & Whiteside, T.C.D. 1984. Binocular cues in the perception of distance of a point source of light. *Perception*, 13, 555–566.

Mortier, R., Haddadi, H., Henderson, T., McAuley, D. & Crowcroft, J. 2014. Human-data interaction: The human face of the data-driven society. *Available at* SSRN *2508051*.

Moser, G. & Uzzell, D. 2003. Environmental psychology. In T. Millon & M.J. Lerner (Eds.), *Handbook of Psychology: Personality and Social Psychology* (Vol. 5, pp. 419–445). Hoboken, NJ: Wiley.

Mossey, M. 2013. *Designing a Novel Steering Wheel for Generation-Y, Baby Boomers, and Engineers* (Doctoral dissertation). Retrieved from ProQuest Digital Dissertations. (UMI 1539415)

Mount, F. 2006. Human factors in space flight. In W. Karwowski (Ed.), *International Encyclopedia of Ergonomics and Human Factors* (2nd ed., Vol. 2, pp. 1956–1962). Boca Raton, FL: CRC Press.

Mowbray, G.H. 1960. Choice reaction time for skilled responses. *Quarterly Journal of Experimental Psychology*, 12, 193–202.

Mowbray, G.H. & Rhoades, M.U. 1959. On the reduction of choice-reaction times with practice. *Quarterly Journal of Experimental Psychology*, 11, 16–23.

Moynihan, G.P., Suki, A. & Fonseca, D.J. 2006. An expert system for the selection of software design patterns. *Expert Systems: International Journal of Knowledge Engineering and Neural Networks*, 23, 39–52.

Mrena, R., Savolainen, S., Kuokkanen, J.T. & Ylikoski, J. 2002. Characteristics of tinnitus induced by acute acoustic trauma: A long-term follow-up. *Audiology and Neurotology*, 7, 122–130.

Mulder, T. & Hulstijn, W. 1985. Sensory feedback in the learning of a novel motor task. *Journal of Motor Behavior*, 17, 110–128.

Müller-Gethmann, H., Ulrich, R. & Rinkenauer, G. 2003. Locus of the effect of temporal preparation: Evidence from the lateralized readiness potential. *Psychophysiology*, 40, 597–611.

Mumford, R. 2002. Improving visual efficiency with selected lighting. *Journal of Optometric Visual Development*, 33(3), 1–6.

Mumm, J. & Mutlu, B. 2011, March. Human-robot proxemics: Physical and psychological distancing in human-robot interaction. In *Proceedings of the 6th International Conference on Human-Robot Interaction* (pp. 331–338). New York: ACM.

Münsterberg, H. 1913. *Psychology and Industrial Efficiency*. Boston, MA: Mifflin.

Muriana, C., Piazza, T. & Vizzini, G. 2016. An expert system for financial performance assessment of health care structures based on fuzzy sets and KPIs. *Knowledge-Based Systems*, 97, 1–10.

Murchison, J. 2010. *Ethnography Essentials: Designing, Conducting, and Presenting your Research*. San Francisco, CA: Jossey-Bass.

Murphy, G., Groeger, J.A. & Greene, C.M. 2016. Twenty years of load theory—Where are we now, and where should we go next? *Psychonomic Bulletin & Review*, 23, 1316–1340.

Murray, J.B. 1974. Renewed interest in attention. *Psychological Reports*, 34, 155–166.

Murray, T. 2000. Lateral chest X-rays of no greater value in Dx: ER pediatricians with only frontal view more accurately tabbed pneumonia cases, also fewer false negatives. *Medical Post*, 36(25, July 4), 16.

Murrell, G.A. 1975. A reappraisal of artificial signals as an aid to a visual monitoring task. *Ergonomics*, 18, 693–700.

Murrell, K.F.H. 1969. The Ergonomics Research Society: The Society's lecture. *Ergonomics*, 12, 691–700.

Muybridge, E. 1955. *The Human Figure in Motion*. New York: Dover Press.

Näätänen, R. 1973. The inverted-U relationship between activation and performance. In S. Kornblum (Ed.), *Attention and Performance IV* (pp. 155–174). New York: Academic Press.

Naess, R.O. 2001. *Optics for Technology Students*. Upper Saddle River, New Jersey: Prentice-Hall.

Nairne, J.S. & Kelley, M.R. 2004. Separating item and order information through process dissociation. *Journal of Memory and Language*, 50, 113–133.

Nairne, J.S. & Neath, I. 2013. Sensory and working memory. In A.F. Healy & R.W. Proctor (Eds.), *Experimental Psychology* (Vol. 4. of *Handbook of Psychology*, 2nd ed., pp. 419–445). Hoboken, NJ: John Wiley.

Nakakoji, K. 2005. Special issue on "Computational Approaches for Early Stages of Design." *Knowledge-Based Systems*, 18, 381–382.

Namamoto, K., Atsumi, B., Kodera, H. & Kanamori, H. 2003. Quantitative analysis of muscular stress during ingress/egress of the vehicle. *JSA Review*, 24, 335–339.

Nanda, R. & Adler, G.L. (Eds.) 1977. *Learning Curves: Theory and Application*. Atlanta, GA: American Institute of Industrial Engineers.

Narumi, J., Miyazawa, S., Miyata, H., Suzuki, A., Kohsaka, S. & Kosugi, H. 1999. Analysis of human error in nursing care. *Accident Analysis and Prevention*, 31, 625–629.

Naruo, N., Lehto, M. & Salvendy, G. 1990. Development of a knowledge-based decision support system for diagnosing malfunctions of advanced production equipment. *International Journal of Production Research*, 28, 2259–2276.

NASA 1978. *Anthropometric Source Book*. NASA Reference Publication 1024. Springfield, VA: National Technical Information Service.

NASA 1995. *Man–Systems Integration Standards*. NASA-STD-3000 (Revision B). Houston, TX: NASA.

National Academy of Engineering. 2004. *The Engineer of 2020*. Washington, DC: The National Academies Press.

National Academy of Engineering. 2005. *Educating the Engineer of 2020*. Washington, DC: The National Academies Press.

National Center for Employee Ownership. 2016. The employee ownership 100. Downloaded on August 19, 2006, from http://www.nceo.org/library/eo100.html

National Transportation Safety Board. 2010. NTSB determines engineer's failure to observe and respond to red signal caused 2008 Chatsworth accident; recorders in cabs recommended (Press release). Retrieved 19 January 2015.

Navon, D. & Gopher, D. 1979. On the economy of the human information processing system. *Psychological Review*, 86, 214–255.

Navon, D. & Miller, J. 1987. Role of outcome conflict in dual-task interference. *Journal of Experimental Psychology: Human Perception and Performance*, 13, 435–448.

Navon, D. & Miller, J. 2002. Queuing or sharing? A critical evaluation of the single-bottleneck notion. *Cognitive Psychology*, 44, 193–251.

Neath, I. & Brown, G.D.A. 2007. Making distinctiveness models of memory distinct. In J.S. Nairne (Ed.), *The Foundations of Remembering: Essays in Honor of Henry L. Roediger, III* (pp. 125–140). New York: Psychology Press.

Neely, G. & Burström, L. 2006. Gender differences in subjective responses to hand-arm vibration. *International Journal of Industrial Ergonomics*, 36, 135–140.

Neerincx, M.A., Ruijsendaal, M. & Wolff, M. 2001. Usability engineering guide for integrated operation support in space station payloads. *International Journal of Cognitive Ergonomics*, 5, 187–198.

Nelson, W.R. 1988. Human factors considerations for expert systems in the nuclear industry. *Proceedings of the IEEE Conference on Human Factors and Power Plants* (pp. 109–114). Piscataway, NJ: IEEE.

Nelson, W.R. & Blackman, H.S. 1987. Experimental evaluation of expert systems for nuclear reactor operators: Human factors considerations. *International Journal of Industrial Ergonomics*, 2, 91–100.

Nemri, A. & Krarti, M. 2006. Analysis of electrical energy savings from daylighting through skylights. *Proceedings of the 2005 International Solar Energy Conference* (pp. 51–57). New York: American Society of Mechanical Engineers.

Neves, D.M. & Anderson, J.R. 1981. Knowledge compilation: Mechanisms for the automatization of cognitive skills. In J.R. Anderson (Ed.), *Cognitive Skills and Their Acquisition* (pp. 57–84). Hillsdale, NJ:

Newell, A. 1990. *Unified Theories of Cognition*. Cambridge, MA: Harvard University Press.

Newell, A. & Rosenbloom, P.S. 1981. Mechanisms of skill acquisition and the law of practice. In J.R. Anderson (Eds.), *Cognitive Skills and Their Acquisition* (pp. 1–55). Hillsdale, NJ: Lawrence Erlbaum.

Newell, A. & Simon, H.A. 1972. *Human Problem Solving*. Englewood Cliffs, NJ: Prentice-Hall.

Newell, K.M. 1976. Knowledge of results and motor learning. *Exercise and Sport Sciences Reviews*, 4, 195–227.

Newell, K.M. & Carlton, M.J. 1987. Augmented information and the acquisition of isometric tasks. *Journal of Motor Behavior*, 19, 4–12.

Newell, K.M., Morris, L.R. & Scully, D.M. 1985a. Augmented information and the acquisition of skill in physical activity. In R.L. Terjung (Ed.), *Exercise and Sport Sciences Reviews* (pp. 235–261). New York: Macmillan.

Newell, K.M., Sparrow, W.A. & Quinn, J.T. 1985b. Kinetic information feedback for learning isometric tasks. *Journal of Human Movement Studies*, 11, 113–123.

Newell, K.M. & Walter, C.B. 1981. Kinematic and kinetic parameters as information feedback in motor skill acquisition. *Journal of Human Movement Studies*, 7, 235–254.

Newman, R.L. 1987. Responses to Roscoe, "The trouble with HUDS and HMDS." *Human Factors Society Bulletin*, 30(10), 3–5.

Ng, T.S., Cung, L.D. & Chicharo, J.F. 1990. DESPLATE: An expert system for abnormal shape diagnosis in the plate mill. *IEEE Transactions on Industry Applications*, 26, 1057–1062.

Nichols, R.G. 1962. Listening is good business. *Management of Personnel Quarterly*, 4, 4.

Nickerson, R.S. 2015. *Conditional Reasoning: The Unruly Syntactics, Semantics, Thematics, and Pragmatics of "If"*. New York, NY: Oxford University Press.

Nicoletti, R., Anzola, G.P., Luppino, G., Rizzolatti, G. & Umiltà, C. 1982. Spatial compatibility effects on the same side of the body midline. *Journal of Experimental Psychology: Human Perception and Performance*, 8, 664–673.

Niederberger, U. & Gerber, W.-D. 1999. Human cerebral potentials during motor training under different forms of sensory feedback. *Journal of Psychophysiology*, 13, 234–244.

Nielsen, J, & Loranger, H. 2006. *Prioritizing Web Usability*. Berkeley, CA: New Riders.

Nielsen, J. & Norman, D. 2016. The definition of user experience. Nielsen Norman Group. Downloaded 1/3/2016 from https://www.nngroup.com/articles/definition-user-experience/

Nikata, K. & Shimada, H. 2005. Facilitation of analogical transfer by posing an analogous problem for oneself. *Japanese Journal of Educational Psychology*, 53, 381–392.

Nilsson, N.J. 1998. *Artificial Intelligence: A New Synthesis*. Los Altos, CA: Morgan Kaufmann.

NIOSH 1981. *Work Practices Guide for Manual Lifting*. DHHS/NIOSH Publication No. 81–122. Washington, DC: U.S. Government Printing Office.

Nobre, K. & Kastner, S. (Eds.) 2014. *The Oxford Handbook of Attention*. New York: Oxford University Press.

Norbäck, D. 2009. An update on sick building syndrome. *Current Opinion in Allergy and Clinical Immunology*, 9, 55–59.

Nordby, K., Raanaas, R.K. & Magnussen, S. 2002. The expanding telephone number part 1: Keying briefly presented multiple-digit numbers. *Behaviour and Information Technology*, 21, 27–38.

Norman, D.A. 1968. Toward a theory of memory and attention. *Psychological Review*, 75, 522–536.

Norman, D.A. 1981. Categorization of action slips. *Psychological Review*, 88, 1–15.

Norman, D.A. 2013. *The Design of Everyday Things* (revised and expanded edition). New York: Basic Books.

Norman, D.A. & Bobrow, D.G. 1976. On the analysis of performance operating characteristics. *Psychological Review*, 83, 508–510.

Norman, D.A. & Bobrow, D.G. 1975. On data-limited and resource-limited processing. *Cognitive Psychology*, 7, 44–60.

Norris, E.M. & Lockey, A.S. 2012. Human factors in resuscitation teaching. *Resuscitation*, 83, 423–427.

Noy, Y.I. & Karwowski, W. (Eds.) 2005. *Handbook of Human Factors in Litigation*. Boca Raton, FL: CRC Press.

Noyes, J.M. 2006. Verbal protocol analysis. In W. Karwowski (Ed.), *International Encyclopedia of Ergonomics and Human Factors* (2nd ed., Vol. 3, pp. 3390–3392). Boca Raton, FL: CRC Press.

Noyes, J.M., Haigh, R. & Starr, A.F. 1989. Automatic speech recognition for disabled people. *Applied Ergonomics*, 20, 293–298.

Nunes, M. S., Souza, M. X., Basso, L., Monteiro, C. M., Corrêa, U. C. & Santos, S. 2014. Frequency of provision of knowledge of performance on skill acquisition in older persons. *Frontiers in Psychology*, 5.

Nygard, C.-H. & Ilmarinen, J. 1990. Effects of changes in delivery of dairy products on physical strain of truck drivers. In C.M. Haslegrave, J.R. Wilson, E.N. Corlett & I. Manenica (Eds.), *Work Design in Practice*

Oatis, C. A. 2016. *Kinesiology: The Mechanics and Pathomechanics of Human Movement*. Philadelphia, PA: Lippincott Williams & Wilkins.

Oberauer, K., Weidenfeld, A. & Hornig, R. 2006. Working memory capacity and the construction of spatial mental models in comprehension and deductive reasoning. *Quarterly Journal of Experimental Psychology*, 59, 426–447.

O'Connor, P. & Kleyner, A. 2012. *Practical Reliability Engineering* (5th ed.). Chichester, UK: John Wiley.

Office of the Assistant Secretary of Defense. 2015. *Systems Engineering: Initiatives: Human Systems Integration*. (Retrieved 1/24/15 from http://www.acq.osd.mil/se/initiatives/init_hsi.html)

Ogden, G.D., Levine, J.M. & Eisner, E.J. 1979. Measurement of workload by secondary tasks. *Human Factors*, 21, 529–548.

Ohlsson, S. 2008. Computational models of skill acquisition. In R. Sun (Ed.), *The Cambridge Handbook of Computational Psychology* (pp. 359–395). New York: Cambridge University Press.

Ohta, N. & Robertson, A.R. 2005. *Colorimetry: Fundamentals and Applications*. Hoboken, NJ: Wiley.

Oldham, G. R., & Fried, Y. 2016. Job design research and theory: Past, present and future. *Organizational Behavior and Human Decision Processes*, 136, 20–35.

Oleari, C. (2016). *Standard colorimetry: Definitions, algorithms and software*. Hoboken, NJ: John Wiley.

Oliver, K. 2002. *Psychology in Practice: Environment*. Abingdon: Hodder & Stoughton.

Öllinger, M., Jones, G., & Knoblich, G. 2014. The dynamics of search, impasse, and representational change provide a coherent explanation of difficulty in the nine-dot problem. *Psychological Research*, 78, 266–275.

Olsen, S.O., Rasmussen, A.N., Nielsen, L.H. & Borgkvist, B.V. 1999. The acoustical reflex threshold: Not predictive for loudness perception in normally-hearing listeners. *Audiology*, 38, 303–307.

Olson, J.R. & Olson, G.M. 1990. The growth of cognitive modeling in human–computer interaction since GOMS. *Human–Computer Interaction*, 5, 221–265.

O'Neil, W.M. 1957. *Introduction to Method in Psychology*. Carlton, Australia: Melbourne University Press.

Ono, H. & Wade, N.J. 2005. Depth and motion in historical descriptions of motion parallax. *Perception*, 34, 1263–1273.

O'Regan, J.K., Rensink, R.A. & Clark, J.J. 1999. Change-blindness as a result of "mudsplashes." *Nature*, 398, 34.

O'Regan, J.K., Deubel, H., Clark, J.J. & Rensink, R.A. 2000. Picture changes during blinks: Looking without seeing and seeing without looking. *Visual Cognition*, 7, 191–211.

O'Reilly, C.A., III 1980. Individuals and information overload in organizations: Is more necessarily better? *Academy of Management Journal*, 23, 684–696.

O'Reilly, R.C., & Munakata, Y. 2003. Psychological function in computational models of neural networks. In M. Gallagher, R.J. Nelson (Eds.), *Handbook of Psychology: Biological Psychology, Vol. 3* (pp. 637–654). Hoboken, NJ: John Wiley.

Ortega, J. 2001. Job rotation as a learning mechanism. *Management Science*, 47, 1361–1370.

OSHA 2003. *OSHA: Employee Workplace Rights* (rev.). Washington, DC: Department of Labor.

Osherson, D.W., Smith, E.E. & Shafir, E.B. 1986. Some origins of belief. *Cognition*, 24, 197–224.

Osman, A., Lou, L., Muller-Gethmann, H., Rinkenauer, G., Mattes, S. & Ulrich, R. 2000. Mechanisms of speed–accuracy tradeoff: Evidence from covert motor processes. *Biological Psychology*, 51, 173–199.

Ostry, D., Moray, N. & Marks, G. 1976. Attention, practice and semantic targets. *Journal of Experimental Psychology: Human Perception and Performance*, 2, 326–336.

Owens, D.A. & Leibowitz, H.W. 1983. Perceptual and motor consequences of tonic vergence. In C.M. Schor & K.J. Cuiffreda (Eds.), *Vergence Eye Movements: Basic and Clinical Aspects* (pp. 23–97). Boston, MA: Butterworth-Heinemann.

Owens, D.A. & Wolfe-Kelly, K. 1987. Nearwork, visual fatigue, and variations of oculomotor tonus. *Investigative Ophthalmology and Visual Science*, 28, 745–749.

Oyama, T. 1987. Perception studies and their applications to environmental design. *International Journal of Psychology*, 22, 447–451.

Paap, K.R. & Ogden, W.G. 1981. Letter encoding is an obligatory but capacity-demanding operation. *Journal of Experimental Psychology: Human Perception and Performance*, 7, 518–528.

Paay, J. 2008. From ethnography to interface design. In J. Lumsden & J. Lumsden (Eds.), *Handbook of Research on User Interface Design and Evaluation for Mobile Technology (Vols. 1 and 2)* (pp. 1–15). Hershey, PA, US: Information Science Reference/IGI Global.

Pachella, R.G. 1974. The interpretation of reaction time in information-processing research. In B.H. Kantowitz (Ed.), *Human Information Processing: Tutorials in Performance and Cognition* (pp. 41–82). Hillsdale, NJ: Lawrence Erlbaum.

Pack, M., Cotten, D.J. & Biasiotto, J. 1974. Effect of four fatigue levels on performance and learning of a novel dynamic balance skill. *Journal of Motor Behavior*, 6, 191–197.

Packer, O. & Williams, D.R. 2003. Light, the retinal image, and photoreceptors. In S.K. Shevell (Ed.), *The Science of Color*. Amsterdam: Elsevier.

Paepcke, A., Wang, Q., Patel, S., Wang, M. & Harada, S. 2004. A cost-effective three-in-one personal digital assistant input control. *International Journal of Human–Computer Studies*, 60, 717–736.

Paik, J., & Ritter, F.E. 2016. Evaluating a range of learning schedules: Hybrid training schedules may be as good as or better than distributed practice for some tasks. *Ergonomics*, 59, 276–290.

Paillard, A.C., Quarck, G., Paolino, F., Denise, P., Paolino, M., Golding, J.F., & Ghulyan-Bedikian, V. 2013. Motion sickness susceptibility in healthy subjects and vestibular patients: Effects of gender, age and trait-anxiety. *Journal of Vestibular Research: Equilibrium & Orientation*, 23(4–5), 203–210.

Paivio, A. 1986. *Mental Representations: A Dual Coding Approach*. New York: Oxford University Press.

Palanica, A., & Itier, R.J. 2015. Eye gaze and head orientation modulate the inhibition of return for faces. *Attention, Perception, & Psychophysics*, 77, 2589–2600.

Palmer, A.R. 1995. Neural signal processing. In B.C.J. Moore (Ed.), *Hearing* (pp. 75–121). San Diego, CA,

Palmer, J. 1986. Mechanisms of displacement discrimination with and without perceived movement. *Journal of Experimental Psychology: Human Perception and Performance*, 12, 411–421.

Palmer, S.E. 2003. Visual perception of objects. In A.F. Healy & R.W. Proctor (Eds.), *Experimental Psychology* (pp. 179–211), Vol. 4 in I.B. Weiner (editor-in-chief) *Handbook of Psychology*. Hoboken, NJ: Wiley.

Pandey, S. 1999. Role of perceived control in coping with crowding. *Psychological Studies*, 44(3), 86–91.

Pang, N., Foo, S., Raamkumar, A.S., Zhang, X., & Vu, S. 2015, July. Object recognition-based mnemonics mobile app for senior adults communication. In 2015 *6th International Conference on Computing, Communication and Networking Technologies (ICCCNT)* (pp. 1–6). Piscataway, NJ: IEEE.

Parasuraman, R. 1979. Memory load and event rate control sensitivity decrements in sustained attention. *Science*, 205, 924–927.

Parasuraman, R. & Davies, D.R. 1976. Decision theory analysis of response latencies in vigilance. *Journal of Experimental Psychology: Human Perception and Performance*, 2, 569–582.

Parasuraman, R. & Nestor, P.G. 1991. Attention and driving skills in aging and Alzheimer's disease. *Human Factors*, 33, 539–557.

Parisi, T. 2016. *Learning Virtual Reality: Developing Immersive Experiences and Applications for Desktop, Web, and Mobile*. Sebastopol, CA: O'Reilly Media.

Park, K.Y. 1987. *Human Reliability: Analysis, Prediction, and Prevention of Human Errors*. Amsterdam: Elsevier.

Park, M.-Y. & Casali, J.G. 1991. A controlled investigation of in-field attenuation performance of selected insert, earmuff, and canal cap hearing protectors. *Human Factors*, 33, 693–714.

Park, S.H. & Woldstad, J.C. 2000. Multiple two-dimensional displays as an alternative to three-dimensional displays in telerobotic tasks. *Human Factors*, 42, 592–603.

Parsaye, K. & Chignell, M. 1987. *Expert Systems for Experts*. New York: Wiley.

Parsons, K.C. 2000. Environmental ergonomics: A review of principles, methods, and models. *Applied Ergonomics*, 31, 581–594.

Parsons, S.O., Seminara, J.L. & Wogalter, M.S. 1999. A summary of warnings research. *Ergonomics in Design*, 7(1), 21–31.

Parush, A., Nadir, R. & Shtub, A. 1998. Evaluating the layout of graphical user interface screens: Validation of a numerical computerized model. *International Journal of Human–Computer Interaction*, 10, 343–360.

Pashler, H. 1989. Dissociations and dependencies between speed and accuracy: Evidence for a two-component theory of divided attention in simple tasks. *Cognitive Psychology*, 21, 469–514.

Pashler, H. 1994. Dual-task interference in simple tasks: Data and theory. *Psychological Bulletin*, *16*, 220–224.

Pashler, H. 1998. Introduction. In H. Pashler (Ed.), *Attention* (pp. 1–11). Hove, UK: Psychology Press

Pashler, H. & Baylis, G. 1991. Procedural learning: 2. Intertrial repetition effects in speeded choice tasks. *Journal of Experimental Psychology: Learning, Memory, and Cognition*, 17, 33–48.

Pashler, H. & Johnston, J.C. 1989. Chronometric evidence for central postponement in temporally overlapping tasks. *Quarterly Journal of Experimental Psychology*, 41A, 19–45.

Pashler, H.E. 1998. *The Psychology of Attention*. Cambridge, MA: MIT Press.

Passchier-Vermeer, W. 1974. Hearing loss due to continuous exposure to steady state broad-band noise. *Journal of the Acoustical Society of America*, 56, 1585–1593.

Paté-Cornell, E. 2002. Finding and fixing systems weaknesses: Probabilistic methods and applications of engineering. *Risk Analysis*, 22, 319–334.

Patel, V.L., Arocha, J.F. & Zhang, J. 2005. Thinking and reasoning in medicine. In K.J. Holyoak & R.G. Morrison (Eds.), *Cambridge Handbook of Thinking and Reasoning* (pp. 727–750). New York: Cambridge University Press.

Patterson, R.D. 1982. Guidelines for auditory warning systems on civil aircraft: The learning and retention of warnings. MRC Applied Psychology Unit, Civil Aviation Authority Contract 7D=S=0142.

Patterson, R., Winterbottom, M.D., & Pierce, B.J. 2006. Perceptual issues in the use of head-mounted visual displays. *Human Factors*, 48, 555–573.

Paulignan, Y., MacKenzie, C., Marteniuk, R. & Jeannerod, M. 1990. The coupling of arm and finger movements during prehension. *Experimental Brain Research*, 79, 431–436.

Pauwels, L., Vancleef, K., Swinnen, S.P., & Beets, I.M. 2015. Challenge to promote change: Both young and older adults benefit from contextual interference. *Frontiers in Aging Neuroscience*, 7.

Payne, D. & Altman, J. 1962. An index of electronic equipment operability: Report of development. Report AIR-C-43–1=62. Pittsburgh, PA: American Institutes of Research.

Pearson, K. & Gordon, J. 2000. Locomotion. In E.R. Kandel, J.H. Schwartz & T.M. Jessell (Eds.), *Principles of Neural Science* (4th ed.), pp. 737–755). New York: McGraw-Hill.

Peebles, D., Derbinsky, N., & Laird, J.E. 2013. Effective and efficient forgetting of learned knowledge in Soar's working and procedural memories. *Cognitive Systems Research*, 24, 104–113.

Peirce, C.S. 1940. Abduction and induction. In J. Buchler (Ed.), *The Philosophy of Peirce: Selected Writings* (pp. 150–156). London: Routledge & Kegan Paul.

Percival, K.A., Martin, P.R., & Grünert, U. 2013. Organisation of koniocellular-projecting ganglion cells and diffuse bipolar cells in the primate fovea. *European Journal of Neuroscience*, 37, 1072–1089.

Pereira, A., Wachs, J. P., Park, K., & Rempel, D. 2015. A user-developed 3-D hand gesture set for human–computer interaction. *Human Factors*, 57, 607–621.

Perrow, C. 1999. *Normal Accidents: Living with High-Risk Technologies*. Princeton, NJ: Princeton University Press.

Peterman, T.K., Jalongo, M.R. & Lin, Q. 2002. The effects of molds and fungi on young children's health: Families' and educators' roles in maintaining indoor air quality. *Early Childhood Education Journal*, 30, 21–26.

Peters, G.A. 1977. Why only a fool relies on safety standards. *Hazard Prevention*, 14(2).

Peters, G.A. & Peters, B.J. 2006. *Human Error: Causes and Control*. Boca Raton, FL: CRC Press.

Peterson, B., Stine, J.L. & Darken, R.P. 2005. Eliciting knowledge from military ground navigators. In H. Montgomery, R. Lipshitz & B. Brehmer (Eds.), *How Professionals Make Decisions* (pp. 351–364). Mahwah, NJ: Lawrence Erlbaum.

Peterson, D.R., & Bronzino, J.D. (Eds.) 2014. *Biomechanics: Principles and Practices*. Boca Raton, FL: CRC Press.

Peterson, L.R. & Gentile, A. 1963. Proactive interference as a function of time between tests. *Journal of Experimental Psychology*, 70, 473–478.

Peterson, L.R. & Peterson, M.J. 1959. Short-term retention of individual verbal items. *Journal of Experimental Psychology*, 58, 193–198.

Petros, T.V., Bentz, B., Hammes, K. & Zehr, H.D. 1990. The components of text that influence reading times and recall in skilled and less skilled college readers. *Discourse Processes*, 13, 387–400.

Pew, R.W. 2008. More than 50 years of history and accomplishments in human performance model development. *Human Factors*, 50, 489–496.

Pew, R.W., & Mavor, A.S. (Eds.) 2007. *Human-System Integration in the System Development Process: A New Look*. Washington, DC: National Academies Press.

Pew, R.W. & Mavor, A.S. (Eds.) 1998. *Modeling Human and Organizational Behavior*. Washington, DC: National Academy Press.

Pheasant, S. & Haslegrave, C.M. 2006. *Bodyspace: Anthropometry, Ergonomics, and the Design of Work*. Boca Raton, FL: CRC Press.

Philp, R.B., Fields, G.N. & Roberts, W.A. 1989. Memory deficit caused by compressed air equivalent to 36 meters of seawater. *Journal of Applied Psychology*, 74, 443–446.

Pickles, J.O. 1988. *An Introduction to the Physiology of Hearing* (2nd ed.). San Diego, CA: Academic Press.

Pilcher, J.J., Nadler, E. & Busch, C. 2002. Effects of hot and cold temperature exposure on performance: A meta-analytic review. *Ergonomics*, 45, 682–698.

Pilczuk, D., & Barefield, K. 2014. Green ergonomics: Combining sustainability and ergonomics. *Work*, 49, 357–361.

Piñango, M.M., Zurif, E. & Jackendoff, R. 1999. Real-time processing implications of enriched composition at the syntax-semantics interface. *Journal of Psycholinguistic Research*, 28, 395–414.

Pinto, M., Cavallo, V., & Saint-Pierre, G. 2014. Influence of front light configuration on the visual conspicuity of motorcycles. *Accident Analysis and Prevention*, 62, 230–237.

Pisoni, D.B. 1982. Perceptual evaluation of voice response systems: Intelligibility, recognition, and understanding. *Workshop of Standardization for Speech I/O Technology* (pp. 183–192). Gaithersburg, MD: National Bureau of Standards.

Pisoni, D.B. & Remez, R.E. (Eds.) 2005. *The Handbook of Speech Perception*. Malden, MA: Blackwell.

Plack, C.J. 2005. *The Sense of Hearing*. Mahwah, NJ: Lawrence Erlbaum.

Plateau, J.A.F. 1872. Sur la mesure des sensations physiques, et sur la loi que lie l'intensité de ces sensations à l'intensité de la cause excitante. *Bulletins de l'Académie Royal des Sciences, des Lettres, et des Beaux-Arts de Belgique*, 33, 376–388.

Plomp, R. 2002. *The Intelligent Ear: On the Nature of Sound Perception*. Mahwah, NJ: Lawrence Erlbaum.

Pockett, S. 2006. The neuroscience of movement. In S. Pockett, W.P. Banks & S. Gallagher (Eds.), *Does Consciousness Cause Behavior?* (pp. 9–24). Cambridge, MA: MIT Press.

Pohlman, L.D. & Sorkin, R.D. 1976. Simultaneous three-channel signal detection: Performance and criterion as a function of order of report. *Perception & Psychophysics*, 20, 179–186.

Poldrack, R.A. 2010. Subtraction and beyond: The logic of experimental designs for neuroimaging. In. S. J. Hanson, M. Bunzl, S. J. Hanson, & M. Bunzl (Eds.), *Foundational Issues in Human Brain Mapping* (pp. 147–159). Cambridge, MA: MIT Press.

Pollack, I. 1952. The information of elementary and auditory displays. *Journal of the Acoustical Society of America*, 24, 745–749.

Pollack, S.R. 1990. Tech wrecks. *Detroit Free Press*, September 1, 1990, 1C.

Polson, P.G. 1988. The consequences of consistent and inconsistent interfaces. In R. Guindon (Ed.), *Cognitive Science and its Applications for Human–Computer Interaction* (pp. 59–108). Hillsdale, NJ: Lawrence Erlbaum.

Pomerantz, J.R. 1981. Perceptual organization in information processing. In M. Kubovy & J.R. Pomerantz (Eds.), *Perceptual Organization* (pp. 141–180). Hillsdale, NJ: Lawrence Erlbaum.

Ponds, R.W.H.N., Brouwer, W.B. & van Wolffelaar, P.C. 1988. Age differences in divided attention in a simulated driving task. *Journal of Gerontology*, 43, P151–P156.

Pongratz, H., Vaic, H., Reinecke, M., Ercoline, W. & Cohen, D. 1999. Outside-in vs. inside-out: Flight problems caused by different flight attitude indicators. *Safe Journal*, 29, 7–11.

Poock, G.K. 1969. Color coding effects in compatible and noncompatible display-control arrangements. *Journal of Applied Psychology*, 53, 301–303.

Poon, J.M.L. 2004. Effects of performance appraisal politics on job satisfaction and turnover intention. *Personnel Review*, 33, 322–334.

Poon, L.W., Walsh-Sweeney, L. & Fozard, J.L. 1980. Memory skill training for the elderly: Salient issues on the use of imagery mnemonics. In L.W. Poon, J.L. Fozard, L.S. Cermak, D. Arenberg & L.W. Thompson (Eds.), *New Directions in Memory and Aging*. Hillsdale, NJ: Lawrence Erlbaum.

Poredos, P. 2016. Raynaud's Syndrome: A neglected disease. *International Angiology: A Journal of the International Union of Angiology*, 35, 117–121.

Porta, M. 2007. Human–Computer input and output techniques: An analysis of current research and promising applications. *Artificial Intelligence Review*, 28, 197–226.

Posner, M.I. 1986. Overview. In K.R. Boff, L. Kaufman & J.P. Thomas (Eds.), *Handbook of Perception and Human Performance* (Vol. II, pp. V-3–V-10). New York: Wiley.

Posner, M.I. & Boies, S.J. 1971. Components of attention. *Psychological Review*, 78, 391–408.

Posner, M.I., Klein, R., Summers, J. & Buggie, S.C. 1973. On the selection of signals. *Memory & Cognition*, 1, 2–12.

Posner, M.I., Nissen, M.J. & Ogden, W.C. 1978. Attended and unattended processing modes: The role of set for spatial location. In H.L. Pick & I.J. Saltzman (Eds.), *Modes of Perceiving and Processing Information*. Hillsdale, NJ: Lawrence Erlbaum.

Post, T.R. & Brennan, M.L. 1976. An experimental study of the effectiveness of a formal vs. an informal presentation of a general heuristic process on problem solving in tenth grade geometry. *Journal for Research in Mathematics Education*, 7, 59–64.

Postman, L. & Underwood, B.J. 1973. Critical issues in interference theory. *Memory & Cognition*, 1, 19–40.

Poulton, E.C. 1957. On prediction in skilled movements. *Psychological Bulletin*, 54, 467–478.

Poulton, E.C. 1974. *Tracking Skill and Manual Control*. San Diego, CA: Academic Press.

Povel, D.-J. & Collard, R. 1982. Structural factors in patterned finger tapping. *Acta Psychologica*, 52, 107–123.

Preczewski, S.C. & Fisher, D.L. 1990. The selection of alphanumeric code sequences. *Proceedings of the Human Factors Society 34th Annual Meeting* (pp. 224–228). Santa Monica, CA: Human Factors Society.

Preece, A.D. 1990. DISPLAN: Designing a usable medical expert system. In D. Berry & A. Hart (Eds.), *Expert Systems: Human Issues* (pp. 25–47). Cambridge, MA: MIT Press.

Prentzas, J., & Hatzilygeroudis, I. 2016. Assessment of life insurance applications: An approach integrating neuro-symbolic rule-based with case-based reasoning. *Expert Systems: International Journal of Knowledge Engineering and Neural Networks*, 33, 145–160.

Prinzmetal, W. & Banks, W.P. 1977. Good continuation affects visual detection. *Perception & Psychophysics*, 21, 389–395.

Proctor, R.W. 2011. Playing the Simon game: Use of the Simon task for investigating human information processing. *Acta Psychologica*, 136, 182–188.

Proctor, R.W. & Capaldi, E.J. 2001. Improving the science education of psychology students: Better teaching of methodology. *Teaching of Psychology*, 28, 173–181.

Proctor, R.W. & Capaldi, E.J. 2006. *Why Science Matters: Understanding the Methods of Psychological Research*. Malden, MA: Blackwell.

Proctor, R.W. & Cho, Y.S. 2006. Polarity correspondence: A general principle for performance of speeded binary classification tasks. *Psychological Bulletin*, 132, 416–442.

Proctor, R.W. & Lu, C.-H. 1999. Processing irrelevant location information: Practice and transfer effects in choice-reaction tasks. *Memory & Cognition*, 27, 63–77.

Proctor, R.W. & Proctor, J.D. 2012. Sensation and perception. In G. Salvendy (Ed.), *Handbook of Human Factors and Ergonomics* (4th ed., pp. 59–94). Hoboken, NJ: Wiley.

Proctor, R.W., & Schneider, D.W. 2018. Hick's law for choice reaction time: A review. *Quarterly Journal of Experimental Psychology*, 71.

Proctor, R.W. & Vu, K.-P.L. 2006a. Laboratory studies of training, skill acquisition, and retention of performance. In K.A. Ericsson, N. Charness, P.J. Feltovich & R.R. Hoffman (Eds.), *Cambridge Handbook of Expertise and Expert Performance* (pp. 265–286). Cambridge: Cambridge University Press.

Proctor, R.W. & Vu, K.-P.L. 2006b. Location and arrangement of displays and control actuators. In W. Karwowski (Ed.), *Handbook of Standards and Guidelines in Ergonomics and Human Factors* (pp. 309–409). Mahwah, NJ: Lawrence Erlbaum.

Proctor, R.W. & Vu, K.-P.L. 2006c. *Stimulus–Response Compatibility Principles: Data, Theory, and Application*. Boca Raton, FL: CRC Press.

Proctor, R.W. & Vu, K.-P.L. 2006d. The cognitive revolution at age 50: Has the promise of the human information-processing approach been fulfilled? *International Journal of Human–Computer Interaction*, 21, 253–284.

Proctor, R.W., & Vu, K.-P.L. 2010. Cumulative knowledge and progress in human factors. *Annual Review of Psychology*, 61, 623–651.

Proctor, R.W., Vu, K.-P.L., & Schultz, E. 2009. Human factors in information security and privacy. In J. N. D. Gupta & S. K. Sharma (Eds.), *Handbook of Research on Information Security and Assurance* (pp. 402–414). Hershey, PA: Information Science Reference.

Proctor, R.W., & Vu, K.-P.L. 2011. Complementary contributions of basic and applied research in human factors and ergonomics. *Theoretical Issues in Ergonomics Science*, 12, 427–434.

Proctor, R.W., & Vu, K.-P.L. 2012. Human information processing: An overview for human-computer interaction. In J. Jacko (Ed.), *The Human-Computer Interaction Handbook* (3rd ed., pp. 21–40). Boca Raton, FL: CRC Press.

Proctor, R. W., & Vu, K. L. 2016. Principles for designing interfaces compatible with human information processing. *International Journal of Human-Computer Interaction*, 32, 2–22.

Proctor, R.W. & Wang, H. 1997. Differentiating types of set-level compatibility. In B. Hommel & W. Prinz (Eds.), *Theoretical Issues in Stimulus–Response Compatibility* (pp. 11–37). Amsterdam: North-Holland.

Proctor, R.W., Wang, H. & Vu, K.-P.L. 2002. Influences of conceptual, physical, and structural similarity on stimulus–response compatibility. *Quarterly Journal of Experimental Psychology*, 55A, 59–74.

Proctor, R.W., Vu, K.-P.L., Najjar, L.J., Vaughan, M.W. & Salvendy, G. 2003. Content preparation and management for e-commerce Web sites. *Communications of the ACM*, 46(12), 289–299.

Proctor, R.W., Wang, D.-Y.D. & Pick, D.F. 2004. Stimulus–response compatibility with wheel-rotation responses: Will an incompatible response coding be used when a compatible coding is possible? *Psychonomic Bulletin & Review*, 11, 811–847.

Proctor, R.W., Young, J.P., Fanjoy, R.O., Feyen, R.G., Hartman, N.W. & Hiremath, V.V. 2005. Simulating glass cockpit displays in a general aviation flight environment. *Proceedings of the 13th International Symposium on Aviation Psychology* (pp. 481–484). Oklahoma City, OK.

Proctor, R.W., & Xiong, A. 2015. Polarity correspondence as a general compatibility principle. *Current Directions in Psychological Science*, 24, 446–451.

Proffitt, D.R., & Caudek, C. 2013. Depth perception and the perception of events. In A.F. Healy & R.W. Proctor (Eds.), *Experimental Psychology* (pp. 212–235). Volume 4 in I.B. Weiner (Editor-in-Chief), *Handbook of Psychology* (2nd. ed.). Hoboken, NJ: John Wiley.

Propst, R.L. 1966. The action office. *Human Factors*, 8, 299–306.

Puel, J.-L., Ruel, J., Guitton, M. & Pujol, R. 2002. The inner hair cell afferent/efferent synapses revisited: A basis for new therapeutic strategies. In D. Felix & E. Oestreicher E (Eds.), *Rational Pharmacotherapy of the Inner Ear. Advances in Otorhinolaryngology* (Vol. 59, pp. 124–130). Basel, Switzerland: Karger.

Pulaski, P.D., Zee, D.S. & Robinson, D.A. 1981. The behavior of the vestibulo-ocular reflex at high velocities of head rotation. *Brain Research*, 222, 159–165.

Pynnönen, M., Ritala, P., & Hallikas, J. 2011. The new meaning of customer value: A systemic perspective. *Journal of Business Strategy*, 32, 51–57.

Radwin, R.G., Vanderheiden, G.C. & Li, M.-L. 1990. A method for evaluating head-controlled computer input devices using Fitts' law. *Human Factors*, 32, 423–438.

Rahman, M.M., Sprigle, S. & Sharit, J. 1998. Guidelines for force-travel combinations of push button switches for older populations. *Applied Ergonomics*, 29, 93–100.

Rakubutu, T., Gelderblom, H., & Cohen, J. 2014, September. Participatory design of touch gestures for informational search on a tablet device. In *SAICSIT 2014: Proceedings of the Southern African Institute for Computer Scientist and Information Technologists Annual Conference* 2014 (pp. 276–285). New York: ACM.

Ramachandran, V.S. 1988. Perception of shape from shading. *Nature*, 33, 163–166.

Ramachandran, V.S. 1992. Filling in gaps in perception: I. *Current Directions in Psychological Science*, 1, 199–205.

Ramkumar, A., Stappers, P. J., Niessen, W. J., Adebahr, S., Schimek-Jasch, T., Nestle, U., & Song, Y. 2016. Using GOMS and NASA-TLX to Evaluate Human-Computer Interaction Process in Interactive Segmentation. *International Journal of Human–Computer Interaction*, 33, 123–134.

Ramsey, C.L. & Schultz, A.C. 1989. Knowledge representation for expert systems development. In J. Liebowitz & D.A. De Salvo (Eds.), *Structuring Expert Systems: Domain, Design, and Development* (pp. 273–301). Englewood Cliffs, NJ: Yourdon Press.

Ramsey, J.D. 1995. Task performance in heat: A review. *Ergonomics*, 38, 154–165.

Randle, R. 1988. Visual accommodation: Mediated control and performance. In D.J. Oborne (Ed.), *International Reviews of Ergonomics* (Vol. 2, pp. 207–232).

Ranney, T.A., Simmons, L.A. & Masalonis, A.J. 2000. The immediate effects of glare and electrochromatic glare-reducing mirrors in simulated truck driving. *Human Factors*, 42, 337–347.

Ranzini, M., Lisi, M., & Zorzi, M. 2016. Voluntary eye movements direct attention on the mental number space. *Psychological Research*, 80, 389–398.

Rash, C.E., Verona, R.W. & Crowley, J.S. 1990. Night flight using thermal imaging systems. *The Visual Performance Group Technical Newsletter*, 12(3), 1–7.

Rasmussen, J. 1982. Human errors: A taxonomy for describing human malfunction in industrial installations. *Journal of Occupational Accidents*, 4, 311–333.

Rasmussen, J. 1983. Skills, rules, and knowledge; signals, signs, and symbols, and other distinctions in human performance models. *IEEE Transactions on Systems, Man, and Cybernetics, SMC-13*, 257–266.

Rasmussen, J. 1985. The role of hierarchical knowledge representation in decision making and system management. *IEEE Transactions on Systems, Man, and Cybernetics, SMC-15*, 234–243.

Rasmussen, J. 1986. *Information Processing and Human–Machine Interaction: An Approach to Cognitive Engineering*. Amsterdam: North-Holland.

Rasmussen, J. 1987. Cognitive control and human error mechanisms. In J. Rasmussen, K. Duncan & J. Leplat (Eds.), *New Technology and Human Error* (pp. 53–61). New York: Wiley.

Ratcliff, R., Smith, P. L., Brown, S. D., & McKoon, G. (2016). Diffusion decision model: Current issues and history. *Trends in cognitive sciences*, 20, 260–281.

Raugh, M.R. & Atkinson, R.C. 1975. A mnemonic method for learning a second-language vocabulary. *Journal of Educational Psychology*, 67, 1–16.

Rautaray, S.S., & Agrawal, A. 2015. Vision based hand gesture recognition for human computer interaction: A survey. *Artificial Intelligence Review*, 43, 1–54.

Raymond, M.W. & Moser, R. 1995. Aviators at risk. *Aviation, Space, and Environmental Medicine*, 66, 35–39.

Reason, J. 1987. The psychology of mistakes: A brief review of planning failures. In J. Rasmussen, K. Duncan & J. Leplat (Eds.), *New Technology and Human Error* (pp. 45–52). New York: Wiley.

Reason, J. 1990. *Human Error*. Cambridge: Cambridge University Press.

Reason, J. 2013. *A Life in Error: From Little Slips to Big Disasters*. Burlington, VT: Ashgate.

Rebenitsch, L., & Owen, C. 2016. Review on cybersickness in applications and visual displays. *Virtual Reality*, 20, 101–125.

Reed, C.M., Rabinowitz, W.M., Durlach, N.I., Braida, L.D., Conway-Fithian, S. & Schultz, M.C. 1985. Research on the Tadoma method in speech communication. *Journal of the Acoustical Society of America*, 77, 247–257.

Reed, C.M., Durlach, N.I., Delhorne, L.A., Rabinowitz, W.M. & Grant, K.W. 1989, September. Research on tactual communication of speech: Ideas and findings. *Volta Review*, 91(5), 65–78.

Reeve, T.G. & Proctor, R.W. 1984. On the advance preparation of discrete finger responses. *Journal of Experimental Psychology: Human Perception and Performance*, 10, 541–553.

Reeve, T.G., Dornier, L. & Weeks, D.J. 1990. Precision of knowledge of results: Consideration of the accuracy requirements imposed by the task. *Research Quarterly for Exercise and Sport*, 61, 284–291.

Refinetti, R. 2016. *Circadian physiology*. Boca Raton, FL: CRC Press.

Reid, G.B., Shingledecker, C.A. & Eggemeier, F.T. 1981. Application of conjoint measurement to workload scale development. *Proceedings of the Human Factors Society 25th Annual Meeting* (pp. 522–526). Rochester, NY: Human Factors Society.

Reinach, S. & Viale, A. 2006. Application of a human error framework to conduct train accident/incident investigations. *Accident Analysis and Prevention*, 38, 396–406.

Reiner, M., & Hecht, D. 2009. Behavioral indications of object-presence in haptic virtual environments. *Cyberpsychology & Behavior*, 12, 183–186.

Reingold, E.M., Loschky, L.C., McConkie, G.W. & Stampe, D.M. 2003. Gaze-contingent multiresolutional displays: An integrative review. *Human Factors*, 45, 307–328.

Remington, L.A. 2012. *Clinical Anatomy and Physiology of the Visual System* (3rd ed.). St. Louis, MO: Butterworth-Heinemann.

Rendon-Velez, E., van Leeuwen, P. , Happee, R., Horváth, I., van der Vegte, W.F., & de Winter, J.F. 2016. The effects of time pressure on driver performance and physiological activity: A driving simulator study. *Transportation Research Part F: Traffic Psychology and Behaviour*, 41(Part A), 150–169.

Rensink, H.J.T. & van Uden, M.E.J. 2006. The development of a human factors engineering strategy in petrochemical engineering and projects. In W. Karwowski (Ed.), *International Encyclopedia of Ergonomics and Human Factors* (2nd ed., Vol. 3, pp. 2577–2583). Boca Raton, FL: CRC Press.

Rensink, R. A. 2014. Limits to the usability of iconic memory. *Frontiers in Psychology*, 5.

Rensink, R.A., O'Regan, J.K. & Clark, J.J. 2000. On the failure to detect changes in scenes across brief interruptions. *Visual Cognition*, 7, 127–145.

Reuter, T. 2011. Fifty years of dark adaptation 1961–2011. *Vision Research*, 51, 2243–2262.

Reynolds, D.D. & Angevine, E.N. 1977. Hand-arm vibration. Part II: Vibration transmission and characteristics of the hand and arm. *Journal of Sound and Vibration*, 51, 255–265.

Reynolds, L.A. & Tansey, E.M. (Eds.) 2003. *The MRC Applied Psychology Unit* (Vol. 16: *Wellcome Witnesses to Twentieth Century Medicine*). London: The Wellcome Trust.

Rice, L.M. 2005. Improving lower beam visibility range. In *Lighting Technology and Human Factors* (SP-1932, pp. 17–25). Warrendale, PA: Society of Automotive Engineers.

Richardson, L. 2003. Fire alarm evacuation—Are you ready? *NSFP Journal*, Online exclusive, September%20 03. Downloaded on September 20, 2006, from http://www.nfpa.org/categoryList.asp?categoryID =717&URL=Publications/NFPA%20Journal®/September%20/%20October%202003/Online%20 Exclusives&cookie%5Ftest=1.

Richeson, A., Bertus, E., Bias, R.G., & Tate, J. 2011. Determining the value of usability in Web design. In K.-P.L. Vu & R.W. Proctor (Eds.), *Handbook of Human Factors in Web Design* (2nd ed., pp. 753–764). Boca Raton, FL: CRC Press.

Rieman, J. 1996. A field study of exploratory learning strategies. *ACM Transactions on Computer–Human Interaction*, 3(3), 189–218.

Riggio, L., Gawryszewski, L.G. & Umiltà, C. 1986. What is crossed in crossed-hand effects? *Acta Psychologica*, 62, 89–100.

Rips, L.J. 1989. Similarity, typicality, and categorization. In S. Voisniadou & A. Ortony (Eds.), *Similarity, Analogy, and Thought*. New York: Cambridge University Press.

Rips, L.J. 2002. Reasoning. In H. Pashler & D. Medin (Eds.), *Stevens' Handbook of Experimental Psychology* (3rd ed., Vol. 2: *Memory and Cognitive Processes*, pp. 363–411). Hoboken, NJ: Wiley.

Rips, L.J. & Marcus, S.L. 1977. Supposition and the analysis of conditional sentences. In M.A. Just & P.A. Carpenter (Eds.), *Cognitive Processes in Comprehension*. Hillsdale, NJ: Lawrence Erlbaum.

Riva, G., & Wiederhold, B.K. 2015. The new dawn of virtual reality in health care: Medical simulation and experiential interface. *Annual Review of Cybertherapy and Telemedicine*, 1, 33–36.

Rizzolatti, G, Bertoloni, G. & Buchtel, H.A. 1979. Interference of concomitant motor and verbal tasks on simple reaction time: A hemispheric difference. *Neuropsychologia*, 17, 323–330.

Robbins, S.R. 1983. *Organizational Theory: The Structure and Design of Organizations*. Englewood Cliffs, NJ: Prentice-Hall.

Roberts, J.W., Bennett, S.J., Elliott, D., & Hayes, S.J. 2015. Motion trajectory information and agency influence motor learning during observational practice. *Acta Psychologica*, 159, 76–84.

Robertson, M.M. 2006. Analysis of office systems. In W. Karwowski (Ed.), *International Encyclopedia of Ergonomics and Human Factors* (Vol. 2, pp. 1528–1535). Boca Raton, FL: CRC Press.

Robinette, K.M. & Daanen, H. 2003. Lessons learned from CAESAR: A 3-D anthropometric survey. In *International Ergonomics Association 2003 Conference Proceedings*.

Robinette, K.M. & Hudson, J.A. 2006. Anthropometry. In G. Salvendy (Ed.), *Handbook of Human Factors and Ergonomics* (3rd ed., pp. 322–339). Hoboken, NJ: Wiley.

Robinson, C.P. & Eberts, R.E. 1987. Comparison of speech and pictorial displays in a cockpit environment. *Human Factors*, 29, 31–44.

Robinson, J.O. 1998. *The Psychology of Visual Illusion*. Mineola, NY: Dover.

Rock, I. & Palmer, S. 1990. The legacy of Gestalt psychology. *Scientific American*, 263(6), 84–90.

Rockway, M. & Franks, P. 1959. Effects of variations in control backlash and gain on tracking performance (Report No. 58-553). Wright Air Development Center: U.S. Air Force.

Rockway, M.R. 1957. Effects of variations in control deadspace and gain on tracking performance. (AF WADC TR 57-326). Wright Air Development Center: U.S. Air Force.

Roebuck, J.A., Jr. 1995. *Anthropometric Methods: Designing to Fit the Human Body*. Santa Monica, CA: Human Factors and Ergonomics Society.

Rogers, J. 1970. Discrete tracking performance with limited velocity resolution. *Human Factors*, 12, 331–339.

Rogers, K.F. 1987. Ergonomics today: Interviews with Julien Christensen, Harry Davis, Kate Ehrlich, Karl Kroemer, Rani Lueder, and Gavriel Salvendy. In K.H. Pelsma (Ed.), *Ergonomics Sourcebook: A Guide to Human Factors Information*. Lawrence, KS: Ergosyst Associates.

Rogers, R.O., Boquet, A., Howell, C., & DeJohn, C. 2010. A two-group experiment to measure simulator-based upset recovery training transfer. *International Journal of Applied Aviation Studies*, 10, 153–168.

Rogers, S.P., Asbury, C.N. & Haworth, L.A. 2001. Evaluations of new symbology for wide-field-of-view HMDs. *Proceedings of SPIE—The International Society for Optical Engineering*, 4361, 213–224.

Rojas, R. (Ed.) 2001. *Encyclopedia of Computers and Computer History*. Chicago, IL: Fitzroy Dearborn.

Roscoe, S.N. 1987. The trouble with HUDS and HMDS. *Human Factors Society Bulletin*, 7(10), 1–3.

Roscoe, S.N., Eisele, J.E. & Bergman, C.A. 1980. Information and control requirements. In S.N. Roscoe (Ed.), *Aviation Psychology* (pp. 33–38). Ames, IA: Iowa State University Press.

Rose, N.S., Buchsbaum, B.R., & Craik, F.M. 2014. Short-term retention of a single word relies on retrieval from long-term memory when both rehearsal and refreshing are disrupted. *Memory & Cognition*, 42, 689–700.

Rosenbaum, D.A. 2002. Motor control. In H. Pashler & S. Yantis (Eds.), *Stevens' Handbook of Experimental Psychology*, Vol. 1: *Sensation and Perception* (pp. 315–339). New York: Wiley.

Rosenbaum, D.A. 2005. The Cinderella of psychology: The neglect of motor control in the science of mental life and behavior. *American Psychologist*, 60, 308–317.

Rosenbaum, D.A. 2010. *Human motor control* (2nd ed.). San Diego, CA: Academic Press.

Rosenbaum, D.A., Augustyn, J.S., Cohen, R.G. & Jax, S.A. 2006b. Perceptual-motor expertise. In K.A. Ericsson, N. Charness, P. Feltovich & R.R. Hoffman (Eds.), *The Cambridge Handbook of Expertise and Expert Performance* (pp. 505–520). New York: Cambridge University Press.

Rosenbaum, D.A., Cohen, R.G., Meulenbroek, R.G.J. & Vaughan, J. 2006a. Plans for grasping objects. In M. Latash & F. Lestienne (Eds.), *Motor Control and Learning*. New York: Springer.

Rosenbaum, D.A., Marchak, F., Barnes, H.J., Vaughan, J., Slotta, J.D. & Jorgensen, M.J. 1990. Constraints for action selection: Overhand versus underhand grips. In M. Jeannerod (Ed.), *Attention and Performance XIII* (pp. 321–342). Hillsdale, NJ: Lawrence Erlbaum.

Rosenbaum, D.A., Meulenbroek, R.J., Vaughan, J. & Jansen, C. 2001. Posture-based motion planning: Applications to grasping. *Psychological Review*, 108, 709–734.

Rosenbloom, P.S. 1986. The chunking of goal hierarchies: A model of stimulus–response compatibility and practice. In J. Laird, J., P. Rosenbloom & A. Newell (Eds.), *Universal Subgoaling and Chunking: The Automatic Generation and Learning of Goal Hierarchies* (pp. 133–282). Boston, MA: Kluwer Academic.

Rosenbloom, P.S. & Newell, A. 1987. An integrated computational model of stimulus–response compatibility and practice. In G.H. Bower (Ed.), *The Psychology of Learning and Motivation* (Vol. 21, pp. 1–52). New York: Academic Press.

Ross, K.G., Lussier, J.W. & Klein, G. 2005. From the recognition primed decision model to training. In T. Betsch & S. Haberstroh (Eds.), *The Routines of Decision Making* (pp. 327–341). Mahwah, NJ: Lawrence Erlbaum.

Ross, K.G., Shafer, J.L. & Klein, G. 2007. Professional judgments and "naturalistic decision making." In K.A. Ericsson, N. Charness, P.J. Feltovich & R.R. Hoffman (Eds.), *Cambridge Handbook of Expertise and Expert Performance* (pp. 403–420). Cambridge: Cambridge University Press.

Ross, S. & Aines, A.A. 1960. Human engineering—1911 style. *Human Factors*, 2, 169–170.

Rosson, M.B. & Mellon, N.M. 1985. Behavioral issues in speech-based remote information retrieval. In L. Lerman (Ed.), *Proceedings of the Voice I/O Systems Applications Conference '85*. San Francisco, CA: AVIOS.

Roswarski, T.E. & Proctor, R.W. 2000. Auditory stimulus–response compatibility: Is there a contribution of stimulus-hand correspondence? *Psychological Research*, 63, 148–158.

Rothstein, A.L. & Arnold, R.K. 1976. Bridging the gap: Application of research on videotape feedback and bowling. *Motor Skills: Theory into Practice*, 1, 35–62.

Rothwell, J.C., Traub, M.M., Day, B.L., Obeso, J.A., Thomas, P.K. & Marsden, D. 1982. Manual motor performance in a deafferented man. *Brain*, 105, 515–542.

Rothwell, W.J. 1999. On-the-job training. In D.G. Landon, K.S. Whiteside & M.M. McKenna (Eds.), *Intervention Resource Guide: 50 Performance Improvement Tools*. San Francisco, CA: Jossey-Bass.

Rouse, W.B. 1979. Problem solving performance of maintenance trainees in a fault diagnosis task. *Human Factors*, 21, 195–203.

Rouse, W.B., & Boff, K.R. 2012. Cost/benefit analysis for human systems investments: Predicting and trading off economic and noneconomic impacts of human factors and ergonomics. In G. Salvendy (Ed.), *Handbook of Human Factors and Ergonomics* (4th ed., pp. 1122–1138). Hoboken, NJ: John Wiley.

Rouse, W.B. & Cody, W.J. 1988. On the design of man–machine systems: Principles, practices and prospects. *Automatica*, 24, 227–238.

Rouse, W.B. & Cody, W.J. 1989. Designers' criteria for choosing human performance models. In G.R., McMillan, D. Beevis, E. Salas, M.H. Strub, R. Sutton & L. Van Breda (Eds.), *Applications of Human Performance Models to System Design* (pp. 7–14). New York: Plenum Press.

Rouse, W.B., Cody, W.J. & Boff, K.R. 1991. The human factors of system design: Understanding and enhancing the role of human factors engineering. *International Journal of Human Factors in Manufacturing*, 1, 87–104.

Rubinstein, M.F. 1986. *Tools for Thinking and Problem Solving*. Englewood Cliffs, NJ: Prentice-Hall.

Rugg, M.D. & Coles, M.G.H. (Eds.) 1995. *Electrophysiology of Mind: Event-Related Brain Potentials and Cognition*. Oxford: Oxford University Press.

Rühmann, H.-P. 1984. Basic data for the design of consoles. In H. Schmidtke (Ed.), *Ergonomic Data for Equipment Design* (pp. 15–144). New York: Plenum Press.

Rumelhart, D.E. & McClelland, J.L. 1986. *Parallel Distributed Processing: Explorations in the Microstructure of Cognition* (Vol. 1: *Foundations* and Vol. 2: *Psychological and Biological Models*). Cambridge, MA: MIT Press.

Rumelhart, D.E. & Norman, D.A. 1982. Simulating a skilled typist: A study of skilled cognitive-motor performance. *Cognitive Science*, 6, 1–36.

Rumelhart, D.E. & Norman, D.A. 1988. Representation in memory. In R.C. Atkinson, R.J. Herrnstein, G. Lindzey & R.D. Luce (Eds.), *Stevens Handbook of Experimental Psychology* (2nd ed., pp. 511–587). New York: Wiley.

Rundell, O.H. & Williams, H.L. 1979. Alcohol and speed–accuracy tradeoff. *Human Factors*, 21, 433–443.

Runeson, R. & Norbäck, D. 2005. Associations among sick building syndrome, psychosocial factors, and personality traits. *Perceptual and Motor Skills*, 100, 747–759.

Ruotolo, F., Senese, V. P., Ruggiero, G., Maffei, L., Masullo, M., & Iachini, T. 2012. Individual reactions to a multisensory immersive virtual environment: The impact of a wind farm on individuals. *Cognitive Processing*, 13(Suppl 1), S319–S323.

Russo, J.E. 1977. The value of unit price information. *Journal of Marketing Research*, 14, 193–201.

Russo, J.E., Johnson, E.J. & Stephens, D.L. 1989. The validity of verbal protocols. *Memory & Cognition*, 17, 759–769.

Rutherford, M.D & Brainard, D.H. 2002. Lightness constancy: A direct test of the illumination-estimation hypothesis. *Psychological Science*, 13, 142–149.

Ruthruff, E., Johnston, J.C. & Van Selst, M. 2001. Why practice reduces dual-task interference. *Journal of Experimental Psychology: Human Perception and Performance*, 27, 3–21.

Ryan, N.S., Rossor, M.N., & Fox, N.C. 2015. Alzheimer's disease in the 100 years since Alzheimer's death. *Brain: A Journal of Neurology*, 138, 3816–3821.

Ryan, S.J., Schachat, A.P., Wilkinson, C.P., Hinton, D.R., Sadda, S.R., & Wiedemann, P. 2013. *Retina* (5th ed., 3 Vols.). Philadelphia: W. B. Saunders.

Sackett, G.P., Ruppenthal, G.C. & Gluck, J. 1978. Introduction: An overview of methodological and statistical problems in observational research. In G.P. Sackett (Ed.), *Observing Behavior*, Vol. II: *Data Collection and Analysis Methods* (pp. 1–14). Baltimore, MD: University Park Press.

Sackett, P.R. 2000. Personnel selection: Techniques and instruments. In A.E. Kazdin (Ed.), *Encyclopedia of Psychology* (Vol. 6, pp. 152–156). Washington, DC: American Psychological Association.

Sackett, R.S. 1934. The influences of symbolic rehearsal upon the retention of a maze habit. *Journal of General Psychology*, 10, 376–395.

Sagan, C. 1990. Why we need to understand science. *Skeptical Inquirer*, 14, 263–269.

Salas, E., Dickinson, T.L., Converse, S.A. & Tannenbaum, S.I. 1992. Toward an understanding of team performance and training. In R.W. Swezey & E. Salas (Eds.), *Teams: Their Training and Performance* (pp. 3–29). Norwood, NJ: Ablex.

Salas, E., Fiore, S.M., & Letsky, M.P. 2012. *Theories of Team Cognition: Cross-Disciplinary Perspectives*. New York: Routledge.

Salas, E., Wilson, K.A., Priest, H.A. & Guthrie, J.W. 2006. Design, delivery, and evaluation of training systems. In G. Salvendy (Ed.), *Handbook of Human Factors and Ergonomics* (3rd ed., pp. 472–512). Hoboken, NJ: Wiley.

Salmoni, A.W., Schmidt, R.A. & Walter, C.B. 1984. Knowledge of results and motor learning: A review and critical reappraisal. *Psychological Bulletin*, 95, 355–386.

Salthouse, T.A. 1984. Effects of age and skill in typing. *Journal of Experimental Psychology: General*, 113, 345–371.

Salthouse, T.A. 1986. Perceptual, cognitive, and motoric aspects of transcription typing. *Psychological Bulletin*, 99, 303–319.

Salvendy, G. (Ed.) 2012. *Handbook of Human Factors and Ergonomics* (4th ed.). Hoboken, NJ: Wiley.

Salvucci, D.D. & Lee, F.J. 2003. Simple cognitive modeling in a complex cognitive architecture. *CHI Letters*, 5, 265–272.

Sanchez, J. I., & Levine, E. L. 2012. The rise and fall of job analysis and the future of work analysis. *Annual Review of Psychology*, 63, 397–425.

Sandal, G.M. 2001. Crew tension during a space station simulation. *Environment and Behavior*, 33, 134–150.

Sanders, A.F. 1998. *Elements of Human Performance*. Mahwah, NJ: Lawrence Erlbaum.

Sanders, A.F. & Lamers, J.M. 2002. The Eriksen flanker effect revisited. *Acta Psychologica*, 109, 41–56.

Sanders, M.S. & McCormick, E.J. 1987. *Human Factors in Engineering and Design* (5th ed.). New York: McGraw-Hill.

Sanders, M.S. & McCormick, E.J. 1993. *Human Factors in Engineering and Design* (7th ed.). New York: McGraw-Hill.

Sanderson, P.M. 2003. Cognitive work analysis. In J.M. Carroll (Ed.), *HCI Models, Theories, and Frameworks: Toward a Multidisciplinary Science* (pp. 225–264). San Francisco, CA: Morgan Kaufman.

Sanli, E.A., Patterson, J.T., Bray, S.R., & Lee, T.D. 2013. Understanding self-controlled motor learning protocols through the self-determination theory. *Frontiers in Psychology*, 3, 611. doi:10.3389/fpsyg.2012.00611.

Saunders, J.A. & Knill, D.C. 2004. Visual feedback control of hand movements. *Journal of Neuroscience*, 24, 3223–3234.

Sawchuk, M., Linville, M., Cornish, N., Pollock, A., Lubin, I., Gagnon, M., & Stinn, J. 2014. *The Essential Role of Laboratory Professionals: Ensuring the Safety and Effectiveness of Laboratory Data in Electronic Health Record Systems*. Atlanta, GA: Center for Surveillance, Epidemiology and Laboratory Services, Centers for Disease Control and Prevention; May 2014. Retrieved from http://www.cdc.gov/labhit/paper/Laboratory_Data_in_EHRs_2014.pdf

Sayer, J.R. & Mefford, M.L. 2004. High visibility safety apparel and night time conspicuity of pedestrians in work zones. *Journal of Safety Research*, 35, 537–546.

Sayers, B.A. (Ed.) 1988. *Human Factors and Decision Making*. New York: Elsevier Applied Science.

Schall, J.D. & Thompson, K.G. 1999. Neural selection and control of visually guided eye movements. *Annual Review of Neuroscience*, 22, 241–259.

Scharf, B., Quigley, S., Aoki, C., Peachey, N. & Reeves, A. 1987. Focused auditory attention and frequency selectivity. *Perception & Psychophysics*, 42, 215–223.

Schendel, J.D. & Hagman, J.D. 1982. On sustaining procedural skills over a prolonged retention interval. *Journal of Applied Psychology*, 67, 605–610.

Schilling, R.F. & Weaver, G.E. 1983. Effect of extraneous verbal information on memory for telephone numbers. *Journal of Applied Psychology*, 68, 559–564.

Schlauch, R.S. 2004. Loudness. In J.G. Neuhoff (Ed.), *Ecological Psychoacoustics* (pp. 317–345). San Diego, CA: Elsevier Academic Press.

Schlittenlacher, J., & Ellermeier, W. 2015. Simple reaction time to the onset of time-varying sounds. *Attention, Perception, & Psychophysics*, 77, 2424–2437.

Schmidt, R.A. 1975. A schema theory of discrete motor skill learning. *Psychological Review*, 82, 225–260.

Schmidt, R.A. 1989. Unintended acceleration: A review of human factors contributions. *Human Factors*, 31, 345–364.

Schmidt, R.A. & Bjork, R.A. 1992. New conceptualizations of practice: Common principles in three paradigms suggest new concepts for training. *Psychological Science*, 3, 207–217.

Schmidt, R.A. & Lee, T.D. 2011. *Motor Control and Learning: A Behavioral Emphasis* (5th ed.). Champaign, IL: Human Kinetics.

Schmidt, R.A. & Young, D.E. 1991. Methodology for motor learning: A paradigm for kinematic feedback. *Journal of Motor Behavior*, 23, 13–24.

Schmidt, R.A., Young, D.E., Swinnen, S. & Shapiro, D.C. 1989. Summary knowledge of results for skill acquisition: Support for the guidance hypothesis. *Journal of Experimental Psychology: Learning, Memory and Cognition*, 15, 352–359.

Schmidt, R.A., Heuer, H., Ghodsian, D. & Young, D.E. 1998. Generalized motor programs and units of action in bimanual coordination. In M. Latash (Ed.), *Progress in Motor Control* (Vol. 1, pp. 329–360). Champaign, IL: Human Kinetics.

Schmidt, R.A., Zelaznik, H.N., Hawkins, B., Frank, J.S. & Quinn, J.T. 1979. Motor-output variability: A theory for the accuracy of rapid motor acts. *Psychological Review*, 84, 415–451.

Schmorrow, D.D. (Ed.) 2005. *Foundations of Augmented Cognition*. Mahwah, NJ: Lawrence Erlbaum.

Schmuckler, M.A. 2004. Pitch and pitch structures. In J.G. Neuhoff (Ed.), *Ecological Psychoacoustics* (pp. 271–315). San Diego, CA: Elsevier Academic Press.

Schmuckler, M.A. & Bosman, E.L. 1997. Interkey timing in piano performance and typing. *Canadian Journal of Experimental Psychology*, 51, 99–111.

Schneider, D.W. 2015. Isolating a mediated route for response congruency effects in task switching. *Journal of Experimental Psychology: Learning, Memory, and Cognition*, 41, 235–245.

Schneider, S., Abeln, V., Popova, J., Fomina, E., Jacubowski, A., Meeusen, R., & Strüder, H.K. 2013. The influence of exercise on prefrontal cortex activity and cognitive performance during a simulated space flight to Mars (MARS500). *Behavioural Brain Research*, 236, 1–7.

Schneider, W. & Chein, J.M. 2003. Controlled and automatic processing: Behavior, theory, and biological mechanisms. *Cognitive Science*, 27, 525–559.

Schneider, W. & Fisk, A.D. 1983. Attention theory and mechanisms of skilled performance. In R.A. Magill (Ed.), *Memory and Control of Action* (pp. 119–143). Amsterdam: North-Holland.

Schneider, W. & Shiffrin, R.M. 1977. Controlled and automatic human information processing: I. Detection, search, and attention. *Psychological Review*, 84, 1–66.

Schnell, T., Bentley, K. & Hayes, R.M. 2001. Legibility distances of fluorescent signs and their normal color counterparts. *Transportation Research Record*, 1754, 31–41.

Schoenmarklin, R.W. & Marras, W.S. 1989a. Effects of handle angle and work orientation on hammering: I. Wrist motion and hammering performance. *Human Factors*, 31, 397–412.

Schoenmarklin, R.W. & Marras, W.S. 1989b. Effects of handle angle and work orientation on hammering: II. Muscle fatigue and subjective ratings of body discomfort. *Human Factors*, 31, 413–420.

School, P.J. 2006. Noise at work. In W. Karwowski (Ed.), *International Encyclopedia of Ergonomics and Human Factors* (2nd ed., Vol. 2, pp. 1821–1825). Boca Raton, FL: CRC Press.

Schowengerdt, B.T. & Seibel, E.J. 2004. True three-dimensional displays that allow viewers to dynamically shift accommodation, bringing objects displayed at different viewing distances into and out of focus. *CyberPsychology and Behavior*, 7, 610–620.

Schraagen, J.M., Chipman, S.F. & Shalin, V.L. (Eds.) 2000. *Cognitive Task Analysis*. Mahwah, NJ: Lawrence Erlbaum.

Schultz, E.E. 2012. Human factors and information security. In G. Salvendy (Ed.), *Handbook of Human Factors and Ergonomics* (Vol. 4; pp. 1250–1266). Hoboken, NJ: Wiley.

Schumacher, E.H., Seymour, T.L., Glass, J.M., Fencsik, D.E., Lauber, E.J., Kieras, D.E. & Meyer, D.E. 2001. Virtually perfect time sharing in dual-task performance: Uncorking the central cognitive bottleneck. *Psychological Science*, 12, 101–108.

Schunk, D.W., Bloechle, W.K. & Laughery, R. 2002. Micro Saint modeling and the human element. In E. Yücesan, C.-H. Chen, J.L. Snowden & J.M. Charnes (Eds.), *Proceedings of the 2002 Winter Simulation Conference* (pp. 187–191).

Schutte, J. 2006. AFRL seeks ways to prevent hearing loss in military environments. *Air Force Research Technology Horizons*. Human Effectiveness Directorate, Technical Article HE-H-05–03. Downloaded on August 19, 2006, from http://www.afrlhorizons.com/Briefs/Aug06/HE_H_05_03.html

Schwartz, D., Sparkman, J. & Deese, J. 1970. The process of understanding and judgment of comprehensibility. *Journal of Verbal Learning and Verbal Behavior*, 9, 87–93.

Schwartz, D.R. & Howell, W.C. 1985. Optional stopping performance under graphic and numeric CRT formatting. *Human Factors*, 27, 433–444.

Schwartz, O., Sejnowski, T.J., & Dayan, P. 2009. Perceptual organization in the tilt illusion. *Journal f Vision*, 9(4), 1–20.

Schwartz, S.H. 2010. *Visual Perception: A Clinical Orientation*. New York: McGraw-Hill.

Schweickert, R. 1983. Latent network theory: Scheduling of processes in sentence verification and in the Stroop effect. *Journal of Experimental Psychology: Learning, Memory, and Cognition*, 9, 353–383.

Schweickert, R., Fisher, D. L., & Goldstein, W. M. 2010. Additive factors and stages of mental processes in task networks. *Journal of Mathematical Psychology*, 54, 405–414.

Schweickert, R., Fisher, D.L. & Proctor, R.W. 2003. Steps toward building mathematical and computer models from cognitive task analyses. *Human Factors*, 45, 77–103.

Scialfa, C.T., Garvey, P.M., Gish, K.W., Deering, L.M., Leibowitz, H.W. & Goebel, C.C. 1988. Relationships among measures of static and dynamic visual sensitivity. *Human Factors*, 30, 677–687.

Seagull, F.J. & Gopher, D. 1997. Training head movement in visual scanning: An embedded approach to the development of piloting skills with helmet-mounted displays. *Journal of Experimental Psychology: Applied*, 3, 163–180.

See, J.E., Howe, S.R., Warm, J.S. & Dember, W.N. 1995. Meta-analysis of the sensitivity decrement in vigilance. *Psychological Bulletin*, 117, 230–249.

Seibel, R. 1963. Discrimination reaction time for a 1,023 alternative task. *Journal of Experimental Psychology*, 66, 215–226.

Seidl, A. & Bubb, H. 2006. Standards in anthropometry. In W. Karowski (Ed.), *Handbook of Standards and Guidelines in Ergonomics and Human Factors* (pp. 169–196). Mahwah, NJ: Lawrence Erlbaum.

Seidl, A., Trieb, R., Wirsching, H.J., Smythe, A., & Guenzel, T. 2016. SizeNorthAmerica—The new North American anthropometric survey: Conceptual design, implementation and results. In R. Goonetilleke & W. Karowski, *Advances in Physical Ergonomics and Human Factors* (pp. 457–468). Berlin Heidelberg: Springer International.

Selye, H. 1973. The evolution of the stress concept. *American Scientist*, 61, 692–699.

Senders, J.W. 1964. The human operator as a monitor and controller of multi-degree of freedom systems. *IEEE Transactions on Human Factors and Electronics*, HFE-5, 2–5.

Senders, J.W. & Moray, N.P. 1991. *Human Error: Cause, Prediction, and Reduction*. Hillsdale, NJ: Lawrence Erlbaum.

Seo, N.J., Armstrong, T.J., & Drinkaus, P. 2009. A comparison of two methods of measuring static coefficient of friction at low normal forces: A pilot study. *Ergonomics*, 52, 121–135.

Seow, S.C. 2005. Information theoretic models of HCI: A comparison of the Hick–Hyman law and Fitts' law. *Human–Computer Interaction*, 20, 315–352.

Serber, H. 1990. New developments in the science of seating. *Human Factors Society Bulletin*, 33(2), 1–3.

Serniclaes, W., Ventura, P., Morais, J. & Kolinsky, R. 2005. Categorical perception of speech sounds in illiterate adults. *Cognition*, 98, B35–B44.

Servant, M., White, C., Montagnini, A., & Burle, B. 2015. Using covert response activation to test latent assumptions of formal decision-making models in humans. *The Journal of Neuroscience*, 35, 10371–10385.

Sesto, M.E., Radwin, R.G. & Richard, T.G. 2005. Short-term changes in upper extremity dynamic mechanical response parameters following power hand tool use. *Ergonomics*, 48, 807–820.

Seta, J.J., Paulus, P.B. & Schkade, J.K. 1976. Effects of group size and proximity under cooperative and competitive conditions. *Journal of Personality and Social Psychology*, 34, 47–53.

Sevilla, J.A.M. 2006. Tactile virtual reality: A new method applied to haptic exploration. In M.A. Heller & S. Ballesteros (Eds.), *Touch and Blindness: Psychology and Neuroscience* (pp. 121–136). Mahwah, NJ: Lawrence Erlbaum.

Shadbolt, N. & Burton, M. 1995. Knowledge elicitation: A systematic approach. In J.R. Wilson & E.N. Corlett (Eds.), *Evaluation of Human Work: A Practical Ergonomics Methodology* (2nd ed., pp. 406–440). Philadelphia, PA: Taylor & Francis.

Shaffer, L.H. & Hardwick, J. 1968. Typing performance as a function of text. *Quarterly Journal of Experimental Psychology*, 20, 360–369.

Shaffer, M.T., Shafer, J.B. & Kutch, G.B. 1986. Empirical workload and communication: Analysis of scout helicopter exercises. In *Proceedings of the Human Factors Society 30th Annual Meeting* (pp. 628–632). Santa Monica, CA: Human Factors Society.

Shaffer, V.A., Probst, C.A., Merkle, E.C., Arkes, H.R., & Medow, M.A. 2013. Why do patients derogate physicians who use a computer-based diagnostic support system? *Medical Decision Making*, 33, 108–118.

Shafir, E.B., Smith, E.E. & Osherson, D.N. 1990. Typicality and reasoning fallacies. *Memory & Cognition*, 18, 229–239.

Shang, H. & Bishop, I.D. 2000. Visual thresholds for detection, recognition and visual impact in landscape settings. *Journal of Environmental Psychology*, 20, 125–140.

Shanks, D.R. & Cameron, A. 2000. The effect of mental practice on performance in a sequential reaction time task. *Journal of Motor Behavior*, 32, 305–313.

Shannon, C.E. 1948. A mathematical theory of communication. *The Bell System Technical Journal*, 27, 379–423.

Sharpe, L.T. & Jägle, H. 2001. Ergonomic consequences of dichromacy: I used to be color blind. *Color Research and Application*, 26, S269–S272.

Sharps, M.J. & Price-Sharps, J.L. 1996. Visual memory support: An effective mnemonic device for older adults. *Gerontologist*, 36, 706–708.

Shatenstein, B., Kergoat, M.-J. & Nadon, S. 2001. Anthropometric indices and their correlates in cognitively-intact and elderly Canadians with dementia. *Canadian Journal on Aging*, 20, 537–555.

Shaughnessy, J.J., Zechmeister, J.S. & Zechmeister, E.B. 2014. *Research Methods in Psychology* (10th ed.). Boston, MA: McGraw-Hill.

Shaw, E.A.G. 1974. The external ear. In W.D. Keidel & W.D. Neff (Eds.), *Handbook of Sensory Physiology*, Vol. 5: *Auditory System* (pp. 455–490). New York: Springer.

Shaw, M.E. 1981. *Group Dynamics: The Psychology of Small Group Behavior* (3rd ed.). New York: McGraw-Hill.

Shea, C.H., Lai, Q., Black, C. & Park, J.-H. 2000. Spacing practice sessions across days benefits the learning of motor skills. *Human Movement Science*, 19, 737–760.

Shea, C.H. & Wulf, G. 2005. Schema theory: A critical appraisal and reevaluation. *Journal of Motor Behavior*, 37, 85–101.

Sheedy, J.E. & Bailey, I.L. 1993. Vision and motor vehicle operation. In D.G. Pitts & R.N. Kleinstein (Eds.), *Environmental Vision: Interactions of the Eye, Vision, and the Environment* (pp. 351–357). Boston, MA: Butterworth-Heinemann.

Sheedy, J.E., Bailey, I.L., Burl, M. & Bass, E. 1986. Binocular vs. monocular task performance. *American Journal of Optometry and Physiological Optics*, 63, 839–846.

Sheik-Nainar, M.A., Kaber, D.B. & Chow, M.-Y. 2005. Control gain adaptation in virtual reality mediated human–telerobot interaction. *Human Factors and Ergonomics in Manufacturing*, 15, 259–274.

Shelhamer, M. 2015. Trends in sensorimotor research and countermeasures for exploration-class space flights. *Frontiers in Systems Neuroscience*, 9, 1–5.

Shelley-Tremblay, J. & Mack, A. 1999. Metacontrast masking and attention. *Psychological Science*, 10, 508–515.

Shepard, R. & Metzler, J. 1971. Mental rotation of three-dimensional objects. *Science*, 171, 701–703.

Shepherd, A. & Marshall, E. 2005. Timeliness and task specification in designing for human factors in railway operations. *Applied Ergonomics*, 36, 719–727.

Sheridan, T.B. 2016. Human–robot interaction status and challenges. *Human Factors*, 58, 525–532.

Sherman, S.M., & Guillery, R.W. 2013. *Functional Connections of Cortical Areas: A New View from the Thalamus*. Cambridge, MA: MIT Press.

Sherrick, C.E. & Cholewiak, R.W. 1986. Cutaneous sensitivity. In K.R. Boff, L. Kaufman & J.P. Thomas (Eds.), *Handbook of Perception and Human Performance* (Vol. I: *Sensory Processes and Perception*) pp. 12–1–12–58). New York: Wiley.

Sherrington, C.S. 1906. *Integrative Action of the Nervous System*. New Haven, CT: Yale University Press.

Sherwood, D.E. & Lee, T.D. 2003. Schema theory: Critical review and implications for the role of cognition in a new theory of motor learning. *Research Quarterly for Exercise and Sport*, 74, 376–382.

Shevell, S. (Ed.) 2003. *The Science of Color*. Amsterdam: Elsevier.

Shibata, T. 2002. Head mounted display. *Displays*, 23, 57–64.

Shiffrin, R.M. 1988. Attention. In R.C. Atkinson, R.J. Herrnstein, G. Lindzey & R.D. Luce (Eds.), *Stevens' Handbook of Experimental Psychology* (2nd ed., Vol. 2, pp. 739–811). New York: Wiley.

Shiffrin, R.M. & Schneider, W. 1977. Controlled and automatic human information processing: II. Perceptual learning, automatic attending, and a general theory. *Psychological Review*, 84, 127–190.

Shilling, R.D. & Shinn-Cunningham, B. 2002. Virtual auditory displays. In K.M. Stanney (Ed.), *Handbook of Virtual Environments: Design, Implementation, and Applications* (pp. 65–92). Mahwah, NJ: Lawrence Erlbaum.

Shinar, D. & Acton, M.B. 1978. Control-display relationships on the four-burner range: Population stereotypes versus standards. *Human Factors*, 20, 13–17.

Shingledecker, C.A. 1980. Enhancing operator acceptance and noninterference in secondary task measures of workload. *Proceedings of the 24th Annual Meeting of the Human Factors Society* (pp. 674–677). Santa Monica, CA: Human Factors Society.

Shiri, R., & Falah-Hassani, K. 2015. Computer use and carpal tunnel syndrome: A meta-analysis. *Journal of the Neurological Sciences*, 349, 15–19.

Shore, D.B., Sheng, Z., Cortina, J.M., & Yankelevich Garza, M. 2015. Personnel selection: A primer. In D.A. Boehm-Davis, F.T. Durso, & J.D. Lee (Eds.), *APA Handbook of Human Systems Integration* (pp. 485–500). Washington, DC: American Psychological Association.

Showalter, M. 2006. Now she can escort others down road to rehabilitation. *Lafayette Journal and Courier*, March 26.

Shulman, H.G. 1970. Encoding and retention of semantic and phonemic information in short-term memory. *Journal of Verbal Learning and Verbal Behavior*, 9, 499–508.

Siddique, C.M. 2004. Job analysis: A strategic human resource management practice. *International Journal of Human Resource Management*, 15, 219–244.

Siegel, A. & Wolf, J. 1969. *Man–Machine Simulation Models*. New York: Wiley.

Siegel, A.I. & Crain, K. 1960. Experimental investigations of cautionary signal presentations. *Ergonomics*, 3, 339–356.

Siegel, A.I., Schultz, D.G. & Lanterman, R.S. 1963. Factors affecting control activation time. *Human Factors*, 5, 71–80.

Silber, B.Y., Papafotiou, K., Croft, R.J., Ogden, E., Swann, P. & Stough, C. 2005. The effects of dexamphetamine on simulated driving performance. *Psychopharmacology*, 179, 536–543.

Silverman, B.G. 1992. Modeling and critiquing the confirmation bias in human reasoning. *IEEE Transactions on Systems, Man, and Cybernetics*, 22, 972–982.

Simunovic, M.P. 2010. Colour vision deficiency. *Eye*, 24, 747–755.

Simon, H.A. 1957. *Models of Man*. New York: Wiley.

Simon, H.A. & Gilmartin, K. 1973. A simulation of memory for chess positions. *Cognitive Psychology*, 5, 29–46.

Simon, J.R. 1969. Reactions toward the source of stimulation. *Journal of Experimental Psychology*, 81, 174–176.

Simon, J.R. 1990. The effects of an irrelevant directional cue on human information processing. In R.W. Proctor & T.G. Reeve (Eds.), *Stimulus–Response Compatibility: An Integrated Perspective* (pp. 31–86). Amsterdam: North-Holland.

Simons, D.J. 2000. Attentional capture and inattentional blindness. *Trends in Cognitive Sciences*, 4, 147–155.

Simons, D.J. & Ambinder, M.S. 2005. Change blindness: Theory and consequences. *Current Directions in Psychological Science*, 14, 44–48.

Simons, D.J. & Levin, D.T. 1998. Failure to detect changes to people during a real-world interaction. *Psychonomic Bulletin & Review*, 5, 644–649.

Simpson, C.A. & Williams, D.H. 1980. Response time effects of alerting tone and semantic context for synthesized voice cockpit warnings. *Human Factors*, 22, 319–330.

Simpson, C.A., McCauley, M.E., Roland, E.F., Ruth, J.C. & Williges, B.H. 1985. System design for speech recognition and generation. *Human Factors*, 27, 115–141.

Sinclair, M.A. 2005. Participative assessment. In J.R. Wilson & N. Corlett (Eds.), *Evaluation of Human Work* (3rd ed., pp. 83–111). Boca Raton, FL: CRC Press.

Singleton, W.T. (Ed.) 1978. *The Analysis of Practical Skills*. Baltimore, MD: University Park Press.

Singley, M.K. & Anderson, J.R. 1989. *The Transfer of Cognitive Skill*. Cambridge, MA: Harvard University Press.

Sirois, S., & Brisson, J. 2014. Pupillometry. *Wires Cognitive Science*, 5, 679–692.

Sivak, M. 1987. Human factors and road safety. *Applied Ergonomics*, 18, 289–296.

Sivak, M. & Olson, P.L. 1985. Optimal and minimal luminance characteristics for retro-reflective highway signs. *Transportation Research Record*, 1027, 53–57.

Sivak, M., Olson, P.L. & Zeltner, K.A. 1989. Effect of prior headlighting experience on ratings of discomfort glare. *Human Factors*, 31, 391–395.

Skipper, J.H., Rieger, C.A. & Wierwille, W.W. 1986. Evaluation of decision-tree rating scales for mental workload estimation. *Ergonomics*, 29, 383–399.

Slee, S. J., & Young, E. D. 2014. Alignment of sound localization cues in the nucleus of the brachium of the inferior colliculus. *Journal of Neurophysiology*, 111, 2624–2633.

Slatter, P.E. 1987. *Building Expert Systems: Cognitive Emulation*. Chichester, England: Ellis Horwood.

Sluchak, T.J. 1990. Human factors: Added value to retail customers. In *Proceedings of the Human Factors Society 34th Annual Meeting* (pp. 752–756). Santa Monica, CA: Human Factors Society.

Smart, K. 2001. Human factors and life support issues in crew rescue from the International Space Station. *Human Performance in Extreme Environments*, 5, 2–6.

Smetacek, V. & Mechsner, F. 2004. Making sense: Proprioception: Is the sensory system that supports body posture and movement also the root of our understanding of physical laws? *Nature*, 432(7013), 21.

Smiley, A., MacGregor, C., Dewar, R.E. & Blamey, C. 1998. Evaluation of prototype highway tourist signs for Ontario. *Transportation Research Record*, 1628, 34–40.

Smith, A. 2015. U.S. *Smartphone Use in 2015*. Report, Pew Research Center, Washington, DC, April 1. http://www.pewinternet.org/2015/04/01/us-smartphone-use-in-2015/

Smith, E.E. 1989. Concepts and induction. In M.I. Posner (Ed.), *Foundations of Cognitive Science* (pp. 501–526). Cambridge, MA: MIT Press.

Smith, E.E. & Medin, D.L. 1981. *Categorization and Concepts*. Cambridge, MA: Harvard University Press.

Smith, E.E., Shoben, E.J. & Rips, L.J. 1974. Structure and process in semantic memory: A feature model for semantic decision. *Psychological Review*, 81, 214–241.

Smith, M.J. 1987. Occupational stress. In G. Salvendy (Ed.), *Handbook of Human Factors*. New York: Wiley.

Smith, S.D. 2006. Seat vibration in military propeller aircraft: Characterization, exposure assessment, and mitigation. *Aviation, Space, and Environmental Medicine*, 77, 32–40.

Smith, S.W. & Rea, M.S. 1978. Proofreading under different levels of illumination. *Journal of the Illuminating Engineering Society*, 8, 47–52.

Snook, S.H. 1999. Future directions of psychophysical studies. *Scandinavian Journal of Work Environment and Health*, 25(Suppl. 4), 13–18.

Snook, S.H. & Ciriello, V.M. 1991. The design of manual tasks: Revised tables of maximum acceptable weights and forces. *Ergonomics*, 34, 1197–1213.

Snyder, H. 1976. Braking movement time and accelerator-brake separation. *Human Factors*, 18, 201–204.

Snyder, H.L. & Taylor, G.B. 1979. The sensitivity of response measures of alphanumeric legibility to variations in dot matrix display parameters. *Human Factors*, 21, 457–471.

So, J.C.Y., Macrowski, L.M., Dunston, P.S., Proctor, R.W. & Goodney, J.E. 2016. Operator training transfer from simulator to real machine for hydraulic excavators. In J.L. Perdomo-Rivera, A. González-Quevedo, C. López del Puerto, F. Maldonado-Fortunet, & O.I. Molina-Bas (Eds.), *Proceedings of 2016 Construction Research Congress* (pp. 1968–1977). American Society of Civil Engineers.

So, J.C.Y., Proctor, R.W., Dunston, P.S., & Wang, X. 2013. Better retention of skill operating a simulated hydraulic excavator after part-task than after whole-task training. *Human Factors*, 55, 449–460.

Söderholm, A., Öhman, A., Stenberg, B., & Nordin, S. 2016. Experience of living with nonspecific building-related symptoms. *Scandinavian Journal of Psychology*, 57, 406–412.

Solomon, S.S. & King, J.G. 1997. Fire truck visibility. *Ergonomics in Design*, 5(2), 4–10.

Sommer, R. 1967. Sociofugal space. *American Journal of Sociology*, 72, 654–659.

Sommer, R. 2002. Personal space in a digital age. In R.B. Bechtel & A. Churchman (Eds.), *Handbook of Environmental Psychology* (pp. 647–660). Hoboken, NJ: Wiley.

Sonnentag, S. & Frese, M. 2003. Stress in organizations. In W.C. Borman, D.R. Ilgin & R.J. Klimoski (Eds.), *Industrial and Organizational Psychology* (Vol. 12: *Handbook of Psychology*) pp. 453–491. Hoboken, NJ: Wiley.

Soegaard, M., & Dam, R. F. (Eds.) 2016. *The Encyclopedia of Human-Computer Interaction* (2nd ed.). Denmark: Interaction Design Foundation. https://www.interaction-design.org/literature/book/the-encyclopedia-of-human-computer-interaction-2nd-ed

Sorkin, R.D. 1987. Design of auditory and tactile displays. In G. Salvendy (Ed.), *Handbook of Human Factors* (pp. 549–576). New York: Wiley.

Sorkin, R.D. 1989. Why are people turning off our alarms? *Human Factors Society Bulletin*, 32(4), 3–4. Sorkin, R.D., Kantowitz, B.H. & Kantowitz, S.C. 1988. Likelihood alarm displays. *Human Factors*, 30, 445–459.

Sorkin, R.D., Wightman, F.L., Kistler, D.S. & Elvers, G.C. 1989. An exploratory study of the use of movement-correlated cues in an auditory head-up display. *Human Factors*, 31, 161–166.

Sosnik, R., Flash, T., Sterkin, A., Hauptmann, B., & Karni, A. 2014. The activity in the contralateral primary motor cortex, dorsal premotor and supplementary motor area is modulated by performance gains. *Frontiers in Human Neuroscience*, 8.

Southall, D. 1985. The discrimination of clutch-pedal resistances. *Ergonomics*, 28, 1311–1317.

Soranzo, A., Galmonte, A., & Agostini, T. 2009. Lightness constancy: Ratio invariance and luminance profile. *Attention, Perception, & Psychophysics*, 71, 463–470.

Soranzo, A., Lugrin, J., & Wilson, C.J. 2013. The effects of belongingness on the simultaneous lightness contrast: A virtual reality study. *Vision Research*, 86, 97–106.

Space Station Human Productivity Study, Vols. I–V 1985. Lockheed Missiles and Space Company, Inc., Man–Systems Division, NASA, Lyndon B. Johnson Space Center.

Spalton, D.J., Hitchings, R.A. & Hunter, P.A. 2005. *Atlas of Clinical Ophthalmology* (3rd ed.). Philadelphia, PA: Elsevier Mosby.

Spector, P.E. 2006. *Industrial and Organizational Psychology* (4th ed.). Hoboken, NJ: Wiley.

Spector, P.E., 2017. Industrial and Organizational Psychology: Research and Practice (7th ed.). Hoboken, NJ: John Wiley.

Spektor, T., Nikolic, N., Lekakh, O., & Gaynes, B.I. 2015. Efficacy of ScripTalk automated prescription label reader and veterans with visual impairments. *Journal of Visual Impairment & Blindness (Online)*, 109(5), 412.

Spencer, K.M., Dien, J. & Donchin, E. 1999. A componential analysis of the ERP elicited by novel events using a dense electrode array. *Psychophysiology*, 36, 409–414.

Sperling, G. 1960. The information available in brief visual presentations. *Psychological Monographs*, 74, 1–29.

Sperling, L., Dahlman, S., Wikström, L., Kilbom, Å. & Kadefors, R., 1993. A cube model for the classification of work with hand tools and the formulation of functional requirements. *Applied Ergonomics*, 24, 212–220.

Spieth, W., Curtis, J.F. & Webster, J.C. 1954. Responding to one of two simultaneous messages. *Journal of the Acoustical Society of America*, 26, 391–396.

Spitz, G. 1990. Target acquisition performance using a head mounted cursor control device and a stylus with a digitizing table. *Proceedings of the 34th Annual Meeting of the Human Factors Society* (pp. 405–405). Santa Monica, CA: Human Factors Society.

Spurgin, A. 2010. *Human Reliability Assessment: Theory and Practice*. Boca Raton, FL: CRC Press.

St. John, M., Cowen, M.B., Smallman, H.S. & Oonk, H.M. 2001. The use of 2D and 3D displays for shape-understanding versus relative-position tasks. *Human Factors*, 43, 79–98.

Stolzberg, D., Salvi, R. J., & Allman, B. L. 2012. Salicylate toxicity model of tinnitus. *Frontiers in Systems Neuroscience*, 6doi:10.3389/fnsys.2012.00028

Stammers, R.B. 2006. The history of the Ergonomics Society. *Ergonomics*, 49, 741–742.

Stanney, K.M. (Ed.) 2002. *Handbook of Virtual Environments: Design, Implementation, and Applications*. Mahwah, NJ: Lawrence Erlbaum.

Stanney, K.M. 2003. Virtual environments. In J.A. Jacko & A. Sears (Eds.), *The Human–Computer Interaction Handbook: Fundamentals, Evolving Technologies and Emerging Applications* (pp. 621–634). Mahwah, NJ: Lawrence Erlbaum.

Stanney, K., Winslow, B., Hale, K., & Schmorrow, D. 2015. Augmented cognition. In D.A. Boehm-Davis, F.T. Durso, & J.D. Lee (Eds.), *APA Handbook of Human Systems Integration* (pp. 329–343). Washington, DC: American Psychological Association.

Stanton, N.A. 2006a. Error taxonomies. In W. Karwowski (Ed.), *International Encyclopedia of Ergonomics and Human Factors* (2nd ed., Vol. 1, pp. 706–709). Boca Raton, FL: CRC Press.

Stanton, N.A. 2006b. Hierarchical task analysis: Developments, applications, and extensions. *Applied Ergonomics*, 37, 55–79.

Stanton, N.A. 2006c. Speech-based alarm displays. In W. Karwowski (Ed.), *International Encyclopedia of Ergonomics and Human Factors* (2nd ed., Vol. 1, pp. 1257–1259). Boca Raton, FL: CRC Press.

Stanton, N.A. & Baber, C. 2005. Task analysis for error identification. In N.A. Stanton, A. Hedge, K. Brookhuis, E. Salas & H.W. Hendrick (Eds.), *Handbook of Human Factors and Ergonomics Methods* (pp. 38-1–38-9). Boca Raton, FL: CRC Press.

Stanton, N.A., & Salmon, P.M. 2009. Human error taxonomies applied to driving: A generic driver error taxonomy and its implications for intelligent transport systems. *Safety Science*, 47, 227–237.

Stanton, N.A., Salmon, P.M., Rafferty, L.A., Walker, G.H., Baber, C., & Jenkins D.P. 2013. *Human Factors Methods: A Practical Guide for Engineering and Design* (2nd ed.). Burlington, VT: Ashgate.

Stark, L. & Bridgeman, B. 1983. Role of corollary discharge in space constancy. *Perception & Psychophysics*, 34, 371–380.

Statler, S.M. 2005. Preventing 'accidental' injury: Accountability for safer products by anticipating product risks and user behaviors. In Y.I. Noy & W. Karwowski (Eds.), *Handbook of Human Factors in Litigation* (pp. 25-1–25-14). Boca Raton, FL: CRC Press.

Steeneken, H.J.M. & Houtgast, T. 1999. Mutual dependence of the octave-band weights in predicting speech intelligibility. *Speech Communication*, 28, 109–123.

Steiner, L.J. & Vaught, C. 2006. Work design: Barriers facing the integration of ergonomics into system design. In W. Karwowski (Ed.), *International Encyclopedia of Ergonomics and Human Factors* (2nd ed., Vol. 2, pp. 2479–2483). Boca Raton, FL: CRC Press.

Steingrimsson, R., Luce, R.D., & Narens, L. 2012. Brightness of different hues is a single psychophysical ratio scale of intensity. *American Journal of Psychology*, 125, 321–333.

Steinhauer, K. & Grayhack, J.P. 2000. The role of knowledge of results in performance and learning of a voice motor task. *Journal of Voice*, 14, 137–145.

Stenzel, A.G. 1962. Erfahrungen mit 1000lx in einer Kamerafabrik (Experience with a 1000 lx leather factory). *Lichttechnik*, 14, 16–18.

Stenzel, A.G. & Sommer, J. 1969. The effect of illumination on tasks which are largely independent of vision. *Lichttechnik*, 21, 143–146.

Stephanidis, C. & Akoumianakis, D. 2011. A design code of practice for universal access: Methods and techniques. In K.-P.L. Vu & R.W. Proctor & (Eds.), *Handbook of Human Factors in Web Design* (2nd ed., pp. 359–370). Mahwah, NJ: Lawrence Erlbaum.

Stephens, E.C., Carswell, C.M. & Dallaire, J. 2000. The use of older adults on preference panels: Evidence from the Kentucky Interface Preference Inventory. *International Journal of Cognitive Ergonomics*, 4, 179–190.

Sternberg, S. 1966. High-speed scanning in human memory. *Science*, 153, 652–654.

Sternberg, S. 1969. The discovery of processing stages: Extensions of Donders' method. In W.G. Koster (Ed.), *Attention and Performance II* (pp. 276–315). Amsterdam: North-Holland.

Sternberg, S. 2016. In defence of high-speed memory scanning. *Quarterly Journal of Experimental Psychology*, 69, 2020–2075.

Stevens, S. S. (1961). To Honor Fechner and Repeal His Law. *Science*, 133(3446), 80–86.

Stevens, J.C., Okulicz, W.C. & Marks, L.E. 1973. Temporal summation at the warmth threshold. *Perception & Psychophysics*, 14, 307–312.

Stevens, J.K., Emerson, R.C., Gerstein, G.L., Kallos, T., Neufeld, G.R., Nichols, C.W. & Rosenquist, A.C. 1976. Paralysis of the awake human: Visual perceptions. *Vision Research*, 16, 93–98.

Stevens, S.S. 1975. *Psychophysics: Introduction to Its Perceptual, Neural, and Social Prospects*. New York: Wiley.

Stevenson, F., Carmona-Andreu, I., & Hancock, M. 2013. The usability of control interfaces in low-carbon housing. *Architectural Science Review*, 56, 70–82.

Stewart, D.W. 2012. Secondary analysis and archival research: Using data collected by others. In H. Cooper, P.M. Camic, D.L. Long, A.T. Panter, D. Rindskopf, K.J. Sher (Eds.), *APA Handbook of Research Methods in Psychology, Vol 3: Data Analysis and Research Publication* (pp. 473–484). Washington, DC: American Psychological Association.

Steyvers, M., Griffiths, T.L. & Dennis, S. 2006. Probabilistic inference in human semantic memory. *Trends in Cognitive Sciences*, 10, 327–334.

Stohr, E.A. & Viswanathan, S. 1999. Recommendation systems: Decision support for the information economy. In K.E. Kendall (Ed.), *Emerging Information Technologies: Improving Decisions, Cooperation, and Infrastructure* (pp. 21–44). Thousand Oaks, CA: Sage.

Stokes, D.E. 1997. *Pasteur's Quadrant: Basic Science and Technological Innovation*. (p. 196). Washington, DC: Brookings.

Straker, L.M. 2006. Visual display units: Positioning for human use. In W. Karwowski (Ed.), *International Encyclopedia of Ergonomics and Human Factors* (2nd ed., Vol. 2, pp. 1742–1745). Boca Raton, FL: CRC Press.

Strange, B.A., Duggins, A., Penny, W., Dolan, R.J. & Friston, K.J. 2005. Information theory, novelty and hippocampal responses: Unpredicted or unpredictable? *Neural Networks*, 18, 225–230.

Sträter, O. 2005. *Cognition and Safety*. Burlington, VT: Ashgate.

Streeter, L., Laham, D., Dumais, S. & Rothkopf, E.Z. 2005. In A.F. Healy (Ed.), *Experimental Cognitive Psychology and Its Applications* (pp. 31–44). Washington, DC: American Psychological Association.

Stroop, J.R. 1935/1992. Studies of interference in serial verbal reactions. *Journal of Experimental Psychology: General*, 121, 15–23.

Strybel, T.Z. 2011. Task analysis for the design of Web applications. In K.-P.L. Vu & R. W. Proctor (Eds.), *Handbook of Human Factors in Web Design* (2nd ed., pp. 483–507). Boca Raton, FL: CRC Press.

Sturr, F., Kline, G.E. & Taub, H.A. 1990. Performance of young and older drivers on a static acuity test under photopic and mesopic luminance conditions. *Human Factors*, 32, 1–8.

Styles, E.F. 2006. *The Psychology of Attention* (2nd ed.). Philadelphia, PA: Psychology Press.

Sullman, M.M. 2012. An observational study of driver distraction in England. *Transportation Research Part F: Traffic Psychology and Behaviour*, 15, 272–278.

Sun, B.C., Hsia, R.Y., Weiss, R.E., Zingmond, D., Liang, L.J., Han, W., & Asch, S.M. 2013. Effect of emergency department crowding on outcomes of admitted patients. *Annals of Emergency Medicine*, 61, 605–611.

Sun, H.M., Li, S.P., Zhu, Y.Q., & Hsiao, B. 2015. The effect of user's perceived presence and promotion focus on usability for interacting in virtual environments. *Applied Ergonomics*, 50, 126–132.

Suss, J., & Ward, P. 2015. Predicting the future in perceptual-motor domains: Perceptual anticipation, option generation, and expertise. In R.R. Hoffman, P.A. Hancock, M.W. Scerbo, R. Parasuraman, & J.L. Szalma (Eds.), *The Cambridge Handbook of Applied Perception Research, Vol. II* (pp. 951–976). New York: Cambridge University Press.

Sutton, H. & Porter, L.W. 1968. A study of the grapevine in a governmental organization. *Personnel Psychology*, 21, 223–230.

Svaetchin, A. 1956. Spectral response curves of single cones. *Acta Psychologica*, 1, 93–101.

Swain, A.D. & Guttman, H.E. 1983. *Handbook of Human Reliability Analysis with Emphasis on Nuclear Power Plant Applications*. Albuquerque, NM: Sandia National Laboratories.

Swets, J.A. & Pickett, R.M. 1982. *Evaluation of Diagnostic Systems: Methods from Signal Detection Theory*. New York: Academic Press.

Swinnen, S.P. 1990. Interpolated activities during the knowledge-of-results delay and post-knowledge-of-results interval: Effects on performance and learning. *Journal of Experimental Psychology: Learning, Memory, and Cognition*, 16, 692–705.

Swinnen, S.P., Schmidt, R.A., Nicholson, D.E. & Shapiro, D.C. 1990. Information feedback for skill acquisition: Instantaneous knowledge of results degrades learning. *Journal of Experimental Psychology: Learning, Memory, and Cognition*, 16, 706–716.

Szabo, R.M. 1998. Carpal tunnel syndrome as a repetitive motion disorder. *Clinical Orthopedics and Related Research*, 351, 78–89.

Szalma, J. L. 2009. Individual differences in human–technology interaction: Incorporating variation in human characteristics into human factors and ergonomics research and design, *Theoretical Issues in Ergonomics Science*, 10, 381–397.

Szalma, J.L., & Hancock, P.A. 2015. Psychophysical methods and signal detection: Recent advances in theory. In R.R. Hoffman, P.A. Hancock, M.W.Scerbo, R. Parasuraman, & J. Szalma (Eds.), *The Cambridge Handbook of Applied Perception Research* (Vol. I., pp. 22–39). New York: Cambridge University Press.

Szilagyi, A.D., Jr., & Wallace, M.J., Jr. 1983. *Organizational Behavior and Performance* (3rd ed.). Glenview, IL: Scott, Foresman.

Taatgen, N.A. & Lee, F.J. 2003. Production compilation: A simple mechanism to model complex skill acquisition. *Human Factors*, 45, 61–76.

Tagliabue, M., Zorzi, M., Umiltà, C. & Bassignani, F. 2000. The role of LTM links and STM links in the Simon effect. *Journal of Experimental Psychology: Human Perception and Performance*, 26, 648–670.

Tan, H., Lim, A. & Traylor, R. 2000. A psychophysical study of sensory saltation with an open response paradigm. *Proceedings of the ASME Dynamic Systems and Control Division, 69-2*, 1109–1115.

Tanaka, Y. 2015. Spinal reflexes during postural control under psychological pressure. *Motor Control*, 19, 242–249.

Tang, S.K. 1997. Performance of noise indices in air-conditioned landscaped office buildings. *Journal of the Acoustical Society of America*, 102, 1657–1663.

Taub, E. & Berman, A.J. 1968. Movement and learning in the absence of sensory feedback. In S.J. Freedman (Ed.), *The Neuropsychology of Spatially Oriented Behavior* (pp. 173–192). Homewood, IL: Dorsey Press.

Taylor, F.W. 1911/1967. *The Principles of Scientific Management*. New York: W.W. Norton.

Taylor, J. R. 2016. *Human Error in Process Plant Design and Operations: A Practitioner's Guide*. Boca Raton, FL: CRC Press.

Teichner, W.H. 1962. Probability of detection and speed of response in simple monitoring. *Human Factors*, 4, 181–186.

Teichner, W.H. & Krebs, M.J. 1972. Laws of the simple visual reaction time. *Psychological Review*, 79, 344–358.

Teichner, W.H. & Krebs, M.J. 1974. Laws of visual choice reaction time. *Psychological Review*, 81, 75–98.

Telford, C.W. 1931. Refractory phase of voluntary and associative responses. *Journal of Experimental Psychology*, 14, 1–35.

Theeuwes, J., Alferdinck, J.W.A.M. & Perel, M. 2002. Relation between glare and driving performance. *Human Factors*, 44, 95–107.

Theise, E.S. 1989. Finding a subset of stimulus–response pairs with a minimum of total confusion: A binary integer programming approach. *Human Factors*, 31, 291–305.

Thomas, M., Gilson, R., Ziulkowski, S. & Gibbons, S. 1989. Short term memory demands in processing synthetic speech. *Proceedings of the Human Factors Society 33rd Annual Meeting* (Vol. 1, pp. 239–241). Santa Monica, CA: Human Factors Society.

Thomson, D.R., Besner, D., & Smilek, D. 2016. A critical examination of the evidence for sensitivity loss in modern vigilance tasks. *Psychological Review*, 123, 70–83.

Thompson, E.R., Iyer, N., Simpson, B.D., Wakefield, G.H., Kieras, D.E., & Brungart, D.S. 2015. Enhancing listener strategies using a payoff matrix in speech-on-speech masking experiments. *Journal of the Acoustical Society of America*, 138, 1297–1304.

Thompson, R.J., Payne, S.C., & Taylor, A.B. 2015. Applicant attraction to flexible work arrangements: Separating the influence of flextime and flexplace. *Journal of Occupational and Organizational Psychology*, 88, 726–749.

Thompson, S., Slocum, J. & Bohan, M. 2004. Gain and angle of approach effects on cursor-positioning time with a mouse in consideration of Fitts' law. *Proceedings of the Human Factors and Ergonomics Society 48th Annual Meeting* (pp. 823–827). Santa Monica, CA: Human Factors and Ergonomics Society.

Thörn, Å. 2006. Emergence and preservation of a chronically sick building. *Journal of Epidemiology and Community Health*, 54, 552–556.

Thorn, A.C., & Page, M.A. 2009. Current issues in understanding interactions between short-term and long-term memory. In A.C. Thorn & M.A. Page (Eds.), *Interactions between Short-Term and Long-Term Memory in the Verbal Domain* (pp. 1–15). New York: Psychology Press.

Thorndike, E.L. 1906. *Principles of Teaching*. New York: A.G. Seiler.

Thorndyke, P.W. 1984. Applications of schema theory in cognitive research. In J.R. Anderson & S.M. Kosslyn (Eds.), *Tutorials in Learning and Memory*. New York: W.H. Freeman.

Tichauer, E.R. 1976. Biomechanics Sustains Occupational Health and Safety, Industrial Engineering Magazine, February, 8(2), 46–56.

Tiesler-Wittig, H., Postma, P. & Springer, B. 2005. Brightness to the very limit—Headlighting sources with high luminance—Mercury free Xenon HID. In *Lighting Technology and Human Factors* (SP-1932, pp. 145–149). Warrendale, PA: Society of Automotive Engineers.

Tillman, B. 1987. Man–systems integration standards (NASA-STD-3000). *Human Factors Society Bulletin*, 30 (6), 5–6.

Tolhurst, D.J. & Thompson, P.G. 1975. Orientation illusions and aftereffects: Inhibition between channels. *Vision Research*, 15, 967–972.

Tombu, M. & Jolicœur, P. 2005. Testing the predictions of the central capacity sharing model. *Journal of Experimental Psychology: Human Perception and Performance*, 31, 790–802.

Topmiller, D., Eckel, J. & Kozinsky, E. 1982. *Human Reliability Data Bank for Nuclear Power Plant Operators*, Vol. 1: *A Review of Existing Human Reliability Data Banks* (NUREG=CR-2744=1 of 2). Washington, DC: Nuclear Regulatory Commission.

Torenvliet, G.L. & Vicente, K.J. 2006. Ecological interface design—Theory. In W. Karwowski (Ed.), *International Encyclopedia of Ergonomics and Human Factors* (2nd ed., Vol. 1, pp. 1083–1987). Boca Raton, FL: CRC Press.

Townsend, J.T. 1974. Issues and models concerning the processing of a finite number of inputs. In B.H. Kantowitz (Ed.), *Human Information Processing* (pp. 133–185). Hillsdale, NJ: Lawrence Erlbaum.

Townsend, J.T. & Roos, R.N. 1973. Search reaction time for single targets in multiletter stimuli with brief visual displays. *Memory & Cognition*, 1, 319–332.

Treisman, A.M. 1960. Contextual cues in selective listening. *Quarterly Journal of Experimental Psychology*, 12, 242–248.

Treisman, A.M. 1964a. The effect of irrelevant material on the efficiency of selective listening. *American Journal of Psychology*, 77, 533–546.

Treisman, A.M. 1964b. Verbal cues, language, and meaning in selective attention. *American Journal of Psychology*, 77, 206–219.

Treisman, A.M. 1986. Features and objects in visual processing. *Scientific American*, 255, 114–125.

Treisman, A.M., Squire, R. & Green, J. 1974. Semantic processing in dichotic listening? A replication. *Memory & Cognition*, 2, 641–646.

Treisman, A.M., Sykes, M. & Gelade, G. 1977. Selective attention and stimulus integration. In S. Dornic (Ed.), *Attention and Performance VI* (pp. 333–361). Hillsdale, NJ: Lawrence Erlbaum.

Trudel, R., Murray, K.B., Kim, S., & Chen, S. 2015. The impact of traffic light color-coding on food health perceptions and choice. *Journal of Experimental Psychology: Applied*, 21, 255–275.

Tsang, P.S. & Vidulich, M.A. 2006. Mental workload and situation awareness. In G. Salvendy (Ed.), *Handbook of Human Factors and Ergonomics* (3rd ed., 243–268). Hoboken, NJ: Wiley.

Tseng, M.M., Law, P.-H. & Cerva, T. 1992. Knowledge-based systems. In G. Salvendy (Ed.), *Handbook of Industrial Engineering* (2nd ed., pp. 184–210). New York: Wiley.

Tufano, D.R. 1997. Automotive HUDs: The overlooked safety issues. *Human Factors*, 39, 303–311.

Tullis, T.S. 1983. The formatting of alphanumeric displays: A review and analysis. *Human Factors*, 25, 657–682.

Tullis, T.S. 1986. *Display Analysis Program* (Version 4.0). Lawrence, KS: The Report Store.

Tulving, E. 1999. Episodic vs semantic memory. In F. Keil & R. Wilson (Eds.), *The MIT Encyclopedia of the Cognitive Sciences* (pp. 278–280). Cambridge, MA: MIT Press.

Tulving, E. & Donaldson, W. (Eds.) 1972. *Organization of Memory*. New York: Academic Press.

Tulving, E. & Pearlstone, Z. 1966. Availability versus accessibility of information in memory for words. *Journal of Verbal Learning and Verbal Behavior*, 5, 381–391.

Tulving, E. & Thomson, D.M. 1973. Encoding specificity and retrieval processes in episodic memory. *Psychological Review*, 80, 352–373.

Turley, A.M. 1978. Acoustical privacy for the open office. *The Office*. Report in *Space Planning*, Office of the Future. Pasadena, CA: Office Technology Research Group.

Tvaryanas, A.P., Thompson, W.T. & Constable, S.H. 2006. Human factors in remotely piloted aircraft operations: HFACS analysis of 221 mishaps over 10 years. *Aviation, Space, and Environmental Medicine*, 77, 724–732.

Tversky, A. 1969. Intransitivity of preferences. *Psychological Review*, 76, 31–48.

Tversky, A. 1972. Elimination by aspects: A theory of choice. *Psychological Review*, 79, 281–299.

Tversky, A. & Kahneman, D. 1973. Availability: A heuristic for judging frequency and probability. *Cognitive Psychology*, 5, 207–232.

Tversky, A. & Kahneman, D. 1974. Judgment under uncertainty: Heuristics and biases. *Science*, 211, 453–458.

Tversky, A. & Kahneman, D. 1980. Causal schemas in judgments under uncertainty. In M. Fishbein (Ed.), *Progress in Social Psychology* (pp. 49–72). Hillsdale, NJ: Lawrence Erlbaum.

Tversky, A. & Kahneman, D. 1981. The framing of decisions and the psychology of choice. *Science*, 211, 453–458.

Tversky, A. & Kahneman, D. 1983. Extensional versus intuitive reasoning: The conjunction fallacy in probability judgment. *Psychological Review*, 90, 293–315.

Tversky, A., Sattath, S. & Slovic, P. 1988. Contingent weighting in judgment and choice. *Psychological Review*, 95, 371–384.

Tversky, B.G. 1969. Pictorial and verbal encoding in a short-term memory task. *Perception & Psychophysics*, 6, 225–233.

Tyrrell, R.A. & Leibowitz, H.W. 1990. The relation of vergence effort to reports of visual fatigue following prolonged nearwork. *Human Factors*, 32, 341–357.

Tziner, A., Murphy, K.R. & Cleveland, J.N. 2005. Performance appraisal: Evolution and change. *Group and Organization Management*, 30, 4–5.

U.S. Department of Transportation 1998. Center brake lights prevent crashes, save millions in property damage. *Briefing NHTSA*, 16–98.

U. S. Environmental Protection Agency. 2017. Introduction to indoor air quality. https://www.epa.gov/indoor-air-quality-iaq/introduction-indoor-air-quality

Ullén, F., Hambrick, D.Z., & Mosing, M.A. 2016. Rethinking expertise: A multifactorial gene–environment interaction model of expert performance. *Psychological Bulletin*, 142, 427–446.

Ulrich, R., Schröter, H., Leuthold, H., & Birngruber, T. 2015. Automatic and controlled stimulus processing in conflict tasks: Superimposed diffusion processes and delta functions. *Cognitive Psychology*, 78, 148–174.

Umiltà, C. & Liotti, M. 1987. Egocentric and relative spatial codes in S–R compatibility. *Psychological Research*, 49, 81–90.

Umiltà, C. & Nicoletti, R. 1990. Spatial stimulus–response compatibility. In R.W. Proctor & T.G. Reeve (Eds.), *Stimulus–Response Compatibility: An Integrated Perspective* (pp. 89–116). Amsterdam: North-Holland.

Usher, M. & McClelland, J.L. 2001. The time course of perceptual choice: The leaky competing accumulator model. *Psychological Review*, 108, 550–592.

Usher, M., Olami, Z. & McClelland, J.L. 2002. Hick's law in a stochastic race model with speed–accuracy tradeoff. *Journal of Mathematical Psychology*, 46, 704–715.

USNRC 2000. Technical Basis and Implementation Guidelines for Technique for Human Event Analysis (ATHEANA). NUREG-1624, Rev. 1.

Uttal, W.R. & Gibb, R.W. 2001. On the psychophysics of night vision goggles. In R.R. Hoffman & A.B. Markman (Eds.), *Interpreting Remote Sensing Imagery: Human Factors* (pp. 117–136). Boca Raton, FL: Lewis Publishers.

Valencia, G. & Agnew, J.R. 1990. Evaluation of a directional audio display synthesizer. *Proceedings of the Human Factors Society* (Vol. 1, pp. 6–10). Santa Monica, CA: Human Factors Society.

van Bommel, W.M. 2006. Non-visual biological effect of lighting and the practical meaning for lighting for work. *Applied Ergonomics*, 37, 461–466.

Van Cott, H.P. 1980. Civilian anthropometry data bases. In *Proceedings of the Human Factors Society 24th Annual Meeting* (pp. 34–36). Santa Monica, CA: Human Factors Society.

van den Anker, F.W.G. & Schulze, H. 2006. Scenario-based design of ICT-supported work. In W. Karwowski (Ed.), *International Encyclopedia of Ergonomics and Human Factors* (2nd ed., Vol. 3, pp. 3348–3353). Boca Raton, FL: CRC Press.

van der Zee, K.I., Bakker, A.B. & Bakker, P. 2002. Why are structured interviews so rarely used in personnel selection? *Journal of Applied Psychology*, 87, 176–184.

van Tilburg, M. & Briggs, T. 2005. Web-based collaboration. In R.W. Proctor & K.-P.L. Vu (Eds.), *Handbook of Human Factors in Web Design* (2nd ed., pp. 551–569). Mahwah, NJ: Lawrence Erlbaum.

Van Wanrooij M.M. & Van Opstal, A.J. 2005. Relearning sound localization with a new ear. *Journal of Cognitive Neuroscience*, 25, 5413–5424.

Vandenbos, G.R. (editor-in-chief) 2015. *APA Dictionary of Psychology* (2nd ed.). Washington, DC: American Psychological Association.

Vanderheiden, G.C. 2005. Access to Web content by those operating under constrained conditions. In R.W. Proctor & K.-P.L. Vu (Eds.), *Handbook of Human Factors in Web Design* (2nd ed., pp. 267–283). Mahwah, NJ: Lawrence Erlbaum.

VanLehn, K. 1998. Analogy events: How examples are used during problem solving. *Cognitive Science*, 22, 347–388.

Várhelyi, A., Hjälmdahl, M., Hydén, C. & Draskóczy, M. 2004. Effects of an active accelerator pedal on driver behaviour and traffic safety after long-term use in urban areas. *Accident Analysis and Prevention*, 36, 729–737.

Vecera, S.P., Vogel, E.K. & Woodman, G.F. 2002. Lower region: A new cue for figure-ground assignment. *Journal of Experimental Psychology: General*, 131, 194–205.

Veitch, J.A. & McColl, S.L. 2001. A critical examination of perceptual and cognitive effects attributed to full-spectrum fluorescent lighting. *Ergonomics*, 44, 255–279.

Verwey, W.B. 2000. On-line driver workload estimation. Effects of road situation and age on secondary task measures. *Ergonomics*, 43, 187–209.

Vicente, K.J. 2002. Ecological interface design: Progress and challenges. *Human Factors*, 44, 62–78.

Vicente, K.J. & Rasmussen, J. 1992. Ecological interface design: Theoretical foundations. *IEEE Transactions on Systems, Man, and Cybernetics, SMC-22*, 589–606.

Vidulich, M.A., Tsang, P.S., & Flach, J.M. 2016. Aviation psychology: Optimizing human and system performance. In M.A. Vidulich, P.S. Tsang, & J.M. Flach, *Advances in Aviation Psychology* (pp. 3–16). New York: Routledge.

Vidulich, M.A., Ward, G.F. & Schueren, J. 1991. Using the subjective workload dominance technique (SWORD) technique for projective workload assessment. *Human Factors*, 33, 677–691.

Vidulich, M.A., Nelson, W.T. & Bolia, R.S. 2006. Speech-based controls in simulated air battle management. *International Journal of Aviation Psychology*, 16, 197–213.

Vieira, M.M., Ugrinowitsch, H., Gallo, L.G., Pinto Carvaiho, M.S., Fonseca, M.A., & Benda, R.N. 2014. Effects of summary knowledge of results of motor skill acquisition. *Revista De Psicología Del Deporte*, 23, 9–14.

Village, J., Salustri, F. A., & Neumann, W.P. 2013. Cognitive mapping: Revealing the links between human factors and strategic goals in organizations. *International Journal of Industrial Ergonomics*, 43, 304–313.

Vilnai-Yavetz, I., Rafaeli, A. & Yaacov, C.S. 2005. Instrumentality, aesthetics, and symbolism of office design. *Environment and Behavior*, 37, 533–551.

Vischer, J.C., & Wifi, M. 2017. The effect of workplace design on quality of life at work. In G. Fleury-Bahi, E. Pol, & O. Navarro (Eds.), *Handbook of Environmental Psychology and Quality of Life Research* (pp. 241–274). Switzerland: Springer.

Volk, F., Pappas, F., & Wang, H. 2011. User research: User-centered design methods for designing web interfaces. In K.-P.L. Vu & R.W. Proctor (Eds.), *Handbook of Human Factors in Web Design* (2nd ed., pp. 417–438). Boca Raton, FL: CRC Press.

Volpentesta, A.P. 2015. A framework for human interaction with mobiquitous services in a smart environment. *Computers in Human Behavior*, 50, 177–185.

von Bismarck, W.-B., Bungard, W., Maslo, J. & Held, M. 2000. Developing a system to support informal communication. In M. Vartiainen, F. Avallone & N. Anderson (Eds.), *Innovative Theories, Tools, and Practices in Work and Organizational Psychology* (pp. 187–203). Ashland, OH: Hogrefe & Huber.

von Winterfeldt, D. & Edwards, W. 1986. *Decision Analysis and Behavioral Research*. New York: Cambridge University Press.

Voss, J.F. & Post, T.A. 1988. On the solving of ill-structured problems. In M.T.H. Chi, R. Glaser & M.J. Farr (Eds.), *The Nature of Expertise* (pp. 261–285). Hillsdale, NJ: Lawrence Erlbaum.

Vu, K.-P.L., & Chiappe, D. 2015. Situation awareness in human systems integration. In D.A. Boehm-Davis, F.T. Durso, & J.D. Lee (Eds.), *APA Handbook of Human Systems Integration* (pp. 293–308). Washington, DC: American Psychological Association.

Vu, K.-P.L., Kiken, A., Chiappe, D., Strybel, T.Z., & Battiste, V. 2013. Application of part-whole training methods to evaluate when to introduce NextGen air traffic management tools to students. *American Journal of Psychology*, 126, 433–447.

Vu, K.-P.L., & Proctor, R.W. (Eds.) 2011. *Handbook of Human Factors in Web Design* (2nd ed.). Boca Raton, FL: CRC Press.

Vu, K.-P.L., Proctor, R.W., & Garcia, F.G. 2012. Web site design and evaluation. In G. Salvendy (Ed.), *Handbook of Human Factors and Ergonomics* (4th ed., pp. 1323–1353). Hoboken, NJ: John Wiley.

Vu, K.-P.L, Zhu, W., & Proctor, R.W. 2011. Evaluating Web usability. In K.-P.L. Vu & R.W. Proctor (Eds.), *Handbook of Human Factors in Web Design* (2nd ed., pp. 439–460). Boca Raton, FL: CRC Press.

Vuckovic, A., Kwantes, P.J., Humphreys, M., & Neal, A. 2014. A sequential sampling account of response bias and speed–accuracy tradeoffs in a conflict detection task. *Journal of Experimental Psychology: Applied*, 20, 55–68.

Wachtler, T., Dohrmann, U. & Hertel, R. 2004. Modeling color percepts of dichromats. *Vision Research*, 44, 2843–2855.

Wade, N., & Swanston, M. 2013. *Visual Perception: An Introduction* (3rd ed.). New York: Psychology Press.

Wade, N.J. & Heller, D. 2003. Visual motion illusions, eye movements, and the search for objectivity. *Journal of the History of the Neurosciences*, 12, 376–395.

Walker, B.N. & Kramer, G. 2006. Auditory displays, alarms, and auditory interfaces. In W. Karwowski (Ed.), *International Encyclopedia of Ergonomics and Human Factors* (2nd ed., Vol. 1, pp. 1021–1025). Boca Raton, FL: CRC Press.

Wallace, R.J. 1971. S–R compatibility and the idea of a response code. *Journal of Experimental Psychology*, 88, 354–360.

Wallace, S.A. & Carlson, L.E. 1992. Critical variables in the coordination of prosthetic and normal limbs. In G.E. Stelmach & J. Requin (Eds.), *Tutorials in Motor Behavior II* (pp. 321–341). Amsterdam: Northe-Holland.

Wallach, H. 1972. The perception of neutral colors. In T. Held & W. Richards (Eds.), *Perception: Mechanisms and Models: Readings from Scientific American* (pp. 278–285). San Francisco, CA: Freeman.

Waller, W.S. & Zimbelman, M.F. 2003. A cognitive footprint in archival data: Generalizing the dilution effect from laboratory to field settings. *Organizational Behavior and Human Decision Processes*, 91, 254–268.

Walsh, V. & Kulikowski, J. (Eds.) 1998. *Perceptual Constancy: Why Things Look the Way They Do*. Cambridge: Cambridge University Press.

Walters, M.L., Dautenhahn, K., Te Boekhorst, R., Koay, K.L., Syrdal, D.S., & Nehaniv, C.L. 2009. An empirical framework for human-robot proxemics. *Proceedings of New Frontiers in Human-Robot Interaction*.

Wang, F., Davies, H., Du, B., & Johnson, P.W. 2016. Comparing the whole body vibration exposures across three truck seats. In *Proceedings of the Human Factors and Ergonomics Society Annual Meeting* (Vol. 60, No. 1, pp. 933–936). Thousand Oaks, CA: Sage.

Wann, J. & Mon-Williams, M. 1996. What does virtual reality NEED? Human factors issues in the design of three-dimensional computer environments. *International Journal of Human–Computer Studies*, 44, 829–847.

Wardell, R. 2005. Product liability for the human factors practitioner. In Y.I. Noy & W. Karwowski (Eds.), *Handbook of Human Factors in Litigation* (pp. 29-1–29-6). Boca Raton, FL: CRC Press.

Wargocki, P., Wyon, D.P., Baik, Y.K., Clausen, G. & Fanger, P.O. 1999. Perceived air quality, sick building syndrome (SBS) symptoms and productivity in an office with two different pollution loads. *Indoor Air*, 9, 165–179.

Warm, J.S. 1984. An introduction to vigilance. In J.S. Warm (Ed.), *Sustained Attention in Human Performance* (pp. 1–14). New York: Wiley.

Warm, J.S., Finomore, V.S., Vidulich, M.A., & Funke, M.E. 2015. Vigilance: A perceptual challenge. In R.R. Hoffman, P.A. Hancock, M.W. Scerbo, R. Parasuraman, & J.L. Szalma (Eds.), *The Cambridge Handbook of Applied Perception Research* (Vol. 1, pp. 241–283). New York: Cambridge University Press.

Warm, J.S., Parasuraman, R., & Matthews, G. 2008. Vigilance requires hard mental work and is stressful. *Human Factors*, 50, 433–441.

Warren, R.M. 1970. Perceptual restoration of missing speech sounds. *Science*, 167, 392–393.

Warrick, M.J. & Turner, L. 1963. Simultaneous activation of bimanual controls. Aerospace Medical Research Laboratories Technical Documentary Report No. AMRL-TDR-63-6.

Warrick, M.J., Kibler, A.W. & Topmiller, D.A. 1965. Response time to unexpected stimuli. *Human Factors*, 7, 81–86.

Wason, P. 1969. Regression in reasoning. *British Journal of Psychology*, 60, 471–480.

Wasserman, D.E. 2006. Human exposure to vibration. In W. Karwowski (Ed.), *International Encyclopedia of Ergonomics and Human Factors* (2nd ed., Vol. 2, pp. 1800–1801). Boca Raton, FL: CRC Press.

Waterhouse, J.M., Minors, D.S., Åkerstedt, T., Reilly, T. & Atkinson, G. 2001. Rhythms of human performance. In J.S. Takahashi, F.W. Turek & R.Y. Moore (Eds.), *Handbook of Behavioral Neurology* (Vol. 12: *Circadian Clocks*, pp. 571–601). New York: Kluwer Academic/Plenum Press.

Waterhouse, J., Reilly, T., Atkinson, G., & Edwards, B. 2007. Jet lag: Trends and coping strategies. *Lancet*, 369, 1117–1129.

Waters, T.R., Putz-Anderson, V., Garg, A. & Fine, L.J. 1993. Revised NIOSH equation for the design and evaluation of manual lifting tasks. *Ergonomics*, 36, 749–776.

Waters, T.R., Putz-Anderson, V. & Garg, A. 1994. Applications manual for the revised NIOSH lifting equation. DHHS (NIOSH) Publication No. 94–110. Washington, DC: National Institute for Occupational Safety and Health.

Watson, A.B. 1986. Temporal sensitivity. In K.R. Boff, L. Kaufman & J.P. Thomas (Eds.), *Handbook of Human Perception and Performance* (Vol. I: *Sensory Processes and Perception*, pp. 6-1–6-43). New York: Wiley.

Watson, A. B., & Yellott, J. I. 2012. A unified formula for light-adapted pupil size. *Journal of Vision*, 12(10), 1–16.

Waugh, N.C. & Norman, D.A. 1965. Primary memory. *Psychological Review*, 72, 89–104.

Web Institute for Teachers 2002. *Krazy Keyboarding for Kids*. Downloaded on February 28, 2006, from http://webinstituteforteachers.org/~gammakeys/teachingguide.html

Webb, J.D. & Weber, M.J. 2003. Influence of sensory abilities on the interpersonal distance of the elderly. *Environment and Behavior*, 35, 695–711.

Weber, E.H. 1846/1978. Der tastsinn und das gemeingefühl. In H.E. Ross & D.J. Murray (D.J. Murray, Trans.), *E.H. Weber: The Sense of Touch*. London: Academic Press.

Webster, J.C. & Klumpp, R.G. 1963. Articulation index and average curve-fitting methods of predicting speech interference. *Journal of the Acoustical Society of America*, 35, 1339–1344.

Weigelt, M., Kunde, W., & Prinz, W. 2006. End-state comfort in bimanual object manipulation. *Experimental Psychology*, 53, 143–148.

Weinstein, A.S., Twerski, A.D., Piehler, H.R. & Donaher, W.A. 1978. *Product Liability and the Reasonably Safe Product*. New York: Wiley.

Welford, A.T. 1952. The 'psychological refractory period' and the timing of high speed performance— A review and a theory. *British Journal of Psychology*, 43, 2–19.

Welsh, M.C. & Huizinga, M. 2005. Tower of Hanoi disk-transfer task: Influences of strategy knowledge and learning on performance. *Learning and Individual Differences*, 15, 283–298.

Weltman, G. & Egstrom, G.H. 1966. Perceptual narrowing in novice divers. *Human Factors*, 8, 499–505.

Wertheim, A.H. 2010. Visual conspicuity: A new simple standard, its reliability, validity and applicability. *Ergonomics*, 53, 421–442.

Wertheim, A.H., Hooge, I.C., & Smeets, P.M. 2011. Conspicuity or visibility: What may cause an object to draw attention? *Food Quality and Preference*, 22, 602.

West, L.J. & Sabban, Y. 1982. Hierarchy of stroking habits at the typewriter. *Journal of Applied Psychology*, 67, 370–376.

Westheimer, G. 2005. Anisotropies in peripheral vernier acuity. *Spatial Vision*, 18, 159–167.

Weston, H.C. 1945. The relationship between illuminance and visual efficiency—the effect of brightness and contrast (Industrial Health Research Board Report No. 87). London: Great Britain Medical Research Council.

Wey, W.-M. 2005. An integrated expert system/operations research approach for the optimization of waste incinerator siting problems. *Knowledge-Based Systems*, 18, 267–278.

Weyers, B.W., Bowen, J., Dix, A., & Palanque, P. 2017. *The Handbook of Formal Methods in Human-Computer Interaction*. New York: Springer.

Wheeler, J. 1989. More thoughts on the human factors of expert systems development. *Human Factors Society Bulletin*, 32(12), 1–4.

White, W.J., Warrick, M.J. & Grether, W.F. 1953. Instrument reading: III. Check reading of instrument groups. *Journal of Applied Psychology*, 37, 302–307.

Whitehurst, H.O. 1982. Screening designs used to estimate the relative effects of display factors on dial reading. *Human Factors*, 24, 301–310.

Whiting, H.T.A. 1969. *Acquiring Ball Skill: A Psychological Interpretation*. London: G. Bell.

Whittingham, R.B. 1988. The application of the combined THERP/HCR model in human reliability assessment. In B.A. Sayers (Ed.), *Human Factors and Decision Making* (pp. 126–138). New York: Elsevier Applied Science.

Wickelgren, W.A. 1964. Size of rehearsal group in short-term memory. *Journal of Experimental Psychology*, 68, 413–419.

Wickens, C.D. 1976. The effects of divided attention in information processing in tracking. *Journal of Experimental Psychology: Human Perception and Performance*, 2, 1–13.

Wickens, C.D. 1984. Processing resources in attention. In R. Parasuraman & R. Davies (Eds.), *Varieties of Attention* (pp. 63–102). San Diego, CA: Academic Press.

Wickens, C.D. 2002a. Multiple resources and performance prediction. *Theoretical Issues in Ergonomics Science*, 3, 159–177.

Wickens, C.D. 2002b. Situation awareness and workload in aviation. *Current Directions in Psychological Science*, 11, 128–133.

Wickens, C.D. & Andre, A.D. 1990. Proximity compatibility and information display: Effects of color, space, and objectness on information integration. *Human Factors*, 32, 61–77.

Wickens, C.D. & Carswell, C.M. 2012. Information processing. In G. Salvendy (Ed.), *Handbook of Human Factors and Ergonomics* (4th ed., pp. 111–149). Hoboken, NJ: Wiley.

Wickens, C.D., Hyman, F., Dellinger, J., Taylor, H. & Meador, M. 1986. The Sternberg memory search task as an index of pilot workload. *Ergonomics*, 29, 1371–1383.

Wickens, C. D., Hutchins, S., Carolan, T., & Cumming, J. 2013. Effectiveness of part-task training and increasing-difficulty training strategies: A meta-analysis approach. *Human Factors*, 55, 461–470.

Wickens, C.D., & McCarley, J.S. 2008. *Applied Attention Theory*. Boca Raton, FL: CRC Press.

Wickens, C.D., Sandry, D.L. & Vidulich, M. 1983. Compatibility and resource competition between modalities of input, central processing, and output. *Human Factors*, 25, 227–248.

Wickens, C.D., & Tsang, P. 2015. Workload. In D.A. Boehm-Davis, F.T. Durso, & J.D. Lee (Eds.-in-Chief), *APA Handbook of Human Systems Integration* (pp. 277–292). Washington, DC: American Psychological Association.

Wickens, D.D. 1972. Characteristics of word encoding. In A.W. Melton & E. Martin (Eds.), *Coding Processes in Human Memory* (pp. 191–215). Washington, DC: Winston.

Wicklund, M.E. & Loring, B.A. 1990. Human factors design of an AIDS prevention pamphlet. *Proceedings of the Human Factors Society* 34th *Annual Meeting* (Vol. 2, pp. 988–992). Santa Monica, CA: Human Factors Society.

Wiegmann, D.A. & Shappell, S.A. 1997. Human factors analysis of postaccident data: Applying theoretical taxonomies of human error. *International Journal of Aviation Psychology*, 7, 67–81.

Wiegmann, D.A. & Shappell, S.A. 2001. Human error perspectives in aviation. *International Journal of Aviation Psychology*, 11, 341–357.

Wiegmann, D.A. & Shappell, S.A. 2003. *A Human-Error Approach to Aviation Accident Analysis: The Human Factors Analysis and Classification System*. Burlington, VT: Ashgate.

Wiegmann, D., Faaborg, T., Boquet, A., Detwiler, C., Holcomb, K. & Shappell, S.A. 2005. Human error and general aviation accidents [electronic resource]: A comprehensive, fine-grained analysis using HFACS. Final report DOT/FAA/AM-05/24. Oklahoma City, OK: Federal Aviation Administration.

Wiener, E.L., 1964. Transfer of training in monitoring. *Perceptual and Motor Skills*, 18, 104.

Wieringa, P.A. & van Wijk, R.A. 1997. Operator support and supervisory control. In T.B. Sheridan & S. Van Luntern (Eds.), *Perspectives on the Human Controller: Essays in Honor of Henk G. Stassen* (pp. 251–260). Mahwah, NJ: Lawrence Erlbaum.

Wierwille, W. 1984. The design and location of controls: A brief review and an introduction to new problems. In H. Schmidtke (Ed.), *Ergonomic Data for Equipment Design* (pp. 179–194). New York:

Wierwille, W.W. & Casali, J.G. 1983. A validated rating scale for global mental workload measurement applications. *Proceedings of the Human Factors Society*, 27, 129–133.

Wightman, D.C. & Lintern, G. 1985. Part-task training for tracking and manual control. *Human Factors*, 27, 267–283.

Wiker, S.F., Langolf, G.D. & Chaffin, D.B. 1989. Arm posture and human movement capability. *Human Factors*, 31, 421–441.

Wilde, G. & Humes, L.E. 1990. Application of the articulation index to the speech recognition of normal and impaired listeners wearing hearing protection. *Journal of the Acoustical Society of America*, 87, 1192–1199.

Williams, C.D. 2003. A novel redesign of food scoops in high volume food service organizations. *Work: Journal of Prevention, Assessment and Rehabilitation*, 20, 131–135.

Williams, J. & Singer, R.N. 1975. Muscular fatigue and the learning and performance of a motor control task. *Journal of Motor Behavior*, 7, 265–269.

Williams, L.J. 1985. Tunnel vision induced by a foveal load manipulation. *Human Factors*, 27, 221–227.

Wilson, G.F. & O'Donnell, R.D. 1988. Measurement of operator workload with the neuropsychological workload test battery. In P.A. Hancock & N. Meshkati (Eds.), *Human Mental Workload* (pp. 63–100). Amsterdam: North-Holland.

Winstein, C.J. & Schmidt, R.A. 1990. Reduced frequency of knowledge of results enhances motor skill learning. *Journal of Experimental Psychology: Learning, Memory, and Cognition*, 16, 677–691.

Winstein, C. & Schmidt, R. 1989. Sensorimotor feedback. In D.H. Holding (Ed.), *Human Skills* (2nd ed., pp. 17–47). New York: Wiley.

Wise, J.A. 1986. The space station: Human factors and habitability. *Human Factors Society Bulletin*, 29(5), 1–3.

Wogalter, M.S., Silver, N.C., Leonard, S.D. & Zaikina, H. 2006. Warning symbols. In M.S. Wogalter (Ed.), *Handbook of Warnings* (pp. 159–176). Mahwah, NJ: Lawrence Erlbaum.

Wohldmann, E.L., Healy, A.F. & Bourne, L.E., Jr. 2007. Pushing the limits of imagination: Mental practice for learning sequences. *Journal of Experimental Psychology: Learning, Memory, and Cognition*, 33, 254–261.

Woldstad, J.C. 2006. Digital human models for ergonomics. In W. Karwowski (Ed.), *International Encyclopedia of Ergonomics and Human Factors* (2nd ed., Vol. 3, pp. 3093–3096). Boca Raton, FL: CRC Press.

Wolfe, J.M., Kluender, K.R., Levi, D.M., Bartoshuk, L.M., Herz, R.S., Klatzky, R.L. & Lederman, S.J., & Merfeld, D.M. 2015. *Sensation and Perception* (4th ed.). Sunderland, MA: Sinauer Associates.

Wolfe, U., & Ali, N. 2015. Dark adaptation and Purkinje shift: A laboratory exercise in perceptual neuroscience. *The Journal of Undergraduate Neuroscience Education*, 13(2), A59–A63

Wolfowitz, P. 2005. Department of defense directive. No. 1000.4, February 12, 2005. Downloaded on August 19, 2006, from http://www.dtic.mil/whs/directives/corres/rtf/d11004x.rtf

Wolpert, L. 2004. Foreword. In R.G. Foster & L. Kreitzman, *Rhythms of Life: The Biological Clocks That Control the Lives of Every Living Thing*. New Haven, CT: Yale University Press.

Wolska, A. 2006a. Human aspects of lighting in working interiors. In W. Karwowski (Ed.), *International Encyclopedia of Ergonomics and Human Factors* (2nd ed., Vol. 2, pp. 1793–1799). Boca Raton, FL: CRC Press.

Wolska, A. 2006b. Lighting equipment and lighting systems. In W. Karwowski (Ed.), *International Encyclopedia of Ergonomics and Human Factors* (2nd ed., Vol. 2, pp. 1810–1816). Boca Raton, FL: CRC Press.

Wood, J.M. 2002. Age and visual impairment decrease driving performance as measured on a closed-road circuit. *Human Factors*, 44, 482–494.

Wood, J.M. & Troutbeck, R. 1994. Effect of visual impairment on driving. *Human Factors*, 36, 476–487.

Wood, N. & Cowan, N. 1995a. The cocktail party phenomenon revisited: Attention and memory in the classic selective listening procedure of Cherry 1953. *Journal of Experimental Psychology: General*, 124, 243–262.

Wood, N. & Cowan, N. 1995b. The cocktail party phenomenon revisited: How frequent are attention shifts to one's name in an irrelevant auditory channel? *Journal of Experimental Psychology: Learning, Memory*, 21, 255–260.

Woods, D.L., Alain, C., Diaz, R., Rhodes, D. & Ogawa, K.H. 2001. Location and frequency cues in auditory selective attention. *Journal of Experimental Psychology: Human Perception and Performance*, 27, 65–74.

Woodson, W.E., Tillman, B. & Tillman, P. 1992. *Human Factors Design Handbook: Information and Guidelines for the Design of Systems, Facilities, Equipment, and Products for Human Use*. New York: McGraw-Hill.

Woodward, A.E., Bjork, R.A. & Jongeward, R.H. 1973. Recall and recognition as a function of primary rehearsal. *Journal of Verbal Learning and Verbal Behavior*, 12, 608–617.

Woodworth, R.S. 1899. The accuracy of voluntary movement. *Psychological Review Monograph Supplements*, 3, 1–119.

Woodworth, R.S. & Sells, S.B. 1935. An atmosphere effect in formal syllogistic reasoning. *Journal of Experimental Psychology*, 18, 451–460.

Woolford, B. & Mount, F. 2006. Human space flight. In G. Salvendy (Ed.), *Handbook of Human Factors and Ergonomics* (3rd ed., pp. 929–944). Hoboken, NJ: Wiley.

Worledge, D.H., Joksimovich, V. & Spurgin, A.J. 1988. Interim results and conclusions of the EPRI operator reliability experiments program. In E.W. Hagen (Ed.), *1988 IEEE Fourth Conference on Human Factors and Power Plants* (pp. 315–322). New York: Institute of Electrical and Electronics Engineers.

Worringham, C.J. & Beringer, D.B. 1989. Operator orientation and compatibility in visual-motor task performance. *Ergonomics*, 32, 387–399.

Worringham, C.J. & Beringer, D.B. 1998. Directional stimulus–response compatibility: A test of three alternative principles. *Ergonomics*, 41, 864–880.

Worthen, J.B., & Hunt, R.R. 2011. *Mnemonology: Mnemonics for the 21st Century*. New York: Psychology Press.

Wright, C.E., Marino, V.F., Chubb, C., & Rose, K.A. 2011. Exploring attention-based explanations for some violations of Hick's law for aimed movements. *Attention, Perception, & Psychophysics*, 73, 854–871.

Wright, G. 1984. *Behavioral Decision Theory*. Beverly Hills, CA: Sage.

Wright, P. 1976. The harassed decision maker: Time pressures, distraction, and the use of evidence. *Journal of Applied Psychology*, 59, 555–561.

Wright, P. 1988. Functional literacy: Reading and writing at work. *Ergonomics*, 31, 265–290.

Wulf, G., Horstmann, G. & Choi, B. 1995. Does mental practice work like physical practice without information feedback? *Research Quarterly for Exercise and Sport*, 66, 262–267.

Wulf, G., & Lewthwaite, R. 2016. Optimizing performance through intrinsic motivation and attention for learning: The optimal theory of motor learning. *Psychonomic Bulletin & Review*, 23, 1382–1414.

Wulf, G., Shea, C.H. & Matschiner, S. 1998. Frequent feedback enhances complex motor skill learning. *Journal of Motor Behavior*, 30, 180–192.

Yamaguchi, M., & Proctor, R. W. 2012. Multidimensional vector model of stimulus–response compatibility. *Psychological Review*, 119, 272–303.

Yamana, N., Kabek, O., Nanako, C., Zenitani, Y. & Saita, T. 1984. The body form of pregnant women in monthly transitions. *Japanese Journal of Ergonomics*, 20, 171–178.

Yang, W., Li, N., Chowdhury, O., Xiong, A., & Proctor, R.W. 2016. An empirical study of password generation strategies. *Proceedings of the 23rd ACM Conference on Computer and Communications Security (CCS 2016)*. New York: Association of Computing Machinery.

Yang, W., Xiong, A., Chen, J., Proctor, R.W., & Li, N. 2017. Use of phishing training to improve security warning compliance: Evidence from a field experiment. In *4th Annual Hot Topics in the Science of Security (HoTSoS) Symposium and Bootcamp*. New York: ACM.

Yantis, S. 1988. On analog movements of visual attention. *Perception & Psychophysics*, 43, 203–206.

Yeh, M. & Wickens, C.D. 2001. Attentional filtering in the design of electronic map displays: A comparison of color coding, intensity coding, and decluttering techniques. *Human Factors*, 43, 543–562.

Yerkes, R.M. & Dodson, J.D. 1908. The relation of strength of stimulus to rapidity of habit-formation. *Journal of Comparative Neurology of Psychology*, 18, 459–482.

Yoshikawa, H. & Wu, W. 1999. An experimental study on estimating human error probability (HEP) parameters for PSA/HRA by using human model simulation. *Ergonomics*, 11, 1588–1595.

Yost, W. A. 2013. In & A.F. Healy R.W. Proctor (Eds.), *Exeperimental Psychology* (pp. 120–151). Volume 4 in I.B. Weiner (Editor-in-Chief) *Handbook of Psychology* (2nd ed.). Hoboken, NJ: John Wiley.

You, H., Simmons, Z., Freivalds, A., Kothari, M.J., Naidu, S.H. & Young, R. 2004. The development of risk assessment models for carpal tunnel syndrome: A case-referent study. *Ergonomics*, 47, 688–709.

Young, E. D. 2007. Physiological acoustics. In T. Rossing (Ed.), *Springer Handbook of Acoustics* (pp. 429–458). New York: Springer.

Young, L.R. 2000. Vestibular reactions to spaceflight: Human factors issues. *Aviation Space and Environmental Medicine*, 71(9, Sect. 2, Suppl.), A100–A104.

Young, M.S., Brookhuis, K.A., Wickens, C.D., & Hancock, P.A. 2015. State of science: Mental workload in ergonomics. *Ergonomics*, 58, 1–17.

Young, M.S. & Stanton, N.A. 2002. Malleable attention resources theory: A new explanation for the effects of mental underload on performance. *Human Factors*, 44, 365–375.

Young, M.S. & Stanton, N.A. 2006. Mental workload: Theory, measurement, and application. In W. Karwowski (Ed.), *International Encyclopedia of Ergonomics and Human Factors* (2nd ed., Vol. 1, pp. 818–821). Boca Raton, FL: CRC Press.

Young, S.L. & Wogalter, M.S. 1990. Comprehension and memory of instruction manual warnings: Conspicuous print and pictorial icons. *Human Factors*, 32, 637–649.

Young, T. 1802. On the theory of light and colours. *Philosophical Transactions of the Royal Society*, 92, 12–48.

Yu, C.-Y. & Keyserling, W.M. 1989. Evaluation of a new workseat for industrial seating operations. *Applied Ergonomics*, 20, 17–25.

Zamanali, J. 1998. Probabilistic-risk-assessment applications in the nuclear-power industry. *IEEE Transactions on Reliability*, 47, 361–364.

Zannin, P.H.T. & Gerges, S.N.Y. 2006. Effects of cup, cushion, headband force, and foam lining on the attenuation of an earmuff. *International Journal of Industrial Ergonomics*, 36, 165–170.

Zarcadoolas, C., Vaughon, W.L., Czaja, S.J., Levy, J., & Rockoff, M.L. 2013. Consumers' perceptions of patient-accessible electronic medical records. *Journal of Medical Internet Research*, 15, 284–300.

Zechmeister, E.B. & Nyberg, S.E. 1982. *Human Memory: An Introduction to Research and Theory*. Monterey, CA: Brooks/Cole.

Zelaznik, H.N., Hawkins, B. & Kisselburgh, L. 1983. Rapid visual feedback processing in single-aiming movements. *Journal of Motor Behavior*, 15, 217–236.

Zelaznik, H.N., Spencer, R.M.C., Ivry, R.B., Baria, A., Bloom, M., Dolansky, L., Justice, S., Patterson, K. & Whetter, E. 2005. Timing variability in circle drawing and tapping: Probing the relationship between event and emergent timing. *Journal of Motor Behavior*, 37, 395–403.

Zennaro, D., Laubli, T., Krebs, D., Krueger, H. & Klipstein, A. 2004. Trapezius muscle motor unit activity in symptomatic participants during finger tapping using properly and improperly adjusted desks. *Human Factors*, 46, 252–266.

Zhang, Y. & Luximon, A. 2006. Voice-enhanced interface. In W. Karwowski (Ed.), *International Encyclopedia of Ergonomics and Human Factors* (2nd ed., Vol. 1, pp. 1357–1360). Boca Raton, FL: CRC Press.

Zhao, Y., Hignett, S., & Mansfield, N.J. 2014. Development and Testing of a New Computerized Link Analysis System. *Human Factors and Ergonomics in Manufacturing & Service Industries*, 24, 479–488.

Zhou, B. & Lukowicz, P. 2017. Textile pressure force mapping. In S. Schneegass & O. Amft (Eds.), Smart Textiles: Fundamentals, Design, and Interaction (pp. 31–47). Switzerland: Springer International.

Zukauskas, A., Shur, M.S & Gaska, R. 2002. *Introduction to Solid-State Lighting*. New York: Wiley.

Zwaga, H.J. 1989. Comprehensibility estimates of public information symbols: Their validity and use. *Proceedings of the Human Factors Society 33rd Annual Meeting* (Vol. 2, pp. 979–983). Santa Monica, CA: Human Factors Society.

Zwicker, E. 1958. Uber psychologische und methodische Grundlagen der Lautheit. *Acustica*, 8, 237–258.

List of Credits

Figure 1.1 Courtesy of the Doug Engelbart Institute dougengelbar.org.

Figure 1.2 Photo courtesy of Karl Van Zandt.

Figure 1.5 Courtesy of Wellcome Library, London. Wellcome Images images@wellcome.ac.uk http://wellcomeimages.org.

Figure 1.6 Courtesy of NDI.

Table 1.1 Courtesy of the Human Factors and Ergonomics Society https://www.hfes.org/web/TechnicalGroups/descriptions.html.

Figure 2.1 From Stokes, D.E. 1997. *Pasteur's Quadrant: Basic Science and Technological Innovation.* (p. 196). Washington, DC: Brookings.

Figure 2.8 From Marras, W.S. and Kroemer, K.H.E., *Human Factors*, 22, 389–399, 1980. Copyright 1980 by the Human Factors Society, Inc. All rights reserved.

Table 2.1 Estimated from Lovasik, J.V. et al., *Human Factors*, 31, 273–289, 1989.

Table 2.4 From Marras, W.S. and Kroemer, K.H.E., *Human Factors*, 22, 389–399, 1980.

Figure 3.2 From D.A. Wiegmann & S.A. Shappell (2003). *A Human Error Approach to Aviation Accident Analysis.* Burlington, VT: Ashgate. Copyright 2003 by Ashgate Publishing.

Table 3.1 From Meister, D. 1989. *Conceptual Aspects of Human Factors.* Baltimore, MD: Johns Hopkins University Press.

Table 3.2 From Berliner, D.C., Angell, D. & Shearer, J.W. 1964. Behaviors, measures and instruments for performance evaluation in simulated environments. Paper presented at a Symposium and Workshop on Quantification of Human Performance, Albuquerque, NM.

Figure B3.1 From R. Schweickert, D.L. Fisher, & R.W. Proctor (2003). Steps toward building mathematical and computer models from cognitive task analyses. Reprinted from *Human Factors*, 45(1), 77–103. Copyright 2003 by the Human Factors and Ergonomics Society, Inc. All rights reserved.

Figure 3.6 From K.R. Boff & J.E. Lincoln (Eds.) (1988). *Engineering Data Compendium: Human Perception and Performance.* Wright-Patterson AFB, OH: AAMRL.

Figure 3.10 From K.R. Boff & J.E. Lincoln (Eds.) (1988). *Engineering Data Compendium: Human Perception and Performance.* Wright-Patterson AFB, OH: AAMRL.

Figure 4.1 From J.T. Townsend & R.N. Roos (1973). *Memory & Cognition,* 1, 319–332. Reprinted by permission of Psychonomic Society, Inc.

Figure 4.3 Adapted from B.H. Kantowitz, The role of human information processing models in system development. Adapted with permission from *Proceedings of the Human Factors Society 33rd Annual Meeting,* 1989, pp. 979–983. Copyright 1989 by the Human Factors Society, Inc. All rights reserved.

Figure B4.1 From D.E. Meyer & D.E. Kieras (1997). A computational theory of executive cognitive processes and multiple-task performance: I. Basic mechanisms. *Psychological Review,* 104(1), 3–65. Published 1997 by the American Psychological Association. Reprinted with permission.

Figure 5.2   Public domain, NIH National Eye Institute.

Figure 5.3   From M. W. Matlin & H. J. Foley (1992). *Sensation and Perception,* Third Edition. Copyright 1992 by Allyn and Bacon. Reprinted with permission.

Figure 5.4   From J.M. Wolfe, K.R. Kluender, D.M. Levi, L.M. Bartoshuk, R.S. Herz, R.L. Klatzky, Lederman, S. J., & Merfeld, D. M. (2015). *Sensation and Perception* (4th ed.) Sunderland, MA: Sinauer Associates. Adapted with permission.

Figure 5.5   From M. W. Matlin & H. J. Foley (1992). *Sensation and Perception,* Third Edition. Copyright 1992 by Allyn and Bacon. Reprinted with permission.

Figure 5.7   From M.W. Matlin (1992). *Sensation and Perception,* Third Edition. Copyright 1992 by Allyn and Bacon. Reprinted with permission.

Figure 5.13   Adapted with permission from A.L. Gilchrist (1977). Perceived lightness depends on perceived spatial arrangement, *Science,* 195, 185–187. Copyright 1977 by the AAAS.

Figure 5.17   From A.B. Watson (1986). Temporal sensitivity. In K.R. Boff & J.P. Thomas (Eds.), *Handbook of Perception and Human Performance.* Copyright 1986 by John Wiley and Sons, Inc. Reprinted with permission.

Table 6.1   From K.R. Boff & J.E. Lincoln (Eds.) (1988). *Engineering Data Compendium: Human Perception and Performance.* Wright-Patterson AFB, OH: AAMRL.

Figure 6.2   From G. Wyszecki (1986). Color appearance. In K.R. Boff & J.E. Lincoln (Eds.) (1988). *Engineering Data Compendium: Human Perception and Performance.* Copyright 1986 Reprinted with permission of John Wiley & Sons, Inc.

Figure 6.3   From K.R. Boff & J.E. Lincoln (Eds.) (1988). *Engineering Data Compendium: Human Perception and Performance.* Wright-Patterson AFB, OH: AAMRL.

Figure 6.6   Provided by NASA.

Figure 6.8   From J. Beck (1966). Effect of orientation and of shape similarity on perceptual grouping. *Perception & Psychophysics,* 1, 300–302. Reprinted by permission of Psychonomic Society, Inc.

Figure 6.10   From W. Prinzmetal & W. P. Banks (1977). Good continuation affects visual detection. *Perception & Psychophysics,* 21, 389–395. Reprinted by permission of Psychonomic Society, Inc.

Figure 6.11   From R. Sekuler & R. Blake (1994). *Perception,* Third Edition. Copyright 1994 Reprinted with permission of the McGraw-Hill Companies.

Figure 6.12   Adapted from Hochberg, 1978.

Figure 6.13   From Victoria and Albert Museum, London.

Figure 6.14   From M. Kubovy(1988). *The Psychology of Perspective and Renaissance Art.* Copyright 1988 by Cambridge University Press. Reprinted by permission of Michael Kubovy.

Figure 6.16   From D.A. Kleffner & V.S. Ramachandran (1992). On the perception of shape from shading. *Perception & Psychophysics,* 52, 18–36. Reprinted by permission of Psychonomics Society, Inc.

Figure 6.19   From B. Julesz (1971). *Foundations of Cyclopean Perception,* University of Chicago Press. Copyright 1971 by Bell Telephone Laboratories, Inc. Reprinted by permission.

Figure 6.20   By Fred Hsu (March, 2005; Wikipedia Commons). Reproduced under the terms of the GNU Free Documentation License.

Figure 8.5　From H.J. Zwaga, Comprehensibility estimates of public information symbols: Their validity and use. Reprinted with permission from *Proceedings of the Human Factors Society 33rd Annual Meeting,* 1989, pp. 979–983. Copyright 1989 by the Human Factors Society, Inc. All rights reserved.

Figure 8.6　From M.E. Wicklund & B.A. Loring, Human factors design of an AIDS prevention pamphlet. Reprinted with permission from *Proceedings of the Human Factors Society 34th Annual Meeting,* 1990, pp. 988–992. Copyright 1990 by the Human Factors Society, Inc. All rights reserved.

Figure 8.7　From M.E. Wicklund & B.A. Loring, Human factors design of an AIDS prevention pamphlet. Reprinted with permission from *Proceedings of the Human Factors Society 34th Annual Meeting,* 1990, pp. 988–992. Copyright 1990 by the Human Factors Society, Inc. All rights reserved.

Figure 8.8　Adapted from R.S. Easterby (1970). The perception of symbols for machine displays. *Ergonomics,* 13, 149–158. Adapted with permission of Taylor & Francis.

Figure 8.9　From J.D. Hitt (1961). An evaluation of five different abstract coding methods – Experiment IV. Reprinted with permission from *Human Factors,* 3(2), 120–130. Copyright 1961 by the Human Factors Society, Inc. All rights reserved.

Figure 8.10　From Hitt, J.D. 1961. An evaluation of five different abstract coding methods— Experiment IV. *Human Factors,* 3, 120–130.

Figure 8.11　From M.G. Helander (1987). Design of visual displays. In G. Salvendy (Ed.), *Handbook of Human Factors* (pp. 507–549). Copyright 1987 Reprinted with permission of John Wiley & Sons, Inc.

Figure 8.14　Photograph by Rama, Wikimedia Commons, Cc-by-sa-2.0-fr.

Figure 8.15　Image courtesy of BAE Systems.

Figure 8.16　From J.C.R. Licklider, D. Bindra, & I. Pollack (1948). The intelligibility of rectangular speech waves. *American Journal of Psychology,* 61, 1–20. Reprinted by permission of University of Illinois Press.

Table 8.1　From B.H. Deatherage (1972). Auditory and other sensory forms of information presentation. In H.P. Van Cott & R.G. Kinkade (Eds.), *Human Engineering Guide to Equipment Design* (revised edition, pp. 123–160). U.S. Government Printing Office.

Table 8.2　From Woodson, W.E., *Human Factors Design,* Handbook, 483. Copyright © 1981 McGraw-Hill, Inc. Reprinted with permission.

Table 8.3　From W.F. Grether & C.A. Baker (1972). Visual presentation of information. In H.P. Van Cott & R.G. Kinkade (Eds.), *Human Engineering Guide to Equipment Design* (revised edition, pp. 123–160). U.S. Government Printing Office.

Table 8.4　From M.G. Helander (1987). Design of visual displays. In G. Salvendy (Ed.), *Handbook of Human Factors* (pp. 507–549). Copyright 1987 Reprinted with permission of John Wiley & Sons, Inc.

Table 8.5　From K.R. Boff & J.E. Lincoln (Eds.) (1988). *Engineering Data Compendium: Human Perception and Performance.* Wright-Patterson AFB, OH: AAMRL.

Figure 9.3　From D. Kahneman (1973). *Attention and Effort.* Copyright 1973 by Prentice Hall, Inc. Reprinted by permission of Daniel Kahneman.

Figure 9.4   From M.I. Posner & S.J. Boies (1971). Components of attention. *Psychological Review*, 78, 391–408. Reprinted by permission of the American Psychological Association.

Figure 9.5   From C.D. Wickens (1984). Processing resources in attention. In R. Parasuraman & R. Davies (Eds.), *Varieties of Attention* (pp. 63–102). Copyright 1984 by Academic Press, Inc. Reprinted by permission.

Figure 9.8   From R.W.H.M. Ponds, W.B. Brouwer, & P.C. Van Wolffelaar (1988). Age differences in divided attention in a simulated driving task. *Journal of Gerontology*, 43, P151–P156. Copyright 1988 by the Gerontological Society of America. Reprinted with permission.

Figure 9.10   From K.R. Boff & J.E. Lincoln (Eds.) (1988). *Engineering Data Compendium: Human Perception and Performance*. Wright-Patterson AFB, OH: AAMRL.

Figure 9.11   From R.J. Lysaght et al. (1989). *Operator Workload: Comprehensive Review and Evaluation of Operator Workload Methodologies*. Technical Report 851, United States Army Research Institute for the Behavioral and Social Sciences.

Figure 9.12   From R.J. Lysaght et al. (1989). *Operator Workload: Comprehensive Review and Evaluation of Operator Workload Methodologies*. Technical Report 851, United States Army Research Institute for the Behavioral and Social Sciences.

Figure 9.13   From J.B. Isreal et al. (1980). The event-related brain potential as an index of display-monitoring workload. Adapted from *Human Factors*, 22(2), 211–224. Copyright 1980 by the Human Factors Society.

Figure 9.14   From W.W. Wierwille & J.G. Casali, A validated rating scale for global mental workload measurement applications. Reprinted with permission from *Proceedings of the Human Factors Society 27th Annual Meeting*, 1983, pp. 129–133. Copyright 1983 by the Human Factors Society, Inc. All rights reserved.

Figure B9.1   From Simons & Chabris (1999); Figure provided by Daniel Simons.

Table 9.1   From K.R. Boff & J.E. Lincoln (Eds.) (1988). *Engineering Data Compendium: Human Perception and Performance*. Wright-Patterson AFB, OH: AAMRL.

Table 9.2   From G.B. Reid, C.A. Shingledecker, & F.T. Eggemeier, Application of conjoint measurement in workload scale development. Reprinted with permission from *Proceedings of the Human Factors Society 27th Annual Meeting*, 1983, pp. 522–526. Copyright 1981 by the Human Factors Society, Inc. All rights reserved.

Table 9.3   From S.G. Hart & L.E. Staveland (1988). Development of NASA-TLX (Task Load Index): Results of empirical and theoretical research. In P.A. Hancock & N. Meshkati (Eds.), *Human Mental Workload*. Reprinted with permission of Elsevier Science Publishers.

Figure 10.3   From Peterson, L.R. and Peterson, M.J., *Journal of Experimental Psychology*, 58, 193–198, 1959.

Figure 10.4   From Sternberg, S., *Science*, 153, 652–654, 1966.

Figure 10.6   From A. Baddeley (2000). The episodic buffer: A new component of working memory? *Trends in Cognitive Sciences*, 4, 417–423. Reprinted by permission of Elsevier.

Figure 10.7   Reprinted by permission from Brooks (1968). Spatial and verbal components of the act of recall. *Canadian Journal of Psychology*, 22, 349–368.

Figure 10.8   From N. Cowan (1988). Evolving conceptions of memory storage, selective attention, and their mutual constraints within the human information processing system.

Dwight Reid Educational Foundation. Published by Heldref Publications, 1319 18th St., N.W., Washington, DC 20036–1802. Copyright 1981.

Figure 14.13   From M.A. Godwin & R.A. Schmidt (1971). Muscular fatigue and discrete motor learning. This article is reprinted with permission from the *Research Quarterly for Exercise and Sport, 42*, 374–383. The Education, Recreation and Dance, 1900 Association Drive, Reston, VA 22091.

Figure 14.14   From T.D. Lee, R.A. Magill, & D.J. Weeks (1985). Influence of practice schedule on testing schema theory predictions in adults. *Journal of Motor Behavior, 17*, 283–299. Reprinted with permission of Helen Dwight Reid Educational Foundation. Published by Heldref Publications, 1319 18th St., N.W., Washington, DC 20036–1802. Copyright 1985.

Figure 14.15   From R.A. Schmidt et al. (1989). Summary knowledge of results for skill acquisition: Support for the guidance hypothesis. *Journal of Experimental Psychology: Learning, Memory, and Cognition, 15*, 352–359. Copyright 1989 by the American Psychological Association. Reprinted by permission.

Figure 14.16   From L.G. Lindahl (1945). Movement analysis as an industrial training method. *Journal of Applied Psychology*

Figure 14.17   From R.A. Schmidt & D.E. Young (1991). Methodology for motor learning: A paradigm for kinematic feedback. *Journal of Motor Behavior, 23*, 13–24. Reprinted with permission of Helen Dwight Reid Educational Foundation. Published by Heldref Publications, 1319 18th St., N.W., Washington, DC 20036–1802. Copyright 1991.

Figure 14.18   From W.R. Carroll & A. Bandura (1982). The role of visual monitoring in observational learning of action patterns: Making the unobservable observable. *Journal of Motor Behavior, 14*, 153–167. Reprinted with permission of Helen Dwight Reid Educational Foundation. Published by Heldred Publications, 1319 18th St., N.W., Washington, DC 20036–1802. Copyright 1982.

Figure 15.2   From D. Howland & M.E. Noble (1953). The effect of physical constants of a control on tracking performance *Journal of Experimental Psychology, 46*, 353–360.

Figure 15.6   From C. R. Kelley (1962). Predictor instruments look into the future. *Control Engineering, 9*, March, 86–90.

Figure 15.7   From R.E. Eberts & W. Schneider (1985). Internalizing the system dynamics for a second-order system. Reprinted with permission from *Human Factors, 27*(4), 371–393. Copyright 1985 by the Human Factors Society. All rights reserved.

Figure 15.8   From J. L. Seminara, W. R. Gonzalez, & S.O. Parsons (1977). *Human Factors Review of Nuclear Power Plant Control Room Design* (EPRI Np-309). Palo Alto, CA: Electric Power Research Institute, March 1977. Reprinted by permission.

Figure 15.9   From D.P. Hunt (1953). The coding of aircraft controls (Report No. 53–221). Wright Air Development Center, U.S. Air Force.

Figure 15.10   From J.V. Bradley (1967). Tactual coding of cylindrical knobs. Reprinted with permission from *Human Factors, 9*(5), 483–496. Copyright 1967 by the Human Factors Society. All rights reserved.

Figure 15.12   From T.G. Moore (1974). Tactile and kinaesthetic aspects of industrial push-buttons. *Applied Ergonomics, 5*, 66–71-38. Copyright 1974 by Butterworth-Heinemann, Ltd. Reprinted by permission.

Figure 15.13   From K.R. Boff & J.E. Lincoln (Eds.) (1988). *Engineering Data Compendium: Human Perception and Performance*. Wright-Patterson AFB, OH: AAMRL.

Figure 16.10	From *Ergonomic Design for People at Work* (Vol. 2). Published by Van Nostrand Reinhold, 1983. Copyright 1983 by Eastman Kodak Company. Reprinted courtesy of Eastman Kodak Company.

Figure 16.11	From D.B. Chaffin & G.B.J. Andersson (1984). *Occupational Biomechanics.* Copyright 1984 by John Wiley & Sons, Inc. Reprinted by permission of John Wiley & Sons, Inc.

Figure 16.13	From E.N. Eklund 7 E.N. Corlett (1986). In E.N. Corlett, J.W. Wilson, & I. Manenica (Eds.), *The Ergonomics of Working Postures* (pp. 319–330). London: Taylor & Francis. Reproduced with permission.

Figure 16.14	From H.P. Rührmann (1984). Basic data for the design of consoles. In H. Schmidtke (Ed.), *Ergonomic Data for Equipment Design* (pp. 115–144). Copyright 1984 by Plenum Publishing Corporation. Reprinted by permission.

Figure 16.15	Courtesy of Mike Pomykacz, FAA William J. Hughes Technical Center.

Figure 16.16	From Design of individual workplaces (1972). In H.P. Van Cott 7 R.G. Kinkade (Eds.), *Human Engineering Guide to Equipment Design* (pp. 381–418). U.S. Government Printing Office.

Figure 16.17	Adapted from C.T. Morgan et al. (Eds.) (1963). *Human Engineering Guide to Equipment Design.* Copyright 1963 by McGraw-Hill.

Table 16.1	From K.H.E. Kroemer, H.J. Kroemer, & K.E. Kroemer-Elbert (1997). *Engineering Physiology,* Third Edition. Copyright 1997 by Van Nostrand Reinhold. Reprinted by permission.

Table 16.2	From E. Tichauer (1978). *The Biomechanical Basis of Ergonomics.* Copyright 1978 by John Wiley & Sons. Reprinted by permission of John Wiley & Sons, Inc.

Table 16.3	From T.J. Armstrong et al. (1982). Investigation of cumulative trauma disorders in a poultry processing plant. *American Industrial Hygiene Association Journal*, 42, 103–116. Reprinted by permission.

Table 16.4	Cal/OSHA Consultation Service and NIOSH, *Easy Ergonomics: A Guide to Selecting Non-Powered Hand Tools*, DHHS (NIOSH) Publication No. (2004–164), National Institute for Occupational Safety and Health, Cincinnati, OH, 2004.

Table 16.5	From E.N. Corlett (2005). The evaluation of industrial seating. In J.R. Wilson & E.N. Corlett (Eds.), *Evaluation of Human Work,* Third Edition, pp. 729–742. Reprinted by permission of CRC Press.

Figure B16.1	From http://differentialdiagnosisoftos.weebly.com/carpal-tunnel-syndrome.html.

Figure 17.1	From W.H. Cushman & B. Christ (1987). Illumination. In G. Salvendy (Ed.), *Handbook of Human Factors* (pp. 670–695). Copyright 1987 Reprinted with permission of John Wiley & Sons, Inc.

Figure 17.2	From W.H. Cushman & B. Christ (1987). Illumination. In G. Salvendy (Ed.), *Handbook of Human Factors* (pp. 670–695). Copyright 1987 Reprinted with permission of John Wiley & Sons, Inc.

Figure 17.3	From P.R. Boyce (2014). *Human Factors in Lighting,* Third Edition. Boca Raton, FL: CRC Press. Reprinted by permission.

Figure 17.4	Reprinted with permission from S.W. Smith & M.S. Rea (1978). Proofreading under different levels of illumination. *Journal of Illuminating Engineering Society*, 8, 47–52.

Figure 17.5 From W.H. Cushman & B. Christ (1987). Illumination. In G. Salvendy (Ed.), *Handbook of Human Factors* (pp. 670–695). Copyright 1987 Reprinted with permission of John Wiley & Sons, Inc.

Figure 17.6 Provided by Casella USA, Amherst, NH.

Figure 17.8 From L.L. Beranek (1989). Balanced noise-criterion (NCB) curves. *Journal of the Acoustical Society of America*, 86, 650–664. Reprinted by permission.

Figure 17.9 From J.D. Miller (1974). Effects of noise on people. *Journal of the Acoustical Society of America*, 56, 729–764.

Figure 17.10 From W. Taylor et al. (1965). Study of noise and hearing in jute weaving. *Journal of the Acoustical Society of America*, 38, 113–120. Reprinted by permission.

Figure 17.11 From *Ergonomic Design for People at Work* (Vol. 1). Published by Van Nostrand Reinhold, 1983. Copyright 1983 by Eastman Kodak Company. Reprinted courtesy of Eastman Kodak Company.

Figure 17.13 From M.-Y. Park & J.G. Casali (1991). A controlled investigation of in-field attenuation performance of selected insert, earmuff, and canal cap hearing protectors. Reprinted with permission from *Human Factors*, 33(6), 693–714. Copyright 1991 by the Human Factors Society, Inc. All rights reserved.

Figure 17.14 Extracts of ISO 5349:1986 are reproduced with the permission of the International Organization for Standardization (ISO). Copies of the complete standard may be obtained from ISO, C.P. 56, 1211 Geneva 20, Switzerland or, in the U.S., from ANSI, 11 West 42nd Street, 13th Floor, New York, NY 10036.

Figure 17.15 From *Ergonomic Design for People at Work* (Vol. 1). Published by Van Nostrand Reinhold, 1983. Copyright 1983 by Eastman Kodak Company. Reprinted courtesy of Eastman Kodak Company.

Figure 17.16 From G.R.J. Hockey (1986). Changes in operational efficiency as a function of environmental stress, fatigue, and circadian rhythms. In K.R. Boff, L. Kaufman, & J.P. Thomas (Eds.), *Handbook of Perception and Human Performance. Volume II: Cognitive Processes and Performance* (pp. 44-1–44-49). Copyright 1986 Reprinted with permission of John Wiley & Sons, Inc.

Figure 17B.1 From A.J. Louviere & J.T. Jackson (1982). Man-machine design for space flight. In T.S. Cheston & D.L. Winter (Eds.), *Human Factors of Outer Space Production,* Westview Press.

Table 17.1 Reprinted from J.E. Flynn (1977). A study of subjective responses to low energy and nonuniform lighting systems. *Lighting Design and Application*, 7(2), 6–15. Published by the Illuminating Engineering Society of North America.

Table 17.2 From L.L. Beranek, W. Blazier, & J.J. Figwer (1971). Preferred noise criterion (PNC) curves and their application to rooms. *Journal of the Acoustical Society of America*, 50, 1223–1228. Reprinted by permission.

Table 17.3 From K.R. Boff & J.E. Lincoln (Eds.) (1988). *Engineering Data Compendium: Human Perception and Performance.* Wright-Patterson AFB, OH: AAMRL.

Figure 18.1 Reprinted by permission from L.N. Jewell & M. Siegall (1990). *Contemporary Industrial/Organizational Psychology,* Second edition. West Publisher.

Figure 18.2 From G.R.J. Hockey (1986). Changes in operational efficiency as a function of environmental stress, fatigue, and circadian rhythms. In K.R. Boff, L. Kaufman, & J.P. Thomas (Eds.),

Handbook of Perception and Human Performance. Volume II: Cognitive Processes and Performance (pp. 44-1–44-49). Copyright 1986 Reprinted with permission of John Wiley & Sons, Inc.

Figure 18.3   From R.C. Browne (1949). The day and night performance of teleprinter switchboard operators. *Journal of Occupational Psychology*, 23, 121–126.

Figure 18.4   From S. F. Folkard, Diurnal variation, In G.R.J. Hockey (Ed.). *Stress and Fatigue in Human Performance* (pp. 245–272). Copyright 1983 by John Wiley & Sons, Inc. Reprinted with permission of John Wiley & Sons, Inc.

Figure 18.7   From R.G. Gifford (2002). *Environmental Psychology: Principles and Practice,* Third Edition. Colville, WA: Optimal Books.

Figure 18.8   From R.L. Propst (1966). The action office. Reprinted with permission from *Human Factors*, 8(4), 299–306. Copyright 1966 by the Human Factors Society. All rights reserved.

Figure 18.9   From Mayor Michael Bloomberg's "Bullpen" office in City Hall, New York City (2013).

Figure 18.10   From M.J. Brookes & A. Kaplan (1972). The office environment: Space planning and affective behavior. Reprinted with permission from *Human Factors,* Vol. 14, No. 5, pp. 373–391, 1972. Copyright 1962 by the Human Factors Society. All rights reserved.

Figure 18.11   From C.D. Dumesnil (1987). Office case study: Social behavior in relation to the design of the environment. Reprinted with permission from the *Journal of Architectural and Planning Research*, 4, 7–13. Copyright Locke Science Publishing Company.

Figure 18.12   Reprinted by permission from L.N. Jewell & M. Siegall (1990). *Contemporary Industrial/Organizational Psychology,* Second edition. West Publisher.

Figure 18.13   Reprinted by permission from L.N. Jewell & M. Siegall (1990). *Contemporary Industrial/Organizational Psychology,* Second edition. West Publisher.

Table 18.1   Reprinted by permission from L.N. Jewell & M. Siegall (1990). *Contemporary Industrial/Organizational Psychology,* Second edition. West Publisher.

Table 18.2   From E.J. McCormick, P.R. Jeanneret, & R.C. Mecham (1972). A study of job characteristics and job dimensions as based on the Position Analysis Questionnaire (PAQ). *Journal of Applied Psychology*, 56, 347–368.

Table 18.3   Excerpted from W.O. Galitz (1984). *Humanizing Office Automation: The Impact of Ergonomics on Productivity.* Copyright 1984 by the QED Publishing Group. Reprinted with permission.

Table 19.1   Based on D. Meister (1989). *Conceptual Aspects of Human Factors.* The Johns Hopkins University Press, Baltimore/London.

Table 19.2   From S. K. Card, P. P. Moran, & A. P. Newell, The model human processor. In K. R. Boff, L. Kaufman, & J. P. Thomas (Eds.), *Handbook of Perception and Human Performance* (pp. 45–1 –45–35). Copyright © 1986 by John Wiley and Sons, Inc. Adapted with permission.

Table 19.3   From B. E. John & A. Newell (1990). Toward an engineering model of stimulus-response compatibility. In R. W. Proctor & T. G. Reeve (Eds.), *Stimulus-Response Compatibility: An Integrated Perspective.* Copyright © 1990 by Elsevier Science Publishers, BV. Reprinted by permission.

Figure 19.1   From M. E. J. van Uden & H. J. T. Rensink (2006). An up-front engineering "level of protection" through human factor design. In W. Karwowski (Ed.), *International handbook of ergonomics and human factors* (2nd ed., Vol. 2: pp. 2455–2465). Boca Raton, FL: CRC Press.

Figure 19.3 From S. Baron, D. S. Kruser, & B. M. Huey (Eds.) (1990). *Quantitative Modeling of Human Performance in Complex, Dynamic Systems*, National Academy Press.

Figure 19.4 From R. Harris, R., H. P. Iavecchia, & A. O. Dick (1989). The human operator simulator (HOS-IV). In G. R. McMillan, D. Beevis, E. Salas, M. H. Strub, R. Sutton, & L. Van Breda (Eds.), *Applications of human performance models to system design* (pp. 275–280). Copyright © 1989 by Plenum Press. Reprinted by permission.

Chapter 8   Suggestions

Figure 8.14 A head-up display.

Figure 8.15 A helmet-mounted display (HMD). If BAE systems has this photo without the shopped glow over the eyepiece that would be great.

Figure 15–17 "Soft" button interface of a digital synthesizer.

Chapter 18   Suggestion

Figure 18.9 Mayor Michael Bloomberg's "Bullpen" office in City Hall, New York City (2013).

Index

Printed in the United States
by Baker & Taylor Publisher Services